T0180239

Microwave, Radar & RF Engineering

Prakash Kumar Chaturvedi

Microwave, Radar & RF Engineering

With Laboratory Manual

Prakash Kumar Chaturvedi
Department of Electronics &
Communication Engineering
SRM University, NCR Campus
Ghaziabad, Uttar Pradesh
India

ISBN 978-981-13-4030-7 ISBN 978-981-10-7965-8 (eBook)
https://doi.org/10.1007/978-981-10-7965-8

© Springer Nature Singapore Pte Ltd. 2018
Softcover re-print of the Hardcover 1st edition 2018
This work is subject to copyright. All rights are reserved by the Publisher, whether the whole or part of the material is concerned, specifically the rights of translation, reprinting, reuse of illustrations, recitation, broadcasting, reproduction on microfilms or in any other physical way, and transmission or information storage and retrieval, electronic adaptation, computer software, or by similar or dissimilar methodology now known or hereafter developed.
The use of general descriptive names, registered names, trademarks, service marks, etc. in this publication does not imply, even in the absence of a specific statement, that such names are exempt from the relevant protective laws and regulations and therefore free for general use.
The publisher, the authors and the editors are safe to assume that the advice and information in this book are believed to be true and accurate at the date of publication. Neither the publisher nor the authors or the editors give a warranty, express or implied, with respect to the material contained herein or for any errors or omissions that may have been made. The publisher remains neutral with regard to jurisdictional claims in published maps and institutional affiliations.

Printed on acid-free paper

This Springer imprint is published by the registered company Springer Nature Singapore Pte Ltd. part of Springer Nature
The registered company address is: 152 Beach Road, #21-01/04 Gateway East, Singapore 189721, Singapore

Preface

Unprecedented growth in the application of microwave has taken place during the last two to three decades, especially in mobile communication, TV transmission, industrial/domestic applications, satellite communication, telemetry, RADAR/navigational aids, etc. This rapid development in the microwave and that of digital technologies has synergised, leading to further accelerated growth rate. This has created an increased demand for trained engineers, in civilian as well as in defence organizations.

This book meets the complete need of the students of engineering courses, i.e. B.S., B.E., M.S., M.Tech., M.Sc., in various countries. However, the students need to have prior knowledge of electromagnetics. **The special features of this book are as follows**: (**a**) It contains fundamental concepts and principles behind microwave engineering explained in a student-friendly lucid language, keeping a balance between physical and analytical approaches, (**b**) contains a large number of solved and unsolved problems after every chapter, for developing practical knowledge, (**c**) has around 400 figures, with special effort put in, for giving realistic numerical values in the graphs/dimensions of the components and devices, for getting a real feel and visualisation of that device, which is missing in many text books, and (**d**) has 15 important experiments, giving full theory, procedures, precautions, and sample readings and observations as expected in each experiment, followed by quiz/viva questions for the benefit of the students and the instructors.

This book consists of **12 chapters** along with an annexure giving related constants and finally an index. Chapter 1 introduces the subject along with the behaviour of lumped components L, C, and R at microwave frequencies, microwave heating mechanism, concept of CW/pulsed signals, decibel, anechoic chamber, EMI/EMC, radiation hazards, etc. Chapter 2 summarizes the basics of wave propagation in different modes/cut off frequencies, etc., in transmission lines and waveguides. It also touches a bit on microstrip lines. Impedance matching covering Smith chart using single and double stub has also been given. Chapter 3 covers cavity resonators of various types and their tuning and coupling. Chapter 4 describes a variety of components like T, directional couplers, isolators used in microwave circuits. Chapter 5 firstly covers limitations of conventional tubes and then various microwave tubes, e.g. klystron tube, magnetron, TWT, used as oscillators and amplifiers. Chapter 6 covers six types of transistors and eight types of diodes, with their working as oscillators and amplifiers. Chapter 7 presents measurement techniques of important parameters and related instruments. Chapter 8 covers theory of microwave propagation in space and through a microwave antenna. Chapter 9 deals with working of various types of RADARs. Chapter 10 introduces filter design theory and design techniques giving emphasis on microstrip line filters, which can be verified by using standard software, e.g. ADS. Chapter 11 gives basic concepts of RF amplifiers, oscillators, and mixers using Smith charts for stability considerations. Chapter 12 presents 15 simple laboratory experiments of the undergraduate level, along with conventional guidelines for students.

I am very thankful to my life partner and children, who have been a source of inspiration, having provided congenial atmosphere even in odd hours and having given useful suggestions while writing the manuscripts of this book.

I am also thankful to authors of various books (as per the references given), which I have referred to during the long period I have been teaching this subject in various institutions.

Finally, I convey my thanks to Springer for their painstaking effort for bringing the book in standard and excellent form.

However, if you notice any mistake, error, and discrepancy, it would be highly appreciated if you bring it to my notice along with your feedback for improvement of the book.

New Delhi, India Prof. Prakash Kumar Chaturvedi

Contents

About the Author

Dr. Prakash Kumar Chaturvedi is currently a Professor in SRM University (NCR Campus), Modinagar, UP. After completing his M.Tech. from BITS, Pilani in 1969, he did his Ph.D. from CEERI, Pilani in 1974. He holds an MBA degree from the University of Stirling, Scotland, which was sponsored by the Government of India.

He started his teaching career from IIT Madras. Thereafter, he joined the Ministry of Information Technology, Government of India. Meanwhile, he continued as a Visiting Professor in DIT (now NSIT), University of Delhi. He retired as Director from the Ministry in 2004, and thereafter, he has been Director, Amity University; Director, GITM, Gurgaon. He carries an experience of 38 years in teaching and research, besides experience of techno-management. He has published 18 research papers in international and Indian journals like IEEE Transactions on Electron Devices, Solid-State Electronics, Microelectronics Journal, Physica Status Solidi. He has also been Project Management Chief in a number of Government of India projects like Digital TV, CODIN, Technology Development for Indian Languages (TDIL). He has travelled widely and has represented India as leader/member of various technical/trade negotiation teams to many European countries (like the UK, France, Germany, Italy, Switzerland, Belgium, Romania, Poland), the USA, Thailand, etc.

Introduction to Microwaves

1

Contents

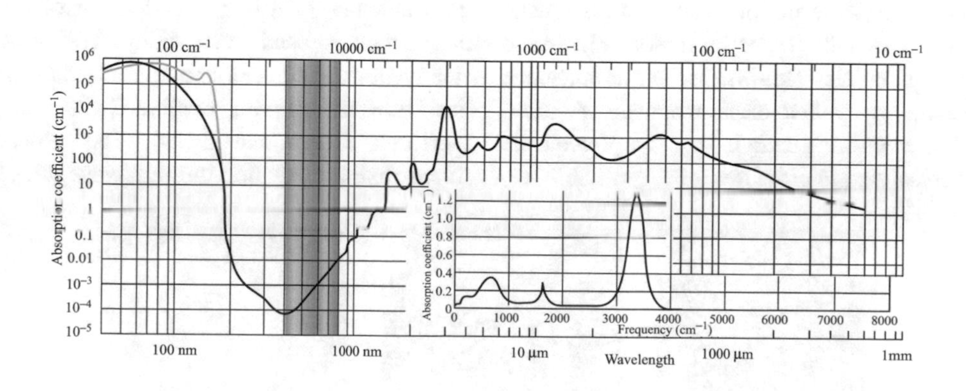

© Springer Nature Singapore Pte Ltd. 2018
P. K. Chaturvedi, *Microwave, Radar & RF Engineering*,
https://doi.org/10.1007/978-981-10-7965-8_1

1.1 Introduction

Microwave is a descriptive term used to identify electromagnetic waves in the frequency spectrum ranging approximately from 1 GHz (wavelength $\lambda = 30$ cm) to 300 GHz ($\lambda = 1$ mm). For wavelength from 1.00 to 0.3 mm, i.e. for frequencies 300–1000 GHz, the EM waves are called millimetre waves (Fig. 1.1) and sub-millimetre waves.

Microwaves are so called as they are normally defined in terms of their wavelength. In fact beyond audio waves, all are electromagnetic waves having E-vector and H-vector which are perpendicular to each other.

These microwaves have several interesting and unusual features, not found in other portions of the electromagnetic frequency spectrum. These features make microwaves uniquely suitable for several useful applications.

Since the wavelengths are small, the phase varies rapidly with distance in the guided media; therefore, the techniques of circuit analysis and design, measurements of power generation and amplification at these frequencies are different from those at lower frequencies.

Analysis based on Kirchhoff's laws and Ohm's law (voltage–current) concepts is not easily possible for describing the circuit's behaviour at microwave frequencies. It is necessary to analyse the circuit or the component in terms of electric and magnetic fields associated with it. For this reason, microwave engineering is also known as electromagnetic engineering or applied electromagnetic. A background of electromagnetic theory is a prerequisite for understanding microwaves.

The complete spectrum of electromagnetic waves is given in Figs. 1.1, 1.2, and 1.3 giving frequency and corresponding wavelength. It also gives names of different frequency bands (e.g. IEEE band, millimetre band, sub-millimetre of UHF and VHF), different applications, guided media of application, etc. The IEEE-defined band is also given separately in Table 1.1.

1.2 History of Microwaves

One of the first attempts to deduce the fundamental law of electromagnetic action in terms of an electric field propagating at finite velocity was done by Karl Friedrick Gauss (1777–1855), a German mathematician. However, the genesis of microwave and electromagnetic waves in general can be taken from Michael Faraday's (1848) experiments on propagation of magnetic disturbance (EM waves), which later got theoretical formulation by James Clerk Maxwell (1865), popularly known as Maxwell's Field Equations. Thereafter, Marconi and Hertz in their experiments (1888) proved Maxwell's theory of RF signal being an EM wave and travel with the velocity of light ($c = \lambda \cdot f = 3 \times 10^8$ m/s). In 1885, J. C. Bose developed a circuit for generating microwave power and in 1898 developed horn antenna, polariser, and detector of RF signal, which is used even today. The slow but steady development in the area of transmission line, transmitters, etc., continued till 1930, but thereafter it got accelerated. The genesis of microwave propagation through waveguides was

Michael Faraday

JC Maxwell

J.C. Bose

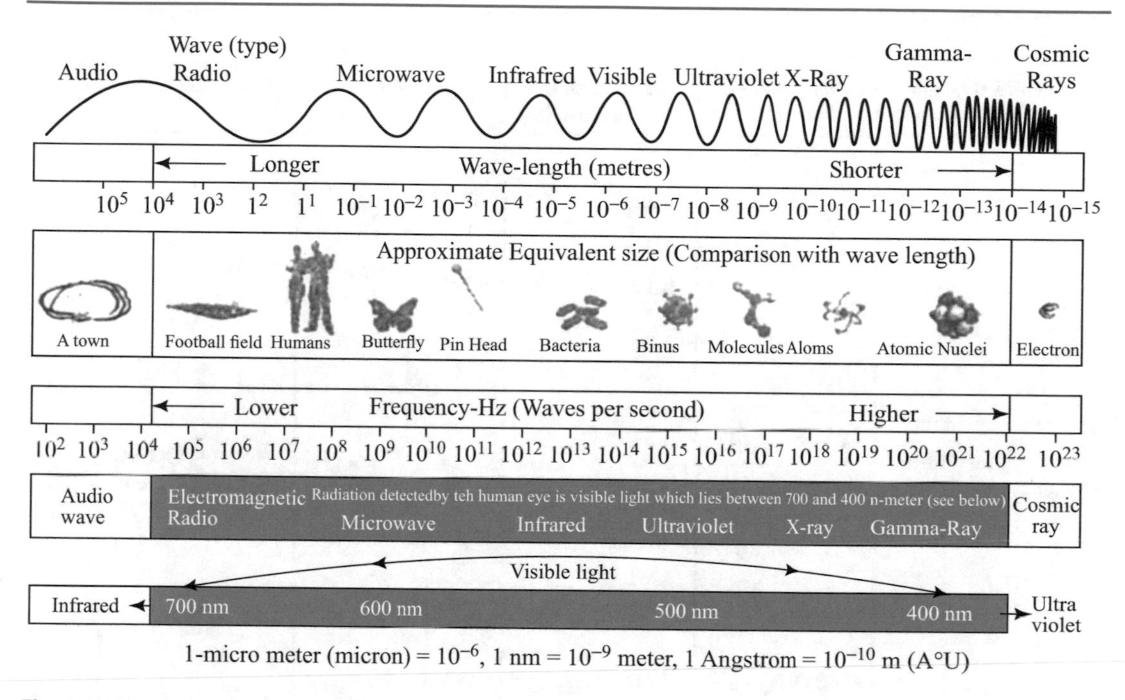

Fig. 1.1 Comparative visualisation of the complete spectrum of EM wavelengths and its frequencies

from the success of Dr. Southworth (1933) of AT and T Labs, USA, when he was able to transmit signal through metal pipe of 4″ diameter. Thereafter, the requirements of World Wars I and II further boosted through the development of microwave tubes—Klystron by Varian brothers (1936) of Stanford University, magnetron by Randel and Boots of UK (1939), Radar by Henry Tizard (UK) during 1940, etc. Thereafter, also the development continued and ferrite devices, TWT, etc., came in 1950s. In 1960s, Solid State Microwave sources, e.g. Gunn diodes, avalanche diodes, microwave transistors came in full swing, which takes very small space and has very low dc power requirements for generating microwave power. Now application of microwaves has entered all the segments of Communication and Telemetry control (audio, video, text, and data), whether it is for use in civilian systems or in defence systems or for space applications. It has other applications also, e.g. heating (in industrial processes or domestic appliances or cancer treatment), microwave spectroscopy, radio astronomy, satellite communication.

Today for microwave power requirements below 5 W, we can use sources of semiconductor devices like IMPATT diode, while for higher power requirements, we use microwave tubes like klystron, magnetron, Travelling Wave Tube (TWT).

1.3 Characteristic Features and Advantages of Microwaves

Unique features leading to advantages of microwave over low-frequency signal are as:

1. **Increased band width availability**: It has large band width because of high frequency. Normally the maximum bandwidth can be 10% of the base signal. A 10% band width at 3 GHz implies availability of 300 MHz band width and hence much more information can be transmitted.
2. **Lower fading and reliability**: Fading effect is high at low frequency, while in microwave due to line-of-sight propagation and high frequency, there is less fading effect and hence microwave communication is more reliable.

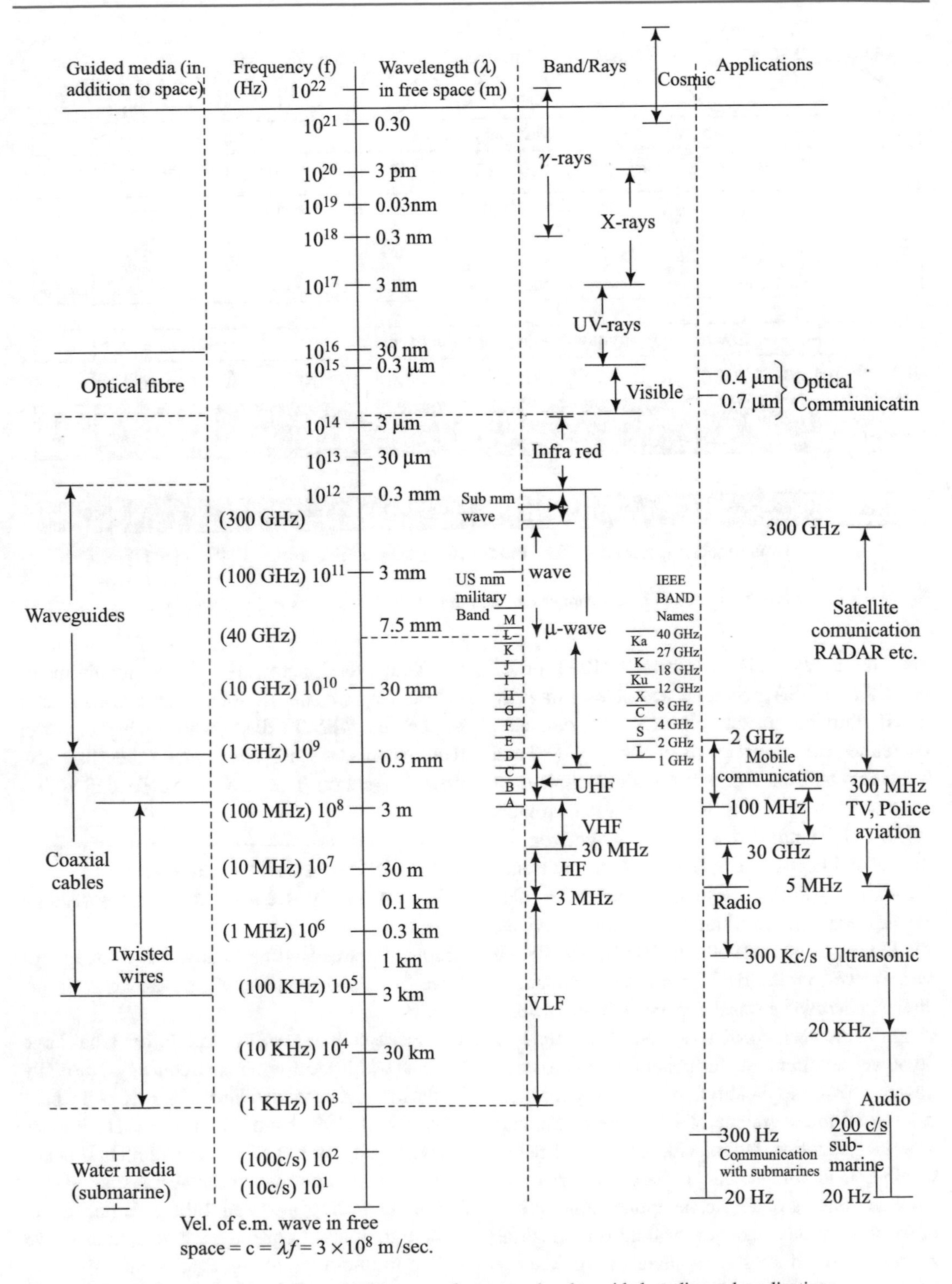

Fig. 1.2 Frequency spectrum: audio and EM waves: frequency, bands, guided media, and applications

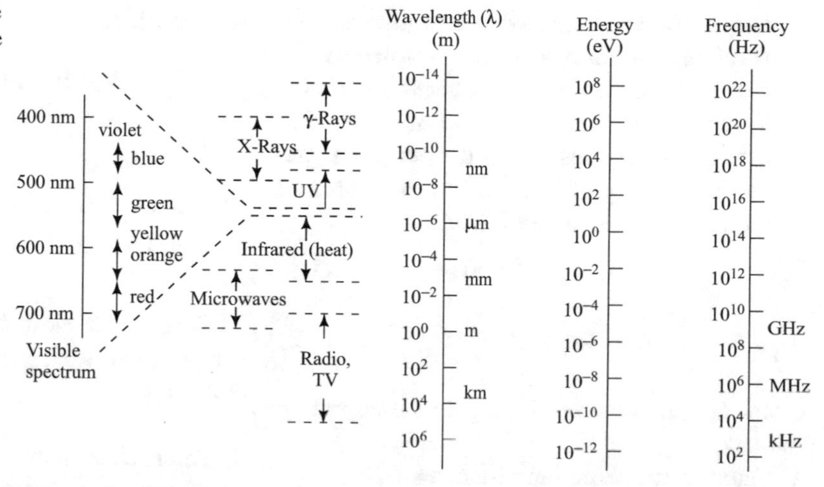

Fig. 1.3 Energy of the wave increases with frequency. The spectrum of electromagnetic radiation.
1 eV = 1.6×10^{-19} J. E_g (eV) = $h\nu = hc/\lambda = 1.24/\lambda$ (μm); $h = 6.55 \times 10^{-34}$ J s

Table 1.1 Microwave frequency band IEEE names

Frequency	Band (old name)	Band (new name)
3–30 MHz	HF	HF
30–300 MHz	VHF	VHF
0.3–1.0 GHz	UHF	C
1–2 GHz	L	D
2–3 GHz	S	E
3–4 GHz	S	F
4–6 GHz	C	G
6–8 GHz	C	H
8–10 GHz	X	I
10–12.4 GHz	X	J
12.4–18.6 GHz	Ku (upper to K)	J
18–20 GHz	K	J
20–26.5 GHz	K	K
26.5–40 GHz	Ka (after K)	K
40–300 GHz	Millimetre	Millimetre
>300.00 GHz	Sub-millimetre	Sub-millimetre

3. **Transparency property**: It has transparency property, i.e. it can easily propagate through air, space, even through an ionised layer surrounding the earth and atmosphere, leading to important applications like:

   - Astronomical research of space.
   - Duplex communication between ground station and speed vehicles.

The only 58–60 GHz frequency band which is used less due to molecule resonance (H_2O and O_2) and hence absorption. Above 400 GHz, some frequencies are blocked by ozone in the atmosphere due to similar reasons.

4. **Low-power requirements**: The dc power required by the transmitter and receiver at microwave frequency is quite low as compared to low-frequency operations especially due to its directivity and low attenuation in space as well as in any guided media like wave guides.

5. **Higher power radiated and higher gain of receiving antenna, at higher frequencies**: In radar system, the power radiated P_r from a tower antenna and the gain G of a receiving antenna for signals reflected from the target, are high, being proportional to the square of the frequency as given below:

$$P_r = m_0 \pi^2 I_0^2 l^2 f^2 / c^2 \quad (1.1)$$

$$G = 4\pi \rho_a \cdot A \cdot f^2 / c^2 \quad (1.2)$$

where I_0 = ac current; l = length of transmitting antenna,
A = area of the receiving dish antenna,
ρ_a = antenna aperture efficiency.

As $E = hv$, the energy of the wave increases with frequency (v); Fig. 1.3 may be referred for this.

6. **Directivity**: As the frequency increases, dispersive angle decreases; hence, directivity increases and beam width angle decreases. This property leads to further less requirement of microwave power in the directions where we want to send signal.

Beam width (θ_B) for parabolic reflector is given by,

$$\theta_B = 140^\circ \times (\lambda/D) \quad (1.3)$$

Therefore,

$$D = 140^\circ \times (\lambda/\theta_B) \quad (1.4)$$

where $\theta_B \rightarrow$ Beam width (degrees), $\lambda \rightarrow$ Wavelength (cm), $D \rightarrow$ Diameter of antenna (cm).

For example, for the same beam width requirement of 1°, smaller antenna is required at microwave frequency, for example:

At 30 GHz (Microwave):

$$\lambda = 1.0 \text{ cm and } \theta_B = 1^\circ$$

So, $D = 140^\circ \times 1.0/1^\circ \rightarrow 140$ cm (which is practical)

At 100 MHz:

$$\lambda = 300 \text{ cm and } \theta_B = 1^\circ$$

So,

$$D = 140^\circ \times 300/1^\circ \rightarrow 42{,}000 \text{ cm}$$
$$= 420 \text{ m (which is not easy to make or use)}$$

Hence, it is clear that antenna size is much smaller and easier to handle at microwave frequency.

7. **Interaction with metal attenuation, penetration, and reflection**: Microwave incident on the metal walls of the oven behaves similar to visible light hitting a silver mirror. The microwaves are absorbed very effectively, since the electric fields of the waves interact very strongly with the nearly free electrons of the metal. In a simple model, the electron undergoes damped forced oscillation and absorbs energy partly. These accelerated electrons re-radiate electromagnetic waves at the same frequency and in phase, hence a major part of the microwave is perfectly reflected. Microscopically, this behaviour is described by the complex dielectric constant $\varepsilon(\omega)$, which is the square of the complex refractive index $(\mu), \varepsilon_d = \varepsilon_r + i\varepsilon_i = (\mu_r + i\mu_i)^2$.

The refractive index (μ) of many metals gives reflectivities close to 100% at low frequencies. The penetration depth of electromagnetic waves of wavelength A is given by

$$\delta = \lambda / 4\pi\mu_i.$$

For example, for microwave with λ = 12.2 cm incident on aluminium, $\delta = 1.2$ μm. These are similar to skin depths, i.e. the attenuation depths of alternating currents of frequency ω in metals (see Fig. 1.4). **This explains why microwaves do not cross metals, e.g. a cell phone kept in a metal enclosure/almirah does not receive signal.**

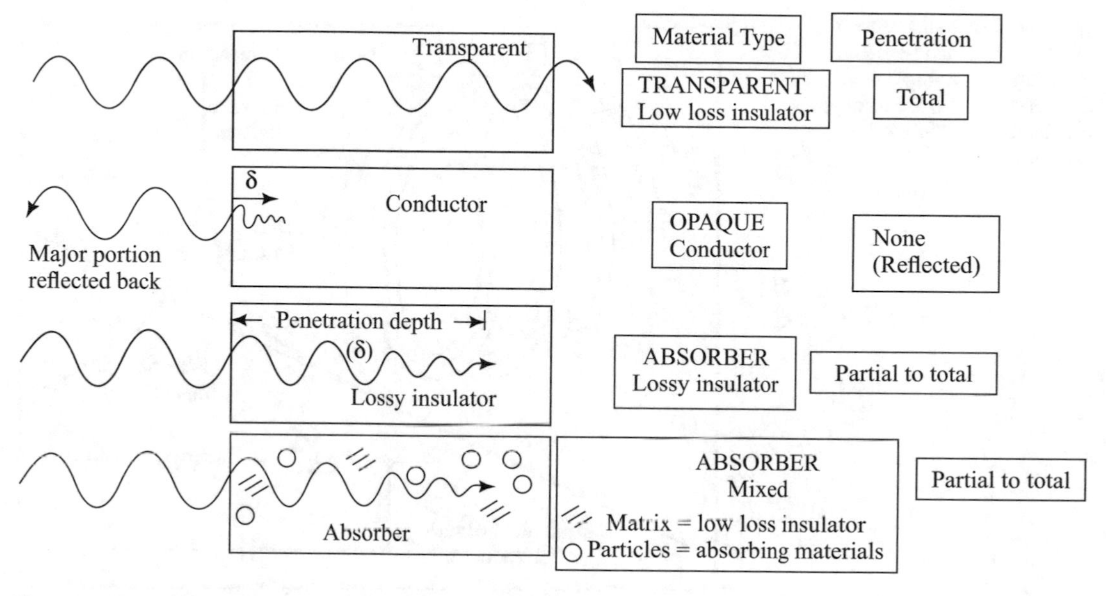

Fig. 1.4 Interaction of microwaves with different materials

8. **Passive lumped components at microwave frequencies**

(a) **Wire and Resistor**

From conventional ac circuit analysis, we know that resistance R is independent of frequency. Also that a low value of capacitor ($C = 1$ pt) and inductor ($L = 1$ pH) have very large reactance $(X_C = 1/\omega c = 3.18 \times 10^9\ \Omega \approx \infty)$ and very low reactance $(\omega_L = 3.14 \times 10^{-7}\ \Omega \approx 0)$ at 50 Hz (Fig 1.5).

As is well known that in a conductor, ac current density is highest at the surface and falls as we go close to the core. This current density falls to $1/\sqrt{2}$ value of the surface current density at a depth called skin depth (δ), which is a function of frequency, conductivity, and permeability as:

$$\delta = 1\Big/\sqrt{\pi f \mu\sigma}$$

With ac charge flowing in the wire establishes an ac magnetic field around, which induces ac

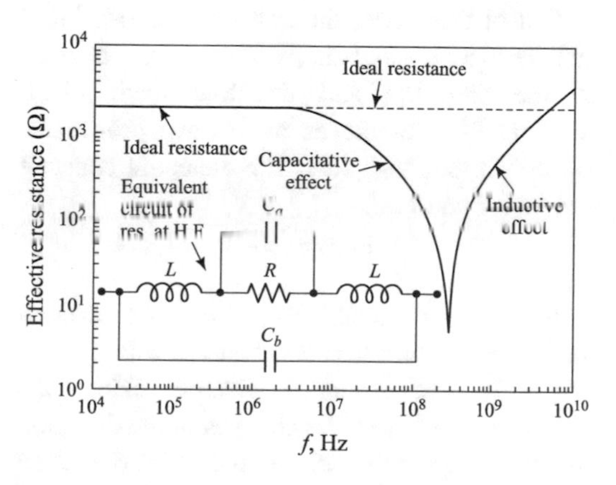

Fig. 1.5 Equivalent circuit of a 2000 Ω resistor and its effective impedance as a function of frequency. At μw frequencies, it behaves like an inductance

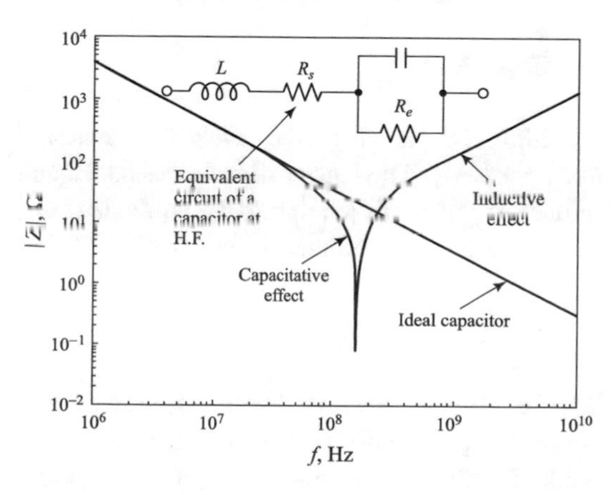

Fig. 1.6 Equivalent circuit of a capacitor and its effective impedance as a function of frequency. At μw frequencies it also behaves more like an impedance

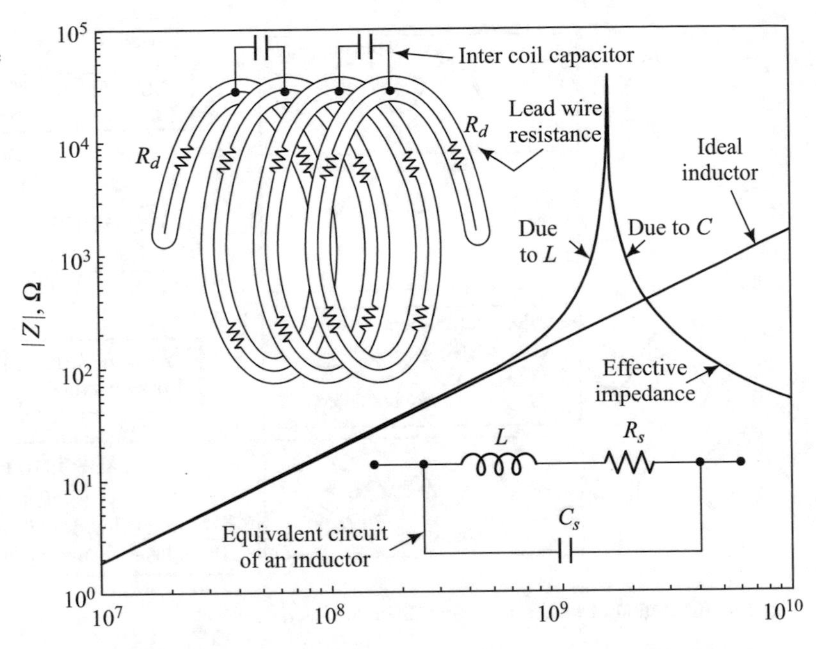

Fig. 1.7 Equivalent circuit of an inductor and its effective impedance as a function of frequency. At μ*w* frequency, it behaves more like a capacitor

electric field (as per Faraday's law), which in turn induces an ac current in opposite direction. This effect is strongest at the core ($r = 0$), thereby increasing the impedance significantly. As a result, the current tends to reside at the outer perimeter called skin effect. This skin depth is $\delta \approx 0.7$ mm at 10 kHz for Cu, which reduces to 2.2 μm at 1 GHz, and it starts behaving like an inductance of value $L = (R_{dc} \cdot r)/(2\delta\omega)$ and resistance of value $R = (R_{dc}r)/(2\delta)$.

Thus, the equivalent circuit of a resistor and its behaviour with frequency are given in Fig. 1.4.

(b) **Capacitor**

At high frequencies, the dielectric material becomes lossy. The leads will have resistance and inductance as well, leading to an equivalent circuit and frequency behaviour as per Fig. 1.6.

(c) **Inductor**

At high frequencies, inter-coil capacitance will come into play, with wire behaving as resistor as well. Therefore, the inductor equivalent circuit and its frequency behaviour are shown in Fig. 1.7.

Because of the above complex behaviour, lumped components are not used in microwaves. Application can be cavity resonator (Chap. 3) and filters (Chap. 10).

9. **Microwave heating mechanism:** Any dielectric (including water or food having water) get heated due to

   (i) Dipole relaxation loss/orientation loss
   (ii) Conduction loss.

Out of these two, the first one is most dominant in bipolar molecule, which is there in most of the dielectric and therefore they orient/oscillate like dumble as microwave propagates through it (see Fig. 1.8). The dielectric constant can be written as:

$$\varepsilon_d = \varepsilon_r + \varepsilon_i$$

Here ε_i is responsible for the dipole relaxation loss or orientation loss and hence heating:

(a) In low-frequency electric fields, the dipoles easily follow the changes in the direction and amplitude of the field so that their orientation changes in phase with the field. (b) At higher frequencies ≥ 0.3 GHz, the inertia of the

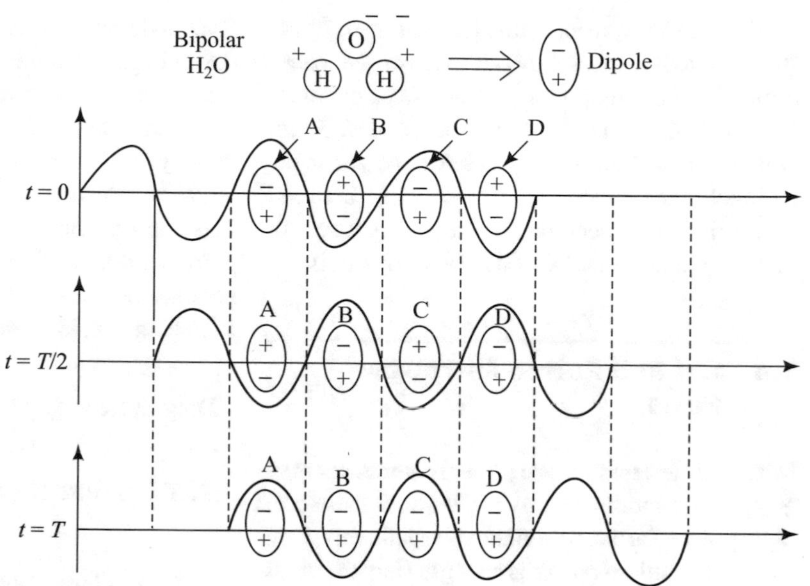

Fig. 1.8 Stationary water bipolar molecules A, B, C, D rotate/oscillate like dumbles with the E-vector, as the wave propagates through water/food, causing friction between bipolar molecules (like water) and hence get heat in this process μw power is lost. Figure shows the alignments of A, B, C, D, at time t = 0. $T/2$ and T, while waves propagate to the right

molecules and their interactions with neighbours make changing orientation more difficult and the dipoles lag behind the field and the electric field absorbs power from the field. This is known as dielectric loss due to dipole relaxation. (c) Finally, at very high frequencies of 1–10 THz the molecules do not respond to the electric field.

Thus, in a microwave oven, the electrically dipolar water molecules absorb most of the microwave energy.

The average microwave power absorbed per unit volume may be written as:

$$P = \omega \varepsilon_0 \varepsilon_i E_{\text{eff}}^2 \text{ w/cc} \quad (1.4a)$$

where E_{eff} is the average electric field intensity of microwave inside the dielectric.

Thus, we see that microwave power absorption in a dielectric is an increasing linear function of frequency [Eq. (1.4a)] while depth of penetration reduces with frequency. Figure 1.9a gives the ε_r and ε_i with frequency, and Fig. 1.9b gives the

(i) power absorption coefficient α and
(ii) penetration depth δ in cm in water and food as dielectric.

Here it may be noted that salt (NaCl) also being bipolar adds to the heating of food. The depth of penetration is given as inverse of absorption coefficient: $l/\alpha = \delta = \lambda/2\pi\mu_i$.

Where μ_i is imaginary part of refractive index μ.

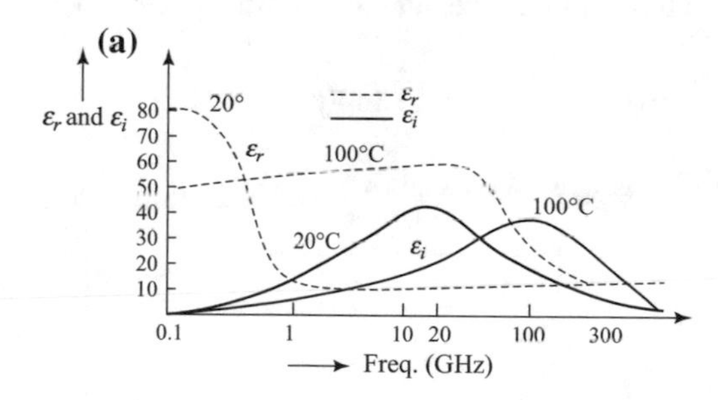

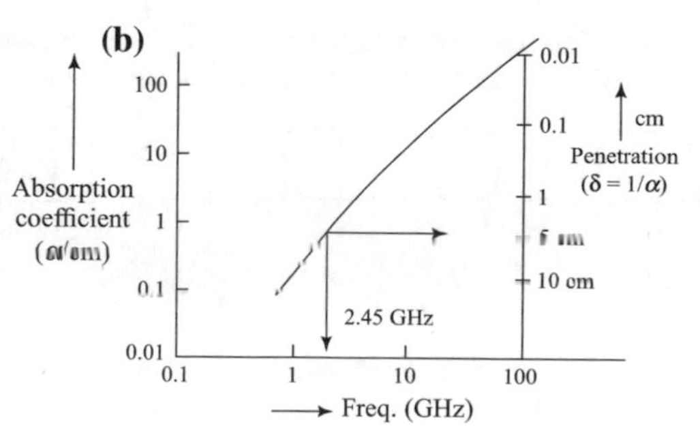

Fig. 1.9 **a** Real (ε_r) and imaginary (ε_i) parts of dielectric constant as a function of frequency (GHz) at two temperatures (20 and 100 °C) of dielectric (water/food), ε_i is maximum near 20 GHz at 20 °C. **b** Absorption coef α/cm and the penetration depth (cm) in water/food. α keeps on increasing up to 300 GHz but penetration falls with frequency. At 2.45 GHz penetration is 5 cm or so

We see that ε_i responsible for heating at 20 °C shows maxima near 20 GHz, where the penetration in water/food will be nearly 0.02 cm only. For food, the depth δ should be at least 5 cm, which is near 2.45 GHz. Therefore as per international convention, as a balance between absorption and penetration, frequency used in microwave oven is 2.45 GHz (λ = 12.2 cm).

1.4 CW and Pulsed Microwave Power

Many a times, microwave power requirement may not be of continuous wave (CW) and therefore, high power for small durations repetitively may suffice, i.e. pulsed power is enough. Here the input power supply to the oscillator has to be rectangular pulses, so that the average dc power requirement reduces and **the oscillator gives bursts of microwave power during the period of the dc pulse** (Fig. 1.10). In this way, even the tiny TRAPATT diode, which can supply just 1–5 W in CW mode near 1 GHz, can supply around 1 kW power in pulsed mode with 0.1% duty cycle. In the case of a magnetron tube which gives 3 kW in CW mode can deliver 5 MW in pulsed mode with 0.1% duty cycle. The average of pulsed power will be 5 kW, which is slightly higher than the CW power. These are possible because for 99.9% of the dc pulse period, the diode gets cooling time to dissipate the heat and gets cooled. In this reference, following should be noted from Fig. 1.10:

$$\text{Duty cycle} = t_{\text{on}}/T \text{ and Duty cycle}(\%) = (t_{\text{on}}/T) \times 100 \quad (1.5)$$

$$\text{PRT i.e. Pulse Repetitive Time Period} = T = t_{\text{on}} + t_{\text{off}} \quad (1.6)$$

$$\text{PRF i.e. Pulse Repetitive Frequency} = 1/T = l/(t_{\text{on}} + t_{\text{off}}) \quad (1.7)$$

$$\text{Average RF power} = \text{duty cycle} \times \text{peak RF power.} \quad (1.8)$$

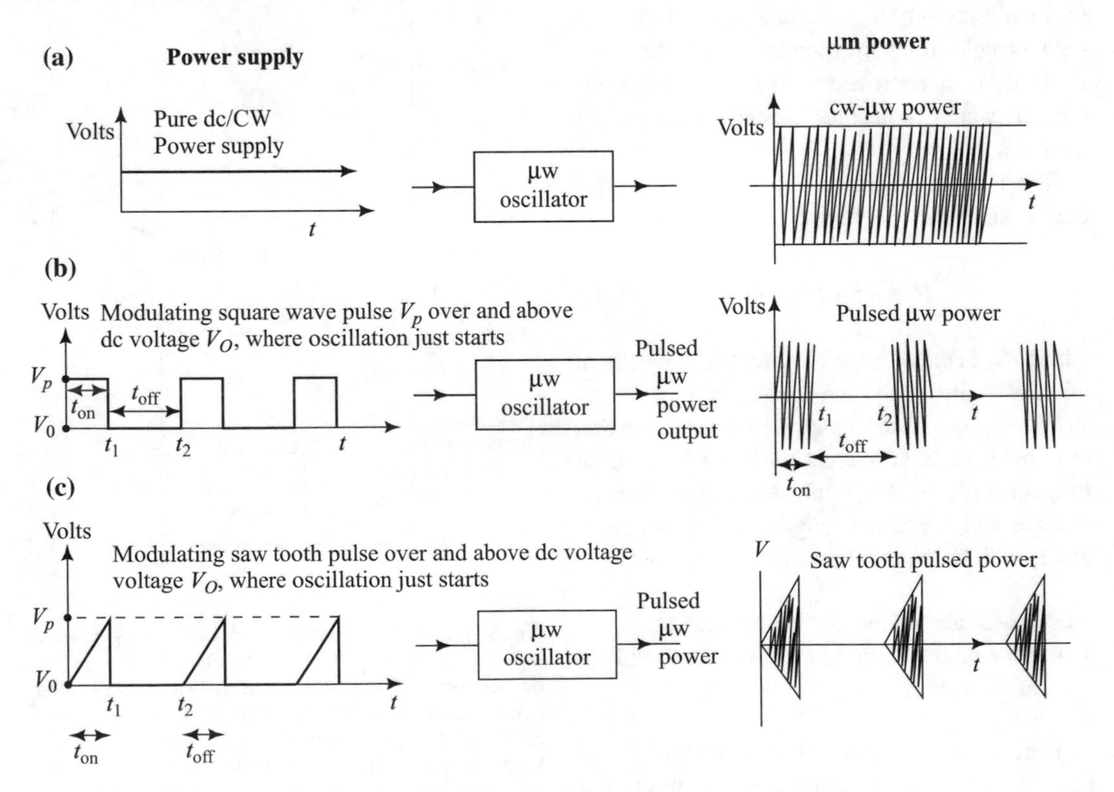

Fig. 1.10 CW and pulsed power supply and the corresponding microwave power

This t_{on} becomes the pulse width of the microwave power supplied. Some of the sources can be designed to supply both CW and pulsed power as per need. Ideally, these sources have the CW power to pulsed power ratio of 10 dB. These sources have another flexibility of being used as oscillators or as amplifiers. Inside the power supply, this pulse is superimposed on the dc supply. **In reflex klystron tube, these pulse modulations are given on the reflector voltage and not on the beam voltage.**

1.5 Decibel—A Unit to Measure Relative Power, Voltage Level, Etc.

This unit of decibel was given by Bell Laboratories, USA, in honour of Alexander Graham Bell for measuring relative values of two powers, which may be electrical, sound, etc. The decibel (abbreviated as dB), a unit 1/10th of bel, was first started in 1923, for measuring sound attenuation in telephone lines. Now it is widely used also for electronic signal power, voice, optical communication, signal-to-noise ratio measurement, etc. The basic definition for decibel for power P_0 with P_{ref} as the reference is:

$$\mathrm{dB} = 10\ \log(P_0/P_{ref}) \qquad (1.9)$$

where one bel = $\log(P_0/P_{ref})$.

The requirement of such a unit in log scale came when it was noted in sound and voice that, ten times higher power, gives a feel to the human ear of just having got doubled, e.g. from 10 to 100 W in loudspeaker our ear perceives sound as just two times loud. The lowest level of sound which a young adult can just hear (ear sensitivity) has a pressure level of 20 mPa (having power equivalent of **10^{-12} W = 1 pW/met^2) is taken as reference power P_{ref} and this was taken as zero dB, as 10log (P_{ref}/P_{ref}) = 0 dB.**

We normally talk at 60 dB level, i.e. 1 μW/m^2 power. The industrial standard for a maximum noise level inside an industry is ≤ 85 dB for 8-h shift. The nightclub high-volume music level is 100 dB, i.e. around 10 mW/m^2, beyond which our ear drum can rupture.

In electronics, we use dB for gain of amplifier and for noise figure of amplifiers/oscillators. In an audio or video system, noise figures of 60, 40, and 30 dB normally qualify it as excellent, good, and poor, respectively. In microwaves, the noise figure of 15 and 20 dB may be treated as just ok and good, respectively.

The noise factor and noise figure are defined as:

$$\text{Noise Factor } (F) = \frac{\text{signal power/noise power at input}}{\text{signal power/noise power at output}} \qquad (1.10)$$

$$\text{Noise figure} = 10\log(\mathrm{F})\ \mathrm{dB} \qquad (1.11)$$

$$= 20\ \log(\mathrm{F})\ \mathrm{dB},\ \text{if F is in voltage terms} \qquad (1.12)$$

In microwaves, the unit dB is normally used for comparing powers (P_0), voltages (V_0), electric field strength (E_0), etc., with respect to their base references

$$P_{ref}, V_{ref} \text{ and } E_{ref}$$

$$\mathrm{dB} = 10\log_{10}\left(\frac{P_0}{P_{ref}}\right), \mathrm{dB} = 20\log_{10}\left(\frac{V_0}{V_{ref}}\right), \mathrm{dB} = 20\log_{10}\left(\frac{E_0}{E_{ref}}\right) \qquad (1.13)$$

Figure 1.11 gives the conversion curve and a table between the ratios of the corresponding

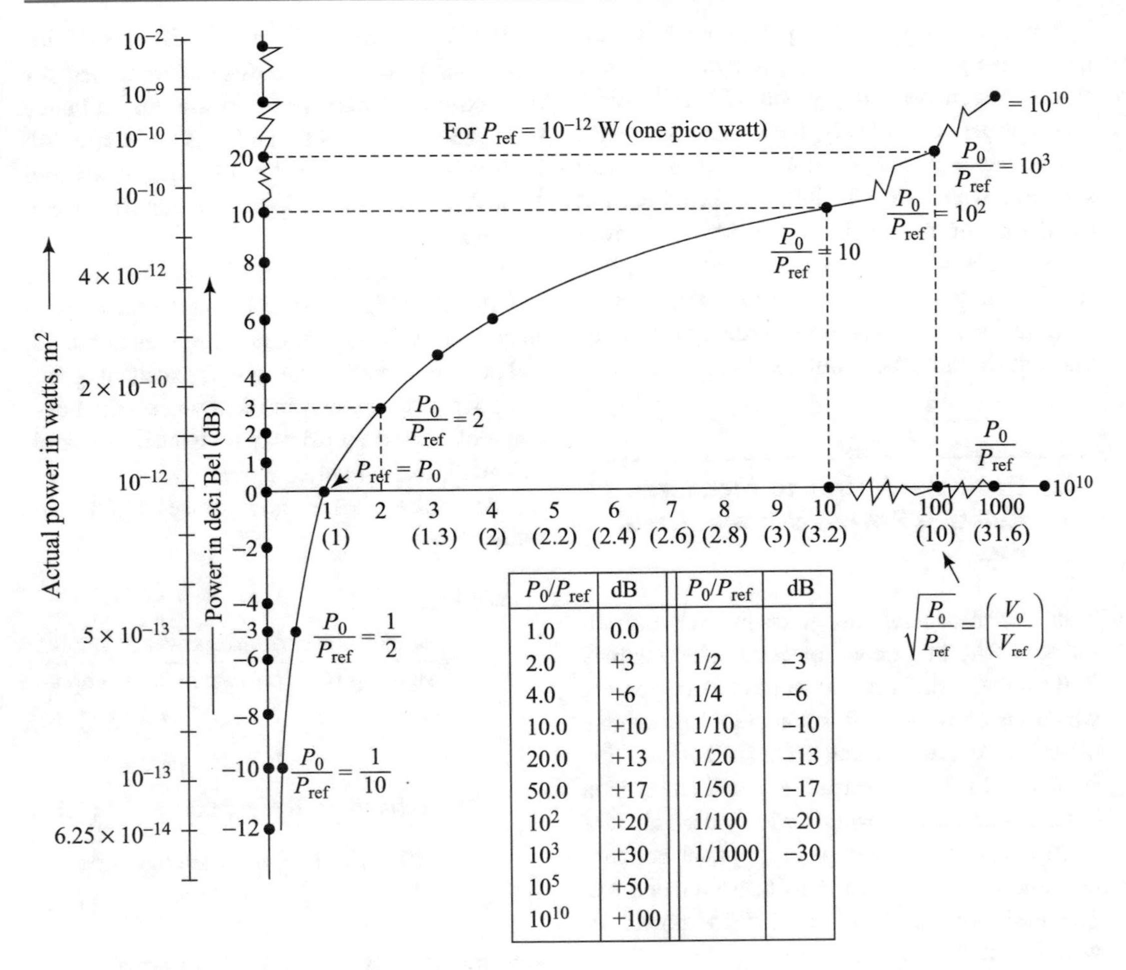

P_0/P_{ref}	dB	P_0/P_{ref}	dB
1.0	0.0		
2.0	+3	1/2	−3
4.0	+6	1/4	−6
10.0	+10	1/10	−10
20.0	+13	1/20	−13
50.0	+17	1/50	−17
10^2	+20	1/100	−20
10^3	+30	1/1000	−30
10^5	+50		
10^{10}	+100		

Fig. 1.11 Conversion graph and table between P_0/P_{ref} and corresponding power in dB and watts *y*-scale. Corresponding V_0/V_{ref} is also given

value of dB. Normally the following terms are used for different base references as given in Eq. (1.13):

(a) dB: For relative power above 1 pW ($P_{ref} = 1 \times 10^{-12}$ W)
(b) dBW: For relative power above 1 W ($P_{ref} = 1$ W)
(c) dBm: For relative power above 1 mW ($P_{ref} = 10^{-3}$ W)
(d) dBV: For relative voltage above 1 V ($V_{ref} = 1$ V)
(e) dBμV: For relative voltage above 1 μV ($V_{ref} = 10^{-6}$ V/m)
(f) dBμV/m: For relative elect. field above 1 μV/m ($E_{ref} = 10^{-6}$ V/m)

Now we can easily use the following conversion equations:

$$\text{dBW} = -30 + \text{dBm} = -60 + \text{dB}\mu\text{W} \quad (1.14)$$

$$\text{dBV} = -60 + \text{dBmV} = -120 + \text{dB}\mu\text{V} \quad (1.15)$$

1.6 Anechoic Chamber: The EM Radiation Free Area

For testing certain microwave devices like cell phones, which work in microwave frequencies, an area is required which is free from any EM signal, e.g. radio, TV, cell/satellite signal. Therefore, a large test room is made, with all its six sides/walls/floor/roof are made with anechoic (free from EM echo) material. This material has high conducting/absorption coefficient for EM waves. If the chamber lining is of conducting material and is earthed, then its inside environment is nearly a free space without any EM waves, as no EM wave can enter inside nor any wave being generated inside (while testing) will get echoed or go out.

In India, we have six Government Laboratories, in Delhi-1, Kolkata-1, Mumbai-1, Chandigarh-1, and Bengaluru-2, where this facility is there. These laboratories are under Ministry of Electronics and IT, Govt. of India. In the private sector also, there are around ten such facilities with HCL, Wipro, L & T, etc. Nowadays, vendors can give a $3' \times 3' \times 3'$ size tiny anechoic chamber also for very small test applications.

1.7 Electromagnetic Interference (EMI) and Electromagnetic Compatibility (EMC)

Because of proximity of circuit components, electromagnetic radiation of one component may induce some field or voltage into the other component, causing interference. These type of E or H fields can be reduced to a great extend, by shielding the circuits by some metal cover, which has to be grounded.

EMC is the ability of an equipment to be able to operate properly, despite EM interference due to radiations existing in the environment, e.g. satellite in space, where the solar EM radiation is not stopped by the ionosphere, etc. On earth also, EM radiations of signals radio, TV, cell phone, etc., are always there. Therefore, for designing an instrument with EMC and then testing it in an EM-free environment, anechoic chamber is used.

1.8 Radiation Hazards for Human Body/Birds, Etc.

We all are exposed to radiations of signals from radio, TV, cell phone, spark ignition of automobiles, etc., all the 24 h. The army weapons as well as explosive also produce EM waves. All these produce harmful biological effect on human being.

All the parts of body—blood, muscles, fat, bone, brain, heart—behave like conductive dielectric. As we know that microwave heating takes place due to vibration of bipolar (dipole) compounds, e.g. water, humidity. This vibration along with heat generated by μ waves damages the body parts. For example, if we expose eyes to μ waves, then blood circulatory system is unable to provide sufficient flow of blood for cooling, then it causes cataract. In addition to cancer and cataract, it can cause blood disorder, leukaemia, birth disorders, sterility in men, interference with pace maker/heart functioning, etc. It is fatal for birds also. They can sense μw power and do not come close to μw towers.

The industrial standard for safe limits of EM radiation (frequency range of 10–100 GHz) for human being are:

- 10 μW/cm^2 (i.e. 80 dB) for 24 h. (General public)
- 10 mW/cm^2 (i.e. 140 dB power) for 8-h working shift
- 100 mW/cm^2 (i.e. 150 dB power) for 2-h continuous working
- 1 W/cm^2 (i.e. 160 dB power) for 15–20 min maximum.

For working in higher radiation environments, e.g. radars, one has to wear microwave absorptive suit made out of stainless steel woven into a fire-retardant fibre. This suit produces attenuation up to 20 dB for 250–500 MHz, 20–35 dB for 0.5–1.0 GHz, 35–40 dB for 1.0–10 GHZ range. Here we may note that 20 dB attenuation means power reduction from 1 W/cm^2 to 10 MW/cm^2.

In laboratory experiments, danger is less as we use coaxial waveguides. If the input from klystron is 20 mW power in a wave guide, it will radiate only with open end and at 10 cm distance from open waveguide flange, power = 20 2.5/47π × (10)^2 = 0.04 mW/cm^2, which is within safe limits.

1.9 Application Areas of Microwaves

Long-distance line propagation through ionosphere with low attenuation is possible only by using microwaves, e.g., in:

1. **Communication**: Microwave band is widely used for communication in cell phone, telephone networks, and TV broadcast. Other important communication applications are equipment used by police, railways, defence services, etc.
2. **RADAR systems**: Microwaves are extensively used in radio detecting and ranging (RADAR) systems, capable of detecting and locating planes, ships, missiles, and other moving objects within its range. Other RADAR-type applications include air traffic control (ATC), burglar alarm, garage door openers, vehicle speed detectors by traffic police, etc.
3. **Radio astronomy**: Sensitive microwave receivers are used in radio astronomy to detect and study the electromagnetic radiation, which lie in microwave frequency region of spectrum.
4. **Remote sensing applications**: The microwave radiometers are used to map atmospheric temperature profiles, moisture conditions in soils, crops, and for other remote sensing applications.
5. **Nuclear research**: Microwaves are used in linear accelerators to produce high energy beams of charged particles for use in atomic and nuclear research.
6. **Industrial and medical application using its property of heating**: These applications are due to its heating property, as it can heat any bipolar (dipole) compound like H_2O (humidity), which is free for rotational oscillations (like a dumble) due to its electric field which is changing direction at microwave frequencies. This property is used in a number of applications.
   - **Industrial applications:** Food processing (drying potato chips, pre-cooked food, etc.), plastic, rubber industries, textile industries for drying clothes.
   - **Mining:** Breaking rocks, tunnel boring, drying, breaking of concrete blocks, curing of cement, etc.
   - **Biomedical applications:** Diathermy for superficial and deep heating in cancer treatment, lungs water detection, heart beat monitoring, etc.

Some more applications are summarised in Table 1.2.

Table 1.2 Applications of microwaves

S. no.	Applications	Frequency band (GHz)
1.	Civil and defence—television, satellite communication, RADAR, point-to-point communication, altimeter, air- and shipborne RADAR, mobile communication, cell phone	0.3–30
2.	Industry and research—food industry, textile industry for dyeing clothes, basic research, microwave spectroscopy, space research, nuclear physics, biomedical applications like monitoring of heart beat, lung water detection, body imaging	1.0–300

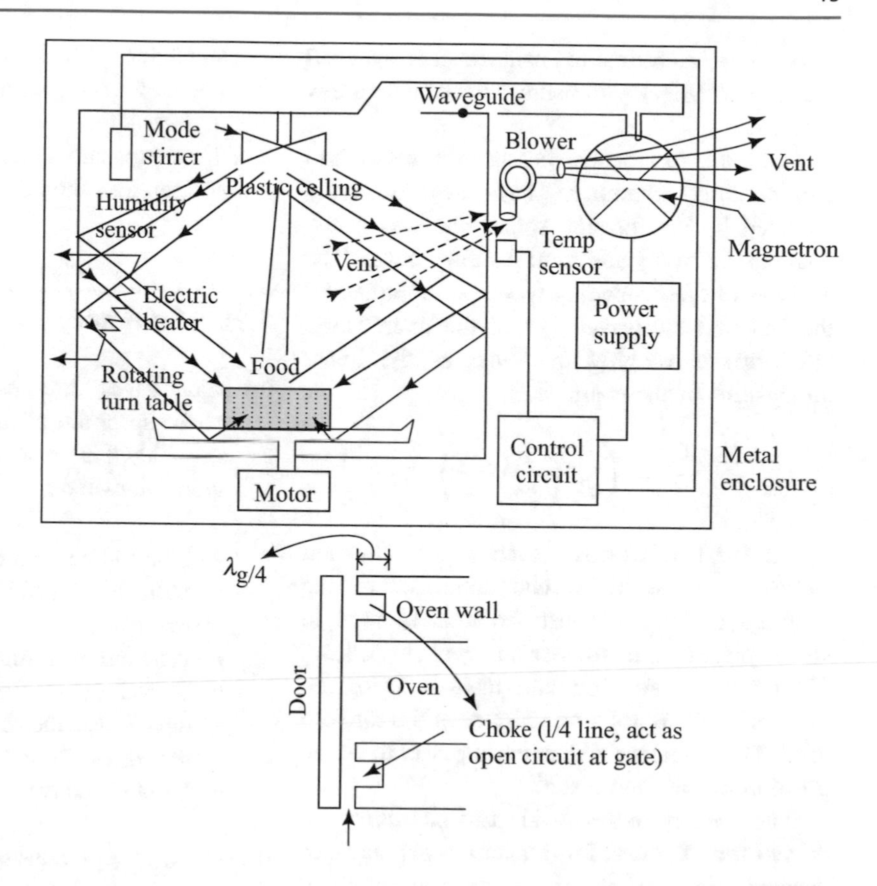

Fig. 1.12 Schematic diagram of a typical microwave oven

7. **Electronics warfare:** This is used in electronic counter measures (ECM) system, spread spectrum system (SSS), etc.
8. **Electronics devices:** A number of devices are made, e.g. switches, phase shifter, tuning element.
9. **Microwave oven:** Having studied the microwave heating mechanism in Sect. 1.3, we now study the microwave oven as application. Following international conventions, microwave ovens at home or in restaurants operate at frequencies of about 2.45 GHz, i.e. $\lambda = 12.23$ cm. The reason for this has been explained in item No. 9 of Sect. 1.3.

Figure 1.12 depicts a typical microwave oven. Microwaves are generated in a magnetron which feeds via a waveguide into the cooking chamber. Most microwaves cook food on a rotating turntable in this chamber and have a rotating reflector, acting as a stirrer. Expensive models may include thermometers, addition conventional cooking facilities such as grills, oven heaters, and even refrigeration. The size of the chamber is normally as:

$$a = L_x = 28\text{–}35 \text{ cm}, \quad b = L_y = 27.33 \text{ cm}, \quad c = L_z = 17\text{–}21 \text{ cm}.$$

Magnetrons allow either continuous or pulsed microwave generation with power up to megawatts and frequencies between 1 and 40 GHz. Efficiencies are around 80% and lifetime about 5000 h.

The magnetrons in domestic microwave ovens emit microwaves at 2.45 GHz ($\lambda = 12.2$ cm) (repeatable, each time the magnetron is switched on, to ±10 MHz) with bandwidths of only a few MHz.

Once the microwave has been coupled into the cooking chamber, they are effectively reflected by the metallic walls. The waves resonate in the cavity and form standing waves. The analysis of these standing waves is simplified by the fact that the wavelength of the microwaves (12.2 cm) is roughly the same as the linear dimensions of the chamber.

$$\frac{1}{\lambda^2} = \left(\frac{1}{a^2} + \frac{1}{b^2} + \frac{1}{c^2}\right)$$

An ideal microwave oven cooks all food evenly, but the nodes and antinodes of the standing waves can cause the food to burn in some places, but to remain cool in others. Therefore, (a) rotating turn table and (b) the mode stirrer (i.e. rotating reflector at the top) are used. They together give uniform power to all the position of the food inside.

The energy losses inside the chamber are due to four factors. First, microwaves may exit through the housing of the microwave oven. Safety regulations ensure that this contribution is negligibly small. **Second,** losses occur due to absorption in the walls; **third** due to absorption in the food in the cooking chamber (the desired mechanism), and **fourth,** there is the chance that microwaves are coupled back into the magnetron. The latter mechanism may play a role if the oven is used empty, and it should be avoided in order to ensure a long lifetime of the magnetron. Hazards in μw oven can be listed as:

1. Closed container, e.g. eggs can explode due to increased steam pressure inside.
2. Metals (including metal painted ceramic) if kept inside the oven, it will act as antenna, resulting into spark due to dielectric breakdown of air above 3 mV/m. Here air forms conducting plasma (spark) along with formation of ozone plus nitrogen oxide, and both are unhealthy.
3. Direct exposure to human body is injurious and therefore should be avoided.

1.10 Summary

(a) Microwaves are electromagnetic waves, whose spectrum of frequency ranges from 1 to 300 GHz and have the features/ applications like:

- Line-to-line communication.
- More band width possible being higher frequency.
- Penetration or transparency property.
- Directivity—As the frequency increases, directivity increases and beam width decreases. $\theta_B = 140° \times (\lambda/D)$.
- Used in industrial and commercial fields.

(b) Microwave presents special properties in generation transmission as well as in circuit design, not found at low frequencies. Another problem is that high radiation energy causes cancer, etc. It also leads to death of birds, etc. Birds can sense high radiation levels and therefore do not come close to microwave towers, etc.

Review Questions

1. Name the different frequency ranges in microwave spectrum.
2. Give the range of frequencies of radio waves, microwaves, mm waves, infrared, optical and ultraviolet waves. Are all these waves electromagnetic waves?
3. Give the application of microwaves.
4. Give the advantages of microwaves.

5. Explain the decibel (dB) used in microwave measurement of power and electric fields. What are dBm, dB μW/m? Please explain.
6. What do you mean by EM-free chamber?
7. While working with microwave sources and waveguides in the laboratory, what safety precautions should be taken as far as radiation to the body is concern?
8. How does heating takes place in a microwave oven? How rolls of clothes are dried instantly (after dyeing) by microwaves? Please explain.
9. Explain heating mechanism in microwaves.

2 Transmission Lines, Waveguides, Strip Lines, and Stub Matching by Smith Chart

Contents

© Springer Nature Singapore Pte Ltd. 2018
P. K. Chaturvedi, *Microwave, Radar & RF Engineering*,
https://doi.org/10.1007/978-981-10-7965-8_2

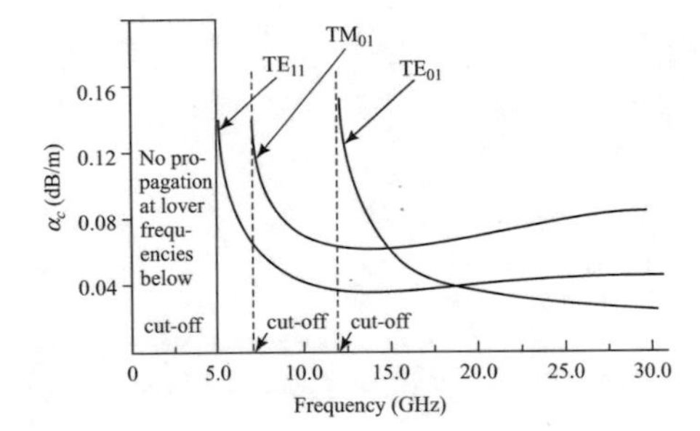

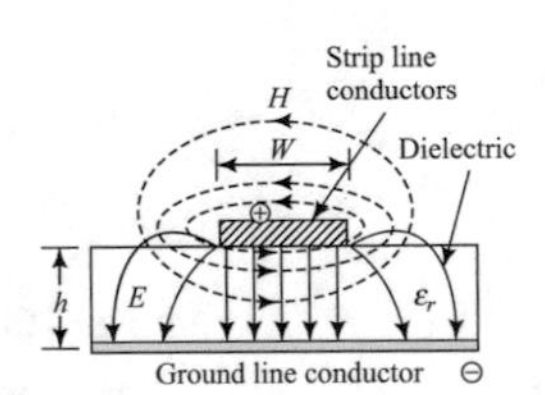

2.1 Introduction

Above 1 GHz frequency, conventional 2-wire cables and coaxial lines become very much lossy lines and very high power (mega watt range) is not possible at all. These losses are:

(a) *Dielectric loss*: Between two lines.
(b) *Copper loss*: At higher frequencies current gets more and more restricted to the surface of conductor (skin effect), increasing the current density over there and hence also the heat dissipated losses at the surface.
(c) *Radiation loss*: At higher frequencies the wire acts as antenna and energy gets transferred from line to the space itself.
(d) *Reflections*: In two-wire or coaxial lines, reflections will take place from non-matched load points or joint or uneven location, and full energy does not get transferred to the destination.

Therefore above 1 GHz frequency, waveguides (rectangles or circular hollow tube) are more useful. **These waveguides can be termed as single-wire line**. The waves travel through these hollow tubes by multiple normal reflections from surfaces just like optical waves travel in fibre-optic cables through total internal reflections (optical property). When microwaves get reflected from inner surface of the guide, then current gets induced due to magnetic field loops along the surface. This leads to power loss, which is minimised by coating the inner surface by,

(a) Highly conducting metal like silver and sometimes gold
(b) Having better/smooth surface of 25 μ polish.

If we compare waves travelling in the waveguides and in 2 wire transmission line, then it has:

(i) Dissimilarities: As given in the Table 2.1).
(ii) Similarities: As per listed below

1. **Phase velocity**: Both have phase velocity, and wave attenuates with distance.
2. **Load mismatch**: Unless the load impedance at the end of a waveguide is matched to absorb the wave power, reflections take place. Same is true in coaxial cable also for **frequency >10 MHz, when λ approaches the usable length of 2-wire line i.e. <30 m.**

Table 2.1 Comparison of waves travelling in waveguide and 2-wire transmission lines

S. no.	Property	2-Wire transmission line (coaxial cable)	Waveguide
1.	Current	Conduction current with a return path for current	Wave current by multiple reflections from inner surface
2.	Cut-off frequency (f_c)	It depends on the diameter of the wire and the dielectric between them. Losses increases with frequency and near 1 GHz; it becomes so high that it cannot be used	There are two modes TE and TM, and the lowest cut-off frequency is of TE_{10} and TM_{11} given by a general equation. It acts as high pass filter $f_{cmn} = \frac{c}{2}\sqrt{\left(\frac{m}{a}\right)^2 + \left(\frac{n}{b}\right)^2}$ $\lambda_{cmn} = \frac{2ab}{\sqrt{(mb)^2 + (na)^2}}$
3.	One line or two lines	Two-conductor line	Body acts as ground line while in the hollow region waves travel; hence, it is a one line system
4.	Propagation velocity	Same as in a cable ($v = c$) at the velocity of light	Wave velocity along the length is less than c and is given by group velocity: $v_g = c\sqrt{1 - (\lambda_0/\lambda_c)}^2$
5.	Characteristic impedance Z_0 and wave impedance	Z_0, the characteristic impedance depends upon the size of the line. Concepts of Z_0 starts from frequency $f > 10$ MHz i.e. $\lambda < 30$ m	The wave impedance (Z_w) in the guide is similar to wave impedance in free space of 377 Ω. This Z_w depends on the geometry of waveguide/mode $Z_{TM} = Z_0 \cdot \sqrt{1 - (\lambda_0/\lambda_c)^2} < Z_0$ $Z_{TE} = Z_0/\sqrt{1 - (\lambda_0/\lambda_c)^2} > Z_0$
6.	Theory for analysis	Conventional circuit theory is applicable	Field theory using Maxwell equation is applicable
7.	Reflections	Concept of reflection starts above 10 MHz (where $\lambda < 30$ m) and so does the concept of Z_0	If the load is not matched or line is shorted at the end, reflection starts and standing wave is formed. Signal bounces back to form standing waves
8.	Maximum power handling capacity	Near 1 GHz up to 500 W CW power and pulse power of 10 kW is possible	With large waveguide size (1.3″ × 2.3″ = $b \times a$), peak pulse power of 3 MW near 3 GHz is possible while at higher frequency, 20 GHz with narrow waveguide ($a \times b$ = 0.4″ × 0.2) only 50 kW peak pulse power is possible

3. **Irregularities** in a waveguide or transmission line also produce reflections.
4. **Standing waves**: When both incident and reflected waves are present, it forms standing wave pattern.

2.1.1 Polarisation of Waves: Circular or Elliptical or Linear

An EM wave is said to be polarised in the direction of the E-vector of the wave. This vector does not remain in single direction all the time while propagating. The three types of polarisation (e.g. circular, elliptical, and linear) of the wave is when the tip of the *E*-vector follows a locus (path) (in *x*–*y* plane with angular velocity *ω*, while propagating in the *z*-direction), which is a circle or ellipse or a straight line as in Fig. 2.1. This *ω* is equal to the frequency of the wave propagation in the *z*-direction), which is a cycle or ellipse or a straight line as in Fig. 2.1. This *ω* is equal to the frequency of the wave.

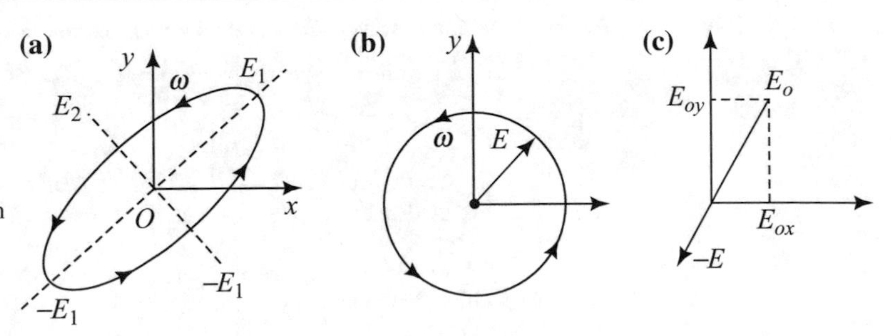

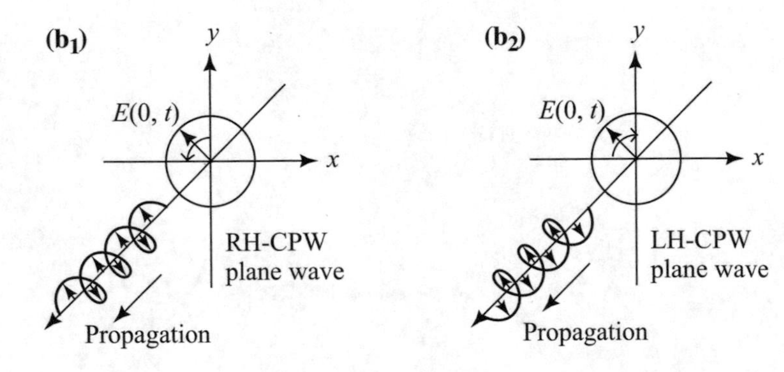

Fig. 2.1 EM wave can be polarized **a** Elliptically **b** circularly, **c** linearly according to the movement (locus) of the tip of E-vector while propagating in z-direction. The circularly can be of right hand (clockwise) or left hand (anticlockwise) (Fig. $\mathbf{b}_1$, $\mathbf{b}_2$)

Analytical representation of these polarisation

1. **Elliptical polarisation**: The TEM waves propagating in z-direction have two components; amplitude of each of them changes with time while propagating along z-direction as:

$$E_x(t) = E_{ox} \cdot \sin(\omega t) \tag{2.1a}$$

$$E_y(t) = E_{oy} \cdot \sin(\omega + \theta) \tag{2.1b}$$

with the resultant

$$\begin{aligned} E(t) &= \sqrt{E_x^2 + E_y^2} \\ &= \sqrt{E_{ox}^2 \cdot \sin^2(\omega t) + E_{oy}^2 \cdot \sin^2(\omega t + \theta)} \\ &\sqrt{a^2 + b^2} \end{aligned} \tag{2.1c}$$

i.e.

$$(E_{ox}/E)^2 \sin^2(\omega) + (E_{oy}/E)^2 \sin^2(\omega + \theta) = 1 \tag{2.1d}$$

This represents an ellipse, and the wave is called elliptically polarised wave.

2. **Circular polarisation**: When $\theta = \pi/2$ with $E_{ox} = E_{oy} = E_o$, then the above equation of ellipse becomes a circle:

$$(E_x/E_0)^2 + (E_y/E_0)^2 = 1 \tag{2.1e}$$

Thus the tip of E describes a circle of radius E_0 with angular velocity ω, while propagating, and therefore the wave is called circularly polarised wave.

The elliptical polarisation and circular polarisation can be of clockwise or anticlockwise rotation, called right-hand polarised wave (RPW) or left-hand polarised wave (LPW), respectively. For circular case corresponding abbreviations RCPW and LCPW are depicted in Fig. 2.1.

3. **Linear polarisation**: When $\theta = 0$, with $E_{ox} = E_{oy} = E_0$, then the components will be:

$E_x = E_0 \cdot \sin(\omega t)$; $E_y = E_0 \cdot \sin(\omega t)$, with resultant

$$\begin{aligned} E &= \sqrt{E_x^2 + E_y^2} \\ &= \sqrt{2} E_0 \sin(\omega t) \end{aligned}$$

Thus the tip of E oscillates in a straight line from, with frequency ω, the frequency of the wave. The wave is called horizontally or vertically polarised wave, if the plane of polarisation is horizontal or vertical with earth surface.

These two types of waves can be generated by waves emanating from dipole antenna placed vertically or horizontally as in Fig. 2.2.

4. **Plane wave and spherical waves**: All the above three types of waves when emitted from a point source will propagate in an expanding spherical surface; which becomes plane surface at infinity called plane wave.

2.2 Propagation of Waves in the Transmission Line

When a RF or microwave gets transmitting through two open conductors or transmission line, energy is lost due to radiation. This loss is reduced by transmission through guided structures like coaxial line or strip line or waveguides. There are some special features in microwaves travelling through these guided structures, which are not there at lower frequencies. These special features are constants like attenuation a (Np/m), phase-β (rad/m), propagation constant $\gamma = \alpha + j\beta$, characteristic impedance z_0, reflection coefficient ρ, voltage standing wave ratio (VSWR)-S.

At *RF* and microwave frequencies, the line parameters are distributed along the line in the direction of propagation (see Fig. 2.3). The equivalent symmetric T network of the line consists of R, L, C, and G per unit length (e.g. Ω/m, H/m, F/m, and ℧/m, respectively). The line voltage and current per unit length which are not in phase decrease as z increases and are written as:

$$\frac{dV}{dz} = -(R + j\omega L) \cdot I = -Z \cdot I \qquad (2.1f)$$

$$\frac{dI}{dz} = -(G + j\omega c) \cdot V = -Y \cdot V \qquad (2.1g)$$

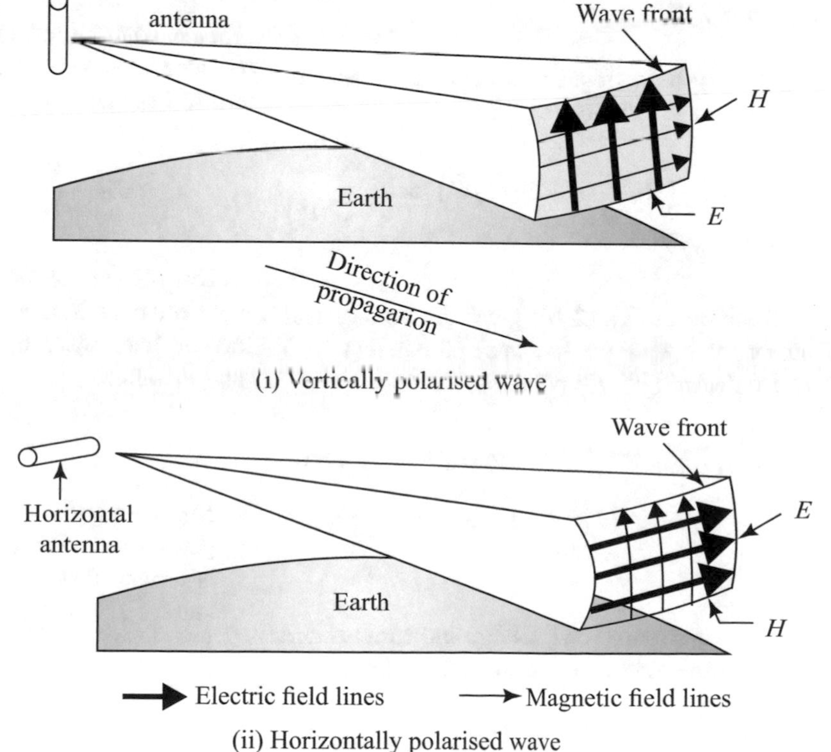

Fig. 2.2 Generation of **i** vertical polarised wave and **ii** horizontal polarised wave

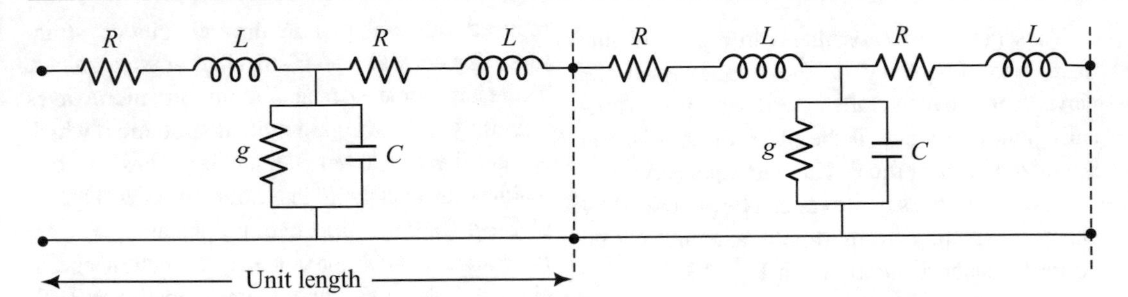

Fig. 2.3 Distributed impedance in a waveguide

Here

$$Z = (R + j\omega L) \text{ and } Y = (G + j\omega C) \quad (2.1h)$$

are the series impedance and shunt admittance per unit length of the line. Eliminating I and V in above two equations by inter-substitution, we get

$$\frac{d^2V}{dz^2} = \gamma^2 V \quad \text{and} \quad \frac{d^2I}{dz^2} = \gamma^2 I \quad (2.1i)$$

where

$$\gamma = \sqrt{Z \cdot Y} = \sqrt{(R + j\omega L)(G + j\omega C)} = \alpha + j\beta \quad (2.1j)$$

At very high frequency, where the losses in the line are small, above gives:

$$\alpha = \left(\frac{1}{2}\right)\left(R\sqrt{C/L} + G\sqrt{L/C}\right) \text{ and } \beta = \omega\sqrt{LC} \quad (2.1k)$$

Solution of Eq. (2.1d) gives the voltage and current at any point z, in terms of forward (V^+, I^+) and reflected (V^-, I^-) parts as

$$V(z) = V^+ \cdot e^{-\gamma z} + V^- \cdot e^{\gamma z} \quad (2.1l)$$

$$I(z) = I^+ \cdot e^{-\gamma z} + I^- \cdot e^{\gamma z} \quad (2.1m)$$

where

V^+, I^+ are source-end voltage and currents and
V^-, I^- the load-end voltage and currents.

Using above two equations, we now compute the special feature constants.

1. **Characteristic impedances**: Dividing the two Eqs. (2.1f) and (2.1g), we get

$$\frac{dV}{dI} = \left(\frac{R + j\omega L}{G + j\omega C}\right) \cdot \frac{I}{V}$$

$$\therefore \quad V dV = \left(\frac{R + j\omega L}{G + j\omega C}\right) \cdot I \cdot dI$$

A simple integration gives

$$\frac{V^2}{2} = \frac{R + j\omega L}{G + j\omega C} \cdot \frac{I^2}{2} + K$$

This constant of integration = $K = 0$ when reflected wave = 0; then, we get Z_0 the characteristic impedance of the line as:

$$Z_0 = \frac{V(z)}{I(z)} = \sqrt{\frac{R + j\omega L}{G + j\omega C}}$$

This Z_0 can be taken as impedance of the input end of an infinite lossless line. For low-loss lines at microwave frequencies $R \ll \omega L$, $G \ll$ and therefore,

$$Z_0 = \sqrt{LC}$$

2. **Input impedance**: The ratio $V(z)/I(z)$ in general is the input impedance of the line. As we know that the reflected voltage V^- and current I^- are out of phase, one can write $I^+ = V^+/Z_0$ and $I^- = -V^-/Z_0$; therefore, using Eqs. (2.1l) and (2.1m) we get

$$Z_{\text{in}} = \frac{(V^+.e^{-\gamma z} + V^-.e^{\gamma z})}{(V^+.e^{-\gamma z} - V^-.e^{-\gamma z})/Z_0} \quad (2.1n)$$

As

$$\begin{aligned}\Gamma &= \text{Reflection/Forward Signal} \\ &= V^-.e^{\gamma z}/V^+.e^{-\gamma z}\end{aligned}$$

the above equation becomes

$$\begin{aligned}Z_{\text{in}} &= Z_0 \frac{e^{-\gamma z} + [(Z_L - Z_0)/(Z_L + Z_0)]e^{\gamma z}}{e^{-\gamma z} - [(Z_L - Z_0)/(Z_L + Z_0)] \cdot e^{\gamma z}} \\ &= Z_0 \cdot \left[\frac{Z_L(e^{-\gamma z} + e^{\gamma z}) + Z_0(e^{\gamma z} - e^{-\gamma z})}{Z_0(e^{-\gamma z} + e^{\gamma z}) + Z_L(e^{\gamma z} - e^{-\gamma z})}\right] \\ \therefore \quad Z_{\text{in}} &= Z_0 \cdot \frac{Z_L \cos h(\gamma l) + Z_0 \sin h(\gamma l)}{Z_0 \cos h(\gamma l) + Z_L \sin h(\gamma l)}\end{aligned} \quad (2.1o)$$

For lossless line $\alpha = 0$, $\gamma = j\ \beta$ therefore:

For short end $Z_L = 0$ $\quad \therefore Z_{\text{in}\,SC} = Z_0 \tan h(\gamma l)$
$$= jZ_0 \tan(\beta l) \quad (2.1p)$$

For open end $Z_L = \infty$ $\quad \therefore Z_{\text{in}\,OC} = Z_0 \cot h(\gamma l)$
$$= -jZ_0 \cot(\beta l) \quad (2.1q)$$

3. **Voltage standing wave ratio (VSWR) and reflection coefficient**: The ratio of maximum voltage to minimum voltage is called VSWR (S).

$$\begin{aligned}S &= V_{\max}/V_{\min} = (V^+ + V^-)/(V^+ - V^-) \\ &= (1 + V^-/V^+)/(1 - V^-/V^+) \\ &= (1 + |\Gamma|)/(1 - |\Gamma|) \\ \therefore \quad S &= (1 + |\Gamma|)/(1 - |\Gamma|)\end{aligned} \quad (2.1r)$$

At the point of $V_{\max}$, the reflected and the incident voltages add together being in phase, while at $V_{\max}$ they substract being out of phase. The ratio of forward voltage to reflected voltage is called reflection coefficient.

$$\begin{aligned}\Gamma &= (V^- \cdot e^{\gamma}e)(V^+ \cdot e^{-\gamma l}) \\ &= (V^-/V^+)e^{2\gamma l} \\ &= |\Gamma| \cdot e^{2\gamma l}\end{aligned} \quad (2.1s)$$

Therefore Eq. (2.1n) can be rewritten as

$$Z_{\text{in}} = Z_0 \left(\frac{1+\Gamma}{1-\Gamma}\right)$$

At the load point also this is true

$$\begin{aligned}\therefore \quad Z_{\text{in}} &= Z_L = Z_0 \left(\frac{1+\Gamma}{1-\Gamma}\right) \\ \therefore \quad \Gamma &= \frac{Z_L - Z_0}{Z_L + Z_0}\end{aligned} \quad (2.1t)$$

Therefore as per Fig. 2.4, following can be seen.

At $Z_L = 0$ (short); $|\Gamma| = 1$ and $\Gamma = -1$ i.e. V^+ and V^- are out of phase
At $Z_L = \infty$ (open); $|\Gamma| = 1$ and $\Gamma = +1$ i.e. V^+ and V^- are in phase
At $Z_L = R_L + jX_L$, $\Gamma = \Gamma_r + j\ \Gamma_i$ = complex term

4. **Impedance at $V_{\min}$ and $V_{\max}$ points**: The input impedance is purely real at the two points where voltage is minimum or maximum, with current as maximum and minimum, respectively.

At $V_{\min}$:

$$Z_{\text{in}} = Z_{\min} = V_{\min}/I_{\max} = Z_0 \cdot \left[\frac{1-|\Gamma|}{1+|\Gamma|}\right] = \frac{Z_0}{S} \quad (2.1u)$$

At $V_{\max}$:

$$\begin{aligned}Z_{\text{in}} &= Z_{\max} = V_{\max}/I_{\min} \\ &= Z_0 \cdot \left[\frac{1+\Gamma}{1-|\Gamma|}\right] = Z_0 \cdot S\end{aligned} \quad (2.1v)$$

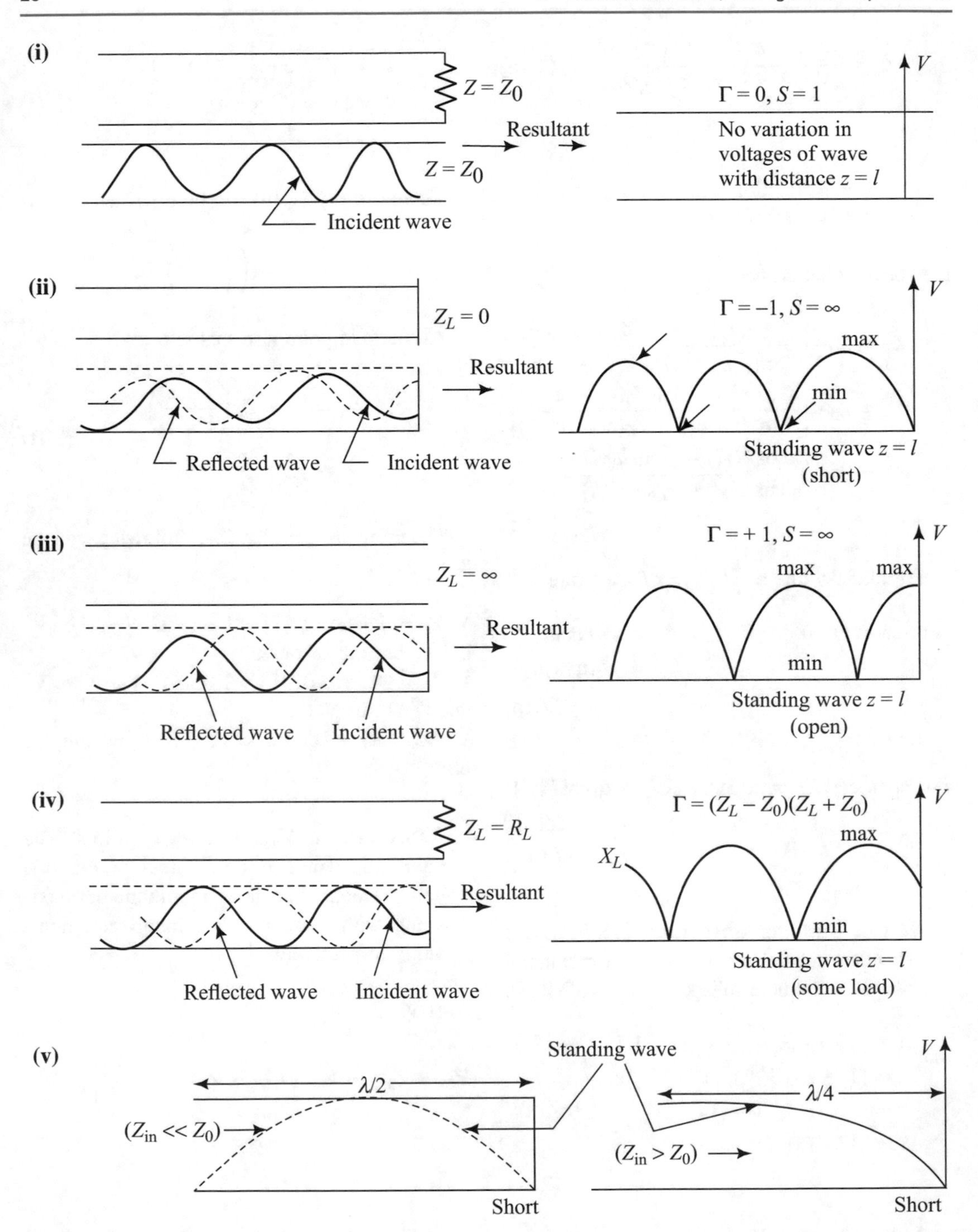

Fig. 2.4 Incident wave, reflected wave, and standing wave with **i** matched load ($\Gamma = 0$), VSWR = 1; **ii** short circuit load, $\Gamma = -1$, $\psi = \pi$, VSWR = ∞; **iii** open circuit load, $\Gamma = +1$, $\psi = 0$, VSWR = ∞; **iv** regular load, $Z_L = R_L + jX_L$; $\Gamma = (Z_L - Z_0)\ (Z_L + Z_0)$; ψ, VSWR = some value; **v** Z_{in} for shorted $\lambda/2$ line and $\lambda/4$ line sections

5. **Losses due to mismatch in transmission lines**: If a transmission line is not terminated by a matched load, there is bound to be a mismatch and reflection of wave between the input termination and output termination of a lossy transmission line. There are various losses like attenuation loss, reflection loss, transmission loss, return loss, and insertion loss due to this mismatch.
These losses have been discussed in Sects. 7.4.5 and 2.7.1 in detail as well.
6. **Short stub termination for matching**: The reflection and hence the amplitude of the reflection wave (V^-) can be reduced to zero by putting a matched load as termination, which should be equal to Z_0, so that VSWR $\rightarrow$ 1 and $\Gamma \rightarrow 0$. Thus losses are minimum, and whole of the microwave power is delivered to the line. Similarly methods for impedance matching is by using short at the end. The end line length has to be varied between $\lambda/4$ and $\lambda/2$, by moving the short stub, and it acquires all the values of reactance required for matching with the line impedance Z_0. This is possible because:

$$\text{at } l = \lambda/2,\ Z_{\text{in}} = Z_L < Z_0 \text{ (minimum value)} \tag{2.2}$$

and at $l = \lambda/4$,

$$Z_{\text{in}} = Z_0^2/Z_L > Z_0 \text{ (maximum value)} \tag{2.3}$$

where Z_L is the effective impedance of the shorted line.

Thus a waveguide short stub termination can be moved between $\lambda/2$ and $\lambda/4$ to get minimum VSWR, i.e. best matching, so that Z_{in} becomes = Z_0 (Fig. 2.4v).

7. **Shunt stub for matching**: Normally the load end is at a fixed distance, and the above method will not work. Therefore single/double stub are used for impedance matching which is given in detail in Sect. 2.7.

2.3 Waveguides: Circular and Rectangular

At microwave frequencies the best media of transmission is waveguides, which are of circular and rectangle types. In general the waves which travel in the waveguides with multiple reflections (Fig. 2.5) are either of the following modes (Fig. 2.6):

TE mode: Transverse electric field mode where $E_z = 0$, $H_z \neq 0$
TM mode: Transverse magnetic field mode where $E_z \neq 0$, $H_z = 0$
Thus for TE mode, electric field along the direction of propagation is not there (i.e. $E_z = 0$) and for TM mode magnetic field along the z-direction is not there ($H_z = 0$).
TEM mode: TEM mode is not possible in waveguides and is only possible in coaxial cable and in open space (see Sect. 2.4.2).

2.4 Propagation of Waves in Rectangular Waveguide

As the wave travel in the guide by multiple reflections from inner surface, the total travel length increases (Fig. 2.5) and therefore effective velocity called group velocity (v_g) is less than 'c',

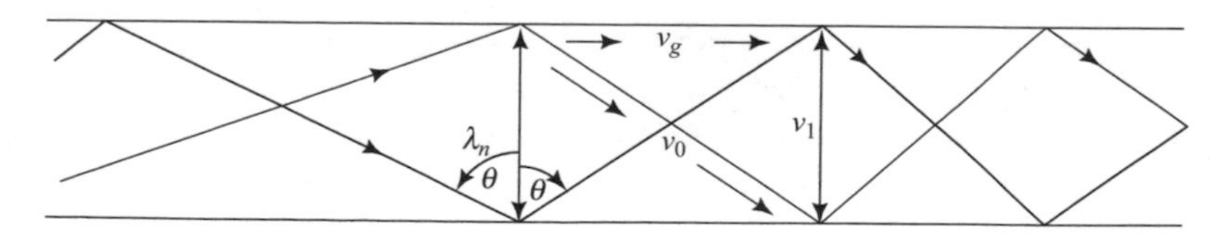

Fig. 2.5 Wave travelling with multiple reflections inside a waveguide

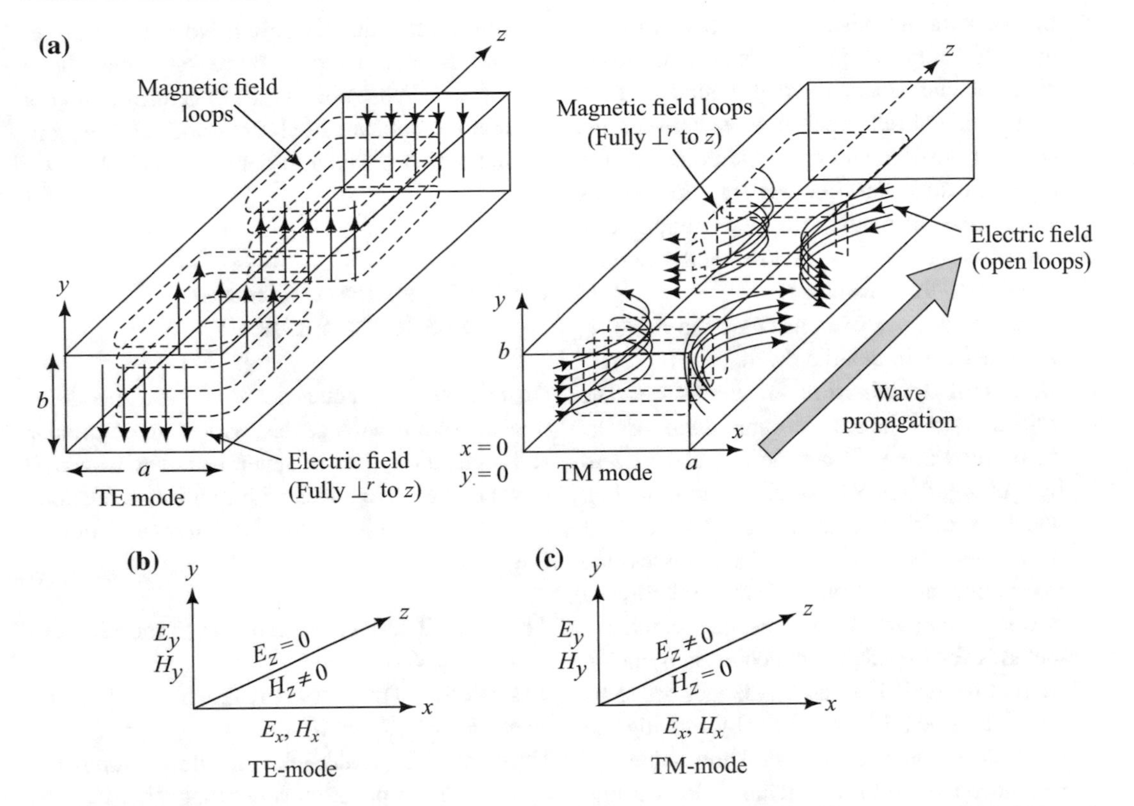

Fig. 2.6 **a** E–H-fields in the TE and TM modes. In TE mode wave, E-field is fully ⊥ʳ to z and not H. In TM mode wave, H-field is fully ⊥ʳ to z and not E. **b** Wave propagation in a waveguide along z

the velocity of light. Also the plane wave in the guide can be resolved into two components.

(i) Standing wave (λ_n) in the direction normal to the reflecting surface which decides the longest wave possible (i.e. cut-off freq.)
(ii) Travelling wave (λ_p) parallel to the reflecting surface. This also means that we have the corresponding electric and magnetic field components also along these two directions. Therefore we define group velocity as $v_g = v_0 \sin\theta$; θ being the angle of reflection of the wave in the waveguides (Fig. 2.5).

For a wave travelling in the waveguide, boundary conditions applied (Fig. 2.6b) **with magnetic field lines as closed loops and electric field lines as open loops are given below**:

(a) The electric field lines are never parallel to the surfaces, but are line starting from one point of the conducting surface and terminating at another point of the same surface/opposite surface as open loops.
(b) Electric field goes through the *H*-loops.
(c) Thus electric field has to be zero at the surface of the waveguide i.e.

$$E(x) = 0 \text{ at } x = 0, a$$
$$E(y) = 0 \text{ at } y = 0, b$$

(d) Magnetic field is always parallel to the surface of the conductor and is in closed loops.
(e) Electric and magnetic fields have to be perpendicular to each other (Fig. 2.7).
(f) Static charge as well as conduction current (*J*) along the surface of the waveguide is zero for a lossless (R = 0) line.

Based on the above boundary value conditions the analysis of wave propagation in the waveguide is done by the following four Maxwell's equations on time and space varying electromagnetic fields:

$$\nabla \cdot D_1 = q_1 \quad (2.4)$$

$$\nabla \cdot B_1 = 0 \quad (2.5)$$

$$\nabla \times E_1 = -\partial B_1/\partial t \quad (2.6)$$

$$\nabla \times H_1 = \partial D_1/\partial t + J \quad (2.7)$$

where

D_1 = $\varepsilon\, E_1$ = electric flux density
B_1 = $\mu\, H_1$ = magnetic flux density
J = $\sigma\, E_1$ = conduction current density
σ_1 = conductivity and
q_1 = Charge density

Here we note that time variation of B_1 leads to space variation in E_1 [Eq. (2.6) and vice versa (Eq. (2.7)].

σ_1, E_1, and H_1 vary with time with σ, E, and H as the peak value as:

$$\sigma_1 = \sigma \cdot e^{j\omega t},\ E_1 = E \cdot e^{j\omega t} \text{ and}$$
$$H_1 = H \cdot e^{j\omega t}$$

Conduction current inside the waveguide $J = 0$.

Substituting these time-varying conditions, the above wave equation will have E and H variables as functions of x, y, and z only:

$$\nabla \times \boldsymbol{H} = j\omega\varepsilon \boldsymbol{E} \text{ (1st Maxwell's Eq.)} \quad (2.8)$$

$$\nabla \times \boldsymbol{E} = -j\omega\varepsilon \mathbf{H} \text{ (2st Maxwell's Eq.)} \quad (2.9)$$

$$\nabla \cdot \boldsymbol{E} = \boldsymbol{0} \quad (2.10)$$

$$\nabla \cdot \boldsymbol{H} = \boldsymbol{0} \quad (2.11)$$

By taking curl Eq. (2.9), we get:

$$\begin{aligned} \nabla \times (\nabla \times E) &= +\nabla \times (-j\omega\mu H) \\ &= -(\nabla \times H) j\omega\mu \\ \therefore\ (\nabla \times \nabla \times E) &= -(+j\omega\mu E)(j\omega\mu) \\ &= -(\omega^2 \cdot \mu\varepsilon)E \end{aligned}$$

α = attenuation constant (Np/m)
β = phase constant,
i.e., phase shift(rad/m)

As per vector formula

$$\begin{aligned} \nabla \times (\nabla \times E) &= \nabla \cdot (\nabla \cdot E) - \nabla^2 E \\ &= -\nabla^2 E \end{aligned}$$

As

$$\nabla \cdot E = q = 0 \,(\text{charge} = 0);$$
$$\Delta \cdot H = 0$$

$$\therefore\ -\nabla^2 \boldsymbol{E} = -\omega^2 \varepsilon\mu \boldsymbol{E} \text{ (Helmholtz Wave Eq.)} \quad (2.12)$$

Similarly

$$\nabla^2 \boldsymbol{H} = -\omega^2 \varepsilon\mu \boldsymbol{H} \text{ (Helmholtz Wave Eq.)} \quad (2.13)$$

These two equations are also called Maxwell's modified equations.

Now for getting the components E_x, E_y, Ez and H_x, H_y, H_z we use the Maxwell's Eqs. (2.8) and (2.9) directly. $\nabla \times H$, in the first Maxwell's equation we get

$$\text{i.e.,} \begin{vmatrix} \hat{i} & \hat{j} & \hat{k} \\ \frac{\partial}{\partial x} & \frac{\partial}{\partial y} & \frac{\partial}{\partial z} \\ H_x & H_y & H_z \end{vmatrix} = j\omega\varepsilon\left[\hat{i}E_x + \hat{j}E_y + \hat{k}E_z\right]$$

($\hat{i}$, $\hat{j}$, $\hat{k}$ are direction vectors)

As the wave is travelling along z-direction and varying as $e^{-\gamma_z}$, γ being the propagation constant = $\alpha + j\ \beta$. Therefore we can replace the operator $\left(\frac{\partial}{\partial z}\right)$ by $-\gamma\left(\text{as}\,\frac{\partial E}{\partial z} = -\gamma E \text{ and } \frac{\partial H}{\partial z} = -\gamma H\right)$ as all E and H vary as $e^{-\gamma_z}$ in line with V and I see Eqs. 2.1p and 2.1q.

$$\begin{vmatrix} \hat{i} & \hat{j} & \hat{k} \\ \frac{\partial}{\partial x} & \frac{\partial}{\partial y} & -\gamma \\ H_x & H_y & H_z \end{vmatrix} = j\omega\varepsilon\left[\hat{i}E_x + \hat{j}E_y + \hat{k}E_z\right]$$

Equating coefficients of $\hat{i}$, $\hat{j}$ and $\hat{k}$ (after expanding) we get

$$\frac{\partial H_z}{\partial y} + \gamma H_y = +j\omega\varepsilon E_x \tag{2.14}$$

$$\frac{\partial H_z}{\partial x} + \gamma H_x = -j\omega\varepsilon E_y \tag{2.15}$$

$$\frac{\partial H_y}{\partial x} - \frac{\partial H_x}{\partial y} = +j\omega\varepsilon E_z \tag{2.16}$$

Now expanding $\nabla \times E$, in the second Maxwell's equation we get

$$\text{i.e.,} \begin{vmatrix} \hat{i} & \hat{j} & \hat{k} \\ \frac{\partial}{\partial x} & \frac{\partial}{\partial y} & \frac{\partial}{\partial z} \\ E_x & E_y & E_z \end{vmatrix} = \begin{vmatrix} \hat{i} & \hat{j} & \hat{k} \\ \frac{\partial}{\partial x} & \frac{\partial}{\partial y} & -\gamma \\ E_x & E_y & E_z \end{vmatrix} = -j\omega\mu\left[\hat{i}H_x + \hat{j}H_y + \hat{k}H_z\right]$$

Expanding the LHS and equating coefficients of $\hat{i}$, $\hat{j}$ and $\hat{k}$ we get

$$\frac{\partial E_z}{\partial y} + \gamma E_y = -j\omega\mu H_x \tag{2.17}$$

$$\frac{\partial E_z}{\partial x} + \gamma E_x = +j\omega\mu H_y \tag{2.18}$$

$$\frac{\partial E_y}{\partial x} - \frac{\partial E_x}{\partial y} = -j\omega\mu H_z \tag{2.19}$$

Use Eqs. (2.14) and (2.17) to eliminate H_y, for getting an expression for E_x as follow.

$$H_y = \frac{1}{j\omega\mu}\frac{\partial E_z}{\partial x} + \frac{\gamma}{j\omega\mu}E_x$$

Substituting this H_y in Eq. (2.14), we get

$$\frac{\partial H_z}{\partial y} + \frac{\gamma}{j\omega\mu}\frac{\partial E_z}{\partial x} + \frac{\gamma^2}{j\omega\mu}E_x = j\omega\varepsilon E_x$$

or

$$E_x\left[j\omega\varepsilon - \frac{\gamma^2}{j\omega\mu}\right] = \frac{\gamma}{j\omega\mu}\frac{\partial E_z}{\partial x} + \frac{\partial H_z}{\partial y}$$

Multiplying by $j\ \omega\mu$, we get

$$E_x\left[-\omega^2\mu\varepsilon - \gamma^2\right] = \gamma\frac{\partial E_z}{\partial x} + j\omega\mu\frac{\partial H_z}{\partial y}$$

or

$$E_x\left[-\left(\gamma^2 + \omega^2\mu\varepsilon\right)\right] = \gamma\frac{\partial E_z}{\partial x} + j\omega\mu\frac{\partial H_z}{dy}$$

where $\gamma^2 + \omega^2\mu\varepsilon = h^2$ (Let another constant).

Dividing by h^2 throughout, we get general equation true for both TE and TM modes:

$$\boldsymbol{E_x = \frac{-\gamma}{h^2}\frac{\partial E_z}{\partial x} - \frac{j\omega\mu}{h^2}\frac{\partial H_z}{\partial y}} \tag{2.20}$$

$$\boldsymbol{E_y = \frac{-\gamma}{h^2}\frac{\partial E_z}{\partial y} - \frac{j\omega\mu}{h^2}\frac{\partial H_z}{\partial x}} \tag{2.21}$$

Similarly

$$\boldsymbol{H_x = \frac{-\gamma}{h^2}\frac{\partial H_z}{\partial x} + \frac{j\omega\mu}{h^2}\frac{\partial E_z}{\partial y}} \tag{2.22}$$

and

$$\boldsymbol{H_y = \frac{-\gamma}{h^2}\frac{\partial H_z}{\partial y} - \frac{j\omega\varepsilon}{h^2}\frac{\partial E_z}{\partial x}} \tag{2.23}$$

These equations give a general relationship for field components within a waveguide (TE and TM). Using E, H field equations for TE and TM modes will be calculated later with different conditions.

As in waveguides TE and TM modes are only possible, the two Eqs. (2.12) and (2.13) become new Helmholtz wave equation.

$$\nabla^2 E_z = (-\omega^2 \mu\varepsilon)E_z (\text{For TM Mode } H_z = 0) \quad (2.24)$$

$$\nabla^2 H_z = (-\omega^2 \mu\varepsilon)H_z (\text{For TE Mode } E_z = 0) \quad (2.25)$$

Expanding these two equations we get:

$$\frac{\partial^2 E_z}{\partial x^2} + \frac{\partial^2 E_z}{\partial y^2} + \frac{\partial^2 E_z}{\partial z^2} = (-\omega^2 \mu\varepsilon)E_z \quad (2.26)$$

(for TM Mode)

$$\frac{\partial^2 H_z}{\partial x^2} + \frac{\partial^2 H_z}{\partial y^2} + \frac{\partial^2 H_z}{\partial z^2} = (-\omega^2 \mu\varepsilon)H_z \quad (2.27)$$

(for TE Mode)

As the wave propagation in the z-direction only, where its fields E_z, H_z vary as $e^{-\gamma_z}$, the differential operator $\partial/\partial z$ on field will bring—γ term out of it and $\partial^2/\partial z^2$ will bring γ^2 out of it. Therefore we can replace operators $\partial/\partial z$ by—γ and $(\partial^2/\partial z^2)$ by γ^2 to get from Eqs. (2.26) and (2.27) as:

$$\therefore \quad \frac{\partial^2 \boldsymbol{E}_z}{\partial \boldsymbol{x}^2} + \frac{\partial^2 \boldsymbol{E}_z}{\partial \boldsymbol{y}^2} + (\boldsymbol{h}^2)\boldsymbol{E}_z = 0$$

(**To be used for TM wave where** $H_Z = 0$) (2.28)

$$\therefore \quad \frac{\partial^2 \boldsymbol{H}_z}{\partial \boldsymbol{x}^2} + \frac{\partial^2 \boldsymbol{H}_z}{\partial \boldsymbol{y}^2} + \boldsymbol{h}^2\boldsymbol{H}_z = 0$$

(**To be used for TM wave where** $E_Z = 0$) (2.29)

where

$$h^2 = (\gamma^2 + \omega^2 \mu\varepsilon) \quad (2.30)$$

2.4.1 TE Waves in Rectangular Waveguides Electrical Field and Magnetic Field Equations

For TE waves we will use the Eq. (2.29) i.e.

$$\frac{\partial^2 H_z}{\partial x^2} + \frac{\partial^2 H_z}{\partial y^2} + h^2 H_z = 0, \quad (2.31)$$
$$\text{where } h^2 = \gamma^2 + \omega^2 \mu E$$

This is a partial differential equation, which can be solved for the EM field components E_x, E_y, H_x, H_y, and H_z by separable variables method, by assuming.

$H_z = P \cdot Q$ [A product of two pure functions of x and y, respectively i.e. $P(x)$ and $Q(y)$]. Putting $H_z = P \cdot Q$ in Eq. (2.29), we get

$$Q \cdot \mathrm{d}^2 P/\mathrm{d}x^2 + P \cdot \mathrm{d}^2 Q/\mathrm{d}y^2 + h^2 PQ = 0$$
$$\therefore \quad (1/P)(\mathrm{d}^2 P/\mathrm{d}x^2) + (1/Q)(\mathrm{d}^2 Q/\mathrm{d}y^2) + h^2 = 0$$

As $(1/P)$ $(\mathrm{d}^2P/\mathrm{d}x^2)$ + $(1/Q)$ $(\mathrm{d}^2Q/\mathrm{d}y^2)$ are pure functions of x, y but sum of them is constant as h^2 is a constant. Therefore each of them also has to be separately constant say $-A^2$ and $-B^2$, so that

$$(1/P)(\mathrm{d}^2 P/\mathrm{d}x^2) = -A^2 \quad (2.32)$$

$$(1/Q)(\mathrm{d}^2 Q/\mathrm{d}y^2) = -B^2 \quad (2.33)$$

$$\therefore \quad (-A^2) + (-B^2) + h^2 = 0 \quad (2.34)$$

Here A and B will be seen later to be phase constants per unit length.

Equations (2.32) and (2.33) are second-order differential equations, and therefore solution for P and Q will be of the type as given below, with C_1, C_2, C_3, and C_4 yet other constants:

$$P = C_1 \cos(Bx) + C_2 \sin(Bx) \quad (2.35)$$

$$Q = C_3 \cos(Ay) + C_4 \sin(Ay) \quad (2.36)$$

$\therefore$

$$\begin{aligned} Hz &= P \cdot Q \\ &= [C_1 \cos(Bx) + C_2 \sin(Bx)] \\ &\quad \cdot [C_3 \cos(Ay) + C_4 \sin(Ay)] \end{aligned} \quad (2.37)$$

Now Eq. (2.14) with $E_z = 0$ becomes simple: i.e.

$$E_x = (-j\omega\mu h^2)\partial H_z/\partial y \quad (2.38)$$

Now by putting H_z from Eq. (2.37) we get:

$$\begin{aligned} E_x = (-j\omega\mu h^2)&[C_1 \cos(Bx) + C_2 \sin(Bx)] \\ &\cdot [-AC_3 \sin(Ay) + AC_4 \cos(Ay)] \end{aligned} \quad (2.39)$$

For getting the values of C_1, C_2, C_3, and C_4 boundary conditions are used.

For the TE mode, fields along the four sides of the waveguide are equal to zero but exist in between and also: $E_z = 0$, along the direction of $z = 0$ to ∞ (i.e. direction of propagation).

Now apply four boundary conditions in the sequence (i) (ii) (iii) (iv) as in Fig. 2.8 **on the TE mode Eq.** (2.38) **of** E**, along with general Eqs.** (2.20)–(2.23) **for getting** E_y, H_x, H, **and y** H_z.

(i) **Bottom surface condition ($E_x = 0$ at $y = 0$)**

Applying $E_x = 0$ at $y = 0$ in Eq. (2.38), we get

$$0 = (j\omega\mu h^2)[C_1 \cos(Bx) + C_2 \sin(Bx)](0 + AC_4)$$

$$\therefore \; A \cdot C_4[C_1 \cos(Bx) + C_2 \sin(Bx)] = 0$$

$$\therefore \; C_4 = 0 \; \{\text{as } A \neq 0 \text{ and } [C_1 \cos(Bx) + C_2 \sin(Bx)] = P \neq 0\}$$

$\therefore$ Equations (2.36) and (2.38) becomes:

$$H_z = [C_1 \cos(Bx) + C_2 \sin(Bx)] \cdot C_3 \cos(Ay)$$

$$E_x = (-j\omega\mu/h^2) [C_1 \cos(Bx) + C_2 \sin(Bx)] \quad [-AC_3 \sin(Ay)]$$

(ii) **Apply, left side wall condition ($E_y = 0$ at $x = 0$)**

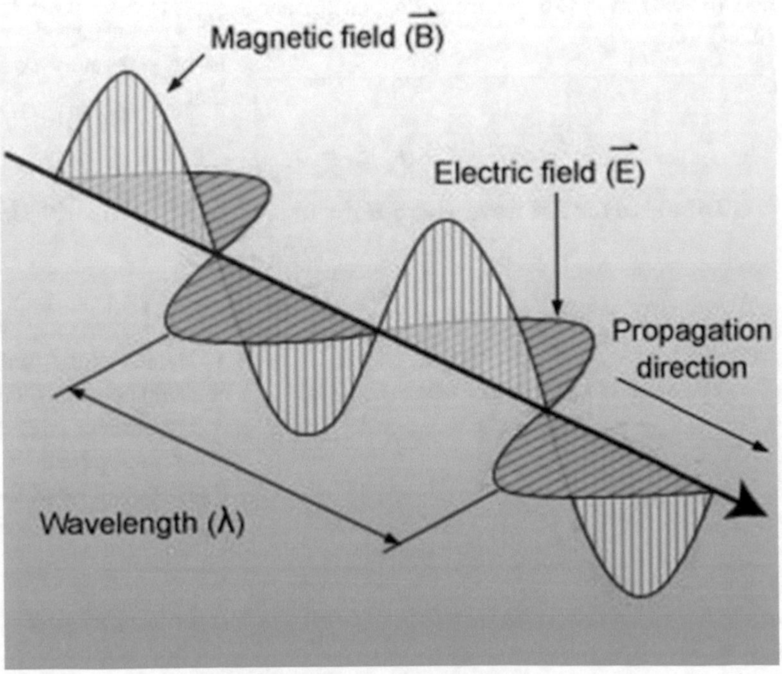

Magnetic field (B)
Electric field (E)
Propagation direction
Wavelength (λ)

Fig. 2.7 Four λ/2 of EM wave in free space propagating in z-direction E-field and H-fields are perpendicular to each other

Substitution H_z of Eq. (2.40) in Eq. (2.21) we get E_y as

$$E_y = (j\omega\mu h^2) \cdot (-C_1 B \sin Bx + C_2 B \cos Bx)(C_3 \sin Ay) \quad (2.42)$$

Putting $E_y = 0$ at $x = 0$ it gives

$$0 = (j\omega\mu/h^2) \cdot (C_2 B C_3 \sin Ay)$$

$$\therefore \quad C_2 = 0\,(as\, C_3 \neq 0;\; B \neq 0;\; \sin Ay \neq 0)$$

$\therefore$ From Eq. (2.40), H_z becomes:

$$H_z = C_1 C_3 \cos Bx \cos Ay \quad (2.43)$$

(iii) **Apply top surface condition (E_x = 0 at y = b)**

$$E_x = 0 \text{ at y} = b \text{ in Eq.(2.14) with } E_z = 0$$

$0 = (-j\omega\mu/h^2)(\partial H_z/\partial y)$, now we put H_z from Eq. (2.43) to get

$$0 = (j\omega\mu/h^2)\partial/\partial y(C_1 C_3 \cos Bx \cos Ay)$$
$$0 = (j\omega\mu h^2) \cdot (-C_1 C_3 A \cos Bx \sin Ay)$$

As $\cos Bx \neq 0$, $\therefore$ $\sin Ay = 0$ (with $y = b$) i.e. $\sin Ab = 0$

$$\therefore \quad Ab = n\omega(n = 0, 1, 2, \ldots, \infty)$$
$$A = (n\pi b) \quad (2.44)$$

(iv) **Apply right wall condition (E_y = 0 at x = a)**

From Eq. (2.21) we get with $E_z = 0$ and H_z from Eq. (2.40)

$$0 = (j\omega\mu/h^2)(\partial H_z/\partial x)$$
$$0 = (j\omega\mu/h^2)\partial/\partial x[C_1 C_3 \cos Bx \cos Ay]$$
$$0 = (j\omega\mu/h^2)[C_1 C_3(-B \sin Bx) \cos Ay]$$

As none of the term can be zero except $\sin Bx$,

$$\therefore \quad \sin Bx = 0 \text{ (with } x = a)$$
$$\therefore \quad Ba = m\pi\,(m = 0, 1, 2, 3, \ldots, \infty)$$
$$B = (m\pi a) \quad (2.45)$$

Thus we get three conditions $\boldsymbol{C_2 = C_4 = 0}$, $\boldsymbol{A = n\pi/b}$, and $\boldsymbol{B = m\pi/a}$, with $\boldsymbol{A}$ and $\boldsymbol{B}$ as phase constants (Rad/m).

$\therefore$ Equation (2.43), becomes

$$H_z = C_1 C_3 \cos(m\pi x/a) \cdot \cos(n\pi/yb)$$

Let C = another constant $= C_1 \cdot C_3$ we get

$$\boldsymbol{H_z = C \cdot cos(m\pi/xa)cos(n\pi/yb)} \quad (2.46)$$

Now put $C_2 = 0$ in Eq. (2.41) gives

$$E_x = \left(-\frac{j\omega\mu}{h^2}\right) \cdot C_1 \cos Bx(-AC_3 \cdot \sin Ay)$$
$$= \left(\frac{j\omega\mu}{h^2}\right) \cdot C_1 C_3(\cos Bx)A(\sin Ay)$$
$$= \left(\frac{j\omega\mu}{h^2}\right) CA \cos(Bx) \sin(Ay)$$
$$\boldsymbol{E_x = C \cdot \left(\frac{j\omega\mu}{h^2}\right)(n\pi/b)cos(m\pi/ax)sin(n\pi/by)} \quad (2.47)$$

Putting H_z of Eq. (2.46) in Eq. (2.21) with $E_z = 0$

$$E_y = \left(\frac{j\omega\mu}{h^2}\right) \cdot [-C_1 B \sin(Bx) C_3 \cos(Ay)]$$
$$\boldsymbol{E_y = C \cdot \left(-\frac{j\omega\mu}{h^2}\right) \cdot (m\pi/a)sin(m\pi x/a)cos(n\pi y/b)} \quad (2.48)$$

we know that

$$\boldsymbol{E_z} = 0 \quad (2.49)$$

$\therefore$ H_x and H_y can be obtained from Eqs. (2.22) and (2.23) with $E_z = 0$ and using H_z of Eq. (2.46).

$$\boldsymbol{H_x} = \left(\frac{-\gamma}{h^2} \cdot C\frac{m\pi}{a}\right) \sin(m\pi x/a) \times \cos(n\pi y/b) \tag{2.50}$$

$$\boldsymbol{H_y} = \frac{-\gamma}{h^2} \cdot C\left(\frac{n\pi}{b}\right)^2 \cos(m\pi x/a) \times \sin(n\pi y/b) \tag{2.51}$$

Thus we get the six components $\boldsymbol{E_x}$, $\boldsymbol{E_y}$, $\boldsymbol{E_z}$, $\boldsymbol{H_x}$, $\boldsymbol{H_y}$, and $\boldsymbol{H_z}$ as above, in Eqs. (2.46)–(2.51), which need to be multiplied by $e^{-\gamma z}$, for getting the phase and attenuation part also as per Table 2.2, as all $\boldsymbol{E}$ and $\boldsymbol{H}$ vary as $e^{-\gamma z}$ in line with $\boldsymbol{V}$ and $\boldsymbol{I}$ (see Eqs. 2.1l and 2.1m).

2.4.2 Non-existence of TEM Mode in Waveguides

TEM means the electric and magnetic fields both are perpendicular to the direction of propagation, i.e. $E_z = H_z = 0$ which will mean all field components = 0 in Eqs. (2.20)–(2.23) i.e. *EM* wave cannot exist. It means also that the current flows in the direction of propagation of wave, but the waveguide is hollow. Which will mean that the current cannot be conduction current but it can be displacement current and for this electric field also has to be along the direction of propagation for forcing the current to flow. But by TEM mode rules the electric field has to be perpendicular to the direction of propagation. Therefore because of this conflicting condition, TEM mode cannot exist in a hollow metal pipe whether it is a circular or rectangular waveguide.

2.4.3 TM Waves in Rectangular Waveguide: Electric and Magnetic Field Equation

In TM waves also we apply $E_z = 0$ at all the four walls (as in TE case) and $H_z = 0$ for all cases. With similar calculations we can show that the fields are as follows.

$$E_x = \left(\frac{-\gamma}{h^2} \cdot Cm\pi/a\right) \cos(m\pi x/a) \sin(n\pi y/b) \tag{2.52}$$

$$E_y = \left(\frac{-\gamma}{h^2} \cdot Cn\pi/a\right) \sin(m\pi xa) \cos(n\pi y/b) \tag{2.53}$$

$$E_z = C \cdot \sin(m\pi x/a) \sin(n\pi y/b) \tag{2.54}$$

$$H_x = \frac{\gamma}{h^2}(Cn\pi/a) \sin(m\pi x/a) \cos(n\pi y/b) \tag{2.55}$$

$$H_y = \left(\frac{j\omega\varepsilon}{h^2} C\right) \cos(m\pi x/a) \sin(n\pi y/b) \tag{2.56}$$

$$H_z = 0$$

For getting these expressions with phase and attenuation part also, all the 6 equations each for TE and TM modes have to be multiplied by the phase factor term ($e^{-\gamma z}$), where γ = propagation factor = $a + j\beta$ and these are summarised in Table 2.2, for lossless lines ($\alpha = 0$).

Table 2.2 Summary of results for rectangular waveguide (lossless line, $\alpha = 0$)

TE_{mn} mode	TM_{mn} mode
$E_z = 0$	$E_z = E_{0z} \sin\frac{m\pi x}{a} \sin\frac{n\pi y}{b} e^{-j\beta z}$
$H_z = H_{0z} \cos\frac{m\pi x}{a} \cos\frac{n\pi y}{b} e^{-j\beta z}$	$H_z = 0$
$E_x = E_{0x} \cos\frac{m\pi x}{a} \sin\frac{n\pi y}{b} e^{-j\beta z}$	$E_x = E_{0x} \cos\frac{m\pi x}{a} \sin\frac{n\pi y}{b} e^{-j\beta z}$
$E_y = E_{0y} \sin\frac{m\pi x}{a} \cos\frac{n\pi y}{b} e^{-j\beta z}$	$E_y = E_{0y} \sin\frac{m\pi x}{a} \cos\frac{n\pi y}{b} e^{-j\beta z}$
$H_x = H_{0x} \sin\frac{m\pi x}{a} \cos\frac{n\pi y}{b} e^{-j\beta z}$	$H_x = H_{0x} \sin\frac{m\pi x}{a} \cos\frac{n\pi y}{b} e^{-j\beta z}$
$H_y = H_{0y} \cos\frac{m\pi x}{a} \sin\frac{n\pi y}{b} e^{-j\beta z}$	$H_y = H_{0y} \cos\frac{m\pi x}{a} \sin\frac{n\pi y}{b} e^{-j\beta z}$

2.4.4 Cut-off Frequencies of Dominant Modes and Degenerate Modes in TE/TM Wave

As proved in Eqs. (2.44) and (2.45), $A = m\pi/a$, $B = n\pi/b$, therefore by Eq. 2.33:

$$h^2 = A^2 + B^2 = (m\pi/a)^2 + (n\pi/b)^2$$

but $h^2 = r^2 + \omega^2\mu\varepsilon$ by Eq. 2.30 therefore:

$$h^2 = \gamma^2 + \omega^2\mu\varepsilon = A^2 + B^2 = (m\pi/a)^2 + (n\pi/b)^2$$

i.e. $$\gamma^2 = (m\pi/a)^2 + (n\pi/b)^2 - \omega^2\mu\varepsilon$$

$$\gamma = \sqrt{(m\pi/a)^2 + (n\pi/b)^2 - \omega^2\mu\varepsilon} = (\alpha + j\beta)$$
= propagation constant.

Here α and β are the attenuation in Nepers and phase shift in radians per unit length of waveguide. Now we consider some special cases.

(a) **At lower frequencies, ($\gamma < 0$)**

$$\omega^2\mu\varepsilon < (m\pi/a)^2 + (n\pi/b)^2$$

$\therefore$ γ is real and positive, and $\gamma = \alpha$ = attenuation constant i.e. with phase constant as zero ($\beta = 0$). Therefore the wave gets completely attenuated and hence does not propagate in a waveguide at all.

(b) **At higher frequencies ($\gamma > 0$)**

$$\omega^2\mu\varepsilon < (m\pi/a)^2 + (n\pi/b)^2$$

and therefore γ is pure imaginary i.e. the wave propagates with $\alpha - 0$, $\gamma = j\beta$.

(c) **At cut-off frequency ($\gamma = 0$)**

The frequency at which γ just becomes zero (i.e. at transition, the propagation just starts) is defined as the cut-off frequency for TE and TM modes.

i.e. at $f = f_c$, at $\gamma = 0$ or $\omega = 2\pi f = 2\pi f_c = \omega_c$

$$0 = (m\pi/a)^2 + (n\pi/b)^2 - \omega^2\mu\varepsilon$$

or

$$\omega_c = \frac{1}{\sqrt{\mu\varepsilon}}\left[(m\pi/a)^2 + (n\pi/b)^2\right]^{\frac{1}{2}}$$

or

$$f_c = \frac{1}{2\pi\sqrt{\mu\varepsilon}}\left[(m\pi/a)^2 + (n\pi/b)^2\right]^{\frac{1}{2}}$$

$$f_c = \frac{C}{2\pi}\left[(m\pi/a)^2 + (n\pi/b)^2\right]^{\frac{1}{2}} \quad \because c = 1/\sqrt{\mu\varepsilon}$$

$$\boxed{= (/2)\sqrt{()^2 + ()^2}} \tag{2.57}$$

The lowest TM mode for which E and H are $\neq 0$ is TM_{11} (m = 1, n = 1) therefore dominant TM mode frequency will be $f_{c11} - (c/2)(1/a^2 + 1/b^2)^{1/2}$.

The cut-off wavelength (λ_c) in general for TE and TM modes is

$$\lambda_c = \frac{c}{f_c} = \frac{c}{(c/2)\sqrt{(m/a)^2 + (n/b)^2}}$$

$$\therefore \boxed{\lambda_{cmn} = \frac{2ab}{\sqrt{m^2b^2 + n^2a^2}} = \frac{2}{\sqrt{(m/a)^2 + (n/b)^2}}} \tag{2.58}$$

Cut-off frequency (f_c) and the corresponding standard size (a, b) of rectangular waveguide as recommended by Electronics Industry Association (EIA*) are given in Table 2.3 for TE_{10} mode of wave propagation. Here X-band is highlighted as in academic laboratories all the equipments which are used, work in X-band only.

All wavelengths greater than λ_c (i.e. at $f < f_c$) get attenuated exponentially as $e^{-\alpha_x}$, and those wavelengths less than λ_c (i.e. $f > f_c$) are allowed to propagate inside the waveguide as in Fig. 2.9.

Thus TM_{mn} denotes the general mode of wave propagation. With different values of m, n of the mode, the corresponding frequency and wavelength can be known. For TE_{mn} mode m denotes the number of half wavelength of electric field in the x-direction (x = 0 to x = a), while n is the number of half wavelength of E-field vector in y-direction (y = 0 to b). The lowest allowed

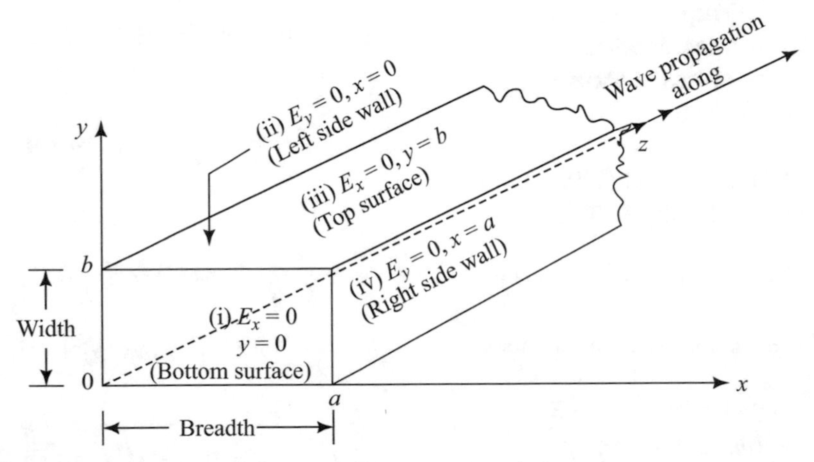

Fig. 2.8 Electric field on the walls of the waveguide is zero and not in between

mode frequency is known as cut-off frequency (f_c). Only frequency above f_c can propagate in a waveguide, which in turn depends on its dimensions. **Therefore waveguide acts as high pass filter**.

The modes which exist as dominant and degenerate modes: In TE propagation TE_{10}, TE_{01}, TE_{11}, TE_{12} etc. exists as the E- and H-field components for these are $\neq 0$. **The cut-off frequency whichever is lowest out of these modes is called dominant mode**. Here f_{10} being lowest cut-off frequency TE_{10} is dominant TE mode. Similarly in TM propagation TM_{11} is the *dominant mode*. See Figs. 2.10 and 2.11 for *E*- and *H*-fields of some of the TM/TE modes.

Sometimes two allowed modes can have the same cut-off frequency for a given dimension of the waveguide; then, these modes are called degenerate mode; e.g. if $a = b$ in a rectangular waveguide, then $f_{12} = f_{21}$, $f_{32} = f_{23}$ etc.; then TE_{12} and TE_{21} modes are called degenerate mode and so are TE_{23} and TE_{32}, etc.

TE_{00}/TM_{00} TM_{10}/TM_{01} modes do not exist: With $m = 0$, $n = 0$ in TM mode Eqs. (2.51)–(2.55) all the electric and magnetic components will be = 0. Similarly in the TE_{00} mode all the electric components = 0. Therefore TM_{00} and TE_{00} do not exist. Similarly we see that TM_{10} and TM_{01} do not exist.

The dimensions of the waveguide are always chosen from Table 2.3 such that the desired mode (dominant modes TE_{10} and TM_{11}) propagates along with their higher modes. Using Eq. (2.58), for $m = 1$, $n = 0$, we get TE_{10} mode wavelength as:

$$\lambda_{c10} = (2ab/b) = 2a \tag{2.59}$$

and with $m = 1$, $n = 1$, we get the TM_{11} mode wavelength as:

$$\gamma_{c11} = 2ab/\sqrt{a^2 + b^2} \tag{2.60}$$

Similarly for TE_{20} and TE_{21} modes:

$$\lambda_{20} = \frac{2ab}{\sqrt{(2b)^2 + 0^2}} = a; \quad \lambda_{21} = \frac{2ab}{\sqrt{4b^2 + a^2}} \tag{2.61}$$

2.4.5 Mode Excitation in Rectangular Waveguides

The waveguide modes are usually excited from a signal source through a coaxial cable. Outer conductor of the cable is connected to the body of the waveguide and the central conductor is projected inside the waveguide, either as a small probe or a loop as in Fig. 2.12. For the TE mode the excitation is done from the broad side of the waveguide, while for TM mode the cable is connected at the shorted end side of the length as in Fig. 2.12. TE_{10} needs only one cable input of $\lambda/2$ distance from shorted end, while TE_{20} and

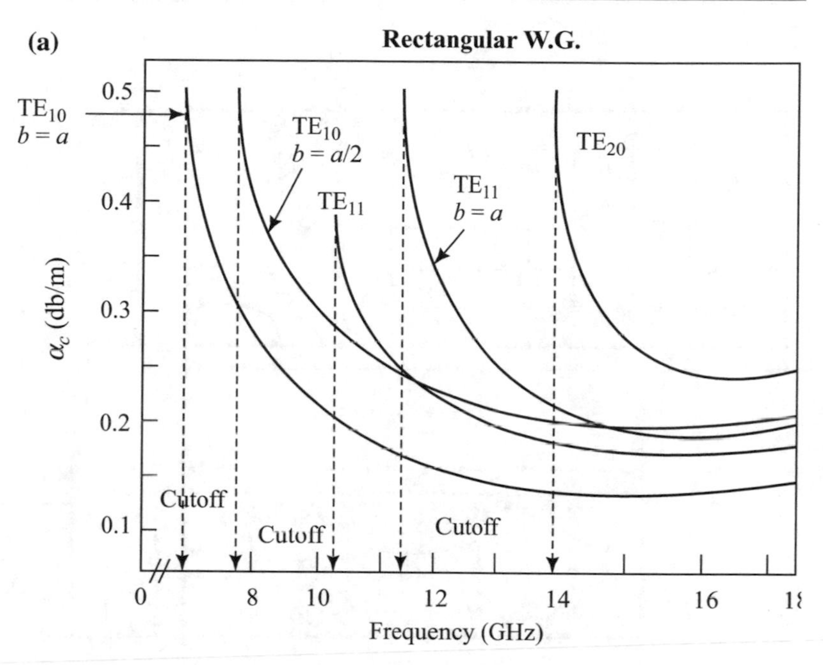

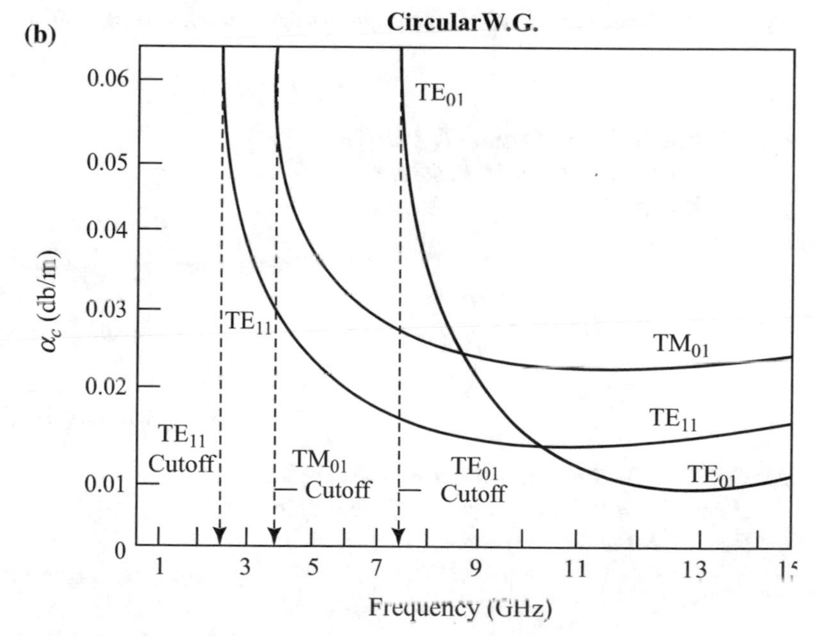

Fig. 2.9 Attenuation versus frequency curve, with cut in **a** rectangular waveguide and **b** circular waveguide, for some typical size of the guides f_c (TE_{10}) is lowest in rectangular and TE_{11} is lowest cut-off frequency in circular guides. Therefore these are dominant modes

TM_{21} need two cables inputs separated by $\lambda/2$ and at a distance $\lambda/2$ from shorted end. Instead of probe, loop is also used (Figs. 3.1, 3.8, and 3.15). **Loop excites the magnetic field, while probe excites electric field. As the E- and H-fields are inseparable, the EM wave gets excited in both the cases**:

$$\lambda_{21} = \frac{2ab}{\sqrt{(2b)^2 + a^2}} = \frac{2ab}{\sqrt{4b^2 + a^2}} \tag{2.62}$$

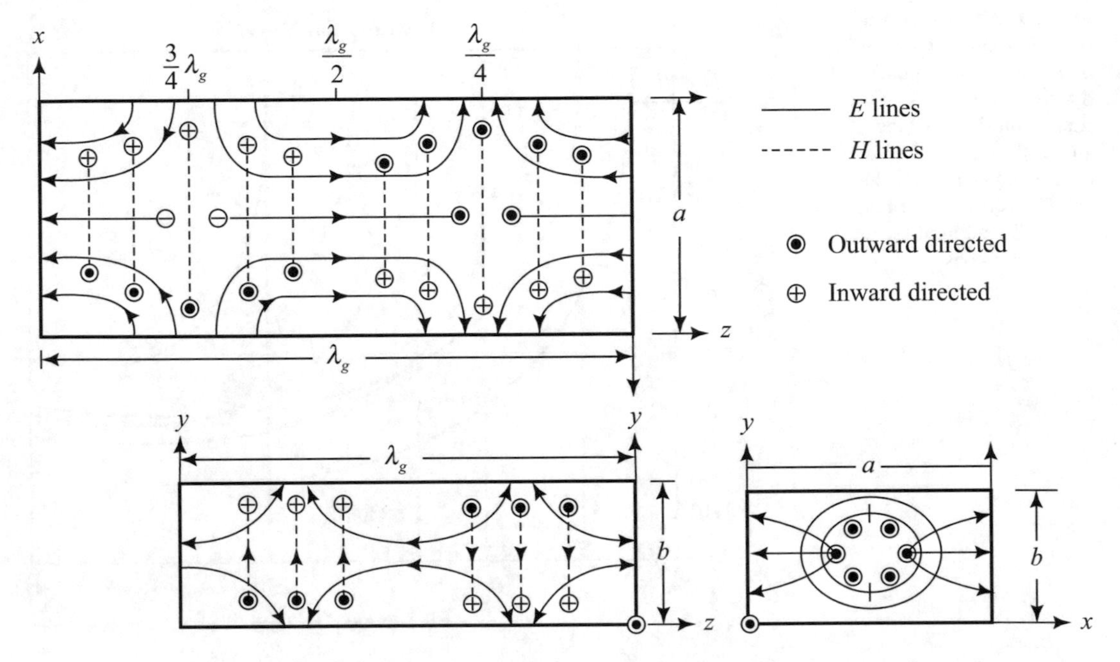

Fig. 2.10 Radiation pattern for TM_{11} waves in rectangular waveguide (H_z = 0) in x–z, y–z and x–y planes

2.4.6 Wave Impedance (Z_w) in TM and TE Waves in lossless Lines

$$
\begin{aligned}
Z_w &= E_x/H_y = -E_y/H_x \\
&= \sqrt{E_x^2 + E_y^2}/\sqrt{H_x^2 + H_y^2}
\end{aligned}
$$

(a) **For TM waves in a lossless waveguide**: Here $H_z = 0$ and $\gamma = j\ \beta$; therefore, Eqs. (2.20) and (2.23) can be used to give

$$
\begin{aligned}
Z_{\text{TM}}\left(E_x H_y\right) &= \frac{-\frac{\gamma}{h^2}\left(\frac{\partial E_z}{\partial x}\right) - \frac{j\omega\mu}{h^2}\cdot\left(\frac{\partial H_z}{\partial y}\right)}{\frac{-\gamma}{h^2}\cdot\left(\frac{\partial H_z}{\partial y}\right) - \frac{j\omega E}{h^2}\cdot\left(\frac{\partial E_z}{\partial x}\right)} \\
&= \frac{-j\beta\cdot\left(\frac{\partial E_z}{\partial x}\right)}{-j\omega\varepsilon\cdot\left(\frac{\partial E_z}{\partial x}\right)} = \frac{\beta}{\omega\varepsilon}
\end{aligned}
$$

As

$$
\begin{aligned}
\beta &= \sqrt{\omega^2\mu\varepsilon - \omega_c^2\mu\varepsilon} \\
&= \omega\sqrt{\mu\cdot\varepsilon}\cdot\sqrt{1-(\omega_c/\omega)^2} \\
&= \omega\sqrt{\mu\cdot\varepsilon}\cdot\sqrt{1-(f_c/f)^2}
\end{aligned}
\tag{2.63}
$$

$$
\begin{aligned}
\mathbf{Z_{TM}} &= \sqrt{\boldsymbol{\mu}\cdot\boldsymbol{\varepsilon}}\times\sqrt{\mathbf{1}-(\boldsymbol{f_c}/\boldsymbol{f})^2} = \sqrt{\boldsymbol{\mu}\cdot\boldsymbol{\varepsilon}}\times\sqrt{\mathbf{1}-(\boldsymbol{\lambda_0}/\boldsymbol{\lambda_c})^2} \\
&= \mathbf{Z_f}\sqrt{\mathbf{1}-(\boldsymbol{\lambda_0}/\boldsymbol{\lambda_c})^2}
\end{aligned}
\tag{2.64}
$$

As $\sqrt{\mu\varepsilon} = Z_f$ = Free space impedance

(b) **Free space impedance**:

In free space $\lambda_c = \lambda_0$

$$
\begin{aligned}
\therefore\quad Z_{TM} &= Z_f = \sqrt{\mu\varepsilon} = \sqrt{\mu_0\varepsilon_0} \\
&= 120\pi = 377\Omega
\end{aligned}
$$

Table 2.3 Characteristics of standard rectangular waveguides recommended by EIA[a]

Inner physical dimensions		Cut-off frequency for air-filled waveguide in GHz (f_c)	Recommended frequency range for TE_{10} mode in GHz
Width (a) cm (in.)	Height (b) cm (in.)		
24.765 (9.750)	12.383 (4.875)	0.606	0.76–1.15
19.550 (7.700)	9.779 (3.850)	0.767	0.96–1.46
16.510 (6.500)	8.255 (3.250)	0.909	1.14–1.73
12.954 (5.100)	6.477 (2.500)	1.158	1.45–2.20
10.922 (4.300)	5.461 (2.150)	1.373	1.72–2.61
8.636 (3.400)	4.318 (1.700)	1.737	2.17–3.30
7.214 (2.840)	3.404 (1.340)	2.079	2.60–3.95
5.817 (2.290)	2.908 (1.145)	2.579	3.22–4.90
4.755 (1.872)	2.215 (0.872)	3.155	3.94–5.99
4.039 (1.590)	2.019 (0.795)	3.714	4.64–7.05
3.485 (1.372)	1.580 (0.622)	4.304	5.38–8.17
2.850 (1.122)	1.262 (0.497)	5263	6.57–9.99
2.286 (0.900)	**1.016 (0.400)**	**6.562**	**8.20–12.50**[b]
1.905 (0.750)	0.953 (0.375)	7.874	9.84–15.00
1.580 (0.622)	0.790 (0.311)	9.494	11.90–18.00
1.295 (0.510)	0.648 (0.255)	11.583	14.50–22.00
1.067 (0.420)	0.432 (0.170)	14.058	17.60–26.70
0.864 (0.340)	0.432 (0.170)	17.361	21.10–33.00
0.711 (0.280)	0.356 (0.140)	21.097	26.40–40.00
0.569 (0.224)	0.284 (0.112)	26.362	32.90–50.10
0.478 (0.188)	0.239 (0.094)	31.381	39.20–59.60
0.376 (0.148)	0.188 (0.074)	39.894	49.80–75.80
0.310 (0.122)	0.155 (0.061)	48.387	60.50–91.90
0.254 (0.100)	0.127 (0.050)	59.055	73.80–112.00
0.203 (0.080)	0.102 (0.040)	73.892	92.90–140.00

[a]Electronics Industry Association;
[b]X-band used in academic institution

Thus the wave impedance in free space = 377 Ω.

As λ_0 is always less than λ_c for wave propagation $\therefore Z_{TM} < Z_f$.

This shows that in TM wave the wave impedance inside a waveguide is always less than that in free space.

(c) **For TE waves in a lossless waveguide**: Here $E_z = 0$ and $\gamma = j\,\beta$, $\alpha = 0$; therefore, same two Eqs. (2.20) and (2.23) can be used to give

$$Z_{TM} = (E_x/H_y) = \frac{-\frac{\gamma}{h^2}\left(\frac{\partial E_z}{\partial x}\right) - \frac{j\omega\mu}{h^2}\cdot\left(\frac{\partial H_z}{\partial y}\right)}{\frac{-\gamma}{h^2}\cdot\left(\frac{\partial H_z}{\partial y}\right) - j\omega\varepsilon\cdot\left(\frac{\partial E_z}{\partial x}\right)}$$

$$= \frac{-j\omega\mu\cdot\left(\frac{\partial H_z}{\partial y}\right)}{-\gamma\left(\frac{\partial H_z}{\partial y}\right)} = \frac{-j\omega\mu}{-j\beta} = \frac{\omega\mu}{\beta}$$

As from Eq. (2.63), $\beta = \omega\sqrt{\mu\cdot\varepsilon}\cdot\sqrt{1-(f_c/f_0)^2} = \omega\sqrt{\mu\cdot\varepsilon}\cdot\sqrt{1-(\lambda_0/\lambda_c)^2}$

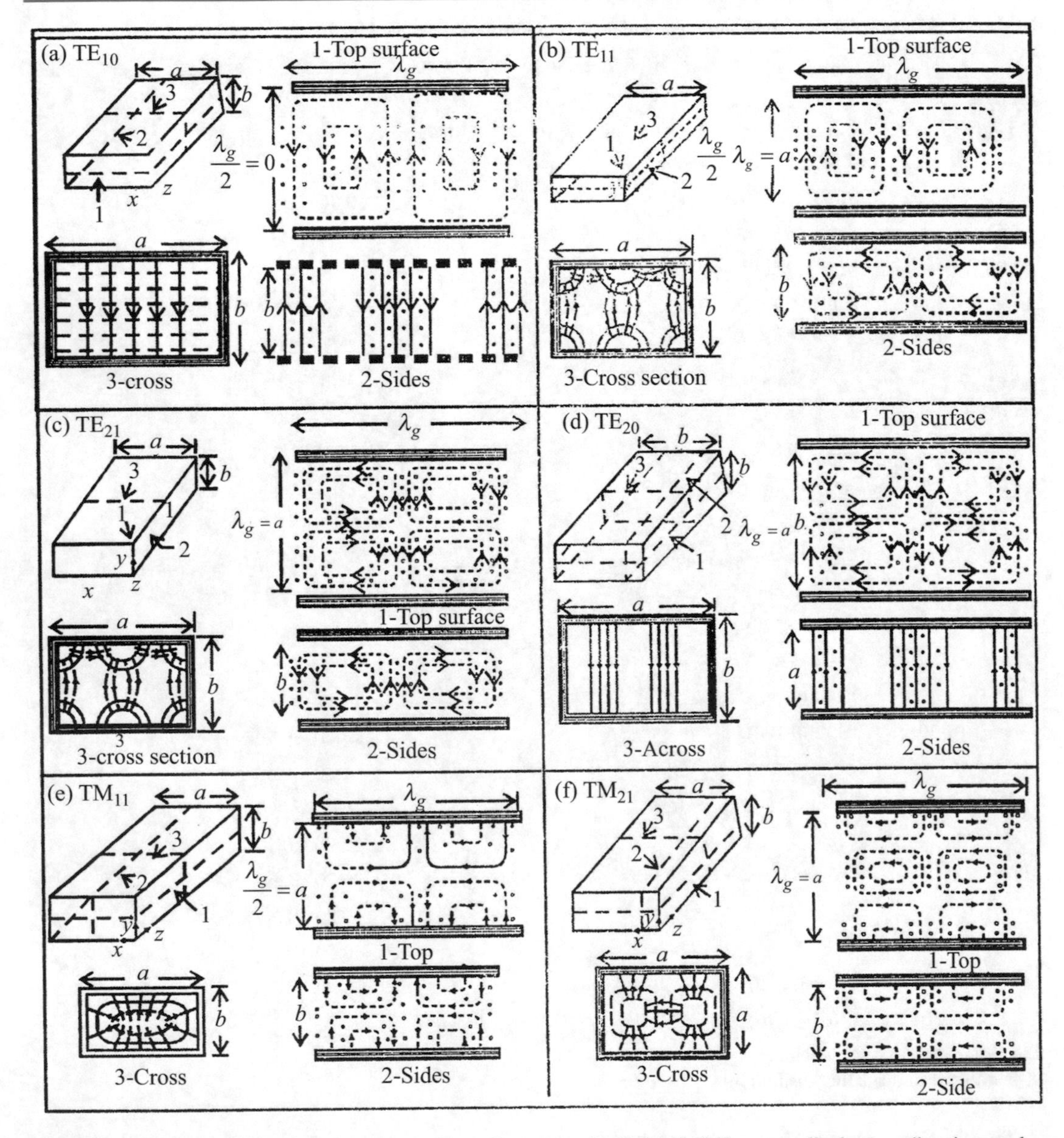

Fig. 2.11 Electric (solid line) and magnetic (dash line) lines of force (i.e. fields) of some of the lower TE and TM modes in rectangular waveguides. Here we note that in TE mode, E-field is fully perpendicular to z-directions and so is H-field in TM mode

$$\therefore Z_{TE} = \frac{\omega\mu}{\omega\sqrt{\mu\varepsilon}.\sqrt{1-(\lambda_0/\lambda_c)^2}} = \sqrt{\frac{\mu}{\varepsilon}}\cdot\frac{1}{\sqrt{1-(\lambda_0/\lambda_c)^2}} \quad (2.65)$$

$$\therefore \mathbf{Z_{TE} = Z_f/\sqrt{1-(\lambda_0/\lambda_c)^2} = Z_f/\sqrt{1-(f_c/f_0)^2}}$$

$Z_{TE} > Z_0$ as $\lambda_0 < \lambda_c$ for wave propagation.

This shows that wave impedance Z_{TE} for a TE wave is always greater than free space impedance (Z_f), while $Z_{TM} < Z_f$.

(d) **For TEM waves between two parallel plans or ordinary parallel wire or coaxial transmission lines**: The cut-off frequency f_c is zero, and wave impedance for TEM wave is the free space impedance itself.

i.e.

$$Z_{TEM} = Z_f \quad (2.66)$$

Thus,

$$\mathbf{Z_{TEM} = Z_f;\quad Z_{TE} > Z_f;\quad Z_{TM} < Z_f} \quad (2.67)$$

2.4.7 Power Transmission and Losses in Waveguide

The power transmitted through a waveguide (where no reflections are there i.e. as it has matched load/termination or is an infinite line, but has lossless dielectric) is given by:

$$P = \oint 1/2\,(E\times H)\,\mathrm{d}s\,\mathrm{W/cc}$$

As

$$Z_g = E_x/H_y = -E_y/H_x \text{ and } |E|^2 = |E_x|^2 + |E_y|^2;\ |H|^2 = |H_x|^2 + |H_y|^2$$

$$\therefore\quad P = 1/(2Z_g)\int |E|^2 \mathrm{d}a = (Z_g/2)\int |H|^2 \mathrm{d}a$$

Here this equation of power can be compared with V^2/R and I^2R.

For TE_{MN} and TM_{MN} modes, the average power transmitted will be (Z_f being the free space impedance)

$$P_{TE} = \frac{\sqrt{1-(f_c/f_0)^2}}{2Z_f}\int_0^b\int_0^a \left(|E_x|^2 + |E_y|^2\right)\mathrm{d}x\cdot\mathrm{d}y \quad (2.68)$$

and

$$P_{TM} = \frac{1/(2Z_f)}{\sqrt{1-(f_c/f_0)^2}}\int_0^b\int_0^a \left(|E_x|^2 + |E_y|^2\right)\mathrm{d}x\cdot\mathrm{d}y \quad (2.69)$$

Power loss in waveguide is due to

(i) Very high attenuation if $f < f_c$ as $\gamma = \alpha$ (only) full attenuation
(ii) Waveguide wall dissipation
(iii) Waveguide dielectric losses

Let us consider the cases of $f < f_c$ and $f > f_c$:

(a) **For $f < f_c$:** Propagation constant $\gamma = \alpha + j\,\beta = \alpha$ (real part only) (i.e. very high attenuation) z.

Here the phase factor $\beta = 2\pi/\lambda_g$ with $\lambda_g = (\lambda/\cos\theta)$ from Fig. 2.5 as $f_c > f$

$$\therefore\quad \lambda_g = \lambda\sqrt{1-(f_c/f)^2} = \lambda/j\cdot\sqrt{(f_c/f^2-1)^2}$$

$$\therefore\quad \beta = 2\pi/\lambda_g = (2\pi/\lambda)\cdot j\sqrt{(f_c/f)^2-1} = j\frac{2\pi\cdot f}{c}\sqrt{(f_c/f)^2-1} = j\frac{2\pi f_c}{c}\sqrt{1-\left(\frac{f}{f_c}\right)^2} = j\alpha_1 \quad \text{(Let)}$$

$\therefore\ \gamma = \alpha + j\,\beta = (\alpha - \alpha_1) = -\alpha_2$ (approx) taking original $\alpha \approx 0$.

∴ The cut-off attenuation constant α_2 causes very high attenuation, with full reflection of the wave (see Fig. 2.9).

$$\alpha_2 = \alpha_1 \frac{54.6}{\lambda_c} \sqrt{1 - (f/f_c)} \text{ dB/length} \quad (2.70)$$

In rectangular waveguides the attenuation is higher than in circular waveguides (Fig. 2.9). (1 dB = 0.115 Np; 1 Np = 8.686 dB)

(b) **For $f > f_c$**. The waveguide offers very low loss; however, these losses are of two types e.g. (*i*) the dielectric loss and (*ii*) conductor loss from the guide walls.

Dielectric loss: When the dielectric is imperfect and non-magnetic (μ_r = 1) in the waveguide, the attenuation constants for TE and TM modes are as:

$$\alpha_{\text{TE}} = \left(\sigma_d . Z_f/2\right) / \sqrt{1 - (f_c/f)^2} \text{ Np/m} \quad (2.71)$$

$$\alpha_{\text{TM}} = \left(\sigma_d \cdot Z_f/2\right) \cdot \sqrt{1 - (f_c/f)^2} \text{ Np/m} \quad (2.72)$$

where σ_d = conductivity of the dielectric

Z_f = intrinsic impedance of free space (377 Ω).

Conductor loss: The attenuation constant due to the imperfect conducting walls both for TE and TM waves is given by

$$\alpha_c = 8.686 \frac{R_S}{b \cdot z_f} \frac{1 + \frac{2b}{a}\left[\frac{f_c}{f}\right]^2}{\sqrt{1 - (f_c/f)^2}} \text{ dB/m} \quad (2.73)$$

R_S = sheet resistivity in Ω/m^2 = $1/(\sigma\delta_S)$ = $\sqrt{\pi f \mu_r \mu_0/\sigma}$

where f = frequency, μ_r = relative permeability of walls being non-magnetic = 1, μ_0 = permeability of free space (4 μ × 10^{-7} H/m, σ = conductivity of the walls of waveguide in mhos), δ_S = skin depth in metres = $1/\sqrt{\pi f \mu_r \mu_0 \sigma}$ Np/m = 8.686 dB/m.

2.4.8 Breakdown Power—Power Handling Capacity in Rectangular Waveguide

(a) **The breakdown power** in an air-filled rectangular waveguide for the dominant mode TE_{10} can be computed from the fact that microwave breakdown takes place when the electric field at the middle of the broad side is of the order of 30 kV/cm:

$$P_{bd\text{TE}_{10}} = 597\ ab\left[1 - (\lambda_0 2a)^2\right]^{\frac{1}{2}}, \text{K.W.} \quad (2.74)$$

where a, b, and λ_0 are in cm.

(b) **Maximum pulse power for dielectric-filled waveguide** in terms of dielectric field strength $E_{\max}$ is:

$$P_{bd\text{TE}_{10}} = P_{\max} = \left(6.63 \times 10^{-4}\right) E_{\max} (a \times b) \sqrt{1 - (\lambda_0 2a)^2} \text{ W} \quad (2.75)$$

where

$E_{\max}$	Maximum field strength of the dielectric in V/cm
a, b	Waveguide sides in cm
λ_0, λ_g	Waveguide lengths in cm in free space and waveguide

Typical peak pulse power handling capacities of lower- and higher-frequency range are as given in Table 2.4 for air as dielectric.

2.4.9 Guide Wavelength, Group Velocity, and Phase Velocity

These three parameters are same for TE and TM modes of propagation. Guide wavelength λ_g is the distance travelled by the wave, during which phase shift of 2π radians takes place. This is

Table 2.4 Typical peak pulse power handling capacities in waveguides (S, X and K bands)

Freq. range (band) (GHz)	$b \times a$ (in.)	P_{max} (Peak pulse power) (kW)
2.6–3.9 (S-band)	1.3 × 2.6	2000–3000
8.2–12.4 (X-band)	0.4 × 0.9	200–300
18.0–26.7 (K-band)	0.17 × 0.42	40–60

shown by Fig. 2.13. With β as the phase shift in radians per unit distance, therefore (Fig. 2.7),

$$\lambda_g = \frac{2\pi}{\beta} \quad (2.76)$$

Thus this guide wavelength λ_g is longer than the wavelength λ_0 in free space and is related to the cut-off wavelength λ_C by $= \frac{1}{\lambda_g^2} = \left(\frac{1}{\lambda_0^2} - \frac{1}{\lambda_C^2}\right)$ (to be proved later).

$$\therefore \lambda_g = \lambda_0 / \sqrt{1 - (\lambda_0/\lambda_C)^2} \text{ (with air dielectric)} \quad (2.77)$$

For waveguide with dielectric constant (ε_r), the wavelength will be

$$\lambda_{gD} - \lambda_0 / \left[\sqrt{\varepsilon_r - (\lambda_0/\lambda_C)^2}\right] \text{ (with a dielectric)} \quad (2.78)$$

As $\varepsilon_r > 1 \therefore \lambda_{g\ \text{dielectric}} < \lambda_{g\ \text{air}}$ and hence cut-off frequency also reduces and therefore lower frequencies can pass through the same guide.

General wavelength for TE_{MN} mode will be:

$$\lambda_{gm,n} = \frac{\lambda_0}{\sqrt{1 - (\lambda_0/\lambda_{Cmn})^2}} \quad (2.79)$$

Thus we see that $\lambda_g < \lambda_0 <$ and $\lambda_g < \lambda_{Cmn}$, and from Eq. (2.60),

(i) If $\lambda_0 \ll \lambda_C$, then from Eq. (2.57), $\lambda_g - \lambda_0$
(ii) If $\lambda_0 = \lambda_C$ then $\lambda_g = \infty$
(iii) If $\lambda_0 > \lambda_C$ then λ_g is imaginary i.e. no propagation in waveguide.

It is clear that if $\lambda_0 \ll \lambda_C$ the denominator in Eq. (2.79) is approximately equal to '1'. As λ_0 increases and approaches λ_C then λ_g increases and reaches infinity at $\lambda_0 = \lambda_C$. When $\lambda_0 > \lambda_C$, it is evident than λ_g is imaginary and no

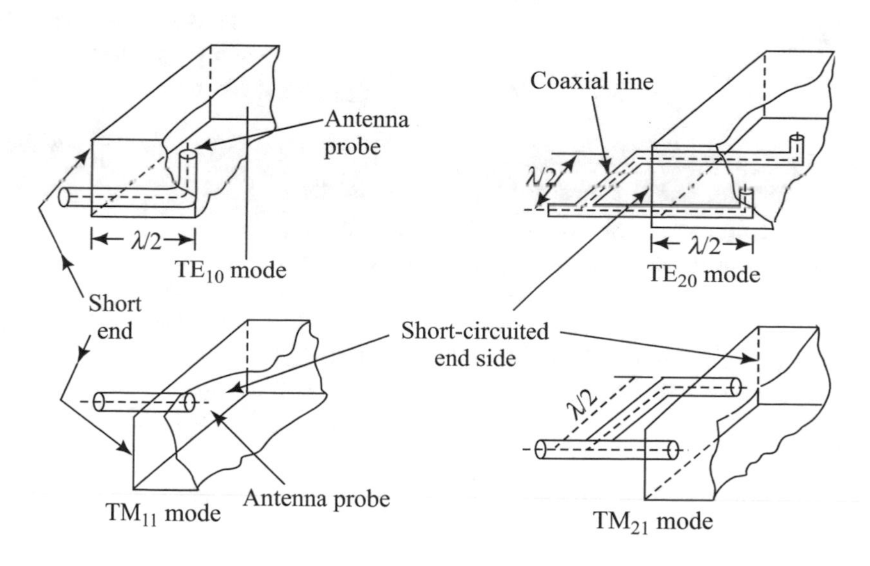

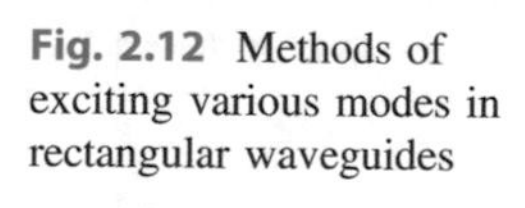
Fig. 2.12 Methods of exciting various modes in rectangular waveguides

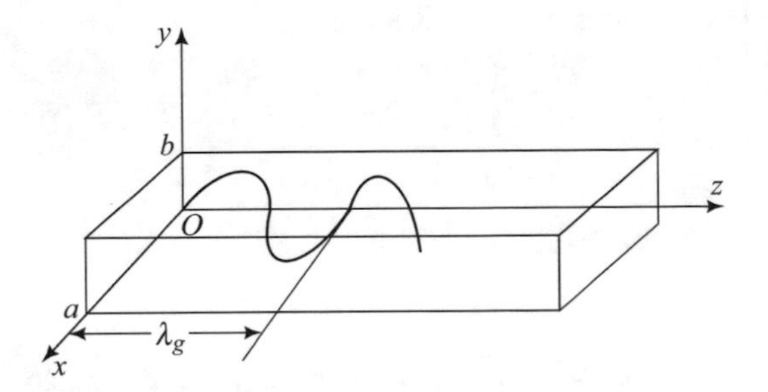

Fig. 2.13 Waves travel in z-direction in TE or TM mode in a waveguide with guide wavelength λ_g along z-direction

propagation is there in the waveguide due to high attenuation.

Phase Velocity V_p and Group Velocity (V_g).

Velocity of any fixed phase point (say Π/3 point) is called the phase velocity V_p of the wave in the waveguide. Therefore $V_p = \lambda_g \cdot f$, f being the frequency of the wave. As $\lambda_g > \lambda_0$ and therefore comparing the velocity of light given by $c = \lambda_0 \cdot f$, we get $V_p > c$. Therefore, **phase velocity is greater than light**, as it is not a velocity of matter, but the rate of change of phase of the guide wave.

Rewriting

$$V_p = \lambda_g \cdot f = \frac{2\pi f \cdot \lambda_g}{2\pi} = \frac{2\pi f}{2\pi \lambda_g} = \frac{\omega}{\beta} \tag{2.80}$$

we get

$$\boldsymbol{V_p = \omega/\beta} \tag{2.81}$$

where, $\omega = 2\pi f$, $\beta = 2\pi \lambda g$.

The Eq. (2.81) can also be proved from the propagation constant term $e^{(j\omega t - \gamma z)}$, which will be constant for two same phase points on the wave $(j\ \omega t - \gamma z)$ = constant = $[j\ \omega t - (\alpha + j\ \beta)z]$

$$\therefore\ (\omega t - \beta z) = \text{constant for a lossless line}\ (\alpha = 0) \tag{2.82}$$

∴ differentiating we get

$$\omega \cdot dt - \beta \cdot dz = 0$$
$$\therefore\ \frac{dz}{dt} = V_p = \frac{\omega}{\beta} \tag{2.83}$$

Also from Fig. 2.5, we know that

$$V_g = V_0 \sin\theta = c \sin\theta \tag{2.84}$$

As wave velocity α (1/wavelength), above becomes

$$\lambda_g = \lambda_0 \sin\theta \tag{2.85}$$

Also we know that $\boldsymbol{\lambda_g = \lambda_p}$ but $\boldsymbol{V_g \neq V_p}$.

∴ In Eq. (2.80) by putting λ_g from (2.84) we get

$$V_p = \lambda_p f = \lambda_g f = \frac{\lambda_0}{\sin\theta} \cdot f = \frac{c}{\sin\theta} \tag{2.86}$$

∴ By (2.84) and (2.86) we get

$$\boldsymbol{V_p . V_g = c^2} \tag{2.87}$$

If there is a modulating signal on the carrier, the modulated carried envelope actually travels at a velocity slower than that of carrier alone and of course slower than that of light. **The velocity of modulation envelope is called the group velocity V_g which is less than the velocity of light**, as the modulation goes on slipping backward with respect to the carrier.

Thus it can be termed as the rate at which the wave propagates through the waveguide and is given by

$$\boldsymbol{V_g = \frac{d\omega}{d\beta}\ \text{while}\ V_p = \frac{\omega}{\beta}} \tag{2.88}$$

(i) **Expression for V_p:** From Eq. (2.80)–(2.87) we know that $V_p = \frac{\omega}{\beta}$

Also,
$$h^2 = \gamma^2 + \omega^2 \mu\varepsilon = A^2 + B^2$$
$$= (m\pi/\alpha)^2 + (n\pi/b)^2$$

And
$$\gamma = \alpha + j\beta$$

For wave propagation, $\gamma = j\beta$ ($\because$ attenuation, $\alpha = 0$)

$$\gamma^2 = (j\beta)^2 = (m\pi/a)^2 + (n\pi/b)^2 - \omega^2\mu\varepsilon \tag{2.89}$$

At $f = f_c$, $\omega = \omega_c$, $\gamma = 0$

$$\omega_c^2\,\mu\varepsilon = (m\pi/a)^2 + (n\pi/b)^2 \tag{2.90}$$

Putting this Eq. (2.89), we get:

$$\gamma^2 = (j\beta)^2 = \omega_c^2\mu\varepsilon - \omega^2\mu\varepsilon$$
$$\therefore \quad \beta^2 = \left(\omega^2\mu\varepsilon - \omega_c^2\mu\varepsilon\right)$$
$$\therefore \quad \beta = \sqrt{\mu\varepsilon\left(\omega^2 - \omega_c^2\right)} = \sqrt{\mu\varepsilon}\sqrt{\left(\omega^2 - \omega_c^2\right)}$$
$$V_p = \frac{\omega}{\beta} = \frac{\omega}{\sqrt{\mu\varepsilon}\sqrt{\omega^2-\omega_c^2}} = \frac{1}{\sqrt{\mu\varepsilon}}\frac{1}{\sqrt{1-(\omega_c/\omega)^2}}$$
$$V_p = c/\sqrt{1-(f_c/f)^2} \tag{2.91}$$

Also $f = c\lambda_0$ and $f_c = c\lambda_c$ where λ_0 is free space wavelength λ_c is cut-off wavelength

$$\boldsymbol{V_P = c/\sqrt{1-(\lambda_0/\lambda_c)^2}} \tag{2.92}$$

(ii) **Expression for V_g**: From Eq. (2.88) we know that $V_g = \frac{d\omega}{d\beta}$

But from Eq. (2.91), $\beta = \sqrt{\mu\varepsilon\left(\omega^2 - \omega_c^2\right)}$. Therefore differentiating β w.r.t. ω, we get

$$\frac{d\beta}{g\omega} = \frac{1}{2\sqrt{\left(\omega^2-\omega_c^2\right)\mu\varepsilon}} \cdot 2\omega\mu\varepsilon$$
$$\therefore \quad \frac{d\beta}{d\omega} = \frac{\sqrt{\mu\varepsilon}}{\sqrt{1-(\omega_c/\omega)^2}} = \frac{\sqrt{\mu\varepsilon}}{\sqrt{1-(f_c/f)^2}}$$

or

$$V_g = \frac{d\omega}{d\beta} = \frac{\sqrt{1-(f_c/f)^2}}{\sqrt{\mu\varepsilon}} = \frac{\sqrt{1-(\lambda_0/\lambda_c)^2}}{\sqrt{\mu\varepsilon}}$$

or

$$\boldsymbol{V_g = c.\sqrt{1-(\lambda_0/\lambda_c)^2}} \tag{2.93}$$

(iii) **Consider the product of V_p and V_g**: From Eqs. (2.92) and (2.93) we get:

i.e.,

$$V_p \cdot V_g = \frac{c}{\sqrt{1-(\lambda_0/\lambda_c)^2}} \cdot c \cdot \sqrt{1-(\lambda_0/\lambda_c)^2}$$

$$\boldsymbol{V_p \cdot V_g = c^2} \tag{2.94}$$

Here again we note that as $V_g < c \therefore V_p > c$.

2.5 Propagation in Circular Waveguides

A circular waveguide is basically a hollow tubular, circular conductor as shown in Fig. 2.14 with inner radius $r = a$ and length l. The general properties of the modes in the circular waveguides are similar to those for rectangular waveguides. **The unique property of circular waveguide is that attenuation falls with increasing frequency** and therefore useful for long low-loss communication links.

For field equations, cylindrical coordinates are used. Here Φ varies from 0 to 2 π, radius r varies from 0 to a, and l varies along z-axis (Fig. 2.14).

2.5.1 TE Waves in Circular Waveguide E- and H-Field Equations

For a TE wave we know $E_z = 0$ and $H_z \neq 0$. Therefore Maxwell's modified equation i.e. [Eq. (2.25)] (Helmholtz wave equation):

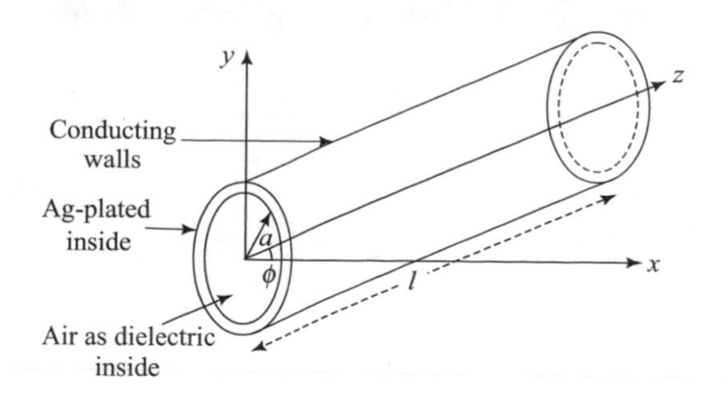

Fig. 2.14 Circular waveguide

$$\nabla^2 H_z = -\omega^2 \mu\varepsilon H_z \tag{2.95}$$

Expanding $\nabla^2 Hz$ in cylindrical coordinates.

$$\frac{\partial^2 H_z}{\partial r^2} + \frac{1}{r}\frac{\partial H_z}{\partial r} + \frac{1}{r^2}\frac{\partial^2 H_z}{\partial \phi^2} + \frac{\partial^2 H_z}{\partial z^2} = -\omega^2 \mu\varepsilon H_z \tag{2.96}$$

As H_z varies along the z-direction as $e^{-\gamma z}$, we can substitute the differentials $\frac{\partial H_z}{\partial z}$ by $-\gamma H_z$ and $\frac{\partial^2 H_z}{\partial z^2}$ by $\gamma^2 H_z$, here we can substitute the operators $\frac{\partial^2}{\partial z^2}$ by γ^2 and $\frac{\partial}{\partial z}$ by $-\gamma$ $\therefore$ Eq. (2.96) becomes

$\therefore$

$$\frac{\partial^2 H_z}{\partial r^2} + \frac{1}{r}\frac{\partial H_z}{\partial r} + \frac{1}{r^2}\frac{\partial^2 H_z}{\partial \phi^2} + (\gamma^2 + \omega^2 \mu\varepsilon) H_z = 0 \tag{2.97}$$

Using $\gamma^2 + \omega^2 \mu\varepsilon = h^2$

$$\frac{\partial^2 H_z}{\partial r^2} + \frac{1}{r}\frac{\partial H_z}{\partial r} + \frac{1}{r^2}\frac{\partial^2 H_z}{\partial \phi^2} + h^2 H_z = 0 \tag{2.98}$$

Being a partial differential equation, it can be solved by variable separable method by assuming $H_Z = P(r) \cdot Q(\phi)$ and the field equations for TE_{nm} modes in circular waveguides can be solved by Bessel's function equation to give:

$$\text{with } h = P'_{nm}/a \tag{2.99}$$

$$E_r = E_{0\rho} J_n\left(\frac{P'_{nm}}{a} r\right) \sin n\phi (e^{-\gamma z}) \tag{2.100}$$

$$E_\phi = E_{0\phi} J'_n\left(\frac{P'_{nm}}{a} r\right) \cos n\phi (e^{-\gamma z}) \tag{2.101}$$

$$E_z = 0 \tag{2.102}$$

$$H_r = -H_{0\rho} J'_n\left(\frac{P'_{nm}}{a} r\right) \cos n\phi (e^{-\gamma z}) \tag{2.103}$$

$$H_\phi = H_{0\phi} J_n\left(\frac{P'_{nm}}{a} r\right) \sin n\phi (e^{-\gamma z}) \tag{2.104}$$

$$H_z = H_0 J_n\left(\frac{P'_{nm}}{a} r\right) \cos n\phi (e^{-\gamma z}) \tag{2.105}$$

where $Z_Z = \frac{E_{0r}}{H_{0\phi}} = \frac{-E_{0\phi}}{H_{0r}}$, the wave impedance in the circular guide with $n = 0, 1, 2, 3, \ldots$ and $m = 1, 2, 3, 4, \ldots$ and $h = P'_{nm}/a$.

2.5.2 TM Modes in Circular Waveguide: E- and H-field Equations

For a TM wave to propagate in a circular waveguide $H_Z = 0$ and $E_Z \neq 0$, the Maxwell's modified i.e. Helmholtz wave equation is [Eq. (2.106)]

$$\nabla^2 E_Z = -\omega^2 \mu\varepsilon E_Z \tag{2.106}$$

Solving this equation on the same lines as TE waves, we get E_Z as a variable of Bessel's function:

$$E_Z = E_{0Z} J_n(rh) \cos n\phi \cdot e^{-\gamma_z} \tag{2.107}$$

By TM mode boundary conditions i.e. $E_Z = 0$ at $r = a$, we get the complete solution by Bessel's function as: (with $h = P_{nm}/a$)

$$E_r = E_{0\rho} J'_n\left(\frac{P_{nm}}{a} \cdot r\right) \cos n\phi (e^{-\gamma z}) \tag{2.108}$$

$$E_\phi = E_{0\phi} J_n\left(\frac{P_{nm}}{a} r\right) \sin n\phi (e^{-\gamma z}) \tag{2.109}$$

$$E_z = E_{0z} J_n \left(\frac{P_{nm}}{a} r \right) \cos n\phi (e^{-\gamma z}) \quad (2.110)$$

$$H_r = \frac{E_{0\phi}}{Z_Z} J_n \left(\frac{P_{nm}}{a} r \right) \sin n\phi (e^{-\gamma z}) \quad (2.111)$$

$$H_\phi = \frac{E_{0\phi}}{Z_Z} J_n' \left(\frac{P_{nm}}{a} r \right) \cos n\phi (e^{-\gamma z}) \quad (2.112)$$

$$H_z = 0 \quad (2.113)$$

All the 12 equations of the TE and TM modes can be rewritten in the following simple shapes also (without the propagation factor) i.e. without $e^{-\gamma_z}$ factor.

2.5.3 Cut-off Wavelength in Circular Waveguide, Dominant and Degenerate Modes

Just like rectangular waveguides. The cut-off wavelength λ_c corresponds to that frequency f_0, below which attenuation is very large and wave does not propagate. Therefore the cut-off λ_c is given by

$$\gamma = \alpha + j\beta \text{ with } \beta = \sqrt{\omega^2 \mu\varepsilon - h^2} \quad (2.114)$$

where, $h = P'_{nm}/a$ for TE wave and $h = P_{nm}/a$ for TM waves. Therefore for TE wave, the cut-off wavelength is given by

$$\lambda_{CTE} = \frac{2\pi}{h} = \frac{2\pi}{(P'_{nm} a)} = \frac{2\pi a}{(P'_{nm})} \quad (2.115)$$

Here λ_C will be maximum if P'_{nm} is minimum. This minimum value of P'_{nm} is known to be 1.841 (Table 2.5) for n = 1 and m = 1. Hence TE_{11} is the **dominant mode in circular waveguide**, with cut-off wavelength (Tables 2.6 and 2.7):

$$\boldsymbol{\lambda_{CTE11} = 2\pi a/1.841 \text{ cm}} (\boldsymbol{a} = radiusincm) \quad (2.116)$$

and

Table 2.5 Values of P'_{nm} for TEnm modes in circular waveguide

n	m		
	1	2	3
0	3.832	7.016	10.173
1	1.841	5.331	8.536
2	3.054	6.706	9.969
3	4.201	8.015	11.346

Note P'_{nm} is dimensionless and P'_{nm} = 1.841 given the lowest i.e. dominant TE mode frequency

Table 2.6 Values of P'_{nm} for TEnm modes in circular waveguide

n	m		
	1	2	3
0	2.405	5.520	8.645
1	3.832	7.106	10.173
2	5.135	8.417	11.620
3	6.380	9.761	13.015

Note P_{nm} is dimensionless

Table 2.7 Fields in circular waveguide without propagation factor $e^{-\gamma z}$

TE_{nm} mode	TM_{nm} = modes
$E_z = o$	$E_r = E_{oz} \cdot J_n(P_2) \cdot \cos(n\phi)$
$H_z = H_{oz} \cdot J_n(P_1) \cdot \cos(n\phi)$	$H_z = o$
$E_r = E_{or} \cdot J_n(P_1). \sin(n\phi)$	$E_r = E_{or} \cdot J_n(P_2) \cdot \cos n\phi$
$E_\phi = E_{o\phi} \cdot J_n(P_1) \cdot \cos(n\phi)$	$E_\phi = E_{o\phi} \cdot J_n(P_2). \sin n\phi$
$H_r = H_{or} \cdot J_n(P_1) \cdot \cos(n\phi)$	$H_r = H_{or} \cdot J_n(P_2) \cdot \sin n\phi$
$H_\phi = H_{o\phi} \cdot J_n(P_1) \cdot \sin(n\phi)$	$H_\phi = H_{o\phi} \cdot J_n(P_2) \cdot \sin n\phi$
Where $J_n(P_1)$ is Bessel's function of first kind	Where $J_n(P_2)$ is Bessel's function of second kind
$P_1 = P'_{nm} \cdot r/a$	$P_2 = P_{nm} \cdot r/a$

$$\begin{aligned} f_{CTE11} &= c/\lambda = cP'_{nm}/2\pi a \\ &= 8.79/a \text{ GHz} \, (a = radiusincm) \end{aligned} \tag{2.117}$$

This cut-off frequency is given in Table 2.8, for standard size of diameters $2a$ of circular waveguides.

Similarly for TM waves:

$$\lambda_{CTM} = 2\pi/h = 2\pi a/P_{nm}$$

Here the lowest frequency will be with highest value of λ_{CTM}, which is with $n = 0$, $m = 1$ i.e. $P_{01} = 2.405$ (Table 2.6) as the dominant mode as

$$\begin{aligned} \lambda_{CTM01} &= 2\pi a/2.405 \\ &= 2.611\, a \text{ cm}(with a\ incm)\ldots(2.88) \end{aligned} \tag{2.118}$$

and

$$\begin{aligned} f_{CTM01} &= c/\lambda \\ &= (11.49/a)\text{GHz}(a = radiusincm) \end{aligned} \tag{2.119}$$

Degenerate Modes in Circular Waveguides

As clear from above, the dominant TE mode in circular waveguide is TE_{11}. Also, we see from Tables 2.5 and 2.6 that $P'_{0m} = P_{1m}$ and hence all the TM_{0m} and TE_{1m} modes are degenerate (i.e. have same wavelength (λ_g) etc.) in a uniform circular waveguide. The field configuration in a circular waveguide is shown in Fig. 2.15.

2.5.4 Phase Velocity, Group Velocity, Guide Wavelength, and Wave Impedance in Circular Waveguides

The expressions for phase velocity, group velocity, and guide wavelength are same as in the case of a rectangular waveguide both for TE and TM modes.

i.e.

$$V_g = \frac{\omega}{\beta} = V_c \cdot \sqrt{1 - (\lambda_0/\lambda_c)^2} \tag{2.120}$$

$$V_p = c^2/V_g \tag{2.121}$$

$$\lambda_g = \lambda_0/\sqrt{1 - (\lambda_0/\lambda_c)^2} \tag{2.122}$$

$$Z_{TE} = Z_0/\sqrt{1 - (\lambda_0/\lambda_c)^2} \tag{2.123}$$

$$Z_{TM} = Z_0/\sqrt{1 - (\lambda_0/\lambda_c)^2} \tag{2.124}$$

where $\lambda = V_p/f$ and $Z_0 = \sqrt{\mu/\varepsilon}$.

Table 2.8 Standard circular waveguide sizes

Inside diameter cm (in.) (2a)	Cut-off frequency for air-filled waveguide (GHz)	Recommended frequency range for TE_{11} mode (GHz)
25.184 (9.915)	0.698	0.80–1.10
21.514 (8.470)	0.817	0.94–1.29
18.377 (7.235)	0.957	1.10–1.51
15.700 (6.181)	1.120	1.29–1.76
13.411 (5.280)	1.311	1.51–2.07
11.454 (4.511)	1.534	1.76–2.42
9.787 (3.853)	1.796	2.07–2.83
8.362 (3.292)	2.102	2.42–3.31
7.142 (2.812)	2.461	2.83–3.88
6.104 (2.403)	2.880	3.31–4.54
5.199 (2.047)	3.381	3.89–5.33
4.445 (1.750)	3.955	4.54–6.23
3.810 (1.500)	4.614	5.30–7.27
3.254 (1.281)	5.402	6.21–8.51
2.779 (1.094)	6.326	1.27–9.97
2.383 (0.938)	**7.377**	**8.49–11.60**[a]
2.024 (0.797)	**8.685**	**9.97–13.70**[a]
1.748 (0.688)	10.057	11.60–15.90
1.509 (0.594)	11.649	13.40–18.40
1.270 (0.500)	13.842	15.90–21.80
1.113 (0.438)	15.794	18.20–24.90
0.953 (0.375)	18.446	21.20–29.10
0.833 (0.828)	21.103	24.30–33.20
0.714 (0.281)	24.620	28.30–38.80
0.635 (0.250)	27.683	31.80–43.60
0.556 (0.219)	31.617	36.40–49.80
0.478 (0.188)	36.776	42.40–58.10
0.437 (0.172)	40.227	46.30–63.50
0.358 (0.141)	49.103	56.60–77.50
0.318 (0.125)	55.280	63.50–87.20
0.277 (0.109)	63.462	72.70–99.70
0.239 (0.094)	73.552	84.80–116.00

[a]Normally X-band frequency is used in laboratory instruments in academic institutions

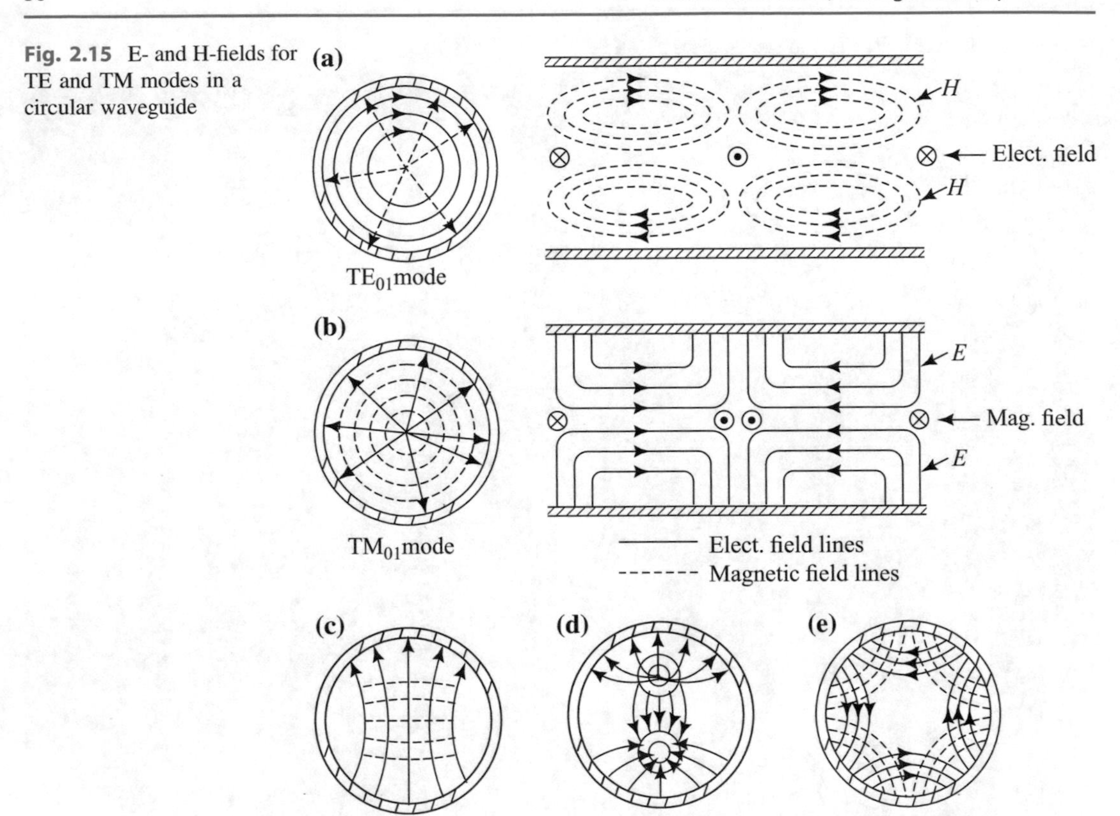

Fig. 2.15 E- and H-fields for TE and TM modes in a circular waveguide

2.5.5 Power Transmission and Attenuation Loss in Circular Waveguide

As in a rectangular waveguide, attenuation in a circular waveguide for TE and TM modes in an air-filled circular waveguide is due to finite conductivity of the guides walls and is given by

$$\alpha = \frac{\text{Power loss/unit length}}{2(\text{Average power transmitted})}$$

Average power transmitted with Z_z as wave impedance is:

$$(P_{nm})_{\text{av}} = \frac{1}{2Z_z}\int_0^{2\pi}\int_0^a \left[|E_\phi|^2 + |E_\rho|^2\right] r\mathrm{d}\phi$$

$$\frac{Z_z}{2}\int_0^{2\pi}\int_0^a \left[|H_\phi|^2 + |H_r|^2\right] r\,\mathrm{d}r\mathrm{d}\phi$$

For power transmitted in TE_{nm} mode [the above becomes with Eq. (2.123)]

$$(P_{nm})_{\text{TE}} = \frac{\sqrt{1-(f_c/f)^2}}{2Z_f}\int_0^{2\pi}\int_0^a \left[|E_\phi|^2 + |E_r|^2\right] r\mathrm{d}p$$

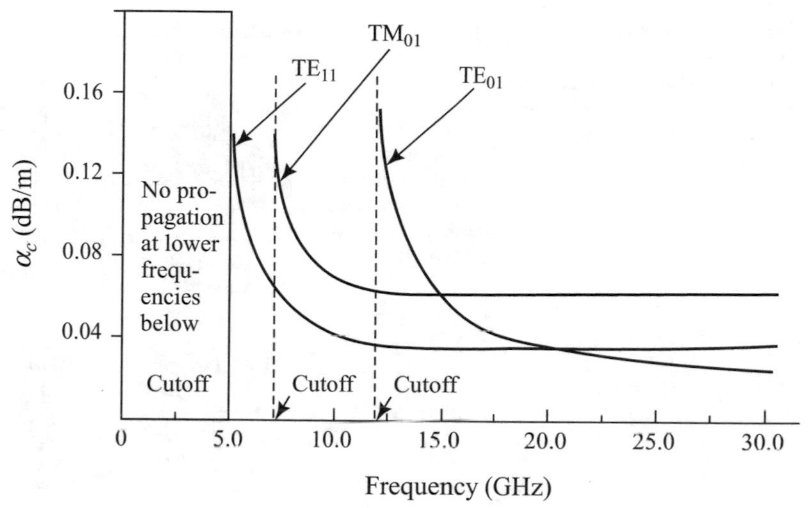

Fig. 2.16 Attenuation in an air-filled circular copper waveguide with inner diameter $D = 3.0$ cm. In circular waveguides guides TE_{11} is the lowest cut-off frequency therefore dominant TE mode. In TM mode TM_{01} is dominant mode

Power transmitted in TM_{nm} modes becomes with Eq. (2.124):

$$(P_{nm})_{TM} = \frac{1}{2Z_f\sqrt{1-(f_c f)^2}}\int_0^{2\pi}\int_0^{a}\left[|E_\phi|^2+|E_r|^2\right] r\,\mathrm{d}r\mathrm{d}\phi$$

The power loss/unit length along the walls of the waveguide will be:

$$P_L = \frac{R_S}{2}\oint \overline{J_S}\cdot\overline{J_S}^{*}\mathrm{d}l$$

where

$Z_f =$ Intrinsic impedance of free space $\sqrt{\mu/\varepsilon}$
$R_s =$ Sheet resistance of waveguide
$= \sqrt{\omega\mu/(2\sigma)}\sqrt{\mu f \mu/\sigma}$
$JS =$ Current on the lossless walls

Using all above, the attenuation constant α for TE and TM modes can be shown to be

$$\alpha_{TE} = \frac{R_s}{aZ_f\sqrt{1-(f_c/f)^2}}\left[\left(\frac{f_c}{f}\right)^2 + \frac{n^2}{(P'^{\,2}_{nm}-n^2)}\right] \text{ Np/m} \tag{2.125}$$

and

$$a_{TM} = \frac{R_s}{aZ_f\sqrt{1-(f_c/f)^2}} \text{ Np/m} \tag{2.126}$$

As a specific case for TE_{01} mode, attenuation falls very fast with frequency as:

$$\alpha = \frac{R_s}{aZ_f}\cdot\frac{f_c^2}{f\sqrt{(f^2-f_c^2)}} \text{ Np/m} \tag{2.127}$$

This shows that the attenuation is lower at higher frequencies of TE_{01} mode and hence this property is useful for long low-loss waveguide communication links. **The attenuation loss in circular waveguides are far less than in the rectangular waveguides (Figs.** 2.9 **and** 2.16).

2.5.6 Power Handling Capacity and Breakdown Power Limits in Circular Waveguide

Maximum field strength $E_{max} = 30$ kV/cm is possible in a waveguide. Therefore maximum power flows in the dominant TE_{11} mode and TE_{01} modes (minimum attenuation mode) for a

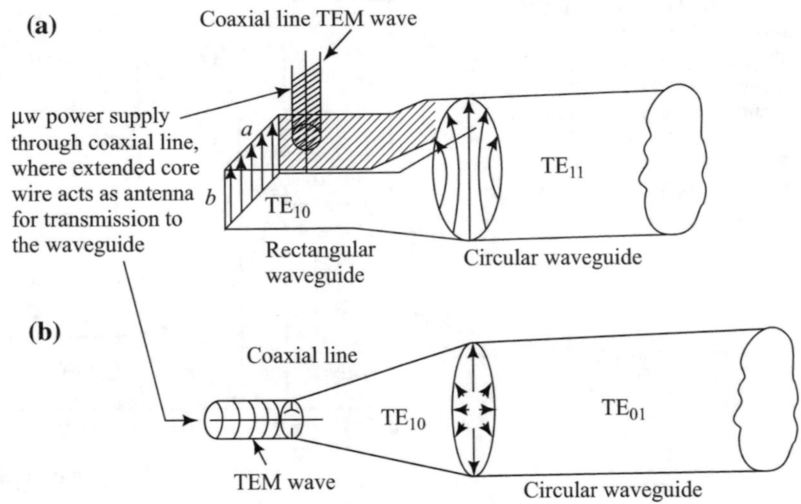

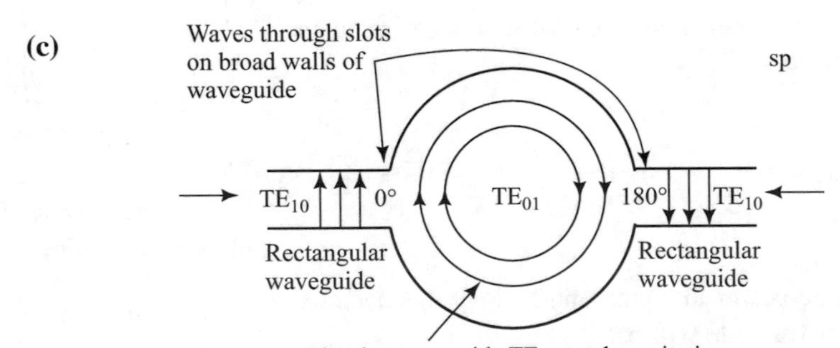

Fig. 2.17 Methods of excitation of TE and TM modes in circular waveguides by coaxial line **a**, **b**, and **c** by two rectangular waveguides directly

circular guide and the corresponding breakdown power with f_{c11} and f_{c10}, as their cut-off frequencies are

Peak pulse power:

$$P_{\max} = 0.498 \times E_{\max}^2 . d^2 (\lambda_0 / \lambda_g) \quad (2.128)$$

Breakdown power:

$$(P_{bd})_{\mathrm{TE11}} = 1790 a^2 \sqrt{\left[1 - (f_{c11}/f)^2\right]} \text{ kW} \quad (2.129)$$

Breakdown power:

$$(P_{bd})_{\mathrm{TE01}} = 1805 a^2 \sqrt{\left[1 - (f_{c01}/f)^2\right]} \text{ kW} \quad (2.130)$$

2.5.7 TEM Wave in Circular Waveguide Do Not Exist

As discussed in Sects. 2.3 and 2.4.2, TEM mode cannot exist in a hollow waveguide and it can exist only in the two parallel plate conductor system or coaxial line or in free space wave propagation. It can be proved [from Eqs. (2.20) to (2.23)] that all the field components in a circular waveguide vanish for $E_z = H_z = 0$ and therefore TEM mode cannot exist.

2.5.8 Excitation of Modes

The electric field generated is perpendicular to the electric current, and its magnetic field is perpendicular to a loop of electric current. Therefore various TE and TM modes can be generated in circular waveguide by coaxial line probes and loops, as shown in Fig. 2.17. In Fig. 2.17a coaxial line probe excites the TE_{10} dominant mode in rectangular waveguide which is converted to TE_{11} dominant mode in the circular waveguide through the transition length in between. In Fig. 2.17b, longitudinal coaxial line probe directly excites the symmetric TM_{10} mode. TE_{01} mode is excited by means of two out of phase signals (in TE_{10} mode each), through two diametrically opposite slots along in length. These two out of phase (TE_{10}) signals are fed through two waveguides with matching slots along the broad side in the conducting sheets on the faces of these waveguides Fig. 2.17c.

2.5.9 Advantages, Disadvantages, and Applications of Circular Waveguides

(a) **Advantages**

(1) The circular waveguides are easier to manufacture than rectangular waveguides and are easier to be connected.
(2) The TM_{01} and TE_{01} modes are rotationally symmetrical, and hence rotation of polarisation can be overcome.
(3) Further TE_{01} mode in circular waveguide has the lowest attenuation per unit length of waveguide and hence suitable for long-distance waveguide transmission, as proved by Eq. (2.127) and Fig. 2.16.

(b) **Disadvantages**

(1) The circular waveguides occupy more space compared to a rectangular system and hence not suitable for some applications. For the same cut-off frequency the cross-sectional area in a circular waveguide is double that of rectangular waveguide (refer solved problem no. 15 of this chapter for proving it).
(2) The plane polarisation rotates, when wave travels through the circular waveguide due to roughness or discontinuities in the circular cross section. Polarisation changes affect the received signal resulting in reflection of the transmitted signal and hence losses.
(3) Due to infinite number of modes existing in a circular waveguide, it becomes very difficult to separate these modes, resulting in interference with the dominant mode.
(4) Propagation in rectangular waveguides is easier to visualise and analyse, than in circular waveguides.

(c) **Applications**
The applications of circular waveguides include:

(1) Rotating joints in radars to connect the horn antenna feeding a paraboloid reflector (which must rotate for tracking).
(2) Long-distance waveguide transmission above 10 GHz in TE_{01} mode.
(3) Short- and medium-distance broadband communication, and it could replace/share coaxial line and microwave links.

2.6 Strip Lines and Microstrip Lines

Because of large size and cost of the waveguide, an extension approach of the 2-wire transmission line was being thought of during 1965. Later this leads to the strip lines, which was called microwave integrated circuit (MIC). Besides size and

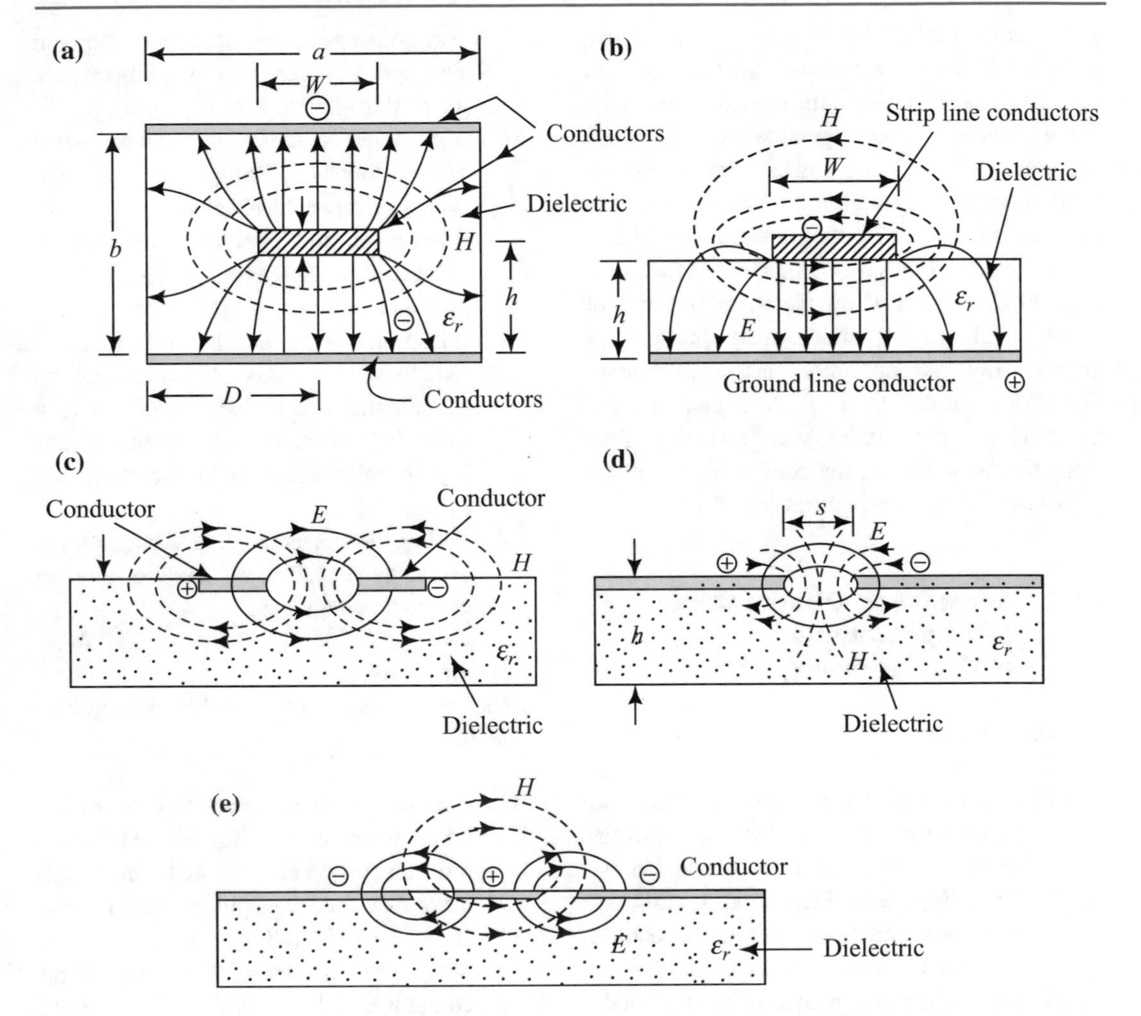

Fig. 2.18 Various strip lines, the electric field (with transient charges) and magnetic field directions along with current direction of a moment. All these fields and charges will reverse after T/2 **a** strip line (strip and two planes); **b** microstrip line (strip and one plane); **c** coplanar strip line (two strips); **d** slotted line (two planes and a slot); and **e** coplanar-double slot (two planes and a strip)

cost advantage, the characteristic impedance can be controlled by the geometry of the strip line. These can be used for frequency much higher than in the coaxial lines. These are of the following types (Fig. 2.18).

(a) **Strip lines**: It has a metal strip between two broad conducting planes, separated by some dielectric.
(b) **Microstrip lines**: It has a strip metal and one broad conducting plane separating by a dielectric.
(c) **Coplanar strip line**: It has two conducting strips on a dielectric surface separated by a distance.
(d) **Slot line**: Here over a conducting plane which is pasted on a dielectric, a slot is made to make two lines.
(e) **Coplanar double slot line**: Here on the same plane on the dielectric surface, two parallel planes and a strip line in between are there. All these are shown in Fig. 2.18; out of all above (a) and (b) are mostly used, and we will discuss these in detail.

(a) **Strip line**: As in Fig. 2.18a it is a metal strip symmetrically between two ground planes separated by low-loss dielectric. The electric field and magnetic field are concentrated near the central conductor.

$$\text{For lossess line } \alpha = g = 0; \quad V_P = \omega/\beta = v_0\sqrt{\lambda\varepsilon} \tag{2.131}$$

where

V_P Phase velocity along the strip line
γ Propagation constant = (α + j β)
v_0 Velocity of electromagnetic wave in free space = c
ε_r Relative permittivity of the dielectric
C Capacitance/length of the strip line between strip and the ground planes

A simple expression for Z_0 is:

$$Z_0 = \frac{60}{\sqrt{\varepsilon_r}} \text{ in} \left(\frac{4b}{\pi d}\right) \tag{2.132}$$

where d = equivalent diameter of the flat central conductor.

The three design conditions for better performance are.

1. $W_p > 5$ h: As can be seen from the fields in Fig. 2.19, the dominant mode of wave propagation is TEM, with fields contained within the transmission line for no radiation losses. For this the width of the ground plane has to be at least 5 times of the spacing between them i.e. $W_p > 5$ h.
2. $\boldsymbol{h < \lambda_g/2}$: For stopping and not allowing the higher order of non-TEM modes (as they create interference with the main dominant mode), distance between two ground planes has to be less than $\lambda_g/2$. i.e. $h < \lambda_g/2$, λ_g being the wavelength in the dielectric media of the line.
3. Low $\mathbf{Z_0}$: Z_0 to be kept as low as possible using the design curve of Fig. 2.19 for lower attenuation/losses. Biggest limitations of strip line are (*i*) higher cost of material and the fabrication and (*ii*) difficult to mount component e.g. diodes capacitor.

(b) **Microstrip lines**: As seen in Fig. 2.18b, it has one strip line and one ground plane separated by a dielectric. The advantages over the simple strip line are as follows:

(i) Fabrication cost and material cost are low.

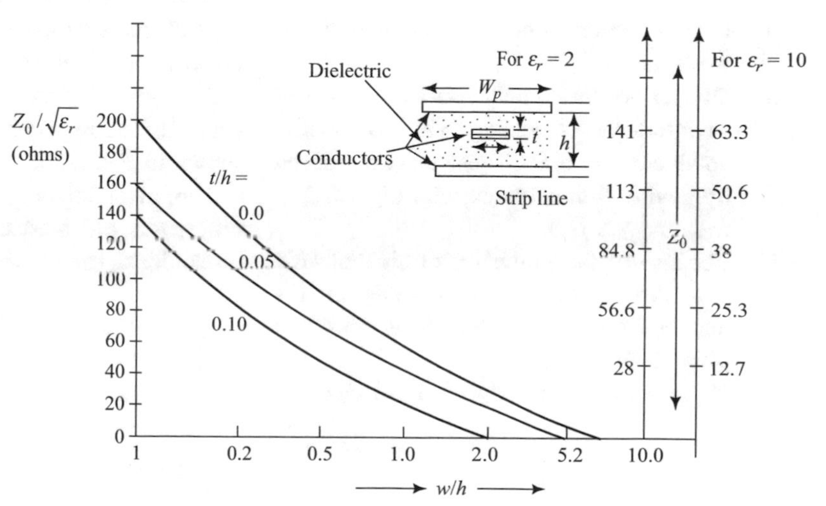

Fig. 2.19 $Z_0/\sqrt{\varepsilon_r}$ versus w/h for different values of t/h in a strip line. The right scale gives corresponding z_0 for $\varepsilon_r = 2$ and 10

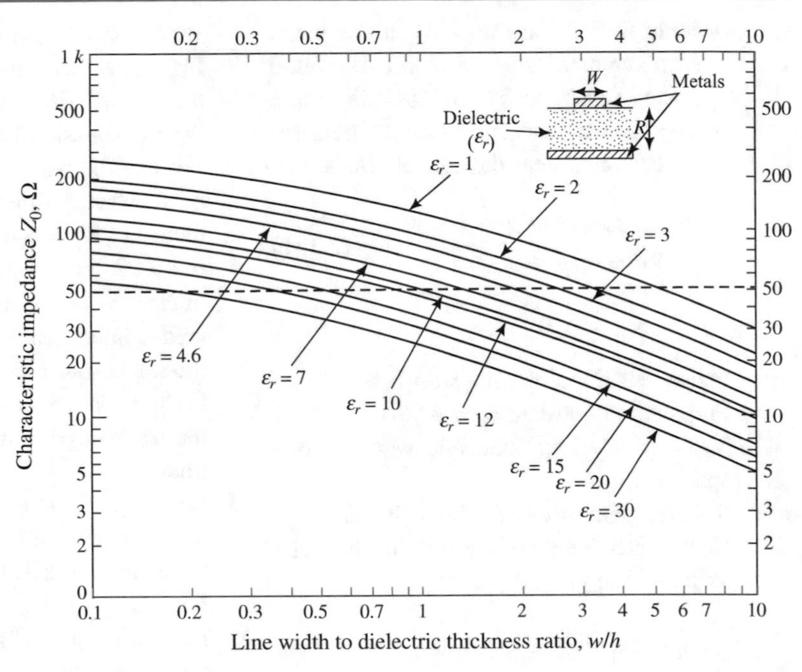

Fig. 2.20 Microstrip characteristic impedance as a function of w/h ratio for different dielectric constant of the substrate

(ii) Designing is easy for frequencies of 1–30 GHz.
(iii) Semiconductor chip can easily be mounted at any location.
(iv) Easy for probing any point for testing, etc.
(v) It is the most popular line today.

The limitations are also there:

(i) Higher attenuation as compared to waveguides.
(ii) Due to openness, radiation losses as well as interference/coupling with the nearby conductor are there. These can be reduced by having thin substrate with high dielectric constant (ε_r).
(iii) Because of the proximity of air dielectric with the conductor, discontinuity in 'E'- and H-field is generated. This makes it difficult to analyse it.
(iv) Highest frequency range is 1–50 GHz.
(v) We can see that there is a concentration of fields below the microstrip element. The electric flux crossing the air–dielectric boundary is small, and therefore a pure TEM mode cannot exist. Hence a small deviation from TEM mode does exist, which can be neglected for approximate analysis.

The characteristic impedance of a microstrip depends quits on its geometry i.e. line width (w), thickness (t), and the distance between the line and the ground plane (h). In fact, the variation of characteristic impedance in terms of w/h ratio is shown in Fig. 2.20.

Empirical relation between Z_0 and w/h for different ε_r has been developed by a number of researchers. They look like the following type:

$$Z_0 = f(w/h, \varepsilon_r) \text{ or } w/h = f(Z_0, \varepsilon_r)$$

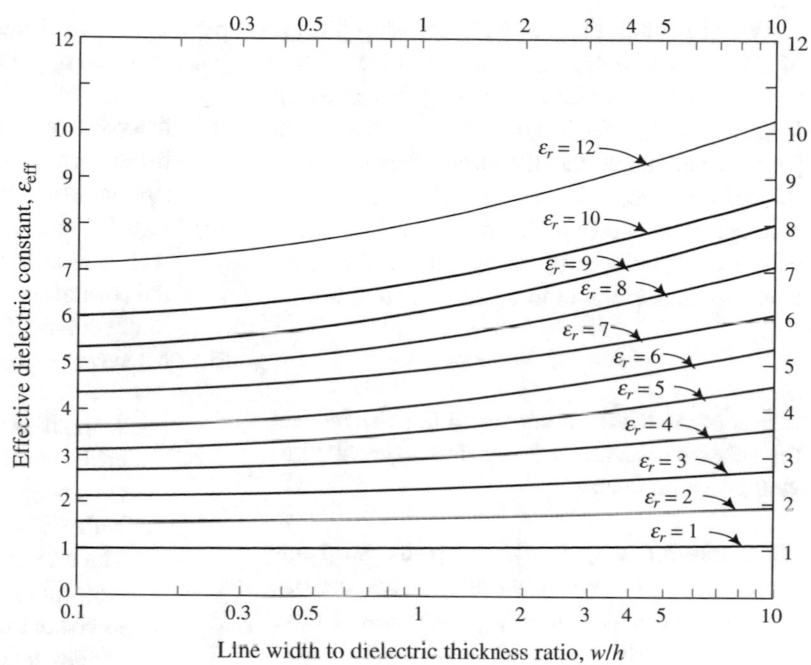

Fig. 2.21 Effective dielectric constant of the microstrip line as a function of w/h ratio for different dielectric constants of the substrate

In microstrip line filter design, etc., Z_0 is obtained by a number of techniques, and finally using its value, the line width (w/h) is computed by the second type of relation given above. This second type of formula is given below for two conditions of $w/h > 2$ or $w/h < 2$ (Fig. 2.20). At $w/h \approx 2$ error in the discontinuity is less than 0.5%.

For $\boldsymbol{w/h} \leq 2$ (Narrow lines) ('*A*' formula).

This is normally true for $Z_0 > 50\ \Omega$ of the line, but not always true as it depends on ε_r also.

$$\frac{w}{h} = \frac{8e^A}{e^{2A} - 2} \tag{2.133}$$

where

$$A = 2\pi \frac{Z_0}{Z_f} \sqrt{\frac{\varepsilon_r + 1}{2}} + \frac{\varepsilon_r - 1}{\varepsilon + 1}\left(0.23 + \frac{0.11}{\varepsilon_r}\right)$$

$$Z_f = \sqrt{\mu_0/\varepsilon_0} = 120\pi$$
$$= 376.8\ \Omega = \text{wave impedance of free space} \tag{2.134}$$

For $\boldsymbol{w/h} \geq 2$ wider lines ('B' formula):

This is normally for $Z_0 < 50\ \Omega$, but varies with the values of ε_r.

$$w/h = \frac{2}{\pi}\left\{ \begin{matrix} (B-1) - \ln(2B-1) \\ + \dfrac{\varepsilon_r - 1}{2\varepsilon_r}\left[\ln(B-1) + 0.39 - \dfrac{0.61}{\varepsilon_r}\right] \end{matrix} \right\} \tag{2.135}$$

$$\text{where } B = Z_f \pi (2z_0\sqrt{\varepsilon_r}) \tag{2.136}$$

Also ε_r gets modified due to the fringe field of the microstrip line as well with w/h rates (Fig. 2.21).

$$\varepsilon_{eff} = \frac{\varepsilon_r + 1}{2} + \frac{\varepsilon_r - 1}{2}\left(1 + 12\frac{h}{w}\right)^{-\frac{1}{2}} \tag{2.137}$$

and

$$\left.\begin{aligned} \lambda_g &= \lambda_0/\sqrt{\varepsilon_{eff}} = \text{wavelength in the microstrip line} \\ \lambda_p &= c/\sqrt{\varepsilon_{eff}} = \text{ phase velocity} \\ \lambda_0 &= c/f = \text{wavelength in the free space} \end{aligned}\right\} \tag{2.138}$$

from Eqs. (2.133) to (2.135), $Z_0 = f(w/h)$ also can be derived.

Power handling capacity of microstrip lines is of a few watts only; however, it has the advantage of miniaturization. For long transmission lengths, they suffer from excessive attenuation per unit length and the attenuation depends upon the electric properties of the substrate and the conductors as well on the frequency. The attenuation constant α is given by sum of dielectric loss, conductor ohmic loss, and radiation loss:

$$\alpha = \alpha_d + \alpha_c + \alpha_r$$

Radiation loss of a microstrip line depends on the substrate thickness and dielectric constant as well as its geometry.

(c) **Coplanar strip lines**: A coplanar strip line consists of two conducting strips on one substrate with one strip grounded as shown in Fig. 2.18. It has the advantages of easy manufacturing and convenient connections, as the cost of high radiation loss, therefore not preferred.

(d) **Slot line and (e) Coplanar double slot line**: Two other types of transmission lines used are known as slot line and coplanar waveguide as shown in Fig. 2.18. A slot line consists of a slot or gap in a conducting coating on a dielectric substrate and here the fabrication process is identical to that of microstrip lines. A coplanar waveguide consists of a strip of thin metallic film deposited on the surface of a dielectric slab, with two ground electrodes running adjacent and parallel to the strip, on same surface. Here the radiation loss is less than coplanar strip line but much greater than microstrip line, hence also not preferred. Therefore microstrip line is the most preferred line today at microwave frequencies.

2.6.1 Microwave Component Using Strip Lines

Most of the microwave devices and component can be designed using strip lines and can be operated in several frequency bands as 1, 5, and 10 GHz. These applications are:

(i) **Microwave receivers**: With noise figures better than 16 dB and very little conversion losses, circuits can be fabricated.

(ii) **Crystal modulators**: Coaxial transition with crystal holder as an integral part of the coaxial line with VSWR, less than 1.5 in the 4.4–5 GHz frequency band.

(iii) **Microwave components**:

(a) **Transitions**: Strip line to coaxial cable transition equivalent to coaxial to waveguide transition can be made with VSWR's as low as 1.2 at 5 GHz. The reverse transition of waveguide to strip line or crossbar feed waveguide to coaxial transitions is also used.

(b) **Magic tees**: Extremely low VSWR, balanced response and negligible radiation design, is possible in strip line magic tee.

(c) **Termination attenuator pads and loads**: Microstrip line coated with a lossy dielectric or graphite paint of appropriate characteristics can be made which can be tapered to get proper matching. Variable attenuation 0–15 dB is obtained by rotation of the flap, which adjusts the length of the dielectric run with respect to the strip line.

(d) **Directional couplers**: Filter elements and antennas can also be fabricated.

(e) **Antenna and arrays**: Antennas and arrays are convenient to make (see Chap. 7).

2.6.2 Microwave IC (MIC) and Monolithic Microwave IC (MMIC)

MIC and MMIC has grown very fast over the last 25 years. **MIC are basically microstrip line with passive components (normally also made**

out of microstrip line) as well as active devices, mounted and connected with the circuit. The area grew very fast, as at mm waves, very thin waveguide of size $a \times b$ = (0.2 cm × 0.1 cm) becomes very expensive. **Thus this area has grown due to the advantages e.g. (i) on these circuits, passive and active device can be mounted (ii) Large-scale production is possible (iii) Cost is lower than waveguide in large-scale production.**

In MIC the dielectric/base/substrate normally used is alumina on which the microstrip line is fabricated. Now some new materials have also been tried as substrate.

The continued efforts and success towards miniaturisation and then to integrate active and passive components on the same substrate, which should be a dielectric insulator as well, have led to MMIC. **Here in MMIC the substrate used is a semiconductor of very high resistivity close to an insulator**, (e.g. Intrinsic GaAs, has $\rho_{\text{GaAs}} = 2 \times 10^9\,\Omega - \text{cm}$, $\rho_{\text{glass}} = 10^{10}\,\Omega - \text{cm}$, $\rho_{\text{porcelain}} = 10^{11}\,\Omega - \text{cm}$). On this semiconductor itself as substrate, the active and passive devices can be fabricated or grown. MESFET is the most commonly used active device on GaAs.

In MMIC devices, successful operation has been possible in mm wave range up to 100 GHz using GaAs substrate and up to 200 GHz using InP substrate. As far as power is concern, 30 dBm using GaAs/InP and up to 40 dBm using GaN/SiC substrates have been achieved, but the latter is not becoming popular due to high cost.

2.7 Impedance Matching

For maximum power transfer from the source to the load without reflections, the required conditions are:

(a) The resistance of the load should be equal to that of source ($R_L = R_S$).
(b) The reactance of the load should be equal to that of source but opposite in sign (i.e. $jX_L = -jX_S$), for cancelling it. Thus if the source is inductive, then the load has to be capacitive and vice versa.

As the mismatch of load leads to power loss, several ways of matching are used e.g. quarter wave line transformer, single-stub/double-stub/triple-stub matching.

2.7.1 Power Losses Due to Impedance Mismatch

If the line impedance is not matched with the load, then part the energy gets reflected back and standing waves are formed, leading to power loss where VSWR = $S > 1$ and reflection coefficient $\Gamma > 0$.

In this situation, the voltage maxima and minima take place, where at these locations on the line, the forward and reflected voltages add in phase and substract being out of phase, respectively.

$$\therefore\ V_{\max} = |V_{\text{for}}| + |V_{\text{ref}}| = I_{\text{ref}} \cdot \left(\frac{Z_L - Z_0}{2}\right) \cdot (1 + |\Gamma|) \quad (2.139)$$

$$V_{\min} = |V_{\text{for}}| - |V_{\text{ref}}| = I_{\text{ref}} \cdot \left(\frac{Z_L - Z_0}{2}\right) \cdot (1 - |\Gamma|) \quad (2.140)$$

$$S = V_{\max} V_{\min} = (1 + |\Gamma|)(1 + |\Gamma|) \quad (2.141)$$

The total power delivered to the load is then given by:

$$P_L = |V_{\max}| \cdot |V_{\min}|/Z_0 = \left(|V_{\text{for}}|^2 - |V_{\text{ref}}|^2\right)/Z_0 \quad (2.142)$$

$$P_L = (P_S - P_{\text{ref}}) = (\text{source power} - \text{reflected power})$$

The reflected power as the power loss and therefore the fractional power lost is given by $|\Gamma|^2 = [(S-1)/(S+1)]^2$. Therefore for $S = 1$, $\Gamma = 0$ (i.e. no reflections) power loss → 0 as depicted in Fig. 2.22.

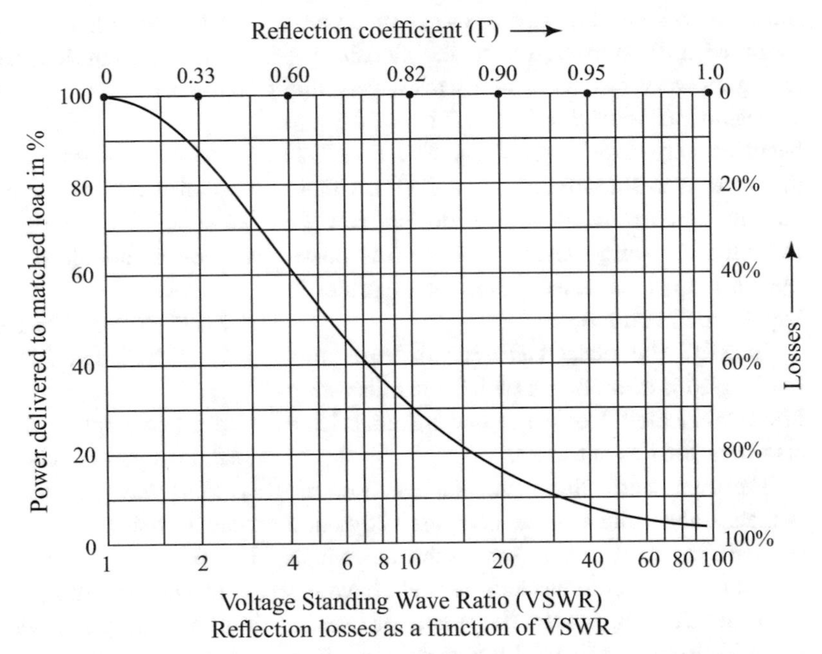

Fig. 2.22 Variation of power to the load and losses as a function of VSWR > 1, due to mismatch and reflection coefficient (Γ)

2.7.2 Quarter Wave Transformer for Impedance Matching

A quarter wave line length (or an odd multiple of $\lambda/4$) has the special property of behaving like an impedance inverter i.e. transforms high impedance (at its output end) to low impedance (as seen from its input end) (Fig. 2.23).

For impedance matching of load Z_L with source impedance Z_S, we have to choose a $\lambda/4$ line of impedance Z'_0, such that $Z_S = (Z'_0)^2/Z_L$ i.e. $Z'_0 = \sqrt{Z_S \cdot Z_L}$.

Therefore a short end load ($Z_L = 0$), and $Z_S = \infty$ (open at input end), behaves like a parallel *LC* circuit as seen from input *AB* side. While open ended load ($Z_L = \infty$), as $Z_S =$ i.e. input side looks like a series *LC* circuit.

Other cases when $l > \lambda/4$ or $l < \lambda/4$ with open and short load are depicted in Fig. 2.24.

To avoid the tedious mathematical methods of matching in transmission line, Philips H Smith devised a simple graphical method, for computing various parameters as well as for impedance matching, called the Smith chart.

2.7.3 Smith Chart and Its Applications

The Smith chart is a polar impedance diagram which is used quite frequently to study the impedance or admittance transformation in a transmission line or waveguide.

Since Γ and Z_L are complex quantities, we may write

$$\Gamma = \Gamma_r + i\Gamma_i \text{ and } \frac{Z_L}{Z_0} = r + jx$$
$$= \text{normalized impedance} \quad (2.143)$$

As we know that:

$$\frac{Z_L}{Z_0} = r + jx = \frac{1+\Gamma}{1-\Gamma} = \frac{1+\Gamma_r + j\Gamma_i}{1-\Gamma_r - j\Gamma_i}$$

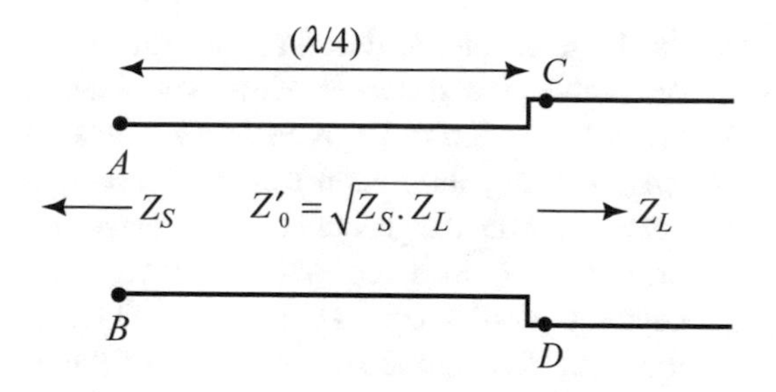

Fig. 2.23 A λ/4 line acts as impedance transformer

l	Short ended load	Open ended load
$= \lambda/4$	$\lambda/4$, A, B, S.C. ⟹ A, B; BSF	$\lambda/4$, A, B, O.C. ⟹ A, B; BPF
$> \lambda/4$	$> \lambda/4$, A, B, S.C. ⟹ A, B; LPF	$> \lambda/4$, A, B, O.C. ⟹ A, B; HPF
$< \lambda/4$	$< \lambda/4$, A, B, S.C. ⟹ A, B; HPF	$< \lambda/4$, A, B, S.C. ⟹ A, B; LPF

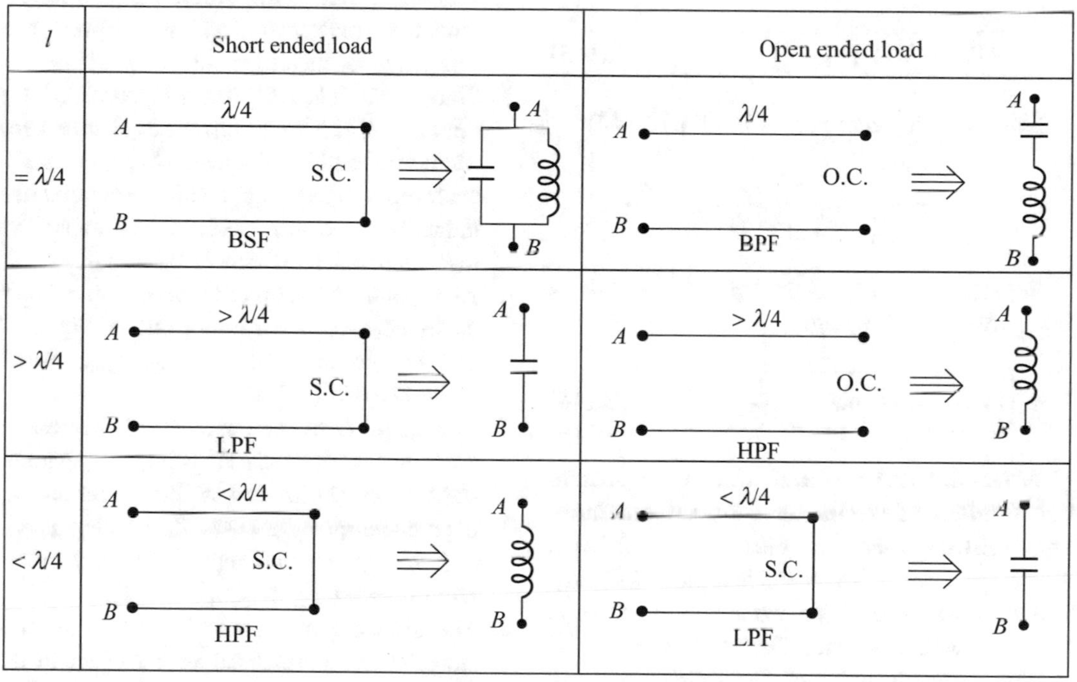

Fig. 2.24 λ/4 line as impedance transformer with l = λ/4, l > λ/4, l < λ/4 with short and open end loads, acting as band stop filter (BSF) or low pass filter (LPF) or high pass filter (HPF)

∴ After rationalisation of denominator we get:

$$\therefore \quad r + jx = \frac{1 + 2j\Gamma_i - \Gamma_r^2 - \Gamma_i^2}{(1 - \Gamma_r)^2 + (\Gamma_i)^2} \tag{2.144}$$

Equating imaginary parts of Eq. (2.144):

$$x = \frac{2\Gamma_i}{(1 - \Gamma_r)^2 + (\Gamma_i)^2}$$

$$(\Gamma_r - 1)^2 + (\Gamma_i)^2 = 2\Gamma_i/x$$

$$\therefore \boxed{(\Gamma_r - 1)^2 + \left(\Gamma_i - \tfrac{1}{x}\right)^2 = \left(\tfrac{1}{x}\right)^2} \tag{2.145}$$

Now equating real part of the Eq. (2.144):

$$r = \frac{1 - \Gamma_r^2 - \Gamma_i^2}{(1 - \Gamma_r)^2 + (\Gamma_i)^2}$$

Separating Γ_r and Γ_i in separate groups, we can prove the following:

$$\boxed{\left(\Gamma_r - \tfrac{r}{1+r}\right)^2 + \Gamma_i^2 = \left(\tfrac{1}{1+r}\right)^2} \tag{2.146}$$

Equations (2.145) **and** (2.146) **represent two families of circles for constant reactance and constant resistance** (Figs. 2.25 and 2.26) in the complex reflection coefficient plane ($|\Gamma|$, θ_L). Superimposition of these two circles is shown in Fig. 2.27 which is called the Smith chart.

The characteristics of the Smith chart are described below in terms of the normalised impedance $Z_L/Z_0 = r + jx$, the normalised admittance $Y_L/Y_o = g + jb$, and the normalised length l/λ that are given below:

1. Equation (2.145) represents the constant reactive (x) circles with radius $1/x$ and centres at (1, $1/x$), for $-\infty \leq x \leq \infty$.
2. Equation (2.146) represents the constant resistance (r) circle with radius $1/(1 + r)$ and centres at $[1/(1 + r), 0]$ on the Γ_r, axis for $0 \leq r \leq \infty$.
3. As we know that the line impedance (Z) and reflection coefficient (Γ) change as we move on the transmission line. Therefore for measurements, the distances along the line is given in the Smith Chart in terms of wavelengths along its circumference. These distances towards the generator is clockwise and towards the load as anticlockwise. One complete rotation covers a distance $\lambda/2$ along the line; the impedance and the reflection coefficient repeat themselves at these intervals, as we will come back to the same point after $\lambda/2$ on the chart.
4. The upper half of the Smith chart represents inductive reactance (jx) and lower half capacitive reactance ($-jx$).
5. Since admittance is the reciprocal of the impedance, the Smith chart can also be used for normalised admittance, where the resistance scale reads the conductance and the inductive reactance scale reads capacitive reactance and vice versa (Fig. 2.27).
6. At a point of maximum voltage, line impedance $(Z_L/Z_0)_{max} = S$ along OK of Fig. 2.27 and at minimum voltage, $(Z_L/Z_0)_{min} = 1/S$, along LO of Fig. 2.27.
7. The centre O of the chart ($S = 1$) represents the matched impedance point, the extreme right of the horizontal radius represents an open circuit point ($S = \infty$, $Z_L/Z_0 = \infty$, $\Gamma = 1$) and the extreme left represents short circuit point ($S = \infty$, $Z_L/Z_0 = 0$, $\Gamma = -1$).
8. The distances measured along the transmission line are normalised with respect to the wavelength and are measured towards the generator and also towards the load along the periphery or unit circle ($|\Gamma| = 1$) in the Smith chart (Fig. 2.27).
9. Any circle drawn with O as the centre will be constant VSWR circle (Fig. 2.28).

Smith charts are used for the following.

1. Plotting a complex impedance z on the chart and drawing a circle with O as centre and radius = OZ.
2. Finding admittance to it, as a diagonal opposite point.

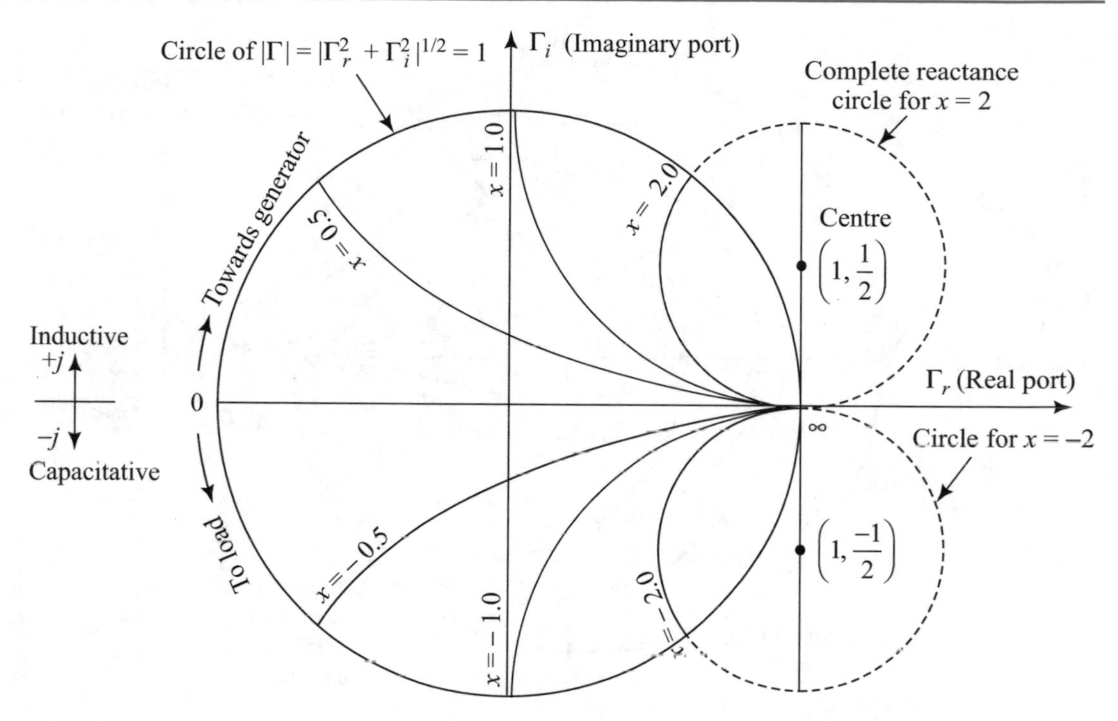

Fig. 2.25 Normalised constant reactance (x) circles on the Γ_r, Γ_i plane like conventional x–y plane. Upper half represents ($+j$) region (i.e. inductive) while lower half represents ($-j$) region (i.e. capacitive), for $x = \pm 0.5$, ± 1.0, and ± 2.0 only

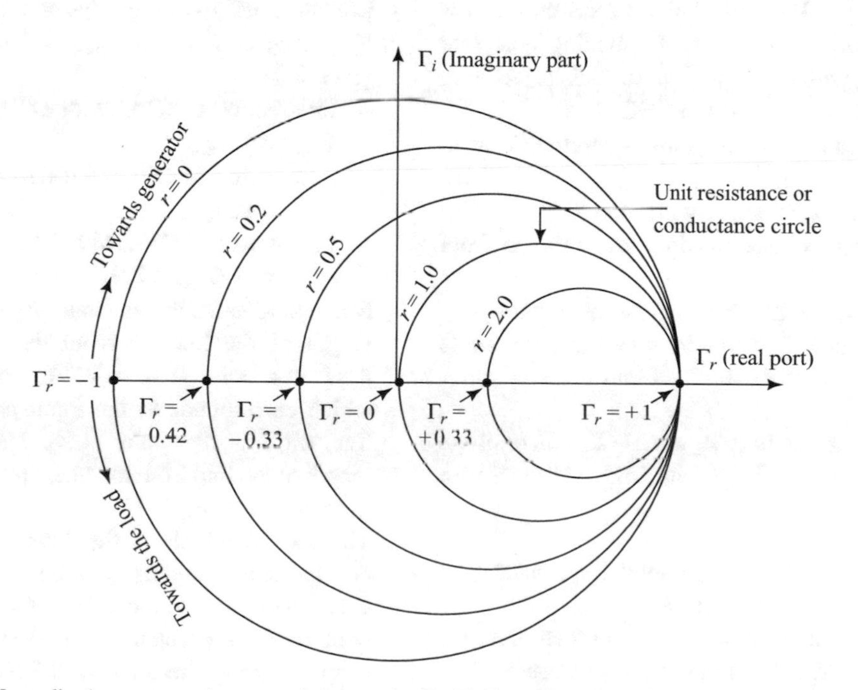

Fig. 2.26 Normalised constant resistance r circles on the Γ_r, Γ_i plane like conventional x–y plane for r = 0, 0.4, 0.5, 1.0, and 2.0 only. The values of Γ_i for a particular value of r are from Eq. 2.146 with $\Gamma_i = 0$

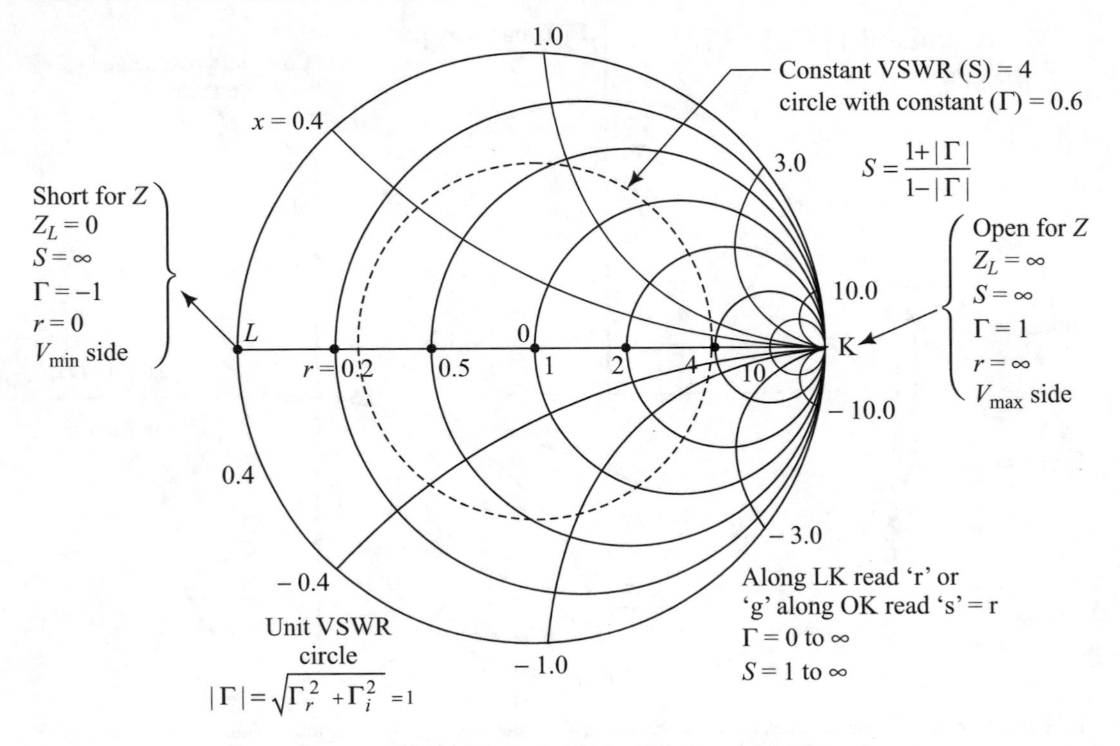

Fig. 2.27 Superimposition of r and x—circles i.e. the Smith chart for some value of x and r

3. Finding VSWR of a given load as the distance of Z point from 0, making complete circle with 'O' as centre, then reading VSWR = r, on the line OK.
4. Find input admittance of a shorted or open load.
5. Finding Z_{in} of a line with load Z_L.
6. Locating voltage maxima and minima from load.
7. Matching of line by single/double stub.
8. Phase shift by a line by measuring the angle of rotation at its start and end points.

Steps for finding Z_{in}, VSWR, Γ, voltage maxima and voltage minima using Smith chart

1. Locate the normalised load impedance z_l/z_o on the Smith chart at, say A.
2. Join centre point O and A, and then draw the circle with OA as radius and O as centre. This is the constant VSWR circle (CVC). Extend OA up to the periphery to cut it at B. Intersection of this constant VSWR circle with LK-axis lines gives the value and locations of V_{min}, r_{min} at E (V_{max}, r_{max} and S at G) as given below:
   - Intersection of CVC with LO line at E is V_{min} with r_{min}
   - Intersection of CVC with OK line at G is V_{max} with r_{max}
   - Intersection of CVC with OK line at G has $r = S = 5$ (Fig. 2.29)
3. Rotation towards the generator by normalised length of the line l/λ from the load point A gives a point D on CVC the outer circle which corresponds to the input point of the line, and z at that point is Z_{in}. Here l is the length of the line from load i.e. arc AD, with l/λ measured along B–C.
4. The movement along the transmission line and the corresponding rotation on the Smith chart, the reflection coefficient and VSWR both remains constant in lossless line. For example along the arc AD VSWR = 5 and correspondingly, with reflection coefficient as $\Gamma = (S - 1)/(S + 1) = 0.66$.

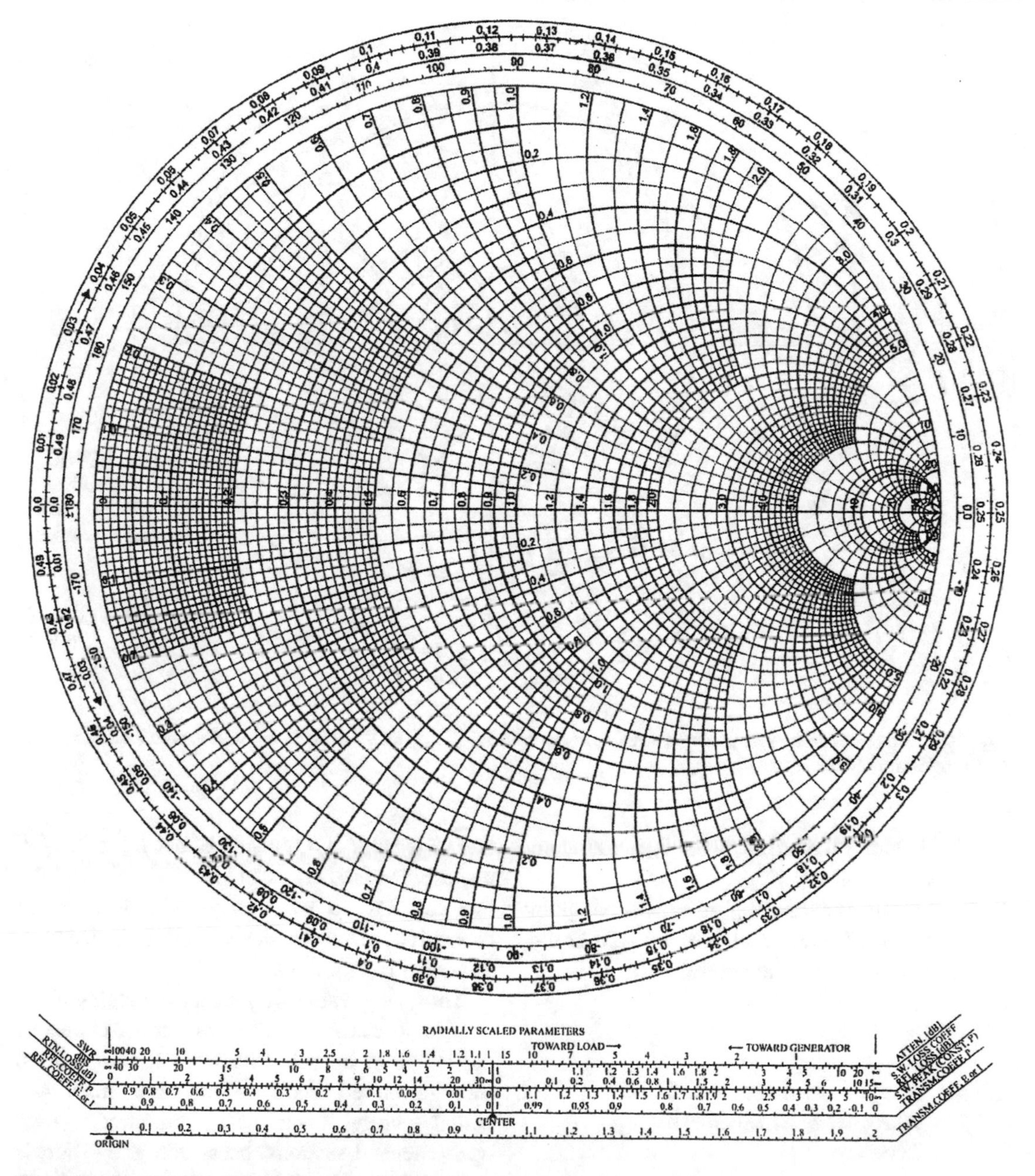

Fig. 2.28 Smith chart: Constant r and x circles in the reflection coefficient plane

5. A circular arc is drawn with O as the centre that passes through A and intersects the line OC at D and the reading of impedance of D gives the required normalised input impedance (Z_{in}/Z_o) of the transmission line as $(0.4 - 2.0\,j)$.

6. The voltage minima (point E) and voltage maxima (point G) in Fig. 2.29, are the intersection points of constant VSWR (S=5) circle (i.e. arc ADEG) with the horizontal line, which is a pure resistance line (reactance = 0). Therefore their distances from load in terms

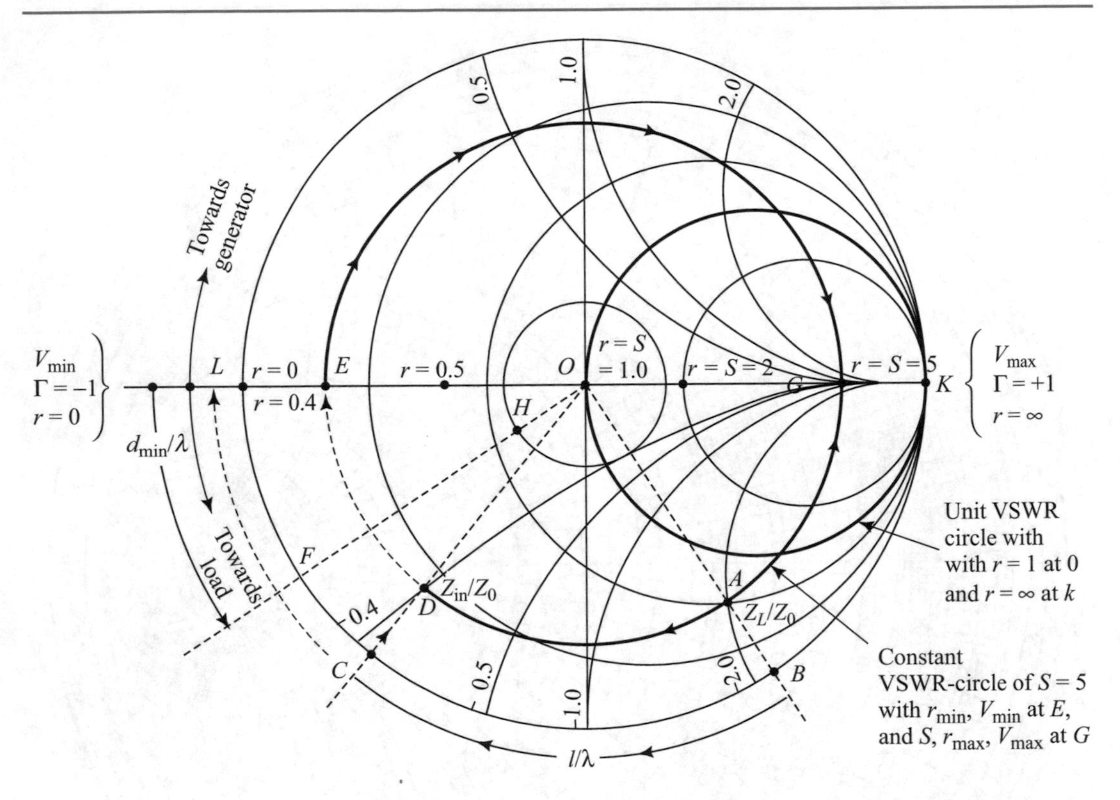

Fig. 2.29 Determination of input impedance (Z_{in}) value and V_{max}, V_{min} locations in a lossless line with load Z_L at point A, using Smith chart

of λ is along the arc BCL for V_{min} and along the arc BCLC for V_{max}.

7. The voltage maxima V_{max} is intersection point of constant VSWR circle i.e. are ADEG distance in terms of λ at point G is V_{max}.

2.7.4 Single- and Double-Stub Matching in Lossless Lines

For a lossless line the Eq. (2.1p) gives:

$$\gamma = j\beta \text{ and } Z_{in} = Z_0\left(\frac{Z_i + Z_0 \tan h(\gamma l)}{Z_0 + Z_L \tan h(\gamma l)}\right)$$

$$= Z_0\left[\frac{Z_i + Z_0 \tan(\beta l)}{Z_0 + Z_L \tan(\beta l)}\right]$$

Therefore the impedance of an open-circuited lines $Z_L = \infty$ or a short-circuited $Z_L = 0$ lossless line of length l is a function of cot $\beta\, l$ or tan $\beta\, l$, respectively [see Eqs. (2.1q) and (2.1r)], and the line can have a wide range of inductive and capacitive reactances, depending on the length l, as shown in Fig. 2.30.

Thus, any value of reactance ranging from $-\infty$ to ∞ can be obtained by proper choice of the length of open circuit or short circuit lines. **Short-circuited sections of such lines are called stubs** and are preferred, over open-circuited sections, because a good short is easily achieved in all kinds of transmission lines.

(a) Single-Shunt Stub Matching in a Lossless Line

When a complex load $Z_L = R_L \pm jX_L$ is required to be matched using a single stub, where the load impedance and the stub impedance appear in parallel, as shown in Fig. 2.31, the matching concepts can be explained better in

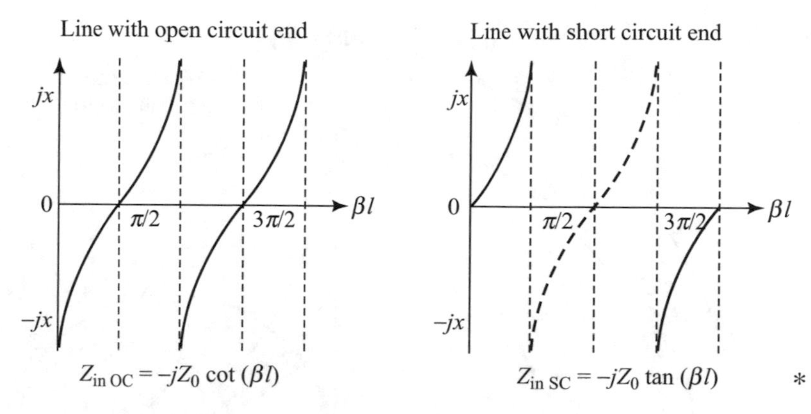

Fig. 2.30 Variation of reactance (x) of open- and short-circuited line with line length, e.g. for $\beta l = (2n + 1)\pi/2$, $Z_{in\ oc} = x = 0$, and $Z_{in\ sc} = x = \infty$

terms of admittances $y_L = \frac{Y_L}{Y_0} = g_L \pm Jb_L$, normalised by Y_0. The generator end is assumed to be matched to the characteristic impedance.

To study the value of Y_{in} at the location of V_{max} and V_{min}, we move away from the complex load towards the generator along line; there are voltage minima at which the reflection coefficient is negative real quantity. Here the input admittance is pure conductance of value $y_{in} = (1 - \Gamma)/(1 + \Gamma) = g = S$, where S is the load side VSWR i.e. AL line. Similarly there are voltage maxima at which the admittance is pure conductance of value $y_{in} = (1 - \Gamma)/(1 + \Gamma) = 1/S$. **Therefore, in between these two points (Fig.** 2.31**), there must be a point *B*, at a certain distance *d* from the load, where the real part of the normalised admittance is unity**; i.e. line impedance there at A is Z_0, so that $y_{in} = 1 \pm jb$ (Fig. 2.31). The reactive component of this admittance can be cancelled by an equal and opposite susceptance $\mp jb$ of parallel single stub of length 1, located at that point,

$$\therefore \quad y_{tot} = 1 \pm jb \mp \cot \beta l = 1; \text{ with } \cot \beta l = b \tag{2.147}$$

Therefore the following steps can be used in the design of single stub with help of the Smith chart of Fig. 2.32.

(i) Let us have the load impedance of $Z_L = (46.9 + 16.7j)\ \Omega$ with 50 Ω line. Therefore the load in normalised form will be $z_e = Z_L/50 = (0.94 + 0.37j)$ and correspondingly normalised load admittance will be $y_e = 1/z_e = (1.9 - 0.75j)$.

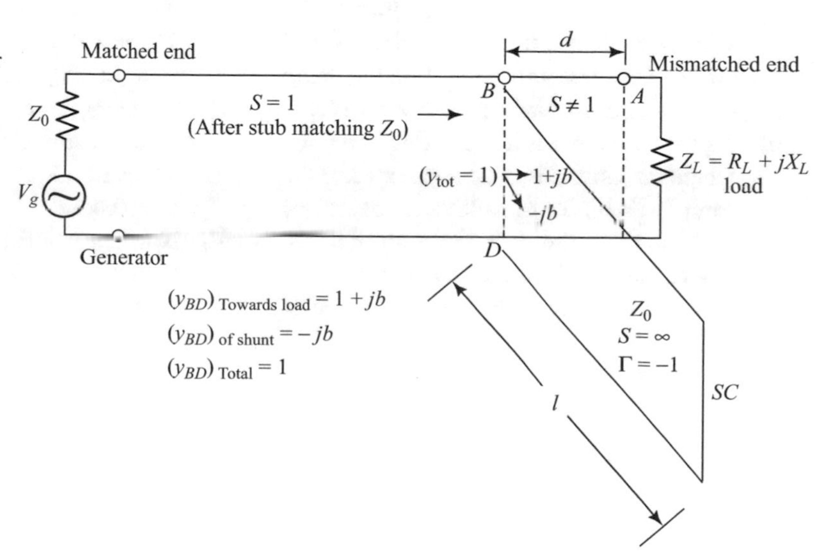

Fig. 2.31 Single-stub matching, when the generator end is a matched Z_0 line

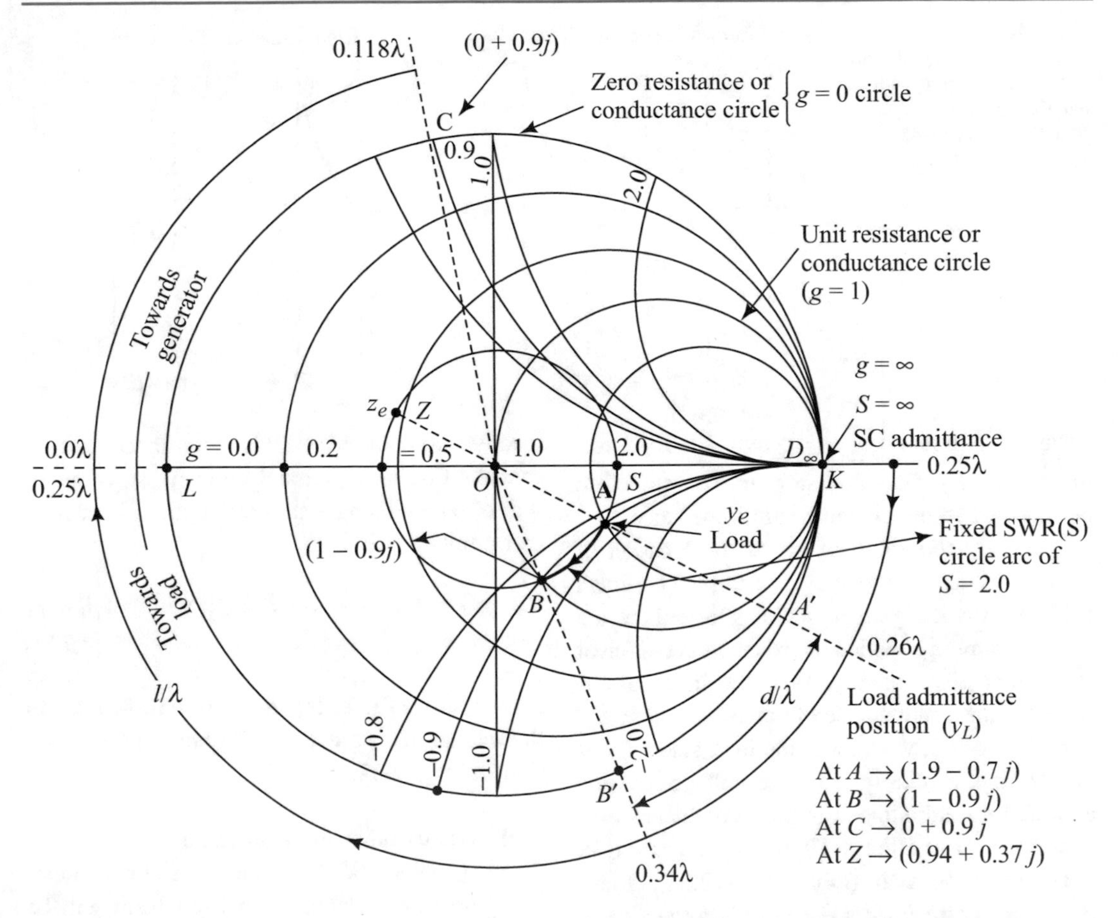

Fig. 2.32 Single-stub design using the Smith chart Note (i) Value of VSWR = 2 is read from intersection of constant VSWR circle between O and K. (ii) Value of g is read from intersection of constant VSWR circle between LK

This y_e may be plotted on the Smith chart at '*A*'. Now join *A* with *O* and draw an arc with *O* as centre to read VSWR = *S* = 2 on line *OK*. The load point *A* on the Smith chart outermost circle is 0.26 λ position.

(ii) As the short stub is to be placed at a point where the normalised admittance is of the form $(1 \pm jx)$ for cancelling the reactance part $(\mp jx)$ by the stub, draw an arc with *OA* radius clockwise to cut the circle of unit conductance at *B*, which we read to normalised admittance of $(1 - 0.9i)$ with real part as unit conductance.

Therefore *B* is the position of the short stub in Fig. 2.32 at a distance of *AB* wavelength i.e. A′(0.26 λ on Smith chart outer most circle) to *B*′(0.34 λ). Therefore '*d*' of Fig. 2.31 will be $d = (0.34\lambda - 0.26\lambda) = 0.08\ \lambda$.

(iii) Now the short stub at *B* has to be pure susceptance $(0 + 0.9j)$. For finding its length, stub being short we start from right most point K on the Smith chart clockwise on the outermost circle to reach point C $(0 + 0.9j)$. Therefore its length is K–C i.e. K–L and L, C i.e. stub length $l = (0.5–0.25\ \lambda) + 0.118\ \lambda = 0.368\ \lambda$.

(iv) For 9 GHz, $\lambda_0 = 3.33$ cm $\therefore l = 0.368\ \lambda = 1.2$ cm and $d = 0.08\ \lambda = 0.2$ cm.

(b) Double-Stub Matching in a Lossless Line

As per previous section, single stub must be located at a fixed point near the load, where the real part of the line admittance is Y_0. In microwave transmission lines, such as coaxial lines, or waveguides, it is not practical to find such an exact position to place the stub as seen in the example of single-stub case $d = 0.2$ cm. Also, this position changes for every load. Therefore matching in coaxial cables can be better achieved by the use of two stubs, placed at a fixed distance (d) from load and each separated by normally $3\lambda/8$ distance, but each having variable lengths as shown in Fig. 2.33, with the assumption that the main line and the stub lines have the same characteristic impedance.

Double-stub matching is conveniently performed by use of the Smith chart from the knowledge of the Z_L and its VSWR.

The key points of double-stub matching can be explained as (see Fig. 2.33):

(a) The final admittance at C as seen from generator should be $= Z_0$ i.e. normalised impedance $y_C = 1$, with stub-2.

(b) Prior to putting stub-2, the impedance across C should be

$$y_C = 1 \pm jb_c$$

so that the stub-2 cancels the reactance part as:

$$y_{S2} = \mp jb_c$$

(c) The line between the two stubs i.e. $l_{be} = 3\lambda/8$ acts as $3\lambda/8$ transformer, such that all admittance (y) of unit conductance circle at point C lies on the unit conductance circle rotated anticlockwise by $3\lambda/8$ (i.e. towards the load side) i.e. for point B.

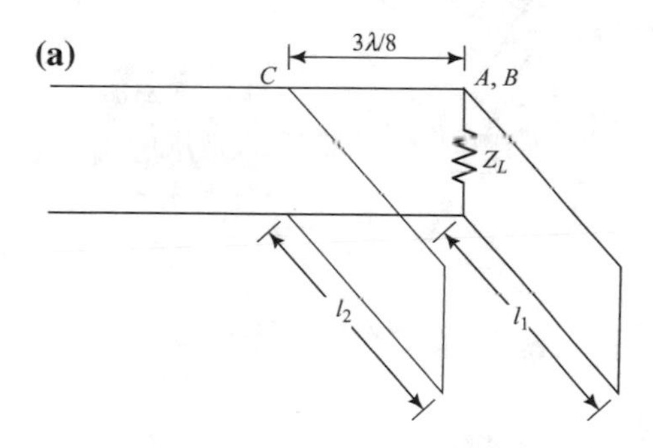

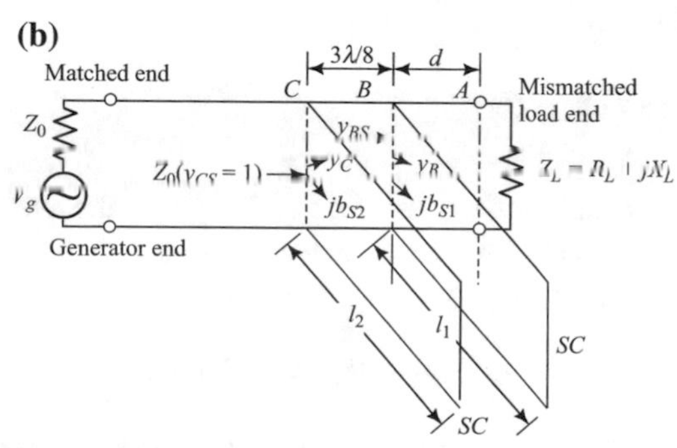

Fig. 2.33 Possible double-stub matching: **a** first stub at load end being not practical is not used at all, **b** first stub at a distance d from load is the normally used situation

An example of double-stub matching is shown in Fig. 2.34, where it is desired that the input admittance at C should be y_0 corresponding to a point C on the unit circle ($g = 1$) of the Smith chart, having its centre at O, before the addition of the second stub. **A movement towards the load from C by 3 λ /8 makes all points of the unit conductance circle to rotate by the same amount, so that in effect, the entire unit conductance circle rotates**. Stub-1 adds susceptance to the load such that the resulting admittance lies on the rotated circle, and consequently, a $3\lambda/8$ movement towards the generator makes the admittance at C to lie on $g = 1$ circle. The second stub then cancels susceptance ib_B and provides matching. The following steps are involved with symbols like y_B and y_C as admittance at points B and C without stub, while y_{Bs} and y_{CS} as admittance at points B and C, with stub. The susceptances of the two stubs are indicated as y_{s1} and y_{s2}. Steps are summarised on the Smith chart.

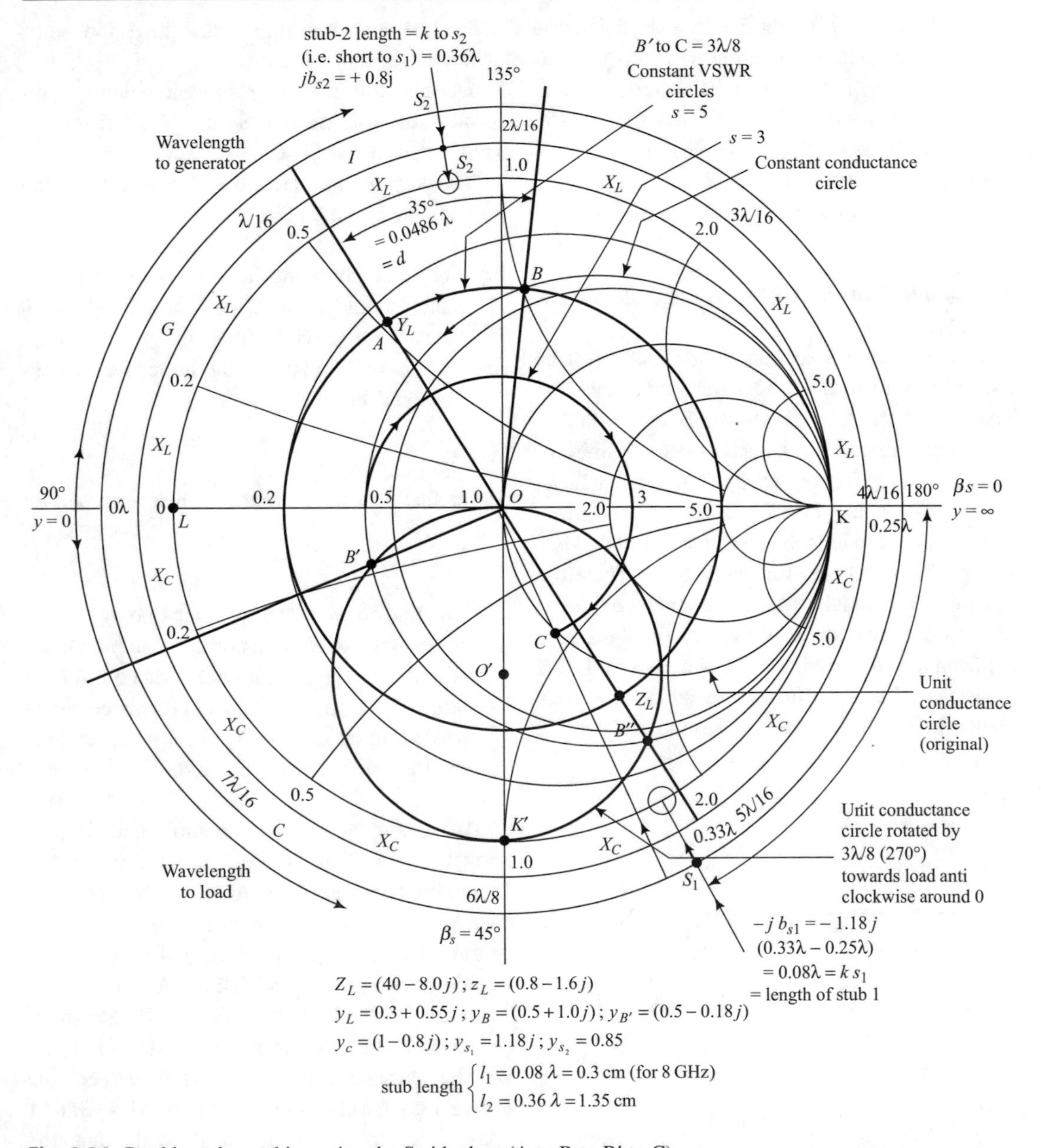

Fig. 2.34 Double-stub matching using the Smith chart (A to B to B′ to C)

1. The normalised load admittance (y_L = 0.3 + 0.55j) is located at A, diametrically opposite normalised to load impedance $Z_L \backslash Z_0$, on the constant VSWR circle 1 of the diameter Z_LA (Fig. 2.34).
2. From the load admittance ($y_A = g_A + jb_A$) point A, a rotation towards generator on the constant VSWR circle to B by an arbitrary distance, say, d = 0.0486 λ = 35° gives the position B of the first stub, where the normalised line admittances are found to be as image of y_L = 0.3 + j0.55 as:

$$y_B = g_B + jb_B = (0.5 + \mathrm{j}1.0)$$

3. The unit conductance circle (g = 1) is rotated around the Smith chart centre O_1 towards the load by $3\lambda/8$ (i.e. 270°), where the new centre

of the circle unit conductance becomes O′ on the vertical line OK′.

4. Reading of the point of intersection (i.e. B′) between rotated unit conductance circle and constant conductance circle g_B becomes the value of y_{BS} of Fig. 2.33. Now with OB′ as radius, we get the constant VSWR circle of segment of line B′C (Fig. 2.33). Read $y_B{}' = y_{BS} = (0.5 - 0.18j)$
As

$$y_{B'} = y_{BS} = y_B + y_{s1}, \quad \text{where } y_{s1} = jb_{s1}$$

$$\therefore \quad y_{Si} = y_{BS} - y_B = (0.5 - 0.18j) - (0.5 + 1.0j) = -1.18j$$

Thus, b_{s1} the susceptance of the first stub is determined, as we know y_{BS} and y_B from the Smith chart. Hence the length l_1/λ of stub is obtained from the Smith chart, starting from short position K.

5. From the point B′, rotate towards generator by $3\lambda/8$ on the constant VSWR circle of line B′C (Fig. 2.33), to reach the point C on the circle (Fig. 2.34). The total normalised admittance without stub-2 at C for the original line and first stub is

$$y_C = 1 + jb_C = (1 - 0.8j) \tag{2.148}$$

6. For matching, the total admittance at C with stub should be

$$y_{CS} = 1.0$$

This is achieved by adjusting the length l_2 of the second stub to provide an equal and opposite susceptance of $b_{S2} = -b_C = 0.8j$ corresponding to point C in Fig. 2.33 on the Smith chart.

7. **The stub lengths are found by rotating towards the generator** around the unit reflection coefficient circle, starting from the short point K of the Smith chart (i.e. $Y = \infty$ the right-end short termination of short-circuited stubs), clockwise up to the point of corresponding stub susceptance read on the chart. Therefore,

For short stub S_1 of $jb_{S1} = -1.18j$; $l_1 = 0.08\ \lambda$
For short stub S_2 of $jb_{S2} = +0.8j$; $l_2 = 0.36\ \lambda$.
Actual lengths for 8 GHz (i.e. $\lambda = 3.75$ cm); $l_1 = 0.3$; $l_2 = 1.35$ cm

Summary of Steps for Double-Stub Matching (A to B to B′ to C)

1. Plot Y_L at A on the Smith chart (Fig. 2.34).
2. Rotate WLTG along its constant VSWR ⊙ of S = 5 up to length d = AB of our choice of convenience, to reach the point B of stub-1. This VSWR is of RHS of stub no. 1 (segment AB of Fig. 2.33). Here the impedance is called image load of Z_L at A.
3. Rotate WLTL on the constant conductance ⊙ of image load, so as to intersect the rotated unit ⊙ at B′.
4. Note this new constant VSWR ⊙ of S = 3 of B′, and this VSWR is of the left-side position of point B of stub-1 i.e. segment BC (Fig. 2.33).
5. Now rotate WLTG by $3\lambda/8$ (i.e. distance between stubs 1 and 2) along this second, constant VSWR of S = 3 to reach C (the stub-2), the intersect on original unit conductance circle (Fig. 2.34).
6. Read the values of reactances of point B, B′ (stub-1) and point C (stub-2).
7. Using normal Smith chart get the lengths of short stubs 1 and 2 as distance measured from short load point K by moving WLTG corresponding to their reactances. K–S_1 for stub-1 and K–S_2 for stub-2 (Fig. 2.34).
Here
WLTG = Wavelengths towards generator
WLTL = Wavelengths towards load

Note: When we move from B over the constant VSWR circle S = 5 to meet the rotated unit conductance circle, we get first intersection at B′ and second intersection at B″; therefore, B″ also gives second solution for the two stubs.

Limitation

The double-stub system cannot match all impedance. If the normalised load conductance exceeds $1/\sin^2(\beta \cdot s)$, double stub cannot be used, since the matching condition will not be satisfied for those values of g_L, where s is the separation distance between the stubs. Also alternative, if $g_L > 2$, there will be no point of intersection between the rotated circle and g = 1 circle, and no solution will be possible. Therefore $g_L = 2$ circle becomes prohibited region.

Solved Examples

Problem 1 A rectangular waveguide 1 cm × 3 cm operates at 10 GHz in the dominant mode in TE_{10} mode. Calculate the maximum pulse power it can handle if the maximum potential gradient is 2 kV/cm.

Solution

$$\lambda_{c10} = 2a = 2 \times 3 = 6 \text{ cm}$$

$$\lambda_0 = c/f = 3 \times 10^{10}/(10 \times 10^9) = 3 \text{ cm}$$

$$\therefore \quad \lambda_g = \frac{\lambda_0}{\sqrt{1 - \left(\frac{\lambda_0}{\lambda_{c10}}\right)^2}} = \frac{3}{\sqrt{1 - \left(\frac{3}{6}\right)^2}}.$$

$$= \frac{3 \times 2}{\sqrt{3}} = 2\sqrt{3} = 3.464 \text{ cm}$$

$$\begin{aligned}\text{Max. Pulse Power} &= \left(6.63 \times 10^{-4}\right) \cdot E_{\max}^2 \cdot a \times b\left(\lambda_0 \lambda_g\right) \\ &= 6.63 \times 10^{-4} \times (2 \times 10^3)^2 \times 1 \times 3 \times 3/2\sqrt{3} \\ &= 6.63 \times 2 \times 3\sqrt{3} \times l0^2 \\ &= 6.63 \times 6 \times 1.732 \times 100 \\ &= 6.89 \text{ kW}\end{aligned}$$

Problem 2 A dielectric-filled rectangular waveguide operates at dominant mode TE_{10}. Dielectric constant = ε_r = 9: a × b = 7 cm 3.5 cm. Calculate f_c, v_p and λ_g at 2 GHz.

Solution

$$f_{c10} = \frac{1}{2\sqrt{\mu\varepsilon}}\sqrt{\frac{m^2}{a^2} + \frac{n^2}{b^2}}$$

$$= \frac{c}{2\sqrt{\varepsilon_r}} \cdot \frac{1}{a} = \frac{3 \times 10^8}{2 \times \sqrt{9}} \times \frac{1}{7 \times 10^{-2}}$$

$$= 0.714 \text{ GHz}$$

$$V_p = \frac{\omega}{\beta_g} = \frac{c}{\sqrt{\varepsilon_r} \cdot \sqrt{1 - \left(\frac{f_c}{f}\right)^2}}$$

$$= \frac{3 \times 10^8}{\sqrt{9} \times \sqrt{1 - \left(\frac{0.714}{2}\right)^2}} = 1.07 \times 10^8 \text{ m/s}$$

$$\lambda_g = 2a = 2 \times 7.0 = 14.0 \text{ cm}$$

Problem 3 An air-filled X-band waveguide has a × b = 2.286 cm × 1.016 cm. Find the cut-off frequencies of most dominant TE and TM modes.

Solution In rectangular waveguides the dominant TE mode is TE_{10} and TM mode is TM_{11} and for air filled $\varepsilon_r = 1$.

$$\therefore f_{c10} = \frac{c}{2\sqrt{\varepsilon_r}} \cdot \frac{1}{a} = \frac{c}{2a} = \frac{3 \times 10^8}{2 \times 2.206 \times 10^{-2}}$$

$$= 3.28 \times 10^9 = 3.28 \text{ GHz } (TE_{10})$$

$$f_{c11} = \frac{c}{2}\sqrt{\frac{1}{a^2} + \frac{1}{b^2}} = \frac{3 \times 10^8}{2} \times \sqrt{0.19 + 0.97}$$

$$= 16 \text{ GHz } (TM_{11})$$

Problem 4 In an air-filled waveguide of a = 10 cm, b = 8 cm, find out the modes possible below 4 GHz.

Solution

$$f_0 = 4\text{ GHz};\quad \lambda_0 = c/f_0$$
$$= \left[3\times 10^8/(4\times 10^9)\right] = 0.075\text{ m};$$
$$f_c = \frac{c}{\lambda_c}$$
$$\lambda_{cmn} = 2/\sqrt{(m/a)^2 + (n/b)^2}\ \text{and}$$
$$f_{cmn} = c/\lambda_{cmn}$$

Calculation gives:

Modes	m	n	λ_{Cmn}	f_c[GHz]
TE_{10}	1	0	0.2	1.5
TE_{01}	0	1	0.16	1.875
TE_{11}	1	1	0.125	2.4
TE_{20}	2	0	0.10	3.0
TE_{02}	0	2	0.08	3.75
TE_{21}	2	1	0.087	3.54
TE_{12}	1	2	0.074	4.05 > 4.00
			($<\lambda_0$)	Not possible

The last mode and higher modes will not propagate as $\lambda_{c12} < \lambda_o = 7.5$ cm, and $f_{cmn} > f_o$, while the rest of the modes will propagate as the condition of f < 4 GHz applied.

Problem 5 If a rectangular guide with a:b = 2:1 has waveguide wavelength of 5 cm for operating frequency of 10 GHz, calculate the values of a, b, v_p, and v_{groups}.

Solution The TE_{10} mode is the dominant mode, and for this, we know that $\lambda_c = 2a$. Also we know that

$$\frac{1}{\lambda_0^2} = \frac{1}{\lambda_g^2} + \frac{1}{\lambda_c^2}\quad\text{and}$$
$$\lambda_0 = \frac{c}{f} = \frac{3\times 10^{10}}{10\times 10^9} = 3.0\text{ cm}$$
$$\therefore\ \frac{1}{3^2} = \frac{1}{(2a)^2} + \frac{1}{5^2}$$
$$\therefore\ a = \frac{3\times 5}{4\times 2} = \frac{15}{8} = 1.875\text{ cm}$$

and as per given condition $\therefore b = a/2 \therefore \lambda_c = 2a = 3.74$ cm and

$$V_p = \frac{c}{\sqrt{1-(\lambda_0/\lambda_c)^2}}$$
$$= \frac{3\times 10^{10}\text{ cm/s}}{\sqrt{1-(3/3.74)^2}}$$
$$= 3.74\times 10^{10}\text{ cm/s}$$

and

$$V_{\text{group}} = \frac{c^2}{V_p} = \frac{(3\times 10^{10})^2}{3.74\times 10^{10}}$$
$$= \frac{9}{3.74}\times 10^{10}\text{ cm/s}$$
$$= 2.4\times 10^{10}\text{ cm/s}$$

Ans. a = 1.9 cm, b = 0.95 cm, $V_p = 3.74\times 10^{10}$ cm/s. and

$$V_{\text{group}} = 2.4\times 10^{10}\text{ cm/s}$$

Problem 6 Normally in the laboratories X-band (8.2–12.4 GHz) waveguides are used, which has inner dimension as (2.286 × 1.016) cm × cm. Find the TE modes which are possible.

Solution The dominant mode will be TE_{10}

$$\therefore \quad \lambda_{c10} = 2a = 4.572 \text{ cm}$$

$$f_{10} = \frac{c}{\lambda_{c10}} = \frac{3 \times 10^{10} \text{ cm/s}}{4.572} = 6.56 \text{ GHz}$$

i.e. above this frequency we can use the waveguide.

For TE_{11} mode:

$$\lambda_{c11} = \frac{2ab}{\sqrt{m^2b^2 + n^2a^2}} = \frac{2 \times 2.286 \times 1.016}{\sqrt{(2.286)^2 + (1.016)^2}} = \frac{4.45}{\sqrt{6.250}} = 1.719 \text{ cm}$$

and

$$f_{c11} = \frac{3 \times 10^{10}}{1.779} = 16.86 \text{ GHz}$$

For mode TE_{01}:

$$\lambda_{c01} = 2b = 2.032 \text{ cm} \quad \text{and}$$

$$\therefore f_{01} = \frac{3 \times 10^{10}}{2.032} = 14.8 \text{ GHz}$$

Therefore we see that for:

(1) TE_{01} mode: $\lambda_{c01} = 2.032$, f_{01} = 14.8 GHz
(2) TE_{11} mode: $\lambda_{c11} = 1.779$ cm, f_{11} = 16.86 GHz
(3) TE_{12} mode: $\lambda_{c12} = 1.455$ cm, f_{12} = 20.62 GHz
(4) TE_{21}mode: $\lambda_{c21} = 1.229$ cm, f_{21} = 24.44 GHz

All these higher modes also will be present beside the dominant mode (TE_{10}).

Problem 7 In a waveguide it is given that we want to propagate only $\lambda_{10} = 16$, $\lambda_{11} = 16$ and $\lambda_{21} = 5.6$ cm. If our source has only two frequencies f_A = 3 and f_B = 6 GHz can we have the above modes when either f_A or f_B has been put into the waveguide.

Solution For f_A = 3 GHz:

$$\lambda_{0A} = \frac{c}{f_A} = \frac{3 \times 10^{10}}{3 \times 10^9} = 10 \text{ cm}$$

Here only λ_{10} can propagate as $\lambda_{10} > \lambda_{0A}$ is the condition to propagate, as for other given modes $\lambda_{11}, \lambda_{21} \not> \lambda_{0A}$

For f_B = 6 GHz:

$$\lambda_{0B} = \frac{3 \times 10^{10}}{6} = 5 \text{ cm}$$

As $\lambda_{c10} > \lambda_{0B}$, $\lambda_{c11} > \lambda_{0B}$, $\lambda_{c21} > \lambda_{0B}$, is true all the three can propagate.

Problem 8 In circular waveguides the dominant mode of microwave propagation is TE_{11}. If our source is 6 GHz and the condition of dimension of the waveguide is $f_c = 0.8\ f_0$ for TE_{10} mode, then find the dia of guide and V_g. If it has a dielectric media inside with $\varepsilon_r = 4$, find the new V_g.

Solution We know that the most dominant mode is TE_{11} and for this:

For diameter:

$$\lambda_{c11} = 2\pi r/1.841 \quad \text{and}$$
$$f_c = 0.8 \times 6 \text{ GHz} = 4.8 \text{ GHz}$$

Also

$$\lambda_{c11} = c/f_{cn} = 3 \times 10^{10}/4.8 \times 10^9 = 6.25 \text{ cm}$$

$$\therefore \; 2\pi r = (\lambda_{c11}) \times 1.841 \quad \text{and}$$

$$\text{diameter} = 2r = \frac{6.25 \times 1.841}{3.14} = 3.66 \text{ cm}$$

For V_g:

$$\lambda_0 = c/f_0 = (3 \times 10^{10}/6 \times 10^9) = 5 \text{ cm}$$

$$\therefore \lambda_g = 1/\sqrt{1/\lambda_0^2 - 1/\lambda_{c11}^2}$$

$$= 1/\sqrt{1/(5)^2 - 1/(6.25)^2} = 8.33 \text{ cm}$$

$$V_p = \frac{c}{\sqrt{1 - \lambda_0^2/\lambda_{c11}^2}}$$

$$= \frac{3 \times 10^{10}}{\sqrt{1 - (5)^2/(6.25)^2}} \frac{3 \times 10^{10}}{\sqrt{0.9744}}$$

$$= \frac{3 \times 10^{10}}{0.9935} = 3.0196 \times 10^{10} \text{ cm/s}$$

As

$$V_g = c^2/V_p = 2.981 \times 10^{10} \text{ cm/s}$$

With dielectric (ε_r = 4):

$$V_{\text{gnew}} = \frac{V_g}{\sqrt{\varepsilon_r}} = \frac{2.98 \times 10^{10}}{\sqrt{4}}$$

$$= 1.89 \times 10^{10} \text{ cm/s}$$

$$\lambda_{\text{gnew}} = \lambda_g/\sqrt{\varepsilon_r} = 8.33\sqrt{4} = 4.17 \text{ cm}$$

Problem 9 In a waveguide, only TE_{10} mode is propagating and the slotted waveguide probe detector experiment shows that the distance between maximum and minimum is 4.55 cm. If the waveguide dimension is a × b = 6 × 4 cm, find the frequency of the wave, v_p and v_g.

Solution

$$X_{\max} - X_{\min} = 4.55 \text{ cm} \quad \therefore \; \lambda_g/4 = 4.55 \text{ cm}; \quad \lambda_g = 18.2 \text{ cm}$$

Also

$$\lambda_c = 2a = 2 \times 6 = 12 \text{ cm}$$

$$\therefore \; \frac{1}{\lambda_0^2} = \left(\frac{1}{\lambda_c^2} + \frac{1}{\lambda_g^2}\right)$$

$$= \left(\frac{1}{1.44} + \frac{1}{331.24}\right) = 0.0099034$$

$$\therefore \; \lambda_0^2 = 100.3673$$

$$\therefore \; \lambda_0 = 10.0184 \text{ cm}$$

$$f_0 = \frac{c}{\lambda_0} = \frac{3 \times 10^{10} \text{ cm/s}}{10.0184 \text{ cm}} = 2.994 \text{ GHz}$$

Group velocity is the velocity of wave in the waveguide, which is $<c$.

Also we know

$$v_p \times v_g = c^2 \text{ and } v_p = \frac{c}{\sqrt{1 - \left(\frac{\lambda_0}{\lambda_c}\right)^2}} > c$$

$$\therefore \; V_p = \frac{3 \times 10^{10}}{\sqrt{1 - 100.367/144}}$$

$$= \frac{12 \times 3 \times 10^{10}}{\sqrt{143.63}} = 5.45 \times 10^{10} \text{ cm/s}$$

$$V_g = c^2/V_p = 1.65 \times 10^{10} \text{ cm/s}$$

Ans: f_0 = 2.944 GHz, $V_p = 5.45 \times 10^{10}$ cm/s, $V_g = 1.65 \times 10^{10}$ cm/s.

Problem 10 A hollow rectangular waveguide has dimensions of $a \times b = 1.5 \times 1.0$ cm. Calculate the attenuation factor for the frequency of 6 GHz.

Solution We know that for the dominant mode TE_{10}, $(\lambda_c)_{10} = 2a = 2 \times 1.5 = 3$ cm

$$(f_c)_{10} = \frac{c}{\lambda_{c10}} = \frac{3 \times 10^{10}}{3} = 10 \text{ GHz}$$

Therefore signal below this frequency will not pass. The signal of 6 GHz also will not pass as its attenuation will be very high and it will be:

$$\alpha = \sqrt{(m\pi/a)^2 + (n\pi/b)^2 - \omega^2\mu\varepsilon}$$
$$= \sqrt{(\pi/0.015)^2 + 0 - (2\pi \times 6 \times 10^9)^2 \cdot 4\pi \times 10^{-7} \times 8.854 \times 10^{-12}}$$
$$= 167.5\,\text{Np/m} = 1456.5\,\text{dB/m}(1\,\text{Np} = 8.686\,\text{dB})$$

Problem 11 In a waveguide of size 2.3×1.0 cm, the dominant mode is 10 GHz. Find the

(i) Max. electric field for max power of 746 W
(ii) Breakdown power.

Solution The breakdown power for TE_{10} mode is given below.

We know that:

$$P_{bd} = 597\,\text{ab}\sqrt{1 - (\lambda_0/\lambda_g)^2}\ \text{kW}$$

Here

$$\lambda_0 = (c/10 \times 10^9) = 3\ \text{cm};$$
$$\lambda_g = 2a = 2.3 \times 2 = 4.6\ \text{cm}$$
$$\therefore\ P_{bd} = 597 \times 2.5 \times 10 \times \sqrt{1 - (3/4.6)^2}$$
$$= 1041\ \text{kW}$$

And

$$P_{max} = 746\text{W} = 6.63 \times 10^{-4} E_{max}^2 (\lambda_0/\lambda_g)\text{W}$$
$$746 = 6.63 \times 10^{-4} E_{max}^2 \times (3/4.6)$$
$$E_{max}^2 = \frac{746 \times 4.6 \times 10^4}{3 \times 6.63}$$
$$= \left(\frac{746 \times 4.6}{19.89}\right) \times 10^4 = 172.5 \times 10^4$$
$$\therefore\ E_{max} = 13.1 \times 10^2 = 1310\ \text{V/m}$$

Problem 12 In a waveguide of size 3×1 cm^2 the operating frequency = 9 GHz. Calculate the max power handing capacity, if the maximum electric field (i.e. potential gradient) is 3 kV/cm.

Solution

$$\lambda_0 = c/f_0 = 3 \times 10^{10}/9 \times 10^9 = 3.33\ \text{cm}$$
$$\lambda_{cl0} = 2a = 2 \times 3 = 6\ \text{cm}$$
$$\lambda_g = 1/\sqrt{1/\lambda_0^2 - 1/\lambda_{c10}^2} = 5\ \text{cm}$$
$$P_{max} = (6.63 \times 10^{-4}) E_{max}^2 \cdot ab \cdot (\lambda_0/\lambda_g)\ \text{W}$$
$$= (6.63 \times 10^{-4})(3 \times 10^3)^2$$
$$\times 3 \times 1 \times (3.33/5) = 11.9\ \text{kW}$$

Problem 13 A strip line has distance between the two ground planes of 0.32 cm. If the diameter of the conductor equivalent to the flat core of the stripline is 0.054 cm, find the characteristic impedance (z_0) and velocity of propagation (v_p) if the dielectric constant of the dielectric in between is 2.3.

Solution

$$\varepsilon_r = 2.3,\ b = 0.32\ \text{cm},\ d = 0.054\ \text{cm}$$
$$\therefore\ Z_0 = \frac{60}{\sqrt{\varepsilon_r}} \ln\left(\frac{4b}{\pi d}\right)$$
$$= \frac{60}{\sqrt{2.3}} \ln\left(\frac{4 \times 0.32}{3.14 \times 0.054}\right)$$
$$= 34.6 \times \ln(7.55)$$
$$= 34.6 \times 2.02 \cong 70\,\Omega$$

and

$$v = c/\sqrt{\varepsilon_r} = 3 \times 10^8 \sqrt{2.3} = 1.98 \times 10^8\ \text{m/s}$$

Problem 14 A microstrip line has a strip of width (w) 0.6 mm and thickness t = 0.076 mm. The dielectric is nylon phonobic board (h = 0.5 mm) (ε_r = 4.20). Calculate Z_0 and velocity of propagation of wave. If width of line is 5.0 mm, what will be new Z_0.

Solution

(a) w = 0.6 mm, h = 0.5 mm, t = 0.076 mm, $w/h \approx 1.2$

$$\therefore \quad Z_0 = \frac{87}{\sqrt{\varepsilon_r + 1.41}}, \ \ln\left(\frac{5.98\,h}{0.8w + t}\right)$$
$$= \frac{87}{\sqrt{9.20 + 1.41}} \ln\left(\frac{5.98 \times 0.5}{0.8 \times 0.6 + 0.076}\right)$$
$$= 36.755 \times \ln(6.28)$$
$$= 36.755 \times 1.8776 = 67.5$$

(b) w = 5.0 mm, h = 0.5 $\therefore w/h = 10 \gg 1$

$$\therefore \quad Z_0 = \frac{377}{\varepsilon_r} \cdot \frac{h}{w} = \frac{377}{4.2} \times \frac{0.5}{5} = 9\,\Omega$$

Thus $\boldsymbol{w/h}$ ratio controls the value of $\mathbf{Z_0}$.

Problem 15 Prove that for the same dominant mode cut-off frequency in the standard rectangular and circular waveguide, the cross-sectional areas are in the ratio 1:2 in TE mode and 1:1.5 in TM mode approximately.

Solution Let us use the index of the variables as 'r' for rectangular and 'c' for circular waveguides (WG).

(a) TE mode propagation: The cut-off frequencies for dominant:

Modes TE_{10} (rectangular WG) and TE_{11} (circular WG) are known to be

[Eqs. (2.57) and (2.117)] with usual symbols:

Rectangular WG:

$$f_{cr10} = \frac{c}{2}\sqrt{\frac{m^2}{a^2} + \frac{n^2}{b^2}} = \frac{c}{2}\sqrt{\frac{1}{a^2} + 0} = \frac{c}{2a_r}$$

Circular WG:

$$f_{cc11} = \frac{c \cdot P'_{nm}}{2\pi r} = \frac{c \cdot 1.84}{2\pi r} \text{ (from Table 2.5)}$$

For some cut-off frequency $f_{cr10} = f_{cc11}$:

$$\therefore \quad \frac{c}{2a_r} = \frac{c \cdot 184}{2\pi r} \qquad \therefore \quad \frac{r}{a} = \frac{1.84}{\pi}$$

Area ratio

$$A_c A_r = \pi r^2 (a \cdot b) = \pi r^2 (a \cdot a/2) = 2\pi (r/a)^2$$

(As in standard waveguides b = a/2.)

$$\therefore \quad A_c/A_r = 2\pi (1.84/\pi)^2 = 2 \times (1.84)^2/\pi$$
$$= 2.156 = 2$$

Therefore circular waveguide cross-sectional area is twice, requiring double the copper material to make i.e. becomes expensive.

(b) In TM mode propagation: The cut-off lowest frequency in TM mode may not be of dominant mode. These are TM_{11} mode in standard rectangular waveguide (where b = a/2) and TM_{10} mode in circular waveguide with [$(P'_{nm})_{mm}$ = 2.405 (Table 2.5)] are with usual symbols.

Rectangular WG:

$$\lambda_{c11} = 2/\sqrt{(m/a)^2 + (n/b)^2} = 2/\sqrt{1/a^2 + 1/b^2}$$
$$= 2/\sqrt{1/a^2 + (2/a)^2} = 2a/\sqrt{5}$$
$$f_{cr11} = c/\lambda_{c11} = c\sqrt{5}/(2a)$$

Circular WG:

$$\therefore \quad \lambda_{c10} = 2\pi r/(P_{nm})_{mm} = 2\pi r/2.405$$
$$f_{cc10} = c/\lambda_{c10} = 2.405c/(2\pi r)$$

For same cut-off frequency $f_{cc10} = f_{cr11}$ then:

$$r/a = 2 \times 2.405c/\left(2\pi c\sqrt{5}\right)$$
$$= 2.405/\left(\pi/\sqrt{5}\right) = 0.343$$
$$\text{Area ratio} = A_c/A_r = \pi r^2/(a \times b)$$
$$= 2\pi r^2/(a \times a/2)$$
$$= 4\pi(r/a)^2 = 4\pi(0.343)^2$$
$$= 1.478 = 1.5$$

$\therefore$ Here in TM mode also circular waveguide has larger cross-sectional area.

Review Questions

1. Explain how waveguide and transmission lines differ in frequency range of use, range of characteristic impedance, cost of the line, velocity of wave propagation, and methods of impedance matching. Why waveguides are sometimes called as high pass filter.
2. Derive the wave equation for TM wave and obtain all the field components in a rectangular waveguide. (MDU-2004)
3. What are degenerate modes? Explain why TEM mode cannot exist in metallic waveguide. (UPTU-2002)
4. Show that metal rectangular waveguide is a high pass filter. Derive the necessary formula used to prove it. (UPTU-2003)
5. Describe in brief the methods of excitation of TE_{10} TE_{20}, TM_{11} modes in rectangular metal waveguides. (UPTU-2003)
6. What is a microstrip line? How does its characteristic impedance change with width-to-height ratio? Give a reason for using a substrate having lower dielectric constant at higher microwave frequencies. (UPTU-2003)
7. Calculate the cut-off frequency of the dominant mode in a 2.5-cm-diameter circular waveguide filled with Teflon ($\varepsilon_v = 2.3$). What is its maximum operating frequency if the possibility of higher mode propagation is to be avoided? (UPTU-2003)
8. Show the field configuration of TE_{10} mode in a rectangular waveguide. Does the field direction remain same with time? (MDU-2006)
9. An air-filled rectangular waveguide of cross section $1 \times 2\ cm^2$ is operating in TE_{10} mode at frequency of 12 GHz. What is the maximum power handling capacity of the guide, if dielectric strength of the medium = 3×10^6 V/cm? (UPTU-2003)
10. What do subscripts of modes TE_{mn}, TM_{mn} designate in a rectangular waveguide of size $a \times b$. Write an expression for cut-off frequency.
11. Deduce the expressions for cut-off frequency of a waveguide. Explain the physical significance of cut-off wavelength. (MDU-2003)
12. A rectangular waveguide has $a = 4$, $b = 3$ cm. Find all TE mode which will propagate at 5 GHz. (MDU-2001)
13. An X-band waveguide normally used in the laboratory has $a = 2.286$ and $b = 1.01$ cm. Find the cut-off frequency for TE_{10}, TE_{01}, TM_{11} modes. Calculate guide wavelength (λ_g), phase velocity (v_p), and group velocity (v_g).
14. For Teflon-filled K-band rectangular waveguide of size $a = 1.069$ cm, $b = 0.432$ cm. Find the f_c for TE_{11} mode ($\varepsilon_r = 2.1$). Ans. = (25.82 GHz)
15. Design a waveguide for cut-off frequency for TE mode as $f_{10} = 14, f_{11} = 30$ GHz. (i.e.

find the values of a, b, by solving the two equations of cut-off frequency).

16. Calculate cut-off wavelength (λ_c), guide wavelength (λ_g) for a circular waveguide of internal dia = 12 cm. Ans. = ($\lambda_c = 4.477$ cm, $\lambda_g = 6.865$ cm)
17. Prove that $Z_{TEM} = Z_0$; $Z_{TE} > Z_0$, $Z_{TM} < Z_0$.
18. In a certain size of a circular waveguide TM_{11} is dominant and not TE_{01}, explain why.
19. In a microstrip line with dielectric constant is 10, $h = 0.3$ cm, and d the equivalent diameter of the strips is 0.05 cm, then find the characteristic impedance.
20. Circular waveguides are preferred over the rectangular, for longer transmission, in spite of its higher cost. Explain.
21. Explain the physical significance of phase velocity and group velocity?

Microwave Cavity Resonators

3

Contents

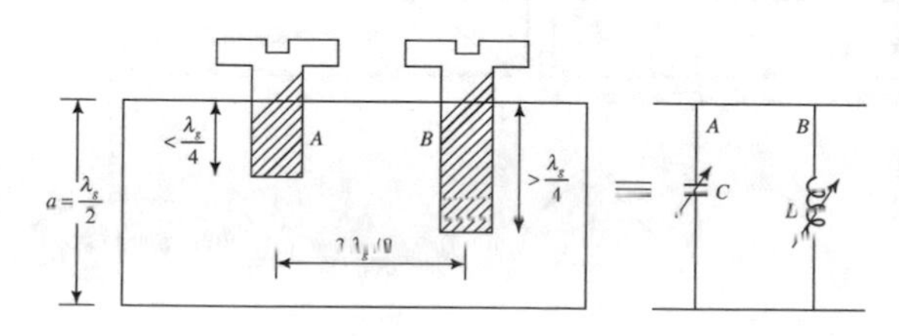

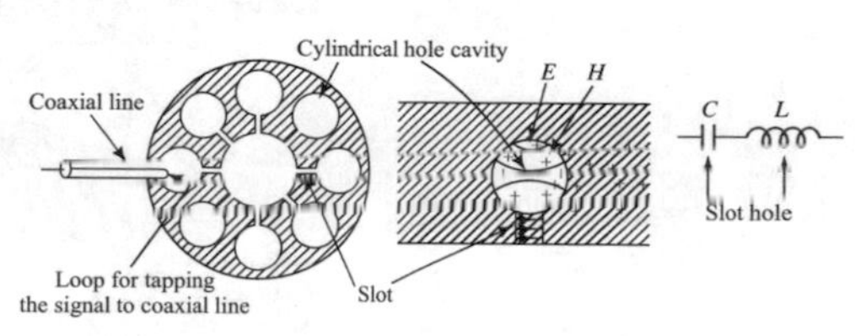

© Springer Nature Singapore Pte Ltd. 2018

P. K. Chaturvedi, *Microwave, Radar & RF Engineering*,
https://doi.org/10.1007/978-981-10-7965-8_3

3.1 Introduction

At low frequencies, the resonant circuit consists of an inductance (L) and capacitance (C) in series or parallel, which has the resonant frequency $f_0 = 1/(2\pi\sqrt{LC})$. For getting higher resonant frequencies, one can keep on reducing the lumped L and C up to a limit only, e.g. half turn of coil and stray capacitance, giving highest frequency of 1 GHz or so. At microwave frequencies, the transmission line (waveguide, strip line, etc.) has distributed L and C, which are of very low value. Once such a line is closed from the two ends, it forms a cavity, acting as having L and C in parallel. In general, a cavity resonator is a metallic enclosure where the electromagnetic energy is confined. The electric and magnetic energies stored inside the cavity determine its equivalent inductance (L) and capacitance (C), while the energy dissipated due to the finite conductivity of the walls determines its equivalent resistance (R).

The **mechanism of resonance** in a cavity can be explained by taking a waveguide having terminations (shorting plates) at a distance *of* $n \cdot \lambda_g/2$ from the source, where it is already shorted at that end (here n = 1, 2, 3 …). When the wave gets reflected back and forth, it leads to resonance type of thing and hence the name of cavity resonator (see Figs. 3.1 and 3.2). A given resonator in general has infinite number of resonant modes, and each mode has a resonant frequency. At these resonant frequencies, maximum amplitude of standing wave occurs. **The mode having the lowest resonant frequency is known as dominant mode**. There are a number of types of cavities, e.g. rectangular circular, coaxial, reflection type, transmission type, etc.

All these resonators are used as tuned circuits in various types as oscillator and amplifiers, e.g. in klystron, magnetron, etc. It is also used for frequency measurement with tuning short plunger in frequency metre.

3.2 Rectangular Waveguide Resonators of Lossless Lines

(a) ***Resonant Frequency***

From our previous chapter, we know that:

$$h^2 = (\gamma^2 + \omega^2\mu\varepsilon) = A^2 + B^2$$
$$= (m\pi/a)^2 + (n\pi/b)^2 \tag{3.1}$$

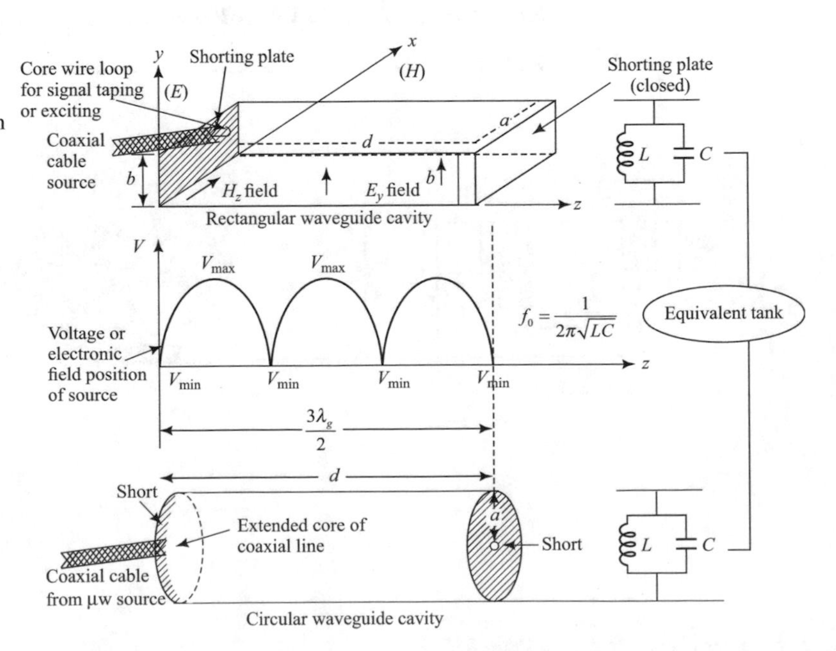

Fig. 3.1 Resonant cavity of rectangular and circular waveguide of length $n\lambda g/2$ (n = 3) with voltage max–min

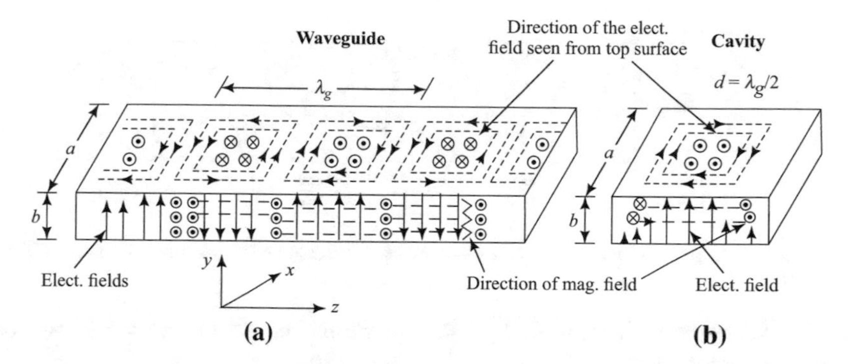

Fig. 3.2 TE-waves in a waveguide and in a cavity. Here on the surface, the magnetic field are in loops while elect. field start from a surface and terminate on another surface. The direction changes every $T/2$, i.e. $1/(2f)$ s

The lossless line has non-attenuated wave propagation, i.e. $\alpha = 0$

$$\therefore \quad \gamma = \alpha + j\beta = j\beta \text{ and hence } \gamma^2 = -\beta^2$$

Therefore, above Eq. (3.1) gives

$$\omega^2\mu\varepsilon = (m\pi/a)^2 + (n\pi/b)^2 + \beta^2 \qquad (3.2)$$

As the phase factor $\beta = 2\pi/\lambda_g$ is in radians phase shift per unit length, phase shift over a distance d is $\beta \cdot d$. This phase shift for the condition of resonance has to be multiple of π radians (i.e. multiple of $\lambda_g/2$)

$$\therefore \quad \beta \cdot d = p \cdot \pi (p = 1, 2, 3, \ldots \text{ any integer}) \qquad (3.2a)$$

$$\therefore \quad \beta = p \cdot \pi/d \text{ and } \omega_0^2 = (2\pi f_0)^2 = 4\pi^2 f_0^2; c = 1/\sqrt{\mu\varepsilon}$$

$$\therefore \text{ Eq. (3.2) gives } 4\pi^2 f_0^2 \mu\varepsilon = \left(\frac{m\pi}{a}\right)^2 + \left(\frac{n\pi}{b}\right)^2 + \left(\frac{p\pi}{d}\right)^2$$

$$\therefore \textbf{For TE/TM modes}: f_0 = \frac{c}{2}\sqrt{\left(\frac{m}{a}\right)^2 + \left(\frac{n}{b}\right)^2 + \left(\frac{p}{d}\right)^2} \qquad (3.3)$$

General mode of propagation in a resonant cavity will be TM_{mnp}, or TE_{mnp}, mode with both frequencies being same. Dominant mode for $b < a < d$ is TE_{101} (Fig. 3.2b).

(b) ***Rectangular Cavity Resonator: Field Expressions for TM_{nmp} and TE_{nmp} Modes***

We know that for TM wave $E_z \neq 0$, $H_z = 0$, and the field equation can be obtained by the wave equation given by:

$$\frac{\partial^2 E_z}{\partial x^2} + \frac{\partial^2 E_z}{\partial y^2} + \gamma^2 E_z = -\omega^2\mu\varepsilon E_z \qquad (3.4)$$

As $(\gamma^2 + \omega^2\mu\varepsilon) = h^2$ above equation becomes:

$$\frac{\partial^2 E_z}{\partial x^2} + \frac{\partial^2 E_z}{\partial y^2} + h^2 E_z = 0 \qquad (3.5)$$

Solving this equation by variable separable method we get:

$$E_z = [C_1\cos(B \cdot x) + C_2\sin(B \cdot x)][C_3\cos(A \cdot y) + C_4\sin(A \cdot y)]$$

For getting the values of C_1, C_2, C_3, and C_4, we apply the boundary conditions (as in Chap. 2) on the four walls of x, y directions with the reflection condition.

$\beta = p\pi/d$ is z direction (shorted ends) of the resonator. Finally, we get the field equation for:

$$\mathbf{TM}_{mnp}\ \textbf{mode}: \mathbf{E}_z = \mathbf{C} \cdot \sin(m\pi x/a) \cdot \sin(n\pi y/b) \cdot \cos(p\pi z/d) \cdot \mathbf{e}^{j\omega t - \gamma z} \qquad (3.6)$$

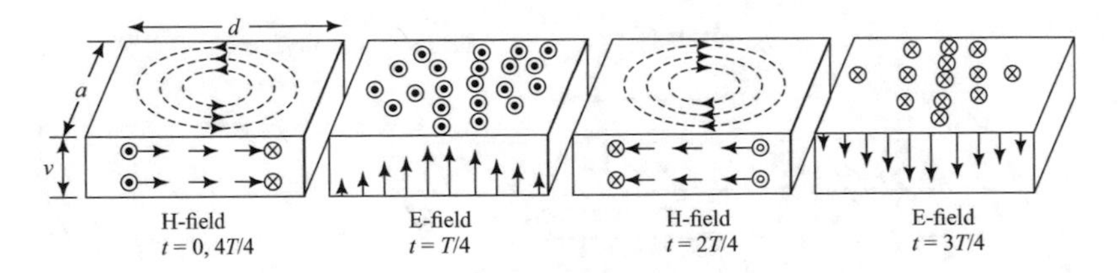

Fig. 3.3 TE_{101} mode field pattern resonating with time: *E*- and *H*-fields shown at time intervals of *T*/4

Similarly, for TE wave with $H_z \neq 0$, $E_z = 0$, we can get the field equations:

$$\mathbf{TE}_{mnp}\ \mathbf{mode} : \boldsymbol{H_z} = \boldsymbol{C}\cos(\boldsymbol{m\pi x/a}) \cdot \cos(\boldsymbol{n\pi y/b}) \cdot \sin(\boldsymbol{p\pi z/d}) \cdot e^{j\omega t - \gamma z} \tag{3.7}$$

Figures 3.3 and 3.4 give the geometry of the resonator with $a \times b \times d$ as its breadth, height, and length using the above Eqs. (3.6) and (3.7); field pattern can be sketched for various modes for a certain moment which reverses after *T*/2. Here, *m*, *n*, and *p* represent the half wave ($\lambda_g/2$) periodicity along *x*, *y*, and *z* directions, respectively.

3.3 Circular Waveguide Resonators of Lossless Line

(a) ***Resonant Frequency***

Here, the two ends are short (Fig. 3.1) with *a* = radius of the waveguide and *d* = length of the waveguide. The condition for resonance is $\beta = p\pi/d$.

In circular waveguide, we know that

$$\begin{aligned} h^2 = (\gamma^2 + \omega^2\mu\varepsilon) = h_{nm}^2 = (P_{nm}/a)^2 \\ \omega^2\mu\varepsilon = (P_{nm}/a)^2 - \gamma^2 \end{aligned} \tag{3.8}$$

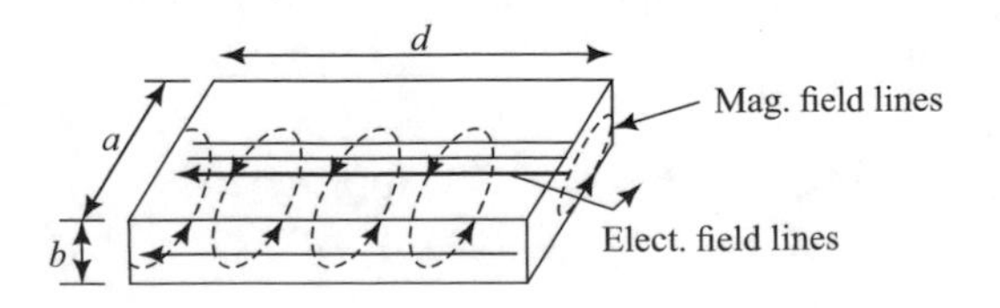

Fig. 3.4 TE_{111} mode of a rectangular cavity: fields at a particular moment, which will reverse after *T*/2

From Eq. (3.1): $\gamma^2 = -\beta^2 = -(p\pi/a)^2$, and $\omega = 2\pi f_0$.

Putting these in Eq. (3.8), we get the TM_{nmp} mode frequency as:

$$\boxed{\mathbf{TM}_{nmp}\ \mathbf{mode}\, f_0 = \frac{c}{2\pi}\left[(P_{nm}/a)^2 + (p\pi/d)^2\right]^{1/2}} \tag{3.9}$$

Similarly, it can be proved that for TE_{nmp} mode frequency:

$$\boxed{\mathbf{TE}_{nmp}\ \mathbf{mode}\, f_0 = \frac{c}{2\pi}\left[(P'_{nm}/a)^2 + (p\pi/d)^2\right]^{1/2}} \tag{3.10}$$

where *n*, *m*, *p* are integers as = 0, 1, 2, 3, … represent $\lambda_g/2$ wavelengths along ϕ, *r*, *z*, directions, respectively, and P_{nm}, P'_{nm} being Bessel's functions (see Chap. 2).

Dominant modes are TE_{111} for $2a \leq d$ and TE_{110} for $2a < d$.

(b) ***Circular Resonator: Field Expressions for TE_{nmp} and TM_{nmp} Modes***

In TE mode ($E_z = 0$), the Helmholtz equation for H_z will be:

$$\nabla^2 H_z = \gamma^2 H_Z \tag{3.11}$$

This can be solved with (*a*) resonant condition $\beta = p\pi/d$ and (*b*) standing wave condition of amplitudes of forward wave (A^+) and reflected wave (A^-) being equal ($A^+ = A^-$) at $z = 0$ to give:

$$\mathrm{TE}_{nmp}\ \text{mode:}\ H_z = C' \cdot J'n(\rho h) \cdot \cos(n\phi) \cdot \sin(p\pi z/d) \cdot e^{j(\omega t - \beta z)} \tag{3.12}$$

Similarly, for TM mode ($H_z = 0$), with the two conditions of, (*a*) resonant condition $\beta = p\pi/d$ (*b*) reflection condition $A^- = A^+$ at $z = 0$, the Helmholtz Eq. (3.11) can be solved to give:

$$\text{TM}_{nmp} \text{ mode} : E_z = C' \cdot J_n'(\rho h) \cdot \cos(n\phi) \cdot \sin(p\pi z/d) \cdot e^{j(\omega t - \beta z)} \quad (3.13)$$

Here, $J_n(\rho h)$ and $J'_n(\rho h)$ are the two Bessel's functions, C, C' are two constants, and n, m, p are integral no. of half cycle variable along, ρ, ϕ, z, the three variables of the cylindrical axis. Using Eqs. (3.12) and (3.13), the field pattern can be sketched for various modes.

The field directions in the cavity keep on changing at the microwave frequency of resonance. For a typical TE_{101} mode, a sample field of a moment is shown in Fig. 3.5. We know that electric field is from a conducting surface (+ve charge) to another conducting surface (–ve charge). In waveguides, these two surfaces get localised transient (very short timed) charges due to the field, which changes direction with the microwave frequency, and hence, the conduction current losses are there.

3.4 Coaxial Line Resonators

There are three basic configurations of coaxial line resonant cavities: (*a*) open-ended quarter-wave coaxial cavity; (*b*) short-ended half-wave coaxial cavity; and (*c*) coaxial cavity with a shortening capacitance, as shown in Fig. 3.6, along with their equivalent circuits.

Method to determine the resonant length of these cavities is by computing the total susceptance of the oscillatory system which becomes

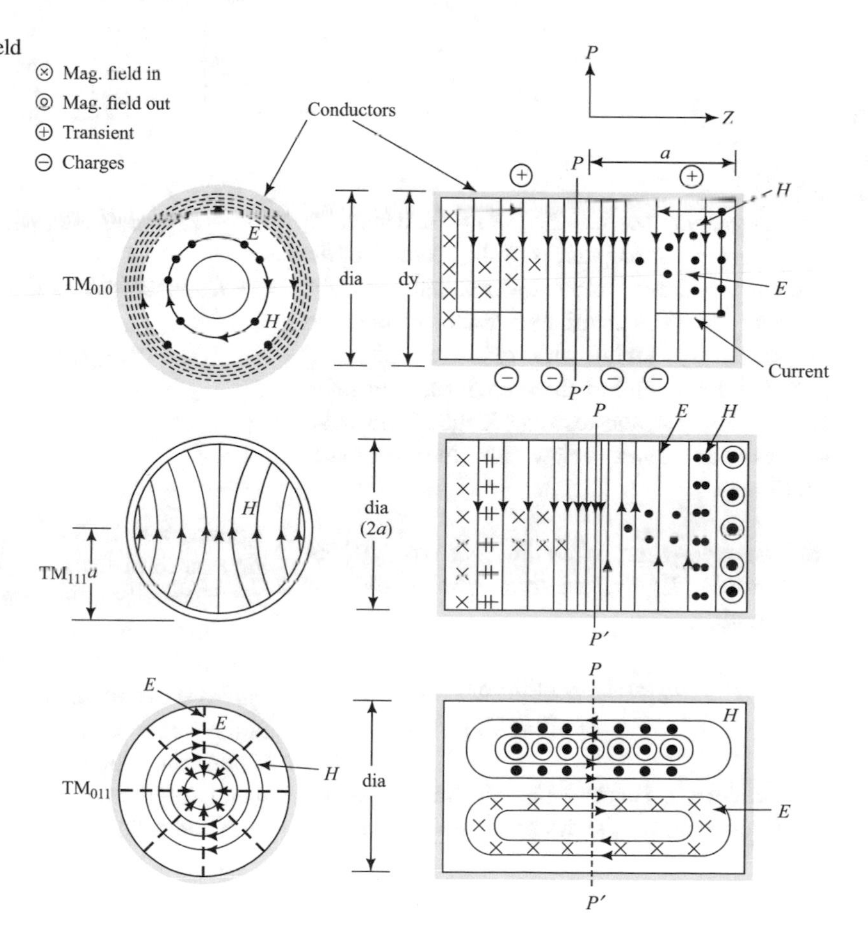

Fig. 3.5 Circular cavity field pattern configurations (left-hand side is the cross-section through PP')

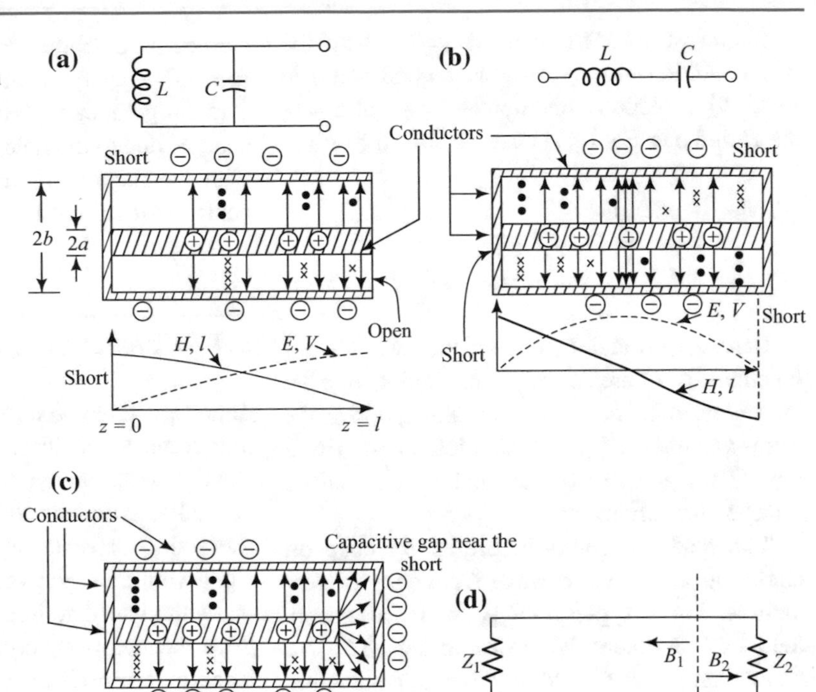

Fig. 3.6 Electric and mag. fields in coaxial cavities with equivalent circuits: **a** quarter-wave cavity **b** half-wave cavity **c** capacitive end cavity **d** all types at resonance condition

zero at resonance, i.e, $X_1 + X_2 = 0$, where, X_1 and X_2 are the susceptances of the sections of the transmission lines looking towards left and right, respectively, from an arbitrary reference plane on the line. By taking the reference plane at right-hand side end of these cavities, with reference to Fig. 3.6, the resonant length l and fields are as per given below for the resonance conditions.

(a) **Quarter-wave Coaxial Cavity (Open Ended)**: Equivalent to a parallel resonant circuit.

$$jZ_0 \tan \beta l = \text{infinite, with,}$$
$$l = (2n - 1)\lambda/4$$

(b) **Half-wave Coaxial Cavity (Short Ended)**: Equivalent to a series resonance circuit.

$$jZ_0 \tan \beta l = 0, \text{ with,} \quad l = (2n - 1)\lambda/2$$

(c) **Capacitive-End Coaxial Cavity**:

$$jZ_0 \tan \beta l = -1/j\omega_r C \text{or,}$$
$$l = \frac{\lambda}{2\pi} \tan^{-1}\left(\frac{1}{Z_0 \omega_r C}\right)$$

Here, $n = 1, 2, 3, \ldots$; Z_0 is the characteristic impedance of the coaxial line, and C is the gap capacitance between the central conductor and the shorting termination of Fig. 3.6c. The diameters a and b of all the coaxial cavities in TEM modes are restricted by the generation of the next higher-order TE and TM mode.

As the diameters of the coaxial cavities determine the power loss in the cavity, Q varies with b/a ratio and attains a maximum value at

$b/a = 3.6$. **Due to microwave radiation from the open end of the quarter-wave coaxial cavity, half-wave coaxial resonators are preferred over quarter-wave sections and are used in microwave resonant wave meters or diode detector tuning portion.**

The coaxial cavity of type (c) (Fig. 3.6) can be tuned either by changing the capacitive gap *d*, by means of a capacitive ridge at a fixed length *l*, or by changing length *l* by a variable shorting plunger at a fixed value of gap length *d*.

3.5 Re-entrant Cavity Resonator

For efficient energy transfer from an electron beam to high *Q* cavity resonators, the electron transit time across the cavity field region must be very small. Consequently, the cavity grids need to be spaced very closely to form re-entrant structure with *a* and *b* as inner and outer radius of the cylindrical structure, as shown in Fig. 3.7. The *E*-field is concentrated in the small gap *g* on this capacitance region of the capacity allows the flow of electrons through this gap, as its surface (grid) is perforated (for a klystron tube). The tuning of the cavity can be accomplished by means of short-circuit plungers, as in klystron tube. For more details, Fig. 5.8 of Chap. 5 can be referred (Figs. 5.8 and 5.13).

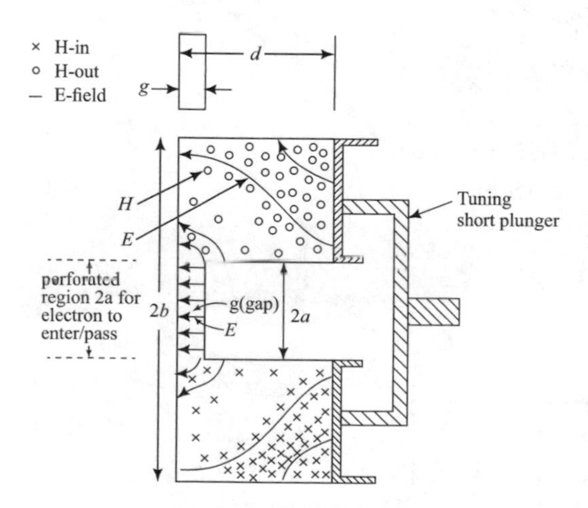

Fig. 3.7 Re-entrant cavity (Also see Figs. 5.8 and 5.13)

The re-entrant cavity of length *d* with a gap thickness $g \ll d$ may be taken as a coaxial line with radii of the inner and outer conductors as *a* and *b*, respectively. The gap capacitance *C* and the coaxial line below the gap provide equal and opposite reactances at the plane of *C* at resonance frequency ω_r with the condition.

$$1/\omega_r C = Z_0 \tan(2\pi d/\lambda_g)$$

$$\text{or,} \quad d = (\lambda_g/2\pi)\tan^{-1}(1/Z_0\omega_r C)$$

where $C = \varepsilon_0 \pi a^2/g$, and $Z_0 = 60 \ln(b/a)$ for air dielectric.

For a small capacitance *C*, $Z_0\ \omega_r\ C \ll 1$, $\beta\ d \approx \pi/2$, i.e. the line is practically a quarter-wave long. For large *C* or small gap (*g*), the length *d* should be shortened. Thus, for a given gap (*g*) or *C*, the resonant length *d* of the cavity can be varied by the short plunger, from multiple of quarter-wavelength to some smaller value for satisfying the resonant condition given above.

By increasing *C*, the electric energy stored in the cavity increases. A corresponding increase in the magnetic energy W_m has to be provided by a larger microwave current in the cavity walls, as ($W_e = W_m$ at resonance) which results in higher dissipative losses due to finite conductivity of the cavity walls. Consequently, the unloaded Q_0 of coaxial cavity decreases.

The cavity modes are: (*a*) TEM type when $d > (b - a)$; $b < \lambda_g/4$. (*b*) TM type if $d < (b - a)$ and $b = \lambda_g/4$, with electric field directed mainly along the cavity axis. The TM mode cavity can be easily tuned by adjusting the capacitive gap *g*, provided that the resonant mode is TM_{010}.

3.6 Cylindrical Hole-and-Slot Cavity Resonator

Microwave resonant cavity of the hole-and-slot type, shown in Fig. 3.8, is **used in multi-cavity magnetron oscillators**. The resonant frequency of the cavities can be determined from the relation:

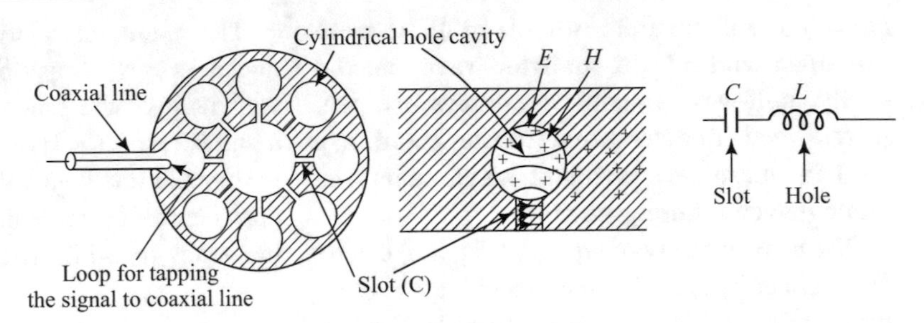

Fig. 3.8 Hole-and-slot cavity and lumped series LC equivalent circuit

$$\omega_r = 1/\sqrt{LC}$$

The function of the lumped inductance L is carried out by the surface of the hole which is equivalent to a loop of metallic band, while the lumped capacitance C is formed by the slot cut through the copper block, **with more or less no tuning possible**.

3.7 Microstrip Line Resonators

In strip lines, the resonator can be formed on the substrate, which could have different geometry, e.g. in the form of

(a) Circular ring.
(b) Circular disc.
(c) Rectangular disc.
(d) Just a gap element on the line.

In all these cases, fringe field extends beyond the physical end of the strip causing a fringe capacitance of the resonator.

(a) **Circular ring resonator**: As in Fig. 3.9, the circular ring resonator is a ring, which resonates when its mean circumference is equal to integral multiple of half the wavelength (a and b being inner and outer radii).

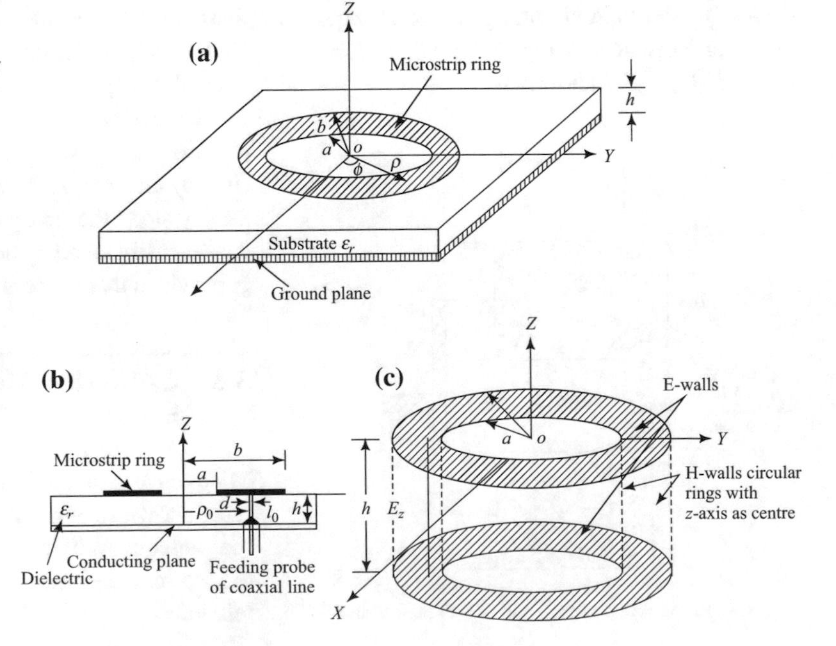

Fig. 3.9 Microstrip ring resonator: **a** isometric view **b** excitation by probe **c** cavity model-electric and magnetic field walls

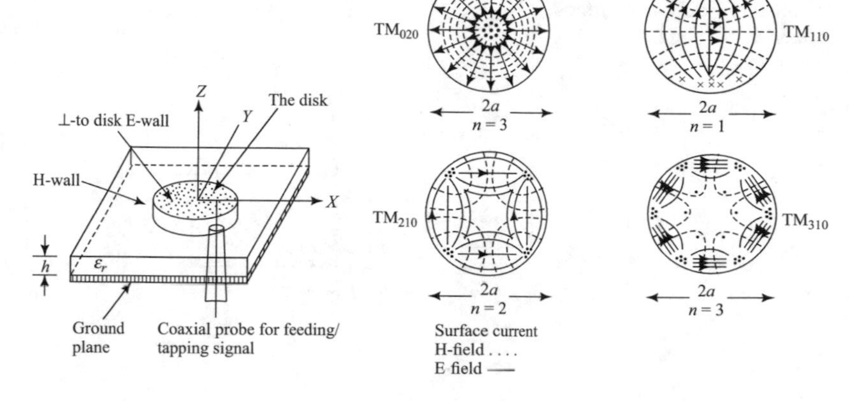

Fig. 3.10 Circular disc microstrip resonator and different mode fields

i.e.

$$(a+b)/2 = n \cdot c/(2f_r\sqrt{\varepsilon_{\text{eff}}}) = n \cdot \lambda/2$$

The region between the ring and the ground plane will act as cavity with magnetic field walls at the edges and electric field terminals at the top and bottom rings. For thin dielectric substrate (as preferred), $h \ll (b-a), \varepsilon_{\text{eff}} \cong \varepsilon_r$.

(b) **Circular disc resonator**: In the ring geometry, if $a = 0$, then it becomes a circular disc resonator, with magnetic field at the fringe end of the disc and electric field is between the disc and the ground plane as in Fig. 3.10. The space between the disc and the ground plane acts as cavity resonator.

(c) **Rectangular disc resonator**: Here, magnetic field is at the periphery making a surrounding H-wall, while the rectangular disc to ground plane forms the electric walls. The electric field inside this cavity with dielectric can be found easily if the dielectric thickness is very much smaller than wavelength ($h \ll \lambda$) (Fig. 3.11).

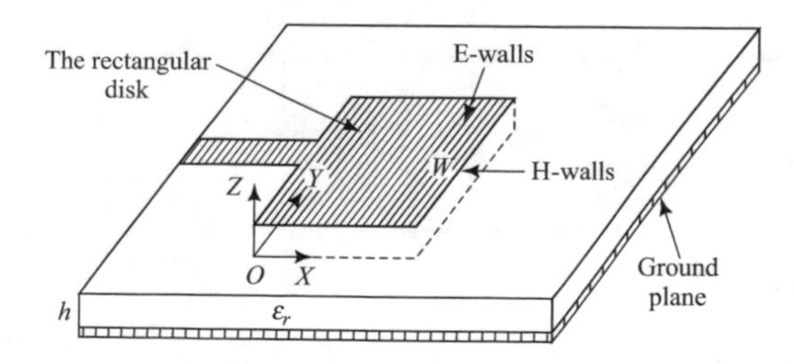

Fig. 3.11 Microstrip rectangular disc resonator

(d) **Microstrip line gap coupled resonator**: As in Fig. 3.12, a half-wavelength open-circuited microstrip line, which is capacitively coupled to an input line, forms a resonator open-circuited end. This is having fringe electric field, resulting in transient accumulation of electric charges for the cavity, at the resonant frequency.

3.8 Coupling of Cavities with the Line: Reflection and Transmission Types

For exciting or for extracting the signal, just like our ordinary LC-resonant circuit, continuous supply of signal energy is required by two cabled wire. In microwave, it is done by some different types of techniques called coupling from the resonator.

This coupling is for injecting signal into the cavity for excitation or for taking out the signal:

1. **Using Another Waveguide**: Which are of two types:

   (a) Reflection cavity.
   (b) Transmission cavity.

2. **Using Coaxial Cable**: Which could be by two methods:

   (a) Probe coupling excites electric field.
   (b) Loop coupling excites magnetic field.

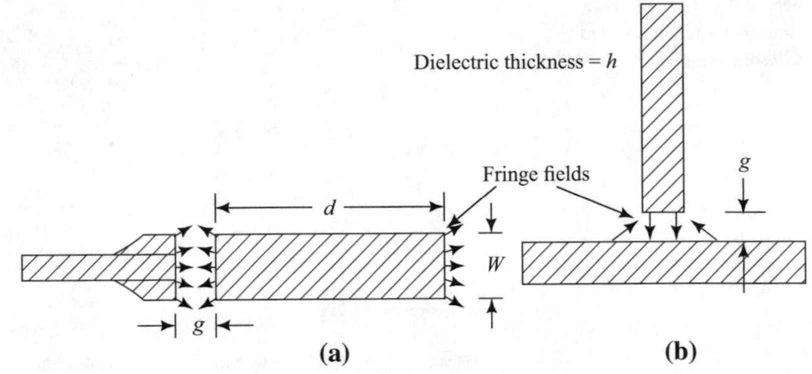

Fig. 3.12 Microstrip resonators—only top surface conductor is shown

Here, it may be noted that E-field leads to H-field and vice versa as they coexist.

All these four types of coupling are—

(1a) **Reflection Cavity Coupling with Waveguides**

This type of cavity resonator is excited by a generator matched to the feeder line impedance z_0, by means of a small aperture (centred hole on the transverse wall) as shown in Fig. 3.13 with its equivalent circuit. The size of the cavity decides the frequency of resonance.

(1b) **Transmission Type Cavities Coupling with Waveguides**

These types of cavities are coupled to both, the generator and load as in Figs. 3.14b, c. Figure 3.14a shows a cavity which is transmission and reflection type together. Figure 3.14d gives their equivalent circuit coupling of these types of cavities, represented by ideal transformers of turn ratios $1{:}n_1$ at the generator side and $n_2{:}1$ at the load end. The transmission–reflection-type cavity as given in Fig. 3.14a is used in wave metre (frequency metre) with tuning slug.

(2a) **Probe Coupling with Coaxial Line**

Here, the coaxial line inner conductor extends inside the waveguide section at the mid-point of one of the wide walls.

As the electric field in the vicinity of the probe is normal to the axis of the probe (as in Fig. 3.15a), higher modes can exist. But these higher modes can be suppressed by proper choice of dimensions of the waveguide section with the probe at the field maxima, which will be at the centre of the length 'l'. For matching the impedance of probe and the waveguide, we have to

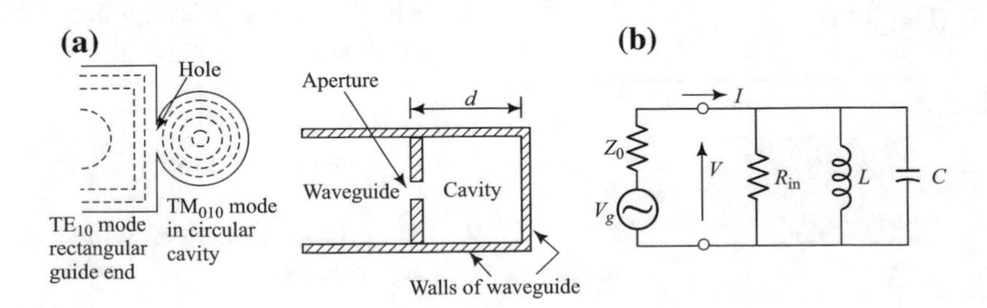

Fig. 3.13 **a** Reflection-type aperture-/hole-coupled cavity, **b** lumped equivalent circuit–parallel LC circuit

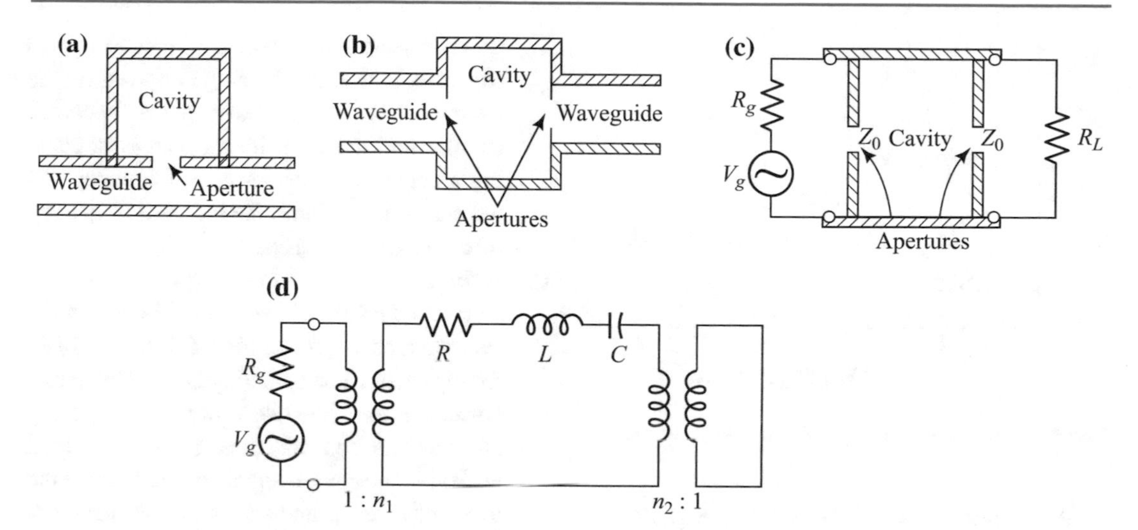

Fig. 3.14 Transmission-type cavity resonators and their series *LC* equivalent circuits. **a** Reflection + transmission type together type used in frequency metres with tuning slug **b** transmission type, **c** part equivalent circuit, **d** full equivalent cct. of (**b**)

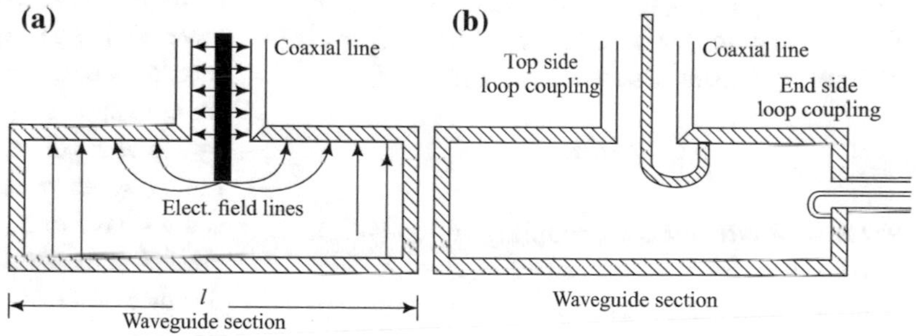

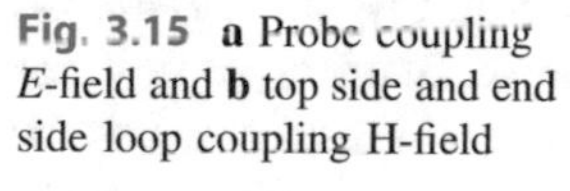
Fig. 3.15 **a** Probe coupling *E*-field and **b** top side and end side loop coupling H-field

adjust the probe length, its position, its flaring and rounding of its end, experimentally.

(2b) **Loop Coupling with Coaxial Line**

Here, the loop should be placed at a position so that it is perpendicular to the magnetic field. If it is parallel to the H-field, it will not excite any current in the loop wire. Therefore, the loop can be rotated around the hole for getting maximum signal output or best coupling.

3.9 Coupling Factor

When we supply microwave power to a cavity or extract power from it, using a coaxial line, etc. (also refer Figs. 2.9 and 2.13), then the coupling between the source cavity or cavity load can be:

(i) Critical coupling ($K = 1$)
(ii) Over coupled ($K > 1$)
(iii) Under coupled ($K < 1$)

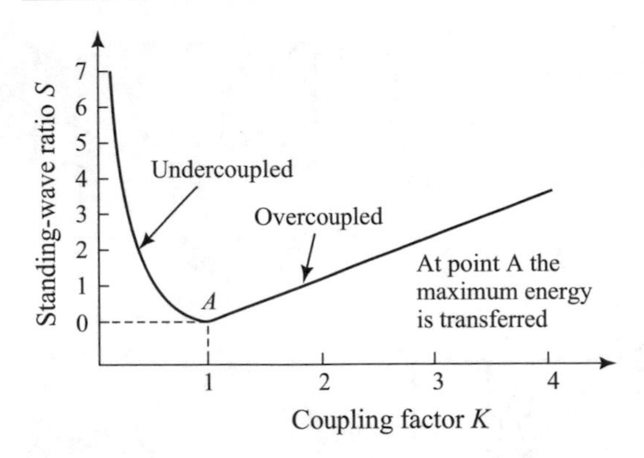

Fig. 3.16 Coupling factor versus standing wave ratio

At coupling factor $K = 1$, $Q_l = Q_0/2$ (i.e. loaded Q = half of unloaded Q) and *VSWR* is lowest, when maximum energy is transferred from the generator or to the load, as the stored energy in the cavity will be at its maximum (Fig. 3.16). Therefore, *VSWR* should be checked for best coupling. The coupling factor coefficient K is related to the source impedance Z_g and cavities equivalent resistance (R) by:

$$K = N^2 Z_g / R$$

where N = turn ratio of coupling in equivalent transformer. In fact, the coupling means how well the energy is supplied to the cavity through the coaxial cable, etc.

3.10 Frequency Tuning of Cavity

The f_0 can be varied by changing the volume mechanically (volume tuning). Varying the distance d of the plate will result in a new resonant frequency, because the change in inductance and the capacitance of the cavity in the circular or rectangular waveguides. If the volume is decreased, the resonant frequency will be higher (Figs. 3.17 and 3.18).

(a) **Capacitive tuning** of a cavity is shown in Figs. 3.17a and 3.18a. An adjustable non-magnetic slug or screw is placed in the area of maximum *E*-lines. The distance d represents the distance between two capacitor plates. As the slug is moved in, the distance between the two plates becomes smaller and the capacitance increases, causing a decrease in the resonant frequency. As the slug is moved out, the resonant frequency of the cavity increases.

(b) **Inductive tuning** is by moving a non-magnetic slug into the area of maximum H-lines on the side wall, as shown in Figs. 3.17b and 3.18b for the cavity. The changing H-lines induces a current in the slug that sets up an opposing H-field. The opposing field reduces the total H-field in the cavity, and therefore reduces the total inductance, and raises the resonant frequency. Similarly, increasing the inductance is done by moving the slug out, which reduces the resonant frequency.

(c) **Inductive and capacitative tuning together**: For wider tuning of a cavity, we can adopt a combination as given in Fig. 3.19. Here, we have two types of mechanical control for tuning a cavity for a certain frequency and for the corresponding λ_g. The capacitive tuning is by screw depth less than $\lambda_g/4$, while inductive tuning by screw depth more than $\lambda_g/4$. These screws have to be placed at the centre of the width b of the waveguide cavity separated by a distance of $3\lambda_g/8$ (Fig. 3.19).

3.11 The Q Factor of a Cavity Resonator

(a) **Simple LC Circuits**

The quality factor Q is a measure of selectivity of frequency of a resonant or anti-resonant circuit, and it is defined by the following equation

$$\begin{aligned} Q &= \frac{\omega_0 W}{P} \\ &= 2\pi f_0 \frac{\text{Maximum energy stored in tank circuit}}{\text{Energy dissipated per cycle}} \end{aligned} \quad (3.14)$$

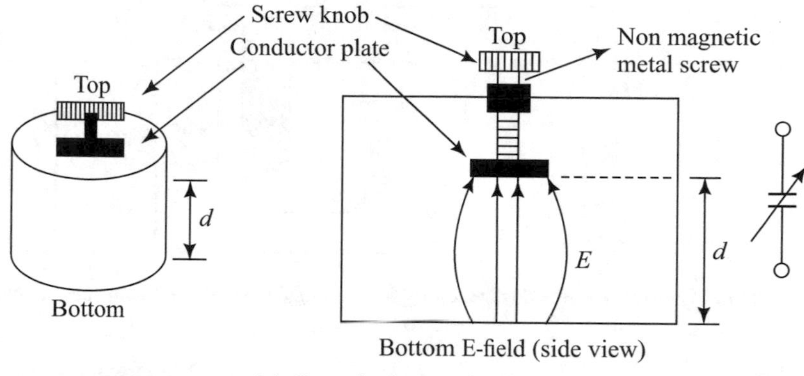

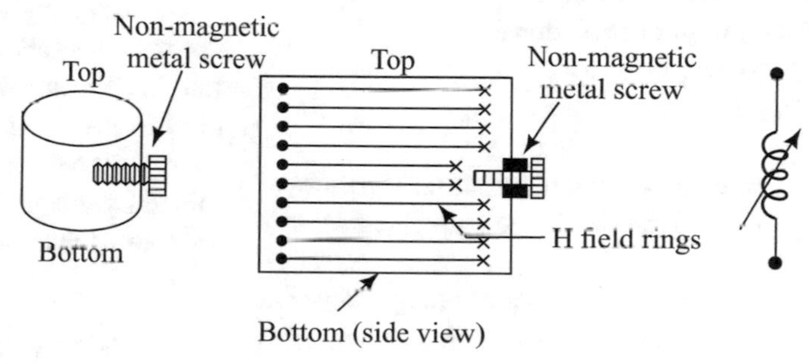

Fig. 3.17 Frequency tuning of circular cavity by changing the resonant frequency of a cavity **a** by varying the capacitance **b** varying the inductance

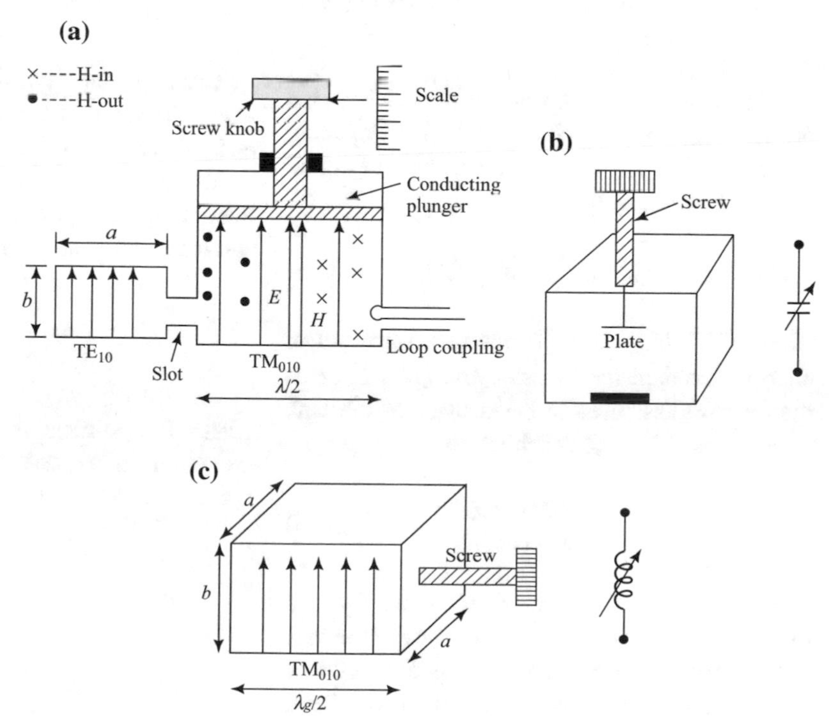

Fig. 3.18 Frequency tuning of rectangular waveguide cavity (**a**) and **b** capacitive, **c** inductive tuning

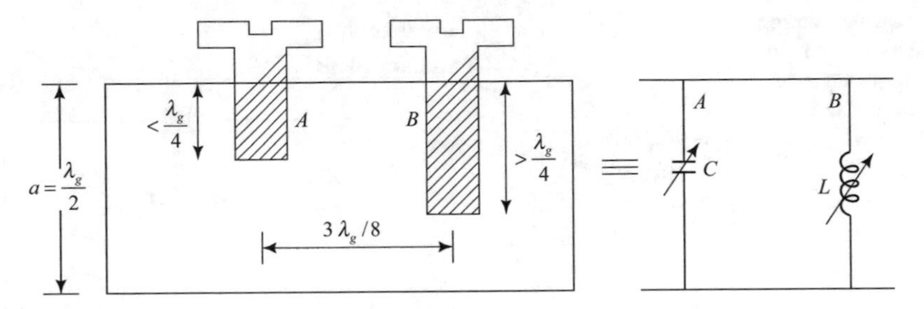

Fig. 3.19 Capacitative + inductive tuning in a waveguide cavity and equivalent circuit

where

- ω_0 Resonant frequency
- W Maximum energy stored
- P Average power loss
- Q Quality factor

From the above-mentioned relation, it is observed that higher the Q factor of a resonant circuit:

(i) the higher is the amount of energy stored in its tank circuit storage elements (such as inductors and capacitors) and

(ii) smaller is the amount of energy dissipating in it (the dissipating elements being wall resistance).

This means that high value of Q must have the product (Lf_0) to be much higher than R. For using this relation of Eq. (3.14), we know that Q of an LCR circuit is given by:

$$Q = 2\pi f_0 \times (1/2)LI_{\max}^2 / \left[(1/2)I_{\max}^2 R\right] \quad (3.15)$$

where (1/2) $I_{\max}^2 \cdot L$ is the maximum amount of energy stored in the inductor and (1/2) $I_{\max}^2 R$ is the energy dissipated in the resistor of the tank circuit. Therefore, Q simplifies to:

$$\begin{aligned} Q &= 2\pi f \times L/R \\ \therefore \quad Q &= \omega L/R \end{aligned} \quad (3.16)$$

where $f = 1/T$ is the resonant frequency and $\omega = 2\pi f$ is angular resonant frequency of the tank circuit. But in microwave, L and R cannot be measured; therefore, magnetic and electric fields become important.

(b) **Microwave Circuits**

At resonant frequency, the magnetic and electric energies are equal, but in time quadrature, i.e. when magnetic energy is maximum, the electric energy is zero and vice versa. The total energy stored in the cavity resonator is determined by integrating the energy density over the volume of the resonator,

$$\begin{aligned} W_e &= \int_V \frac{\varepsilon}{2} |E|^2 \mathrm{d}v = W_m \\ &= \int_V \frac{\mu}{2} |H|^2 \mathrm{d}v = W(\text{Let}) \end{aligned} \quad (3.17)$$

Here, W_e and W_{m} = maximum electric and magnetic stored energy in the volume inside; $|E|$ and $|H|$ are peak values of the field intensities.

The average power loss can be evaluated by integrating the power density over the inner surface area of the resonator. Thus,

$$P = \frac{R_s}{2} \int_S |H_t|^2 \,\mathrm{d}a \quad (3.18)$$

where R_s = surface resistance of the resonator and H_t and H_n = peak value of the tangential and normal magnetic intensity, respectively.

By substituting the value of W and P from the above Eqs. (3.17) and (3.18), respectively, in Eq. (3.14),

$$Q = \frac{\omega\mu \int_V |H|^2 \mathrm{d}v}{R_s \int_S |H_t|^2 \mathrm{d}a} \quad (3.19)$$

The peak value of magnetic intensity is related to its normal and tangential components $|H|^2 = |H_t|^2 + |H_n|^2$. Also, the approximate value of integrals of $|H^2|$ over surface is approximately twice the value of $|H|^2$ integrated over the volume; therefore, Q becomes:

$$Q = \frac{\omega\mu\,(\text{volume})}{2R_s\,(\text{Surface areas})} \tag{3.20}$$

An unloaded resonator can be represented by either a series or a parallel resonant LC-circuit. The resonant frequency (f_0) and the unloaded Q_0 of a cavity resonator are given:

$$f_0 = \frac{1}{2\pi\sqrt{LC}} \text{ and } Q_0 = \frac{\omega_0 L}{R} \tag{3.21}$$

The loaded resonant Q_l is different from Q_0 and will be related to the coupling factor K between the source of signal and the cavity as:

$$Q_l = \frac{\omega_0 L}{R(1+K)} = \frac{Q_0}{1+K}$$

This can also be written as:

$$\frac{1}{Q_l} = \frac{1}{Q_0} + \frac{1}{Q_{\text{ext}}},$$

where Q_{ext} is external Q given by $Q_{\text{ext}} = Q_0/K = \omega_0 L(KR)$.

If the resonator is matched to the generator, i.e. critically coupled cavity $K = 1$, then $Q_l = Q_0/2$. For most practical purpose, the following formula is sufficient.

$$Q_l = f/\Delta f \tag{3.22}$$

Here, Δf is half-power band width (see Chap. 8 also).

3.12 Solved Problems

Problem 3.1 A rectangular cavity resonator has the dimension $a = 2$, $b = 1$, $d = 3$ cm. Find the resonant frequency. If we want that the cavity to be tunable by ±20% of resonant frequency, find the movement of the tuning short slug if $d = 3$ cm is the central position. Also, find the position for $f_0 = 7.2$, 7.5, and 8.0 GHz.

Solution For fixed cavity, lowest resonant frequency f_{101} is

$$f_{101} = \frac{c}{2}\sqrt{\left(\frac{m}{a}\right)^2 + \left(\frac{n}{b}\right)^2 + \left(\frac{p}{d}\right)^2} = \frac{c}{2}\sqrt{\frac{1}{a^2} + \frac{1}{d^2}}$$

$$= \frac{3 \times 10^{10}}{2} \times \sqrt{\frac{1}{2^2} + \frac{1}{3^2}} = 9.01\,\text{GHz} \approx 9\,\text{GHz}$$

For ±20% frequency variation, i.e. (9 ± 1.8), GHz = 10.8 to 7.2 GHz.

For $f_1 = 10.8$ GHz as resonant frequency:

$$10.8 \times 10^9 = \frac{3 \times 10^{10}}{2}\sqrt{\frac{1}{4} + \frac{1}{d_1^2}}$$

$$\therefore \quad d_1 = 1.93\,\text{cm}$$

For $f_2 = 7.2$ GHz as resonant frequency:

$$d_2^2 = -\text{ve} \quad \therefore d_2 = \text{imaginary}$$

∴ Not possible for this frequency to exist.

For other frequencies, we see that

f_2 (GHz)	d_2
7.2	Not possible
7.5	Not possible
7.75	7.8 cm
8.0	5.71 cm
9.0	3 cm (central)
10.8	1.93 cm

Thus, maximum and minimum tuning which is possible is with $d_1 = 1.93$ cm to $d_2 = 7.8$ cm, giving $f_0 = 10.8$ to 7.75 GHz.

Problem 3.2 A circular waveguide has a radius of 3.5 cm. Find the distance between two shorted conducting plates at the two ends, if we want a resonator for 12 GHz for TM_{011} and TE_{011} modes. Given that Bessel's function values $P_{01} = 2.405$ (TM mode), and $P'_{01} = 3.832$ (TE mode).

Solution

$$f_0 = \frac{c}{2\pi}\left[\left(\frac{P_{n,m}}{a}\right)^2 + \left(\frac{p\pi}{d}\right)^2\right]^{\frac{1}{2}} \quad \textbf{for TM}_{mnp}\textbf{ mode}$$

$$= \frac{c}{2\pi}\left[\left(\frac{P'_{n,m}}{a}\right)^2 + \left(\frac{p\pi}{d}\right)^2\right]^{\frac{1}{2}} \quad \textbf{for TE}_{mnp}\textbf{ mode}$$

For TE$_{011}$ mode, $P'_{mn} = 3.832$; therefore, for $p = 1$, $P'_{01} = 3.832$:

$$f_0 = 12 \times 10^9 = \tfrac{3\times10^{10}}{2\times3.14}\sqrt{\left(\tfrac{3.832}{3.5}\right)^2 + \left(\tfrac{3.14}{d_1}\right)^2}$$

$$\therefore \quad \left(\tfrac{3.14}{d_1}\right)^2 = \left[\left(\tfrac{12\times10^9\times2\times3.14}{3\times10^{10}}\right)^2 - \left(\tfrac{3.832}{3.5}\right)^2\right]$$

$$\therefore \quad d_1 = 1.39\,\text{cm}$$

For TM$_{011}$ mode, $P_{mn} = 2.405$; therefore, with $p = 1$ and $P_{01} = 2.405$

$$f_0 = 12 \times 10^9$$

$$= \frac{3 \times 10^{10}}{2 \times 3.14}\sqrt{\left(\frac{2.405}{3.5}\right)^2 + \left(\frac{3.14}{d_2}\right)^2}$$

$$= (6.31 - 1.06) = 5.25d_2$$

$$= 3.14/2.29 = 1.37\text{cm}$$

Problem 3.3 Calculate resonant frequency for a circular waveguide resonator having dia of 6 cm and length = 1.62 cm for the modes TM$_{011}$, TE$_{011}$, TE$_{012}$.

Solution

$$\text{Dia} = 6\,\text{cm radius}$$

$\therefore$ radius (a) = 3 cm; the values of P_{mn} for TM$_{011}$, TE$_{011}$, and TM$_{012}$ are 2.405, 3.832, and 3.832, respectively

$$f_0(\text{TM}_{011}) = \frac{3 \times 10^{10}}{2 \times 3.14}\sqrt{\left(\frac{2.405}{0.3}\right)^2 + \left(\frac{3.14}{1.62}\right)^2}$$

(for $m = 0$, $n = 1$, $p = 1$, $P_{01} = 2.405$)

$$= 0.4777 \times 10^{10}\sqrt{4.39999}$$
$$= 10.02\text{ GHz}$$

$$f_0(\text{TE}_{011}) = \frac{3.00 \times 10^{10}}{2 \times 3.14}\sqrt{\left(\frac{3.832}{3}\right)^2 + \left(\frac{3.14}{1.62}\right)^2}$$

(for $m = 0$, $n = 1$, $p = 1$, $P'_{01} = 3.832$)

$$= 0.4777 \times 10^{10}\sqrt{1.632 + 3.757}$$
$$= 1.11 \times 10^{10}$$
$$= 11.10\,\text{GHz}$$

$$f_0(TE_{012}) = \frac{3.00 \times 10^{10}}{2 \times 3.14}\sqrt{\left(\frac{3.832}{3}\right)^2 + \left(\frac{2 \times 3.14}{1.62}\right)^2}$$

(for $m = 0$, $n = 1$, $p = 2$, $P'_{01} = 3.832$)

$$= 0.4777 \times 10^{10}\sqrt{1.632 + 15.028}$$
$$= 21.097\,\text{GHz}$$

Problem 3.4 If two waveguides are filled with a dielectric material of $\varepsilon_r = 2.5$, find the new resonant frequencies, if the resonant frequencies with air filled are 9 GHz and 10.01 GHz.

Solution Rectangular waveguide: $f_0 = \frac{1}{2\sqrt{\mu\varepsilon}}\sqrt{\left(\frac{m}{a}\right)^2 + \left(\frac{n}{b}\right)^2 + \left(\frac{p}{d}\right)^2}$

In circular waveguide:

$$f_0 = \frac{1}{2\pi\sqrt{\mu\varepsilon}}\sqrt{\left(\frac{P_{mn}}{a}\right)^2 + \left(\frac{p\pi}{d}\right)^2}$$

Therefore, whether it is a circular or rectangular W.G., $1/\sqrt{\mu\varepsilon} = c$ for air-filled waveguides, but for the dielectric filled waveguide: $1/\sqrt{\mu\varepsilon} = c/\sqrt{\varepsilon_r}$

$\therefore$ Both the frequencies get divided by $\sqrt{\varepsilon_r} = 1.58$ as given below:

Air filled (GHz)	Dielectric filled (GHz)
$f_0 = 9$	$f_0 = 9/1.58 = 5.70$
$f_0 = 10.02$	$f_0 = 10.02/1.58 = 6.34$

Problem 3.5 If a rectangular waveguide cavity has the dimension a, b, d, prove that the resonant frequency for TE_{101} mode $f_0 = [c/(2d)]\sqrt{1+(d/a)^2}$.

Solution As $f_0 = \frac{c}{2}\sqrt{\left(\frac{m}{a}\right)^2+\left(\frac{n}{b}\right)^2+\left(\frac{p}{d}\right)^2}$

$\therefore$ for $m = 1$, $n = 0$, $p = 1$, $f_0 = \frac{c}{2}$

$\sqrt{\frac{1}{a^2}+\frac{1}{d^2}} = \frac{c}{2d}\sqrt{\frac{d^2}{a^2}+1} = \frac{c}{2d}\sqrt{1+\left(\frac{d}{a}\right)^2}$

Problem 3.6 A section length of 6 cm has been cut from the X-band waveguide and the two ends closed by metal. Find the TE_{101} mode resonant frequency (for a standard X-band waveguide, $a = 2.286$ cm, $b = 1.01$ cm).

Solution

$$f_0 = \frac{c}{2}\sqrt{\left(\frac{m}{a}\right)^2+\left(\frac{n}{b}\right)^2+\left(\frac{p}{d}\right)^2}$$

$$= \frac{3\times10^{10}}{2}\times\sqrt{\left(\frac{1}{2.286}\right)^2+\left(\frac{0}{1.01}\right)^2+\left(\frac{1}{6}\right)^2}$$

$$= \frac{3\times10^{10}}{2}\times0.468$$

$$= 7.02\,\text{GHz}$$

Problem 3.7 In a rectangular cavity, cut-off wavelength $\lambda_c = 4\,\text{cm}$. If the resonant frequency is 10 GHz in TE_{101} mode, then find the length.

Solution Cut-off wavelength $\lambda_c = 4$ cm $= 2a$ $\therefore$ $a = 2$ cm

$$f_0 = 10\,\text{GHz} = 20\times10^9$$

$$= \frac{c}{2}\sqrt{\left(\frac{m}{a}\right)^2+\left(\frac{n}{b}\right)^2+\left(\frac{p}{d}\right)^2}$$

$$\because\ m = 1, n = 0, p = 1, a = 2, c = 3\times10^{10}\,\text{cm/s}$$

$$\because\ 10\times10^9 = \frac{3\times10^{10}}{2}\sqrt{\frac{1}{4}+\frac{1}{d^2}}\quad \therefore d = 2.27\,\text{cm}.$$

Problem 3.8 A rectangular cavity has dominant resonant frequency = 5 GHz. If $a = 2b = d/2$. Find its dimension.

Solution Resonant mode of frequency in a rectangular cavity is:

$$f_{m,n,1} = \frac{c}{2\sqrt{\varepsilon_r}}\sqrt{\left(\frac{m}{a}\right)^2+\left(\frac{n}{b}\right)^2+\left(\frac{p}{d}\right)^2}$$

Here, the dominant mode is f_{101}, i.e.

$$m = 1, n = 0, p = 1, \varepsilon_r = 1, c = 3\times10^{10}\,\text{cm/s}$$

$$\therefore\ 5\times10^9 = \frac{3\times10^{10}}{2\times1}\sqrt{\frac{1}{a^2}+0+\frac{1}{(2a)^2}}$$

$$\therefore\ a = \tfrac{2}{3}\times\sqrt{5}\,\text{cm} = 3.354\,\text{cm}$$
$$\therefore\ b = \tfrac{a}{2} = 0.745\,\text{cm}$$
$$d = 2a = 2.98\,\text{cm}$$

Problem 3.9 In a circular cavity, its resonant frequency for TE_{101} mode is 5 GHz, if $d = 2a$. Find its dimensions.

Solution The resonant frequency with a = radius and d = length:

$$f_{m,n,p} = \frac{1}{2\pi\sqrt{\mu\varepsilon}}\sqrt{\left(\frac{P'_{mn}}{a}\right)^2+\left(\frac{p\pi}{d}\right)^2}$$

Dominant mode is f_{011}, i.e. for $m = 0$, $n = 1$, $p = 1$ $P'_{mn} = 3.892$ for TE mode

$$\therefore 5\times10^9 = \frac{3\times10^{10}}{2\times3.14}\sqrt{\left(\frac{3.832}{a}\right)^2+\left(\frac{3.14}{2a}\right)^2}$$

Computing this, we get $a = 3.96$ cm and $d = 2a = 7.92$ cm, Dia = 7.92 cm, length = 7.92 cm **Ans.**

Problem 3.10 A microwave source, a 20 dB directional coupler (dc), a lossless waveguide (WG) section, and a resonant cavity are connected in series. Here, source is at port-1 of dc, waveguide at port 2 of dc. The power output at port-3 of dc was noted to be 4 mW and the VSWR in the WG-section as 2.5. Then, find (a) source power (b) power reflected to the port 4 of dc (c) power used by the cavity.

Solution

(a) As it is a 20-dB direction coupler, power at port 3 is 20 dB down (i.e. 1/100) of the power input at port 1.

$$\therefore \quad P_1 = 100 \times P_3 = 100 \times 4 = 400\,\text{mW}$$

$\therefore$ P_{in} to the cavity will be

$$P_{\text{in}} = (P_1 - P_3) = 400 - 4 = 396\,\text{mW}$$

(b) As the VSWR in the WG section = 2.5, the reflected signal from cavity P_{ref} enters the WG section and then port 2 of dc and a 20 dB portion (i.e. 1/100) of P_{ref}, i.e. P_4 reaches port 4, through the dc holes. This P_4 gets absorbed fully at the port 4 absorbant/attenuator.

As

$$\frac{P_{\text{ref}}}{P_{\text{in}}} = (\Gamma)^2 = \left(\frac{S-1}{S+1}\right)^2 \therefore \frac{P_{\text{ref}}}{P_{\text{in}}} = \left(\frac{2.5-1}{2.5+1}\right) = 0.185$$

$$\therefore P_{\text{ref}} = P_{\text{in}} \times 0.185 = 396 \times 0.185 = 73.2 \quad \text{mW}$$

The portion $P_4 = P_{\text{ref}}/100 = 73.2/100 = 0.732$ mW

(c) Power used by the cavity = 396 − 73.2 = 322.8 mW.

Problem 3.11 A rectangular cavity has the dimension $a = 4.755$ cm and $b = 2.215$ cm and is filled with dielectric of $\varepsilon_r = 2.25$. Find the length required for having the lowest resonant frequency = 5 GHz.

Solution Frequency (resonant) at $\text{TE}_{m,\,n,\,l}$ mode is

$$f_{mnl} = \frac{1}{2\sqrt{\mu\varepsilon}}\sqrt{\left(\frac{m}{a}\right)^2 + \left(\frac{n}{b}\right)^2 + \left(\frac{p}{d}\right)^2}$$

Most dominant mode will be TE_{101} when $d > a > b$ and $m = 1$, $n = 0$, $p = 1$.

For dielectric material of $\varepsilon_r \therefore \varepsilon = \varepsilon_0 \cdot \varepsilon_r$ and $c = 1/\sqrt{\mu_0\varepsilon_0}$; $\mu = \mu_0$

$$f_{101} = 5 \times 10^9 = \frac{c}{2\sqrt{\varepsilon_r}} \times \sqrt{\frac{1}{a^2} + \frac{1}{d^2}}$$

$$5 \times 10^9 = \frac{3 \times 10^{10}}{2 \times \sqrt{2.25}}\sqrt{\left(\frac{1}{4.755}\right)^2 + \frac{1}{d^2}}$$

$$\left[0.25 - \left(\frac{1}{4.755}\right)^2\right] = \frac{1}{d^2} \quad \therefore \frac{1}{d} = 0.45362$$

$$\therefore \quad d = 2.2\,\text{cm}$$

Review Questions

1. Explain what is a cavity resonator? If the length of the rectangular cavity is 3 $\lambda_g/2$, will it resonate?
2. Derive the equations for resonant frequencies for a rectangular cavity resonator.
3. Derive the field expressions for a rectangular cavity resonator and plot the fields.
4. Derive the field expressions for a circular cavity resonator and plot the fields.
5. Discuss the quality factor of a cavity resonator and explain the term unload Q, loaded Q, critical coupled Q, under coupled Q and over coupled Q related to a cavity resonator.

6. An air-filled circular waveguide has a radius of 4 cm and is used as a resonator for TM_{101} mode at 8 GHz by placing two perfectly conducting plates at its two ends. Determine the minimum distance between them.
7. A circular waveguide with the radius of 5 cm and is used as a cavity resonator for TM_{011} mode at 8 GHz by inserting two perfectly conducting plates at its two ends. Find their distance.
8. A rectangular cavity is formed by shorting the ends of an X-band waveguide. ($a = 2.28$ cm, $b = 1.01$ cm). The length $d = 6$ cm. Calculate the resonant frequency for the TE_{101} mode.
9. A cylindrical waveguide cavity has a radius of 2 cm. The cavity is turned by means of plunger that allowed to be varied from 5 to 7 cm. Determine the range of resonant frequencies for the TE_{111} mode.
10. A cavity resonator with dimensions $a = 2$ cm, $b = 1$ cm is excited by TE_{101} mode of 20 GHz. Calculate the length of the cavity.
11. Describe the various ways of coupling energy to a resonator, i.e. signal excitation of a cavity, explain with figure
12. A circular waveguide has a radius of 3 cm resonating at 10 GHz in TM_{011} mode. Find its length with (*a*) air as dielectric (*b*) dielectric material filled ($\varepsilon_r = 2.5$).
13. A cubical box of metal has each side of 3 cm. Find its lowest resonant frequency.
14. In a cylindrical waveguide resonator, if the side '*a*' is changed from 2 to 3 cm with $b = 1$ cm, $d = 4$ cm fixed, find the variation of the resonant frequency in TE_{111} mode.
15. Draw the *E*- and *H*-fields in a rectangular waveguide cavity of length $l = 3\lambda/2$ for the four moments of $t = 0$, $T/4$, $2T/4$, $3T/4$, $4T/4$.

Microwave Components and Their Scattering Matrices

4

Contents

© Springer Nature Singapore Pte Ltd. 2018
P. K. Chaturvedi, *Microwave, Radar & RF Engineering*,
https://doi.org/10.1007/978-981-10-7965-8_4

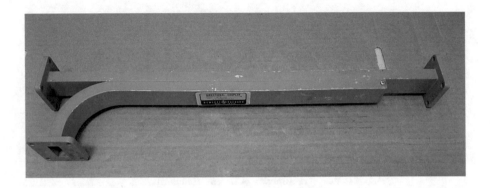

4.1 Introduction

As we have seen that, at microwave frequencies the media of transmission is normally coaxial cable waveguides, which are mostly straight and of finite length. For all these some special types of components are required. At the same time we know that for different frequency range, we need different sizes of the rectangular and circular waveguides and therefore all the components have to be chosen accordingly. Normally in the laboratories, for simple experiments, the size of a rectangular waveguide is (0.4″ × 0.9″) (i.e. 1.016 cm × 2.286 cm), having cut-off frequency of 6.56 GHz which is suitable for frequency band of 8.2−12.5 GHz (X-band), as seen from the table given in Chap. 2 (Table 2.2).

All these components have to be build with low VSWS, low attenuation, low insertion loses, etc. At microwave frequencies voltage, current, and impedance cannot be measured in a direct manner. The direct measurable quantities are:

1. Relative field strength (minimum and maximum), and hence the VSWR and power.
2. Amplitude and phase angle of the wave reflected, relative to the incident wave.

Moreover in most of the microwave devices, the scattered (reflected) wave amplitudes are linearly related to the incident wave amplitude as sum of its fractional parts. **The matrix describing fractional (S_{ij}) relationship between forward wave and reflected wave is called scattering matrix. All these scattering matrix coefficients are smaller than one ($S_{ij} < 1$).** As in fact reflection cannot be made equal to zero fully, therefore all the terms of the scattering matrix exist. The letter '*a*' is used for forward microwave signal amplitude into a system, and '*b*' is used for reflected signal coming in as given in the two-port network in Fig. 4.1.

The reflected signal will be represented by the *S*-parameters as:

$b_1 = S_{11}\,a_1 + S_{12}\,a_2$ = sum of S_{11} and S_{12} of fractions a_1, a_2, respectively.

Table 4.1 List of Important Passive Components used in μω

S. no.	Article	Components	Applications
1	4.2	Coaxial cables and connection	Connecting measuring instrument or oscillator with waveguide
2	4.3	Waveguide junction	
3	4.4	(a) H-plane/tee	For adding/splitting/getting differential output
4	4.5	(b) E-plane/tee	For adding/splitting/getting differential output
5	4.6	(c) E–H-plane tee	As mixer, duplexer, etc.
6	4.7	(d) Hybrid ring (rat-race junction)	As mixer in impedance measurement or in duplexers
7	4.8	Directional coupler	Sampling of power for power flow measurement
8	4.9	(a) Bends, twists	For changing the direction of flow of signal or its plane of polarisation
9		(b) Transitions	For switching to different types of line from rectangular to circular waveguide or vice versa
10	4.10	Attenuators and terminators	For reducing the power or fully terminating the power without reflections
11	4.11	Iris and screws/posts	For introducing susceptance (C or L) in the path or for impedance matching
12	4.12	Probes and loops	For exiting/feeding signal into a waveguide for taking out/tapping the signal
13	4.12	Diode detectors	For measuring $\mu\omega$-power using rectified dc voltage
14	4.13	Cavity wave metres	For frequency measurement
15	4.14	Faraday rotation and ferrite devices	For rotation of polarisation of a wave used in, isolators, gyrators, and circulators
16	4.15	Phase shifters	For phase shifting of a wave for some objective in mind

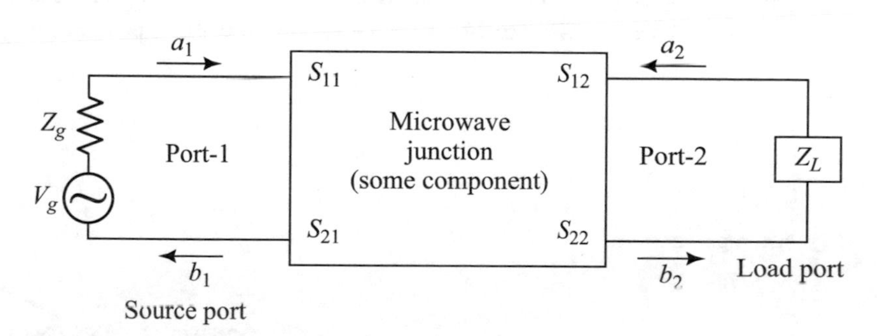

Fig. 4.1 *S*-matrix elements of two-port network

$b_2 = S_{21}\, a_1 + S_{22}\, a_2$ = sum of S_{21} and S_{22} of fractions a_1, a_2, respectively.

If $Z_L = Z_0$ (matched load), then there is no reflection from Z_L, i.e. no signal from Z_L ($a_2 = 0$). Here it may be noted that *S*-coefficients are always <1, i.e. fractions.

$$\therefore S_{11} = \frac{b_1}{a_1}; S_{21} = \frac{b_2}{a_1} \quad \text{(with } a_2 = 0\text{)}$$

If $Z_g = Z_0$ (matched source) and only port 2 is driven by a signal, then ($a_1 = 0$)

$$\therefore S_{12} = \frac{b_1}{a_2}; S_{22} = \frac{b_2}{a_2} \quad (\text{with } a_1 = 0)$$

Thus the four elements can be said to represent:

> **S_{11}, S_{22} → Reflection coefficient of ports 1 and 2, respectively.**
> **S_{21}, S_{12} → Attenuation coefficient of forward and reverse signal, respectively.**

Above equations can be written in a matrix form as

$$\begin{bmatrix} b_1 \\ b_2 \end{bmatrix} = \begin{bmatrix} S_{11} & S_{12} \\ S_{21} & S_{22} \end{bmatrix} \begin{bmatrix} a_1 \\ a_2 \end{bmatrix} \tag{4.1}$$

Here we should remember that the scattering matrix signal elements *a* and *b*, i.e. input wave and reflected wave, are in voltage terms and for getting power equation all the elements have to be squared.

It we consider a *n*-port (Fig. 4.2) network then with '*a*' the forward wave into the network and '*b*' the reflected wave amplitude out of the network, then the scattering matrix equation will be:

$$\underset{\text{Reflected wave}}{\begin{bmatrix} b_1 \\ b_2 \\ \vdots \\ b_n \end{bmatrix}} = \underset{\text{Scattering matrix}}{\begin{bmatrix} S_{11} & S_{12} & \cdots & S_{1n} \\ S_{21} & S_{22} & \cdots & S_{2n} \\ \vdots & \vdots & & \vdots \\ S_{n1} & S_{n1} & \cdots & S'_{nn} \end{bmatrix}} \underset{\text{Incident wave}}{\begin{bmatrix} a_1 \\ a_2 \\ \vdots \\ a_n \end{bmatrix}}$$

Here

- S_{ij} represents the scattering coefficient due to signal input at *i*th port and output taken at *j*th port.
- S_{ii} represents the reflection at *i*th port due to signal at *i*th port itself (for matched load at *i*th port there is no reflection and in this case $S_{ii} = 0$).

Properties of Scattering Matrix [*S*]

1. **Square Matrix Property**: It is always a square matrix [$n \times n$].
2. **Unitary Property**: [*S*] [*S**] = [1], product of a matrix and its complex conjugate = unitary matrix.
3. **Symmetry Property**: It is symmetric: $S_{ij} = S_{ji}$ the input to output and output to input port does not change the transmission properties, being a passive component.

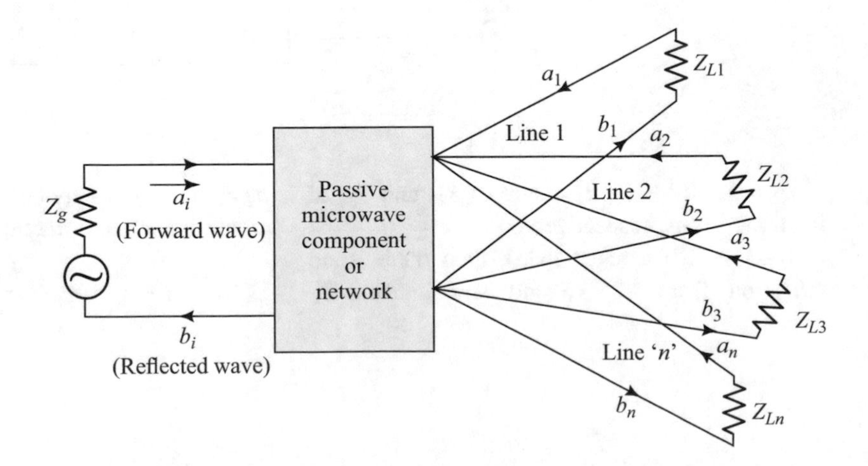

Fig. 4.2 *n*-port network

4. **Complex Conjugate (Column × Row) Property**: The sum of product of a term of any column or row multiplied by the complex conjugate of any other column or row is always =0.

$$\sum S_{ik} S_{jk}^{*} = 0 \text{ for } i \neq j \text{ (Row Case)}$$

$$\text{and } \sum_{i=1}^{n} S_{ik} \cdot S_{ij}^{*} = 0 \text{ for } k \neq j \text{ (Column Case)}$$

Example of a column case

$$\begin{bmatrix} \begin{pmatrix} S_{11} \\ S_{21} \\ \\ S_{n1} \end{pmatrix} & \begin{matrix} S_{12} \\ S_{22} \\ \\ S_{n2} \end{matrix} & \begin{pmatrix} S_{13} \\ S_{23} \\ \\ S_{n3} \end{pmatrix}^{*} & \begin{matrix} \ldots S_{1n} \\ \ldots S_{2n} \\ \\ \ldots S_{nn} \end{matrix} \end{bmatrix} \quad \begin{matrix} \text{where} \\ k = 1 \\ j = 3 \\ i = 1, 2, \ldots n \end{matrix}$$

$$\therefore \left(S_{11} \cdot S_{13}^{*} + S_{21} \cdot S_{23}^{*} + \cdots S_{n1} \cdot S_{n3}^{*}\right) = 0$$

5. **Lossless Network**: A network is lossless if no real power is lost, i.e. for the first row elements of S-matrix = 1:

$$\sum_{\substack{i=1 \\ j=1 \text{ to } N}} \left|S_{ij}\right|^{2} = 1 \qquad (4.2)$$

Example If $S_{11} = 0.2\angle 45°; S_{22} = 0.2\angle 90°$

$$S_{12} = 0.5\angle -90°; S_{21} = 0.5\angle 0°$$

Find whether the network is lossy, symmetric, and reciprocal? Also find the insertion loss.

Solution

(1) $\sum_{\substack{ij \\ j=1,2}} S_{ij} = (0.2)^2 + (0.5)^2 = 0.29 \neq 1.$

Therefore it is lossy.

(2) As $S_{12} = S_{21}$, it is reciprocal.

(3) As $S_{12} = S_{21}$ and $S_{11} = S_{21}$, it is symmetrical.

(4)
$$\begin{aligned} \text{Insertion loss} &= -20 \log_{10} |S_{21}| \text{ dB} \\ &= -20 \log_{10}(0.5) \\ &= -6 \text{ dB} \end{aligned}$$

6. **Shift of reference follows for measuring matrix elements**: As S-matrix relates the magnitude and phase of the travelling wave and also the reflected wave, therefore shift of reference plane along the transmission line by length 'l' causes a shift of 2 βl. Therefore each of the elements has to be multiplied by $e^{-2j\beta l}$, where β is the phase factor/length, i.e.

$$S_{ij}' = S_{ij}\, e^{-2j\beta l}$$

4.2 Coaxial Cables and Connectors

(a) **Coaxial cables**: Coaxial line and connectors are integral part of microwave system, as we require to connect the signal with some measuring/test instrument with cables, requiring connectors at the two ends. The outer conductor of the coaxial line is used to guide the signal through TEM mode and shield the external or internal signal from leakage. The standard characteristic impedance of these cables is normally 50 Ω or 75 Ω.

There are three basic types of coaxial cables with increasing order of shielding, i.e. flexible, semi-rigid, and rigid. The outer conductors of all these types are of knitted wire mesh, while the central conductor is single copper wire. The only difference is in the dielectric in between (Fig. 4.3), which may be solid or flexible. The coaxial cable has low loss solid or foam polythene dielectric. The rigid cable has air dielectric with small dielectric supports/spacers at some regular distance for giving mechanical support (Fig. 4.3b). The frequency range of these coaxial

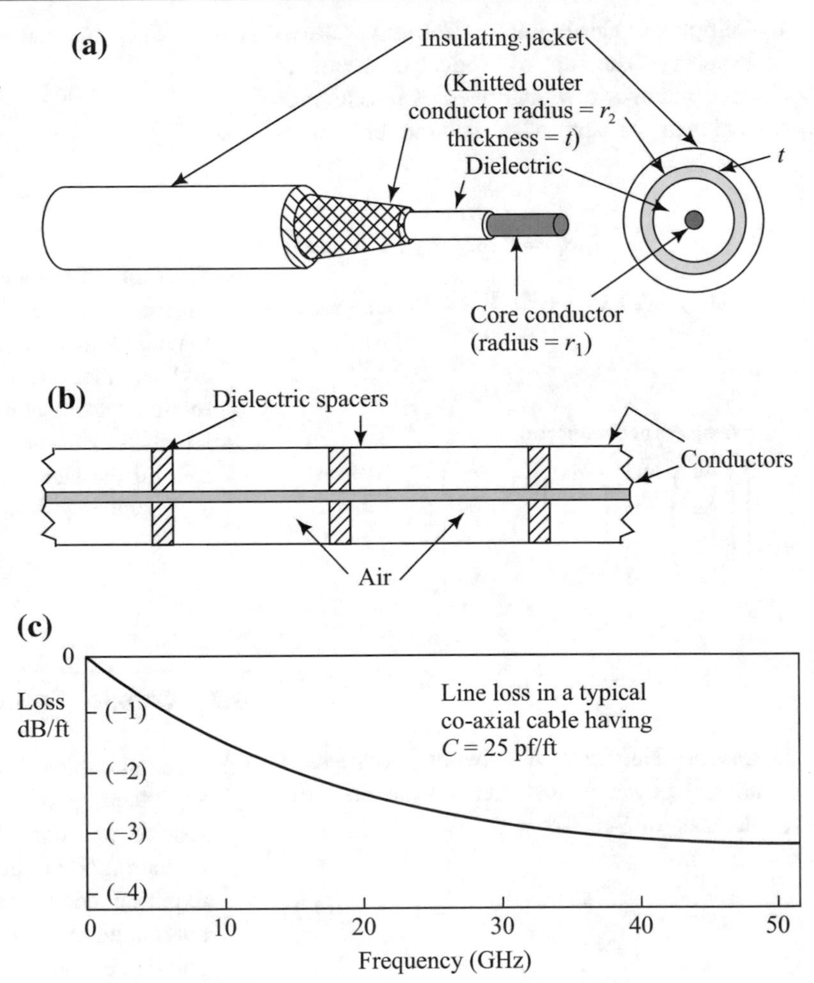

Fig. 4.3 Coaxial cables. **a** Flexible cable. **b** Rigid cable. **c** Typical loss in dB per foot

cables is dc to microwave frequencies. The characteristic impedance of these cables is given by:

$$\text{For lossless cable: } Z_0 = \sqrt{\frac{L}{C}} = \frac{1}{2\pi}\sqrt{\frac{\mu}{\in}} \cdot \ln\left(\frac{r_1}{r_2}\right) = \frac{60}{\in_r} \cdot \ln\left(\frac{r_1}{r_2}\right)\Omega$$

$$\text{For lossy cable: } Z_{0l} = \frac{1}{2\pi \cdot r_2 \sigma t} \cdot \frac{(1+j)(t/\delta)}{\sinh[(1+j)(t/\delta)]} \quad \text{for } t \ll r_1 \ll \delta$$

where r_2 and r_1 are radius of knitted and core conductors (Fig. 4.3), t = knitted conductor thickness, σ its conductivity, and δ, its skin depth. For better shielding the transfer impedance Z_{0l} has to be small, while for better 'σ', both the conductors are to be made of silver plated copper.

(b) **Coaxial cable connectors**: Coaxial cables are terminated or connected to another cable or instruments of components by connectors. These are various types depending upon the frequency range of operation and cable diameter (Fig. 4.4).

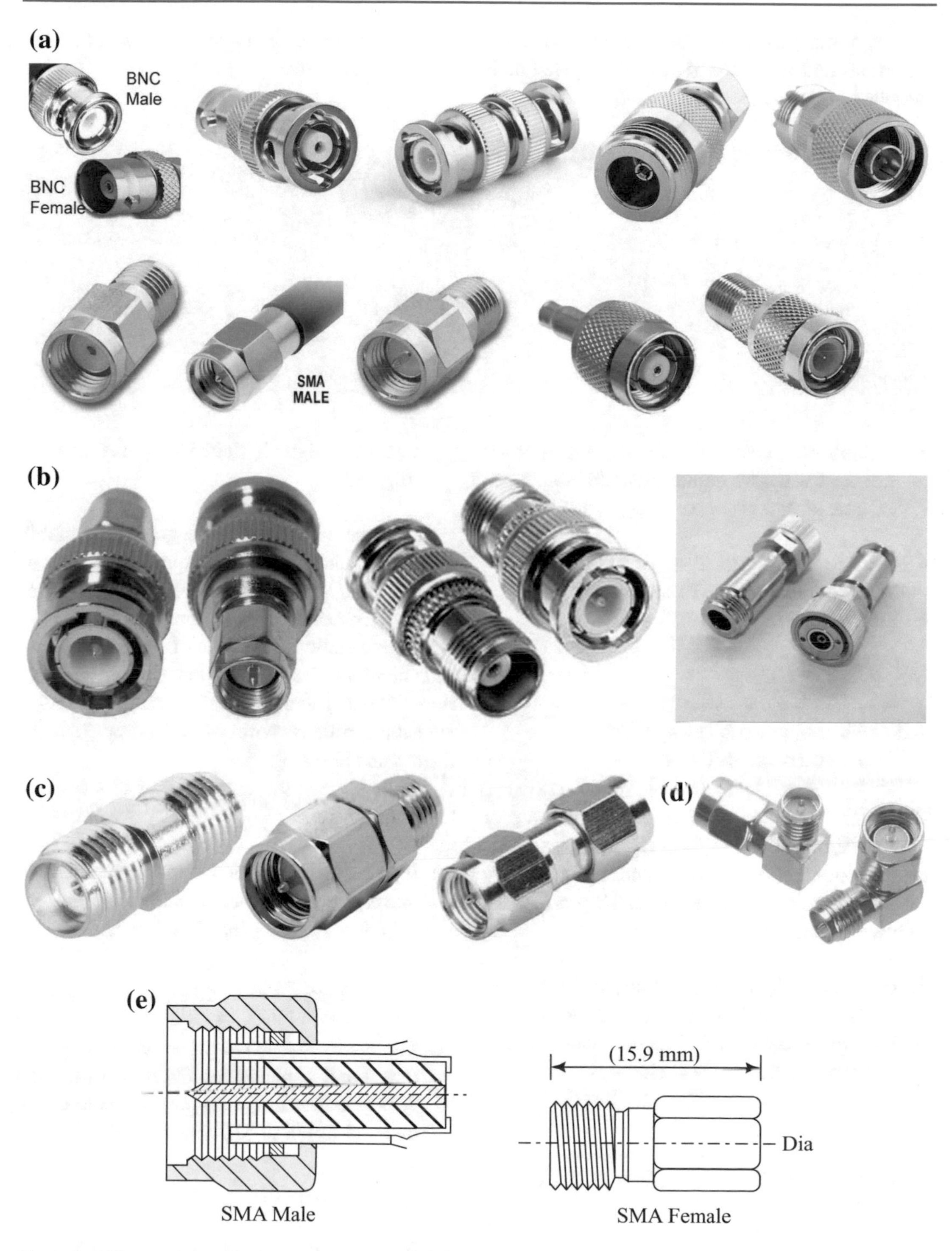

Fig. 4.4 Different types of cable connectors and adaptors. **a** Connectors of male/female to cables. **b** Adapters for connecting different style/types of connectors. **c** Adapters for same class male/female connectors. **d** Adapter elbow bends. **e** Schematic diagram of a SMA-type connector

The commonly used connectors are given in Fig. 4.4a, and the table given below gives their frequency range, etc.

Types/full name	Frequency range (GHz)	Male/female	Dielectric space	Impedance (Ω)
N (Navy)	0–18	M/F	Air	50/75
BNC (Baynet Navy Connector)	0–1	M/F	Solid	50/75
TNC (Threaded Navy Connector)	0–12	M/F	Solid	50/75
SMA (Sub-Miniature-A)	0–24	M/F	Solid	50
APC-7 (Amphenol Precision Connector-7)	0–18	M/F	Air	50
APC-3.5 (Amphenol Precision Connector-3.5)	0–35	M/F	Air	50

(c) **Adapters**: These components (Fig. 4.4b–e) are used when we connect two different types of coaxial lines as given below.

- BNC to SMA type, TNC to APC, etc.
- Male (M) to female (F) or vice versa.
- Elbow bends.

4.3 Microwave Waveguide Junctions: 4-Types

They are intersections or junctions of three or four waveguides in the form of T or waveguide in the form of ring, etc., and are used for branching the microwave signal. They are of the following four types.

1. H-plane tee junction (current junction)
2. E-plane tee junction (voltage junction)
3. E–H-plane tee (hybrid junction/magic tee)
4. Retrace junction (hybrid ring).

4.4 H-Plane Tee Junction (Current Junction)

Here the plane of the axis of the T-joint waveguide is parallel to the:

(a) Plane of the main transmission line, i.e. joint on 'b' side and

(b) Plane of magnetic field loops lines of force (Fig. 4.5).

In all these cases once we get their S-matrix equation, then the input–output cases can be derived straight way. Therefore knowing their S-matrix becomes the most important thing.

The magnetic field divides itself into the arm and therefore also calls power current junction. Here the port 1 and port 2 of the main waveguide are called collinear ports while the port 3, as the H-arm/side arm (Fig. 4.5).

Following are the properties of this tee (see Fig. 4.6a–e).

(a) If power is given at port 3, then it gets divided into equal portions and in phase to ports 1 and 2 (**power splitting into 3 dB each**)

$$\text{i.e. } P_3\angle 0 = P_1\angle 0 + P_2\angle 0 \qquad (4.3)$$

(b) The reverse of above property, i.e. if power (equal and in phase) are fed into arms 1 and 2, than power output at port 3 is sum of the two and in phase (**power adding**):

$$P_1\angle 0 + P_2\angle 0 = P_3\angle 0 \qquad (4.4)$$

(c) If equal and out of phase power are fed into arms 1 and 2, then no output is there in the arm 3, as the electric. Fields will cancel at the junction point (**power canceller**)

$$P_1\angle 0 + P_2\angle 180 = 0 = P_3 \quad (4.5)$$

(d) If power is fed into arm 1 only, then the outputs at arms 2 and 3 are in phase (**power splitting**) along with a portion reflected back to port 1.

$$P_1\angle 0 = P_2\angle 0 + P_3\angle 0 + P_{1\text{ref}}\angle 0 \quad (4.6)$$

Proof

All the above rules and more (Fig. 4.6) can be proved with the help of *S*-matrix (3 × 3). Properties of *S*-matrix are applied on a 3 × 3 matrix of H-plane tee.

(i) Plane of symmetry of H-tee around port 3,

$$S_{13} = S_{23}\left[\begin{array}{r}\text{Whether input is}\\ \text{at port 1 or port 2, result is similar}\end{array}\right] \quad (4.7)$$

(ii) Symmetry property of *S*-matrix,

$$S_{12} = S_{21}, S_{23} = S_{32} = S_{13} = S_{31} \quad (4.8)$$

(iii) Port 3 is perfectly matched with the junction,

$$\text{i.e. } S_{33} = 0 \quad (4.9)$$

Therefore by substituting above equation in Eqs. (4.7), (4.8), (4.9), in the 3 × 3 *S*-matrix we get, with a_1, a_2, a_3 as inputs and b_1, b_2, b_3 as outputs at ports 1, 2, and 3:

$$\begin{bmatrix} b_1 \\ b_2 \\ b_3 \end{bmatrix} = \begin{bmatrix} S_{11} & S_{12} & S_{13} \\ S_{12} & S_{22} & S_{13} \\ S_{13} & S_{13} & 0 \end{bmatrix} \begin{bmatrix} a_1 \\ a_2 \\ a_3 \end{bmatrix} \quad (4.10)$$

Therefore we are left with only four unknown parameters (S_{11}, S_{12}, S_{22}, and S_{13}).

(iv) Appling the unitary property $[S] \cdot [S]^* = [I]$ on the above matrix Eq. (4.10) we get:

$$\begin{bmatrix} S_{11} & S_{12} & S_{13} \\ S_{12} & S_{22} & S_{13} \\ S_{13} & S_{13} & 0 \end{bmatrix} \cdot \begin{bmatrix} S_{11}^* & S_{12}^* & S_{13}^* \\ S_{12}^* & S_{22}^* & S_{13}^* \\ S_{13}^* & S_{13}^* & 0 \end{bmatrix} = \begin{bmatrix} 1 & 0 & 0 \\ 0 & 1 & 0 \\ 0 & 0 & 1 \end{bmatrix} \quad (4.11)$$

For getting the four unknown parameters, we open the above product of matrices and using the fact that $SS^* = |S|^2$, we get:

$$\text{Row 1} \times \text{Col 1 gives: } |S_{11}|^2 + |S_{12}|^2 + |S_{13}|^2 = 1 \quad (4.12)$$

$$\text{Row 2} \times \text{Col 2 gives: } |S_{12}|^2 + |S_{22}|^2 + |S_{13}|^2 = 1 \quad (4.13)$$

$$\text{Row 3} \times \text{Col 1 gives: } S_{13}S_{11}^* + S_{13}S_{12}^* = 0 \quad (4.14)$$

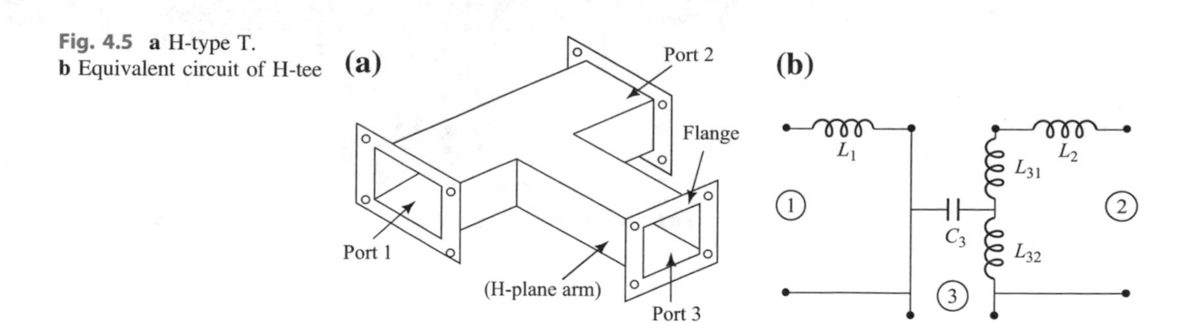

Fig. 4.5 **a** H-type T. **b** Equivalent circuit of H-tee

Fig. 4.6 H-plane tee: the five cases of inputs and outputs with electric, magnetic fields, and power flow directions with phase shown in each case

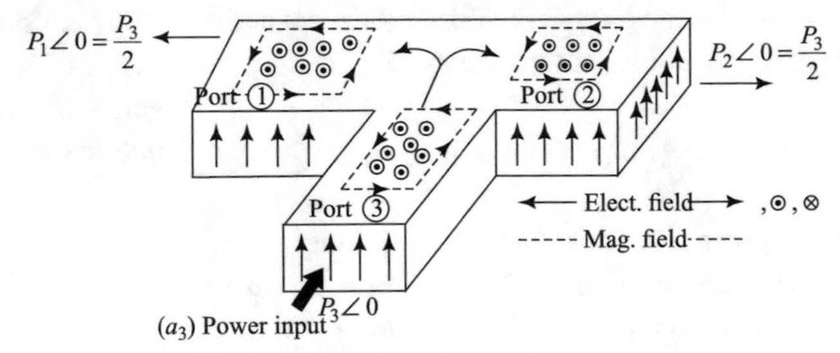

(a) Case 1 power splitter (in phase)

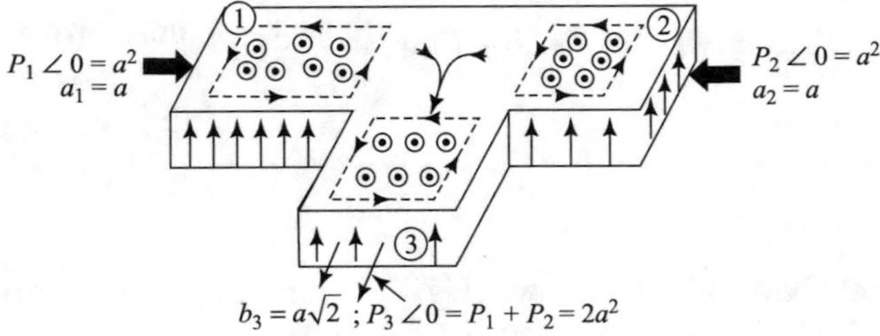

(b) Case 2 power adder (in phase)

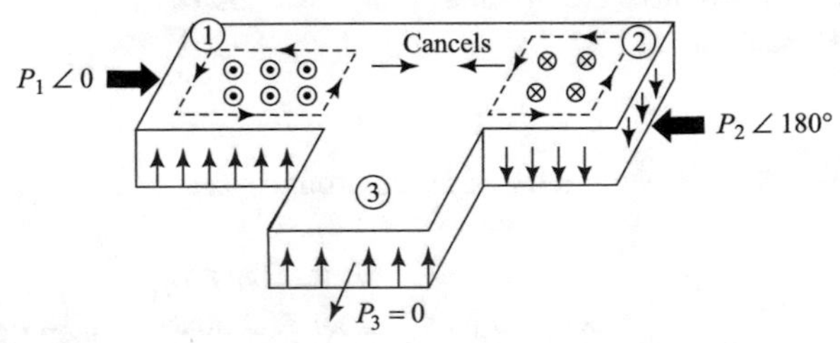

(c) Case 3 power adding out of phase at (1) (2), cancels

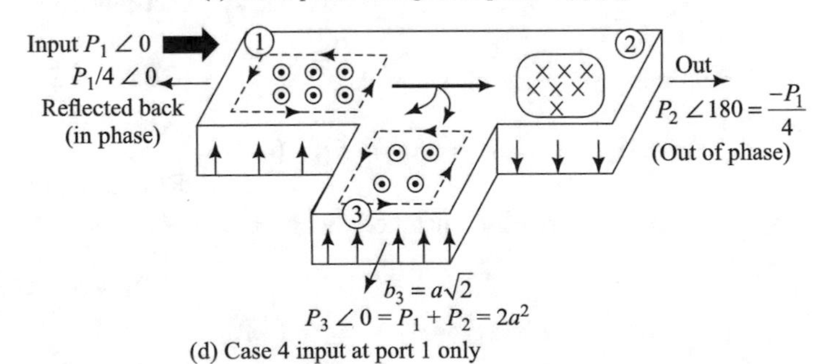

(d) Case 4 input at port 1 only

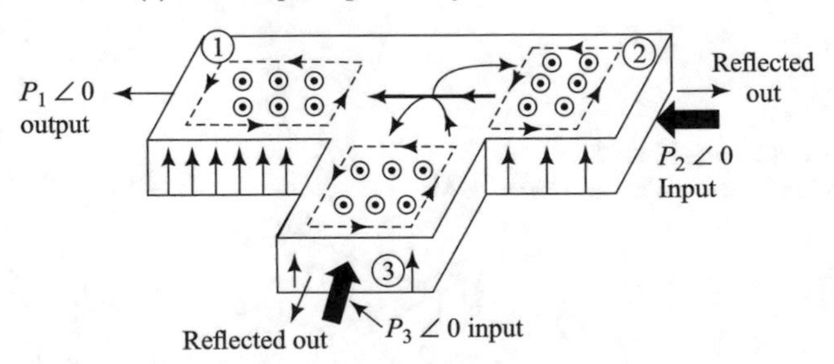

(e) Case 5 power addition in phase [partly reflected back to ports (2, 3)]

Row 3 × Col 3 gives: $|S_{13}|^2 + |S_{13}|^2 = 1$

$$(4.15)$$

∴ Equation (4.15) gives

$$|S_{13}| = \frac{1}{\sqrt{2}} \quad (4.16)$$

∴ Equations (4.12) and (4.13) give

$$|S_{11}|^2 + |S_{12}|^2 + \frac{1}{2} = 1 \quad (4.17)$$

and

$$|S_{12}|^2 + |S_{22}|^2 + \frac{1}{2} = 1 \quad (4.18)$$

$$\text{i.e. } |S_{11}|^2 + |S_{12}|^2 = \frac{1}{2} = |S_{12}|^2 + |S_{22}|^2$$

$$\therefore |S_{11}|^2 = |S_{22}|^2 \quad \text{i.e., } S_{11} = S_{22} \quad (4.19)$$

∴ Equation (4.14) gives

$$\frac{1}{\sqrt{2}} \cdot S_{11}^* + \frac{1}{\sqrt{2}} \cdot S_{12}^* = 0 \text{ i.e., } S_{11} = -S_{12} \quad (4.20)$$

∴ Equation (4.12) gives

$$|S_{11}|^2 + |S_{11}|^2 + \frac{1}{2} = 1 \text{ i.e., } |S_{11}|^2 = \frac{1}{2} \times \frac{1}{2}$$

$$\therefore S_{11} = \frac{1}{2} \quad (4.21)$$

∴ By Eq. (4.20)

$$\therefore S_{12} = -\frac{1}{2} \quad (4.22)$$

∴ S-matrix (Eq. 4.10) relation for H-plane tee will become:

$$\begin{bmatrix} b_1 \\ b_2 \\ b_3 \end{bmatrix} = \begin{bmatrix} 1/2 & -1/2 & 1/\sqrt{2} \\ -1/2 & 1/2 & 1/\sqrt{2} \\ 1/\sqrt{2} & 1/\sqrt{2} & 0 \end{bmatrix} \begin{bmatrix} a_1 \\ a_2 \\ a_3 \end{bmatrix} \quad (4.23)$$

Lossless network: For this H-plane tee, we see that:

$$\sum_{k=1,3} |S_{1k}|^2 = S_{11}^2 + S_{12}^2 + S_{13}^2$$
$$= (1/2)^2 + (-1/2)^2 + (1/3)^2 = 1$$

Therefore by the rule of Eq. (4.2), it is a lossless network.

Let us re-examine the properties of H-type junction given at Fig. 4.5 and Eqs. (4.3)–(4.7).

Expanding the matrix equation we get:

$$b_1 = a_1/2 - a_2/2 + a_3\Big/\sqrt{2} \quad (4.24)$$

$$b_2 = -a_1/2 + a_2/2 + a_3\Big/\sqrt{2} \quad (4.25)$$

$$b_3 = a_1\Big/\sqrt{2} + a_2\Big/\sqrt{2} + 0 \quad (4.26)$$

Now let us examine the power I/O properties also using the above three equations.

Case 1: Input at port 3 only (3 dB *splitter without loss*)

Here $a_3 \neq 0$ and $a_1 = a_2 = 0$.

Therefore Eqs. (4.24), (4.25), and (4.26) give

$$b_1 = a_3\Big/\sqrt{2};\ b_2 = a_3\Big/\sqrt{2};\ b_3 = 0$$

That is, outputs at ports 1 and 3 are equal in magnitude and phase, i.e. same as Eq. (4.3). Therefore power outputs will be square of these voltage relations.

Therefore power output at '1' and '2' will be:

$$P_1 = P_3/2 \text{ and } P_2 = P_3/2$$

The ratio of power at port 1 and port 3 in dB = 10 log is $\left(\frac{P_1}{P_3}\right)$

$$= 10 \log\left(\frac{1}{2}\right)$$
$$= -10 \times 0.3010$$
$$= -3.01 \text{ dB i.e., approximately 3 dB down}$$

Case 2: Equal input at ports 1 and 2 (*power adder*)

Here

$$a_1 = a_2 = a \text{ and } a_3 = 0$$

Using Eqs. (4.24), (4.25), and (4.26), the outputs $b_1 = b_2 = 0; b_3 = \frac{2a}{\sqrt{2}} = a\sqrt{2}$.

This means equal and in phase inputs at ports 1 and 2 give output at port 3 in phase and double of each Fig. 4.6 [power adder Eq. (4.4)].

Case 3: Equal but out of phase inputs at ports 1 and 2 (*power reflector*)

Therefore

$$a_1 = a; a_2 = -a; a_3 = 0$$

Therefore Eqs. (4.24)–(4.26) give

$$\text{Outputs } b_1 = a; b_2 = -a; b_3 = 0$$

This means that, equal and out of phase inputs at port 1 and port 2 gets reflected back as such, with no output at port 3 Eq. (4.5).

Case 4: Input at port 1 only (*splitter with power loss*)

$$\text{Input} \quad a_1 = a; a_2 = 0; a_3 = 0$$

By Eqs. (4.24)–(4.26) outputs are as follows:

$$b_1 = \frac{a_1}{2}; b_2 = -\frac{a_1}{2}; b_3 = \frac{a_1}{\sqrt{2}}$$

Case 5: Equal inputs as ports 2 and 3

$$a_1 = a_3 = a \text{ (in phase)}$$

$$\therefore b_1 = a\left(\frac{1}{\sqrt{2}} - \frac{1}{2}\right)$$

$$b_2 = a\left(\frac{1}{\sqrt{2}} + \frac{1}{2}\right) \quad \text{(reflected back)}$$

$$b_3 = \frac{a}{\sqrt{2}} \quad \text{(reflected back)}$$

These scattering matrix results are summarised in the table given below (Table 4.2).

Table 4.2 Input–output (current and power) in H-plane tee (the -ve sign means out by phase by 180°)

Case no.↓	Voltage equivalent (I) elements of scattering matrix						In power equivalent terms (I^2)					
	Inputs			Outputs			Inputs			Outputs		
Port No. →	1	2	3 (H)	1	2	3 (H)	1	2	3 (H)	1	2	3 (H)
1. (Splitter)	0	0	a	$a/\sqrt{2}$	$a\sqrt{2}$	0	0	0	a^2	$a^2/2$	$a^2/2$	0
2. (Adder)	a	a	0	0	0	$a\sqrt{2}$	a^2	a^2	0	0	0	$2a$
3. (Canceller)	a	$-a$	0	0	0	0	a^2	$-a^2$	0	0	0	0
Aid to memory '1′ + '2′ ↔ '3′; '1′ + '2′∠180°												

4.5 E-Plane Tee Junction (Voltage Junction)

This has (a) tee joint at the broader side (i.e. 'a' side) of the waveguide, and (b) the branching direction is parallel to the E-vector; therefore it is called E-type tee (see Fig. 4.7).

When an input signal is given at port 3, then the output will be from port l and port 2 of equal amplitude but with a phase difference of 180°. Along with this fact, matrix properties can be added as done for H-plane tee.

Proving the properties: The 3 × 3 S-matrix equation as usual is:

$$\begin{bmatrix} b_1 \\ b_2 \\ b_3 \end{bmatrix} = \begin{bmatrix} S_{11} & S_{12} & S_{13} \\ S_{21} & S_{22} & S_{23} \\ S_{31} & S_{32} & S_{33} \end{bmatrix} \begin{bmatrix} a_1 \\ a_2 \\ a_3 \end{bmatrix} \quad (4.27)$$

Now apply two properties of the tee junction and two properties of matrix, for simplifying the terms as follows.

1. Since output at port 1 and port 2 is 180° out of phase, then $S_{23} = -S_{13}$
2. If port 3 is perfectly matched with the junction, then $S_{33} = 0$
3. From symmetric property, $S_{ij} = S_{ji}$, $\therefore$ $S_{12} = S_{21}$, $S_{13} = S_{31}$, $S_{23} = S_{32}$
4. Now using unitary property ($S \cdot S^* = I$): after putting the above results we get

$$\begin{vmatrix} S_{11} & S_{12} & S_{13} \\ S_{12} & S_{22} & -S_{13} \\ S_{13} & -S_{13} & 0 \end{vmatrix} \begin{vmatrix} S_{11}^* & S_{12}^* & S_{13}^* \\ S_{12}^* & S_{22}^* & -S_{13}^* \\ S_{13}^* & -S_{13}^* & 0 \end{vmatrix} = \begin{bmatrix} 1 & 0 & 0 \\ 0 & 1 & 0 \\ 0 & 0 & 1 \end{bmatrix} \quad (4.27a)$$

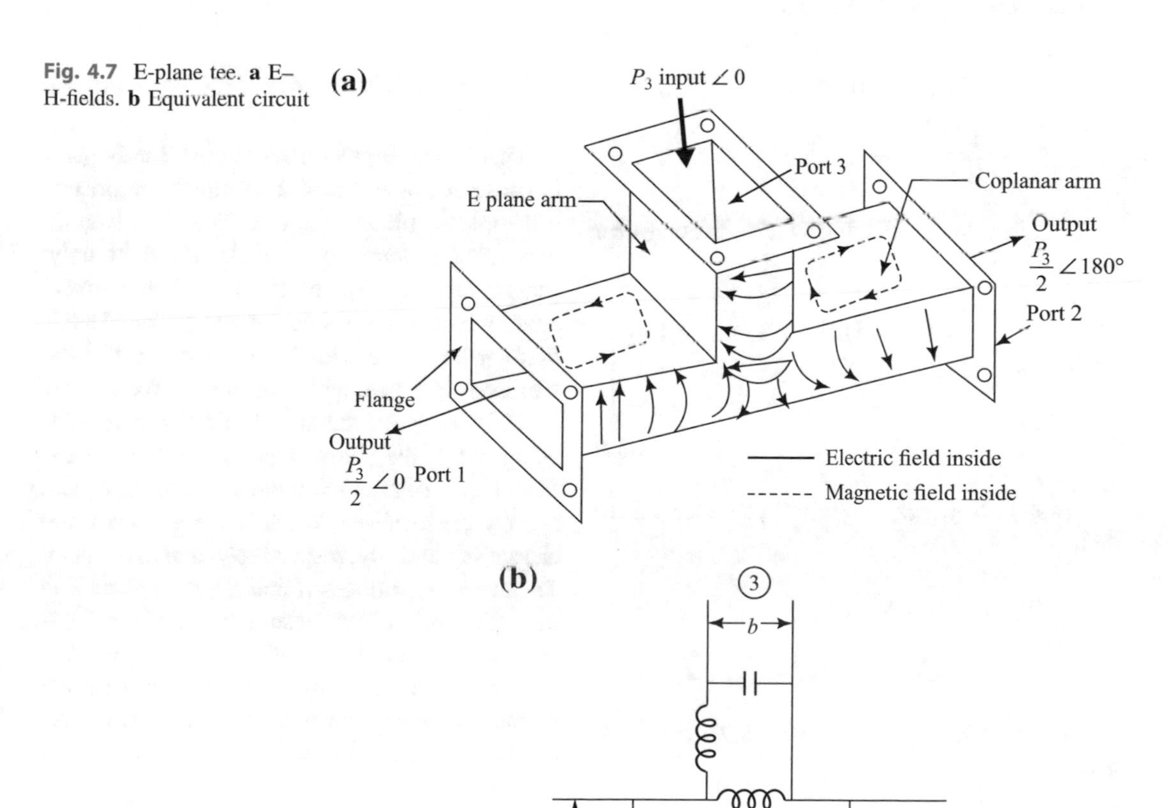

Fig. 4.7 E-plane tee. **a** E–H-fields. **b** Equivalent circuit

Now we expand the matrix to get

$$R_1C_1 : |S_{11}|^2 + |S_{12}|^2 + |S_{13}|^2 = 1 \quad (4.28)$$

$$R_2C_2 : |S_{12}|^2 + |S_{22}|^2 + |S_{13}|^2 = 1 \quad (4.29)$$

$$R_3C_3 : |S_{13}|^2 + |S_{13}|^2 + 0 = 1 \quad (4.30)$$

$$R_3C_1 : S_{13} \cdot S_{11}^* - S_{12}^* S_{13} = 0 \quad (4.31)$$

Subtracting Eq. (4.28) from Eq. (4.29), we get: $S_{11} = S_{22}$.

While Eq. (4.30) gives $S_{13} = 1/\sqrt{2}$ therefore using these two results in Eq. (4.31) we get:

$$S_{13}\left(S_{11}^* - S_{12}^*\right) = 0, \therefore S_{11} = S_{12} = S_{22} \text{ and } S_{13} = 1/\sqrt{2}$$

Therefore Eq. (4.28) becomes:

$$|S_{11}|^2 + |S_{11}|^2 + \frac{1}{2} = 1$$

$$\therefore S_{11} = \frac{1}{2} = S_{12} = S_{22} \text{ and } S_{13} = 1/\sqrt{2}$$

Therefore the matrix Eq. (4.27a) becomes for E-plane tee:

$$\begin{bmatrix} b_1 \\ b_2 \\ b_3 \end{bmatrix} = \begin{bmatrix} 1/2 & 1/2 & 1/\sqrt{2} \\ 1/2 & 1/2 & -1/\sqrt{2} \\ 1/\sqrt{2} & -1/\sqrt{2} & 0 \end{bmatrix} \begin{bmatrix} a_1 \\ a_2 \\ a_3 \end{bmatrix} \quad (4.32)$$

Lossless Network: For this E-plane tee we see that:

$$\sum_{k=1,3} |S_{1k}^2| = S_{11}^2 + S_{12}^2 + S_{13}^2 = (1/2)^2 + (1/2)^2 + (1/\sqrt{2})^2 = 1$$

Therefore by the rule of Eq. (4.2), the tee is lossless

$\therefore$ Expanding the matrix of Eq. (4.32) it gives

$$b_1 = \frac{a_1}{2} + \frac{a_2}{2} + \frac{a_3}{\sqrt{2}} \quad (4.33)$$

$$b_2 = \frac{a_1}{2} + \frac{a_2}{2} - \frac{a_3}{\sqrt{2}} \quad (4.34)$$

$$b_3 = \frac{a_1}{\sqrt{2}} - \frac{a_2}{\sqrt{2}} \quad (4.35)$$

Here a_1, a_2, a_3, and b_1, b_2, b_3 are the parameters of voltage equivalent inputs and outputs. For getting the power parameters we have to square these terms. Following gives the different types of input and output (see Fig. 4.8) relations.

Case 1: Input at port 3 only (3 dB *splitter*)

Therefore by $a_3 \neq 0$; $a_1 = a_2 = 0$ in Eqs. (4.33)–(4.35), we get outputs at ports 1, 2, 3, as

$$b_1 = a_3/\sqrt{2}; b_2 = -a_3/\sqrt{2}; b_3 = 0 \quad (4.36)$$

That is, **one input signal at port 3 only gives outputs at ports 1 and 2 of equal magnitude but out of phase signals**. This is given in Fig. 4.8a, in terms of the electric field only, shown for the TE_{10} mode of a time moment along with the charges on the walls. **The electric fields will reverse after T/2 time along with the transient charges, with microwave frequency**.

The reason for the signal and hence E-fields being out of the phase at ports 1 and 2 is clear from Fig. 4.8a as at $t = 0$ the surfaces P_3, P_1 and Q_3, Q_2 get localised transient (very short time) charge +ve and –ve, respectively, due to the field. Therefore the surfaces R and S get −ve and +ve induced charges. This results into electric fields at ports 1 and 2 to be out of phase (i.e. $b_1 = -b_2$).

The above field directions are for a certain moment and will reverse with microwave frequency after $T/2$ time period.

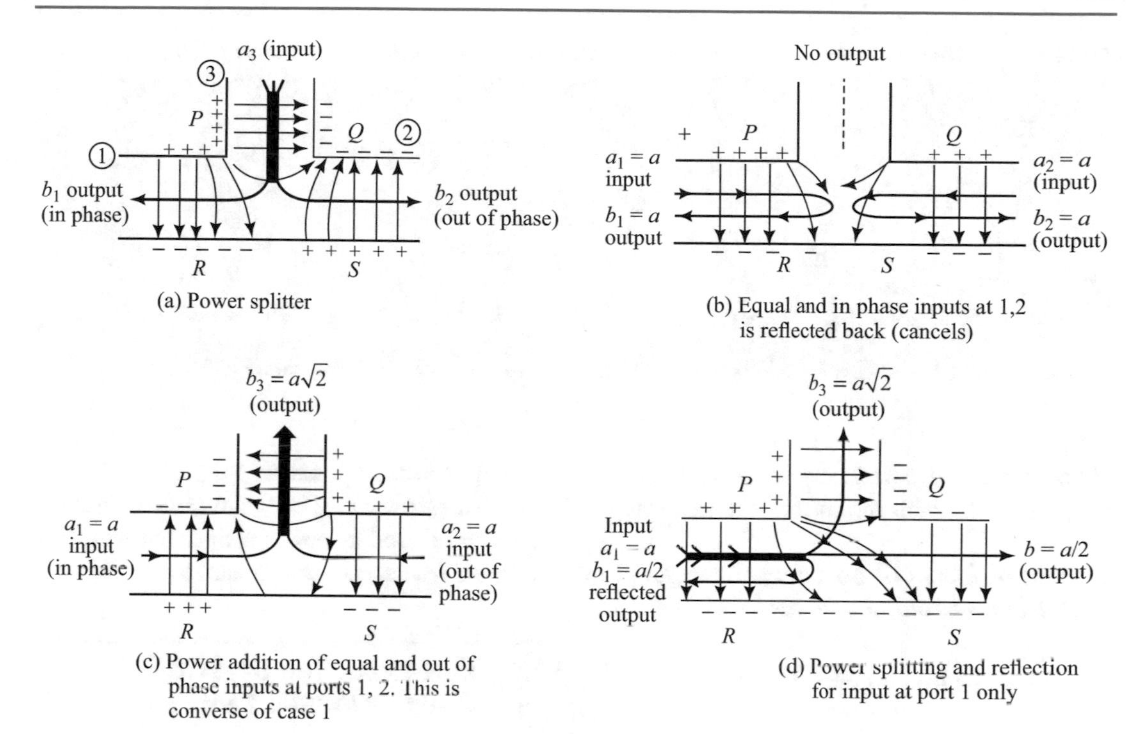

Fig. 4.8 E-plane with four different cases of inputs and their outputs, with electric field (E) power flows and transient charge (q). Shown in each case, E and q will reverse after $T/2$

Case 2: Equal and in phase inputs at port 1 and port 2 (*power reflected back*)

Input signal $a_1 = a_2 = a$; $a_3 = 0$; then, using Eqs. (4.33), (4.34), (4.35), we get the outputs as:

$$b_1 = \frac{a}{2} + \frac{a}{2} = a; b_2 = \frac{a}{2} + \frac{a}{2} = a \text{ and}$$
$$b_3 = \frac{a}{\sqrt{2}} - \frac{a}{\sqrt{2}} = 0$$

That is, equal in phase inputs at port 1 and port 2 give no outputs at port 3, as the fields cancel at port 3. At the port 1 and port 2 signals get reflected back as outputs. This is clear from the field diagram given in Fig. 4.8b. Here we see that at a certain moment ($t = 0$) the surfaces P and Q will become equipotential with no field or voltage over there at port 3 as the surfaces R and S have equal −ve charge.

Case 3: Equal (out of phase) inputs at port 1 and port 2 (*power adder*)

Input signal $a_1 = a$; $a_2 = -a$; $a_3 = 0$; this is converse of case 1.

Then Eqs. (4.33), (4.34), and (4.35) give: $b_1 = 0; b_2 = 0; b_3 = a\sqrt{2}$

That is, equal inputs at ports 1 and 2 appear as addition of both at port 3:

$$P_{b3} = (b_3)^2 = 2a^2 \text{ while } P_{a1} = a^2 \text{ and } P_{a2} = a^2;$$

$\therefore P_{b3} = P_{a1} + P_{a2}$. In this case at a certain moment of the wave cycle, surfaces P and S will be −ve and surfaces Q and R will be +ve charge. These charges then reverses after $T/2$ with the cycle of the microwave frequency. Then the field at port 3 will not be zero but addition of the two

Table 4.3 Input–output (voltage and power) in E-plane tee (the −ve sign means out of phase by 180°)

Case no	Voltage equivalent elements (V) of scattering matrix						In power equivalent terms (V^2)					
	Inputs			Outputs			Inputs			Outputs		
Ports No.→	1	2	3 (E)	1	2	3 (E)	1	2	3 (E)	1	2	3 (E)
1. (Splitter)	0	0	a	$a/\sqrt{2}$	$-a/\sqrt{2}$	0	0	0	a^2	$a^2/2$	$-a^2/2$	0
2. (Reflector)	a	a	0	a	a	0	a^2	a^2	0	a^2	a^2	0
3. (Adder)	a	$-a$	0	0	0	$a\sqrt{2}$	a^2	$-a^2$	0	0	0	$2\,a^2$
Aid to memory '1′ + '2′ ↔ '3′; '1′ + '2′∠180° ↔ 0												

from port 1 and port 2. The field direction will keep changing with the microwave frequency.

Case 4: Input at one port only either port 1 or port 2 (*power splitter with loss*)

$$\text{Here inputs} \quad a_1 = a; a_2 = 0; a_3 = 0$$

$$\text{Then outputs} \quad b_1 = b_2 = \frac{a}{2}; b_3 = \frac{a}{\sqrt{2}}$$

That is, output appears in all the three ports including reflected output at port 1. The charge around surface P remains +ve, which induces −ve charge at surfaces R, S, and Q also. Thus this can also be used as power splitter but with loss. All these results can be summarised in Table 4.3.

4.6 E–H-Plane Tee (Hybrid Junction/Magic Tee)

In this tee the branching is done on both the E- and H-planes and it is called hybrid junction or magic tee (Fig. 4.9). This hybrid junction has the advantage that it has all the four ports matched for TE_{10} mode of propagation. The main

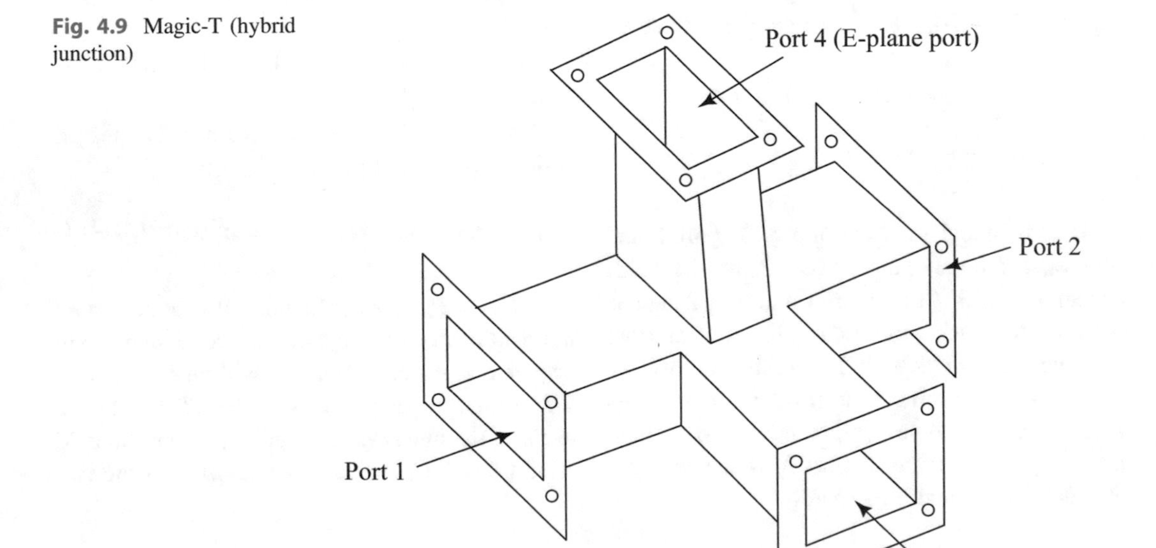

Fig. 4.9 Magic-T (hybrid junction)

application is in mixer section of the microwave radar receivers.

The magic T has the following **special characteristics** for power splitting and in power combining.

(a) **Splitting in phase**: Input at port 3 gives equal and in phase outputs at port 1 and port 2, with no output at port 4.
(b) **Splitting out of phase**: Input at port 4 gives equal and out of phase outputs at port 1 and port 2, with no output at port 3.
(c) **Splitting in phase**: Input at port 1 gives equal and in phase outputs at port 3 and port 4, with no output at port 2.
(d) **Combining two signals in phase**: Converse of case *c*: Two equal and in phase inputs at port 3 and port 4 lead to output as sum at port 1, with no output at port 2.
(e) **Combining two signals in phase**: As converse of '*a*' case two inputs (equal and in phase) at port 1 and port 2 lead to output as sum at port 3, with no output at port 4.

Proof All the above characteristics will be firstly proved by the *S*-matrix, and then the figures can be studied with E–H-fields for these cases, as given below.

Since it has four ports, the *S*-matrix is of 4 × 4 elements size:

$$[S] = \begin{bmatrix} S_{11} & S_{12} & S_{13} & S_{14} \\ S_{21} & S_{22} & S_{23} & S_{24} \\ S_{31} & S_{32} & S_{33} & S_{34} \\ S_{41} & S_{42} & S_{43} & S_{44} \end{bmatrix} \tag{4.37}$$

Now

(i) Because of H-plane tee section symmetry around port 3 [characteristic property (*a*) above],

$$S_{23} = S_{13} \tag{4.38}$$

(ii) Because of E-plane tee section symmetry around port 4 [characteristic property (*b*) above],

$$S_{24} = -S_{14} \tag{4.39}$$

(iii) Because of geometry of the junction, an input at port 3 cannot come out of port 4. Since they are isolated ports and vice versa,

$$S_{34} = S_{43} = 0 \tag{4.40}$$

(iv) From symmetric property, $S_{ij} = S_{ji}$

$$S_{12} = S_{21}, S_{13} = S_{31}, S_{34} = S_{43}, S_{24} = S_{42}, S_{41} = S_{14} \tag{4.41}$$

(v) If ports 3 and 4 are perfectly matched to the junction then

$$S_{33} = S_{44} = 0 \tag{4.42}$$

Now the *S*-matrix becomes as follows with five unknown only:

$$[S] = \begin{bmatrix} S_{11} & S_{12} & S_{13} & S_{14} \\ S_{12} & S_{22} & S_{13} & S_{14} \\ S_{13} & S_{13} & 0 & 0 \\ S_{14} & S_{14} & 0 & 0 \end{bmatrix} \tag{4.43}$$

(vi) Applying the unitary property, i.e. $[S] \cdot [S]* = [I]$.

$$\begin{bmatrix} S_{11} & S_{12} & S_{13} & S_{14} \\ S_{12} & S_{22} & S_{13} & -S_{14} \\ S_{13} & S_{13} & 0 & 0 \\ S_{14} & -S_{14} & 0 & 0 \end{bmatrix} \begin{bmatrix} S_{11}^* & S_{12}^* & S_{13}^* & S_{14}^* \\ S_{12}^* & S_{22}^* & S_{13}^* & -S_{14}^* \\ S_{13}^* & S_{13}^* & 0 & 0 \\ S_{14}^* & -S_{14}^* & 0 & 0 \end{bmatrix} = \begin{bmatrix} 1 & 0 & 0 & 0 \\ 0 & 1 & 0 & 0 \\ 0 & 0 & 1 & 0 \\ 0 & 0 & 0 & 1 \end{bmatrix}$$

Now we expand this to get:

$$R_1C_1 : |S_{11}|^2 + |S_{12}|^2 + |S_{13}|^2 + |S_{14}|^2 = 1 \tag{4.44}$$

$$R_2C_2 : |S_{12}|^2 + |S_{22}|^2 + |S_{13}|^2 + |S_{14}|^2 = 1 \tag{4.45}$$

$$R_3C_3 : |S_{13}|^2 + |S_{13}|^2 = 1 \tag{4.46}$$

$$R_4C_4 : |S_{14}|^2 + |S_{14}|^2 + 1 \tag{4.47}$$

From (4.46) and (4.47)

$$S_{13} = \frac{1}{\sqrt{2}}, S_{14} = \frac{1}{\sqrt{2}} \tag{4.48}$$

Subtracting Eq. (4.25) from (4.44) we get:

$$S_{11} = S_{22} \tag{4.49}$$

Putting Eq. (4.48) we get in (4.44)

$$|S_{11}|^2 + |S_{12}|^2 + \frac{1}{2} + \frac{1}{2} = 1$$
$$\text{i.e., } |S_{11}|^2 + |S_{12}|^2 = 0.$$

This is only possible when:

$$S_{11} = S_{22} = 0$$

From Eq. (4.49), $S_{22} = 0$. Therefore ports 1 and 2 are perfectly matched, when we start with signal at port 3 and port 4 which are perfectly matched.

The magic of magic Tee are (i). If any two ports are perfectly matched to the junction, then the remaining two ports get, automatically matched to the junction and (ii). The signal input at port 1 does not come out of port 2, in spite of being collinear; i.e. ports 1 and 2 are isolated. Similarly ports 3 and 4 are isolated.

After substituting the values of coefficients in Eq. (4.43), *S*-matrix becomes:

$$[S] = \begin{bmatrix} 0 & 0 & \frac{1}{\sqrt{2}} & \frac{1}{\sqrt{2}} \\ 0 & 0 & \frac{1}{\sqrt{2}} & -\frac{1}{\sqrt{2}} \\ \frac{1}{\sqrt{2}} & \frac{1}{\sqrt{2}} & 0 & 0 \\ \frac{1}{\sqrt{2}} & -\frac{1}{\sqrt{2}} & 0 & 0 \end{bmatrix} = \frac{1}{\sqrt{2}} \begin{bmatrix} 0 & 0 & 1 & 1 \\ 0 & 0 & 1 & -1 \\ 1 & 1 & 0 & 0 \\ 1 & -1 & 0 & 0 \end{bmatrix}$$

Therefore with input–output parameters the *S*-matrix equation will be:

Lossless Network: We see that

$$\sum_{k=1,4} S_{1k}^2 = (S_{11}^2 + S_{12}^2 + S_{13}^2 + S_{14}^2)$$
$$= 0^2 + 0^2 + (1/\sqrt{2})^2 + (1/\sqrt{2})^2 = 1$$

Therefore by the rule of Eq. (4.2), the magic tee is also a lossless network.

$$\text{i.e.} \begin{bmatrix} b_1 \\ b_2 \\ b_3 \\ b_4 \end{bmatrix} = \frac{1}{\sqrt{2}} \begin{bmatrix} 0 & 0 & 1 & 1 \\ 0 & 0 & 1 & -1 \\ 1 & 1 & 0 & 0 \\ 1 & -1 & 0 & 0 \end{bmatrix} \begin{bmatrix} a_1 \\ a_2 \\ a_3 \\ a_4 \end{bmatrix} \tag{4.50}$$

Expanding the above matrix equation gives:

$$b_1 = \frac{1}{\sqrt{2}}(a_3 + a_4); \quad b_3 = \frac{1}{\sqrt{2}}(a_1 + a_2) \tag{4.51}$$

$$b_2 = \frac{1}{\sqrt{2}}(a_3 - a_4); \quad b_4 = \frac{1}{\sqrt{2}}(a_1 - a_2) \tag{4.52}$$

Let us now analyse different cases of input (Fig. 4.10) as given in the characteristics earlier and prove it by *S*-matrix Eqs. (4.51) and (4.52).

Case 1: Input at port 3 only: (*in-phase splitting*)

$$a_3 \neq 0, a_1 = a_2 = a_4 = 0$$

$$\therefore \text{output} \quad b_1 = \frac{a_3}{\sqrt{2}}, b_2 = \frac{a_3}{\sqrt{2}}, b_3 = b_4 = 0$$

This is the property of H-plane tee of in-phase splitting.

The reason of zero output at the mouth of port 4 is that the branched signal towards port 1 and port 2 is out of phase, and they cancel at the junction point, with no signal coming out of port 4.

Case 2: Input at port 4 only (*out-of-phase splitting*)

$$a_1 = a_2, = a_3 = 0, a_4 \neq 0$$

Outputs are :

$$b_1 = \frac{a_4}{\sqrt{2}}, b_2 = \frac{-a_4}{\sqrt{2}}, b_3 = b_4 = 0$$

This is the property of *E*-plane tee of out-of-phase splitting.

Just like that in E-type tee junction if a signal is fed into the port 4 of the magic-T, it will get divided into two out of phase components appearing out of port 1 and port 2 with no output at port 3. This is because the electrical vector at the junction point plane is maximum at the centre of the waveguide and becomes = 0 as we approach the two side walls point *P* and *Q* as in Fig. 4.10.

Case 3: Input at port 1 only (in-phase splitting)

$$a_1 = a, a_2 = a_3 = a_4 = 0$$

Then outputs are:

$$b_1 = 0; b_2 = 0; b_3 = \frac{a_1}{\sqrt{2}} = b_4$$

This is the property of power splitting or branching in phase.

For this case (Fig. 4.10c), the signal input of port 1 gets equal divided into two portions giving outputs at port 3 (H-arm) and port 4 (E-arm), with zero output at port 2. Near the junction point of port 2, the two elect. fields (due to branching) cancels being out of phase by 180°, leading to no signal out of port 2.

Similarly input at port 2 will give outputs at port 3 and port 4 with no output at port 1.

Case 4: Equal in-phase inputs at ports 3 and 4 (*power addition*)

$$a_1 = a_2 = 0; a_3 = a_4 = a$$

Then outputs are

$$b_1 = \frac{2a}{\sqrt{2}} = a\sqrt{2}; b_2 = b_3 = b_4 = 0$$

This is the property of power addition and is converse of case 3.

Case 5: Two equal power inputs at ports 1 and 2 (*power adder*)

$$a_1 = a_2; a_3 = a_4 = 0 \,\text{then}\, b_1 = b_2 = b_4 = 0; b_3 = \frac{2a_1}{\sqrt{2}} = a\sqrt{2}$$

This is also power addition property and is converse of case 1 (see Fig. 8.14).

Case 6: Two equal out of phase inputs at port 1 and 2 (*power adder*)

$$a_1 = -a_2 = a, a_3 = a_4 = 0; b_1 = b_2 = b_3 = 0; b_a = 2a$$

This is converse of case 2.

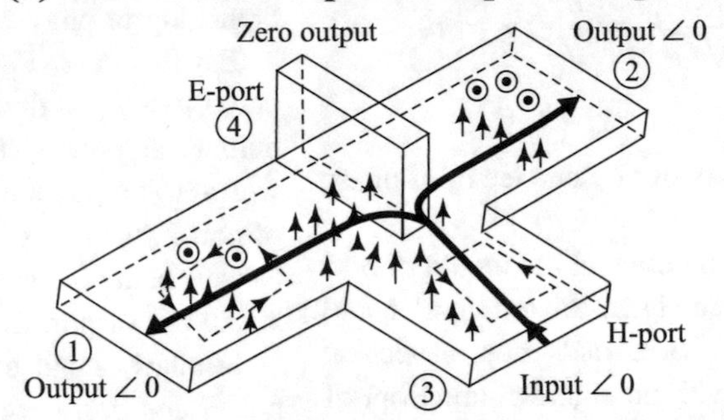

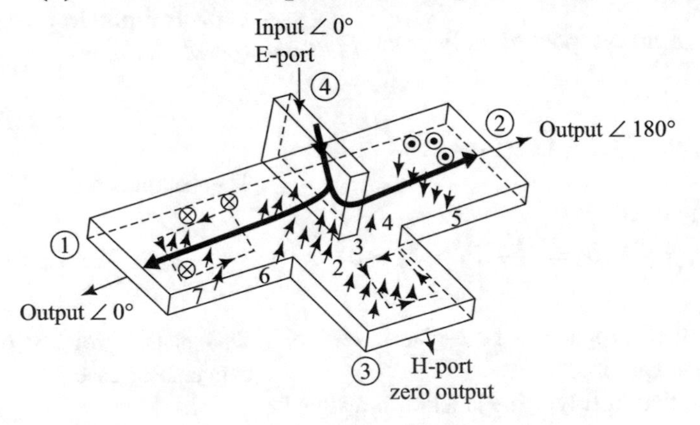

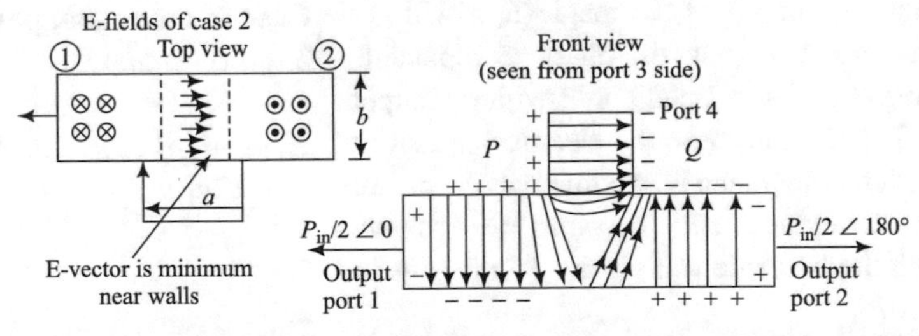

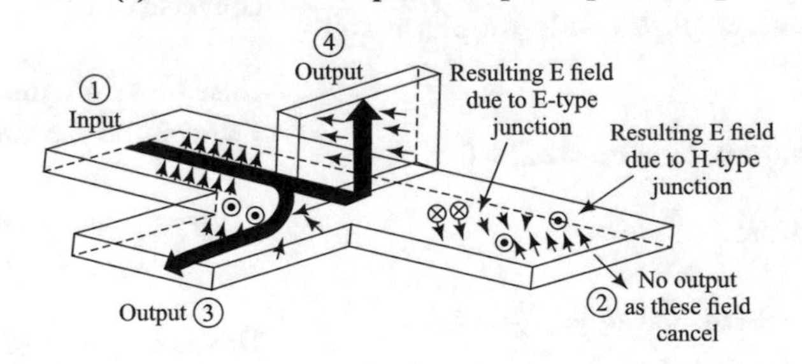

Fig. 4.10 Magic tee and its E–H-fields for three cases with input at. **a** port 3 only. **b** Port 4 only, and **c** Port 1 only

All these cases are summarised in Table4.4 for microwave power points of view.

4.6.1 Applications and Limitations of Magic Tee

Out of a number of applications the main ones are:

1. Impedance measurement on a waveguide: This has been given in detail in the chapter on measurements.
2. As duplexer in trans-receiver system.
3. Discriminator and microwave bridge.
4. Automatic frequency controller for a source like klystron to give a stable signal output for any application.

The main limitation in a magic tee is the reflection due to mismatched junction, leading to power losses. These reflections lead to the following problems.

(i) Whole energy supplied does not reach the load.
(ii) Reflections cause standing waves, which reduce the maximum power handling capacity.

These **reflections can be reduced** at the cost of reduction of power handling capacity by impedance matching at the ports by.

(i) Screw posts on the H-plane of waveguide on the arms 1 and 2.
(ii) Iris at arm 4 for matching in the E-plane as given in Fig. 4.11.

4.7 Hybrid Ring (Rat-Race Junction)

The power limitation problem, the reflection problem, and hence impedance matching problem of magic-T are overcome in hybrid ring. It has four E-plane ports on a circular waveguide having total ring length of $6\lambda_g/4 = 1.5\lambda$ (Fig. 4.12). All consecutive ports are separated from each other by $\lambda_g/4$ or $3\lambda_g/4$. It can be used as a signal splitter or for getting differential output of unequal signals.

1. **As signal splitter:** Input given at port 1 will split equally into two ports, travelling clockwise and anticlockwise (Fig. 4.12a). **When these two waves reach (a) port 3 they cancel being out of phase by 180°**, having path length of odd multiples of $\lambda_g/2$ and

Table 4.4 Summary of input–output power (adders/splitters) in magic tee (−ve power means out of phase power)

SN/Ports	Power inputs				Power outputs			
	Port 1	Port 2	Port 3 (H)	Port 4 (E)	Port 1	Port 2	Port 3 (H)	Port 4 (E)
1. H-port power splitter	0	0	P_i	0	$P_i/2$	$P_i/2$	0	0
2. E-port power splitter	0	0	0	P_i	$P_i/2$	$-P_i/2$	0	0
3. Port 1 power splitter	P_i	0	0	0	0	0	$-P_i/2$	$P_i/2$
4. Port 3, 4 power adder	0	0	P_i	P_i	$2P_i$	0	0	0
5. Port 1, 2 power adder	P_i	P_i	0	0	0	0	$2P_i$	0
6. Port 1, 2 out of phase power adder	P_i	$-P_i$	0	0	0	0	0	$2P_i$
Adder/splitter: Aid to memory $1+2 \leftrightarrow 3$; $1+2\angle 180° \leftrightarrow 4; 3+4 \leftrightarrow 1$								

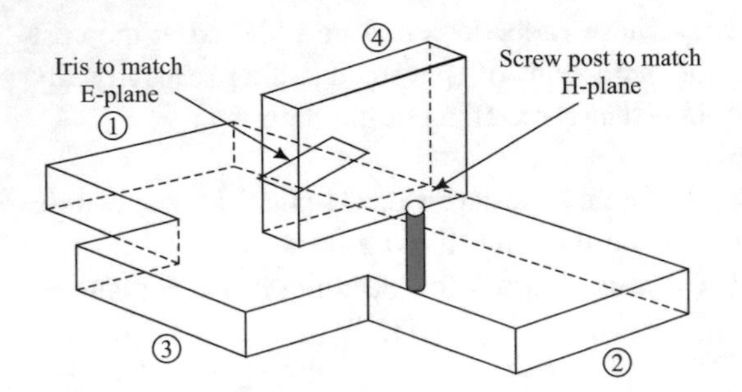

Fig. 4.11 Magic-T impedance matching by iris and screw post

nothing comes out of port 3, **(b) reaches ports 2 and 4, they add** (having path length differential of multiple of λ_g, being in phase) giving equal output at port 2 and 4. However signal at port 4 will be in-phase with input, but at port 2 out of phase. With similar reasonings, we can show that input at port 3 will give equal split outputs at port 2 and port 4, with no output at port 1 and so on.

Thus voltage level input a_1 at port 1 leads to $b_4 = \frac{a_1}{\sqrt{2}}; b_2 = -\frac{a_1}{\sqrt{2}}, b_3 = 0$.

2. **For getting differential output of unequal signal**: As seen two unequal signal inputs at port 1 give output proportion to the sum at port 2 and port 4. The sum at port 2 is out of phase with the input signal while at port 4 it is in phase. The port 3 gives a differential output if signal levels are unequal Fig. 4.12b, but zero if levels are equal.

The scattering matrix of hybrid junction can be written under ideal condition [zero coupling between ports and all the ports matched] as:

$$[S] = \begin{bmatrix} 0 & S_{12} & 0 & S_{14} \\ S_{21} & 0 & S_{23} & 0 \\ 0 & S_{32} & 0 & S_{34} \\ S_{41} & 0 & S_{43} & 0 \end{bmatrix} = \frac{1}{\sqrt{2}} \begin{bmatrix} 0 & -1 & 0 & 1 \\ -1 & 0 & -1 & 0 \\ 0 & -1 & 0 & -1 \\ 1 & 0 & -1 & 0 \end{bmatrix}$$

Here only S_{14} and S_{41} are = +1 and rest are −1 or zero.

4.8 Directional Couplers for Power Sampling/Testing

In general directional couplers are four-port junction for sampling or taping the power flow using secondary auxiliary waveguide, without disturbing the main primary waveguide line power. Therefore **some directional couplers are unidirectional meant for measuring** incident

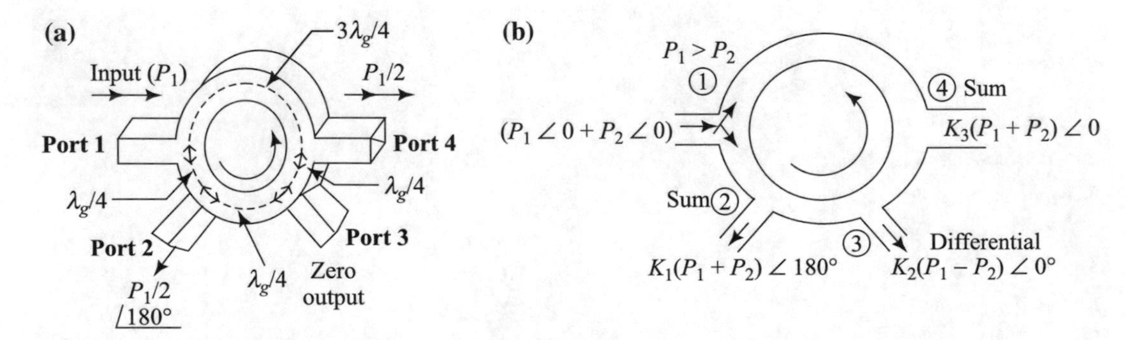

Fig. 4.12 **a** Hybrid ring output of signal available at alternative ports only. Signal power splitting out of phase. **b** Getting sum and differential of two unequal signals

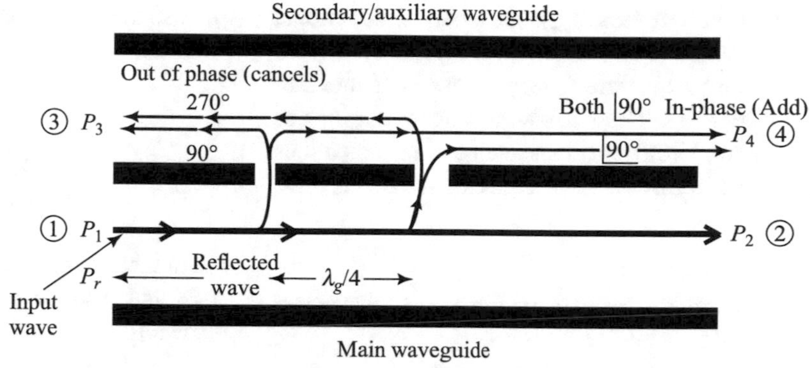

Fig. 4.13 A two-hole four-port directional coupler. At port 4 signals add while at port 3 the two branched signals cancel, and a part reflected back to port 4

power, while some **directional** couplers are **bidirectional** for measuring both incident and reflected power (Fig. 4.13). If power input is given at port 1, then

(i) The main power reaches port 2.
(ii) no power reach port 3 and no reflected wave power (i.e. perfect matched line).
(iii) part power reach port 4.

Thus for P_1 at port 1, the incident power is not coupled to port 3. Similarly the ports 2 and 4 are decoupled.

Performance of a directional coupler is measured in terms of four parameters, like coupling factor (C), directivity (D), isolation (I_{SO}), and insertion loss (I_L), with reference to Fig. 4.13.

(a) **Coupling factor (C)**: It is defined as the ratio of the incident power (I_{ns}) at port 1 to the forward power P_4 at port 4 and is measured in dB.

$$\therefore C = 10\log_{10}\left(\frac{P_1}{P_4}\right)\ \text{dB} \qquad (4.53)$$

This is a measure of how much power is being sampled. Normally directional couplers have $C = 20$ dB or 30 dB or 40 dB; i.e. 1/100 or 1/1000 or 1/10,000 of input power goes to port 4 as a sample to measure continuously the useful power P_2 going to port 2.

(b) **Directivity (D)**: It is the ratio of the sampled forward wave power P_4 (as port 4) to the reflected backward power P_3 (at port 3) and is expressed in dB.

$$\therefore D = 10\ \log_{10}\left(\frac{P_4}{P_3}\right)\ \text{dB} \qquad (4.54)$$

It is a measure of how well the directional coupler distinguishes between the forward and reverse travelling power in the secondary waveguide. Therefore for ideal case $P_3 = 0$, $\therefore D = \infty$.

(c) **Isolation (I_{so})**: It is the ratio of incident at port 1 to backward power at port 3:

$$I_{so} = 10\ \log_{10}\left(\frac{P_1}{P_3}\right)\ \text{dB}$$
$$= 10\ \log_{10}\left(\frac{P_4}{P_3} \times \frac{P_1}{P_4}\right)\ \text{dB} \qquad (4.55)$$
$$\therefore I_{so} = C + D$$

(d) **Insertion loss (I_{ns})**: Generally a directional coupler or any microwave component causes more and more insertion loss at higher frequencies. These losses are due to sampling of signal (P_4) for measurement or reflections due to mismatched load, etc. We define it as:

$$I_{ns} = 10\ \log_{10}[(P_2 + P_3 + P_4)/P_1]\ \text{dB} \quad (4.56)$$

In an ideal case the expression inside capital bracket = 1; therefore, $I_{ns} = 0$. In a well-designed directional coupler $D \geq 30$ to 35 dB. $C > 20$ dB and $P_r = 0$.

Example If tapped power P_4 is (l/100)th of P_1 and reflected backward power P_3 is 1/1000th of the P_4
then

$$P_4 = P_1/100 \text{ and } P_3 = P_4/1000$$

$$\therefore \quad C = 10\ \log_{10}(100) = 10 \times 2 = 20\ \text{dB}$$

$$D = 10\ \log(1000) = 30\ \text{dB}$$

$$I_{so} > (30 + 20) \text{ i.e. } I_{so} > 50\ \text{dB}$$

Scattering Matrix of a Directional Coupler

(a) Directional coupler being a four-port network, the 4 × 4 matrix will be:

$$\text{i.e., } [S] = \begin{bmatrix} S_{11} & S_{12} & S_{13} & S_{14} \\ S_{21} & S_{22} & S_{23} & S_{24} \\ S_{31} & S_{32} & S_{33} & S_{34} \\ S_{41} & S_{42} & S_{43} & S_{44} \end{bmatrix} \quad (4.57)$$

(b) When all four ports are perfectly matched to the junction, then all the diagonal elements will be zero, i.e.

$$S_{11} = S_{12} = S_{33} = S_{44} = 0$$

(c) Also symmetric property of scattering matrix gives:

$$S_{ij} = S_{ji}$$

$$S_{12} = S_{21}, S_{23} = S_{32}, S_{13} = S_{31}, S_{24} = S_{42}, S_{34} = S_{43}, S_{41} = S_{14}$$

(d) As there is no coupling between ports 1 and 3, and also no coupling between 2 and 4, therefore $S_{13} = S_{31} = 0$, i.e. $I_{so} = \infty$ and $S_{24} = S_{42} = 0$.
All the above leads to scattering matrix with four unknown parameters:

$$[S] = \begin{bmatrix} 0 & S_{12} & 0 & S_{14} \\ S_{12} & 0 & S_{23} & 0 \\ 0 & S_{23} & 0 & S_{34} \\ S_{14} & 0 & S_{34} & 0 \end{bmatrix}$$

(e) Now using unitary property on the above matrix, i.e. $[S] \cdot [S]^* = I$, we get:

$$\begin{bmatrix} 0 & S_{12} & 0 & S_{14} \\ S_{12} & 0 & S_{23} & 0 \\ 0 & S_{23} & 0 & S_{34} \\ S_{14} & 0 & S_{34} & 0 \end{bmatrix} \begin{bmatrix} 0 & S_{12}^* & 0 & S_{14}^* \\ S_{12}^* & 0 & S_{23}^* & 0 \\ 0 & S_{23}^* & 0 & S_{34}^* \\ S_{14}^* & 0 & S_{34}^* & 0 \end{bmatrix} = \begin{bmatrix} 1 & 0 & 0 & 0 \\ 0 & 1 & 0 & 0 \\ 0 & 0 & 1 & 0 \\ 0 & 0 & 0 & 1 \end{bmatrix} \quad (4.58)$$

Expanding and writing each equation separately we get (writing $x, x^* = |x|^2$):

$$R_1C_1 : |S_{12}|^2 + |S_{14}|^2 = 1 \quad (4.59)$$

$$R_2C_2 : |S_{12}|^2 + |S_{23}|^2 = 1 \quad (4.60)$$

$$R_3C_3 : |S_{23}|^2 + |S_{34}|^2 = 1 \quad (4.61)$$

$$R_1C_3 : S_{12}S_{23}^* + S_{14}S_{34}^* = 0 \quad (4.62)$$

Subtracting Eq. (4.59) from (4.58) gives

$$S_{14} = S_{23} \tag{4.63}$$

Subtracting (4.59) from (4.60) gives

$$S_{12} = S_{34} \tag{4.64}$$

Assuming that S_{12} is real and +ve = 'p', therefore

$$S_{12} = S_{12}^{*} = S_{34} = p = S_{34}^{*} \tag{4.65}$$

From Eqs. (4.62) and (4.65) $p(S_{23} + S_{14}) = 0$.
But by Eq. (4.62) $S_{14} = S_{23}$, i.e. $p\left(S_{23} + S_{23}^{*}\right) = 0$.
Since $p \neq 0$, $S_{23}^{*} = -S_{23}$, and this is possible only if
S_{23} is complex = jq (Assume). Therefore

$$S_{23} = S_{14} = jq \text{ and } S_{12} = S_{34} = p \tag{4.66}$$

Thus final scattering matrix of the directional coupler will be:

$$[S] = \begin{bmatrix} 0 & p & 0 & jq \\ p & 0 & jq & 0 \\ 0 & jq & 0 & p \\ jq & 0 & p & 0 \end{bmatrix} \tag{4.67}$$

With

$$p^2 + q^2 = 1 \tag{4.68}$$

This relation can be proved by Eq. (4.59) (unit property) which will give

$$S_{12} \cdot S_{12}^{*} + S_{14} \cdot S_{14}^{*} = 1 \tag{4.69}$$

By Eq. (4.65):

$$S_{12} \cdot S_{12}^{*} = p^2 \tag{4.70}$$

and by Eq. (4.66):

$$S_{14} = jq \text{ and } S_{14}^{*} = -jq \therefore S_{14} \cdot S_{14}^{*} = q^2$$

Thus Eq. (4.69) with values of the two terms gives:

$$p^2 + q^2 = 1 \tag{4.71}$$

Thus Eq. (4.2), i.e. $\sum |S_{1k}|^2 = 1$, is satisfied as $|S_{11}|^2 + |S_{12}|^2 + |S_{13}|^2 + |S_{14}|^2 = 0 + p^2 + q^2 + 0 = 1$
by Eq. (4.65). Therefore directional couplers are also lossless network.

4.8.1 Various Types of Directional Couplers (DC)

There is a variety of directional couplers, and we will discuss the following four types only, which are normally used.

(a) One-hole directional coupler, for connecting a probe for sampling the signal (Fig. 4.14) **(Rarely used now)**.
(b) Two-hole directional coupler, used mostly for sampling signal for measurement of VSWR, etc. (Fig. 4.15).
(c) Multihole coupler for sampling transmitted energy as well as the reflected energy (Fig. 4.16) **(mostly used now)**.
(d) Lange hybrid coupler for strip lines (Fig. 4.17).

(a) **Single-hole (Bethe) directional coupler (not used now)**: The theory of directional coupler was first established by Bethe (1942), using a single hole in between. Two rectangular waveguide sections, one kept on the other. Signal coupled through a circular hole on the cavity of the common broad wall. For achieving the directional coupling, the two waveguides are inclined at an angle θ. (see Fig. 4.14). The amount of energy coupling is decided by the frequency to be used and the dimension of the coupling hole.

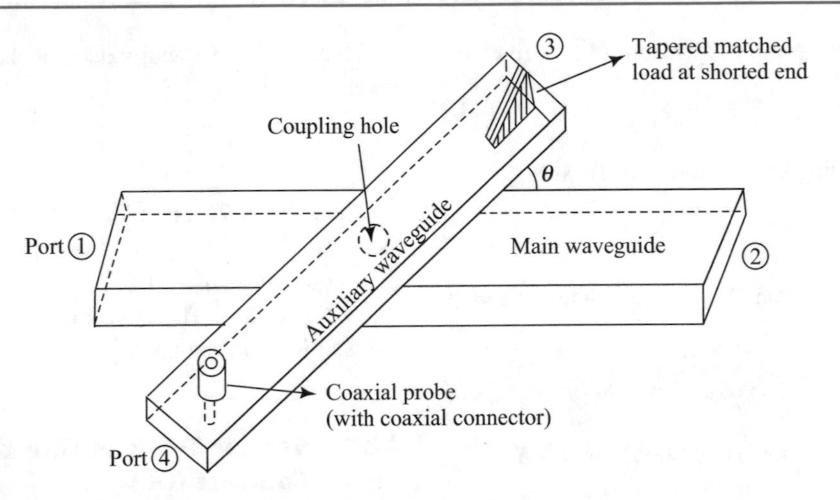

Fig. 4.14 Single hole be the directional coupler (rarely used now)

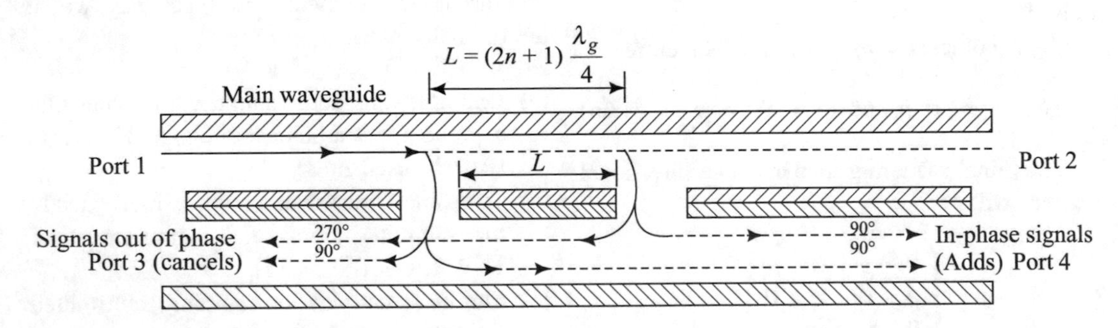

Fig. 4.15 Two-hole waveguide directional coupler

Fig. 4.16 Multihole directional coupler—the most commonly used directional coupler is three-port device

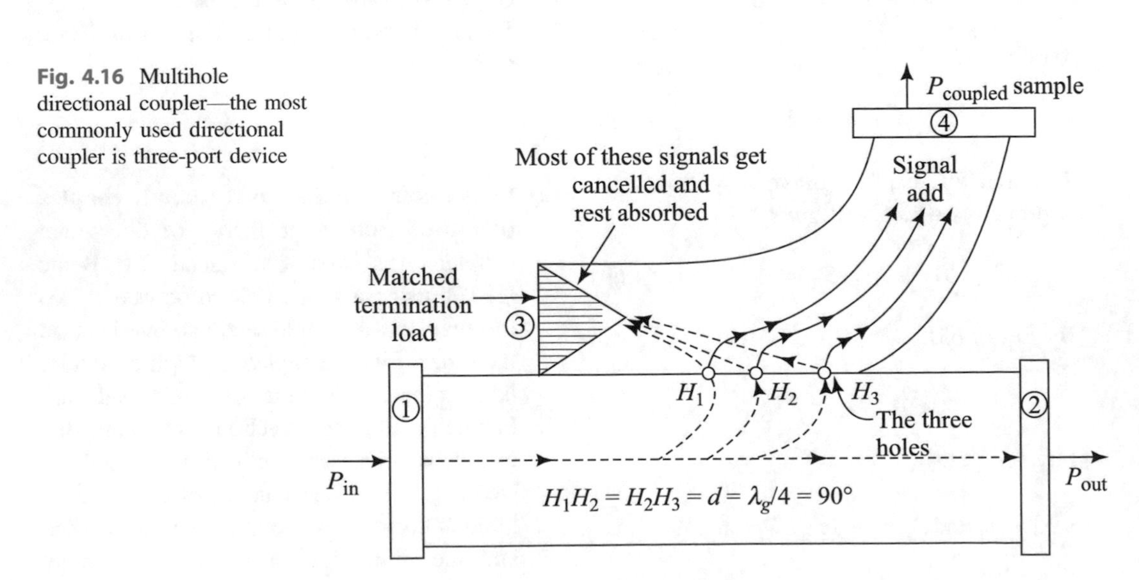

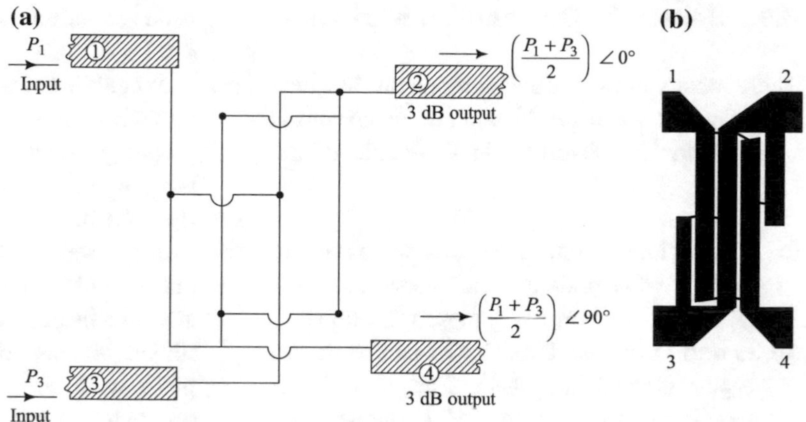

Fig. 4.17 **a** Lange hybrid coupler for microstrip lines. Two power inputs P_1 and P_3 add together and divide equally to ports 2 and 4. If $P_3 = 0$, $P_1 \neq 0$, then also 3 dB power appears at ports 2 and 4 each. **b** Actual design of Lange coupler

A part of the power into the port 1 gets into ports 3 and 4 due to coupling through the hole. The power at port 3 is absorbed by the matched load, while the power at port 4 is tapped by the coaxial probe for any application like VSWR measurement, etc. If $\theta = 0$, then the hole has to be offset and is not at the centre. The TE_{10} mode gives the vertical electric field along the probe as well as through the hole. **Because of bad directivity and high losses, this dc is not used at all.**

(b) **Two-hole directional couplers**: It has two waveguides (Fig. 4.15), the main and the auxiliary waveguides joined on the broad side, having two holes in the common wall at a distance of $\lambda_g/4$ or its odd multiple, i.e. $(2n + 1)\ \lambda_g/4$, being a +ve integer. At port 3, two portions of signal reach with a phase difference corresponding to odd multiplied of $\lambda_g/4$, i.e. out of phase, and therefore cancel. Total phase shift of 270° is due to (a) crossing the hole by 90° ($\equiv\lambda_g/4$) and (b) travelling extra distance $2L = 2\lambda/4$ for the wave turning around second hole from ports 1 to 3, i.e. additional 180° ($\equiv 2 \times \lambda_g/4$). At port 4, the two portions are in-phase and therefore add together. As no impedance matching is required at port 3, this is the most commonly used directional coupler. It has already been dealt in detail in Sect. 4.8.

(c) **Multihole directional coupler**: Multihole directional coupler also works on the same principle as of two-hole coupler (Fig. 4.16) wherein all consecutive holes are separated by $\lambda_g/4$. The size of the hole decides the frequency range over which it can be used. The coupling vs frequency variation is kept at a maximum of ±0.5 dB.

(d) **Lange hybrid coupler for microstrip line**: It consists of four parallel strip lines with alternate lines connected together (Fig. 4.17). This coupler is used for dividing the input power (port 1) equally 3 dB each to two output ports 2 and 4 in quadrature; i.e. signals at ports 2 and 4 have a phase difference of 90°. When this coupler is in same circuit, then the reflections from the circuit back to the output parts (2 or 4) will flow to the port 3 (normally shorted) or cancel at the port 1 (input port). Thus the unwanted reflected signal cannot damage the source. The device is mechanically and electrically symmetric, with high degree of isolation between two input ports and two output ports. Because of this quality it is used in mixers, balanced modulators, balanced amplifiers, etc.

4.9 Bends, Twists, and Transitions

There are a variety of other components as given below, where some specification in dimensions is required for no distortion in the mode of the signal (see Fig. 4.18).

(a) **Transitions**: From rectangular waveguide to circular waveguide and vice versa where the signal can be sent using these transitions.
(b) **E and H bends**: These bends can be of two types: gradual and sharp. The gradual one has a radius of curvature $>2\lambda_g$, while sharp bends are of 45° separated by distance of $\lambda_g/4$ (Fig. 4.18).
(c) **Waveguide twist**: Sometimes one requires rotation of fields in microwave circuits for proper physical joints of power. The twist has to be gradual over a length $>2\lambda_g$ (Fig. 4.18).
(d) **Flexible waveguide**: It may be due to compulsion of less space in some cases where we also use flexible waveguide made of round ribbon of brass with inner surface electroplated with chromium. In these guides the power loss is quite high (Fig. 4.18).

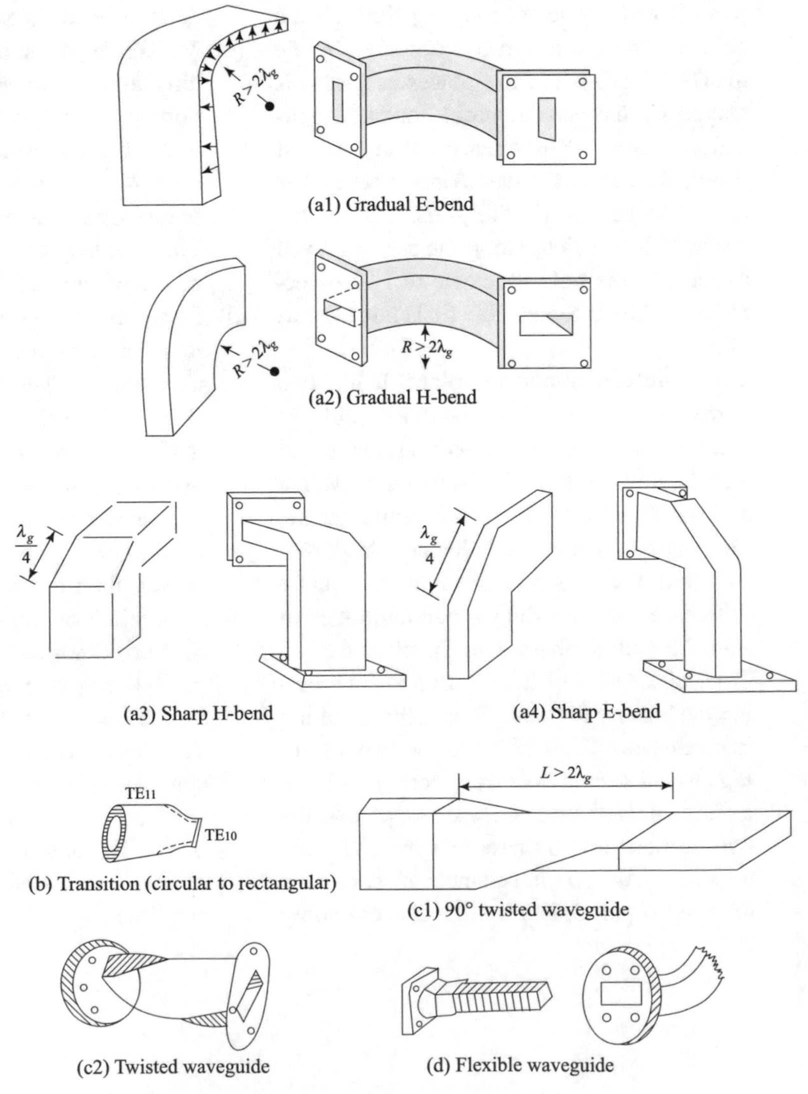

Fig. 4.18 $\mathbf{a}_1$, $\mathbf{a}_2$ *E*- and *H*-plane gradual bends of 90°. $\mathbf{a}_3$, $\mathbf{a}_4$ *E*- and *H*-plane sharp bends. **b** Transition from circular to rectangular. $\mathbf{c}_1$ and $\mathbf{c}_2$ twisted waveguide. **d** Flexible waveguide

4.10 Attenuators and Terminators

(a) **Attenuators** are required for partly absorbing/controlling/reducing the power flow from one point to the another, without introducing reflections (Fig. 4.19). It normally has a power absorbing/dissipating/resistive element or pad in the section of a waveguide, with its plane parallel to the electric field and at the central line, where the electric field is at its maximum, i.e. centre of width a for TE_{10} mode. **Attenuation is a function of frequency**. And a fixed 10 dB attenuator at 5 GHz, may give 13 dB attenuation at 10 GHz. In general attenuation may be defined with input (P_1) and output power (P_0) through a device/circuit/component as:

$$\text{Attenuation} = 10\ \log_{10}(P_i/P_0)\ \text{dB}$$

Some attenuators are used for completely absorbing the power at one end, without any reflection, e.g. port 3 in directional coupler in Figs. 2.14 and 2.16.

Types of attenuators

(i) Fixed for partly or fully absorption
(ii) Variable as per need.

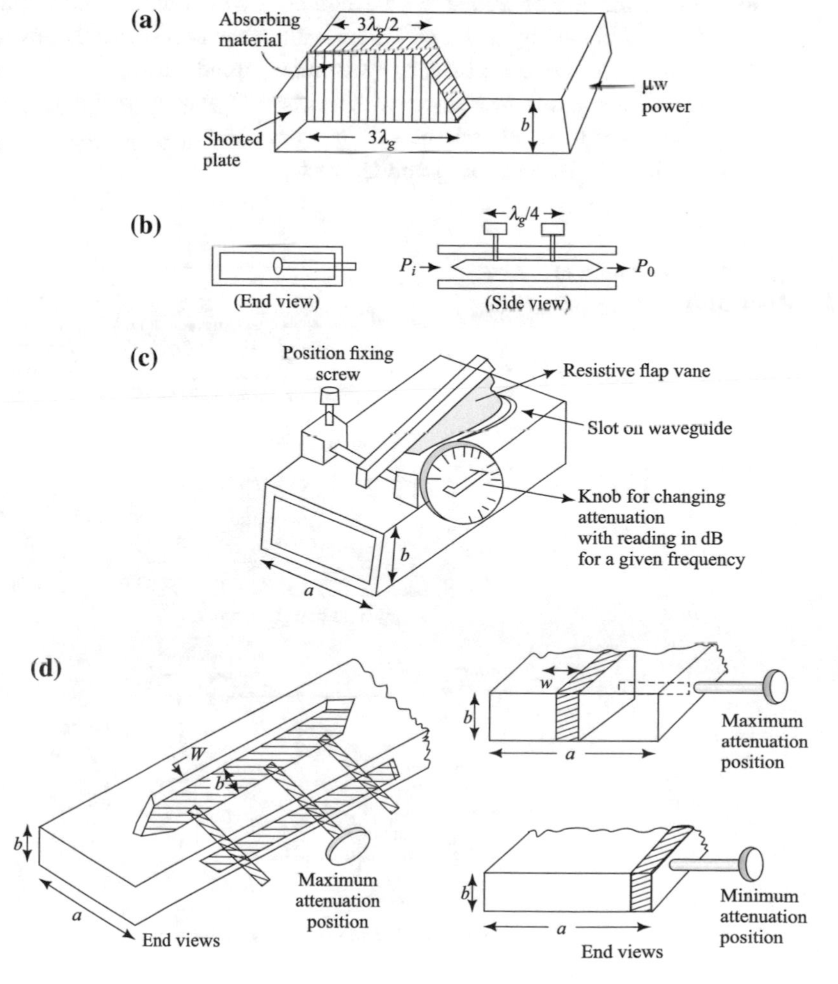

Fig. 4.19 Resistive material attenuators. **a** Fixed type full absorption. **b** Fixed type part absorption. **c** Variable angle vane rotatable. **d** Variable position wane push–pull type

Figure 2.20 gives the fixed and variable attenuators of various types. The fixed ones can be at the end of the waveguide (Fig. 4.19a) or a resistive element at the central line (Fig. 4.19b). The variable types can be rotatable vane (Fig. 4.19c) or variable position vane inside (Fig. 4.19d). Here these vanes are made of some resistive material.

(b) **Terminations:** These terminations are resistive load for fully absorbing power at one end of the waveguide, without reflection, e.g. port 3 in directional coupler (Fig. 4.20). A good termination should lead to VSWR (1.0). An absorbing wedge type of material placed at the central point of the width *a*, along the length, as in Fig. 4.20a, b can also do the job. Instead of wedge the waveguide at that end is closed by metal and filled with mixture of graphite and powder of ceramic sand used for making resistance, which acts as perfect absorber of microwave power (Fig. 4.20c, d). Instead of graphite, metal powder is also used. The tunable short (Fig. 4.20e, f) is the best type of terminations with lowest VSWR.

4.11 Iris and Screw Posts for Impedance Matching/Introducing L or C

These are used for (a) impedance matching and (b) introducing capacitance or inductance: When there is mismatch in the impedance between the source and the load in a waveguide, reflection will be there. The difference in the load impedance and characteristic impedance will be some susceptance capacitance or inductive due to which the reflections are there. This susceptance has to be cancelled by introducing susceptance of opposite sign.

For example, $j\omega L$ (inductive) or $\frac{-j}{\omega c}$ capacitive or vice versa.

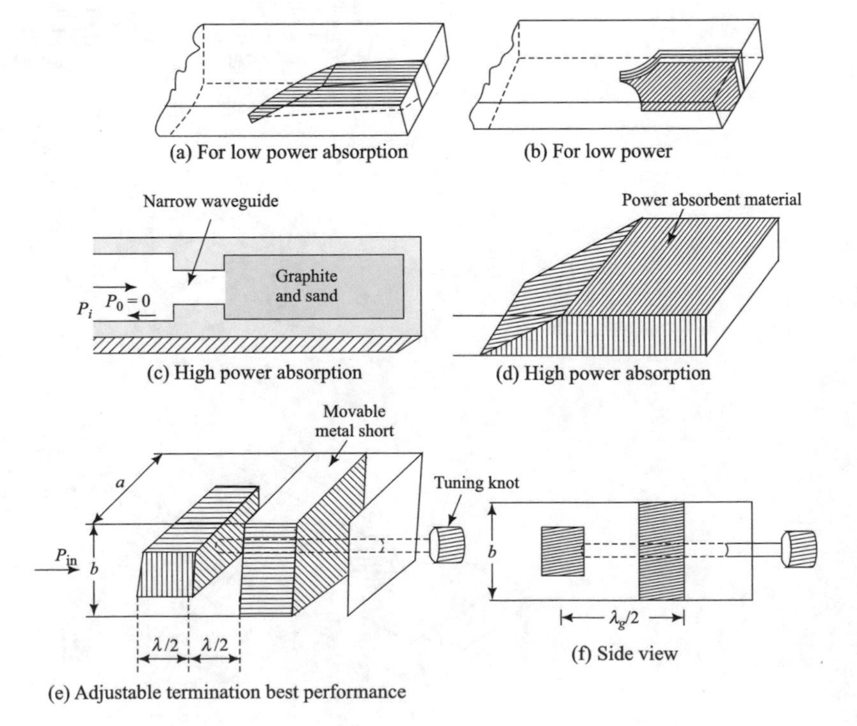

Fig. 4.20 Terminations of various types

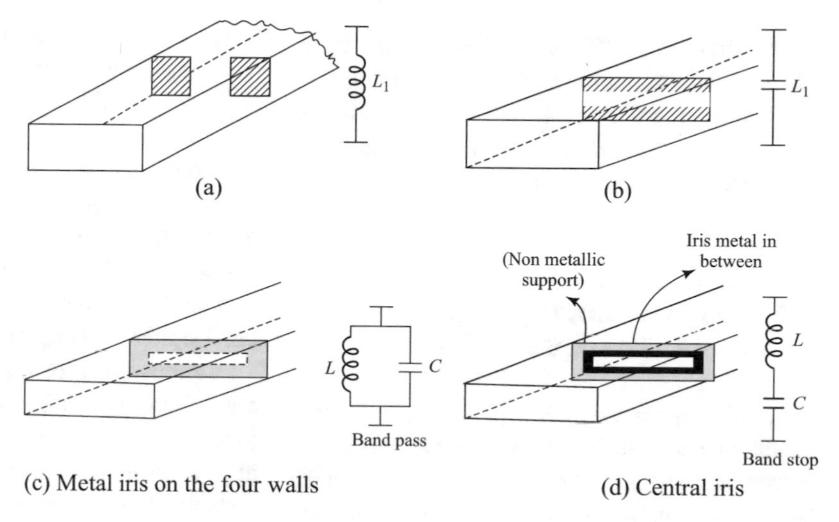

Fig. 4.21 Various types of iris for only *L* or *C* or *LC*-tuned circuit and their equivalent circuit

This is done by introducing (a) iris or (b) posts/screws as in Fig. 4.22 with appropriate size/length of irises or screws, so that the impedance matches and the VSWR comes close to one.

(a) **Irises**: These are metal plates on the path of the waveguide. The value of the inductance or capacitance of the irises depends on its size and design as in Fig. 4.21. In TE_{10} mode the irises plate parallel to elect. field gives inductive and parallel to the magnetic field plane gives capacitative, while a combination of both will give resonant *LC* circuit.

(b) **Posts and screws**: If a metal rod or screws projects inside the broad side of the waveguide, then this gives capacitative or resistive or inductive load depend on how much (l) is the projection inside, i.e. $l < \frac{\lambda_g}{4}$ or $= \frac{\lambda_g}{4}$ or $> \frac{\lambda_g}{4}$ respectively as in Fig. 4.22a, b. The susceptance increases with the projection length inside (l) and also with the dia (D) of the screw. The advantage of screw/post is that its depth can be adjusted for frequency turning. Two screws separated by a distance of $(2n + 1)\frac{\lambda_g}{4}$ and depths of penetration of one of them $>(\lambda_g/4)$ and of the

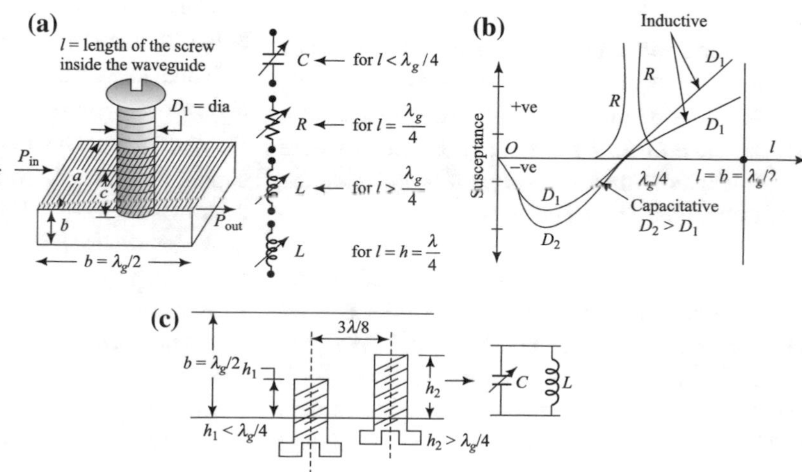

Fig. 4.22 **a** Screws and posts in waveguides introduces capacitative, inductive or resistive load, which depends on depth of insertion into the waveguide. **b** Susceptance as a function of screw depth for diameters D_1 and D_2 of screw. Larger dia. gives higher susceptance. **c** Two screw tuned circuit

second $< \frac{\lambda_g}{4}$ will give load similar to L, C circuit with tuning of L and C (Fig. 4.22c). Using iris and screws, various types of filters, e.g. band stop, band pass, high pass and low pass filters, can be made.

4.12 Signal Tapping/Feeding and Detecting

Normally signal from the generator or cavity resonator needs to be fed to the coaxial line or waveguide or the signal is to be detected from the line. For this some special components are required (e.g. probes and loops) for tapping the signal. For detecting the signal, it has to be rectified by a diode with RF bypass capacitor and sent to VSWR-metre.

4.12.1 Probes and Loops (for Tapping/Exciting/Feeding μω Power into a Waveguide or for Taking Out Microwave Power from the Waveguide)

Normally power is fed into a waveguide using a coaxial cable connected either:

(a) At the shorted end, with its central conductor loop projecting inside up to $\lambda_g/4$.
(b) At the top surface, with a probe placed at a distance of $\lambda_g/4$ from the shorted end where E is maximum end, and projecting inside the waveguide as in Fig. 4.23.

The loop excites the magnetic field, while the probe excites the electric field. As E- and H-fields are inseparable, the EM wave gets excited/transferred to the waveguide in both the cases. The probe in the broad side is also used for sampling signal level (as in slotted waveguide) for measuring SWR (Chap. 8).

4.12.2 Diode Detectors Using Schottky Barrier Diode (SBD)

A diode mounted (as in Fig. 4.24) normally inside a waveguide or coaxial line having a tunable metal short slug with the coaxial cable connecter on its broad side works as a detector. The tuning is required for impedance matching between the detector and the transmission line.

Various types of probe system and detectors are given (Fig. 4.24a–d) which may be with matched load (Fig. 4.24b) or tunable (Fig. 4.24a) at the waveguide itself (Fig. 4.24c) or at the coaxial line side (Fig. 4.24a). The diode (Fig. 4.24g) rectifies the microwave signal, and the filtered dc output goes through the coaxial cable to the metre for measurement of field strength at the position of the probe.

Schottky or point contact diodes which have switching time of picosecond range are used for detection of microwave signal. For input signal power <10 W, the forward I–V characteristic is approximately parabolic and follows the square law, i.e. $I \propto V^2$ as in Fig. 4.24e. If the microwave signal voltage $v = V_0$ cot (ωt) is applied across a diode, the current is given by:

$$\begin{aligned} i &= I_0(\mathrm{e}^{av} - 1) \cong I_0\mathrm{e}^{av} \\ &\approx I_0\left(aV\cos\omega t + a^2V_0^2/4\right)(1 + \cos 2\omega t) \end{aligned}$$

Here $a = 1/(\eta V_T); kT/q$ with $\eta = 1.1$ for Schottky diode and 1.4 for point contact diode.

Thus the rectified dc current of the diode detector reaching the metre is given by $I = I_0a^2V_0^2/4$, which is proportional to V^2 of $\mu\omega$ signal and hence directly proportional to the microwave power. Here the bypass capacitor, near the diode (Fig. 4.24g), by passes out ac ($\mu\omega$) component of the rectified wave.

For larger power input >10 W, the I–V characteristic goes to the linear portion and therefore power needs to be attenuated or a portion to be taken out by directional coupler before reaching the detector for its operation in the square law region.

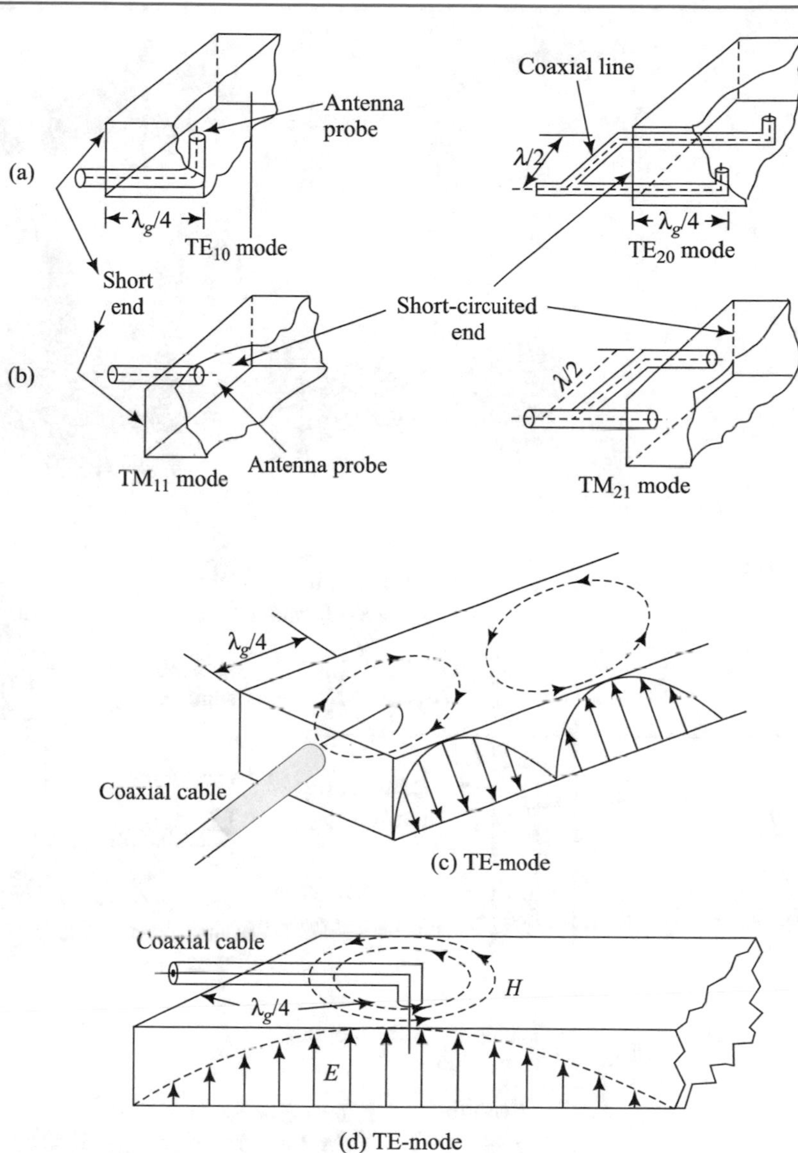

Fig. 4.23 **a** Methods of exciting various TE modes in rectangular waveguides by signal fed from coaxial cable extended core wire as antenna at the base at a distance of $\lambda_g/4$ from shorted end. **b** Exciting TM modes by probes at the shorted side. **c** E- and H-fields in the case of loop feeding at the centre of the magnetic field. **d** E- and H-fields in the case of probe feeding at the locations of maximum electric field

These detectors are very sensitive, operate with $\mu\omega$ signal without any dc bias, and disturb the main signal very less, keeping the VSWR < 1.3 (also see Sect. 6.12).

4.13 Wave Metres/Frequency Metre

A cavity wave metre is a transmission cavity with tuning short plunger coupled through a hole (Fig. 4.25c, d) on the main waveguide carrying the microwave (Fig. 4.25). The tuning plunger has a dial at the other end, where frequency can be read. At resonance condition in the cavity, maximum signal will be taken by the cavity and as a result the power reaching the load P_2 falls (Fig. 4.25e). Here the calibrated position of the plunger gives the frequency of the microwave power flowing in the waveguide.

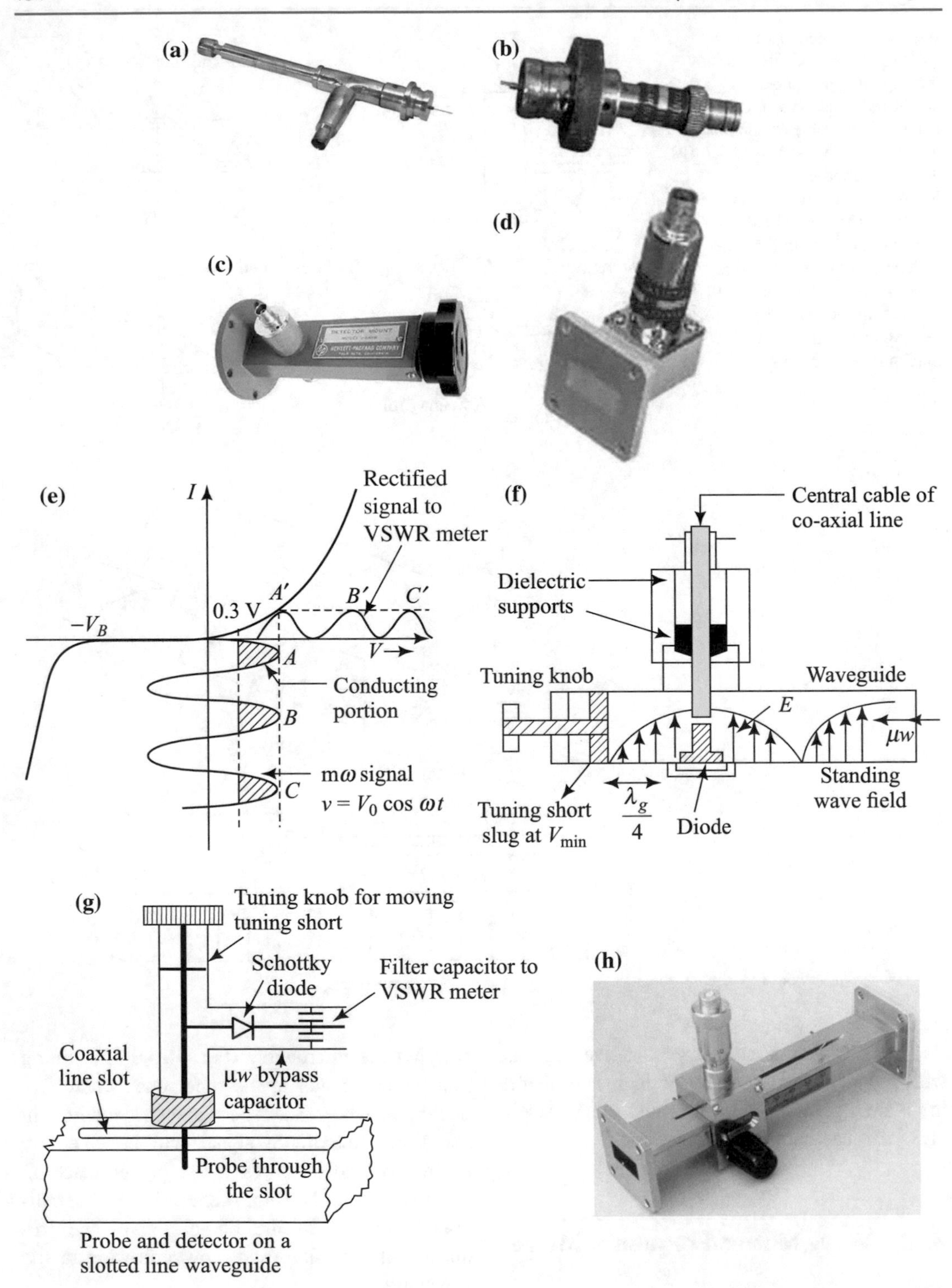

Fig. 4.24 Probe and detectors of various types. **a** Tunable probe for slotted waveguide. **b** Fixed tune for fixed frequency for slotted waveguide. **c** Tunable waveguide Schottky diode detector. **d** Fixed waveguide matched detector mount. **e** I–V characteristic of Si Schottky diode. **f** Inner of a tunable waveguide detector with electric field. **g** Probe and detector on slotted line waveguide

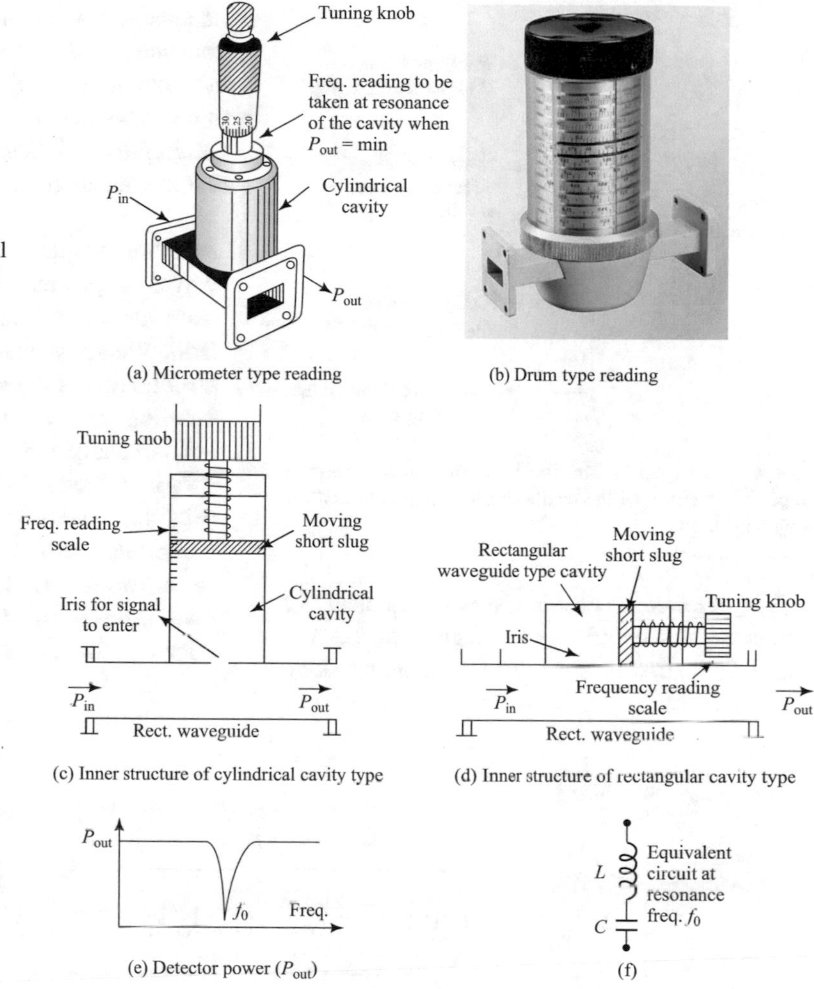

Fig. 4.25 Wave metre, i.e. frequency metres. **a** Micrometre-type reading cylindrical tunable cavity. **b** Drum-type reading cylindrical tunable cavity. **c** Inner structure of (**a**) and (**b**). **d** Rectangular cavity types. **e** Detector power versus frequency curve for all types. **f** Equivalent circuit of all above reflection type cavities

4.14 Faraday Rotation and Ferrite Devices—Isolators, Gyrators, and Circulators

(a) Properties of Ferrite

Ferrite material is ferro or ferric magnetic material consisting of oxides of Mn, Cd, Ni, Zn, Co along with Fe_2O_3. Therefore the generic formula can be written as $M.O.Fe_2O_3$, where M is one of the above di-valent, metal, or their mixture. This mixture is heated at 1100 °C and pressed into required shapes. Ferrites materials have the following five special properties.

1. **Electrical insulator/Low loss**: High resistivity of the order of 10^8 to 10^{10} Ω-m, as compared to copper of 1.7×10^{-8} Ω-m. Because of this Eddy current posses are extremely low.
2. **High dielectric constant** with ε_r = 12–15.
3. **High permeability**: $\mu_r \approx 2000$, i.e. 3000 times than iron which has $\mu_r = 60$.
4. **It can be magnetised like iron** in magnetic field; the spinning electron axis does not align fully, but wobbles around the magnetic field direction as its axis of wobbling, with its resonant frequency, which is in the range of microwaves. In this process it gets magnetised (Fig. 4.26) as explained below.

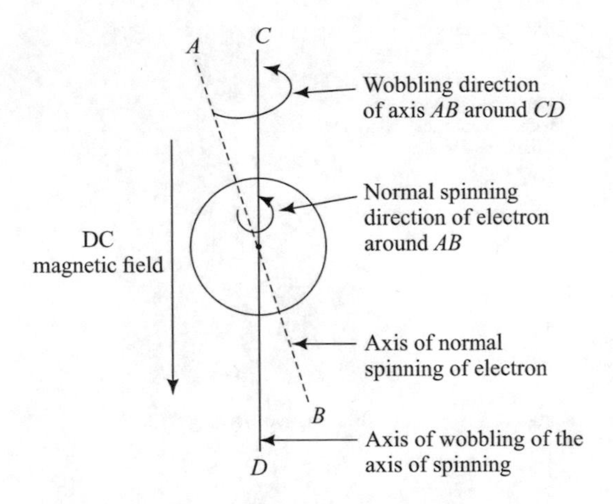

Fig. 4.26 Property of ferrite: Wobbling of spin axis of unpaired electrons of the ferrite around the direction of the magnetic field

The electrons orbiting around nucleus of atoms do not contribute to magnetism, but the spin of unpaired electron does by aligning along the magnetic field applied externally (Fig. 4.26). In magnetic field, the electron tries to align its own spinning axis with the field, but its own binding force tries to stop it. As a result, it does not align fully and therefore starts wobbling. The partial alignment contributes to magnetism.

5. **Another property which is used in microwaves** is that the polarisation plane of EM wave gets rotated, in the presence of magnetic field. We know that the **EM wave has two circularly polarised wave (one clockwise rotating and other anticlockwise)**, called left circularly polarised and right circularly polarised waves. **When LCPW and RCPW** (Fig. 4.27) **pass through ferrite, then its wobbling electrons react differently with these two waves, leading to different permeabilities and different** velocities (also refer Fig. 2.1 and Sect. 2.1.1).

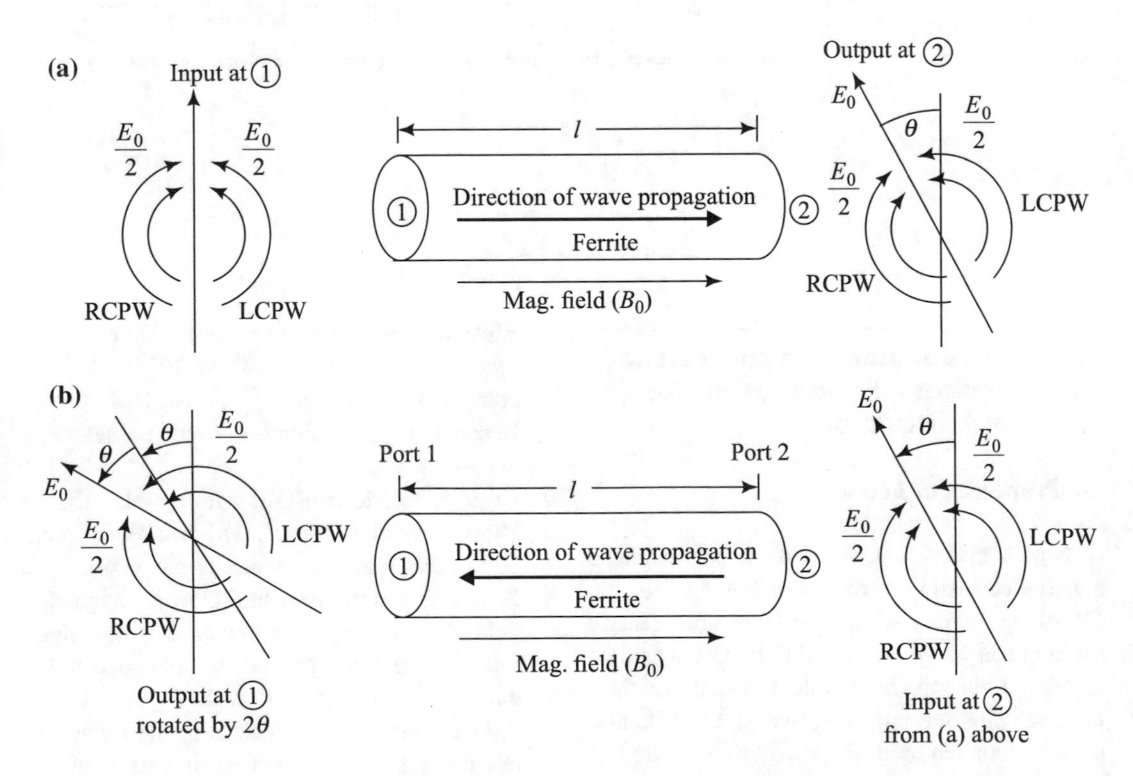

Fig. 4.27 Faraday rotation due to different velocity and hence different phase shift of left and right circularly polarised waves (LCPW and RCPW), direction of rotation is same (anticlockwise as seen from port 1 side), whether the wave goes **a** from port 1 to 2 or **b** from 2 to 1, provided B_0 remains in the same direction (also refer Fig. 2.1)

(b) **Faraday Rotation and its Properties**

If a plane polarised microwave which always has LCPW and RCPW (Fig. 4.27) passes through a ferrite placed in a magnetic field along the direction of the wave, then the plane of polarisation of the resultant wave front rotates. This mechanism is known as, Faraday rotation. It can be explained by its following properties.

(i) **LCPW and RCPW**: We know that any plane-polarised wave can be taken as a vector sum of left and right circularly polarised waves (LCPW and RCPW) (Fig. 4.27).

(ii) **Different properties of LCPW and RCPW in ferrite**: When the plane-polarised wave passes through a ferrite, placed in a magnetic field along the direction of wave propagation, then the properties of LCPW and RCPW are different in terms of their:

(a) Velocities.
(b) Attenuation.
(c) Electrical properties (μ, ε, ρ, phase constant β).

Therefore when they emerge out of the ferrite, then the net resultant, E-vector of the two components LCPW and RCPW may not be in the same direction as original E-vector, resulting into Faraday rotation.

(iii) **Direction of rotation** depends on:

(a) Whether $\omega > \omega_0$ or $\omega < \omega_0$, i.e. clockwise or anticlockwise (ω_0 being the wobbling resonant frequency of ferrite).
(b) Direction of the magnetic field.

Direction of rotation does not depend

(a) On the direction of propagation of the wave.
(b) Magnetic field strength, so far as $B > B_0$ (B_0 = saturated magnetic field of ferrite)

(iv) **The amount of rotation** depends on:

(a) Length of the material and is around 100°/cm.

$$\phi = (\phi_L - \phi_R) = \frac{l}{2}(\beta_L - \beta_R)$$

(here β_L, β_R = phase constants of LCPW and RCPW)

(b) Frequency of the wave.
(c) Not on the magnetic field strength beyond its saturation value B_0.

This property of Faraday rotation is used in some of the microwave devices,
e.g. isolators, gyrator, circulator.

Thus ferrite rotates the signal in the same direction travelling from its end A to D or D to A (as seen from end A), when the magnetic field direction is unchanged (say A to B).

4.14.1 Isolator

In an isolator, signal flows in one direction only from port 1 to port 2 as attenuation from port 2 to 1 is very high for any reflected signal. **This way an isolator isolates the source from the load** (Fig. 4.28). **It consists of three portions** where the propagation mode changes from TE_{10} (rectangular waveguide) to TE_{11} (circular waveguide) to TE_{10} (rectangular waveguide):

(i) A 45° waveguide twist connected to a circular waveguide as transition.
(ii) The circular waveguide position has the tapered ferrite of known length placed in a magnetic field along the direction of wave propagation.
(iii) This circular waveguide in turn is connected by transition to another rectangular waveguide at the other end.

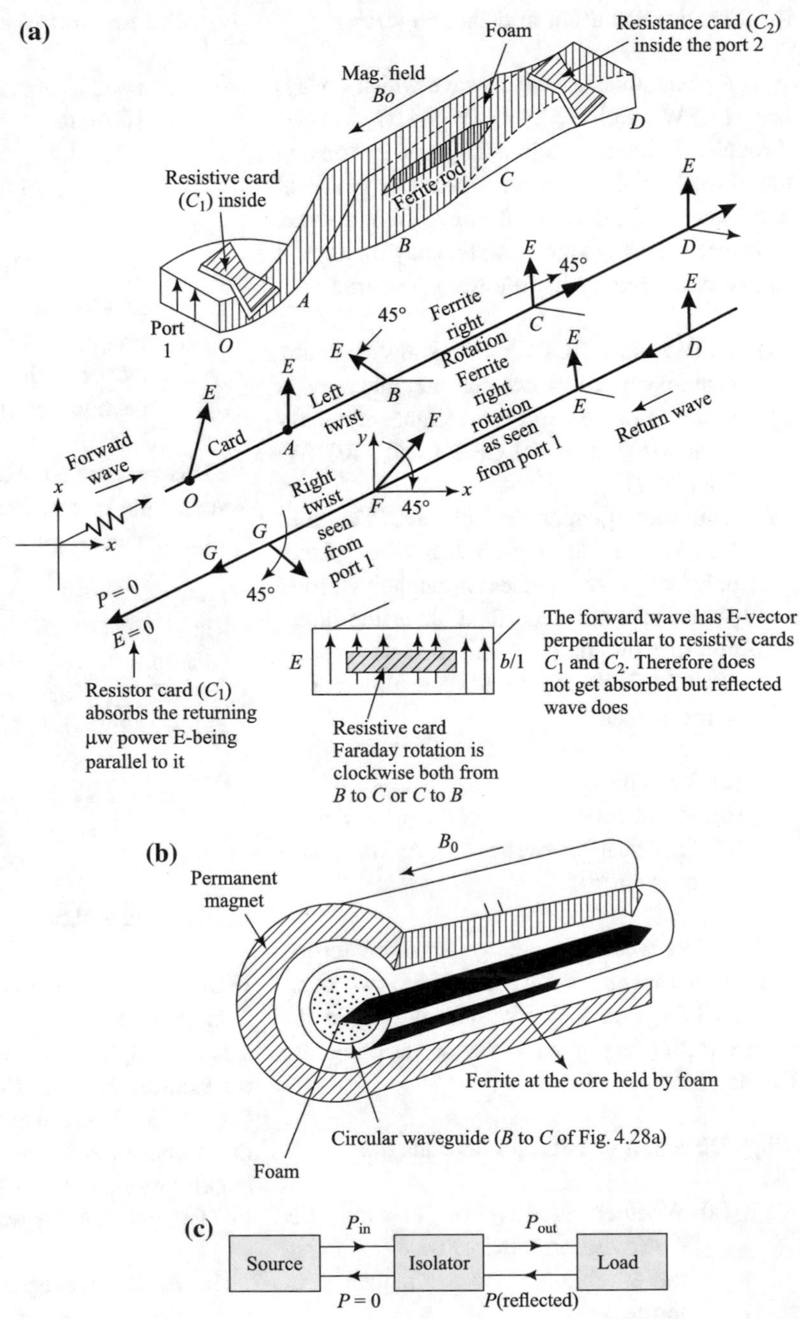

Fig. 4.28 Isolator: **a** Rotation of E-vector of the wave TE_{10}. **b** Circular waveguide at the central portion with ferrite and permanent magnet for ferrite. **c** One-way signal propagation from source due to isolator

Two resistive cards are placed in the rectangular waveguide at the two ends, perpendicular to the electric field lines of the wave (Fig. 4.28) and along the two coplanar portion of rectangular waveguide (i.e. OA and CD).

When the TE_{10} wave propagates from A to B, its vertical E-vector gets rotated anticlockwise by 45° by the twist and then from B to C it rotates anticlockwise by 45°, by ferrite, becoming vertical again. The two resistive cards *P* and *Q* do not absorb the E-vector being perpendicular to its surface. When the wave follows the return path, the vertical E-vector passes unattenuated from D to E, while from E to F it rotates clockwise through the ferrite by 45°. Then from F to G the signal rotated by another 45° clockwise through the twist. **Here we may note the ferrite rotates the signal clockwise (as seen from port 1), whether it is travelling forward or backward,** when magnetic field direction remains same. This makes the E-vector parallel to the resistive card (C_2) and hence gets absorbed fully in the form of heat. Thus in an isolator, wave travels in one direction only and any reflected wave due to mismatched impedance gets absorbed and does not reach the source of signal. Thus the isolator isolates the source and the loads and normally 20–30 dB isolation is there between port 2 and 1. The S-matrix of this port isolator will be: $[S] = \begin{bmatrix} 0 & 0 \\ 1 & 0 \end{bmatrix}$.

4.14.2 Gyrator

As in Fig. 4.29 it is a two-port device to give a phase shift of 180° in the forward direction and zero in the backward direction. It is similar to isolator and also has three segments except the following four differences:

(i) The waveguide twist is of 90°.
(ii) The ferrite is of double the length such that rotation due to it is of 90°.
(iii) The direction of magnetic field is kept opposite to that in isolator, for Faraday rotation of 90° to be in the same direction as of 90° of the twist.
(iv) There are no power absorbing resistive cards.

Because of the above and the fact that Faraday rotation is in the same direction (e.g. anticlockwise as seen from port 1), whether it is travelling forward or backward, following happens to the E-vector (as clear in Fig. 4.29).

(i) **In the forward direction the plane of polarisation rotates by 180°, when is equivalent to phase shift of 180°.**
(ii) In the reverse direction there is no rotation between input and output wave.

The change of modes from TE_{10} to TE_{10} is same as in the isolator.

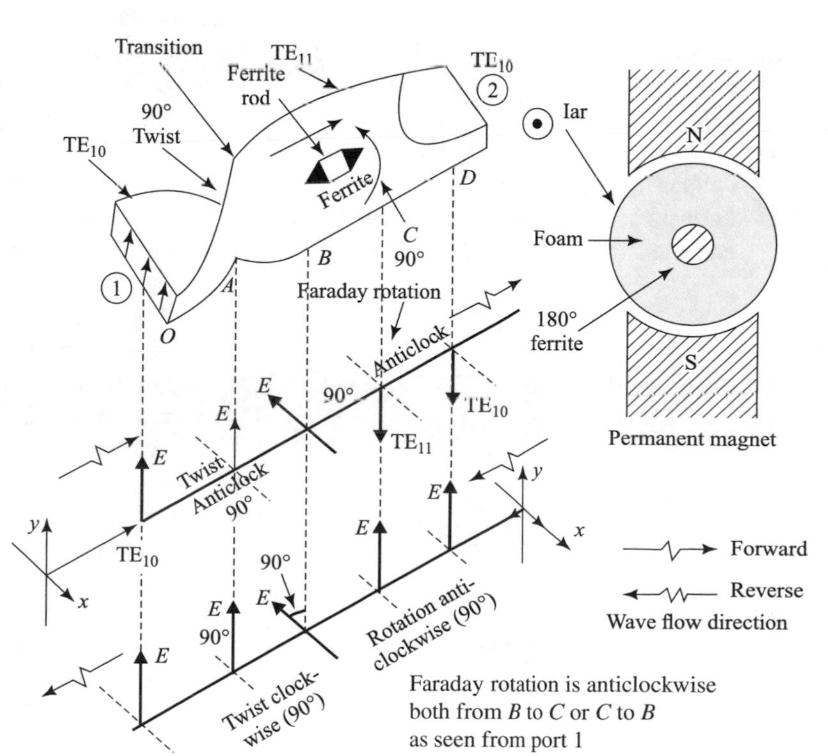

Fig. 4.29 Gyrator: 180° phase shift in forward direction only, with central portion similar to isolator (see Fig. 4.28b)

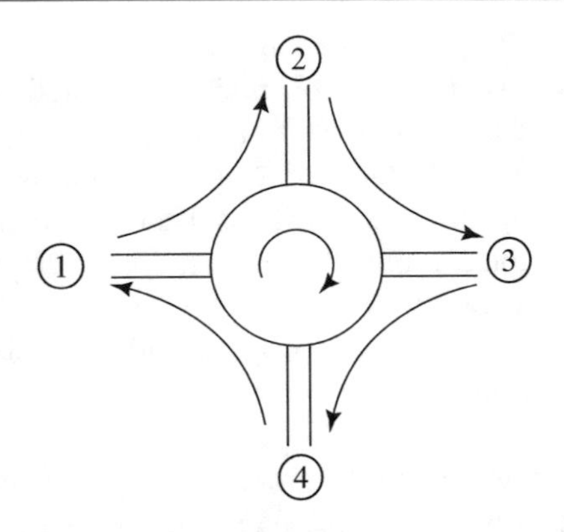

Fig. 4.30 Symbol of a circulator

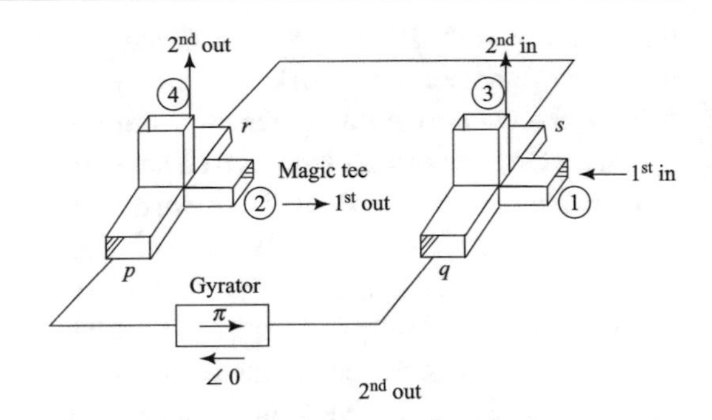

Fig. 4.32 Four-port circulator using two magic tee and a gyrator. The four ports 1, 2, 3, 4 behave like a circulator of Fig. 4.30

4.14.3 Circulators

It is a multiport device such that signal flows from port-'n' to port-(n + 1) only. Generally four-port circulator is common, where input at port-① gives output at port-② and so on as shown in Fig. 4.30. For example, signal moves as $1 \rightarrow 2 \rightarrow 3 \rightarrow 4 \rightarrow 1$.

They are useful in parametric amplifiers, duplexers in radar, etc. There are normally four types of circulators, e.g.

(i) Three-port ferrite circulator (Fig. 4.31).
(ii) Four-port circulator using two magic T and one gyrator (Fig. 4.32).
(iii) Four-port ferrite circulator directional coupler type (Fig. 4.33).
(iv) Four-port circulator using tee branching and transitions (Fig. 4.34).

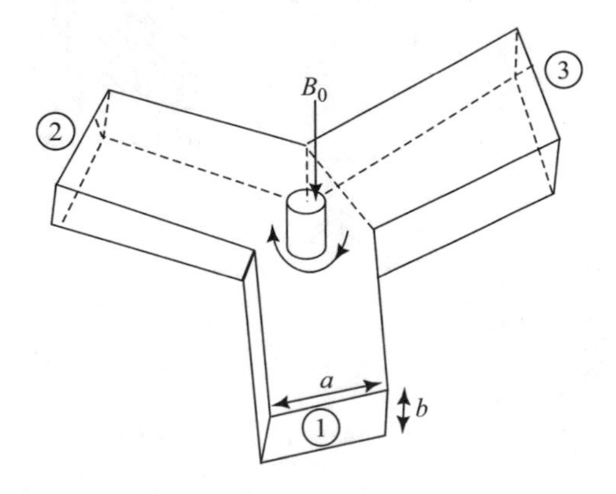

Fig. 4.31 Three-port ferrite circulator

(i) **Three-port ferrite circulators**: It consists of three sections of waveguides connected at 120° along the broad side planes, with a ferrite rod at the centre on which magnetic field applies axially (Fig. 4.31).

The microwave power at port 1 divides into two equal portions circulating clockwise and anticlockwise so that port 2 and port 3 each get two signals. The direction of ferrite and its magnetic field strength is such that the phase shift variation in the two signals at port 2 add together while get subtracted (i.e. cancelled) at port-3 giving zero output similarly signal input at port-2 will reach port-3 only.

(ii) **Four-port circulator using magic tee**: The standard properties of the magic T are used along with a gyrator for phase shifting by 180° in one direction only (Fig. 4.32). The external electric length of propagation has to be same, i.e. (q to p) = (r to s) = (s to q), for keeping the same phase differences. When a wave is incident on port 1, it splits into two equal (in-phase) waves coming out of port-q and port-s, which reach p and r ports in phase; thereby sum comes out from port 2 only with nil at port 4 and port 3. If the wave is incident at port 2, it splits out from port-p and port-r in-phase, but reaches port-q and port-s out of phase, thereby giving sum at port 3 and nil at ports 1 and 4.

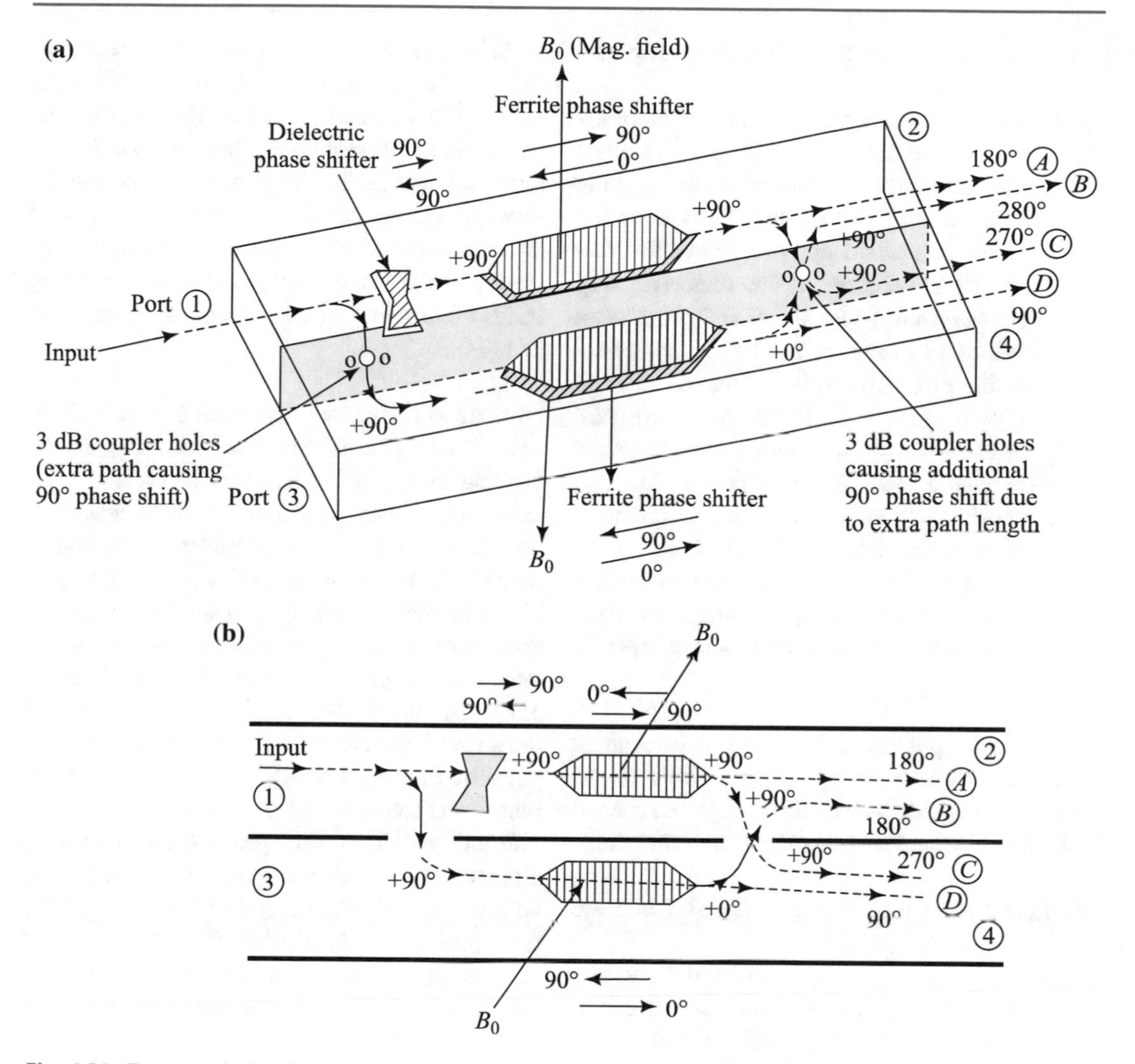

Fig. 4.33 Four-port ferrite-directional coupler type of circulator **a** 3-D figure and **b** 2-D simple representation. The ferrites phase shifters are non-reciprocal in direction and have opposite magnetic fields

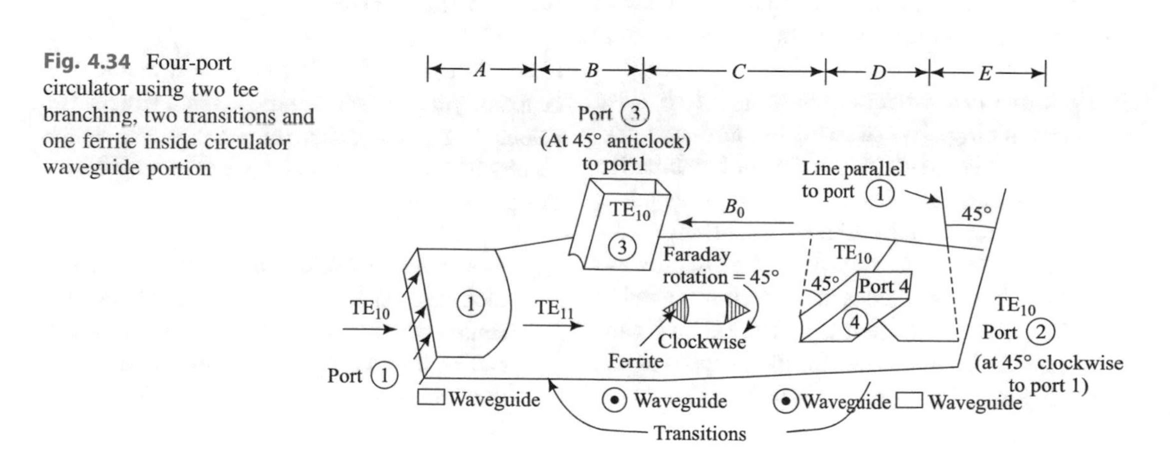

Fig. 4.34 Four-port circulator using two tee branching, two transitions and one ferrite inside circulator waveguide portion

Similarly input at port 3 will reach port 4 only.

(iii) **Four-port ferrite-directional coupler-type circulator**: A simple compact four-port circulator can be made by using a directional coupler has 3 dB coupling holes, **using two non-reciprocal 90° ferrite phase shifters and a dielectric slap (bidirectional 90° phase shifter) placed inside**, as in Fig. 4.33. The magnetic field of the first ferrite is in a direction so as to give a phase shift of 90° in the forward direction only, while on the second ferrite magnetic field is in opposite direction causing 90° phase shift in reverse direction only. The dielectric phase shifter gives phase shifting of 90° in both the directions. **Signal passing through the coupler holes also causes a phase shift of 90°**.

A signal input at port 1 of the first waveguide will get branched through the first hole to the second waveguide as well as through the second hole also. As is clear from Fig. 4.33b that the:

(a) Port 2 will get two waves *A* and *B*, which are in-phase and get added.
(b) Port 4 will get two waves *C* and *D*, with a phase difference of 180° and therefore cancel, being of same amplitude as well.
(c) Port 3 does not get any signal at all.

Similarly we will see that input at port 2 will give output at 3 only and so on.

(iv) **Four-port circulator-using two tee branching, two transitions and one ferrite**: This circulator consists of combination of rectangular and circular waveguides requiring two transitions as in Fig. 4.34. Here we will see that the microwave entering port 1 finds its E-vector aligned to the E-vector in port 2 and at 90° with the rest of the two ports. Similarly input signal at port 2 finds port 3 aligned and so on.

This circulator consists of five segments (Fig. 4.35), e.g. ***A***, ***B***, ***C***, ***D***, and ***E***. These are rectangular waveguide ***A***, transition ***B***, circular waveguide ***C***, transition ***D***, and finally a rectangular waveguide ***E***. There are two rectangular waveguides connected as tee at the two ends of the central circular waveguide. The middle segment ***C*** is the Faraday rotation ferrite part, being inside the circular waveguide with magnetic field as in Fig. 4.34.

Mechanism: The power entering port 1 is in TE_{10} mode and is converted to TE_{11} mode, because of gradual rectangular to circular transition. This power passes port 3 unaffected since the electric field is not significantly cut and is rotated clockwise through 45° due to the ferrite; also this power passes port 4 unaffected for the same reason. Finally it emerges out of port 2 which has its plane of polarisation already tilted clockwise by 45° with respect to port 1. Now if the power input is at port 2 (which already has its plane at clockwise by 45° to port 1), it crosses port 4 unaffected as the electric field is not significantly cut. This wave gets rotated by another clockwise by 45° due to ferrite rod. This power, whose plane of polarisation is tilted through 90°, finds port 3 suitably aligned and emerges out of it. Similarly we can show that port 3 is coupled only to port 4 and port 4 to port 1.

4.15 Phase Shifters

Phase shifters can be of two types—variable type or fixed type. **In fact the dielectric reduces the velocity of propagation of microwave, which results in an increased electric path and hence the phase shift**.

(i) **A simple variable type** is just a resistive dielectric slab vane (Fig. 4.35a) (at the centre slot where E-vector is max), which can be inserted by rotation by a knob, as in the variable angle attenuator (Fig. 4.19c)

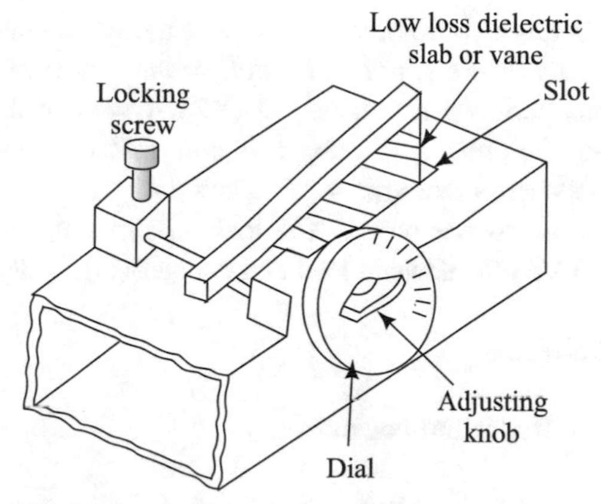

(a) Variable phase shifter—vane type-like attenuator

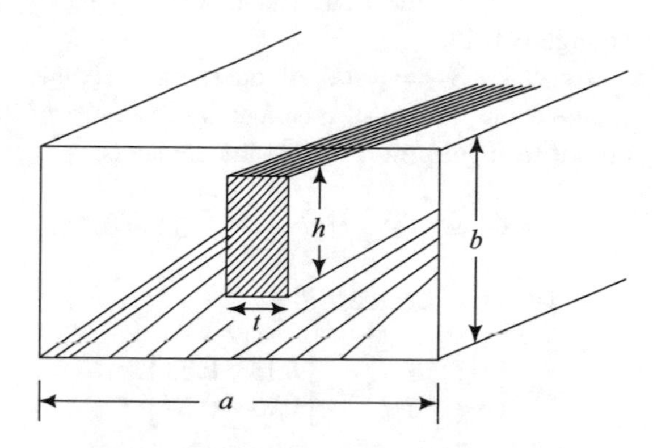

(b) Phase shifter—fixed type using resistive dielectric slab in the centre where the E-field is maximum

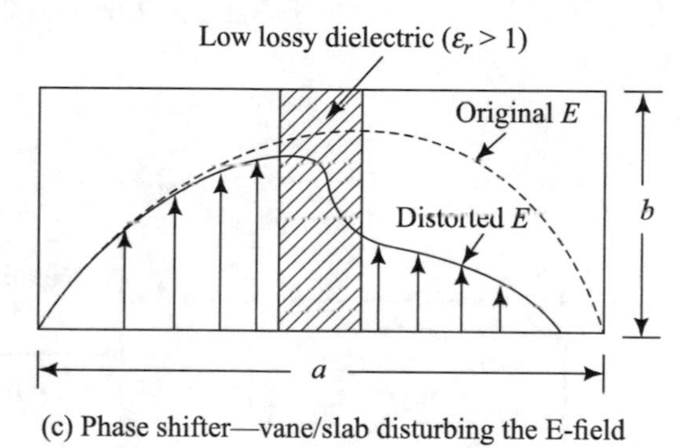

(c) Phase shifter—vane/slab disturbing the E-field

Fig. 4.35 Phase shifters

having resistive vane. When the vane is inserted deeper, the electric field gets disturbed (Fig. 4.35) and phase shift increases as the plane of the vane is parallel to the E-field vector.

(ii) **Fixed dielectric phase shifter** can be made by having a dielectric slab placed at the central line along the length of the waveguide, with its plane parallel to the E-vector (Fig. 4.34). Amount of phase shift depends on the dielectric constant ε_r and its length l. **Typically dielectric constant of 2.5 and of length 1.5 cm can give phase shift of 180° at 10 GHz**. Ferrite phase shifters can also be used as fixed phase shifter in one direction only. The total phase shift is given by:

$$\phi = K_\phi \cdot l \cdot \varepsilon_\mathrm{r}$$

Here K_ϕ is proportionally constant.

Solved Problems

Problem 4.1 When a microwave component is placed in a microwave circuit with a matched load, then VSWR = 1.3. The input power (source) is 150 mW, and the matched load gets only 80 mW. Moreover the component gives the same result if its input–output ports are inter-charged. Then find the S-matrix of the component.

Solution As VSWR = 1.3, therefore reflection coefficient $(\Gamma) = \frac{S-1}{S+1} = \frac{0.3}{2.3} = 0.13$.

$S_{11} = S_{22}$ = reflection coefficient (Γ) of voltage = 0.13.

As $S_{12} = S_{21}$ = Ratio of microwave signals (voltage output/input), therefore by the ratio of output to input power we get the elements;

$$S_{12} = S_{21} = \sqrt{P_{\text{out}}/P_{\text{in}}} = \sqrt{8/150} = 0.73$$

$\therefore$ S-matrix becomes $[S]$ =

$$\begin{bmatrix} S_{11} & S_{12} \\ S_{21} & S_{22} \end{bmatrix} = \begin{bmatrix} 0.13 & 0.73 \\ 0.73 & 0.13 \end{bmatrix}$$

Problem 4.2 A 20 dB directional coupler is used to monitor power delivered to the load and for this two bolometer placed at the port 3 and port 4 shows 2 mW and 8 mW power. The port 3 has perfectly matched load (VSWR = 1), while the bolometer at port 4 is not matched and VSWR on that arm = 2.0. Then find:

(a) source power, (b) load dissipation, and (c) VSWR between load and coupler (Fig. 4.36).

Solution

(1) **Reflection coefficient**

$$\Gamma_4 = \frac{\text{VSWR} - 1}{\text{VSWR} + 1} = \frac{2 - 1}{2 + 1} = \frac{1}{3} = \sqrt{\frac{P_{\text{r4}}}{P_{\text{i4}}}}$$

The bolometer B_4 at port 4 shows 8 mW

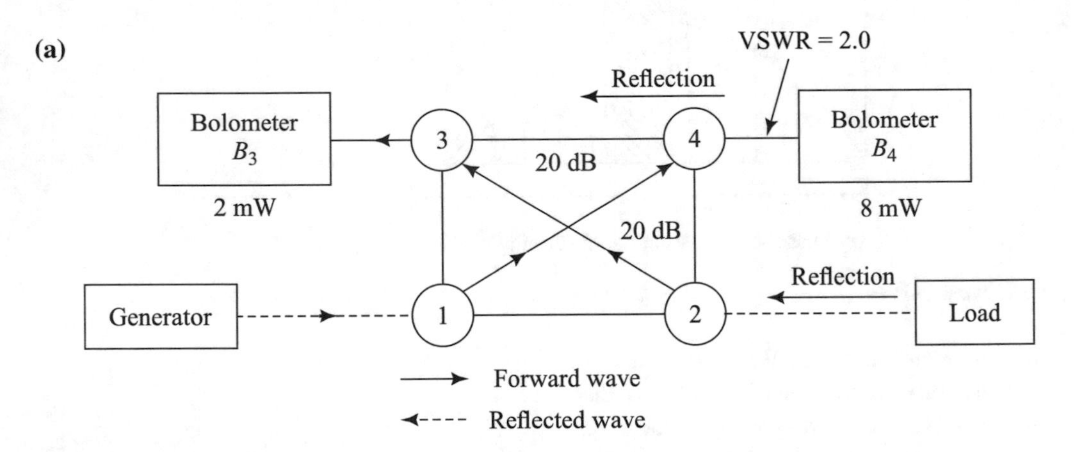

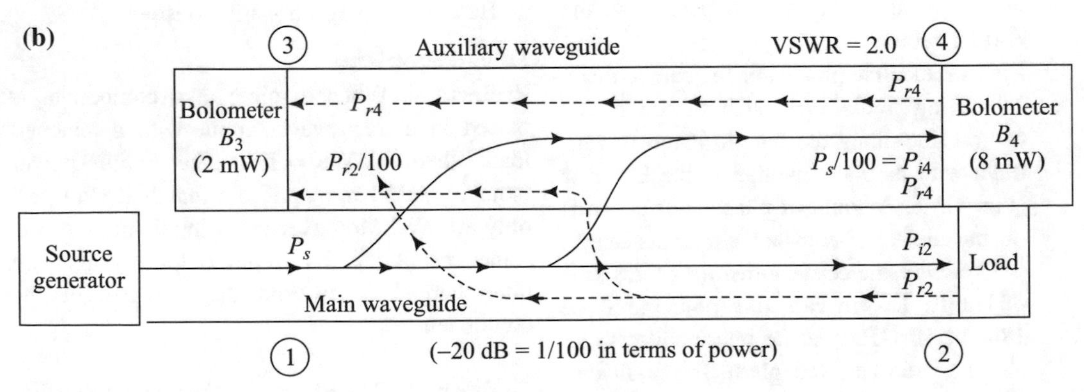

Fig. 4.36 Problem 4.2: Directional coupler having source, load, and two bolometer for power measurement of sampled signal

$$P_{i4} - P_{r4} = 8 \text{ mW}$$
$$P_{i4} = (8 + P_{r4})$$

Above equation becomes:

$$\Gamma_4^2 = \left(\frac{1}{3}\right)^2 = \frac{P_{r4}}{8 + P_{r4}}$$

P_{r4} = 1 mW and P_{i4} = 8 + 1 = 9 mW.

(2) **Calculating Source Power (P_s)**
P_{i4} comes from 20 dB holes of the coupler

$$\therefore \quad \frac{P_{i4}}{P_s} = \frac{1}{100} \left(\text{As } 20 \text{ dB} = \frac{1}{100} \text{ in fraction}\right)$$

$$P_s = P_{i4} \times 100 = 9 \times 100 = 900 \text{ mW}$$

(3) **Calculating Power to Bolometer (B_3)**
Bolometer B_3 gets signal of P_{r4} and P_{r2}

$$\left(P_{r4} + \frac{P_{r2}}{100}\right) = 2 \text{ mW}$$

As $P_{r4} = 1$ mW $\therefore$
$\frac{P_{r2}}{100} = 1 \text{ mW}_2,\ P_{r2} = 1 \times 100 = 100 \text{ mW}$

(4) **Input Power to Load (P_{i2}) and Dissipation**
Out of 900 mW source power, 9 mW, i.e. $P_{i4} = \frac{900}{100}$, goes to the coupling hole to port 4. The remaining 900−9 = 891 goes to load

$$\therefore P_{i1} = 891 \text{ mW and } P_{i2} = 100 \text{ mW}$$

$\therefore$ Power consumed (dissipated) at load = 891 −100 = 791 mW.

(5) **Reflection Coefficient and VSWR in Arm**
Reflection coefficient between load and coupler, i.e. for port 2;

$$\Gamma_2 = \sqrt{\frac{P_{r2}}{P_{i2}}} = \sqrt{\frac{100}{891}} = 0.335 = \frac{\text{VSWR} - 1}{\text{VSWR} + 1}$$

$\therefore$ VSWR on the main waveguide line = 1.335/0.665 = 2.00.

Problem 4.3 In a H-plane tee, microwave power of 100 mW is fed at port 1, find the power at ports 2 and port 3, when they are terminated with matched load (i.e. VSWR = 1).

Solution We know that in the *S*-matrix equation for voltage level signals at the inputs and outputs at ports 1, 2, 3 as a_1, a_2, a_3 and b_1 b_2, b_3, the *S*-matrix equation for H-plane tee is:

$$\begin{bmatrix} b_1 \\ b_2 \\ b_3 \end{bmatrix} = \begin{bmatrix} 1/2 & -1/2 & 1/\sqrt{2} \\ -1/2 & 1/2 & 1/\sqrt{2} \\ 1/\sqrt{2} & 1/\sqrt{2} & 0 \end{bmatrix} \begin{bmatrix} a_1 \\ a_2 \\ a_3 \end{bmatrix}$$

For getting power instead of voltage if the symbols can be $b'_1, b'_2, b'_3, a'_1, a'_2, a'_3$, then equation will have elements each of *S*-matrix as squared:

$$\therefore \quad \begin{bmatrix} b'_1 \\ b'_2 \\ b'_3 \end{bmatrix} = \begin{bmatrix} (1/2)^2 & (-1/2)^2 & (1/\sqrt{2})^2 \\ (-1/2)^2 & (1/2)^2 & (1/\sqrt{2})^2 \\ (1/\sqrt{2})^2 & (1/\sqrt{2})^2 & 0 \end{bmatrix} \begin{bmatrix} a'_1 \\ a'_2 \\ a'_3 \end{bmatrix}$$

As $a'_1 = 100 \text{ mW}, a'_2 = a'_3 = 0.$

$$\begin{bmatrix} b'_1 \\ b'_2 \\ b'_3 \end{bmatrix} = \begin{bmatrix} 1/4 & 1/4 & 1/2 \\ 1/4 & 1/4 & 1/2 \\ 1/2 & 1/2 & 0 \end{bmatrix} \begin{bmatrix} 100 \\ 0 \\ 0 \end{bmatrix}$$

$$\therefore \quad b'_1 = 25 \text{ mW}, b'_2 = 25 \text{ mW and } b'_3 = 50 \text{ mW}$$

Problem 4.4 In a H-plane tee 100 mW is given at port 3 with its matched load. Calculate the power delivered to port 1 and port 2, where loads of 80 Ω and 60 Ω are connected. The characteristic impedance of the line is 50 Ω.

Solution In a H-plan tee, power input at port 3 gets divided equally into port 1 and port 2 and if b_1, b_2, b_3 and a_1, a_2, a_3 are voltage levels reflected and inputs fed, then corresponding power level coefficients will be $a_1^2, a_2^2, a_3^2, b_1^2, b_2^2, b_3^2$, respectively.

Also if voltage reflection coefficients at port 1 and port 2 are Γ_1 and Γ_2, then power reflection coefficient will be Γ_1^2, Γ_2^2.

Power delivered to each of the port 1 and port 2 is:

$$P_{01} = \frac{1}{2}b_1^2 = P_{02} = \frac{1}{2}b_2^2 = 50 \text{ mW}$$

The power reflected by loads at port 1 and port 2 will be:

$$P_{r1} = \Gamma_1^2 \frac{b_t^2}{2}; P_{r2} = \Gamma_2^2 \cdot \frac{b_2^2}{2}$$

Power dissipated at loads 80 and 60 Ω, respectively, will be the differences of signal coming out of the ports 1 and 2 and the signal reflected back inside:

$$P_{D1} = \left(\frac{b_1^2}{2} - \Gamma_1^2 \cdot \frac{b_1^2}{2}\right); P_{D2} = \left(\frac{b_2^2}{2} - \Gamma_1^2 \cdot \frac{b_2^2}{2}\right)$$

For calculating reflection coefficient Γ_1 and Γ_2, we know that $Z_0 = 50\ \Omega$, $Z_1 = 80\ \Omega$, $Z_2 = 60\ \Omega$

$$\therefore |\Gamma_1| = \frac{80-50}{80+50} = \frac{30}{130} = \frac{1}{13} \therefore \Gamma_1^2 = \frac{1}{169}$$

$$\frac{1}{2}b_1^2 = \frac{1}{2}b_2^2 = 50 \text{ mW}$$

$$\therefore |\Gamma_2| = \frac{60-50}{60+50} = \frac{10}{110} = \frac{1}{11} \therefore \Gamma_2^2 = \frac{1}{121}$$

$$P_{D1} = \frac{1}{2}b_1^2\left(1 - |\Gamma_1|^2\right) = 50 \times 10^{-3}\left(1 - \frac{1}{169}\right)$$
$$= 49.7 \text{ mW}$$

$$P_{D2} = \frac{1}{2}b_2^2\left(1 - |\Gamma_2|^2\right) = 50 \times 10^{-3}\left(1 - \frac{1}{121}\right)$$
$$= 49.6 \text{ mW}$$

Problem 4.5 If in a magic tee, the power input to E-arm is 150 mW, then find the power in remaining three arms.

Solution In magic tee, input at E-arm gets divided equally into coplanar arms 1 and 2, with no output at H-arm

$$P_{01} = \frac{150}{2} = 75 \text{ mW (in phase)}$$

$$P_{02} = \frac{150}{2} = 75 \text{ mW (out phase)}$$

$$P_{03} = 0$$

Problem 4.6 In a magic tee the ports 1, 2, and 4 are having load such that the reflection coefficients in these arms are $\Gamma_1 = 0.5$, $\Gamma_2 = 0.6$, $\Gamma_4 = 0.8$. When we feed 10 W into fully matched port 3, find the reflected power in this arm 3 and the power outputs in remaining arms.

Solution The signal (voltage) input (a_1, a_2, a_3, a_4) and output (b_1, b_2, b_3, b_4) in magic tee are related as:

$$\begin{bmatrix} b_1 \\ b_2 \\ b_3 \\ b_4 \end{bmatrix} = \begin{bmatrix} 0 & 0 & 1/\sqrt{2} & 1/\sqrt{2} \\ 0 & 0 & 1/\sqrt{2} & -1/\sqrt{2} \\ 1/\sqrt{2} & 1/\sqrt{2} & 0 & 0 \\ -1/\sqrt{2} & 1/\sqrt{2} & 0 & 0 \end{bmatrix} \begin{vmatrix} a_1 \\ a_2 \\ a_3 \\ a_4 \end{vmatrix}$$

Then squares of a's and b's relate to the power in those arms. Input power at port 3 is 10 W = $|a_3|^2$; therefore $a_3 = \sqrt{10}$, and the remaining three reflected signals into the ports are:

$$a_1 = \Gamma_1 b_2; a_2 = \Gamma_2 b_2; a_4 = \Gamma_4 b_4$$

As port 3 is matched, therefore there is no reflected output at port 3, i.e. $b_3 = 0$.

$$\therefore \quad a_1 = 0.5\, b_1; a_2 = 0.6\, b_2; a_3 = \sqrt{10}\, b_3;$$
$$a_4 = 0.8\, b_4$$

Matrix equation becomes

$$\begin{bmatrix} b_1 \\ b_2 \\ 0 \\ b_3 \end{bmatrix} = \begin{bmatrix} 0 & 0 & 1/\sqrt{2} & 1/\sqrt{2} \\ 0 & 0 & 1/\sqrt{2} & -1/\sqrt{2} \\ 1/\sqrt{2} & 1/\sqrt{2} & 0 & 0 \\ -1/\sqrt{2} & 1/\sqrt{2} & 0 & 0 \end{bmatrix} \begin{vmatrix} 0.5b_1 \\ 0.6b_2 \\ \sqrt{10} \\ 0.8b_4 \end{vmatrix}$$

$$b_1 = \sqrt{10}\big/\sqrt{2} + 0.8b_4\big/\sqrt{2}$$
$$= (2.230 + 0.566b_4)$$

$$b_2 = \sqrt{10}\big/\sqrt{2} + 0.8b_4\big/\sqrt{2}$$
$$= (1.581 + 0.566b_4)$$

$$b_4 = -\frac{0.5b_1}{\sqrt{2}} + \frac{0.6b_2}{\sqrt{2}} = (-0.354b_1 + 0.42b_2)$$

Solving gives $b_1 = 0.657$, $b_2 = 0.758$, $b_4 = 0.089$, and squaring them gives power output as:

$P_1 = 0.431$ W, $P_2 = 0.574$ W, $P_4 = 0.008$ W.

Problem 4.7 Show that for an ideal isolator, the following is the S-matrix

$$[S] = \begin{bmatrix} 0 & 0 \\ 1 & 0 \end{bmatrix}$$

Solution As the isolator is ideal, full signal will go from ports 1 to 2, but nil from port 2 to 1. Therefore $b_1 = 0$ and $b_2 = a_1$

$$b_1 = 0 \;= 0 \cdot a_1 + 0 \cdot a_2$$
$$b_2 = a_1 \;= 1 \cdot a_1 + 0 \cdot a_2$$

$$\begin{bmatrix} b_1 \\ b_2 \end{bmatrix} = \begin{bmatrix} 0 & 0 \\ 1 & 0 \end{bmatrix} \begin{bmatrix} a_1 \\ a_2 \end{bmatrix}$$

Problem 4.8 Show that ideal three-port circulator has the following S-matrix

$$[S] = \begin{bmatrix} 0 & 0 & 0 \\ 1 & 0 & 0 \\ 0 & 1 & 0 \end{bmatrix}$$

as per the property.

Solution Rewriting input–output property of a circulator as $b_2 = a_1$, $b_3 = a_2$, $b_1 = a_3$
we get

$$b_1 = a_3 = 0 \cdot a_1 + 0 \cdot a_2 + l \cdot a_3$$
$$b_2 = a_1 = 1 \cdot a_1 + 0 \cdot a_2 + 0 \cdot a_3$$
$$b_3 = a_2 = 0 \cdot a_1 + 1 \cdot a_2 + 0 \cdot a_3$$

Writing it in matrix form gives

$$\begin{bmatrix} b_1 \\ b_2 \\ b_3 \end{bmatrix} = \begin{bmatrix} 0 & 0 & 1 \\ 1 & 0 & 0 \\ 0 & 1 & 0 \end{bmatrix} \begin{bmatrix} a_1 \\ a_2 \\ a_3 \end{bmatrix}$$

Problem 4.9 A three-port circulator has insertion loss of 0.5 dB, isolation of 20 dB, and VSWR = 2. Find the S-matrix.

Solution In the forward cycle signal 1–2-port loss = 0.5 dB and the same loss is from 2 to 3 and 3 to 1.

In the reverse direction the isolation signal 2–1 loss = 20 dB and the same loss is from 3 to 2 and 1 to 3.

Insertion loss = 0.5 dB = −20 log(S_{21}), and isolation loss = 20 dB = −20 log(S_{12}).

As both are losses, therefore are −ve.

(a) **The insertion losses**

$$-0.5 = 10 \log\left(P_{\text{output}}/P_{\text{input}}\right) = 20 \log(V_{\text{out}}/V_{\text{in}}) = 20 \log(b_2/a_1)$$

$$\therefore -0.5 = 20 \log(S_{21})$$

$$S_{21} = 10^{\frac{-0.5}{20}} = 10^{-0.025} = 0.944$$

Similarly
$S_{32} = S_{21} = S_{13} = 10^{-0.025} = 0.944.$

(b) **The isolation**

$$= 20 \text{ dB} = -20 \log S_{12}; \text{here } S_{12} = b_1/a_2$$

$$\therefore S_{12} = 10^{-20/20} = 10^{-1} = 0.1$$

and $S_{23} = S_{12} = S_{31} = 0.1$

(c) **Reflection coefficient**

$$S_{11} = S_{22} = S_{33} = \frac{\text{VSWR} - 1}{\text{VSWR} + 1} = \frac{2-1}{2+1} = 0.333$$

$$\therefore [S] = \begin{bmatrix} S_{11} & S_{12} & S_{13} \\ S_{21} & S_{22} & S_{23} \\ S_{31} & S_{32} & S_{33} \end{bmatrix} = \begin{bmatrix} 0.333 & 0.1 & 0.944 \\ 0.944 & 0.333 & 0.1 \\ 0.1 & 0.944 & 0.333 \end{bmatrix}$$

Problem 4.10 A directional coupler has insertion loss of 0.5 dB. Its coupling with port 4 is 20 dB and with port 3 is 35 dB. If all the ports are matched, what will be the signal power outputs at the ports 2, 3, and 4. If an input of 50 W is given at port 1, find insertion loss.

Solution The signal reaching port 4 is (−20) dB of input of port 1

$$\therefore (-20) = 10 \log_{10}(P_4/P_{\text{in}}) = 10 \log_{10}(P_4/50)$$

$$\frac{P_4}{50} = 10^{-\frac{20}{10}} = 10^{-2} = \frac{1}{100} \therefore P_4 = \frac{50}{100} = 0.5 \text{ W}$$

The reflected signal from port 4 reaches at port 3 with 35 dB down:

$$\therefore -35 = 10 \log_{10}(P_3/P_4) = 10 \log_{10}(P_3/0.5)$$

$$\therefore 10^{-\frac{3.5}{10}} = \frac{P_3}{0.5} \therefore P_3 = 0.5 \times 10^{-3.5} = \frac{0.5 \times \sqrt{10}}{10^4} = 0.00158$$

$$\therefore P_2 = \text{Power to the load} = 50 - 0.5 - 0.00158 = 49.49842 \text{ W}$$

Ans. $P_1 = 50$ W, $P_2 = 49.49842$ W, $P_3 = 1.58$ mW; $P_4 = 0.5$ W

Review Questions

1. Write the difference between E-plane and H-plane tee, giving their construction and working.
2. What is the use of directional coupler? What does 20 dB directional coupler mean?
3. What is variable attenuator in waveguides? Explain its construction and working.
4. Explain how in a circulator, signal goes to ports in one direction only.
5. What does isolator do and how, explain.
6. Prove that isolator has S-matrix as:

$$[S] = \begin{bmatrix} 0 & 0 \\ 1 & 0 \end{bmatrix}$$

7. Prove that S-matrix of a circulator is:

$$[S] = \begin{bmatrix} 0 & 0 & 1 \\ 1 & 0 & 0 \\ 0 & 1 & 0 \end{bmatrix}$$

8. How much and in what way does the Faraday rotation depend on

   (a) Magnetic field in any direction on a ferrite
   (b) Length of the ferrite material
   (c) Strength of the microwave signal
   (d) Direction of microwave signal

9. What are ferrites and how does the E-vector of the EM wave get rotated when passed through it in the presence of magnetic field on it.
10. What is the difference between vane-type variable phase shifter and attenuator? Explain.
11. Explain the function of hybrid junction.
12. What is the maximum bend allowed in microwaveguides and why?
13. List the differences between gyrator and isolator.
14. In directional couplers when two signals reach port 1 and port 2, their phase shifts are 90° and 270°, explain.
15. In directional couplers when two signals reach from ports 1 and 4, their phase shifts are 90° each, explain.

Microwave Tubes as Microwave Source (Oscillators) and Amplifiers

5

Contents

© Springer Nature Singapore Pte Ltd. 2018

P. K. Chaturvedi, *Microwave, Radar & RF Engineering*,
https://doi.org/10.1007/978-981-10-7965-8_5

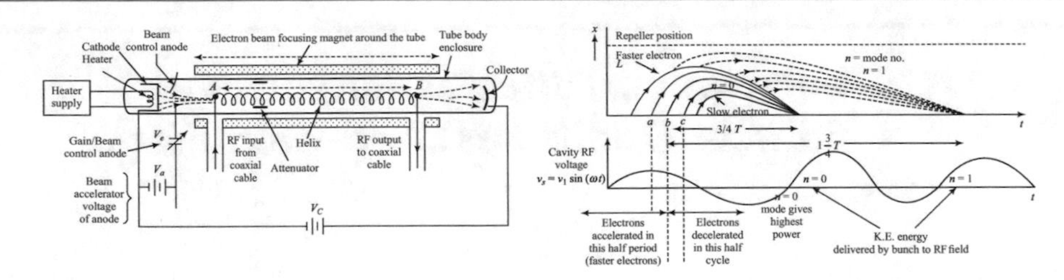

5.1 Introduction

A source of microwave power is essential for any microwave system. All communication and radar system generally use high power in kW range (CW or pulsed) for the transmitter and a few low-power sources for local oscillator/down conversion. The radar transmitter often operates at single frequency in pulsed mode of low duty cycle (<1%) with peak power in the range of 100 kW, having average power of 100 W or so. The electronic warfare system needs tunability over a wide frequency range. Radio astronomy requires low-power local oscillator. The microwave oven needs high-power CW source (700 W) at a single frequency (near 2.5 GHz).

In communication, there is certain frequency range which is avoided due to high absorption loss due to molecular/atomic natural frequency of oscillation; e.g., oxygen has these frequencies as 69 and 122 GHz, while for water it is 23 and 160 GHz (Fig. 5.1). Rain attenuation limits the range to 5 km, while oxygen limits to 1 km at these frequencies.

The requirement of microwave source is met by a variety of microwave solid state devices (SSD) for low-power low frequencies and microwave tubes for high-power high frequencies. Typically, solid state devices (SSD) can generate a maximum CW power of 100 W below 1 GHz and 10 W or so near 10 GHz. A microwave tubes can generate 10 kW to 10 MW even at high frequencies of 100 GHz. Figure 5.1 summarises the power versus frequency performances of sources with the upper limit line given for SSD. For simple laboratory experiments, reflex klystron and Gunn diode sources are used. In industrial applications, solid state devices have completely replaced the tubes.

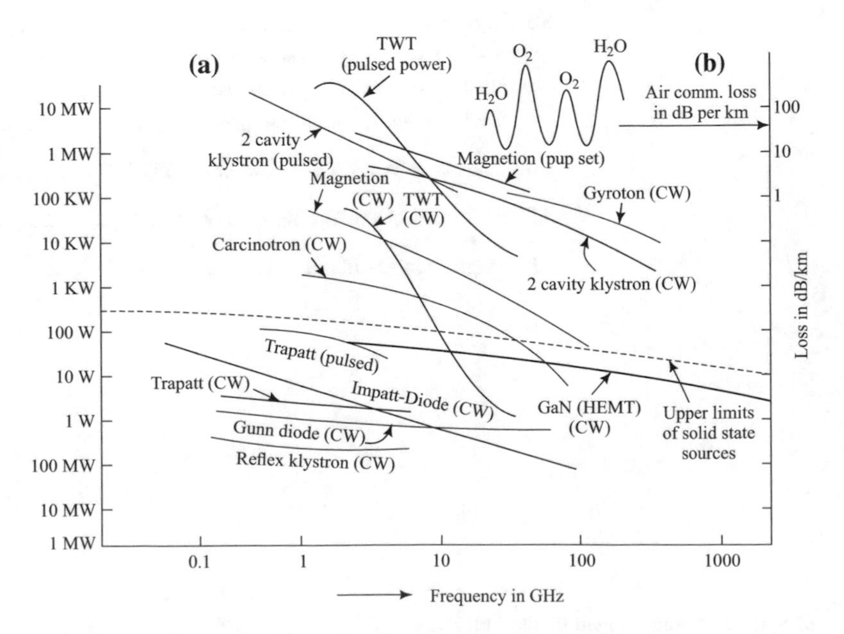

Fig. 5.1 **a** CW and pulsed power versus frequency for various microwave sources (microwave tubes and solid state). **b** Air communication loss in dB/km

The conventional vacuum tubes, which were being used in radio receivers and even computers during 1950s and 1960s, got substituted by transistors/ICs for all low-frequency (<l GHz) operations. These conventional tubes (triode, tetrode, pentode) had lot of limitations and as a result could not be used for higher frequencies and higher powers. Different types of tubes, e.g., klystron, TWT, magnetron, were invented, where the mechanism of electron movement was different (with density modulation and velocity modulation) and not just transit time mechanism of conventional tubes.

5.2 The Conventional Tubes

The normal conventional vacuum tubes used in radio receivers during 1960s were of 2″ to 5″ height with diameter of 0.5″ to 1.5″ (Fig. 5.2), while larger tubes are used even today in radio station transmitters. They are used for frequencies up to 50 MHz and generate over 10 kW power. Figure 5.2a gives the look of an actual small vacuum tube triode along with its functional diagram. It was invented by Lee De Forest in 1907 by placing a grid control element between heated cathode (emitting electron cloud) and anode with +ve voltage attracting the electron. Thus, the basic elements of simple vacuum tubes are:

The filament heater: Which heats the cathode by a small ac voltage supply and placed inside the cathode cylinder. It is like a microimmersion rod with resistance wire covered by ceramic layer.

The cathode: Which covers the heater for getting its full heat. As we know that around a heated metal, electron cloud gets created.

The anode plate: Which is given high +ve voltage for attracting the electrons from cathode, causing an electron current flow.

The grid: Which is wire mesh as screen between anode and cathode. A small −ve voltage on the grid can control the electron flow and hence the plate current to the load. The output voltage can be controlled by grid. Therefore, a small variation (signal) at grid can lead to large variation at the output ac voltage, i.e., acts as amplifier. So as to improve the performance further, tetrode with two grids and pentrode with three grids were also used. Without any grid, the tubes were being used as diode rectifier, but because of cost, size, and heater supply requirements, it got replaced by semiconductor diodes during the 1960s.

5.3 High-Frequency Limitations of Conventional Tubes

Conventional vacuum tubes (triode, tetrode, pentrode) work satisfactorily up to 50 MHz, but beyond this the inter-electrode reactances come into play. This reactance (specially inter-electrode capacitance between cathode, grid, and anode), being parallel to the external circuit, starts shorting the signal at higher frequencies (as $1/\omega C$ reduces). There are other limitations also, because of which we were not able to use the conventional tube, and scientist had to invent new type of tubes. These limitations are:

(i) Inter-electrode capacitance (shorting the signal)
(ii) Lead inductance (impending the signal)
(iii) Transit time effect (being much larger than μw time period)
(iv) Gain band width products (being independent of frequency)
(v) RF losses in wire due to skin effect (signals power loss)
(vi) Dielectric losses (signals power loss)
(vii) Radiation losses (signals power loss)
(viii) I^2R losses caused by charging current (signals power loss).

5.3.1 Inter-Electrode Capacitance–Shorting the Signal

At low frequencies, the large inter-electrode capacitances (C_{gp}, C_{gk}, C_{kp}) are already there but are less effective, while at higher frequencies it affects the performance (Fig. 5.3). For example 1 picofarad of inter-electrode capacitance, the current through it deteriorates the performance

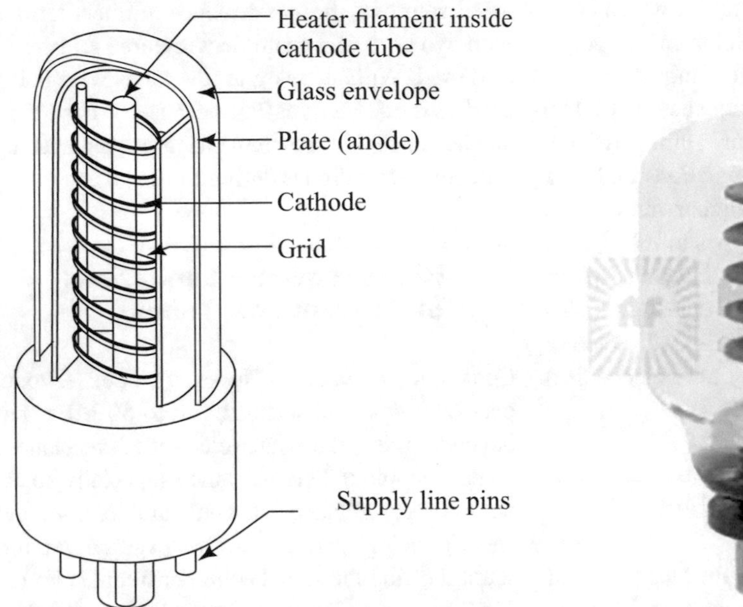

(a) Inner structure of a triode tube

(b) 200 W transmitting tube used during 1950-60

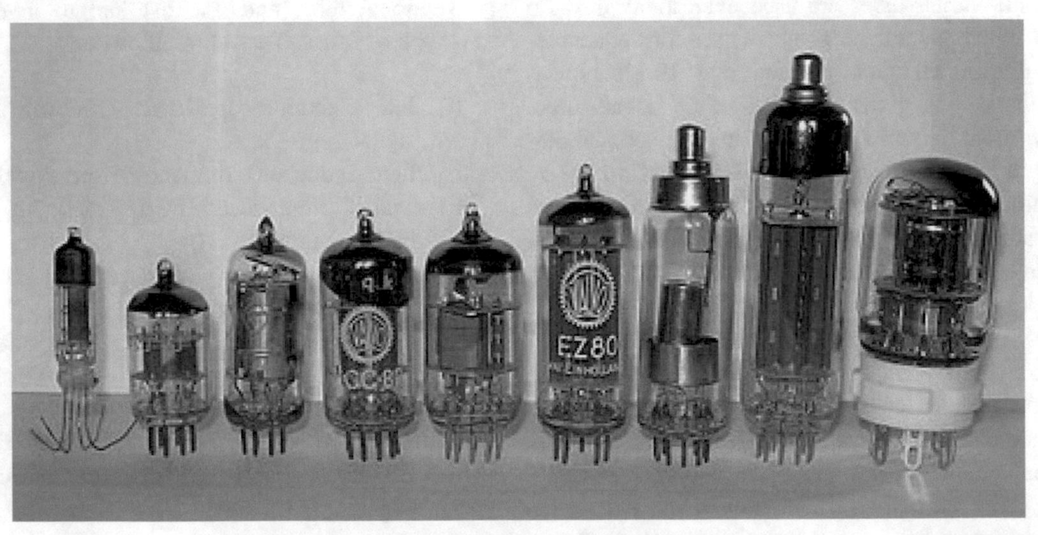

(c) Tubes used in Radio, TV etc. during 1950-60

Fig. 5.2 Actual conventional tubes

drastically at higher frequencies, as clear from Table 5.1.

The effect of inter-electrode capacitance can be reduced by decreasing the area of electrodes (as $C = \varepsilon A/d$) and increasing the distance between electrodes. This in turn reduces the current carrying capacity, power, and frequency of use as well.

5.3.2 Lead Inductance Impeding the Signal

The wires and leads of the tube cannot be straight all the time and have to be bend for connecting the outer circuit. This causes the lead inductance (L_g, L_k, L_p) within the tubes having reactance = $X_L = \omega L$, which increases with

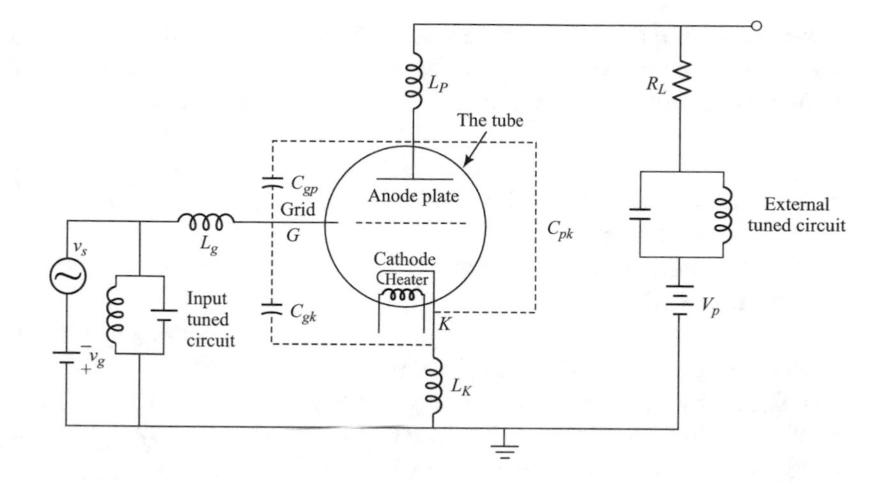

Fig. 5.3 Inter-electrode capacitances and lead inductance of a tuned amplifier

Table 5.1 Effect of inter-electrode capacitance and lead inductance at high frequency for C_{gk} = lpF, L_g = 10 nH, and V_{pp} = 500 V

Frequency of operation	Reactance $X_c = (1/\omega C_{gk})$ (Ω)	Current loss (V_{pp}/X_c) (mA)	$X_L = \omega L$ (Ω)
1 MHz	159,000	3.15	0.06
100 MHz	1590	315	6.14
1 GHz	159	3150	61.4

frequency. A small lead inductance of one nano-Henry has little effect at low frequency but at higher frequency it (a) provides degenerative −ve feedback, reducing the performance of the circuit (Table 5.1), and (b) reduces the voltage reaching the electrodes inside due to series impedance (ω_L). The lead inductance [$L = l/(\mu A)$] can be reduced by (1) reducing the length l, (2) having material of high permeability (μ), and (3) increasing the diameter of the wire for increasing A. All these are possible up to a limit only and that limit is f = 50 MHz or so.

5.3.3 Transit Time Effect Much Larger Than μW Time Period

Transit time is the time taken by electron to travel from cathode to anode and is given by:

$$T = d/v_o,$$

where

v_0 velocity of electron and
d distance between cathode and anode.

Equating kinetic and potential energies of electron:

$$\frac{1}{2}m_0 v_0^2 = eV(V = \text{voltage applied})$$

$$\therefore \quad v_0 = \sqrt{\frac{2eV}{m}}$$

$$\therefore \quad \tau = \frac{d}{v_0} = d\sqrt{\frac{m}{2eV}}$$

This is the transit time of electron (τ) from cathode to anode and is around 10 ns or so. Therefore, at low frequencies, the transit time $\tau \ll T$ (T being the time period of signal) and has no hinderance in its performance. With τ = 10 ns for normal tubes at 1 GHz (T = 1 ns), the signal voltage across the cathode to anode will change 10 times from −ve to +ve during the transit time of transit (τ) of electron. This will cause electron to oscillate back and forth in the cathode grid space and may return back to the cathode as well. This will reduce the efficiency of the conventional tube at microwave region. This limitation is made use, for generating μw signal in μw-tubes. Once the electron crosses the grid, it accelerates towards the anode plate due to high dc voltage.

The effect of transit time can be reduced to a limited extend by reducing τ by way of having lower values of d and high dc plate voltage. **This way these conventional tubes can be used up to 500 MHz and not beyond at all**.

5.3.4 Gain Band width Product Independent of Frequency Becomes Limitations

In a circuit, gain is maximum (A_{max}) at the resonant frequency (f_0) of an amplifier of Fig. 5.3 and it falls on either side of it. The difference between these two side frequencies (f_1 and f_2) where the gain falls to half of the maximum gain is called band width ($f_2 - f_1$) (Fig. 5.4). This gain band width product $A_{max}(f_2 - f_1)$ defines the quality of the amplifier, and it keeps on reducing as we go to higher frequencies in the conventional tubes. A conventional tube amplifier, its functional diagram, and its equivalent circuit are shown in Fig. 5.5. The internal plate resistance of the tube effectively will be parallel to the external LC-tank circuit and the load. If g_m is transconductance, i.e., ratio of plate ac current to input ac voltage, then the output load voltage (ac) will be

$$v_L = \frac{g_m v_g}{\frac{1}{r_p} + \frac{1}{R_l} + j\left(\omega C - \frac{1}{\omega L}\right)} = \frac{g_m v_g}{G + jX};$$

where G = conductivity of output side $= \frac{1}{r_p} + \frac{1}{R_l}$

$$\text{Therefore gain } A = (v_L/v_g) = g_m/(G + jX) \quad (5.1a)$$

will be maximum at resonant frequency $f_0 = 1/(2\pi\sqrt{LC})$ as here

$$X = (\omega L - 1/\omega C) = 0$$

$$\therefore \; A_{max} = (v_L/v_g) = g_m/G$$

As band width is measured at half power point P_1 and P_2 (Fig. 5.4) i.e. at this voltage gain $A_{1/2} = \frac{A_{max}}{\sqrt{2}} = \frac{g}{G\sqrt{2}}$. Therefore comparing this $A_{1/2}$ with Eq. (5.1a) we get $G = X = \left(\omega C - \frac{1}{\omega L}\right)$.

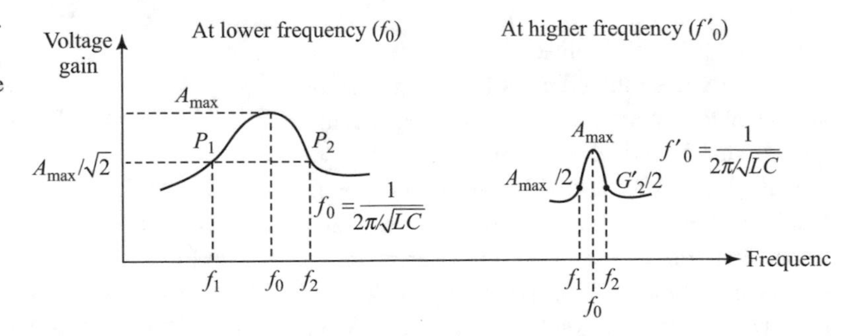

Fig. 5.4 As we go to higher frequencies gain (G) as well as band width (f_2–f_1), the tube amplifier reduces drastically

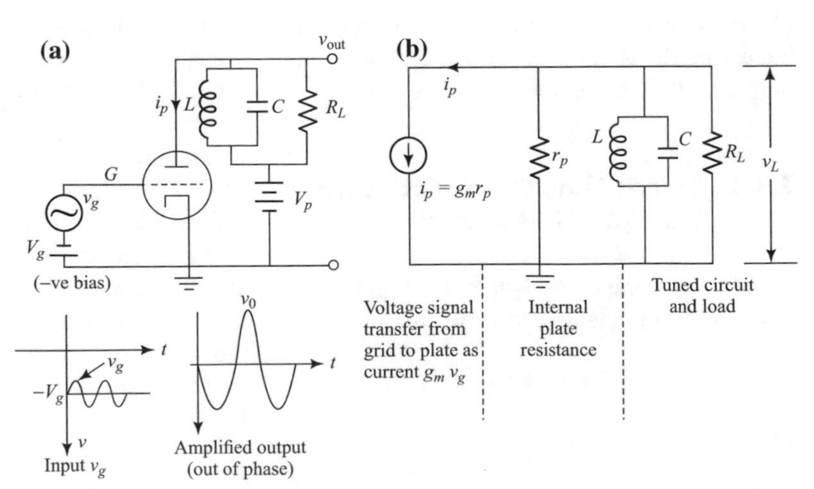

Fig. 5.5 **a** Functional diagram of conventional triode tube amplifier. **b** Its equivalent circuit

This being a quadratic equation in ω gives two values of ω_1 and ω_2 at half power points around f_0 (see Fig. 5.4) for the circuit of Fig. 5.5:

$$\omega_1, \omega_2 = \frac{G}{2C} \pm \sqrt{\left(\frac{G}{2C}\right)^2 + \frac{1}{LC}} = 2\pi f_1,\ 2\pi f_2$$

$\therefore$ Band width (BW) $= (\omega_2 - \omega_1)$

$$= \frac{G}{C}\left(\text{As}\left(\frac{G}{2C}\right)^2 \gg \frac{1}{LC}\right) \tag{5.1b}$$

$\therefore$ Gain band width product of a triode amplifier (Fig. 5.5) by Eqs. (5.1a) and (5.1b) is:

$$A_{\max} \times \text{BW} = \frac{g_m}{G} \times \frac{G}{C} = \frac{g_m}{C}$$
$$= \text{independent of frequency.}$$

Thus in such tubes with resonant tank LC circuit, higher gain can be obtained at the expense of lower band width. In microwave tubes, this limitation is not there and we can have larger band width as well as higher gain together.

5.3.5 RF Losses (I^2R Losses) in Wire and Skin Effect (increasing the resistance)

For ac current in a wire, the current density is higher near the surface and lower at the core, and at higher frequencies, this tendency increases. We define the skin depth (δ), where the current density falls to $1/\sqrt{2}$ and the cross section of that area is defined as A_{eff}, where most of the current flows (Fig. 5.6).

$$\text{Skin depth}\,(\delta) = \frac{1}{\sqrt{\pi f\,\mu\,.\,\sigma}}$$

$\therefore\ \delta \propto 1/\sqrt{f}$ and also because $\delta \propto A_{\text{eff}}$.
$\therefore\ A_{\text{eff}} \propto 1/\sqrt{f}$

$$\therefore\ A_{\text{eff}} = \frac{k}{\sqrt{f}},\ \text{and}\ R = \frac{\rho l}{A_{\text{eff}}} = \frac{\rho l\sqrt{f}}{k}$$

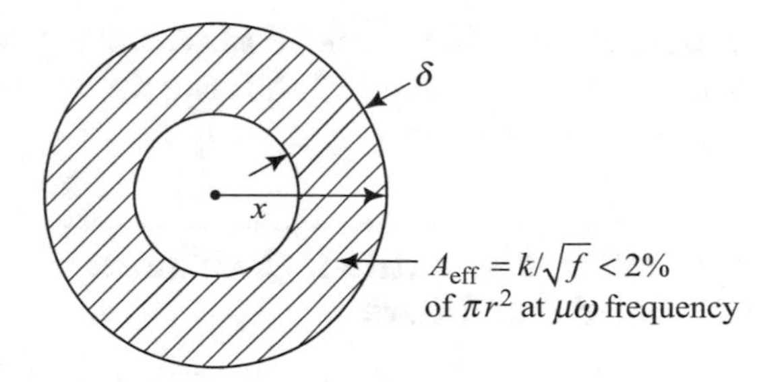

Fig. 5.6 Effective area of current flow in a wire at higher frequencies f and is <2% of cross section at $\mu\omega$ frequencies

Thus, resistance of wire R increases with frequency leading to higher ohmic loss. This can be reduced by having short-length l low-resistivity (ρ) wire of larger area, e.g., planar electrodes, to some extent, but at $\mu\omega$ frequencies, it becomes very large.

Therefore, I^2R power losses for ac current and for charging of inter-electrode capacitances increases very much.

This loss can be reduced by lowering the capacitance and by increasing the number of shunt paths, for the charging current, but this too is possible up to a limited extent up to 100 MHz only.

5.3.6 Dielectric Loss (Signal Power Loss)

Insulating material used in components, e.g., glass, plastic, spaces, causes dielectric loss given by: $P = \pi f V_0^2 \varepsilon_r \tan(\delta)$. Where δ = loss angle of dielectric, ε_r = relative permittivity.

This can be reduced by having smaller surface area of glass and smaller base area, which in turn will increase the inter-electrode capacitance and lead inductance.

5.3.7 Radiation Loss (Signal Power Loss)

At microwave frequencies, the length of the wire in the tubes is close to the wavelength, i.e., cm and mm, and therefore, wire acts as antenna.

Therefore, power is lost due to radiation of signal. This can be reduced by shielding the total circuit including the tube.

5.4 Microwave Tubes, Oscillators, and Amplifiers

So as to overcome the limitation of conventional tubes, special tubes for microwave frequencies were designed. **These tubes make use of the transit time effect. Here, the density modulation of electron (bunching) is made use of profitably, as a blessing in disguise.** In fact, these tubes require large transit time for their operation. Here, the kinetic energy of electrons is converted into RF energy. These tubes (Fig. 5.7) are of two types:

(i) **Linear beam type (0-type)**: dc electric and magnetic fields are parallel to the electron flow. Magnetic field is used just for focusing the beam, e.g., two-cavity klystron, reflex klystron, TWT, and BWO. **They give very high gain (30–70 dB), moderate power output, (<200 W), low noise, and moderate efficiency**.

(ii) **Cross-field type M-type**: Here, the electric and magnetic fields are perpendicular to each other, e.g., magnetron, carcinotron. **These tubes have moderate gain (6–20 dB) but very high power (>1 kW),** high noise, and high efficiency.

Here, in this chapter, we will discuss two-cavity klystron amplifier, reflex klystron oscillator, travelling tube (TWT) amplifier, backward wave oscillator (BWO), and magnetron oscillator.

5.5 Klystrons

The klystron tube can be used as a microwave source, i.e., as an oscillator. The reflex klystron is used as oscillator, while two-cavity/multicavity klystron is used as amplifier (Fig. 5.8).

5.5.1 Two-Cavity Klystron Amplifier

All these klystrons make +ve use of the transit time effect by

(i) Velocity modulation of the electron beam by the RF field of the first cavity.
(ii) Conversion of velocity modulation to density modulation, i.e., electron bunching.
(iii) Transfer of energy to the RF field in second cavity in amplified form.
(iv) Collection of signal by loop tapping of amplified microwave signal to coaxial cable.
(v) Collection of remaining electron by collector anode.

Figure 5.8 shows how the electrons emitted from cathode are forced by the beaming electrode

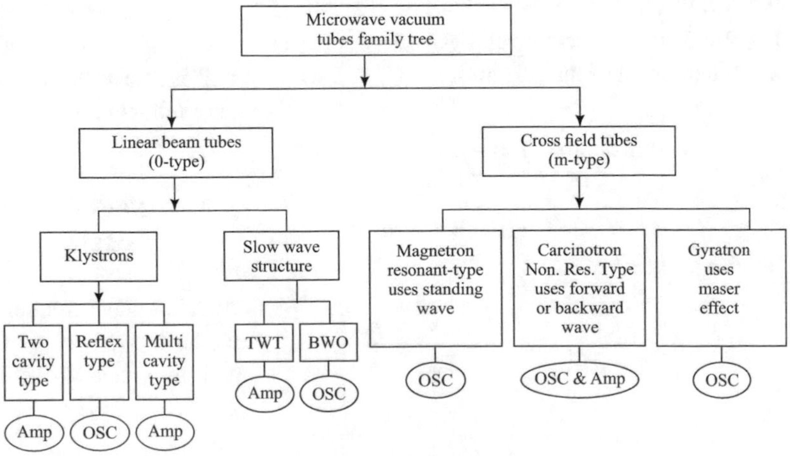

Fig. 5.7 Various types of microwave tubes as oscillators and amplifiers

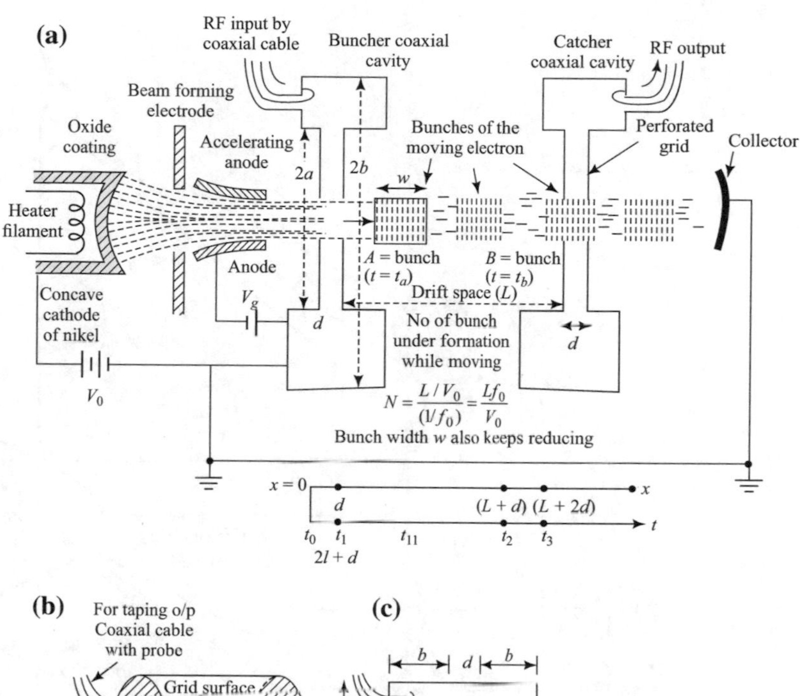

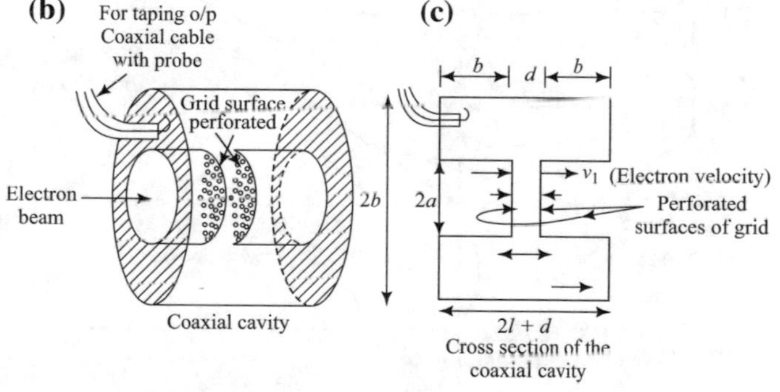

Fig. 5.8 a Two-cavity klystron amplifier—schematic diagram. **b** Structure of the two coaxial re-entrant cavities of it. **c** Functional diagram of cavity

to converge into a beam through the anode, which being at very high potential has attracted them. Due to momentum, the electron keeps moving to the first cavity called buncher cavity, which besides being at high dc potential V_0 is also having ac input voltage V_1, signal. The cavity has a grid gap, through which the electron crosses.

When the electron is first accelerated by high dc voltage V_0, before entering the buncher grids, the velocity is uniform as:

$$v_0 = \sqrt{\frac{2eV_0}{m}} \approx 0.6 \times 10^6 \sqrt{V_0}\ \text{m/s} \qquad (5.2)$$

The buncher dc voltage is slightly higher than that of cathode, for making the electrons to get attracted and allow it to move out through its gaps due to momentum. These electrons are velocity modulated and hence density modulated, forming bunches in the drift space. **Those electrons which enter the first grid when its ac signal is in its −ve phase across the walls of the grid, they slow down. Those electrons which enter the first grid when its ac signal is +ve speed up and catch the slow electron, thus forming a bunch.** This process continues with time as depicted in the Applegate diagram of electron movement (time vs. voltage) (Fig. 5.9). When this pulsating stream of electrons, i.e., bunches, reaches second grid of the catcher cavity and gives its charge to the cavity gate, it increases the grid gap voltage and electric field, which is then tapped out by the loop of coaxial cable from the cavity-2.

Thus, these bunches when reach the cavities, it becomes pulsating beam of electron current (i.e., ac current) which then gives higher amplified voltage across the characteristic impedance of the cavity and the coaxial line.

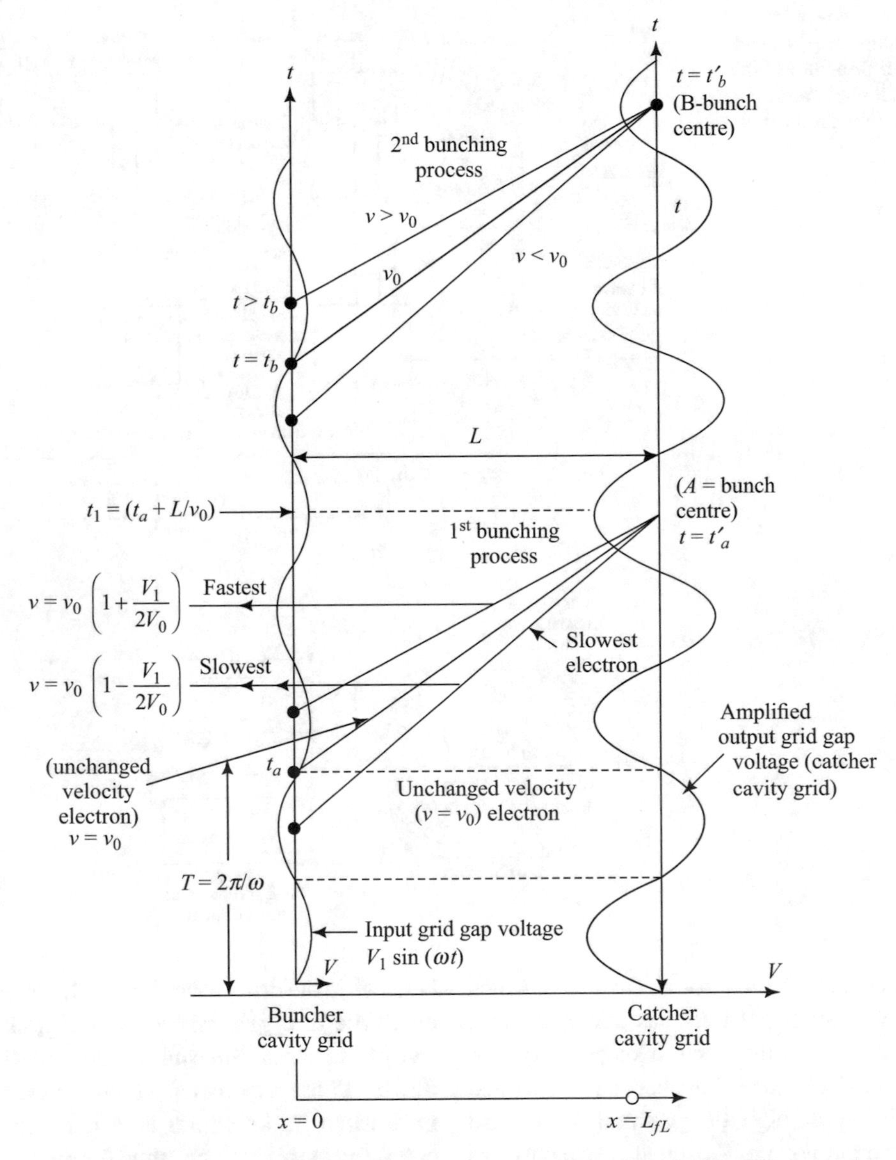

Fig. 5.9 Applegate diagram of electron bunching in the two cavities (buncher and catcher) in two-cavity amplifier. No. of bunches under formation while moving will be = Transit time/wave time period = $(L/v_0)/(1/f_0) = Lf_0/v_0$

Mathematical Analysis

This analysis is carried out with the following assumptions:

1. Electrons leave the cathode with zero velocity, with uniform electron density across the beam.
2. Space charge effect is negligible.
3. The magnitude of the input signal (v_1) is very much smaller than the dc accelerating voltage ($v_1 \ll V_0$).
4. The RF field is fully longitudinal along the electron velocity.
5. Both the cavities are identical and are tuned to the natural frequency of the cavity which is also the input signal frequency.
6. The transit time (t_1) across the gap spacing of the cavity is negligible as compared to the (T) time period of the signal (Fig. 5.6).
7. **The transit time of electron from buncher to catcher (L/v_0) is greater than the time**

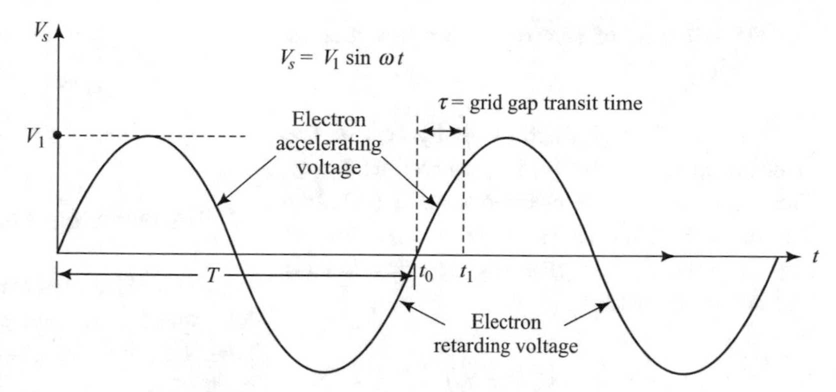

Fig. 5.10 Signal voltage at the buncher cavity grid gap retarding or accelerating the electron bunch

period of signal (T). Therefore, more than one bunch are moving in the drift space (buncher to catcher), and the width of the bunch keeps reducing (Figs. 5.8 and 5.9).

8. Whole of the potential energy given to the electron gets converted to kinetic energy. The ac voltage input signal at the buncher cavity gives gap voltage at the buncher grid (Fig. 5.10) as:

$$V_s = V_1 \sin \omega t$$

where V_1 is the peak amplitude of signal with $V_1 \ll V_0$.

The electron (Figs. 5.8 and 5.9) enters the grid at time $t = t_0$ and then enters the drift space at $t = t_1$ with $(t_1 - t_0)$ as the transit time though the buncher gap is given in:

$$\tau_g = \frac{d}{v_0}$$
$$= (t_1 - t_0)(\text{where } v_0 = \text{electron velocity}) \tag{5.3}$$

(a) **The catcher cavity ac signal V_s**

This cavity-grid-gap transit time τ_g is much smaller than time period T of the signal input (Fig. 5.10). The transit angle (θ_g) will be a fraction of ω.

$$\theta_g = \omega\tau_g = \omega(t_1 - t_0) = \frac{\omega d}{v_0} \tag{5.4}$$

The average voltage $\langle V_s \rangle$ of the buncher grid will be time averaged from t_0 to t_1 (Fig. 5.10).

$$\langle V_s \rangle = \frac{1}{\tau} \int_{t_0}^{t_1} V_1 \sin(\omega t) \mathrm{d}t$$
$$= \frac{-V_1}{\omega t} [\cos(\cos(\omega t_1)) - \cos(\omega t_0)]$$

$\therefore$ using the relation $t_1 = (t_0 + d/v_0)$, we get

$$\therefore \langle V_s \rangle = \frac{+V_1}{\omega t} \left[\cos(\omega t_0) - \cos\left(\omega t_0 + \frac{\omega d}{v_0}\right) \right]$$
$$= \frac{+V_1}{\omega \tau} \left[\left(2 \sin\left(\omega t_0 + \frac{\omega d}{2v_0}\right) \sin\left(\frac{\omega d}{2v_0}\right) \right) \right]$$

As $(\cos A - \cos B) = 2 \sin\left(\frac{A+B}{2}\right) \cdot \sin\left(\frac{B-A}{2}\right)$

$$\therefore \langle V_s \rangle = \frac{V_1 \sin(\theta_g/2)}{(\theta_g/2)}$$
$$\cdot \sin(\omega t_0 + \theta_g/2) \left(\text{As} \frac{\omega d}{v_0} = \theta_g \right)$$
$$V_s = V_1 \beta_i \sin(\omega t_0 + \theta_g/2) \tag{5.5}$$

where $\beta_i = \frac{\sin(\theta_g/2)}{(\theta_g/2)}$ is called as the electron beam coupling coefficient of the

input cavity gap and

V_s = a.c. signal voltage of buncher cavity. (5.6)

(b) **Transit time of electrons from buncher to catcher**

Equation (5.1) gives the velocity (v_0) of the electron up to the cavity, i.e., up to time $t = t_0$. The expression for the electron velocity (v_1) after leaving the cathode (at time $t = t_1$) (as in Figs. 5.8 and 5.9) is given by equating kinetic and potential energies:

$$\frac{1}{2}mv_1^2 = e(V_0 + V_s) \tag{5.7}$$

where v_s is the input ac signal on the electron across the gap grid of the first cavity and V_0 is the dc accelerating voltage on the electron. Therefore, its velocity v_1 at the end of the transit time (t_0−t_1) in the grid gap is:

$$\therefore \quad v_1 = \sqrt{\frac{2e}{m}(V_0 + V_s)}$$

$$= \sqrt{\frac{2eV_0}{m}} \cdot \left(1 + \frac{V_s}{V_0}\right)^{\frac{1}{2}}$$

Expand in binomial, and we retain the first power term only then

$$v_1 \cong v_0\left(1 + \frac{V_s}{2V_0}\right) \text{with } v_0 = \sqrt{\frac{2eV_0}{m}} \tag{5.8}$$

As $\frac{V_s}{2V_0} \ll 1$, in Eq. (5.8), putting v_s of Eq. (5.5) in Eq. (5.8) we get v_1 as:

$$v_1 = v_0\left[1 + \frac{\beta_i V_1}{2V_0}\sin\left(\omega t_0 + \frac{\theta_g}{2}\right)\right] \tag{5.9}$$

From here, the maximum and minimum values of v_1 will be

$$v_{1\text{MAX}} = v_0\left(1 + \frac{\beta_i V_1}{2V_0}\right) \tag{5.10}$$

$$v_{1\text{Min}} = v_0\left(1 - \frac{\beta_i V_1}{2V_0}\right) \tag{5.11}$$

And time taken by faster, slower, and the central electrons to cover the distance L from buncher to catcher cavity with velocities $v_{1\max}$, $v_{1\min}$, and v_1, respectively, is given in (Figs. 5.8 and 5.9):

$$t_f = (t_2 - t_0 - T/4) \tag{5.12}$$

$$t_S = (t_2 - t_0 + T/4) \tag{5.13}$$

$$t_C = (t_2 - t_0) \tag{5.14}$$

(c) **The bunching process**

All the three electrons, i.e., high speed ($v_{1\max}$), low speed $v_{1\min,}$ and central electron (v_1), get bunched after travelling a distance L in the cavity-2 at time $t = t_2$ (Figs. 5.8 and 5.9).

$$\therefore L = v_0(t_2 - t_0) = v_0 \,.\, \Delta t \tag{5.15}$$

As the distance travelled (i.e. velocity X time) by all the three electrons are same, we equate the distance travelled by faster and central electron and put $T = 2\pi/\omega$ in Eq. (5.10–5.14), we get:

$$\therefore \quad L = v_0(t_2 - t_0) = t_f v_{1\max}$$

$$= \left(t_2 - t_0 - \frac{\pi}{2w}\right) \cdot v_0\left(1 + \frac{\beta_i V_1}{2V_0}\right) \tag{5.16}$$

$$\therefore \quad (t_2 - t_0) = \left(t_2 - t_0 - \frac{\pi}{2\omega}\right)\left(1 + \frac{\beta_i V_1}{2V_0}\right)$$

Let $t_2 - t_0 = \Delta t$ and open the brackets:

$$\Delta t = \left(\Delta t - \frac{\pi}{2\omega} + \frac{\beta_i V_1}{2V_0}\Delta t - \frac{\beta_i V_1}{2V_0} \cdot \frac{\pi}{2\omega}\right)$$

$$\Delta t = \frac{-2V_0}{\beta_i V_1}\left(\frac{-\pi}{2\omega} - \frac{\beta_i V_1}{2V_0} \cdot \frac{\pi}{2\omega}\right)$$

$$\therefore \quad \Delta t = \frac{\pi}{2\omega}\left(1 + \frac{2V_0}{\beta V_1}\right)$$

$$= \left(\frac{\pi V_0}{\beta_i \cdot \omega \cdot V_1}\right) \text{ as } \frac{V_0}{V_1} \gg 1 \tag{5.17}$$

The same result can be obtained from equation of slow-moving electron. Now, putting this Δt in Eq. (5.15), we get.

$$L = \left(\frac{\pi V_0 v_0}{\beta_i \omega V_1}\right) \tag{5.18}$$

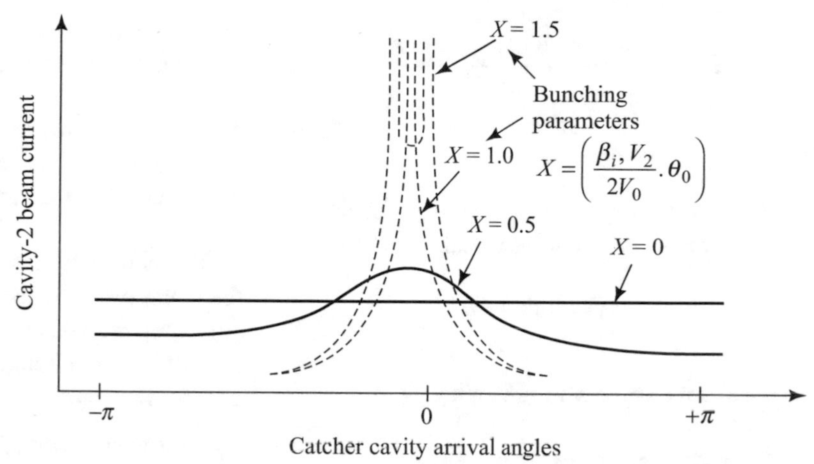

Fig. 5.11 Beam current spikes at catcher cavity versus arrival angles for different bunching

(d) **Catcher cavity current and the number of bunches**

It has been found that the electron beam bunch reaching the catcher cavity gives spikes of current and not sinusoidal signal (Fig. 5.11). The amplitude of the spike and its duration depend on its arrival angle (phase). It is also possible that some times there may be more than one bunch in the drift space, length L [Eq. (5.19)].

The number of bunches N that are getting formed while moving from cavity 1 to 2 is:

$$N = \frac{\text{Transit time of electron in the drift space } L}{\text{One time period of the wave}} = \frac{(L/v_0)}{(1/f)} = \frac{L.f.}{v_0} \quad (5.19)$$

With N number of cycles in the drift space, we now define a bunching parameter,

$$X = \frac{\beta_i V_1}{2V_0} \cdot \theta_0 \quad (5.20)$$

$$\therefore \quad V_1 = 2V_o X/(\beta_i \theta_o) \quad (5.20a)$$

where $\theta_0 = 2\pi N = 2\pi Lf/v_0 = L\omega/v_0$ = drift space transit angle.

Figure 5.11 gives beam current at the catcher cavity versus different arrival angles, for different bunching parameter X.

It has been found by Fourier analysis of the beam current (not covered here) that at $X = 1.841$, the fundamental signal at the catcher cavity is maximum and the corresponding optimum distance between the two cavities is given by the following equation, instead of L of Eq. (5.18):

$$L_{\text{opt}} = \frac{1.173\pi V_0 v_0}{\omega \beta_i V_1} \quad (5.21)$$

(e) **Voltage gain**

If β_0 is beam–cavity coupling coefficient of the output cavity, R_{sh} the output shunt resistance, and I_2 the output RF current, then the output voltage V_2 is:

$$V_2 = \beta_0 \cdot (12 R_{sh}) \quad (5.21a)$$

Also, it can be proved that the peak output current is:

$$I_2 = 2I_0 \cdot j_1(x) \quad (5.21b)$$

where $J_i(x)$ is Bessel's function of x, the bunching parameter. Therefore, by Eqs. (5.20a), (5.21a), and (5.21b), voltage gain will be:

$$A_v = (V_1/V_2) = \frac{\beta^2 \cdot \theta \cdot I_0 \cdot J_1(x) \cdot R_{sh}}{x_0 V_0} = \frac{\beta^2 \cdot \theta \cdot J_1(x) \cdot R_{sh}}{x \cdot R_0} \quad (5.21c)$$

Here, we have taken $\beta_i = \beta_0 = \beta$, for the two cavities to be same and also that:

$$(V_0/I_0) = R_0$$

(f) **Power output and efficiency**

The input power as dc power is:

$$P_{in} = I_0 V_0 \quad (5.22)$$

The RF power (fundamental) at the catcher cavity is

$$P_{out1} = 0.58\, I_0 V_2 \quad (5.23)$$

where V_2 is the fundamental μw RF voltage, close to dc bias V_0 which is >1 kV.

$$\text{Therefore the efficiencies} = \eta = \frac{P_{out1}}{P_{in}} = 0.58 \frac{V_2}{V_0} \quad (5.24)$$

$$\therefore \quad \eta_{max} \approx 0.58 \,(\text{for } V_2 = V_0)$$

In practice, η is always less than 58% due to heat loss, loss in harmonics, etc. It is normally 30–40% only. The power is quite high, e.g., CW power of 500 kW and pulsed power of 30 MW–1.0 GHz. The power gain is 15–70 dB. Size of klystrons ranges from 4″ to 10″ for multicavity designs for higher power requiring water cooling system.

5.5.2 Two-Cavity Klystron Oscillator

A two-cavity klystron amplifier can be converted into an oscillator by +ve feedback as a part of the catcher output into the buncher in phase with the input. For this, the feedback has to be adjusted to give the correct phase and amplitude (frequency and power output tuning) which is done by:

1. Tuning the two resonator cavities together.
2. Adjusting the dc accelerating voltage (V_0).
3. Adjusting the dc grid voltage (V_g).

The two-cavity resonators need to resonate in the some phase for maximum power output. By adjusting the above three, frequency tuning is possible over a range of 10–15% around the central frequency.

Typical Characteristic of two-Cavity Klystron Amplifiers/Oscillators

Beam voltage required depends on the specific klystron, but in general following maximum ranges of specification can be taken (see Fig. 5.1).

1. Beam voltage 1–20 kV
2. Beam current 25 mA–5 A
3. Frequency 0.5–100 GHz
4. Power gain 15–70 dB
5. Noise figure 15–20 dB
6. Efficiency 30–40%
7. Spacing between two cavities 2–5 cm
8. Shunt impedance R_{sh} > 10 kΩ.
9. Power 10–500 kW for CW and 30 MW in pulsed mode at 1.0 GHz.
10. Band width is very small ±10% around the central frequency f_0. Tuning is possible in a small range of frequencies at the cost of gain [$\omega \propto V_0$ as per Eq. (5.18)].

Applications of two-Cavity Amplifier/Oscillator:

1. As a high-power oscillator of the range 5–50 GHz.
2. As high-power amplifier in:

   (a) TV transmitters
   (b) Radar
   (c) Satellite communication
   (d) Tropospheric scanners.

Fig. 5.12 Reflex klystron No. 2K25. **a** Coaxial line and probe are put inside the waveguide hole (through the tube holder) to get supply of μw power. **b** Mechanical tuning knob of re-entrant cavity. **c** and **d** Pins for heater supply voltage line. **e** Pin for accelerator supply line. **f** Repeller supply knob. **g** Central plastic core for stable alignment on the tube holder/socket

5.5.3 Reflex Klystron Oscillator

In the two-cavity klystron if a fraction of output is fed back in phase, i.e., with a phase shift of multiple of 2π, it will oscillate. But normally it is not used due to the problem of (a) tuning two cavity together along with (b) adjustment of feedback path. **In reflex klystron oscillator which has only one cavity, the above problems are removed;** however, it gives low power (less than 2 W) only in the range 1–25 GHz. **The most useful characteristic is its tunability over a wide frequency range, and therefore, it is used in** radar receiver, local oscillator, microwave receiver, variable frequency signal generator, portable microwave links, pump oscillator in parametric amplifier, academic institutions, laboratories, etc.

Figure 5.12 shows the actual reflex klystron (No. 2K25) normally used in the laboratories. The tube is put on the holder/socket fixed on the waveguide to which μw power is supplied.

It consists mainly of

1. Electron cathode gun,
2. Anode,
3. Resonator cavity,
4. Repeller plate,
5. Output coupling,
6. Accelerating dc voltage (250–1000 V) of the anode and the cavity,
7. Repeller −ve dc voltage (100–1000 V).

(a) ***Mechanism of Oscillation***

From the electron gun, electron beam is accelerated towards anode plate due to its high +ve dc potential and crosses through its gap to the cavity resonator. It then crosses the cavity gap mesh grid also (due to momentum) and moves towards the repelled, but never reaches it. It gets repelled back to the cavity due to its high −ve potential (Fig. 5.13). As in any electronics system noise is always present with all frequencies 0 to ∞ and if there is a +ve feedback system which keeps on amplifying only that frequency signal, which corresponds to the cavity resonant frequency, then that frequency signal gets generated. Here, if the repelled electron bunch reaches back to the cavity at the some phase point where it left it, then it is a +ve feedback. That is to say that the best possible time for electrons to return to the gap is at a time when voltage across the gap will apply maximum retardation (see Applegate diagram Fig. 5.14).

Thus, the working of the reflex klystron can be summarised in the following six steps:

1. Electron velocity and its density get modulated (bunched) after crossing the perforated anode cavity due to its RF voltage.
2. Due to momentum, the electron continues to move in the −ve field between cavity and repeller.

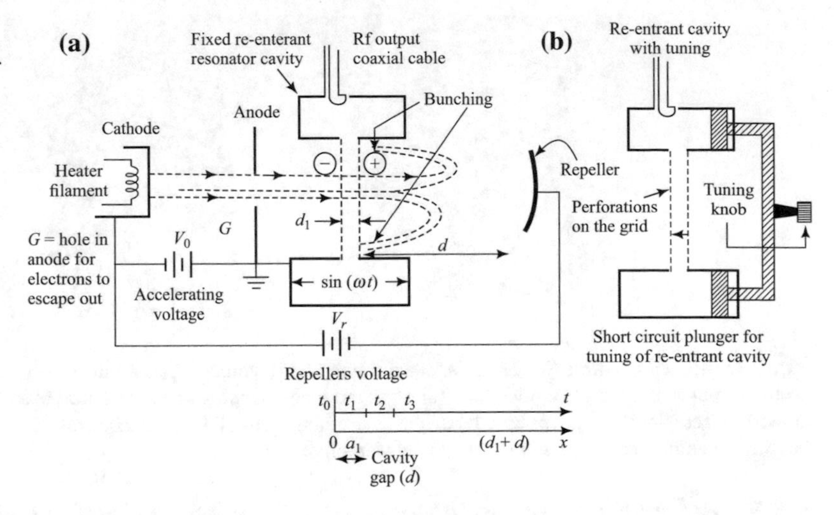

Fig. 5.13 **a** Schematic diagram of a reflex klystron. **b** Mechanical tuning of the re-entrant cavity of reflex klystron

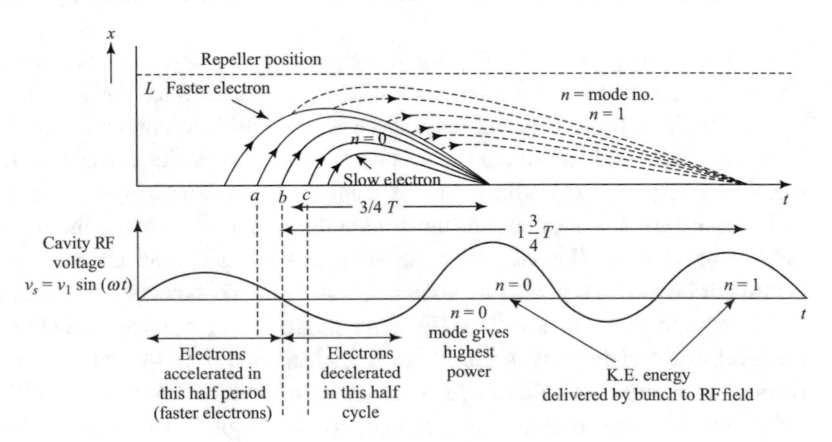

Fig. 5.14 Applegate diagram for a reflex klystron

3. Repeller reduces its velocity and finally repels back to the cavity, and by this time, the modulation (bunching) becomes still more sharper, i.e., denser.
4. **If this bunch of electron (Fig.** 5.13) **returns back at the first wall of the cavity at the moment when it has +ve RF voltage, then it looses energy by transferring to the RF field, which then gets amplified (+feedback process). This leads to sustained RF oscillations.**
5. This amplified signal can be tapped out by probes/loops as usual.
6. Frequency tuning is possible by:

   (a) Mechanical tuning of cavity by rotating the knob by 0–360° (Fig. 5.13b) which can vary frequency up to 1 GHz (i.e., ±5%) around the central frequency in 2 k 25 tubes (Fig. 5.12).

   (b) Electronics tuning by changing the repeller voltage up to 0.4 GHz (i.e., ±2%) around the central frequency in $1\frac{3}{4}$ mode.

The bunching process: The electron '*a*' coming out of the gap is accelerated during its +ve cycle, the electron '*c*' retarded during −ve cycle, while the '*b*' electron comes out with no change in speed when $v_s = V_1 \sin \omega_0 t = 0$. (see Fig. 5.14). The faster electron '*a*' travels deeper in the repeller space as compared to slower electron '*c*' and forms bunches. As the bunch is to be formed exactly at the cavity point, we have to adjust the repeller space and repelled voltage. This way the returning and retarded electron bunch losses its kinetic energy and gives to the RF field in the cavity. When the energy delivered by a certain bunch of electron to the cavity is greater than the energy it had collected (while it was crossing forward), then the oscillation of that frequency signal is sustained.

(b) ***Transit Time and Mode Number***

As seen earlier that for giving up energy by the electron beam and for reinforcing the oscillation in the cavity, the following two conditions should be satisfied:

(i) The reflected back electrons should form bunch just at the time they reach the cavity.
(ii) At this moment, the signal at the cavity is to be in +ve phase.

Thus, the start of the central part of the bunch is when $v_s = V_1 \sin(\omega t) = 0$, and reaching back time is when $v_s = V_1 \sin(\omega t) = V_1$ for optimum efficiency.

Therefore, the time difference between these two events is $(n + 3/4)T$. Hence, $n = 0$ gives the 3/4 mode, $n = 1$ gives the $1\frac{3}{4}$ mode, etc. (Fig. 5.14). With lower reflecting voltage, the electron will travel longer before it is reflected back and therefore corresponds to $n = 1, 2, 3 \ldots$ modes. The same is true for lower accelerating voltage. If we define the mode number as $N_n = (n + 3/4)$, then the repeller (V_R) and accelerator voltages (V_P) are related to it as follows.

$$\frac{V_0 + |V_{R1}|}{V_0 + |V_{R2}|} = \frac{N_1}{N_2} = \frac{n_1 + 3/4}{n_2 + 3/4} \tag{5.25}$$

Figure 5.15 gives (a) power output at different modes and (b) frequency tuning around the cavity resonant frequency f_0, by changing repeller voltage. We will prove a relation [Eq. (5.26)] between f and V_R later.

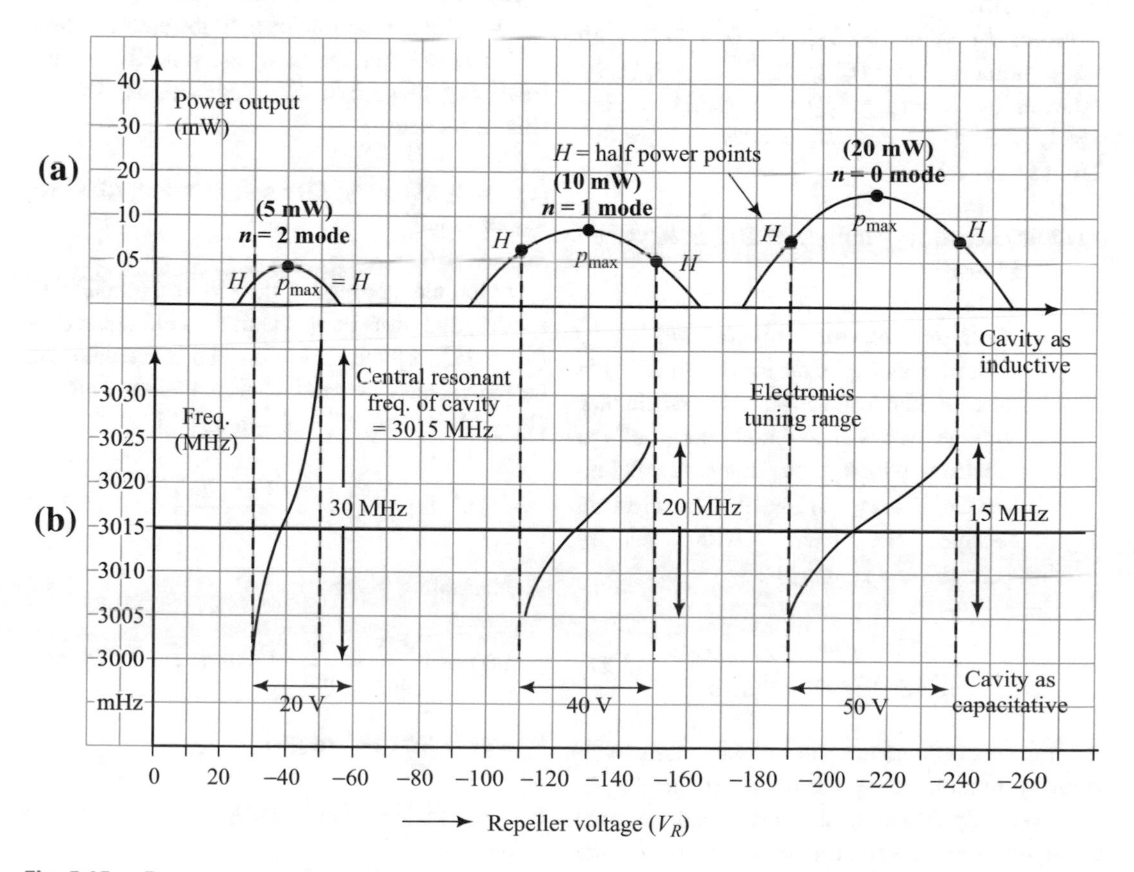

Fig. 5.15 **a** Power output modes. **b** Variation of power output with frequency tuning around the central frequency of the cavity $f_0 = 3015$ MHz by changing repeller voltage, for $V_0 = 750$ V in a typical reflex klystron

The frequency tuning is limited by the half power points of that mode as in Fig. 5.15b. Larger reduction in repeller voltage switches the klystron to the next mode of oscillation range, where the tuning range is less as is clear from Fig. 5.15a. **Besides tuning by (a) repeller voltage, cavity can also be tuned by either (b) short slug in the cavity or (c) changing the grid gap. The latter two methods change the central frequency of oscillation of the cavity. This way there are three methods of tuning the reflex klystron**. A simple and optimal oscillation condition mode '*n*' is given by:

$$f = \frac{(n + \frac{3}{4})(V_R - V_0)}{d} \cdot \sqrt{\frac{e}{8mV_0}} \tag{5.26}$$

This formula will be proved now. It is plotted in Fig. 5.15b.

Note: In some books, $N = (n + 3/4)$ with mode number = n = 0, 1, 2, 3 … might be replaced by $N = (n - 1/4)$ with mode number n = 1, 2, 3, 4 …; both are correct, but are different conventions.

(c) **Mathematical Analysis of Reflex Klystron Oscillator**

(i) *Electron between cathode and cavity anode*: This is similar to that done in the case of klystron earlier. The accelerator voltage V_0 gives potential energy eV_0 to electron which then converts it into kinetic energy $(1/2)mv_0^2$. The electron reaches the cavity with velocity $\left[\text{i.e. } eV_0 = (1/2)mv_0^2\right]$:

$$\therefore \quad v_0 = \sqrt{2eV_0/m} = 0.592 \times 10^6 \sqrt{V_0}\ \text{m/s} \tag{5.27}$$

This electron comes out of the cavity with velocity modulated by the ac signal inside. As in two-cavity klystron, ac signal can be time averaged over the transit time (d_1/v_0) of the electron across the cavity-grid-gap period (see Fig. 5.13). We get the velocity of electron coming out of the grid gap at time t_1 as:

$$v_1(t) = v_0\left[1 + \frac{\beta_i V_1}{2V_0}\sin\left(\omega t_0 + \frac{\theta_g}{2}\right)\right] \tag{5.28}$$

[$\theta_g/2$ being the average (i.e., half) transit angle of the gap].

(ii) ***Electron between cavity and repeller (V_r and V_0 relation and resonant frequency)***

The electron coming out of the cavity at time t_1 has the velocity v_1 (t_1) given by Eq. (5.28), but for all practical purpose we can take

$$v_1 \approx v_0 = \sqrt{2eV_0/m}\ (\text{As } V_1 \ll V_0). \tag{5.28a}$$

The electron comes out at time t_1 and enters cavity-repeller region, where it gets repelled due to −ve field and returns back to the cavity at time t_2. Therefore, for the analysis, **we will use the following three boundary conditions (BC) in this sequence**:

(1) $t = t_1$ at $v = v_1$; (2) $t = t_1$ at $x = 0$; (3) $t = t_2$ at $x = 0$

In the repeller region, the electron accelerates, retards, and turns back due to the field created by $(V_R - V_0)$ voltage. Therefore, its movement can be represented by mass X acceleration = $eE = e(V_R - V_0)/d$. Here, V_R is taken as $|V_R|$.

$$\therefore \quad m\left(\frac{d^2x}{dx^2}\right) = e\left(\frac{V_R - V_0}{d}\right) \tag{5.29a}$$

Integrating gives

$$v(t) = \left(\frac{dx}{dt}\right) = \frac{e}{md}(V_R - V_0)t + c_1 \tag{5.29b}$$

Apply first BC to give

$$v_1 = \left(\frac{e}{md}\right)(V_R - V_0)t_1 + c_1 \tag{5.29c}$$

Put this c_1 of Eq. (5.29c) in Eq. (5.29b) of $\frac{dx}{dt}$:

$$\therefore \quad \frac{dx}{dt} = \left(\frac{e}{md}\right)(V_R - V_0)(t - t_1) + v_1 \quad (5.29d)$$

Integrating it again, we get

$$x = \left(\frac{e}{md}\right)(V_R - V_0)\frac{(t - t_1)^2}{2} + v_1 t + c_2$$

Apply second BC to give $c_2 = -v_1 t_1$, and therefore, the above becomes

$$x = \frac{e}{2md}(V_R - V_0)(t - t_1)^2 + v_1(t - t_1)$$

By applying third BC on this equation, we get the following:

$$\therefore \quad (t_2 - t_1) = \frac{2md\, v_1}{e(V_R - V_0)} \quad (5.29e)$$

For maximum energy transfer by the electron to the cavity, this round-trip time (t_2–t_1) has to be equal to $\left(n + \frac{3}{4}\right) T$, T being the time period of the signal of oscillator, and n the mode of oscillation.

$$\therefore \quad (t_2 - t_1) = \left(n + \frac{3}{4}T\right) = \frac{2md\, V_1}{e(V_R - V_o)} = \frac{2md \cdot \sqrt{2eV_0/m}}{e(V_R - V_0)}$$

$$\therefore \quad \left(n + \frac{3}{4}\right)T = \frac{2m \cdot d}{e(V_R - V_0)} \cdot \sqrt{\frac{2eV_0}{m}} \quad (5.29f)$$

This gives the relation between V_R and V_0 for a given frequency f_0.

As $T = 1/f_0$, therefore, the frequency will be:

$$\boxed{f_0 = \frac{\left(n + \frac{3}{4}\right)(V_R - V_0)}{d} \cdot \sqrt{\frac{e}{8mV_0}}} \quad \boxed{\text{with } \left(n + \tfrac{3}{4}\right) = N} \quad (5.30)$$

When the frequency tuning is done by changing V_R for a given mode, we see that by differentiating above we get electronic and mechanical tuning rate with V_R and d as:

$$\frac{\partial f_0}{\partial V_R} = +\frac{\left(n + \frac{3}{4}\right)}{d}\sqrt{\frac{e}{8m \cdot V_0}} \text{ Hz/V} \quad (5.30a)$$

$$\frac{\partial f_0}{\partial d} = -\frac{\left(n + \frac{3}{4}\right)}{d^2}\sqrt{\frac{e}{8mV_0}} \text{ Hz/m} \quad (5.30b)$$

This gives the central frequency of oscillation of the cavity for mode 'n' ($n = 0, 1, 2, 3, \ldots$) with repeller voltage V_R, accelerator voltage V_0, and 'd' as the gap between repeller and cavity. This gives the equation of the curve given in Fig. 5.15b, and using this, the frequency tuning width (tuning range) can be computed. If the tuning range of V_R or d is given, then we can compute $\frac{\Delta f_0}{\Delta V_R}$ and $\frac{\Delta f_0}{\Delta d}$ as per Eqs. (5.30a) and (5.30b).

(d) ***Frequency Tuning***

- For electronic tuning of frequency by varying the frequency changes around the central frequency of resonance of the cavity (i.e., f_0). Here, $\left(n + \frac{3}{4}\right)$ $(V_R - V_0)$ variation is very small within a mode. This variation of frequency tuning is ±2% only (Fig. 5.15 and Eq. 5.30a).
- For mechanical tuning by varying angle of the knob, which change the cavity size, and hence, this frequency variation is over ±5% (Fig. 5.13 and Eq. 5.30b).

(e) ***Power Output and Efficiency***

Using the charge conservation law, we can compute the fundamental component of the ac current (I_f) induced by bunching pack at the cavity. If $V_0 I_0$ is the input dc power (P_0), then the maximum RF power and efficiency will be (Fig. 5.15b):

$$P_{RF}(\max) = \frac{0.399 I_0 V_0}{n + 3/4} \quad (5.31)$$

$\therefore$ The efficiency will be

$$\eta_n = \frac{P_{RF}}{P_0} = \frac{0.399 \times 100}{n + 3/4} \quad (5.32)$$

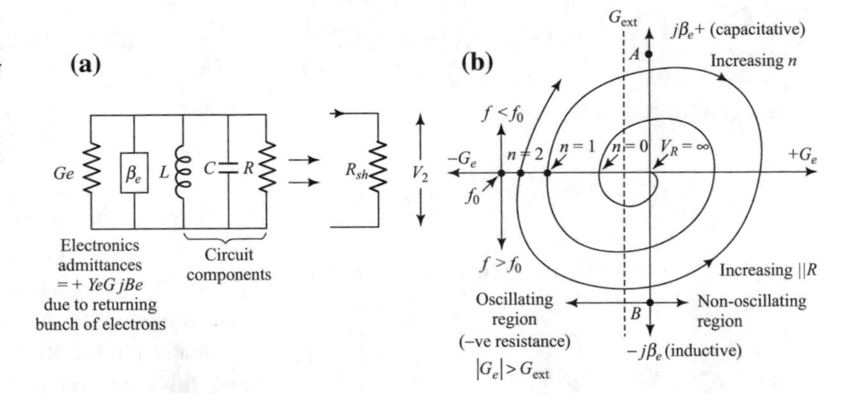

Fig. 5.16 **a** Equivalent circuit of reflex klystron. **b** Electronics admittance of reflex klystron for different modes as a function of V_R

For $n = 0, 1, 2, 3$ …, these modes will give theoretical efficiencies as:

$$\therefore \quad \eta_0 = 53\%,\ \eta_1 = 22.8\%,\ \eta_2 = 14.5\%$$

This is just theoretical power and efficiency. In practice the efficiency is ≈15 to 30% in CW mode.

Practical range of power, frequency, etc.

1. Frequency ……………0.4–200 GHz
2. Power (CW) …………0.1 mW to 3 W
3. Efficiency ……………0.10–30%
4. Tuning range …………..electronic = ±2%, mechanical = ±5%
5. Repeller voltage (V_r) 20–600 V
6. Accelerator (V_0) 250–600 V
7. Beam dc current (I_0) 10–30 mA
8. Spacing between repeller and cavity 0.2–1.0 cm
9. Oscillation modes possible: n = 0–3 (is normal)
10. Size 4″–8″ of different manufacturers.

(e) ***Equivalent Circuit of a Reflex Klystron***

The reflex klystron equivalent circuit as in Fig. 5.16a consists of resonator cavity components L, C, R, and $Y_e = G_e + j\beta_e$, the admittance presented by electron bunches returning to the cavity just before the transit time $T_0 = (n + 3/4)$ T. If the RF beam current lags behind the RF field of cavity, then Y_e is inductive, and if it leads, then Y_e is capacitative. The oscillation condition is when G_e is −ve and greater than external conductance G_{ext}. Fig. 5.16b (i.e., left of AB) gives the admittance plot for $Y_e = G_e + j\beta_e$, in the complex plane. At the origin, $|V_R| = \infty$, and it keeps reducing as we move along the spiral starting from origin. With inductive value of $j\beta_e$ (i.e., +ve $j\beta_e$), frequency of oscillation (f) is above resonant frequency (f_0), while with capacitative $j\beta_e$ (i.e., −ve $j\beta_e$ below x-axis) frequency of oscillation $f < f_0$.

Application of Reflex Klystron: These includes as a:

(i) Pump source in parametric amplifiers.
(ii) Local oscillators in μw-receivers.
(iii) Low-power μw-links
(iv) Source for laboratory experiments.

5.6 Travelling Wave Tube Amplifier (TWTA)

The TWTA was invented in 1944 by Kompfner, when he felt that in two-cavity klystron, full energy of the electron is not getting transferred to the microwave signal in the cavity for amplification, due to interaction in electron beam and RF field being only in the cavity. In TWT the cavity is not there, and continuous interaction between electron beam and RF field is there, by making their velocities some. This is by slowing down RF wave field velocity from c to $cp/\pi D$ (as proved latter in Eq. 5.33a), in the axial direction, by making it to pass through a helical path. The microwave RF signal is pumped through coaxial cable, with its

Fig. 5.17 **a** TWT structure. **b** Helix slow wave structure. **c** Interaction between beam and the increasing amplitude of RF field of the helix by gaining energy from the beam

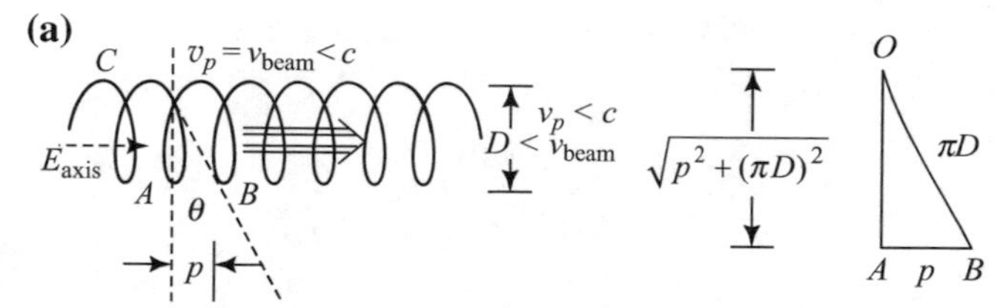

In the same time t_p (1) The beam travels from A to B linear distance p, with velocity V_p

(2) The wave travels from A to B circular + linear distance

$\sqrt{p^2 + (\pi D)^2}$ with velocity C.

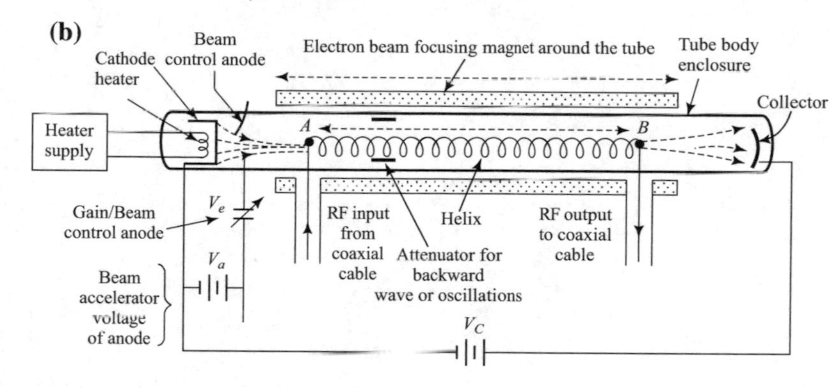

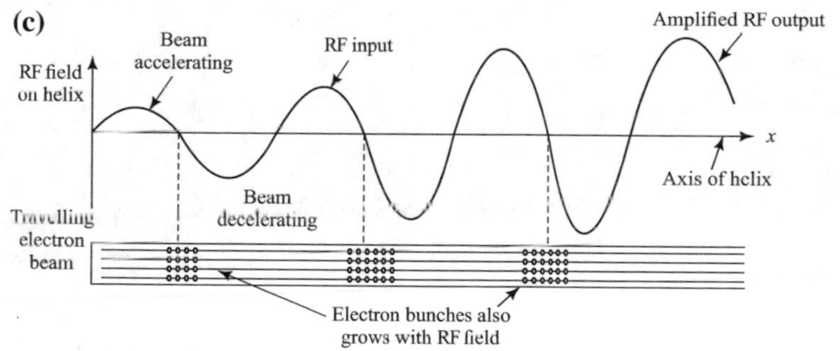

central cable connected to the helical wire through which the current moves with the velocity of light (c) (Fig. 5.17). This current causes an RF field inside the helical wire, and an electron beam is passed through the axis of the helix. This electron beam interacts continuously over the length of the helix (around 12″ or so, instead of just 1″ or so within a resonant cavity as in two-cavity klystron) and transfers energy to the RF field and hence to the beam current of helix. This causes amplification of RF signal when it reaches the other end of output. Thus, we see that **there are some major differences (see Table 5.2) attached between TWTA amplifier and two-cavity klystron amplifier in their operation.**

From Fig. 5.11a, we see that if the electron beam and RF field wave of helix both reach A to B together, then time of movement of both is same ($t_p = t_{beam}$)

$$\therefore \quad t_p = \frac{p}{v_p} = \frac{\sqrt{p^2 + (\pi D)^2}}{c}$$

$$\therefore \quad v_p = \frac{pc}{\sqrt{p^2 + (\pi D)^2}} \approx \frac{pc}{\pi D} \qquad (5.33a)$$

where p = pitch, i.e., distance between two rings of helix, D = its diameter, ϕ = pitch angle, c = velocity of light = 3×10^{10} cm/s, and $\pi D \gg p$. The phase of velocity of the wave v_p is = the velocity of electricity beam v_{beam} that is

Table 5.2 Comparison of TWTA and two-cavity klystron amplifier

Property/function	TWTA	Two-cavity klystron amplifier
1. Resonator	Conventional resonator does not exist and hence has non-resonant *μw* circuit. However, each of the helix can be treated as a cavity	It has two cavities as resonators
2. EM-field and *e*-beam	'*e*'-beam bunches travel in synchronous with field/wave current which is the helix	*e*-beam bunches travel but field wave is stationary in the two cavities
3. Velocities of EM-field current and *e*-beam	'*e*' beam velocity $v_0 = \sqrt{2eV_0/m}$, while velocity of *EM*-field current along the helix = *c* = velocity of high (V_0 = collector voltage)	*e*-beam velocity $v_0 = \sqrt{2eV_0/m}$(V_0 = Anode voltage) as collector places voltage = 0 (grounded)
4. Input and output	The two ends of helix is used for input signal and amplified output signal	The two independent separator cavities are for input and output
5. Energy transfer	Interaction and energy transfer from *e*-beam to *EM*. Wave in helix is continuous over whole of its length of 12″	Energy transfer is only in the two cavities, i.e., cavity-1 to *e*-beam and then *e*-beam to cavity-2 as output
6. Slow wave structures for synchronising the two velocities	The high speed *EM* wave field current in helix (of velocity '*e*') needs to be synchronised. Therefore, its path is increased by having it helical, for keeping its pace with linearly moving *e*-beam	No such requirement here
7. *e*-beam bunching	It takes place all along the length	Also all along the length
8. Tuning and frequency band of a given design or model	Each of the helix can be treated as a cavity. Therefore, it can be used over a very wide frequency band being a non-resonant circuit device	For tuning, the frequency of the two cavities has to change together, which is impossible. Therefore, tuning is ±10% only, that too at the cost of gain
9. Gain	50–60 dB	15–70 dB
10. Efficiency	10–20% (near 3 GHz)	30–40%
11. Power	Very high power up to 5000 kW is possible	Reasonable high power from 10 to 500 kW
12. Life	Very long life up to 50,000 h	Much lower life than TWTA

$$eV_0 = \frac{1}{2}mv_{\text{beam}}^2 \tag{5.33b}$$

∴ by the above two equations, we get

$$D = (pc/\pi)\sqrt{m/2eV_0} \tag{5.33c}$$

(a) **Construction and amplification process in TWT**: Physical construction of a TWT is given in Fig. 5.17a. The electron gun is just like that in klystron, and the electron beam is regulated by a control anode so as to pass through the centre of the long helix. An axial magnetic focusing field prevents the beam from spreading and guides it through the centre of the helix. Finally, the electrons are collected by the concave collector plate.

When the RF signal propagates through the wire of the helix, it produces an RF electric field along the centre of the helix as in Fig. 5.17c. When the velocity of the electron beam is close to the velocity of this axial RF field, then due to the interaction between them, the electron beam delivers energy to the RF wave of the helix.

This leads to more and more amplification of the RF field and the wave of helix, as it keeps moving along its length, with axial velocity of $v_p = pc/(\pi D)$, as given by Eq. (5.33). This phase velocity is helix geometry dependent and therefore a constant. Therefore, a TWT can be used over a wide range of frequencies.

(b) **Velocity modulation and bunching** of electron beam take place along the axis of

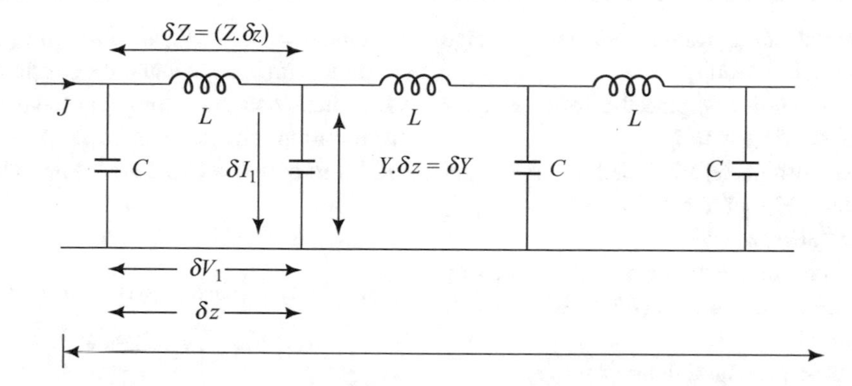

Fig. 5.18 Helix TWT equivalent circuit. Here. δz is the elemental increase in length. While δZ and δV are the addition of impedance and conductance over this δz length. Z is impedance per unit length

helix. Those electrons which move along with the +ve cycle of RF field get accelerated, while those electrons moving along the −ve cycle of RF field get decelerated. Thus, some electrons are moving slower, while the electron behind might be faster, and this leads to some of them catching them up, causing bunch formation. When this bunch encounters retarding field, it delivers energy to the wave resulting into amplification (see Fig. 5.17c).

Here, the RF field causes velocity modulation, which in turn amplifies the RF field/signal, leading to re-generative amplification of each other, as we move along the axis. For better operation, electron beam velocity v_0 is kept slightly greater than the RF field wave velocity v_p, as more electrons face the decelerating field and give energy to the field (i.e., its amplification).

(c) **Mathematical small-signal analysis**: Small-signal analysis is done so as to determine the following:

(i) AC current (I_1) and ac voltage (V_1) relations

(ii) The propagation constants (γ), phase constant (β_e), and attenuation constants (α) of the space charge wave, which exists on the helix along with electron beam and

(iii) The gain of the device, for which a small-signal analysis can be carried out. For this analysis, the following assumption is made:

1. Conductivity of the helix sheath along the wire is infinite and is zero perpendicular to the wire.
2. The travelling wave along with the helix has a longitudinal component of electric field, causing velocity modulation and bunching.
3. The velocity (v_0) of the beam electron is slightly greater than the phase velocity (v_p) of the RF wave, for keeping its phase in advance of RF wave for better transfer of energy.
4. The helix behaves like a lossless transmission line, with series impedance Z per unit length and shunt admittance Y per unit length, with an equivalent circuit as in Fig. 5.18 (i.e., $R = 0$, $G = 0$).

The RF input in the helix induces RF electric field along its axis which causes velocity modulation and bunching of electron beam. This is equivalent to the RF current (I_1) induced in the beam. This RF current in the beam in turn adds to the ac wave voltage (V_1) of helix. This continuous interaction along the length of helix leads to growth of:

(a) Beam current (I_1) and (*b*) RF wave voltage in the helix circuit.

Now, we will compute the following, for understanding it fully.

(i) Beam current (I_1) induced by RF input voltage (V_1) of the helix circuit **(i.e., I_1 V_1 relation)**.

(ii) Helix circuit RF voltage (V_1) induced by the RF beam current (I_1) **(i.e., $V_1 \rightarrow I_1$ relation)**.

(iii) Complex propagation constant (γ).

(iv) Propagation of the three waves (v) gain.

(i) **Beam current (I_1) induced by RF input voltage (V_1) of the helix ($I_1 \rightarrow V_1$ relation)**: This ac space charge current induced (I_1) in the beam is also called convection current and can be computed from the electronic equation on velocity $v(z)$, charge density $q(z)$, beam current $I(z)$, axial field $E(z)$, and ac voltage V_1.

$$v(z) = v_0 + v_1 \cdot e^{(j\omega t - \gamma z)} \tag{5.34}$$

$$q(z) = q_0 + q_1 \cdot e^{(j\omega t - \gamma z)} \tag{5.35}$$

$$J(z) = J_0 + J_1 \cdot e^{(j\omega t - \gamma z)} \tag{5.36}$$

$$E(z) = E_{z1} \cdot e^{(j\omega t - \gamma z)} \tag{5.37}$$

$$V(z) = V_0 + V_1 \cdot e^{(j\omega t - \gamma z)} \tag{5.38}$$

$$\therefore \quad I(z) = I_0 + I_1 \cdot e^{j\omega t - \gamma z} \tag{5.39}$$

Here v_0, γ_0, J_0, V_0, I_0 represent the state dc values and v_1, q_1, J_1, E_{z1}, V_1, I_1 the much smaller time varying peak values of the above six variables.

Here, as dc potential, V_0 is constant $E_0 = \frac{dV_0}{dz} = 0$.

Now, using equation $I = q_1 v_1$ and continuity equation $\Delta J = -\frac{\partial q}{\partial t}$, we will get I_1 in terms of v. Then, using the electron acceleration equation, $md^2x/dt^2 = eE$. We get V_1 in terms of v_1. Then, by eliminating, we get $I_1 \leftrightarrow V_1$ relation.

Here, $\gamma = -\alpha_e + j\beta_e$, the electronic propagation constant with α_e = attenuation constant and $\beta_e = \omega/v_0$ phase constant along the axial wave. The beam current charge density, the beam velocity, and beam cross-sectional area A are related by.

$$q(z) \,.\, v(z) \times A = I(z) \tag{5.40a}$$

$$\therefore \quad I(z) = \left[q_0 v_0 + (q_0 v_1 + q_1 v_0) e^{(j\omega t - \gamma z)} + q_1 V_1 . e^{2(j\omega t - \gamma z)}\right] . A \tag{5.40b}$$

comparing Eqs. (5.39) and (5.40b), we can write

$$I_0 = q_0 v_0 \,.\, A \text{ and } I_1 \cong (q_0 v_1 + q_1 v_0) \cdot A \,.\, e^{j\omega t + \gamma z} \tag{5.40c}$$

As $V_0 \gg V_1$, $t_0 \gg t_1$, we have neglected the last term of Eq. (5.40b).

Comparison of this with Eq. (5.36) gives $J_1 = (q_0 v_1 + q_1 v_0)$.

Putting $J(z)$ and $q(z)$, from Eqs. (5.35) and (5.36) in the continuity equation,

$$\Delta J = (-\partial q/\partial t) \text{ i.e., } (-\partial J/\partial z) = (-\partial q/\partial t)$$

We get $-\gamma J_1 = -j\omega q_1$ where we put $J_1 = I_1/A$

$$\therefore \quad q_1 = -\frac{j\gamma I_1}{\omega A} \tag{5.41}$$

Putting this in Eq. (5.40c) gives

$$I_1 = (q_0 v_1 + v_0 q_1) A = \left(q_0 v_1 - \frac{j\gamma I_1}{\omega A} \cdot v_0\right) A$$

$$\therefore \quad I_1 \left(1 + \frac{j\gamma v_0}{\omega}\right) = q_0 v_1 A \tag{5.42}$$

By putting $\omega = \beta_e \,.\, v_0$

$$\text{we get } I_1 = \frac{\omega q_0 v_1 A}{\omega + j\gamma v_0} = \frac{j\beta_e q_0 A \,.\, v_1}{j\beta_e - \gamma} \tag{5.43a}$$

As mass $\times$ acceleration of electron = force = eE (E is the axial electric field).

$$\therefore \quad m\frac{dv(z)}{dt} = e \,.\, E(z) = e\frac{\partial V(z)}{\partial z} \qquad (5.43b)$$

As $v(z)$ is a function of z and t also, therefore

$$\frac{dv(z)}{dt} = \left(\frac{\partial v_1(z)}{\partial t} + \frac{\partial v_1(z)}{\partial z} \cdot \frac{dz}{dt}\right) = j\omega v_1(z) - \gamma v_1(v_0 + v_1) \qquad (5.43c)$$

Eliminating $dv(z)/dt$ using Eqs. (5.43b) and (5.43c), we get:

$$\frac{e}{m}\frac{\partial V(z)}{\partial z} = j\omega v_1(z) - \gamma v_1(v_0 + v_1)$$

By using V of Eq. (5.38),

$$\eta\gamma V_1 = j\omega v_1(z) - \gamma v_1(v_0 + v_1) \text{ where } \eta = \frac{e}{m}$$

$$\therefore \quad v_1 = \frac{\eta\gamma V_1}{v_0(j\beta_e - \gamma)} \qquad (5.44)$$

As $v_1 \ll v_0$ and $\beta_e = \omega/v_0$. Put this v_1 in Eq. (5.43a) to get the $I_1 \leftrightarrow V_1$ relation:

$$\boldsymbol{I_1} = \frac{\boldsymbol{jI_0 \cdot \beta_e}}{\boldsymbol{2V_0(j\beta_e - \gamma)^2}} \cdot \boldsymbol{V_1}, \qquad (5.45)$$

Here, $I_0 = q_0 v_0 A$, $V_0 = v_0^2/2\eta$(As $eV_0 = \frac{1}{2}mv_0^2$).

This Eq. (5.45) gives the relation between ac beam current (I_1) (convection current) induced by the axial field and the RF voltage (V_1) which has generated the axial field, i.e., $I_1 \leftrightarrow V_1$ relation.

(ii) **Helix circuit RF voltage (V_1) induced by the RF current (I_1) of the beam ($V_1 \leftrightarrow I_1$ relation)**: Now, for getting $V_1 \leftrightarrow I_1$ relation, we assume that the helix is like a lossless transmission line ($R = 0$, $G = \infty$) as per Fig. 5.15, with series z impedance and shunt admittance Y per unit length. From transmission theory, we know:

$$\gamma_0 = j\sqrt{YZ} = \text{propagation constant}$$

$$Z_0 = \sqrt{\frac{Z}{Y}} = \text{characteristic impedance} \qquad (5.46)$$

Multiplying them gives $Z = -j\gamma_0 Z_0$ (5.47)

Dividing them gives $Y = -j\gamma_0 Z_0$ (5.48)

Voltage and current per unit length of Fig. 5.15 will be:

$$\frac{\partial V_1}{\partial z} = -jzI_1 \qquad (5.49)$$

$$\frac{\partial I_1}{\partial z} = J - jV_1 Y \qquad (5.50)$$

Here, J is the impressed current per unit length, due to coupling between electron beam and the circuit:

$$J = \frac{\partial I_1}{\partial z}.$$

Here $I_1 = A\,.\,(q_0 v_1 + v_0 q_1)$ of Eq. (5.40c) (5.51)

As all the ac quantities vary with times and distance by $e^{j(wt-\gamma z)}$, then above Eqs. (5.49) and (5.50) become

$$-\gamma I_1 = -jYV_1 + \gamma I_1 \qquad (5.52)$$

$$\text{And} \quad -\gamma V_1 = -jZI_1 \qquad (5.53)$$

Eliminating I_1, we get

$$-\gamma\left(\frac{-\gamma V_1}{-jZ}\right) = -jYV_1 + \gamma I_1$$

Using Z, Y of Eqs. (5.47) and (5.48), we will get $V_1 \leftrightarrow I_1$ relation

$$\boldsymbol{V_1} = \left[\frac{-\boldsymbol{\gamma} \cdot \boldsymbol{\gamma_0 z_0}}{\boldsymbol{\gamma^2 - \gamma_0^2}}\right] \cdot \boldsymbol{I_1} \qquad (5.54)$$

This gives the helix circuit equation relating RF circuit voltage (V_1) generated by RF beam current (I_1), i.e., $V_1 \leftrightarrow I_1$ relation.

(iii) **The three waves, their complex propagation constant (v), phase constant (β_e) and the gain**:

Equations (5.45) and (5.54) are complimentary to each other; therefore, eliminating V_1 and I_1 by substituting V_1 from Eqs. (5.54) to (5.45), we get:

$$\frac{j\,.\,I_0 Z_0 \gamma_o \gamma^2 \beta_e}{2V_0 (j\beta_e - \gamma)^2 (\gamma_o - \gamma^2)} = 1 \quad (5.55)$$

$$\therefore \quad (\gamma_0 - \gamma^2)(j\beta_e - \gamma)^2 = \frac{j\gamma^2 \gamma_o \beta_e Z_0 I_0}{2V_0} \quad (5.55a)$$

This is a fourth-order equation in γ and therefore has four roots of γ. We also note that $\gamma_0 = j\sqrt{ZY}$ is propagation constant of the helix in the absence of the beam current. While $\gamma = -{}_e + j\beta_e$ is the complex propagation constant in the presence of the beam current, therefore, Eq. (5.55a) has four roots of γ as well as of α_e. **Corresponding to these four roots, four waves also are there in the helix.** For getting an approximate solution of Eq. (5.55a), let us put $\gamma = -\alpha + j\beta$.

$$\frac{-Z_0 I_0 \beta_e^2 (-\beta_e^2 - 2j\beta_e \alpha + \alpha^2)}{2V_0 (2j\beta_e \alpha - \alpha^2)\alpha^2} = 1 \quad (5.56)$$

As we know that $\alpha \ll \beta$, we can neglect α^2 and $2j\beta_e\alpha$ in the numerator while α^4 in the denominator to get:

$$\alpha^3 = \frac{-jZ_0 I_0 \beta_e^3}{4V_0} = -jC^3 \beta_e^3$$
$$\alpha = (-j)^{1/3} C\,.\,\beta_e \quad (5.57)$$

where C is a constant $\left(C = \frac{Z_0 I_0}{4V_0}\right)$ and depends mainly on the dc beam voltage (V_0) and dc beam current (I_0). Eq. (5.57) has three roots of α as:

$$\alpha_1 = \left(\frac{\sqrt{3}}{2} - \frac{j}{2}\right) C\,.\,\beta_e$$
$$\alpha_2 = \left(-\frac{\sqrt{3}}{2} - \frac{j}{2}\right) C\,.\,\beta_e \quad (5.58)$$
$$\alpha_3 = j\,.\,C\,.\,\beta e$$

Corresponding to these three equations, three propagation constants (α_n) will have modified phase constants and attenuation constants as given below:

$$\gamma_1 = -\beta_e \frac{C\sqrt{3}}{2} + j\beta_e \left(1 + \frac{C}{2}\right) = \alpha' + j\beta''(\text{Let}) \quad (5.59a)$$

$$\gamma_2 = \beta_e \frac{C\sqrt{3}}{2} + j\beta_e \left(1 + \frac{C}{2}\right) \alpha'' + j\beta''(\text{Let}) \quad (5.59b)$$

$$\gamma_3 = j\beta_e (j - C) = \alpha''' + j\beta''(\text{Let}) \quad (5.59c)$$

The first two waves have phase constant as

$$\beta'' = \beta' = \beta_e \left(1 + \frac{C}{2}\right) = \frac{\omega}{v_0}\left(1 + \frac{C}{2}\right) = \frac{\omega}{v_{p12}}$$

Therefore, their phase velocity is $v_{p12} = \frac{v_0}{1 + \frac{C}{2}} < v_0$; i.e., they travel slower than the RF wave. The third wave is unattenuated and travels faster than the first two waves, and its phase velocity v_{p3} can be known from equation:

$$\beta'' = \beta_e (1 - C)$$
$$= \frac{\omega}{v_0/(1 - c)}$$
$$= \frac{\omega}{v_{p3}} \text{ and therefore } v_{p3} > v_0$$

(iv) **Propagation of the three waves**: Thus, the amplitudes of the three waves vary with time and space as:

First: $e^{\frac{\sqrt{3}}{2}\beta_e cz}\,.\,e^{j\left[\omega t - \beta_e \left(1 + \frac{C}{2}\right)z\right]}$ (slow-moving growing wave gains energy from electron beam)
Second: $e^{-\frac{\sqrt{3}}{2}\beta_e cz}\,.\,e^{j\left[\omega t - \beta_e \left(1 + \frac{C}{2}\right)z\right]}$ (slow-moving attenuating wave transfers energy to electron beam)
Third: $e^{j[\omega t - \beta_e (1 - C)]}$ (faster wave unattenuated wave)

Thus, we infer that this third wave γ_3 does not transfer any energy to the RF wave and travel unattenuated. The first wave γ_1 has −ve attenuation [Eq. (5.59a)] and therefore keeps growing exponentially as it travels along z by acquiring energy from the electron beam. The second wave γ_2 is a decaying wave (has +ve attenuations), and energy flows from this wave to the electron beam. The approximate solution of Eq. (5.55) has lead to three roots only. The exact solution obtained by numerical method shows that the **fourth wave is a backward wave with $\gamma_4 = -j\beta_e\,(1 - C^3/4)$; i.e., it travels backward with a phase velocity more than v_0. It is attenuated either by an attenuator at the central distance of the helix or by perfect matched structure.**

(v) **Suppression of the backward wave (r_4)/ oscillations by an attenuator**: Even a small mismatch in impedance can cause reflection of waves in the helix, leading to +ve feedback to the input end of the amplified signal and hence oscillation. This undesired backward wave oscillation is stopped by having an attenuator for breaking the feedback path (Fig. 5.17).

This attenuator is just a lossy/conducting material (e.g., wire graphite/aquadag paint) inside the glass wall, placed closer to the input end of the helix. This attenuator absorbs the forward-growing wave also to some extent, and therefore, overall gain reduces a bit.

(vi) **Gain due to first wave only**: As seen above that, sum of the three forward RF signals should have the same amplitude at the input end, being equally split up of the main RF signal input $V(0)$:

$$\therefore \quad V(0) = (V_1 + V_2 + V_3)$$

$$\text{i.e., } V_1 = V_2 = V_3 = \frac{V(0)}{3}$$

Therefore, the starting amplitude of the growing first wave will be

$$V(z) = \frac{V(0)}{3} \cdot e^{\left(\frac{\sqrt{3}}{2}\beta_e \cdot C\right)z} \tag{5.60}$$

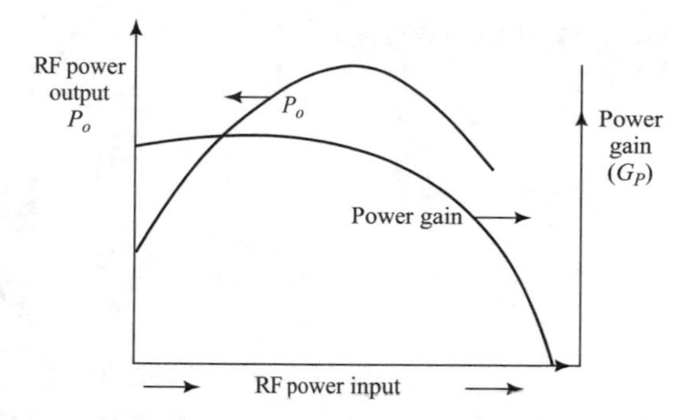

Fig. 5.19 RF power output and gain as a function of input power in TWT amplifier

At the end of the helix, spiral of length $z = L$

$$\therefore \quad V(L) = \frac{V(0)}{3} \cdot e^{\left(\frac{\sqrt{3}}{2}\beta_e C\right)L}$$

$$\begin{aligned}
\therefore \quad \text{Power gain in dB} &= 10\log_{10}\left|\frac{V(L)}{V(0)}\right|^2 \\
&= 10\,\log_{10}\left[\frac{e^{\frac{\sqrt{3}}{2}\beta_e CL}}{3}\right]^2 \\
&= 10\left[\frac{\sqrt{3}}{2}\beta_e CL \,.\, \log_{10}(e) - \log(9)\right] \\
&= [(47.3CL/\lambda_e) - 9.54]\text{dB} \;(\text{As } \beta_e = 2\pi/\lambda_e)
\end{aligned}$$

where constant $C = \frac{I_0 Z_0}{4V_0}$ and the length (L) controls the power gain with the initial loss of gain of 9.54 due to splitting of signal into three waves. Here, λ_c = effective beam wavelength inside helix. The gain and RF power, as a function of input power is depicted in Fig. 5.19.

The gain is maximum when beam velocity v_0 is in synchronism with the axial wave phase velocity (v_p). The lower and higher frequency limits of a given TWT are due to its geometry, i.e., length and diameter of helix. For increasing the frequency limits, the diameter has to be smaller ($f.D = c/\pi$).

(vii) **Number of helicals**: The number of helicals in the helix can be computed by $N_h = 1/\lambda_e = l.f/v_0$, where l = length of helix and v_0 = beam velocity $\sqrt{2eV_0/m} = 0.593 \times 10^6\sqrt{V_0}$ m/s.

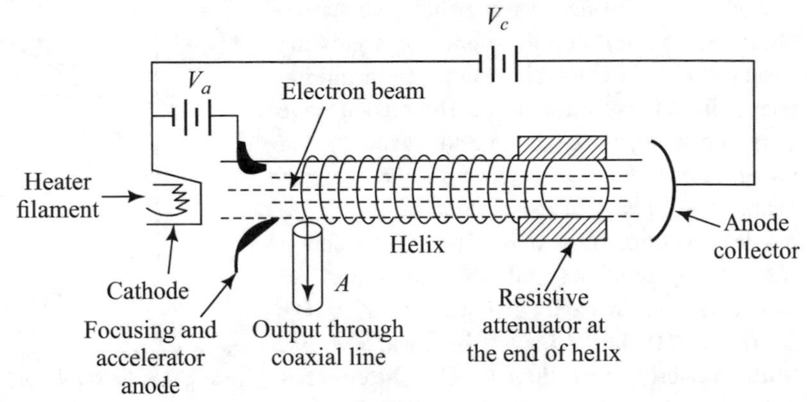

Fig. 5.20 Helix (TWT)-type BWO

(viii) **Frequency of operation**: From above, we see that $f = (N_h/l) \cdot \sqrt{2eV_0/m}$

$$\therefore \quad f = (1/p)\sqrt{2eV_0/m} \qquad (5.60a)$$

Thus, beam voltage and pitch decide the frequency of operation.

ix **Diameter of helix**: By Eq. (5.33c), we get

$$D = (pc/\pi) \cdot \sqrt{m/2eV_0} \qquad (5.60b)$$

$$\therefore \quad f \,.\, D = c/\pi = \text{constant} \qquad (5.60c)$$

Performance Characteristic Range of TWT

1. Collector beam voltage: 1–10 kV
2. Beam current: 10–100 mA
3. Frequency: 5–100 GHz
4. Cut-off power output: 5 MW (10–40 GHz) to 300 kW (3 GHz)
5. Efficiency: 5–20%
6. Band width: Being a non-resonant device, large band width ±30% can be there; e.g., a typical TWTA can give 35 ± 3 dB gain from 2 to 4 GHz.
7. Helix length ≈ 12″: helix diameter = 0.2–0.5 mm
8. Life: 50,000 h, much larger than other tubes
9. Noise: 5 dB (low-power and lower-frequency TWT), 15 dB (high-power and high-frequency TWT)

Application of TWTA

(i) In medium- and high-power satellite transponder, because of its very long life.
(ii) In wideband communication links.
(iii) In CW-RADAR and RADAR jamming on land, aeroplane (air), and ship (water).
(iv) At very higher power and wide tunable [by beam voltage] device.

5.7 Backward Wave Oscillator (BWO)

It is an extension of TWT amplifier with a built-in +ve feedback inside, caused by reflection of wave from the right end of the helix, terminated by a resistive attenuator coating. As usual in an oscillator, out of the noise signal, (which has all frequencies 0 to ∞), an appropriate frequency (suitable to the diameter and length of helix which decides the central frequency of the BWO) will keep on getting amplified with +ve feedback, when the beam reaches back to the point of starting current (Fig. 5.20). Thus, the oscillation generates the signal of that frequency. These BWOs have two types of structures.

(i) Helix TWT-type BWO
(ii) Zigzag line-type BWO.

In both the structures, the forward electron beam interacts with the backward-moving wave

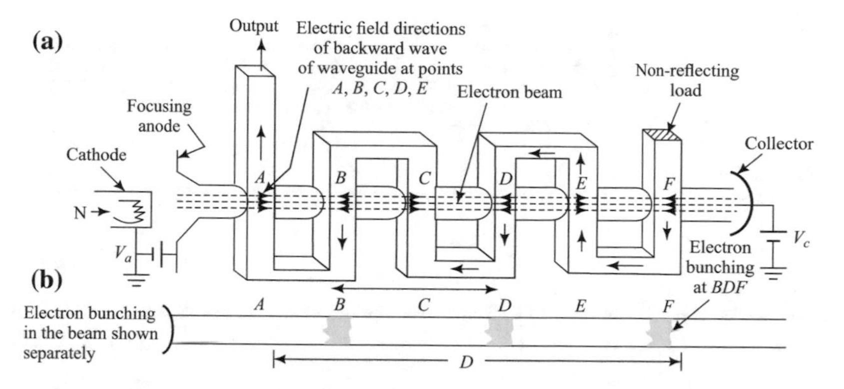

Fig. 5.21 The zigzag line BWO. **a** Structure. **b** Bunching process. Electric field directions of backward wave in the waveguide at the electron beam crossing points A, B, C, D, E, F are shown of a particular moment

through either the spiral or zigzag waveguide line. This leads to the absorption of energy by the wave from the electron beam and hence keeps growing. Both the BWOs give continuous wave (CW) output with very wide range of frequency and tunability up to 40 GHz.

1. **Helix TWT-type BWO**: Structure is just like helical slow wave structure of TWT amplifier and operates on the same principle of interaction between

   (a) RF wave of helix moving backward in BWO unlike TWT and
   (b) the electron beam moving forward inside the helix (Fig. 5.21).

The electron beam is focused by an axial magnetic field and the focusing electrodes. That frequency signal of the noise/transient in the helix, which reaches back with a +ve feedback phase (i.e., $2\pi n$) at the output end A, keeps on getting amplified after acquiring energy from electron bunching formation by the RF wave as in TWT. This leads to oscillation of that frequency signal and hence becomes a source of microwave power. The growing RF wave in the helix has a group velocity in the backward direction, while the beam velocity of electrons in the forward direction.

The best feature of the BWO is that the frequency and amplitude of oscillation can be varied:

(i) Frequency by changing beam accelerating voltage V_a, between gun and focusing anode.
(ii) Amplitude by changing the electron beam current, i.e., by voltage between electron gun and collection anode (V_c).

BWO is also called M-carcinotron. It can be of circular form also, where the electron beam is rotated in a circular path by a magnetic field perpendicular to the beam, for reducing the length/size of the device. The efficiency of these circular BWOs is much higher than the linear BWO.

2. **Zigzag backward line-type BWO**: In this type of BWO, the helix is replaced by a zigzag waveguide, in which the transient reflected wave gets amplified as it moves backward by acquiring energy from the electron beam moving in the forward direction as in Fig. 5.21. The beam moves forward through the waveguide hole, where the backward wave crosses it at 90° in the waveguide and electron beam junction, thereby wave of waveguide acquires energy from the beam.

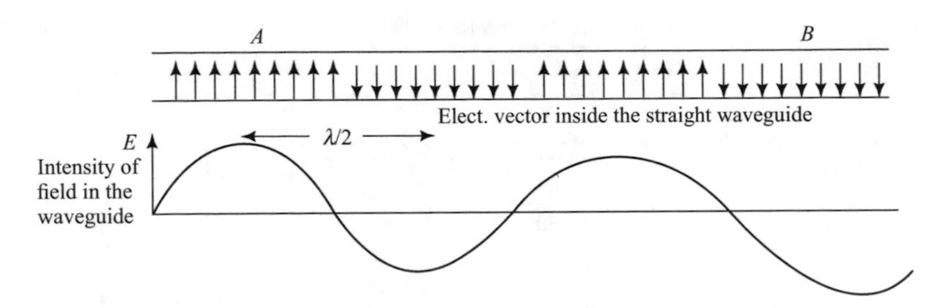

Fig. 5.22 Electric vector of a wave moving in a straight waveguide, where the points *A* and *B* could correspond to that in Fig. 5.21

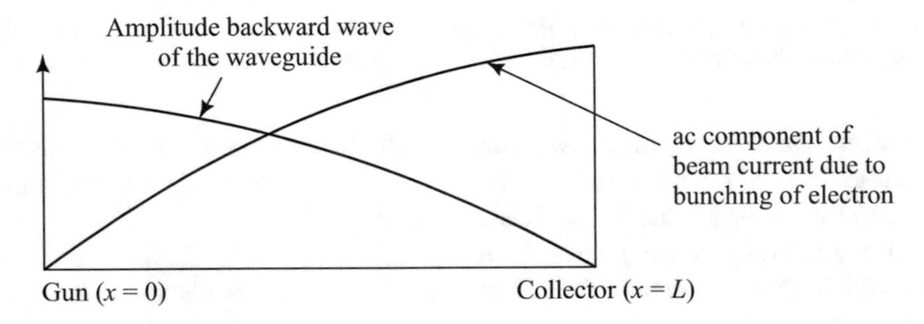

Fig. 5.23 In zigzag growth of amplitudes of wave in the waveguide from collector to gun and ac, component of beam current from gun to collector

In this folded waveguide-type BWO, we make use of the electric vector in the waveguide which is directed from upper wall to lower wall and changes its direction after every λ/2 distance, as it moves with the velocity of light in a backward direction (Fig. 5.21a) and crosses the electron beam at junctions F, E, D, C, B, C, A of the BWO. Inside the zigzag waveguide, the field intensity is as given in Fig. 5.22. At the start of the waveguide near *F*, non-reflection load is put. **Due to electron bunching in the beam current, ac component gets incorporated into it.**

The velocity of electron is such that at A, C, E, it encounters accelerating direction electric field, while at B, D, F, it receives retarding field of the waveguide. Therefore, the electron bunching will take place near B, D, F joints (Fig. 5.21b). As the electron gets retarded, they loose energy and the bunch delivers energy to the wave of the waveguide. The growth of RF wave as it travels inside the waveguide in the reverse direction $x = L - 0$ is given in Fig. 5.23, along with fall of A C component of beam current, by transferring its energy to the RF wave.

3. **Power output frequency of BWO**

(i) Frequency range	1–100 GHz
(ii) CW power output (much lower than TWT)	(1–2 GHz) 200 W; Near 3 GHz: 100 W; 100 GHz: 20 W; 200 GHz: 1 W
(iii) Efficiency	10% for linear BWO, over 30% in circular BWO
(iv) Noise figure	10 dB or so
(v) Tunability	up to 40 GHz by changing collector voltage

4. **Application of BWO**: Because of the beam voltage, the beam current controls the beam velocity as well as the frequency of oscillation, and therefore, wideband tuning is possible. Therefore, it is used as:

   (i) Sweep generator in instruments.
   (ii) Broadband noise source for enemy's RADAR for jamming it.
   (iii) Voltage-tunable band pass amplifier by controlling beam current below threshold oscillation.

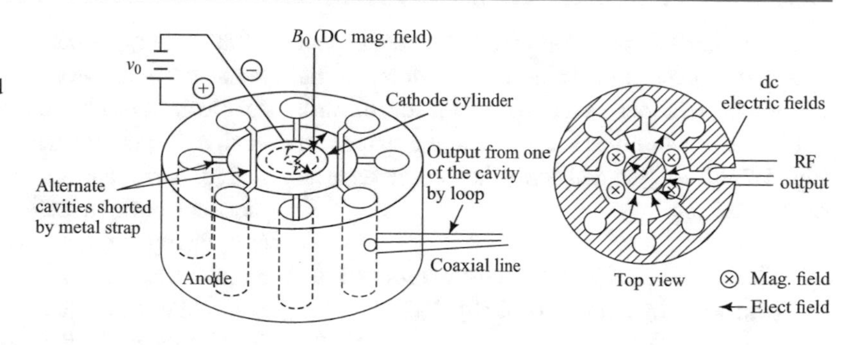

Fig. 5.24 Eight cavity, magnetion with the dc magnetic, and electron field

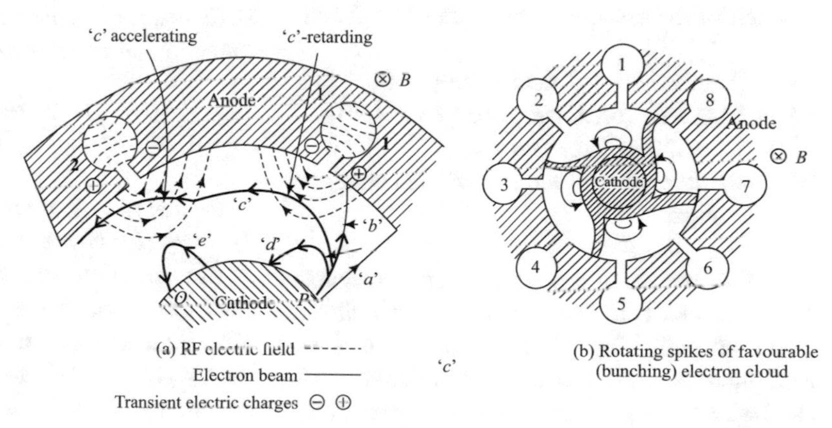

Fig. 5.25 **a** RF field on cavity and moving electron path, **b** rotating spikes of favourable electron cloud bunch with RF (dc magnetic field is into the paper.)

(iv) Amplitude modulation, as by changing the beam current; oscillation can be reduced to zero.

(v) A noiseless oscillator of medium-power source in a number of other devices.

5.8 Magnetron Oscillator

All the tubes discussed so far are called o-type (e.g., klystron, TWT, BWO), as the magnetic field is normally used for focusing the electron beam and is in the same direction as the electric field, except in circular BWO, where the magnetic field is perpendicular to the beam and is used just for rotating the beam for reducing the linear size of BWO. In magnetron also, the magnetic field is perpendicular to the electric field and is therefore called cross-field or M-type. It was invented by Hull in 1921 and improvised by Randall and Boot in 1939. For low-power (i.e., mW) requirement such as laboratory/experimental purpose, klystron, etc., are OK, but for higher requirements, e.g., >100 W power, magnetron is used.

In klystron, the electron beam carrying energy interacts with RF field for a short duration in the cavity grid only, and hence, the efficiency is around 10–20% only. For higher efficiency and power, the electron beam can be made to interact with the RF field for larger duration and distance as in TWT, in a linear path. The same is done in magnetron in a circular path. The difference is that the linear slow wave structure is in TWT, while in magnetron we have multiple cavities in a circular path, reducing the size as well.

The magnetron normally consists of circular anode with eight cavities, with a coaxial cylindrical cathode (Fig. 5.24). The electrons coming out of cathode have to move in the circular path between anode and cathode; therefore, a dc magnetic field perpendicular to the dc electric field has to be there, i.e., into the paper in Fig. 5.25. Now, we will study the various aspects of its working.

(a) **Electron beam path with dc magnetic field (B_0) and dc electric field (by V_0)**

Each of the electron experiences four forces:

(i) Anode dc radial field force ($-eE$),
(ii) Magnetic rotational force [$-e(v \times B)$],
(iii) Centrifugal force (mv^2/r),
(iv) RF field around the slot of the cylindrical cavities.

In the absence of any RF field or dc magnetic field, the electron moves undisturbed radically along '*a*' path (Fig. 5.25). As the dc magnetic field increases, the electron tries to rotate anti-clockwise to path '*b*', than to path '*c*' to path '*d*'. The path '*d*' is when the magnetic field is very high and electron comes back to the cathode and heats the cathode.

(b) **Sustained oscillation and favourable and unfavourable electrons**
In the presence of the above first three forces, if the RF field and the rotating electron have the same frequency, then electron profile mechanism forces to form bunch and moves in spikes (as in Fig. 5.25b). This way it delivers maximum energy to the RF field.

The RF oscillation starts from some noise transient, when the magnetic field strength is such that electron follows the path '*c*'. **The signal with frequency suitable or close to the cyclotron frequency** acquires more and more energy from '*c*' electron and gets sustained. We know that electron accelerates in a −ve electric field and retards in +ve electric field. The electron '*c*' when it goes into the electric field region of cavity '1', then its retards, i.e., transfer its kinetic energy (1/2) mv^2 to the RF field and hence drifts towards the anode. By the time, this '*c*' electron reaches cavity 2, the RF field of this cavity '2' gets reversed (opposite to that shown in Fig. 5.23), and the electron again retards further and again transfers energy to the RF field and finally merges with anode. Therefore, the angular velocity of electron (ω_c) has to be such that while it travels from cavity '1' to cavity '2' [i.e., angle = ($2\pi/N$)), time elapsed has to be ($T/2$) (T being the time period of RF wave of frequency f and angular frequency ω.

$$\therefore \quad \frac{2\pi/N}{\omega_c} = \frac{T}{2} = \frac{1}{2f}$$

Cyclotron frequency = ω_c = ($4\pi f/N$) = ($\pi f/2$) = $\omega/4$ [for N = 8 cavities]

The number of 'c' types of favourable electrons is much more in number than those of 'd' and 'e' type unfavourable electrons; i.e., energy given by '*c*' electrons to RF is more than the energy taken by '*d*' and '*e*' electrons from RF, and therefore, sustained oscillation takes place ($N_c > N_d, N_e$). **These favourable electrons are like a car getting green signal at every crossing (cavity gaps).**

This was for electrons coming out from the region *P* at that moment, when the RF field configuration is as in Fig. 5.25. Another electron *e* emitted at the same moment from region of cathode (opposite to cavity 2) finds an RF electric field in opposite direction and therefore accelerates gaining more energy. As a result, this *e*-electron experiences higher magnetic force [$-e(v \times B)$] to turn back to the cathode. Such electrons constitute 5% of anode power and heat the cathode.

Thus, the electrons following the **'c' path (i.e., favourable electron) form a bunch and therefore form spokes (electron cloud path)** as in Fig. 5.25b, which rotates with angular velocity corresponding to the two poles per cycle. The blank space between the spikes is those electrons which turn back to cathode [like *d* and *e* of Fig. 5.25a].

(c) **Frequency of oscillation**
For the favourable electron (of path ‘c’), centrifugal force and magnetic force have to be equal for equilibrium condition:

$$mv^2/r = ev \,.\, B_0 \quad (5.61)$$

where v is the linear velocity, r the radius of cycloidal path, and B_0 the dc magnetic cross-field.

$$\therefore \text{ Angular velocity} \quad \omega = v/r = eB_0/m$$
$$\text{Period of revolution} \quad T = 2\pi/\omega = 2\pi m/(eB_0) \quad (5.62)$$

For maximum power transfer to RF signal, the frequency sustained oscillations will be when $f = (1/T)$, i.e.,

$$f = 1/T = eB_0/(2\pi m) \quad (5.63)$$

(d) **The π mode**
The RF electric field between the two sidewalls of the entry points of a cavity is as in Fig. 5.25 at a particular moment. This field configuration will change with the frequency of the RF The field shown inside the cavity is of TE_{110} mode in the cylindrical waveguide cavity resonator. In this mode, the phase difference (ϕ) between two adjacent cavities is π and is called π-mode. In higher modes, this difference ϕ has to be $<\pi$, but with the condition that $8\phi = 2n\pi$, i.e., total phase shift in the complete circle has to be multiple of 2π.

(e) **Mode jumping and strapping**
These resonant mode frequencies are very close to each other; there is always a possibility of mode jumping, and purity of signal may not be there all the time. For avoiding this, alternate anode cavities are connected by a metal straps (Fig. 5.24), so that they are in phase. This way the phase of the eight cavities at any moment of time $t = t_0$ will be (ϕ, $\phi + \pi$, $\phi + 2\pi$, $\phi + 3\pi$, …, $\phi + 7\pi$), taking the first cavity RF signal at phase $\phi(t_0)$. **This method of stopping/preventing mode jumping is called strapping, and the magnetron is said to operate in π mode.** Also, we have seen that the condition for maximum power transfer from electron beam to the RF field takes place when the cyclotron (rotating electron) frequency is equal to angular velocity of RF wave (Fig. 5.25), i.e., $f = eB_0/2\pi m$.

(f) **Frequency pushing and pulling**
Just like reflex klystron, resonant frequency can be increased (pushed) by increasing the anode dc voltage and the dc magnetic field. This will change the angular velocity of electron and hence the rate of energy transfer to anode cavity resonator. This causes change of oscillation frequency $f = eB_0/2\pi m$. **The frequency pulling mechanism is the reduction of frequency caused by the change of load, as it causes reflection of signal from load into cavity resonator.** Frequency pushing can be prevented by using stabilised power supply, while the frequency pulling can be stopped by putting an isolator at the output and so that reflected signal does not reach the cavity.

(g) **Mathematical analysis–angular velocity and Hull cut-off magnetic field**

(i) *Computing angular velocity of electron (ω), cathode, and anode*: Let the radius of cathode and anode be a and b. The dc anode field attracts the electron with force eV radically outside, while the cross dc magnetic field B_0 exerts a force perpendicular Fig. 5.26a to electric field and magnetic fields. As a result, the path of electron bends towards the circle and becomes parabolic, as is clear from

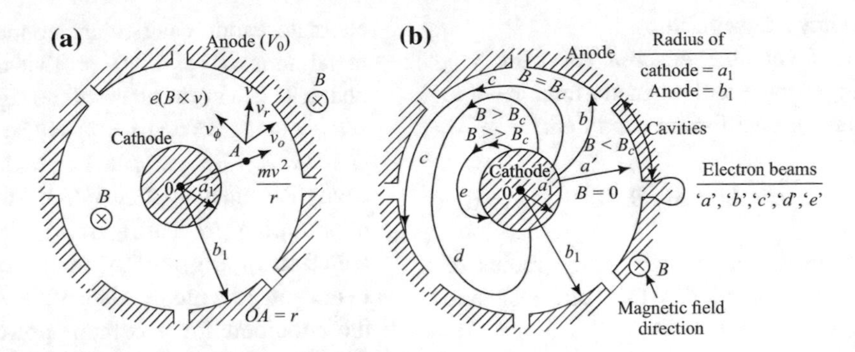

Fig. 5.26 **a** Three forces on the electron due to dc electric, dc magnetic fields, and centrifugal force, **b** electron path for different strength of dc magnetic field above and below critical value *Bc*

Fig. 5.26b. The force exerted by the magnetic field when the electron has velocity v and is at a radial centre of cathode is $F = e(B \times V)$. The torque on the electron due to the magnetic field will be at an angle ϕ from original velocity.

$$T_\phi = r\,.\,F = r\,.\,e\,.\,(B \times v)$$

$$T_\phi = r\,.\,e\,.\,B\,.\,v_r \tag{5.73}$$

where v_r is the radical component of the velocity v and v_ϕ the velocity component perpendicular to v_r. Therefore, $v^2 = v_\phi^2 + v_r^2$ (Fig. 5.26) and

$$v_r = \mathrm{d}r/\mathrm{d}t$$

Also, the angular momentum P is given by the product of angular velocity X moment of inertia.

$$P = (\mathrm{d}\phi/\mathrm{d}t)\,.\,mr^2 \tag{5.74}$$

The torque is the rate of change of angular momentum; therefore, Eqs. (5.74) and (5.73) give

$$\begin{aligned}\frac{\mathrm{d}P}{\mathrm{d}t} &= \frac{\mathrm{d}}{\mathrm{d}t}\left(\frac{\mathrm{d}\phi}{\mathrm{d}t}\,.\,mr^2\right) = rev_rB \\ &= r\,.\,e\,.\,\frac{\mathrm{d}r}{\mathrm{d}t}\cdot B \\ &= eB\,.\,r\,.\,\frac{\mathrm{d}r}{\mathrm{d}t}\end{aligned} \tag{5.75}$$

Integrating w.r.t. time t, we get:

$$\int \frac{\mathrm{d}}{\mathrm{d}t}\left(\frac{\mathrm{d}\phi}{\mathrm{d}t}\,.\,mr^2\right)\mathrm{d}t = \int eBr\frac{\mathrm{d}r}{\mathrm{d}t}\,.\,\mathrm{d}t$$

$$(\mathrm{d}\phi/\mathrm{d}t)mr^2 = e\,.\,B\,.\,r^2/2 + K \tag{5.76}$$

For finding the value of *K*, we apply the first boundary condition that *r* = *a*, i.e., the surface of the cathode, from where the electron has just emerged out with radial velocity $\mathrm{d}r/\mathrm{d}t = V_r$ and the angular velocity $V_\phi = (\mathrm{d}\phi/\mathrm{d}t) = 0$.

Therefore, Eq. (5.76) gives:

$$0 = eB\,.\,a^2/2 + K$$

Putting this value of K in Eq. (5.76), we get:

$$\omega = \frac{\mathrm{d}\phi}{\mathrm{d}t} = \frac{eB}{2m}\left(1 - \frac{a^2}{r^2}\right) \tag{5.77}$$

At the cathode $(\mathrm{d}\phi/\mathrm{d}t) = 0$ and near anode $a \ll r$, therefore:

$$\omega_{\max} = \left(\frac{\mathrm{d}\phi}{\mathrm{d}t}\right)_{\max} = \frac{eB_e}{2m} \tag{5.78}$$

$$\mathrm{d}\phi/\mathrm{d}t = \omega = \omega_{\max}\left(1 - a^2/r^2\right) \tag{5.79}$$

(ii) *Computing cut-off magnetic field (B_c) and voltage (Vc)*

Now, using energy conservation of kinetic energy $(1/2)mv^2$ acquired by electron is fully from the potential energy (eV_0), and therefore:

$$eV_0 = \frac{1}{2}mv^2 = \frac{m}{2}\left(v_r^2 + v_\phi^2\right) \quad (5.80)$$

We also know that $v_r = \frac{\mathrm{d}r}{\mathrm{d}t}$ and $v_\phi = r.\frac{\mathrm{d}\phi}{\mathrm{d}t}$.
$\therefore$ Above Eq. (5.80) becomes:

$$eV_0 = \frac{m}{2}\left[\left(\frac{\mathrm{d}r}{\mathrm{d}t}\right)^2 + r^2.\frac{\mathrm{d}^2\phi}{\mathrm{d}t^2}\right] \quad (5.81)$$

Putting $(\mathrm{d}\phi/\mathrm{d}t)$ from Eq. (5.79), we get

$$eV_0 = \frac{m}{2}\left[\left(\frac{\mathrm{d}r}{\mathrm{d}t}\right)^2 + r^2\omega_{\max}^2\left(1 - \frac{a^2}{r^2}\right)^2\right] \quad (5.82)$$

The second boundary condition is at the anode where the favourable electrons are grazing (i.e., they are parallel to its inner circular anode surface where $r = b$; Fig. 5.26); therefore at $r = b$, $v = (\mathrm{d}r/\mathrm{d}t) = 0$, Eq. (6.55) becomes

$$\therefore \quad \frac{m}{2}\cdot b^2\omega_{\max}^2\left(1 - \frac{a^2}{r^2}\right)^2 = eV_0$$

$$\text{where} \quad \omega_{\max}^2 = eB_c/(2m) \quad (5.83)$$

B_c = cut-off magnetic field for maximum frequency $\omega_{\max}$.

$$\therefore \quad \frac{m}{2}\cdot b^2\cdot\left((eB_c/2m)^2\cdot\left(1 - \frac{a^2}{b^2}\right)\right) = eV_0 \quad (5.84)$$

$$\therefore \quad B_c = \sqrt{8V_0m/e}/\left[b\cdot\left(1 - a^2/b^2\right)\right] \quad (5.85)$$

$$\textbf{For}\ \boldsymbol{a \ll b} \quad \boldsymbol{B_c = \left(\sqrt{8V_0m/e}\right)/b} \quad (5.86)$$

The above is called Hull cut-off magnetic field equation. At $B = B_C$, the electron just grazes the anode (Fig. 5.26b) and merges with it thereafter for a given V_0.

If $B > B_c$, the electron will not reach anode but turn back, i.e., anode current = 0.

If we increase V_0 in Eq. (5.25), keeping the B_c constant, then anode current increases and B_c also has to be increased for electrons to remain favourable, i.e., graze the anode. As a result, a new cut-off magnetic field is seen. Therefore, for a given cut-off field B_c, the cut-off V_0 will be from Eq. (5.85) as:

$$\boldsymbol{V_{OC} = \left(eB_c^2b\right)\left(1 - a^2/b^2\right)/(8\mathbf{m})} \quad (5.87)$$

This is called Hull cut-off voltage equation.

(h) **Typical characteristic of magnetron oscillator**

1. Frequency range: 0.5–75 GHz
2. Power output:

Frequency (GHz)	CW. Power	Pulsed power (duly cycle 0.1%) (MW)
1	10 kW	10
10	100 W	1

3. Efficiency: 40–70%
4. Tunability: As seen in the mathematical analysis that $\omega_{\max}$ given by Eq. (5.83), can be increased by increasing B_c and V_{c0} together. The range of tunability can be made as wide as 0.3–12 GHz, by tuning ring capacitors above the π mode straps, which can be move up and down for changing cavity resonant frequency. Therefore, magnetron is the most versatile high-power source in microwaves.
5. Range of anode voltage V_0: 10–100 kV
6. Anode current (I_0): 10–100 A
7. Cross-magnetic field (B_0): 10–500 $\mathrm{mWb/m^2}$

Applications of Magnetron

It generates reasonable high power in the kW range. The frequency of oscillation is decided by the cavity resonating frequency, and therefore, tunability is very low.

Therefore, applications are:

1. Voltage-tunable magnetron (VTM) used as sweep oscillator in telemetry missile with CW power at 70% efficiency in 0.2–10 GHz range.
2. CW fixed frequency magnetron is used in
   (a) Industrial heating
   (b) Transmitter
   (c) Microwave oven at freq. 2.45 GHz and power = 600–1200 W. The efficiency is 64% or so. Therefore, input dc power requirement is in the range of 1000–2000 W depending on the size.
3. Pulsed power magnetron of peak power of megawatts range is in RADAR.

Solved Problems

Problem 5.1

A two-cavity klystron amplifier has the following parameters:

Input power = 10 MW

Amplifier gain = 20 dB

Effective load of input cavity (R_{in}) = 30 kΩ

Effective load of output cavity (R_o) = 30 kΩ

Load impedance = 50 kΩ

Assuming the klystron to be 100% efficient, find the input RF voltage, output RF voltage, and power delivered to the load.

Solution

$$P_{in} = V_{in}^2/R_i$$

$$\therefore V_{in}^2 = P_{in} \times R_{in}$$

$$= 10 \times 10^{-3} \times 30 \times 10^{+3}$$

$$= 300$$

$$V_{in} = 17.3\text{V}$$

$$\text{Voltage gain} = A_v = 20 \log\left(\frac{V_0}{V_{in}}\right) \text{dB};$$

$$20 = 20 \log\left(\frac{V_0}{17.3}\right)$$

$$V_0 = 10 \times 17.3 = 173\text{V}$$

$$P_{out} = \frac{V_0^2}{R_0} = \frac{(173)^2}{80 \times 10^3} = 382.8\,\text{mW}$$

Problem 5.2

A two-cavity klystron has the following parameters with the assumption of efficiency = 100%

Voltage gain = 15 dB, input power = 15 mW

Input cavity shunt impedance = 30 kΩ, output cavity shunt impedance = 40 kΩ

Load impedance (R_L) = 40 kΩ

Find the input, output RF voltage, and the power to the load.

Solution

$$\text{input voltage} = \sqrt{P_{in}.R_{in}}$$

$$= \sqrt{15 \times 10^{-3} \times 30 \times 10^{+3}}$$

$$= 21.2\,\text{V(rms)}$$

Output voltage = V_{in} × gain = 21.2 × (gain), but gain (dB) = 15 dB

15 dB = 20log (V_0 /V_{in})

$$\therefore \quad 10^{\frac{15}{20}} = V_0/V_{in} = 10^{0.75} = 5.62 \text{ and}$$

$$\therefore \quad V_0 = 5.62 \times 21.2 = 119.4\,\text{V}$$

$$\therefore \quad P_0 = \frac{V_0^2}{R_L} = \frac{(119.4)^2}{40 \times 10^3} = 0.36\text{ W}$$

Problem 5.3

A 100% efficient reflex klystron has the following characteristic:

Beam voltage (V_0) = 900 V, beam current (I_0) = 30 ma, frequency = 8 GHz, gap spacing in the two-cavity (d) = 1 mm, spacing between cavity (L) = 4 cm, effective shunt impedance =

40 kΩ. Calculate the electron velocity (v_0), dc resistance (R_0), the transit time (T_0) from one cavity to the other, input and output voltge V_1, V_2, and voltage gain in dB.

Solution

(1)

$$\text{Electron velocity } (v_0) = \sqrt{2eV_0/m} = 5.93 \times 10^5 \sqrt{V_0}\,\text{m/s} = 5.93 \times 10^5 \cdot \sqrt{900} = 1.779 \times 10^7\ \text{m/s}$$

(2) dc resistance $R_0 = V_0/I_0 = 900/30 \times 10^{-3} = 30\ \text{k}\Omega$

(3)

$$\text{Transit time} = \frac{d+l}{v_0} = \frac{10^{-3} + 4 \times 10^{-2}}{1.779 \times 10^7} = 2.3 \times 10^{-9}\text{s}$$

(4) Input voltage gap angle

$$(\theta_g) = \frac{\omega \cdot d}{v_0} = \frac{2\pi \times 8 \times 10^9 \times 10^{\ 3}}{1.779 \times 10^7} = 2.825\,\text{rad}$$

Coupling phase constant $\beta_0 = \frac{\sin(\theta_g/2)}{(\theta_g/2)} = \frac{\sin 1.41}{1.41} \approx 0.70$

Also transit angle between the two cavity

$$\theta_0 = \omega_0 \times (\text{transit time}) = 2\pi \times 8 \times 10^9 \times 2.3 \times 10^{-9} = 115.6\,\text{rad}$$

Maximum input voltage is given by Eq. (5.20a)

$$V_{1\max} = \frac{2 \cdot V_0 X}{\beta_0 \theta_0} = \frac{2 \times 900 \times 1.841}{0.7 \times 115.61} = 40.9\ \text{V}$$

(5) Gain and voltage output: For maximum power output bunching parameter X = 1.841 for fundamental signal with corresponding Bessel's function $J(X)$ = 0.582.

Therefore by Eq. (5.21c)

$$\text{voltage gain} = \frac{V_0}{V_{1\max}} = \frac{\beta_0^2 \cdot \theta_0 \cdot J_1(X)}{R_0 \cdot X} \cdot R_{sh} = \frac{(0.7)^2 \cdot (115.61) \cdot (0.582)}{30 \times 10^3 \times 1.841} \cdot 40 \times 10^3 = 23.8$$

$\therefore$

$$\text{Output voltage} = V_2 = \text{gain} \times V_{1\max} = 23.8 \times 40.9 = 973.4\ \text{V}$$

$$\text{gain in dB} = 20\ \log_{10}(23.8) = 20(1.3766) \simeq 27.5\ \text{dB}$$

Problem 5.4

A two-cavity klystron amplifier for 3 GHz has drift space of L = 2 cm and beam current I_0 = 25 mA. Catcher voltage is 0.3 times of beam voltage. If the cavity gap length $l \ll L$ the drift space, so that input and output voltage are in phase. Therefore, compute P_0 and efficiency (η), beam voltage, input voltage, and output voltage for maximum power output, for $N = 5\frac{1}{4}$ mode.

Solution

Beam voltage

$$V_0 = \frac{m}{2e}\left(\frac{L \cdot f}{N}\right)^2 = \frac{9.1 \times 10^{-31}}{2 \times 1.6 \times 10^{-19}} \times \left(\frac{2 \times 10^{-2} \times 3 \times 10^9}{5.25}\right)^2 = 371.4\,\text{V}$$

For maximum power output, the bunching parameter $X = 1.84 = \frac{\pi N \times V_1}{V_0}$

$$\therefore \quad \text{Input voltage} \quad V_1 = \frac{1.84 \times 317.4}{3.14 \times 5.25} - 41.4\,\text{V}$$

$$\text{Catcher voltage} \quad V_2 = 0.3 \times V_0 = 111.4\,\text{V}$$

$$\text{Output power} \quad P_0 = J_1(X) \cdot I_0 V_2$$

where

$$J(X) = \text{Bessel's function}$$
$$= 0.582 \text{ for maximum power output}$$
$$P_0 = 0.582 \times 25 \times 10^{-3} \times 111.4 = 1.62\text{W}$$
$$P_{dc} = I_0 V_0 = 25 \times 10^{-3} \times 371.4 = 9.3 \text{ W}$$
$$\text{Efficiency} = J_1(X) \cdot \frac{V_2}{V_0} = 0.582 \times 111.4/371.4 = 17.46\%$$

Problem 5.5

A two-cavity klystron operates at 5 GHz with beam voltage (V_0) of 10 kV and gap of cavity = 2 mm. For a given RF voltage, the cavity gap voltage (V_1) = 100 V (maximum). Calculate the minimum and maximum velocity of electron.

Solution

Beam velocity

$$v_0 = 0.593 \times 10^6 \sqrt{v_0} = 0.593 \times 10^7 \text{ m/s}$$

$$\text{Gap transit time } t_g = \frac{d}{v_0} = \frac{2 \times 10^{-3}}{5.93 \times 10^7} = 3.37 \times 10^{-11}\text{s}$$

$$\text{Gap angle } (\theta_g) = \omega t_g = 2\pi \times 5 \times 10^9 \times 3.37 \times 10^{-11} \text{ rad.} = 1.059 \text{ rad.}$$

$$\text{Beam coupling coef} = \beta_1 = \frac{\sin(\theta_g/2)}{(\theta_g/2)} = 0.9537$$

Velocity of electron leaving the gap

$$v(t) = v_0\left[1 + \frac{\beta_1 V_1}{2V_0}\sin(\omega t + \theta_g/2)\right] = v_0(1 + K \cdot \sin\phi)$$

$$\therefore \text{ The minimum velocity} = v_0(1-K) = 5.93 \times 10^7(1 - 0.00477) = 5.902 \times 10^8 \text{m/s}$$

$$\text{And minimum velocity} = v_0(1+K) = 5.93 \times 10^7(1 + 0.00477) = 5.958 \times 10^8 \text{m/s}$$

Problem 5.6

A 8 GHz, two-cavity klystron amplifier has a beam voltage of 900 V, and the distances between two cavities (L) are 4 cm. Find the number of electron bunches under formation and travelling between cavities 1 and 2.

Solution

$$\text{Electron velocity} = \sqrt{2eV_0/m} = 5.93 \times 10^5 \text{ m/s} = 5.93 \times 10^5 \times \sqrt{900} = 1.779 \times 10^7 \text{ m/s} = 1.779 \times 10^9 \text{ cm/s}$$

Time taken by free electron to travel between the cavities = L/v_0

One time period = $(1/f_0)$

$$\therefore \text{ No. of traveling bunches } (N) = \frac{L/v_0}{(1/f_0)} = \frac{Lf_0}{v_0} = \frac{4 \times 8 \times 10^9}{1.779 \times 10^9} \approx 27$$

Problem 5.7

A reflex klystron oscillator operates in $1\frac{3}{4}$ mode at 10 GHz with a beam voltage of 300 volts and beam current of 20 mA. If the repeller is at a distance of 1.0 mm from the cavity, then find the maximum RF power output, the repeller voltage, and efficiency.

Solution

Given $V_0 = 300$ V, $I_0 = 20 \times 10^{-3}$, $N = 1\frac{3}{4} = 1.75$, $L = 1 \times 10^{-3}$

$$P_{\text{r.f.max}} = 0.399 \times I_0 V_0/N = 0.399 \times 10^{-3} \times 300/1.75 = 1.36 \text{ W}$$

Also, by Eq. (5.30)

$$(|V_R| - V_0) = \frac{d.f_0}{N}\sqrt{8\frac{mV_0}{e}} = \frac{1 \times 10^{-3} \times 10 \times 10^9}{1.75}\sqrt{\frac{8 \times 300}{1.759 \times 10^{11}}} = 642.6 \text{ V}$$

$$\therefore \ |V_R| = 642.6 + V_0 = 642.6 + 300 = 942.6 \text{ V}$$

$$\text{Efficiency}\ \eta_{max} = \frac{XJ_1(X)}{\pi N}$$

$$\frac{2.408 \times 1.252}{3.14 \times 1.75} = 0.399/1.75$$

$$= 22.8\%$$

Problem 5.8

A reflex klystron operates at the higher mode of $N = 1.75$, $n = 1$ with beam voltage $V_0 = 300$ V, beam current $I_0 = 20$ mA, RF output rms $(V_1) = 40$ V. Find the input power, output power, and efficiency, $N = (n + 3/4)$.

Solution

P_0 = Input power $= I_0V_0 = 20 \times 10^{-3} \times 300 = 6$ W
ac voltage $V_1 = 40$ V

$$\therefore \quad P_{r.f.} = J_1(X) \times V_1 I_0$$

where $J_1(X)$ is Bessel's function with

$$X = \frac{V_1}{V_{01}} \pi \cdot N = \frac{40}{300} \times 3.14 \times 1.75 = 0.734$$

$$N = \left(n + \frac{3}{4}\right), \ J(X) = 0.345$$

$$\therefore \quad P_{RF} = 0.345 \times 40 \times \left(10^{-3} \times 20\right)$$

$$= 0.276\ \text{W}$$

$$\eta = 0.276/6$$

$$= 4.6\%$$

Problem 5.9

A reflex klystron has the following operating conditions $V_0 = 600$ V, $f_0 = 8$ GHz, $R_{sh} = 20$ kΩ, $L = 1$ mm, $n = 1$ mode. Find the repeller voltage V_R, beam current (I_0), and efficiency, so that the cavity gap ac voltage is 200 V. $(V_1 = 200)$

Solution

We know that from Eq. (5.30) with $d = L$

$$\frac{V_0}{(V_r - V_0)^2} = \frac{(e/m)}{8L^2 \cdot f^2} \cdot \left(n + \frac{3}{4}\right)$$

$$= \frac{1.759 \times 10^{11} \times 1.75}{8 \times (10^{-3})^2 \times (8 \times 10^9)^2}$$

$$\therefore \quad \frac{600}{(V_R - 600)^2} = 1.05 \times 10^{-3}$$

$$\therefore \quad V_R = \sqrt{600/1.05 \times 10^{-3}} + 600$$

$$\sqrt{6 \times 10^5/1.05} + 600$$

$$= 7.55 \times 10^2 + 600$$

$$= 755 + 600 = 1355\ \text{V}$$

$$\text{Bunching} \quad X = \frac{V_1 \pi}{V_0}\left(n + \frac{3}{4}\right)$$

$$= \frac{200}{600} \times 3.14 \times 1.75$$

$$= 1.83$$

From Bessel's function curve $J(1.83) = 0.58$

$$\therefore \quad I_0 = \frac{V_1}{2J_1(1.83) \times R_{sh}}$$

$$= \frac{200}{2 \times 0.58 \times 20 \times 10^3} = 8.6\ \text{mA}$$

$$\text{Now efficiency}(\eta) = \frac{X.J_1(X)}{\pi\left(n + \frac{3}{4}\right)}$$

$$= \frac{1.8 \times 0.58}{3.14 \times 1.75}$$

$$= 18.999\% \cong 19\%$$

Problem 5.10

A TWT operates with beam voltage $(V_0) = 3000$ V, beam current $(I_0) = 30$ mA, characteristic impedance of helix $(Z_0) = 10$ Ω, frequency = 10 GHz, number of helix = 50. Find gain parameter (g) output power in dB and the velocity of beam electrons.

Solution

Beam electron speed

$$v_0 = \sqrt{\frac{2eV_0}{m}} \text{ here } \sqrt{\frac{2e}{m}} = 0.593 \times 10^6$$

$$\begin{aligned} v_0 &= 0.593 \times 10^6 \sqrt{3000} \\ &= 35.58 \times 10^4 \\ &= 3.558 \times 10^5 \text{m/s} \end{aligned}$$

$$\begin{aligned} \text{Gain parameter } (g) &= [(I_0 Z_0 / 4V_0)]^{1/3} \\ &= 2.9 \times 10^{-2} \end{aligned}$$

$$\begin{aligned} \text{Output power (in dB)} &= (47.3N \cdot g - 9.54)\text{dB} \\ &= (47.3 \times 50 \times 2.9 \times 10^{-2} - 9.54) \\ &= 59.5 \text{ dB} \end{aligned}$$

Problem 5.11

In a linear TWT, the beam voltage (V_0) = 10 kV, helix impedance (Z_0) = 25 Ω, helix length = 20 cm. It operates at 4 GHz. Find the number of helicals, beam speed, gain parameter, and gain in dB.

Solution

Given $f = 4$ GHz, $V_0 = 10$ kV, $I_0 = 500$ mA, $Z_0 = 25\ \Omega$, $l = 20$ cm

$$\begin{aligned} v_0(\text{Beam velocity}) &= \sqrt{\frac{2e}{m}} \cdot \sqrt{V_0} \\ &= 0.593 \times 10^6 \sqrt{V_0} \\ &= 0.593 \times 10^6 \times 10^2 \\ &= 5.93 \times 10^7 \text{m/s} \end{aligned}$$

$$\begin{aligned} N = \text{no. of helicals} = \frac{l}{\lambda_e} = \frac{l.f.}{v_0} &= \frac{0.2 \times 4 \times 10^9}{5.93 \times 10^7} \\ &= 13.49 \end{aligned}$$

$$\begin{aligned} \text{Gain parameter } (g) &= [I_0 Z_0 / 4V_0]^{1/3} \\ &= \left[\frac{500 \times 10^{-3} \times 25}{4 \times 10000}\right]^{\frac{1}{3}} = 0.068 \end{aligned}$$

$$\begin{aligned} \text{Gain in dB} &= (47.3.\text{N.g} - 9.45)\text{dB} \\ &= 47.3 \times 13.49 \times 0.068 - 9.54 \\ &= 33.85 \text{ dB.} \end{aligned}$$

Problem 5.12

A helical TWT has helix of 2 cm length and has 100 turns. The dia = 2 mm. Calculate the axial phase velocity and anode voltage at which it can be operated usefully.

Solution

v_p = Phase velocity = c . pitch/($2\pi r$)

$$\begin{aligned} p = \text{Pitch} &= \text{length/turns} \\ &= 2 \times 10^{-2}/100 = 2 \times 10^{-4} \text{ met./turn} \\ &= \text{linear distance per turn} \end{aligned}$$

$$\begin{aligned} \therefore \quad v_p &= 3 \times 10^8 \times 2 \times 10^{-4} / (2 \times 3.14 \times 10^{-3}) \\ &= 0.955 \times 10^7 \end{aligned}$$

Also

$$eV_0 = \frac{1}{2} \cdot mv_p^2$$

$$\begin{aligned} \therefore \quad V_0 &= \frac{1}{2}\frac{m}{e} \cdot v_p^2 \\ &= \left(\frac{9.1 \times 10^{-31}}{2 \times 1.6 \times 10^{-19}}\right) \cdot (0.955 \times 10^7)^2 \\ &= 25.9 \text{ kV} \end{aligned}$$

Problem 5.13

An X-band cylindrical magnetron has $V_0 = 40$ kV, $I_0 = 100$ A, $B_0 = 0.01$ Wb/m^2, $a = 4$ cm, $b = 8$ cm.

Calculate (a) cyclotron angular frequency, cut-off voltage, and cut-off magnetic flux, and (b) if the pulsed peak power output is 60 MW, with 1% duty cycle, find the efficiency.

Solution

Cut-off magnetic flux

$$\begin{aligned} B_c &= \frac{\sqrt{8V_0 \,.\, m/e}}{n(1 - a^2/b^2)} \\ &= \frac{\sqrt{[8 \times 40 \times 10^3 \times 5.685 \times 10^{-12}]}}{8 \times 10^{-2}(1 - 4^2/8^2)} \\ &= 11.12 \text{ mWb/m}^2 \cong 0.011 \text{ Wb}/M^2 > B_0 \end{aligned}$$

Cut-off voltage for the flux applied will be $V_c = \frac{eB_0^2 b^2}{8m}\left(1 - a^2/b^2\right)^2$

$$\therefore \quad V_c = (1/8)1.759 \times 10^{11}(0.011)^2 \times \left(8 \times 10^{-2}\right)^2 \times \left(1 - 4^2/8^2\right) = 7.92 \text{ kV}$$

Cyclotron angular frequency

$$\omega = \frac{eB_0}{m} = 1.759 \times 10^{11} \times 0.011 = 1.76 \times 10^9 \text{ rad/s}$$

$$\text{As} = \frac{e}{m} = 1.759 \times 10^{11} \text{ C/m}$$

$$P_0 = \text{D.C. power disspation} = V_0 I_0 = 40 \times 10^3 \times 100 = 4 \times 10^6 \text{W} = 400 \text{kW}$$

$$P_{\text{r.f.}} = \text{pulsed r.f.} = 60\text{MW at } 10\% \text{ duty cycle}$$

$$\therefore \quad \text{Average r.f.power} = 60 \times 10^3 \times 0.01 = 600 \text{ W}$$

$$\therefore \quad \text{Efficiency} = \frac{P_{\text{r.f.}}}{P_0} = \frac{600}{4000} = 15\%$$

Review Questions

1. A conventional tube cannot be used at microwave frequencies. Explain with figures the reasons for it.
2. Discuss the types of microwave tubes along with their classification. Differentiate between linear and cross-field devices.
3. What is velocity modulation? How it is different from normal signal modulation in electronics? Explain how the velocity modulation is utilised in klystron amplifier (MDU 2004)
4. What problems are encountered in the conventional multielectrode tubes at microwave frequencies? Describe the principle of operations of reflex klystron oscillator (UPTU-2002, 2003)
5. Derive an expression for efficiency of a two-cavity klystron amplifier (MDU 2003)
6. An identical two-cavity klystron amplifier operates at 4 GHz with $V_0 = 1$ kV, $I_0 = 22$ mA, cavity gap = 1 mm, drift space = 3 cm, effective conductance (G_{sh}) of catcher cavity = 0.3×10^{-4} mhos. Calculate
   (i) Beam coupling coefficient
   (ii) Input cavity voltage for maximum output voltage (UPTU 2003, 2004).
7. Explain the working principle of two-cavity klystron by giving the Applegate diagram (MDU 2006)
8. A reflex klystron operates at 9 GHz with $V_0 = 600$ V for $1\frac{3}{4}$ mode, repeller space = 1 mm, $I_0 = 10$ mA, beam coupling coefficient = 1. Calculate repeller voltage, efficiency, and the output power (UPTU-2008-2009)
9. Draw the schematic diagram of a TWT amplifier and describe its principle of operation. Give the propagation characteristic of different waves generated in the amplifier. Explain how RF power output and gain vary with the change of RF input power (UPTU-2002-2003)
10. With support of eight-cavity magnetron, diagram explains its working. Discuss the role of slow wave structure in TWT (UPTU-2008-2009)
11. How is bunching achieved in a cavity magnetron? Explain the phase-focusing effect. Why does spoke cloud (bunch) of electron rotates and with what angular velocity? (MDU-2004)
12. How is continuous interaction between the electron beams and RF field ensured in a TWT? Show how the favourable interactions are for more in number than the unfavourable interactions resulting into amplification (UPTU-2002, 2004)
13. Explain bunching of electron taking place in (*a*) two-cavity klystron, (*b*) reflex klystron, (*c*) TWT, (*d*) magnetron. Explain each separately and then write differences and similarities.
14. Draw the cross-sectional diagram of a magnetron and derive the expression for cut-off voltage (MDU-2006)

15. A magnetron operates in π mode and has the following specification: $N = 10$, $f = 3$ MHz, $a = 0.4$ cm, $b = 0.9$ cm, $l = 2.5$ cm (anode), $V_0 = 18$ kV, and $B_0 = 0.2$ Wb/m^2; therefore, determine:

    (a) Angular velocity of electron
    (b) Radius of electron movement, for which radial force due to magnetic and electronic fields is equal and opposite (MDU-2003)

16. What is BWO? Explain its principle of working (MDU-2003)
17. Explain the purpose of strapping in magnetron. It is said that it suppresses spurious modes and enhances π mode. Explain.
18. Explain the mechanism of turning, i.e., changing the frequency of RF signal generated in reflex klystron and also magnetron. Which has wider tuning and why?
19. Show that in a magnetron of N-cavities, oscillating at frequency f, the angular velocity of electron is given by: $\omega_c = 4\pi f/N$, where n is integers.
20. In magnetron, explain (*a*) frequency pulling and pushing mechanism, (*b*) phase-focusing effect, (*c*) TT mode, (*d*) strapping, (*e*) mode jumping.
21. Derive an expression for Hull cut-off magnetic field B_c. What happens when magnetic field $B > B_c$ and $B < B_C$.
22. If we need 250 W (CW) microwave power at 1 GHz. Which device should we choose and why?
23. In electronic tuning by V_R, the central frequency f_0 of reflex klystron and cavity size remains same in all the modes, while in mechanical tuning being done by rotating the knob of re-entrant cavity, frequency itself changes. Explain the mechanism.

Microwave Semiconductors Devices: Oscillators, Amplifiers, and Circuit

6

Contents

© Springer Nature Singapore Pte Ltd. 2018
P. K. Chaturvedi, *Microwave, Radar & RF Engineering*,
https://doi.org/10.1007/978-981-10-7965-8_6

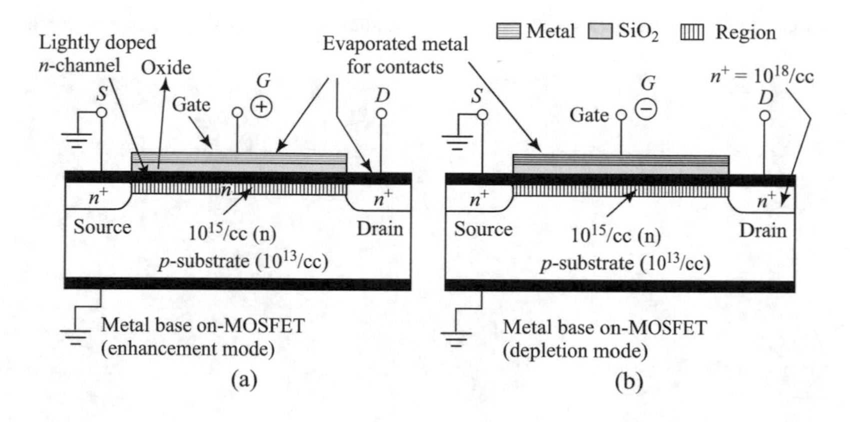

6.1 Introduction

A microwave source is essential for any microwave system, and we have seen that the microwave tubes offer very high-power to very-high frequencies, but at the cost of:

(a) Larger space requirements.
(b) Higher dc power (electric field and magnetic field) requirements.
(c) Filament heater (of cathode) requirements.
(d) Higher cost, etc.

Scientists were on a look out for a simpler source. Therefore, after the invention of transistor, the work on microwave transistor like oscillators/amplifiers, new sources like Gunn diode, IMPATT diodes, TRAPATT diodes, etc., had started. Today we have these semiconductors, i.e. solid-state device as sources, which meet the low-power requirements in microwave. Figure 6.1 gives the comparison of power and frequency performance of solid-state devices and the microwave tubes. Efficiency of the solid-state sources still remains low (10–15%) except (a) Gunn diodes have shown up to 30% efficiency (in the pulsed mode) at low frequencies near 1–2 GHz, (b) CW-IMPATT devices up to 25% near 5–10 GHz, (c) TRAPATT diodes up to 40% near 1–2 GHz in pulsed mode. (d) The most promising upcoming GaN device has shown the highest efficiencies around 90% (because of which it can be used as amplifier giving 100 W also) near *C*-, *X*-bands. The transistor oscillators generally have lower frequency and power capabilities as compared to Gunn and IMPATT, but have more controls on frequency tunability, with lower noise figure and better temperature stability. In this chapter, we will discuss these solid-state devices. Comparison of these devices is given in Table 6.1 with frequency range of its operation along with maximum power possible, applications and special advantages of that device, besides the advantages of lightweight, small in size, easy mountability on the circuit, etc.

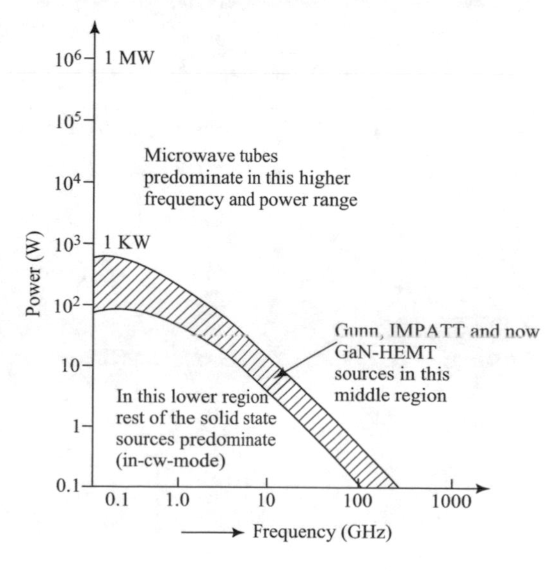

Fig. 6.1 Power versus frequency performance of solid-state sources and the microwave tubes

Table 6.1 Summary of microwave semiconductor devices (diodes and transistors: oscillators, amplifiers, and circuit devices)

Section nos.	Devices	Frequency maximum power range, etc. (microwave ranges)	Applications (as signal oscillator or amplifier)	Special advantage
(A) Devices used as oscillators or amplifiers				
6.4	Junction transistor P^+np^- n^+pn$^-$	f = 1–20 GHz P = (10–1 W) Gain = 30 dB	• Telemetry • RADAR • Communication (as oscillator and amplifier)	• Reliability • Stability • Life longevity • High cw power • high gain
6.5	Unipolar junction FET	f = 1–40 GHz P = (1–0.1 W) Gain ≅ 10 dB	As above	Very low dc power requirements
6.6	MESFET (Si/GaAs)	f = 1–50 GHz P = (1–0.5 W) Gain = 5 dB	As above + Airborne RADAR where GaAs-MESFET is preferred	• Low-power needs • Wide freq. •High cw power • Now preferred over para-amp in airborne radar
6.7	MOSFET	f = 1–30 GHz P = (1–0.1 W) Current gain ≥ 100 (i.e. 20 dB)	As above	1. Large input ac possible than BJT, JFET, MESFET 2. Linear power amplifier, as g_m does not depend on V_g 3. Lowest dc power requirement
6.8	Tunnel diodes	f = 1–20 GHz Gain = 5–15 dB Max ac voltage = 350 mW	Oscillator and amplifier, used in UHF T.V. tuner, CRO, fast-rise pulse generator	Very long life up to 40 years
6.9	(a) TED/Gunn devices (GaAs)	CW: 1–25 GHz (1–0.1 W) pulsed: 1–10 GHz (200–2 W) Efficiency (η) ≈ 2–10%	As microwave source and amplifier	• Reliable • High gain • Wide gain amplifier • Disadvantage is of low frequency and low stability with temperature
	(b) LSA mode	Pulsed 1–20 GHz (500–20 W) Efficiency = 20–25%	Microwave source	Very high pulsed power possible
6.10.1	IMPATT diode (mostly silicon)	CW and pulsed f = 0.5–120 GHz (CW: 10–0.1 W) At pulsed: 100–1 W η = 10–15% band width = ±10% noise = 30 dB	Transmitter for mm wave (source and amplifier)	High CW power, very high frequency possible

(continued)

Table 6.1 (continued)

Section nos.	Devices	Frequency maximum power range, etc. (microwave ranges)	Applications (as signal oscillator or amplifier)	Special advantage
6.10.2	TRAPATT diode $p^{+}nn^{+}$ $n^{+}pp^{+}$ (silicon)	f = 1–20 GHz $P_{\text{pulsed:}}$ = 1000–20 W η = 40–20% Operating = 150–50 V Noise figure 60 dB	Transmitter in phased array radars as source	• Very high peak power • Very high average power • Highest efficiency among microwave devices
6.11	BARITT diode	f = 2–8 GHz P_0 = 20–1 mW η = 2% Noise <10 dB	Local oscillator source	Very low noise lower than IMPATTs/Gunn
(B) Microwave devices used as circuit components/devices				
6.12	Schottky diode	Whole microwave range (0.5–200 GHz) Noise = 4 dB of 2 GHz = 15 dB at 100 GHz	Mainly microwave detector, also as mixer due to very low-noise figure	Very low-noise diode detector of microwave signal
6.13	PIN diode (I = π or v)	f = 0.5–150 GHz (full microwave range) power handling is as per size area. Also 150 kW is possible with several parallel diodes	As a microwave line switch, phase shifter, amplitude modulator, power limiter, etc.	Used as a passive component in the circuit
6.14	Varactor diode (varicap)	Silicon and GaAs diodes Si-up to 25 GHz GaAs-up to 100 GHz Capacitor tuning by bias 10–300 pF	As a variable capacitor in microwave circuit, TV, parametric Also used in harmonic generator, mixer detector, filter, etc.	This is the only non-mechanical variable capacitor in electronics
6.15	Parametric amplifier (as up-convertor, down convertor of frequency)	Same as above	• Low-noise amplifier in long range radar satellite ground station radio astronomy, etc. • In radar −ve resistance parametric amp is preferred as frequency. required by the system is higher than X-band • For low-noise requirements degenerate para-amp is preferred	Very low-noise amplifier as no resistance is involved. However due to complicated circuit, it is now days getting replaced by GaAs-MESFET amplifiers in airborne radar

6.2 Classification of Microwave Semiconductor Devices

In this chapter, we are going to study various semiconductor or solid-state devices (six types of transistors, eight types of diodes, and one para-amplifiers) used in microwaves as oscillator, as amplifier, or as a circuit device. These can be listed in the following two classes:

(a) **Devices used as oscillators and amplifiers**

(i) *Transistors*: Microwave BJT, Jn-FET, MOSFET, MESFET, HEMT, and FINFET.
(ii) *Diodes*: Tunnel diodes, Gunn diodes, IMPATT diodes, TRAPATT diodes, BARITT diodes.

(b) **Diodes used as circuit device for some special applications**

(i) *Schottky diode*: Used as power detector, microwave mixer, etc.
(ii) *PIN diode*: For switching microwave power/phase shifter/Power limiter
(iii) *Varactor diode*: As frequency multiplier/variable capacitor frequency tuning in circuits parametric amplifiers

All these are given in Table 6.1 for comparison of their properties (Table 6.2).

6.3 Microwave Transistors—BJT and FET

Because of lot of advancement in microwave transistors and the fact that a transistor has become the fundamental building block of digital and analog circuits, it has became important in microwaves also. The two properties of the transistor e.g. (a) a small input voltage or current controlling large voltage and current variation and (b) fast response time and accuracy, has found its applications in amplification, switching, modulation and as an oscillator.

There are variety of transistors, but we will be discussing only six types of transistors, e.g. (1) bipolar (BJT), (2) junction field effect transistors (Jn-FET), (3) metal-semiconductor field effect transistors (MESFET), (4) metal-oxide-semiconductor field effect transistor (MOSFET), (5) HEMT, (6) FINFET.

First we will discuss the three FETs and their common properties.

6.3.1 Field Effect Transistors (FETs)

The FETs are called unipolar device, as only one type of carrier current (e.g. electron current for 'n' channel FET) is there. The channel current is controlled by the following three mechanisms of the gates:

Table 6.2 Doping level symbols used for Si, GaAs are as

Symbols	Doping density range
P^{++}, n^{++}	$>10^{19}$/cc (Degenerate)
p^{+}, n^{+}	10^{17}–10^{18}/cc
p, n	10^{15}–10^{16}/cc
p^{-}, n^{-}	10^{13}–10^{14}/cc
$p^{--}(\pi)$, $n^{--}(\nu)$	In Si < 10^{10}/cc (intrinsic)
	In GaAs < 10^{7}/cc (intrinsic)

Note Here it may be noted that the free electron carrier densities of metals are in the range of 10^{22}–10^{23}/cc

(a) pn junction depletion region at the gate in jn FET
(b) Metal–semiconductor jn depletion region (Schottky barrier gate) in MESFET
(c) Capacitative field and charge effect in MOSFET.

Major advantages of FETs over bipolar transistor are:

(i) Low dc power requirement
(ii) FETs are voltage-controlled devices and draw very little power from the dc supply as well as from the input signal
(iii) As no minority carrier is involved, it has more stability
(iv) Both Z_{in} and Z_{out} are very high, therefore do not load either the input side or the output side
(v) Less noisy
(vi) Can be a part of the integrated circuit
(vii) Easy to fabricate than the bipolar Tr.

All the three FETs have the following:

1. **Source**: Through this terminal, the majority carriers enter the channel.
2. **Drain**: Through this terminal, the majority carriers leave the channel.
3. **Gate**: It is used to control the flow of carriers in the channel, by application of a −ve voltage which creates depletion region in the channel, thereby restricting the path and hence current in the channel.
4. **Channel**: The space between drain and source through which the majority carrier current flows.
5. The maximum frequency of oscillation that is possible is:

$$f_{max} = \frac{1}{2\pi\tau} \quad \text{where } \tau = L_g/v_s \qquad (6.1)$$

where τ = transit time of carrier across the gate length (L_g), where the depletion region and its capacitance are formed, v_s being the saturated velocity of carriers which is around 10^7 cm/s for silicon.

6.4 Microwave Bipolar Junction Transistor (BJT)

After the invention of transistor (word derived from transfer of resistor) in 1948 by W. Schockley of Bell Laboratories, lot of development has taken place. Now for microwave low-power applications, silicon bipolar transistor dominates for frequency range from UHF to S-band (i.e. 200 MHz–6 GHz); however, it can give useful power up to 25 GHz.

Silicon bipolar junction transistor (BJT) is less expensive, durable, low noise, integratable in the circuit, and offers higher gain than FET. For higher frequencies, higher temperature, and radiation hardness, GaAs BJT is being used. High-frequency response limit of BJT is determined by the (a) time taken by the carriers injected from emitter to cross the base region and the (b) mobility of the carriers.

As mobilities of electron and holes are 1500 and 450 cm^2/Vs, electron carrier is preferred to be transmitted through the base and therefore the npn-type of BJT. By newer technologies (e.g. ion implantation), base width as lower as 0.05 μ can be achieved, which keeps the hole-electrons-recombination (i.e. carrier losses) at the base also small.

Thus, the μW BJT differs with low-frequency BJT in terms of:

(i) Very low base width (<0.2 μm) and low emitter width (<1 μ).
(ii) High emitter doping (>10^{19}/cc) for reducing base resistance and increase current gain.
(iii) Multifinger emitter and base metalisation contact.

6.4.1 Structure

These transistors are fabricated by the usual planar technology by diffusion of impurities through the strip-type windows formed on the oxide layer as per the design of the masks for that diffusion. These diffusion depths are more for *p*-base diffusion, less for n^+ emitter junction diffusion

(Fig. 6.2a). The p^+ base contact diffusion is done so that the semiconductor has high conductivity and its contact with metal does not form Schottky diode contact but ohmic contact. For the same reason for metal contact at the bottom with the collector (which is n^- epitaxial layer), the substrate is n^+. These depths are controlled by time and temperature of that diffusion. Finally the strip-type windows on the oxide layer are again made at appropriate locations for metallisation contacts for base and emitter. The surface geometry for the diffusions and metallisation can be inter-digited, i.e. multifinger (or some other similar forms, e.g. 'over lay' or 'matrix' form). The objective behind such geometry with alternate emitter and base metallisation strips is to use maximum surface area with lower capacitance for increasing the current and hence higher power capability of the device.

6.4.2 Operation

The bipolar junction transistor is commonly used as amplifier and switch. Normally emitter junction is forward biased and collector junction reversed biased. When both the junctions are reversed biased, it acts as open circuit and when both are forward biased, it is like a short circuit. Out of the three configurations, i.e. common base, common emitter, and common collector, the second one is normally used in microwave circuits.

Figure 6.3 gives the various components of current flow in an npn-BJT under normal bias conditions of collector-Jn reverse biased and emitter-Jn forward biased. As the emitter is forward biased, large number of electrons (majority) gets injected into the base. As the base width is kept very low (e.g. 0.1 μ) (see Fig. 6.2), some of the electrons recombine with the majority (p) of the base, (giving a small current to the base), but most of the electron current diffuses to the collector due to its voltage +ve corresponding electric field attracting them.

The hole current of the emitter and of the collector will be there as minority, and a part will be used at the base for recombining with the electrons coming from emitter. This will constitute the small base current I_B.

$$I_E = I_B + I_C$$

6.4.3 Cut-off Frequency

At microwave frequencies, three parasitic elements come into play. These are (a) inter-electrode bond pad capacitances, (b) inductance, as the current in the lead wires has skin effect, and (c) resistances of the base, emitter, and collector regions in the silicon. All these limit the maximum frequency of operation of the transistor.

Using a simplified equivalent circuit, we can see that the ultimate frequency limitation is due to the following.

(i) *Charging times* τ_{ctc} and τ_{cte} of the

(a) Collector depletion capacitance (C_C)
(b) Emitter junction capacitance (C_E)

$$\therefore \quad \text{Total charging time } \tau_{ct} = \tau_{ctc} + \tau_{cte} \tag{6.2}$$

(ii) *Transit times* τ_{ttb} and τ_{ttc} of

(a) Base (non-depleted part) τ_{ttb}
(b) Collector depletion region τ_{ttc}

$$\therefore \quad \text{Total transit time } \tau_{tt} = \tau_{ttb} + \tau_{ttc} \tag{6.3}$$

Thus, the total delay for the signal to pass from emitter to the collector will be:

$$\tau_T = \tau_{ct} + \tau_{tt} = \tau_{ctc} + \tau_{cte} + \tau_{ttb} + \tau_{ttc} \tag{6.4}$$

This leads to the cut-off frequency as

$$f_T = \frac{1}{2\pi\tau_T} \quad (\beta = 1) \tag{6.5}$$

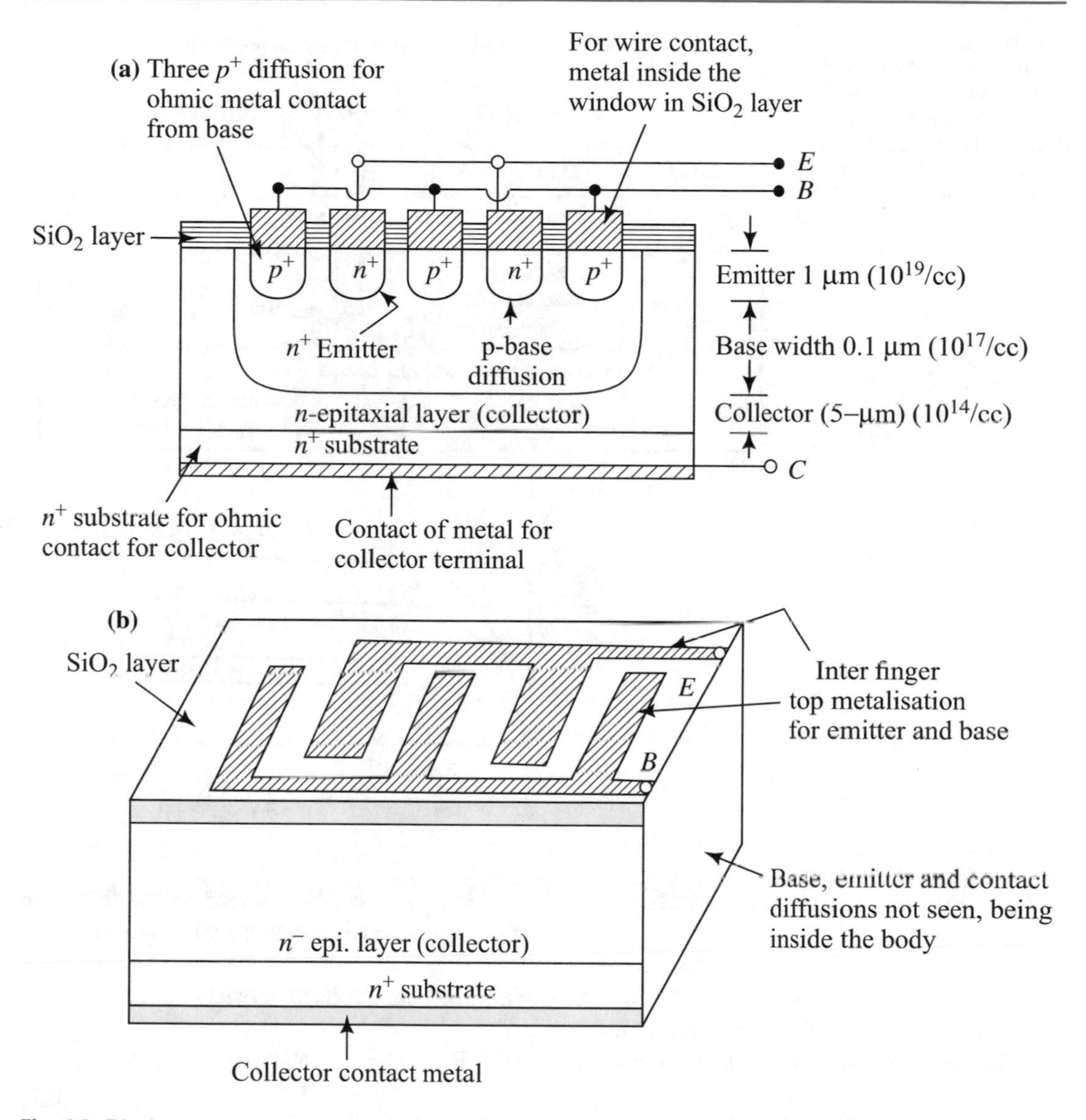

Fig. 6.2 Bipolar npn microwave power transistor **a** cross sections of metallisation and diffusion, with typical measurements and **b** inter-digited (fish bone) geometry of surface metallisation contacts, etc.

Out of the two types of delay given above, the transit time dominates on charging as ($\tau_{tt} \gg \tau_{ct}$). Therefore, the base width and the collector width (Fig. 6.3) have to be made as small as possible for having higher f_T.

At f_T the current gain β falls to unity (i.e. $\beta = 1$ at $f = f_T$) but the power gain (A_P) has not become unity. Therefore, we define f_{max} (which is higher than f_T) where power gain (A_P) falls to unity. These two frequencies are related by the following equation:

$$f_{max} = \sqrt{\frac{f_T}{8\pi R_B C_0}} \quad (A_p = 1) \tag{6.6}$$

where both R_B base spreading resistance and C_0 the collector base depletion layer capacitance are proportional to the width of emitter strip. Reducing these two reduces the power handling capacity but increases f_{max}. Therefore, study of power frequency limitations becomes important.

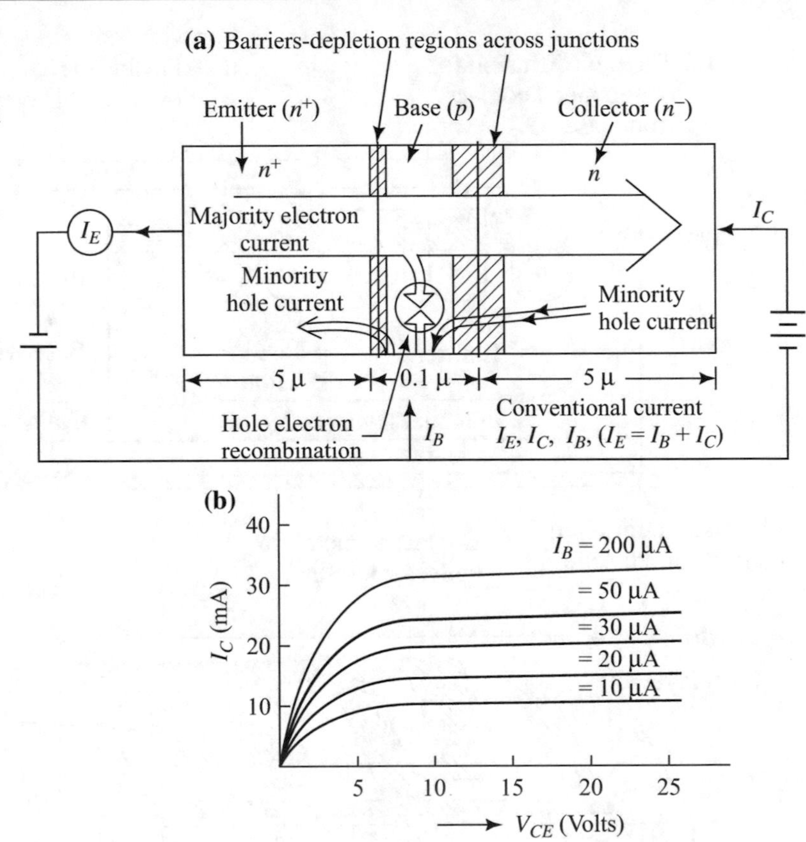

Fig. 6.3 **a** Various components of the current flow in an npn-BJT, under normal biasing condition and **b** a typical DC-operating characteristic of common emitter configuration BJT

6.4.4 Power Frequency Limitation

It has been shown by that product of power (P) and square of frequency f^2 are constant ($Pf^2 = K$) in BJT, MESFET as well as for two terminal devices. These limitations are due to:

(i) Maximum attainable field ($E_m = 2 \times 10^5$ V/cm in Si) in semiconductor without onset of avalanche multiplication.
(ii) Maximum carrier velocity ($v_s = 2 \times 10^7$ cm/s in Si).
(iii) Maximum current a transistor can carry is limited by the base width (L_m).
(iv) The cut-off frequency $f_T = 1/(2\pi\tau')$ where $\tau' = L_m/v_s$ and L_m = emitter collector distance.

These four conditions lead to the following four equations of limitations with frequency:

(a) Maximum voltage allowable (V_m)

$$V_m = E_m L_m = E_m(v_s \tau) = (E_m \cdot v_s)/(2\pi f_T) \tag{6.7}$$

(b) Maximum current possible (I_m)

$$\begin{aligned} I_m &= (V_m/X_C) = E_m \upsilon_s/(2\pi f_T X_C) \\ &= E_m \upsilon_s C_{bc} \end{aligned} \tag{6.8}$$

where X_C = reactance of the collector base capacitance (C_{bc}) at f_T, i.e. $X_C = 1/(2\pi f_T C_{bc})$

(c) *Maximum power*: By multiplying Eqs. (6.6) and (6.7) we get

$$V_m I_m = P_m = [E_m \upsilon_s/(2\pi f_T)]^2/X_C = V_m^2/X_C = V_m^2 \cdot 2\pi \cdot f_T \cdot C_{ac} \quad (6.9)$$

(d) Maximum gain available

$$G_m = [E_m v_s/(2\pi f_T)^2/(V_{th} \cdots V_n) = V_m^2/(V_{th} \cdot V_m) = (V_m/V_{th}) = (V_m \cdot q/kT) = \frac{E_m v_s q}{2\pi f_T kT}$$
$$(\text{where } V_{th} = kT/q = \text{thermal voltage}) \quad (6.10)$$

The voltage limitation, current limitation, power limitation, and gain limitation with frequency can be re-written in the following simple form also:

$$V_m f_T = (I_m X_C) f_T = \left(\sqrt{P_m X_C} \cdot f_T\right) = G_m f_T \cdot \frac{kT}{q} = \frac{E_m v_s}{2\pi}$$

We will now discuss these three FETs in detail.

6.5 Junction Field Effect Transistors (Jn-FET)

The actual structure and the structure used just for explaining the working of n-channel Jn-FET are given in Fig. 6.4a, b, respectively, with Fig. 6.4c giving the symbol of Jn-FET used in the circuits.

The gate junction is reversed based, resulting into a depletion region, which increases with gate reverse voltage. This depletion region being devoid of majority carriers reduces and pinches the conducting portion of the channel and hence reduces the drain–source current. Further increase of −ve gate voltage will spread the depletion layer further and fully pinch the conducting path for I_{ds} current (Figs. 6.4b and 6.5). The characteristic of the I_{ds}-vs-V_{ds} for different values of V_{gs} is given in Fig. 6.5. The depletion region electric field created by V_{gs} controls the I_{ds}, that is how the name field effect transistor.

Thus the pinch-off voltage is the reverse gate voltage that removes all the free charges from the channel and thereafter the channel current saturates (Figs. 6.4 **and** 6.5). The Poisson equation for the voltage in the n-channel in terms of the volume charge density q is given by:

$$\frac{d^2V}{dy^2} = -\frac{q}{\varepsilon_s} = \frac{N_d \cdot e}{\varepsilon_r \varepsilon_0} \quad (6.11)$$

With N_d = electron concentration density (doner) in the n-channel.

ε_s ε_0, ε_r = the permittivity of material, space, and dielectric constant, respectively.

Integrating the Eq. (6.10) once and using the boundary condition of electric field $E = -\frac{dv}{dy} = 0$ at $y = a$ (the channel width), we get

$$\frac{dV}{dy} = -\frac{e \cdot N_d}{\varepsilon_s}(y - a) \text{ volts/metre} \quad (6.12)$$

Integrating again with the boundary condition $V = 0$ at $y = 0$, we get

$$V = \frac{e \cdot N_d}{2\varepsilon_s}(y^2 - 2ay) \quad (6.13)$$

The pinch-off voltage V_p is at $y = a$, therefore Eq. (6.12) gives:

$$V_p = e \cdot N_d \cdot a^2/(2\varepsilon_s) \quad (6.14)$$

Thus, we see that the pinch off is a function of doping concentration N_d and channel width 'a'. With fully pinched-off condition, the FET is said to be in the OFF state, with I_d saturated.

As the drain voltage is increased, the avalanche breakdown across the gate junction takes place, increasing I_d sharply (Fig. 6.5).

Channel resistance can be expressed as

$$R = \frac{qL}{A} = \frac{L}{\sigma A} = \frac{L}{2 \cdot \mu_n e N_d z(a - w)} \quad (6.15)$$

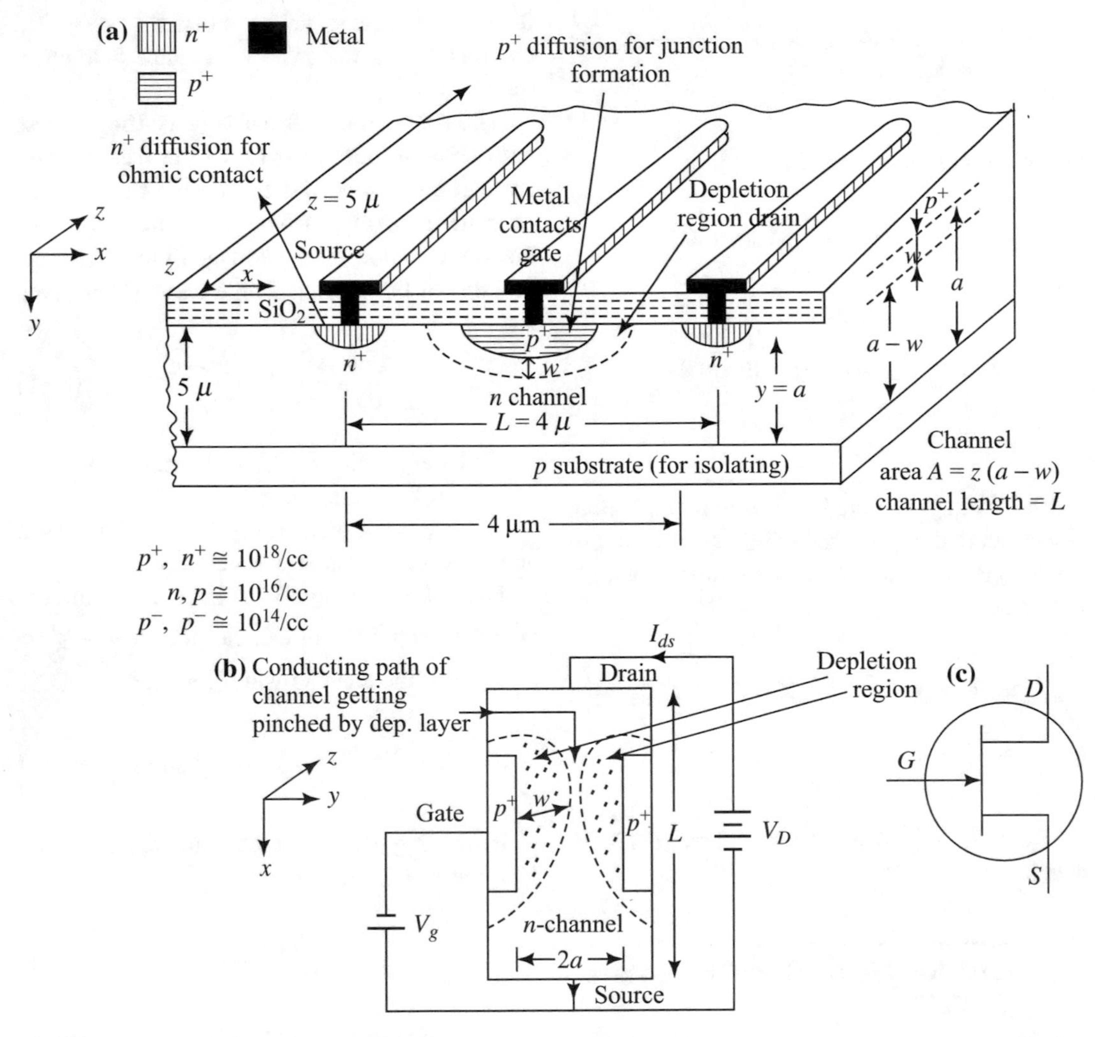

Fig. 6.4 **a** *n*-channel-Jn-FET-actual layout in planar technology giving typical diffusion densities and the measurements of its size. **b** Simplified figure used just for explaining the working of the *n*-channel-Jn-FET. Here diffusion (gate) is shown on both the sides, which is not actual, and **c** circuit symbol of *n*-channel-Jn-FET

μ_n	electron mobility
e	electron charge (coulombs)
L	distance between source and drain (see Fig. 6.4a, b)
z	length of the channel in z direction, i.e. as seen from top surface
a	width of the channel
w	diffusion depth of ohmic-n^+
$\varepsilon_s = (\varepsilon_0 \varepsilon_r)$	permittivity of material Si
A	area of the channel as seen from the right of the chip surface $= z \cdot (a - w)$
σ	conductivity of channel region
ρ	resistivity of the channel region
N_d	doping doner density/cc of the channel.

∴ Using Eqs. (6.13) and (6.14), we can get drain current at pinch off (i.e. saturation current) (as $a \gg w$, therefore $(a - w) \cong a$) as:

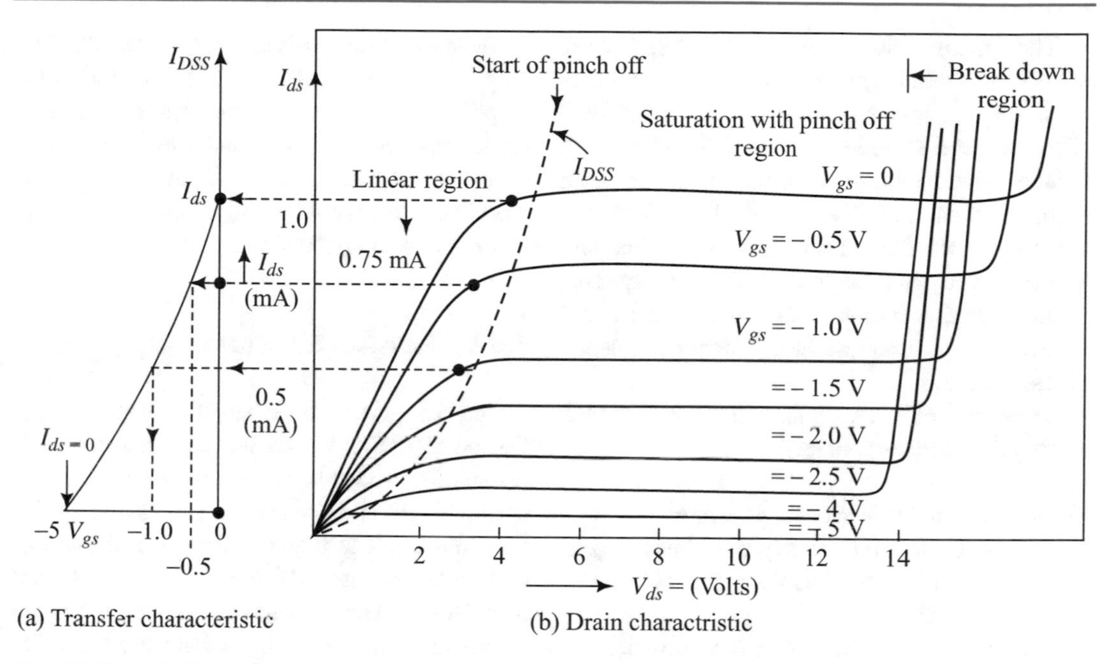

Fig. 6.5 Typical characteristic of a Jn-FET

$$I_{\mathrm{dSS}} = I_{\mathrm{dp}} = \frac{V_{\mathrm{p}}}{R} = \frac{\mu_n e^2 N_{\mathrm{a}}^2 z a^3}{L\,\varepsilon_{\mathrm{s}}} \quad (6.16)$$

It has been found experimentally that general equation of drain current (Shockley equation) is:

$$I_{\mathrm{d}} = I_{\mathrm{dss}}(1 - V_{\mathrm{gs}}/V_{\mathrm{p}})^2 \quad (6.17)$$

Cut-off frequency (i.e. highest frequency possible) is:

$$f_{\mathrm{c}} = \frac{2\mu_n e N_{\mathrm{d}} \cdot a^2}{\pi\,\varepsilon_{\mathrm{s}} L^2} \quad (6.18)$$

By differentiating Eq. (6.16), we get transconductance as

$$\begin{aligned} g_{\mathrm{m}} &= \left(-2I_{\mathrm{DSS}}/V_{\mathrm{p}}\right) \cdot \left(1 - V_{\mathrm{gs}}/V_{\mathrm{p}}\right) \\ &= g_{\mathrm{mo}}(1 - V_{\mathrm{gs}}/V_{\mathrm{p}}) \end{aligned} \quad (6.19)$$

Advantages of Jn-FET over BJT: Refer Sect. 6.3 for all the advantages of FET (See Sect. 6.3).

6.6 Metal–Semiconductor Field Effect Transistor (MESFET)

Instead of forming a rectifying contact of pn junction in Jn-FET, one can form a rectifying gate contact by a contact between lightly doped (n, n^-) semiconductor and metal also called Schottky diode. It may be noted that if the doping is high (n^+) then this junction, instead of Schottky diode, forms ohmic contact. These types of transistors are metal–semiconductor field effect transistor (MESFET). The majority carrier current from drain to source in an n-channel MESFET is controlled by a Schottky metal gate −ve voltage. Just like in Jn-FET, this V_{gs} forms depletion region in the semiconductor, thereby reducing the thickness of the conducting portion of the channel and hence the current I_{DS} reduces.

The only disadvantage of MESFET is the presence of Schottky metal gate, which limits the forward turn-on voltage to <0.7 V for GaAs Schottky diode.

The main advantage of MESFET over MOSFET is the higher mobility of channel carriers. The inversion layer of MOSFET (OFF-MOSFET) which extends into the channel reduces the mobility to half, than the bulk mobility in MESFET. In MESFET the depletion layer separates the carrier from the surface and hence mobility is close to the bulk mobility. This leads to higher current, transconductance, and smaller transit time and hence higher frequency of the device.

Thus use of GaAs rather than Si-MESFET offers additional advantages:

- Electron mobility five times larger.
- Saturated electron velocity two times larger.
- Higher current possible than Si devices.
- Low shot noise.
- Higher electric field before breakdown.
- Operates up to higher temperature than Si.
- Higher frequency than Si.
- Higher μW power output than Si.

Because of these advantages, GaAs-MESFET amplifiers have replaced X-band parametric amplifiers in airborne radar systems, due to less-complicated circuit and less expensive, besides having above-listed advantages. It is also used in microwave IC for high-power, low-noise, and broadband applications.

6.6.1 Physical Structure

Figure 6.6 gives the schematic diagram of a GaAs-MESFET, where we see that two thin layers of n^{-} and n-layers are grown on the thick substrate, either by epitaxial process or by ion implantation. The impurity densities of these n^{-} and n-layers are 10^{14}/cc and 10^{16}–10^{17}/cc, respectively. The n^{-} epitaxial layer of 3 μ is just to isolate the n-channel layer from substrate. The channel layer is very thin (0.15–0.35 μ), on which the metal contacts for gate/source (Au-Ge or Au-Te) on ohmic contact diffusion (n^{+}) region

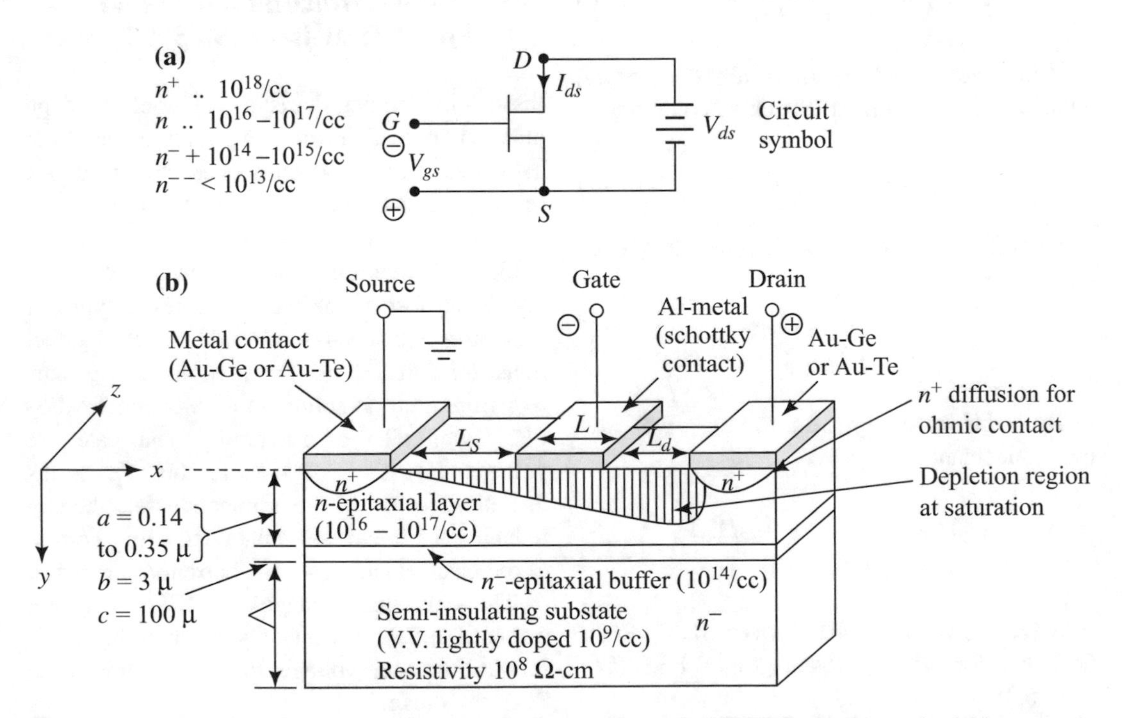

Fig. 6.6 **a** Device symbol in circuit and **b** schematic diagram of a GaAs-MESFET with thin expitaxial layers on the thick semi-insulating substrate. Thickness and impurity densities are given for each of the layer. Here L_s, $L_d \ll L$

and for gate (Al) directly on n-channel layer are put by photolithography method. These two diffusions are also very thin (0.1–0.2 μ). The substrate is doped (n^{--}) very lightly, with Cr and therefore is an insulating thick substrate, from which electrical contact is taken.

Operation: Once the drain-to-source channel is biased with +ve voltage, then the majority carrier current (electron) flows through the channel. This current causes a voltage drop along the channel from drain to source. As the reverse bias between gate and source increases, so does the width of the depletion layer, and finally pinches off the channel against buffer layer near the drain end. This reduces the path for electron flow and hence increasing the channel resistance. This leads to saturation of I_{ds} (as shown in its characteristic in Fig. 6.7), which does not increase after this, not even by increasing V_{ds}.

From Fig. 6.7, we observe that I_{ds} is fully controlled by the field of depletion region created by the gate voltage V_{gs} and hence the name to this FET has been given accordingly.

At pinch off, the −ve gate voltage removes all the free carriers from channel. This pinch-off voltage V_p is derived from Poisson's equation.

$$\frac{d^2V}{dy^2} = \left(-\frac{q}{\varepsilon_s}\right) = \frac{eN_d}{\varepsilon_r\varepsilon_0}$$

Integrating twice with the boundary condition of pinch off $V = V_p$ at $y = a$; we get:

$$V_p = \frac{eN_d \cdot a^2}{2\varepsilon_s} \tag{6.20}$$

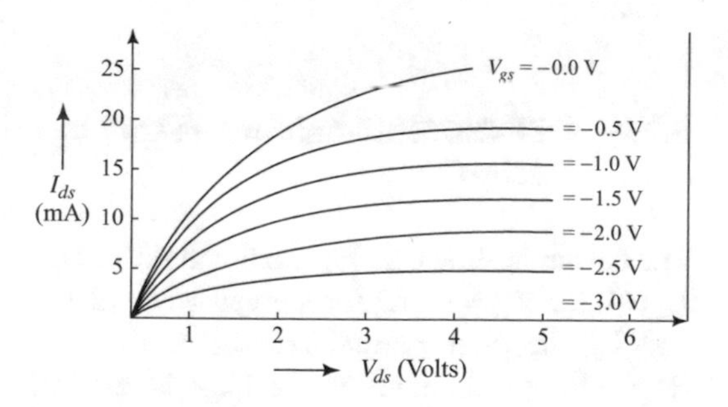

Fig. 6.7 Current–voltage characteristic of a typical n-channel GaAs-MESFET drain–source current-vs-drain–source voltage for different gate–source voltages

The saturation current at pinch off with $V_{gs} = 0$ is

$$I_p = \frac{e \cdot N_d \cdot \mu az \cdot V_p}{3L} \tag{6.20a}$$

where

- a Channel height as in Fig. 6.6 and
- N_d Doping density of channel (free electrons/cc)
- e Electron charge (coulomb)
- q Charge density (coulombs/cc)
- ε_s Permittivity of semiconductor (GaAs)
- L Gate length, i.e. effective channel length (L_s, $L_d \ll L$)
- z Channel length

For $a = 0.1$ μ, $N_d = 10^{18}$/cc; $\varepsilon_r = 13$; $V_p = 7.5$ V.

Thus V_p is dependent on N_d, 'a', and $\varepsilon_s = \varepsilon_0\varepsilon_r$

Drain current and mutual transconductance in a GaAs-MESFET are given by the following in a GaAs-MESFET.

$$I_{ds} = I_{dss}\left(1 + \frac{|V_g|}{V_p}\right)^2; \tag{6.21}$$

$$g_m = \frac{-2I_{dss}}{|V_p|}\left(1 + \frac{|V_g|}{V_p}\right) \tag{6.22}$$

The cut-off frequency (f_{co}) where current gain fall to unity $\beta = 1$ and the maximum frequency of oscillation (f_{max}) where power gain falls to unity, $A_p = 1$, with input–output matched load are given by:

$$f_{co} = \frac{g_m}{2\pi C_{gs}} = \frac{v_s}{4\pi L} = \frac{1}{4\pi\tau} \quad (\text{Here } \beta = 1) \tag{6.23}$$

$$f_{max} = \frac{f_{co}}{2} \cdot \left(\frac{R_d}{R_s + R_g + R_i}\right)^{1/2} \quad (\text{Here } A_p = 1) \tag{6.24}$$

Putting approximate values of these resistances gives:

$$f_{max} = 33000/L = 66\,\text{GHz} \quad (\text{for } L = 0.5\,\mu) \tag{6.25}$$

where R_d, R_s, R_g, R_i are internal device material resistances of drain, source, gate-metallisation and the gate-input side. L is the channel length and τ transit time of the electron to cross L with saturated velocity v_s.

Example A typical GaAs-MESFET has the device parameters as $R_d = 450\ \Omega$, $R_s = 2.5\ \Omega$, $R_g = 3\ \Omega$, $R_i = 2.5\ \Omega$, $g_m = 50$ mhos, $C_{gs} = 0.6$ pF; $L = 10\ \mu$, $a = 0.1$, $z = 500\ \mu m$, $N_d = 10^{18}/$cc, compute f_{co}, f_{max}, I_p and V_p.

Solution Using Eqs. (6.20)–(6.25) we get $f_{co} = 13.3$ GHz; $f_{max} = 49.7$ GHz, $V_p = 7.5$ V; $I_p = 6.6$ mA.

6.6.2 Application of MESFET

Because of so many advantages (as listed earlier), it is used in a number of microwave applications up to 50 GHz.

1. Satellite, receiver, radars, cellular devices, etc.
2. Power amplifier of output stage of microwave links.
3. Power oscillator in a number of applications.
4. Power driver amplifier for high-power transmitters.
5. Low-noise amplifier in microwave receivers etc.

Advantage: All the advantages of Si FET (Sect. 6.3) plus that of GaAs-MESFET.

6.7 Metal Oxide Field Effect Transistor (MOSFET)

All the transistors discussed so far, e.g. bipolar, Jn-FET, MESFET, are three terminal devices, with substrate isolated in Jn-FET and MESFET, while in bipolar transistor, the substrate is the collector itself bonded on the header directly. Thus MOSFET is a four-terminal device where substrate is 4th terminal normally connected to the source and is grounded. Rest of the three terminals being source, drain, and gate. In Jn-FET the p–n junction is at the gate while in MOSFET, there are two p–n junctions at source and drain itself. The MOSFET, because of its simpler structure and lower losses, has superseded the junction transistors (BJT and Jn-FET).

When the gate bias is zero, the two back-to-back pn junctions, between the source and drain, prevent the current flow in either direction. When in a *p*-type substrate MOSFET, a +ve voltage is applied to the gate with respect to source, i.e. v_{gs}, (with substrate and source grounded), then −ve charges are induced in the channel (like a capacitor) and this provides current flow in the channel.

As this MOSFET (Fig. 6.8) is with *p*-substrate, the channel region forms −ve carrier channel for current flow and therefore called *n*-channel MOSFET.

The structure given in Fig. 6.8 also gives the dimensions of the chip and its layers. In practice on a wafer, a large number of such chips are fabricated and chips diced out of it. A MOSFET can be a part of a circuit on a chip also and in such cases the MOSFET is normally surrounded by a thick oxide to isolate it from the adjacent device in a microwave I.C. Two designs of MOSFET are used, e.g. enhancement design (OFF-MOSFET), where n-channel region being very lightly *p*–type doped (10^{13}/cc), it has very less carriers therefore even with V_{ds} bias $I_d = 0$ for $V_g \leq 0$. But by $V_g = +$ve *n* carriers are induced in the channel region, then I_d starts (Fig. 6.8): The other is depletion type depletion design (ON-MOSFET), where *n* type (10^{15}/cc) doping is already done in the channel region, giving enough *n* carriers. Therefore with V_{ds} bias $I_d \neq 0$, whether $V_g \leq 0$ or $V_g \geq 0$ and hence ON-type the name is given.

6.7.1 OFF-MOSFET-Enhancement Design Type

The design is shown as Figs. 6.8 and 6.9, where $I_{ds} = 0$ for $V_{gs} = 0$ and for a given value of V_{gs}, and V_{ds} the drain current I_{ds} will be saturated (Fig. 6.9). A minimum gate voltage is required to induce (i.e. form) the channel, and it is known as threshold voltage V_{gth}. The gate voltage has to be larger than V_{gth}, before a conducting *n*-channel (mobile electrons) is induced. This type

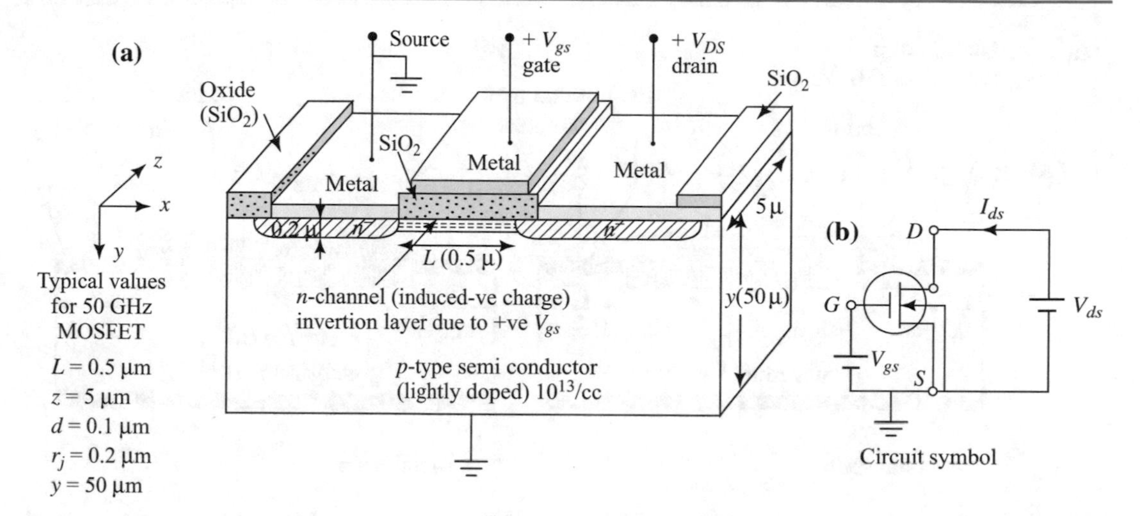

Fig. 6.8 **a** Schematic diagram of n-channel Si-MOSFET with dimensions: Enhancement design type OFF-MOSFET. **b** Circuit symbol

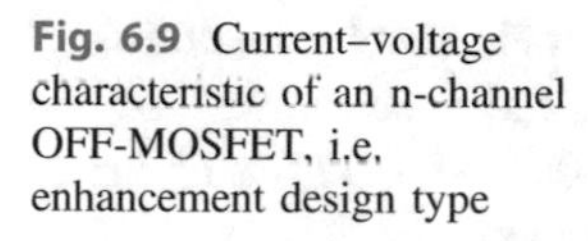

Fig. 6.9 Current–voltage characteristic of an n-channel OFF-MOSFET, i.e, enhancement design type

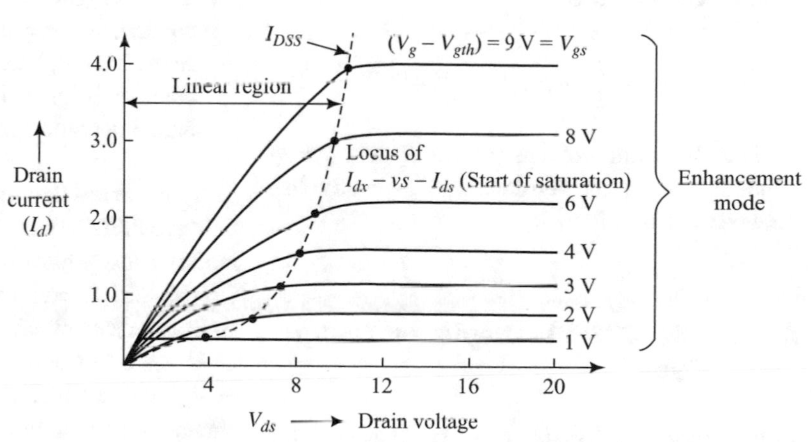

of device design, where the channel region which is p-type material of very lightly doped (10^{13}/cc) semiconductor, is called enhancement design. This device can work only in enhancement mode and not in depletion mode due to very low doping level in the channel and in the substrate area. Therefore $I_{ds} = 0$ for $V_{gs} = 0$ itself (Fig. 6.9), and the drain current can be given by:

(a) **In the linear region**

$$I_{\mathrm{d}} = \frac{Z}{L}\mu_n C_i\left(V_{\mathrm{g}} - V_{\mathrm{gth}}\right) \cdot V_{\mathrm{d}} \quad (\text{for } V_{\mathrm{d}} \ll (V_{\mathrm{g}} - V_{\mathrm{gth}})) \tag{6.27}$$

(b) **In the saturation region**

$$I_{\mathrm{d\,sat}} = \frac{mZ}{L}\mu_n C_{\mathrm{i}}\left(V_{\mathrm{g}} - V_{\mathrm{th}}\right)^2 \tag{6.28}$$

where $m = 0.5$ for low doping and $V_{\mathrm{d\,sat}} = (V_{\mathrm{g}} - V_{\mathrm{gth}})$

Transconductance in saturation region is given by:

$$g_{\mathrm{m}} = \frac{2mZ}{L} \cdot \mu_n C_{\mathrm{i}} V_{\mathrm{d}}\left(V_{\mathrm{g}} - V_{\mathrm{gth}}\right) \tag{6.29}$$

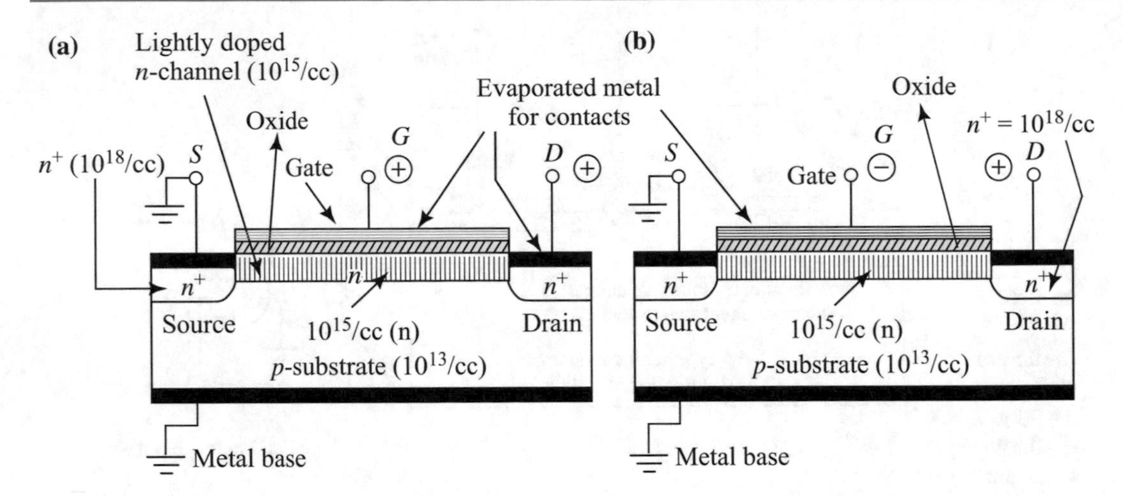

Fig. 6.10 Depletion design structure of *n*-channel ON-MOSFET operated in **a** enhancement mode gate +ve and **b** depletion-mode gate −ve

channel conductance

$$g_c = \frac{Z}{L} \cdot \mu_n C_i (V_g - V_{gth})$$

The threshold voltage (where I_d just starts) depends on doping densities and generally lies between 0.1 and 2 V.

6.7.2 ON-MOSFET—Depletion Design Type

In this design Fig. 6.10a, b we see that $I_d \neq$ for $V_{gs} = 0$ with $V_{ds} > 0$.

This is because of *n*-type of light, doping (10^{15}/cc) is done in the channel region during fabrication (and hence called soft junction, whereas in OFF-MOSFET no such doping is there. Therefore this design can function in the enhancement mode (+ve V_{gs}) as well as in depletion mode (−ve V_{gs}) as the doped n-channel already exists even with $V_{gs} = 0$. The two modes are

(a) **The +ve gate voltage** will induce/enhance more −ve carrier in this *n*-channel (like a capacitor effect due to SiO_2-insulator) and hence called enhancement mode (Figs. 6.10a and 6.11) operation.

(b) **The −ve gate voltage** will induce the +ve carriers in the channel, thereby neutralising/reducting depletes the −ve carriers of the channel (Figs. 6.10a and 6.11), and hence called depletion mode.

Thus we see that the enhancement design, i.e. OFF-MOSFET, can function in enhancement mode only while the depletion design, i.e. ON-MOSFET, can work in depletion mode as well as enhancement mode.

In the enhancement design type, the increase in V_{ds} will not increase the I_{ds} as there is no *n*-channel for conduction, therefore this design type is called OFF-MOSFET. In case of depletion design type, the I_{ds} can increase with V_{ds} even with $V_{gs} = 0$, as the channel carriers are present (due to doped layer) and therefore this design type is called ON-MOSFET.

So far we have discussed the *n*-channel MOSFET only, but all these are true for *p*-channel MOSFET also, with *n*-replaced by *p*.

6.7.3 Applications

MOSFET is generally used as power amplifiers as they have some advantages over BJT, Jn-FET, and MESFET, for example:

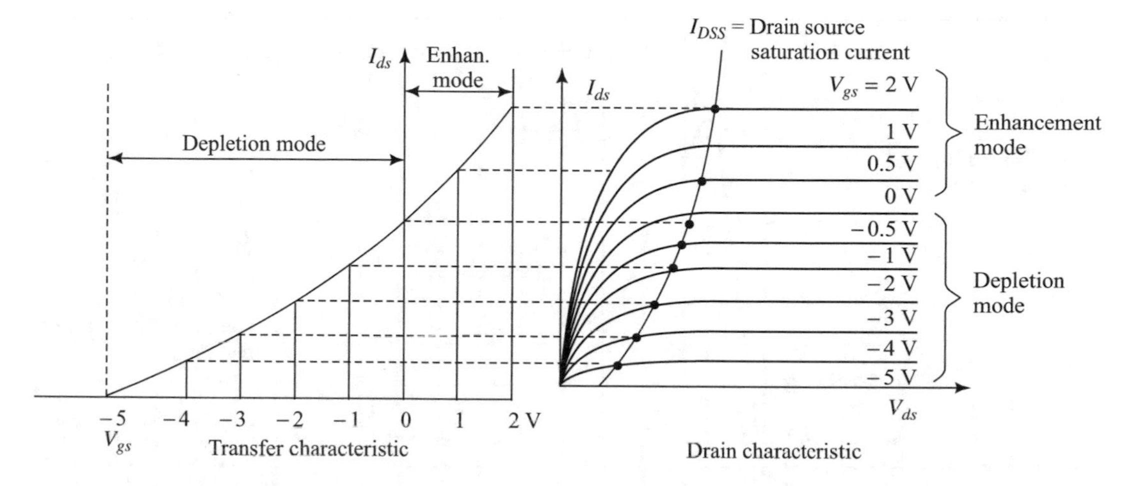

Fig. 6.11 Current characteristic of depletion design of n-channel ON-MOSFET-transfer and drain characteristic

1. It can be linear power amplifier in the enhancement mode as C_{in} and g_{m} do not depend on V_{g}, while c_{out} is independent of v_{ds}.
2. Gate ac input signal can be quite large as n-channel depletion-type ON-MOSFET can operate from depletion-mode region ($-V_{\text{g}}$) to enhancement mode region ($+V_{\text{g}}$).

Advantages: Refer Sect. 6.3 for all the advantages of FET (Sect. 6.3).

6.8 Tunnel Diode Characteristic, and Working Oscillators and Amplifiers

The tunnel diodes are heavily doped (degenerate) pn junction diode that exhibit −ve resistance over a portion of its forward I–V characteristic (Fig. 6.12) and are used as microwave amplifiers or oscillators. Before we study its working, let us consider the following three cases of doping:

1. Simple pn junction has doping density of 10^{14}–10^{15}/cc (around one is 10^{8} atoms), and it exhibits avalanche type of breakdown in the reverse bias with normal turn-on voltage of 0.7 V (in Si) in the forward bias. At high avalanche field $E_{\text{a}} = 400$ kV/cm, electron accelerates and knocks out the outermost electron of Si and multiplication takes place. **This diode is called avalanche diode.**
2. If the impurity doping is 10^{17}–10^{18}/cc, then the reverse breakdown voltage reduces as the depletion width reduces and at the same time the reverse breakdown mechanism become zener, with normal turn-on in the forward bias. As the depletion region is very small, the electron does not have enough distance to accelerate and hence field increases to 4000 kV/cm, i.e. still higher than E_{a} of 400 kV/cm. Then the 'e' of Si-atom is snatched away, as the external electric field is higher than the internal field of the atom, which $E_{\text{z}} \approx 3000$ kV/cm. **This diode is called zener diode.**
3. If the doping density increases very much near 10^{19}–10^{20}/cc (1 part in 10^{3} atoms), then the fermi levels of the p- and n-type semiconductors come very close to the respective band and merge into it. This type of semiconductor is called degenerate semiconductors and the **pn junction becomes tunnel diode**.

In this third case, the depletion layer width is very small of the order of 100 Å (or 10^{-6} cm or 0.01 μ). This diode has nearly zero breakdown voltage in reverse bias and acts as a conductor. In the forward bias, the $I \leftrightarrow V$ characteristic

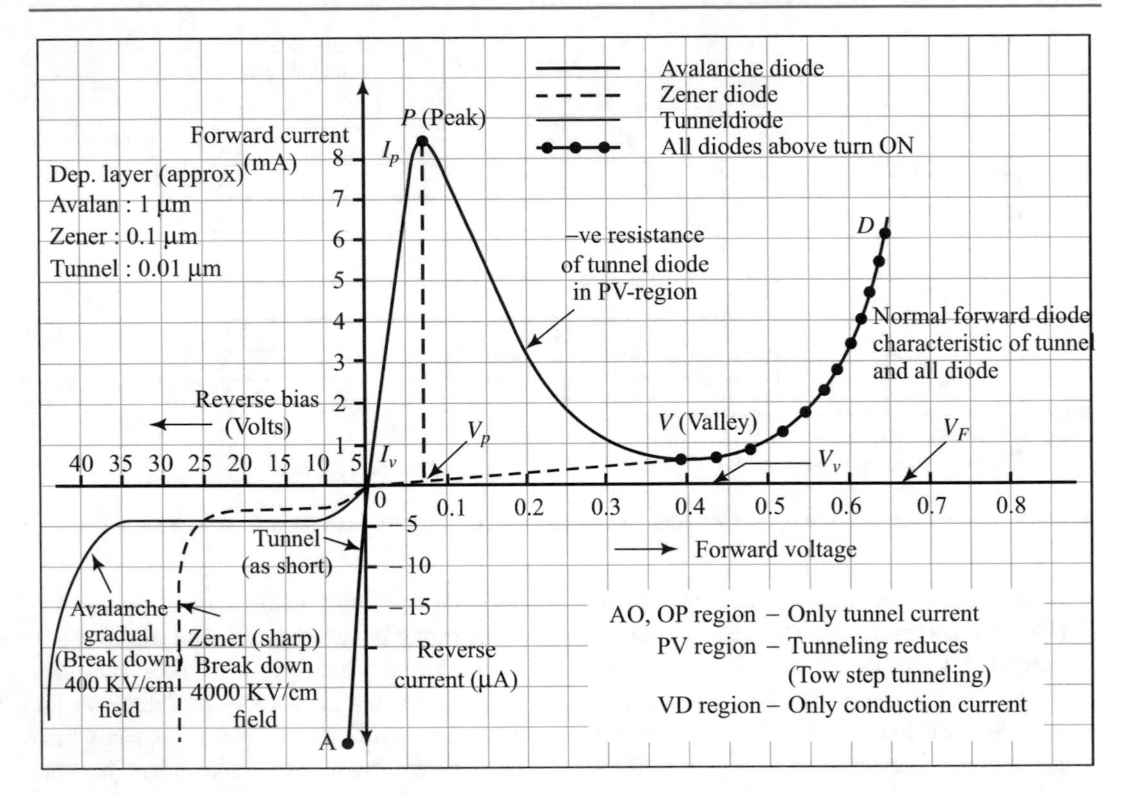

Fig. 6.12 *I–V* characteristic of tunnel diode. Also characteristic of avalanche and zener diode given for comparison for the four regions (AO, OP, PV, and VD) of the tunnel diode characteristic. The −ve resistance along PV is from −30 to −100 Ω depending on the doping level

Table 6.3 Typical tunnel diode parameters

Parameter	Ge	Si	GaAs
I_P/I_V	8	3.5	15 (theoretical value > 50)
V_P (V)	0.055	0.065	0.15
V_V (V)	0.35	0.42	0.50
V_F (V)	0.50	0.7	1.10

exhibits −ve resistance below the normal turn-on voltage of 0.7 (in Si) (Fig. 6.12).

The typical *I–V* characteristic of tunnel diode was first observed by Mr. Esaki in 1958 and therefore also called Esaki diode. This −ve resistance behaviour is due to tunnelling of majority carriers electrons from conduction band of '*n*' type to valence band of *p*-type from *P* to *V* part of the curve (Fig. 6.12). After *V* point, the normal majority current flow starts and the current increases again (Table 6.3).

Ge and GaAs have high I_P/I_V ratio, and therefore these are the material used for tunnel diode and not Si. Due to high mobility of carriers, GaAs has seven additional advantages:

(a) Operating frequency is high as 500 GHz and is controlled by external circuit.

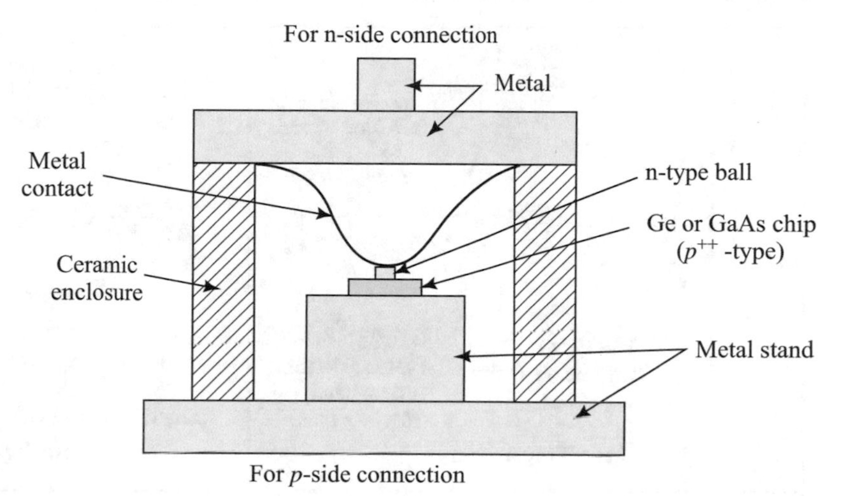

Fig. 6.13 Construction of a tunnel diode

(b) Wide frequency tuning is possible, as tuning is external circuit dependent.
(c) Temperature as high as 340 °C and low as −100 °C is possible for use and its performance is less sensitive to temperature.
(d) High switching speed from ON to OFF due to high doping density and higher mobility of carriers, leading to very small, recombination time.
(e) Because of high doping, tunnel diode is less sensitive to nuclear radiation/space radiation and therefore well suited for space applications.
(f) Very low noise.
(g) Very low-power requirement.

The construction of the diode is given in Fig. 6.13.

We know that the fermi level in a doner (n-type) material is inside the conduction band of energy level of:

$$E_f = E_C - kT \cdot \ln(N_C/N_D) \quad (6.30)$$

where

N_C electrons' energy states in the conduction band $\approx 10^{19}$/cc.
N_D doner level density.

For $N_D > N_C$, i.e. fermi level goes inside the conduction band and the semiconductor is said to become degenerate and diode becomes tunnel diode.

The −vc resistance behaviour is explained by the tunnelling theory of quantum mechanics in Fig. 6.14, by the band diagram of tunnel diode for various biasing conditions. The tunnelling of electrons takes place across the thin junction barrier; and it is defined as cross-band movement of electrons from conduction band of n-side of pn junction to valence band of p-side or vice versa.

(a) **Reverse bias ($V < 0$)**: Tunnelling of electrons of the valence band of 'p' side to the empty states of conduction band of 'n' side takes place. Large current flows as the depletion layer is very thin (100 Å). Here the diode behaves like a short with nearly zero resistance (reverse cross-band tunnelling).
(b) **Small forward bias ($V < V_p$)**: Tunnelling of conduction band electron of 'n' side to the vacant states of valence band of 'p' side, as the energy level of former got raised over the latter. This process increases the current (forward cross-band tunnelling).
(c) **Forward bias below ($V < V_V$)**: Direct band-to-band tunnel got reduced due to level difference but indirect tunnelling continues. Here the electron 'P' of conduction band of 'n' side goes to a local energy state 'Q' of the band gap and then falls to 'R' state in the

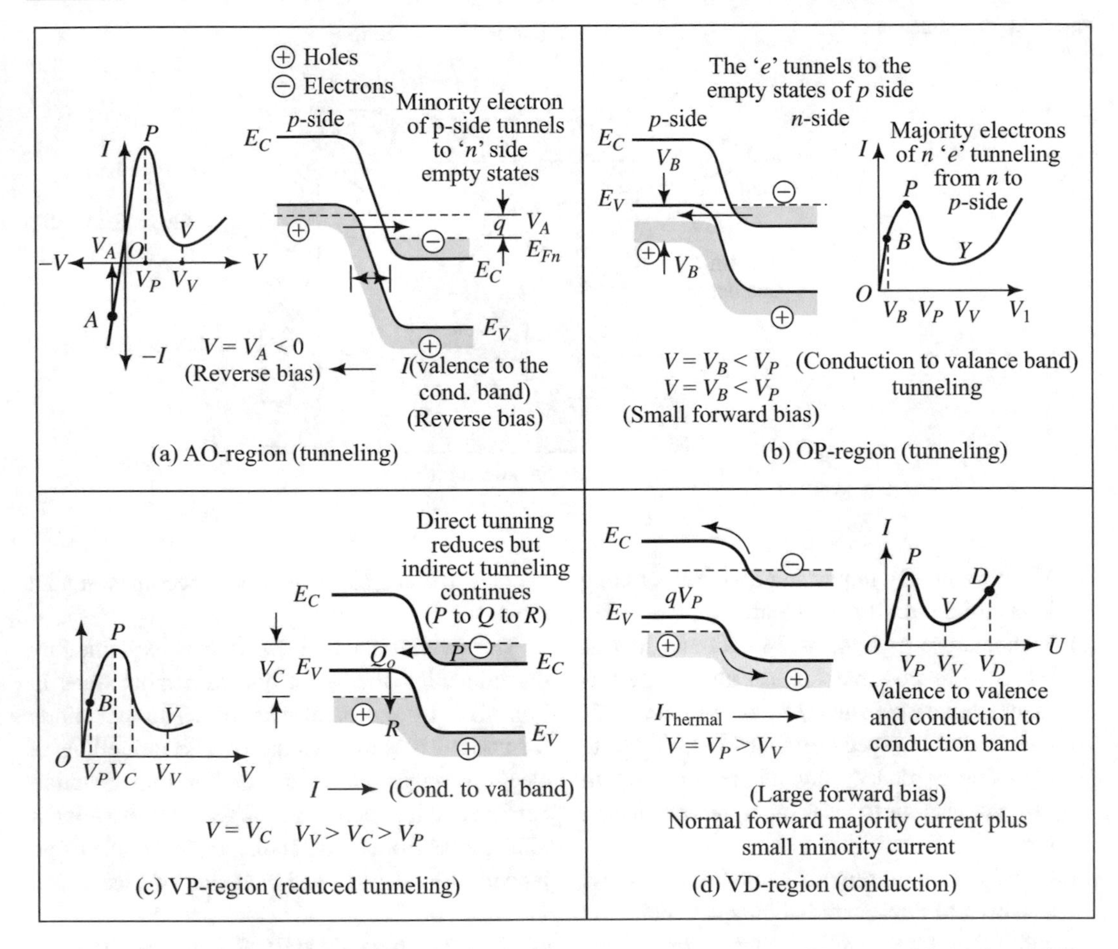

Fig. 6.14 Simplified energy band diagrams of a tunnel diode for various biasing conditions: **a** reverse bias-cross-band tunnelling, **b** small forward bias, cross-band tunnelling reverses, **c** cross-band tunnelling and −ve resistance condition, and **d** large forward bias, which causes minority carrier injection over the barrier as well as the majority carrier flow. No cross-band tunnelling

valence hand of '*p*' side. Thus the current falls, leading to −ve resistance region (cross-band forward tunnelling continues).

(d) **Large forward bias ($V > V_V$)**: Here the tunnelling current tends to zero and (see Fig. 6.12 also) conventional conduction current majority and minority current flows, i.e. conduction band-to-conduction band and valence band-to-valence band, and no cross-band charge flow. Here the current keeps growing exponentially like ordinary diode.

Thus we see that cross-band tunnelling of electron takes places in the reverse bias condition as well in the forward bias up to V_V. Beyond the bias $V = V_V$, normal conduction current (diffusion of majority and drift current of minority) starts flowing (Fig. 6.12).

Tunnel diode characteristic is relatively less sensitive to temperature. I_P increase or decrease with temperature depends on the doping of '*n*' and '*p*' sides, but V_P, V_V, and I_V decrease with temperature very little, e.g. $\partial V_V/\partial T \approx -1$ mV/°C.

The forward bias $I \leftrightarrow V$ characteristic (with refer to Fig. 6.12) can be represented approximately by the equation:

$$I_f = I_p(V/V_p) \cdot e^{(1-V/V_p)} + I_0(e^{V/V_T} - 1)$$
$$\text{(dominant in OPV region)} + \text{(dominant in VD region)} \quad (6.31)$$

Here $V_T = kT/e$. The first term is dominant from $V = 0$ to V_V and second term thereafter.

6.8.1 Tunnel Diode Equivalent Circuit

The equivalent circuit of a tunnel diode is shown in Fig. 6.15. Here R_j and C_j are the diode junction resistance and junction capacitance, while L_S and R_S are the lead inductance and lead resistance of the diode. Typical values of these equivalent components are in Fig. 6.15.

Therefore the impedance of the circuit will be:

$$Z = R_s - \frac{R_j}{1 + (\omega R_j C_j)^2} + jwR_jC_j\left[\frac{L_S}{R_jC_j} - \frac{R_j}{1 + (\omega R_jC_j)^2}\right] \quad (6.32)$$

The condition of oscillation is that each of the real and imaginary parts = 0

i.e. $$R_e(Z) = 0 \quad \text{and} \quad I_m(Z) = 0 \quad (6.33)$$

Correspondingly two important frequencies are given by:

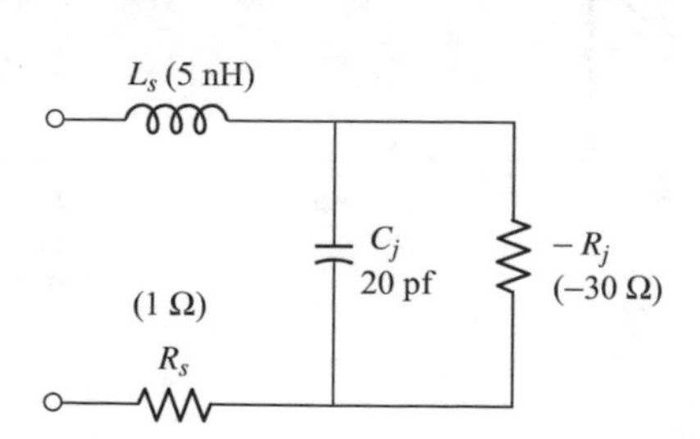

Fig. 6.15 Equivalent circuit of tunnel diode with typical values

Resistive cut off frequency

$$f_T = \frac{1}{2\pi R_jC_j}\sqrt{\left[\frac{R_j}{R_s} - 1\right]} \quad (6.34)$$

Self resonant frequency

$$f_{so} = \frac{1}{2\pi}\sqrt{\left[\frac{1}{L_sC_j} - \frac{1}{(R_jC_j)^2}\right]} \quad (6.35)$$

This f_{so} could be as high as 1000 GHz depending upon the design. At lower frequencies only R_j is effective.

6.8.2 Tunnel Diode Amplifier and Oscillators

(a) Amplifier

Tunnel diode can be used as reflection amplifier by using a circulator to isolate the source from load. Tunnel diode reflection amplifier with circulator is shown in Fig. 6.16. The −ve resistance of zener diode is −30 to −100 Ω, which leads to amplification.

The diode negative resistance R_j must be smaller than that of the characteristic impedance Z_o of the circulator for sustained amplification. At the diode port, the voltage reflection coefficient Γ is given by

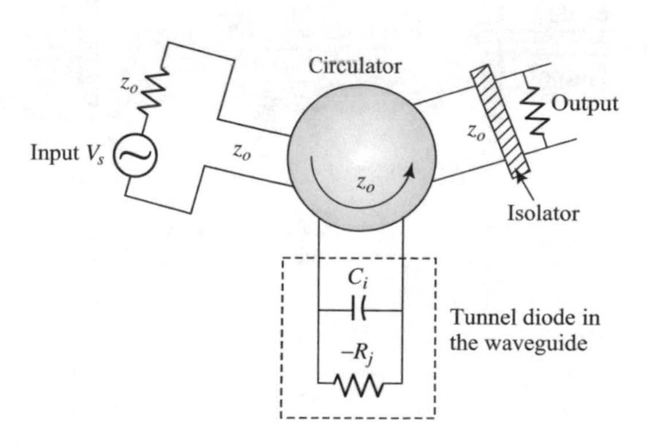

Fig. 6.16 Tunnel diode amplifier using circulator

$$\Gamma = \frac{Z_{in} - Z_o}{Z_{in} + Z_o} \quad (6.36)$$

As C_j is small, it can be neglected in Z_{in} of the diode.

$$\Gamma = \frac{R_j - Z_o}{R_j + Z_o}$$

As power to the load is $P_L = |\Gamma|^2 P_{in}$

$$\therefore \quad \text{Power gain} \quad G_p = |\Gamma|^2 = \left|\frac{R_j - Z_o}{R_j + Z_o}\right|^2 \quad (6.37)$$

(b) Tunnel Diode Oscillator

The diode is used with external tunable circuit, where we use only $(-R_i)$] as $f_0 \ll f_{so}$, where C_j, L_s, R_s are not effective.

Figure 6.17 gives the tunnel diode oscillator and its equivalent circuit. Here the bias of the diode is adjusted at the −ve resistance region by R_B and is isolated from the diode circuit by an inductance L_B. The diode is coupled with the cavity, and for this a dc isolator capacitance C_1 is there. The capacitance C_2 is the tuning screw capacitance of the cavity (or iris in waveguide cavity). L_B of the bias line blocks the μW from reaching the bias port. The equivalent parallel load resistance of the cavity is given by:

$$R_p = 1/\left(\omega^2 \cdot C_2^2 R_1\right) \quad (6.38)$$

For sustained oscillation, $R_p < |R_S|$, and this is achieved by adjusting C_2.

6.8.3 Applications

Because of the seven major advantages described earlier (of wide range of frequency wide tuning range, low biasing, low noise, high switching speed and less sensitive to temperature change or radiation), the GaAs tunnel diodes are used as low-noise amplifier, trigger circuits in CRO, high-speed counters, local oscillator for VHF-TV tuner, memories, etc.

The only disadvantages of this diode are (a) requirement of precise and low dc voltage (<0.5 V) (b) very low μW power output (<1 mW).

6.8.4 Performance Characteristic

1. **Frequency**: Tunnel diode designs can be made for 1–100 GHz frequency using GaAs.
2. **Power**: It gives quite low power in the range of 700 μW even for larger area of the diode.

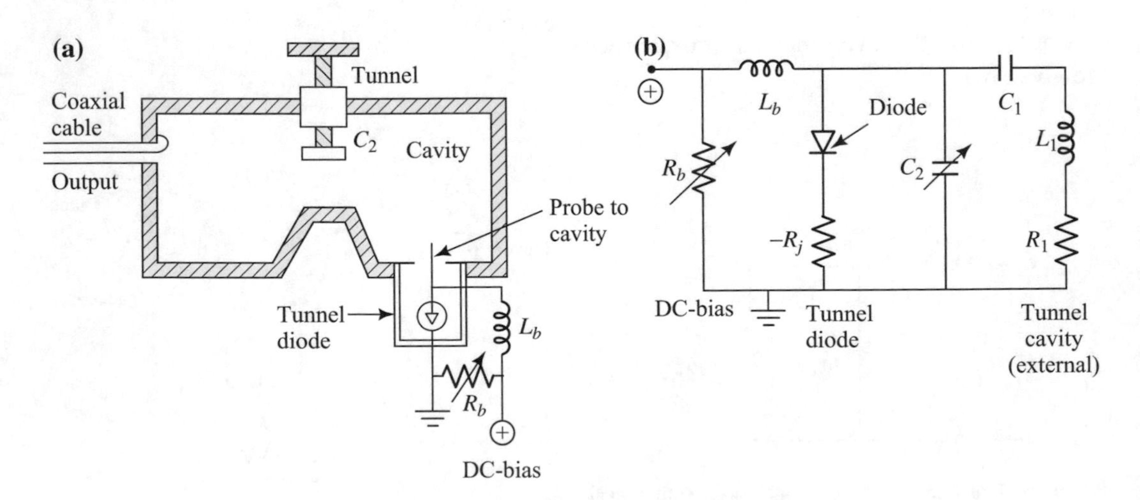

Fig. 6.17 **a** Tunnel diode oscillator and **b** its equivalent circuit with −ve resistance

3. **Tuning**: Frequency tuning of $f_0 \pm 10\%$ is possible. ($f_0 \ll f_{so}$).
4. **Gain**: As an amplifier, gain up to 30 dB is possible.
5. **Biasing range** 0.1–0.4 V, $I \approx 2 - 10$ mA.
6. **Negative resistance** (R_j) = −20 to −100 Ω (depending on doping etc.).
7. **Junction capacitance** (C_j) ≈ 5 pF.
8. **Series resistance** (R_S) ≈ 1 Ω max (depends on the areas).

6.9 Transferred Electron Devices (TED)—Gunn Diodes

6.9.1 Introduction-Bulk Device with No Junction

Over the years, the two-terminal devices have been found to give higher and higher *CW* and pulsed power at higher microwave frequencies 1.0–100 GHz, as compared to these from the best power transistors. The common property of all two-terminal active devices is their negative differential resistance as voltage and current are out of phase leading to power generation ($-i^2R$) instead of power absorption ($+i^2R$). All these devices have some or the other form of junction of two types of material.

Gunn diodes are also two-terminal but a bulk device, without any junction. It is also called diode as it has two terminals. Gunn diodes are also −ve resistance device, normally used as low-power oscillators at microwave frequencies in transmitters, local oscillators of receiver front end.

Gunn (1963) discovered microwave oscillators in GaAs, InP, and CdTe. All these semiconductors have closely spaced two or three energy valleys in the conduction band (Figs. 6.18 and 6.19). At dc voltage and hence low electric field (E_f) in the material, most of the electron will be located in the lowest valley. At higher E_f beyond E_{th} most of the electrons will be transferred to the high-energy satellite valleys, where the effective electron mass is much larger and hence the mobility (μ_2) and velocity (v) are much low than that at lowest valley (μ_1). As the conductivity is proportional to mobility, the conductivity and hence current decreases with

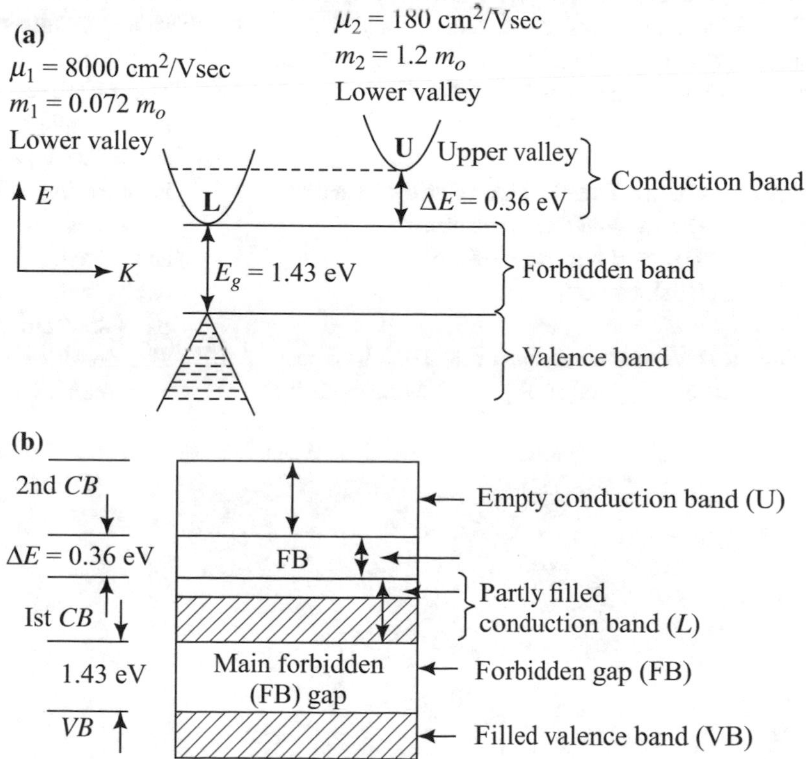

Fig. 6.18 The two-valley model of the *n*-GaAs conduction band at no bias condition. A narrow forbidden gap (ΔE = 0.36 eV) exists between the two valleys in the conduction band

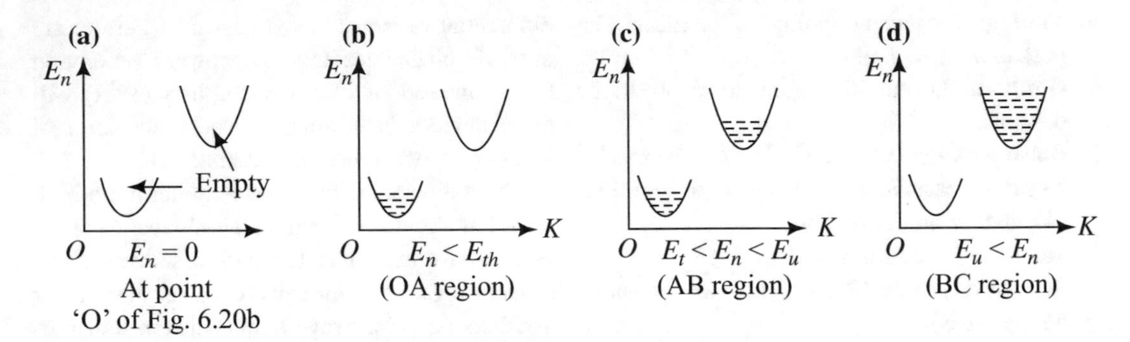

Fig. 6.19 The two valleys of the conduction land and transfer of electrons from valence band to valley one and then to two, valley two with increase of voltage and hence electric field. The fours states *O*, *A*, *B*, *C* are with reference to Fig. 6.20

higher $E_f(E_f > E_{th})$ or voltage ($V > V_{th}$). This is called transferred electron effect, and the device is also called transferred electron device (TED) or Gunn diode Fig. 6.20. The bulk material (without any junction) behaves as a −ve resistance device over a range of applied voltage and therefore used as microwave oscillators.

In the energy (E_n) wave number (*K*) diagram, the valence band will have electrons in valley-like structure, while in the conduction band we will have two valleys, e.g. the lower valley and upper valley (corresponding) to the two conduction bands (CB-1 and CB-2) as in Fig. 6.18. The properties of electrons and energy gap between the two valleys of some of the semiconductors are listed in Table 6.4.

6.9.2 Gunn Effect: Two-Valley Theory (Ridley–Watkins–Hilsum Theory for dc −ve Resistance)

The dc I.V. characteristic of a TED has been found to be as given in Fig. 6.20 along with the drift velocity curve, where we see that for a single value of current, there will be two different fields or two voltages, due to the −ve resistance region appearing in between. This −ve resistance, i.e. Gunn effect, can be explained by the theory of two-valley Ridley–Watkins–Hilsum theory in the conduction band. The basic mechanism in the operation of the bulk *n*-type GaAs device is the transfer of electrons from lower valley (*L*) to upper valley (*U*) in the conduction band. Figure 6.18a, b gives the energy-wave no (*E*–*K*) diagram and band diagram.

For Gunn effect, following three conditions are there for the material:

(i) The effective mass of electrons in the lower valley is smaller than normal ($m_1 = 0.072\ m_0$) and in the upper valley it is larger ($m_2 = 1.2\ m_0$). Accordingly their mobilities are large ($\mu_1 = 8000$ cm²/Vs) and small ($\mu_2 = 180$ cm²/Vs), respectively, while normal electron mobility is $\mu_0 = 8500\,\text{cm}^2/\text{Vs}$.
(ii) The densities of states (allowed slots of electrons) in the upper valley are much

Table 6.4 Band energy gaps, threshold electric field and peak electron velocity in some of the Gunn effect materials

Semiconductors	Eg (eV)	Separation of two valleys (ΔE) (eV)	E_{Th} (kV/cm)	Peak velocity V_P ($\times 10^7$ cm/s)
Ge	0.8	0.18	2.3	1.4
Ga As	1.43	0.36	3.2	2.2
In *P* (it has 3-valleys)	1.33	0.60	10.5	2.3
In Sh	0.16	0.41	0.6	5.0
In As	0.33	1.28	1.60	3.6

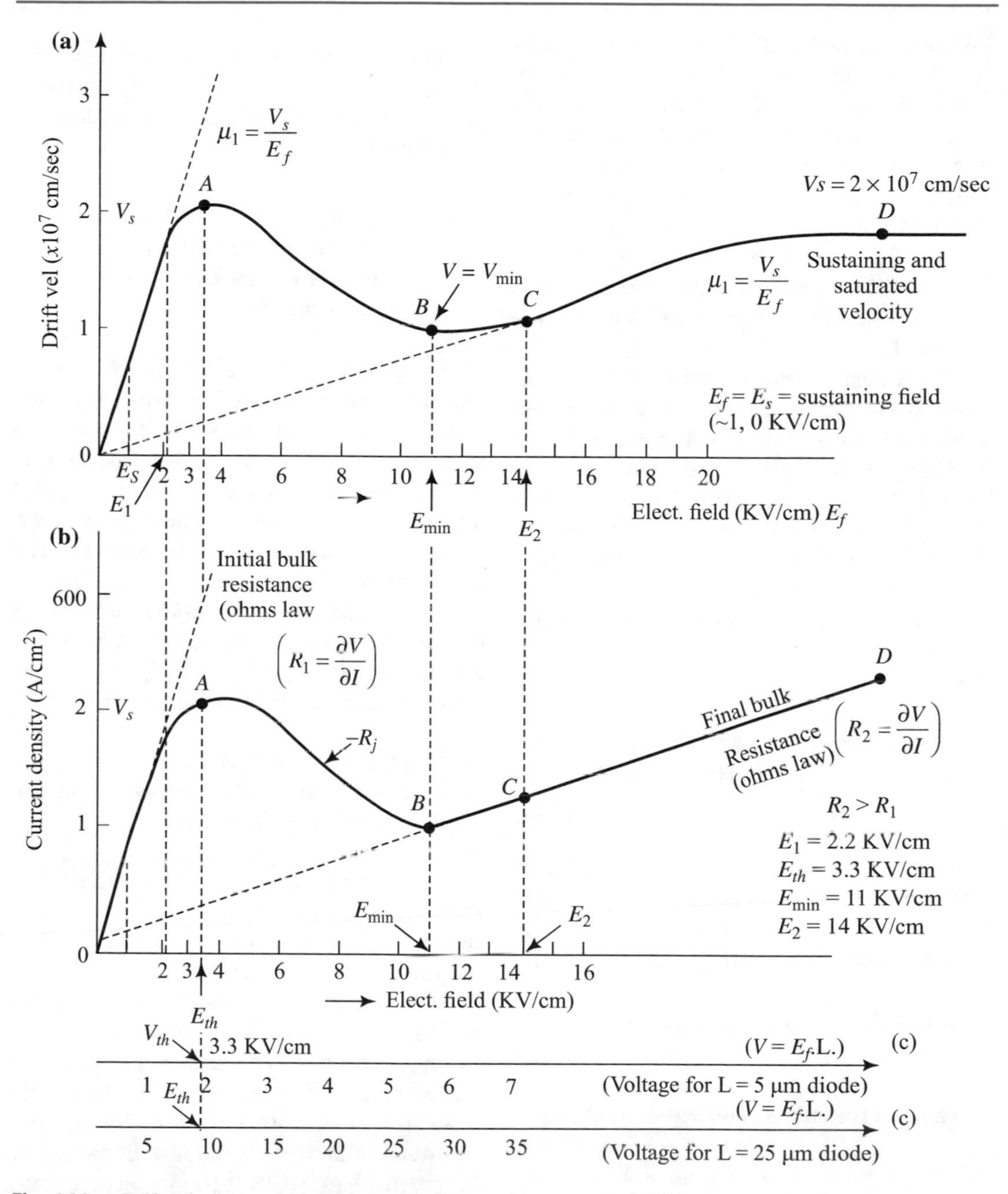

Fig. 6.20 **a** Drift velocity vs elect. field showing lower and upper valley fields (E_1, E_2). Threshold (E_{th}), sustaining velocity field (E_S), valley minimum field E_{min}, **b** current elect. field curve showing the three resistances R_1, R_j, R_2, and **c** corresponding voltage scale for 5 and 25 μm diode

larger (60 times in GaAs) than in lower valley.

(iii) The energy gap (ΔE_n) between valley U and L has to be greater than the thermal energy ($kT = 26$ meV) of electron at room temperature, ($\Delta E_n \gg kT$) or else at room temperature itself electrons will get transferred from 'L' to 'U', without increase of bias voltage.

As the bias voltage and hence the field increase (OA region) from $E_f = 0$ to E_{th} the

I–V curve follow the Ohm's law (Fig. 6.20) with the mobility of lower valley, as the electron keeps moving from valence band to the lower valley of conduction band (Fig. 6.19a, b). Beyond V_{Th} and E_{th}', the electrons start getting transferred from lower to upper valley (Fig. 6.19c) and get fully transferred at $V = V_{min}$ or $E_f = E_{min}$, where higher mobility electron dominates the current. Therefore the resistance R_1 at OA is lower than at BC regions (R_2) (Fig. 6.20).

The current density is given by $J_1 = \sigma \cdot E_f$ when all the carriers (i.e. doping density n_0) are in the lower valley (region OA in Fig. 6.20). As the voltage increases beyond V_{th} and hence field beyond E_{th}, some electrons (n_2) out of n_0 electrons get transferred to upper valley and therefore the new current density for AB region will be:

$$J = \sigma E_f = e(n_1 + n_2).\mu.E_f = en_0\bar{\mu}E_f \quad (6.39)$$

where

$$\bar{\mu} = (n_1\mu_1 + n_2\mu_2)/n_0 = \text{weighted average of the mobility of electrons (with } n_0 = n_1 + n_2) \quad (6.40)$$

Further increase of voltage (up to point B) results into transfer of almost all the electrons from L-valley to V-valley and as a result the current density [in the BC region (Fig. 6.20)] becomes

$$J_2 = \sigma_2 E_f = en_2\mu_2.E_f \approx en_0\mu_2 E_f(n_1 = 0) \quad (6.41)$$

Thus we see that conductivity of GaAs for

$$\left.\begin{array}{ll} \text{For } E_f = 0 \text{ to } E_1 & \sigma_1 = n_1 e\mu_1 = +\text{ve} \\ \text{For } E_f = E_1 \text{ to } E_2 & \sigma_{12} = (n_1 + n_2)e\bar{\mu} = -\text{ve} \\ \text{For } E_f = E_2 \text{ onwards} & \sigma_2 = n_2 e\mu_2 = +\text{ve} \end{array}\right\} \quad (6.42)$$

There σ_{12} is −ve as $J = \sigma E$ relation shows that as $\sigma_{12} = \partial J/\partial E = -\text{ve}$ as J and E_f are out of phase in AB region in this DC −ve conductance (i.e. −ve resistance) region.

6.9.3 Moving High-Field Dipole Domain in the Device and the Phase Difference in ac I and V

In the last section, we saw how the dc −ve differential resistance occurs in a two-valley semiconductor compound like GaAs. Here we will study how the decrease in the drift velocity with electric field can lead to the formation of a high-field dipole domain and hence −ve ac differential resistance for microwave generation and amplification.

In an n-type GaAs, the majority carriers are electron. At low OA region and high (BCD-region) voltages (where $E_f < E_{th}$ and $E_f > E_{min}$), the electric field and conduction current density are:

(a) Proportional to each other.
(b) Uniform throughout the device length and therefore follows Ohm's law.

The current is carried by the free electrons of fixed +ve charge of doners, keeping the net space charge = 0. Density of the doner minus acceptors is equal to the doping density.

High-Field Domain Formation: When applied dc voltage is above V_{th} (with $E_f > E_{th} = 3.3$ kV/cm), then a high-field domain is formed near the cathode. As this region will be the first to experience the inter-valley transfer than the rest, as a result of this (a) electric field in the rest of the material length falls (b) which in turn drops the current to 2/3rd of the maximum value.

The reason for this is obvious, as the applied dc voltage bias is fixed and given by,

$$V = \int_0^L E_f(x)\mathrm{d}x = \text{constant dc bias} \quad (6.42a)$$

Therefore increase of 'E_f' in one region will lead to decrease in rest of the region (Fig. 6.21).

Dipole Domain Formation: In this high-field domain of the device length, electrons from valley 1 gets transferred to valley 2 (Fig. 6.20) and as a result, their velocity drops. The electrons to the right of the domain move out to the anode faster, causing deficiency (region-D_{ep}) of electron (i.e. makes it +ve charged). To the left of the domain, slow-moving valley 2 electrons get accumulated (region A_{cc}), and thus form the dipole charge region around the high-field domain (Fig. 6.21).

This space charge dipole as well as the high field of the domain keeps growing while moving right (towards anode) and exits out of the anode as a current–voltage pulse. Immediately after this, the electric field again grows to a uniform value and the domain formation restarts at the cathode end, i.e. left end. This shows the phase lag of 180° between the current pulse and voltage at output terminal hence the −ve dynamic (ac) resistance. Thus the frequency of the

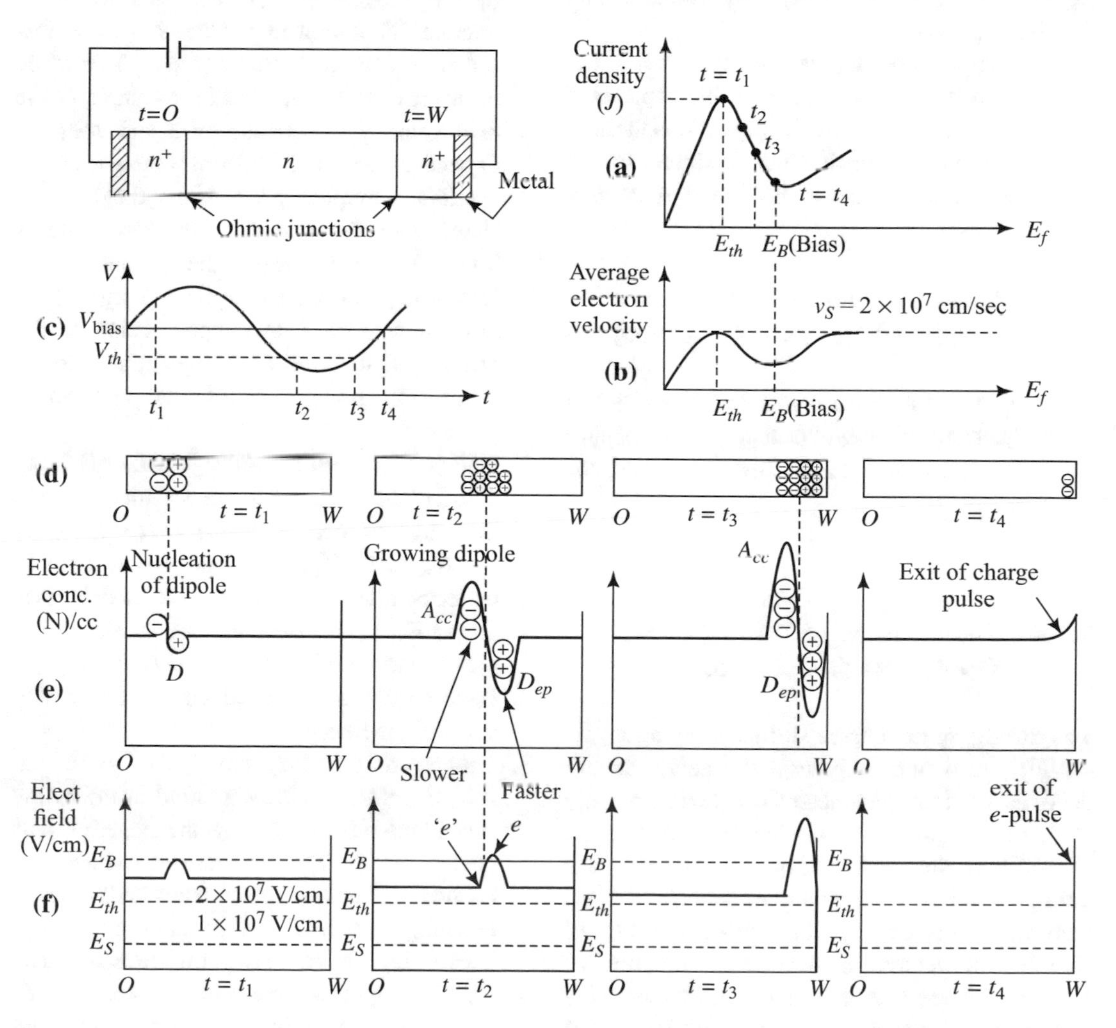

Fig. 6.21 **a** Current density, **b** electron velocity as a function of electric field, **c** changing voltage (DC bias + ac signal) across the diode, for different moments t_1, t_2, t_3 and t_4. Corresponding snap shots of **d** moving *e–h* dipole in the device, **e** electron concentration per cc, **f** varying elect. field, integral of which is voltage given at **c** E_f changes from E_{th} to E_B

current–voltage pulse (i.e. microwave signal) depends on the transit time of the domain across the device length, which in turn depends on the average drift velocity v_d (which is around 10^7 cm/s) during the transit from $x = 0$ to $x = L$.

Therefore the fundamental frequency will be:

$$f = v_d/L = f_{tt}(\text{transit time frequency})$$

The domain has the following properties:

(a) Its charge density is proportional to doping density N_d.
(b) Its width (d) is inversely proportional to the doping density N_d.
(c) It disappears before reaching the anode if the voltage drops below V_{th} and this happens if the device length $L \gg L_{diffusion}$. This thermal diffusion length decides the maximum size of L and hence decides the lower frequency range as:

$$L_{diffusion} = 1\,\mu(\text{for}\, N_d = 10^{16}/\text{cc})$$
$$\text{and} \quad = 10\,\mu, \quad \text{for}(N_d = 10^{14}/\text{cc})$$

(d) Its width will increase with DC bias as it has to absorbs more voltage for keeping Eq. (6.42a) true as E_S = sustaining field, for $V = V_g$

6.9.4 Four Modes of Gunn Device Operation as Oscillator

The strong space charge domain formation, its stability, and transit through the device length depends on four parameters (a) doping density (N_d), (b) length of the device (L), (c) external circuit and the cavity frequency (f), (d) bias voltage applied. As the domain length $d \propto 1/N_d$, therefore the device with similar product of $N_d \cdot L$ will behave similarly as a function of $f \cdot L$ or voltage/L or efficiency. Therefore ($N_d \cdot L$)-vs-($f \cdot L$) graph is very much useful (Fig. 6.22).

Depending upon these four parameters, and because carrier velocity varies with elect. field E, a Gunn oscillator can be made to oscillate in any of the following four frequency modes (Fig. 6.22) within doping range defined by $N_d \cdot L = 10^{12}$–10^{14} and frequency range defined by $f \cdot L = 10^6$–10^8.

1. **Transit time mode ($f \cdot L \approx 10^7$ cm/s)**: When the electron drift velocity (v_d) is equal to the sustaining velocity v_S (Fig. 6.18), the high-field domain is stable and therefore the oscillation period is equal to the transit time ($t_0 = t_{tt} = 1/f_{tt} = (v_S/L)$). For this mode to operate, the circuit impedance has to be low for the domain to continue to stay. This mode is sensitive to the applied bias voltage as the drift velocity v_d depends on it and does not depend on external cavity resonators, as the oscillation frequency $f_{tt} = 1/t_{tt}$ = fixed.
   Power and the efficiency of this mode is lowest ($\approx$10%) because the current is collected only when the high-field charge dipole domain arrives at the anode. Prior to the arrival of domain field, electron velocity v_s is very low and so is the conduction current.
2. **Delayed domain mode $f \cdot L = 10^6$ to 5×10^6cm/s**: Here the transit time is such that domain gets collected at $t = t_{tt}$ while $e < E_{th}$ as in Fig. 6.22, and a new domain collected after a while, when the ac field rises above E_{th} again. Here the oscillation period $t_0 > t_{tt}$ and therefore $f < f_{tt} = 1/f_{tt}$. This mode is controlled by external circuit and cavity and has efficiency around 20%.
3. **Quenched domain mode ($f \cdot L = 10^7$ to 4×10^7 cm/s)**: If the bias field drops below sustaining field E_S during the negative half cycle (Fig. 6.22), the domain which got formed collapses (gets quenched) before reaching anode. The bias field goes back above E_{th}, then only new domain gets formed, repeating the process again. This mode is controlled by the external circuit and the cavity resonator.

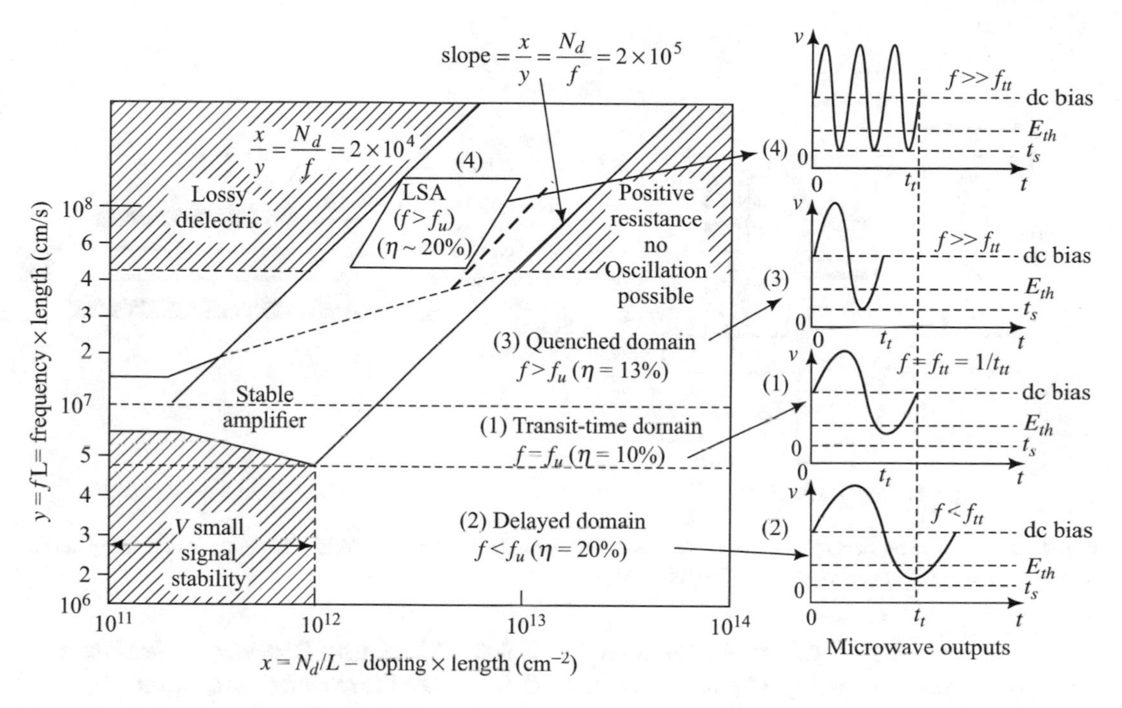

Fig. 6.22 Modes of operation for Gunn diodes, for a given bias, with different design regions in fL-vs-$N_d \cdot L$ plot

The frequency of oscillation is the frequency of cavity resonator and its frequency $f_0 > f_{tt}$. Efficiency of this mode is around 13%.

4. **Limited space charge accumulation (LSA) mode** ($f \cdot L > 4 \times 10^7$ cm/s): Upper and lower boundaries of this mode are between slopes $N_d/f = 1 \times 10^4$ to 2×10^5. This mode has special importance due to its high power and efficiency. In this mode the external circuit, cavity frequency, and the bias are so taken that the frequency is quite high therefore domain does not have sufficient time to form while the field is above E_{th}. Therefore the domains are maintained during the negative resistance state during the large fraction of the microwave voltage cycle. Therefore the internal field will be uniform in the sample and never crosses E_{th}. Therefore the current in the device (J) is proportional to v_d. The frequency of oscillation is independent of transit time and solely depends on external circuit and hence is in our control. The efficiency of this mode is around 20% or so with power output from 6 kW pulsed at 2 GHz to 400 W pulsed at 50 GHz.

The limitations of the LSA modes are:

(a) Very much sensitive to load conditions, temperatures, and doping.
(b) r–f circuit should allow the field to build up very fast, so as to prevent the domain formation.
(c) Power output is proportional to the volume of the device LA. (A = top surface area), but cannot be increased indefinitely due to electrical wavelength, skin depth, thermal dissipation limits, etc. The normal LSA mode device has L = 10–200 μm and A = 5 μm × 20 μm.

Out of the above four modes, only transit time mode has its own frequency of oscillation while in the rest, external circuit decides the frequency.

As an amplifier: The Gunn device can also be used as amplifier with $10^{11} < n_d \cdot L < 10^{12}/\text{cm}^2$, where the negative conductance region is used with limited domain formation and without

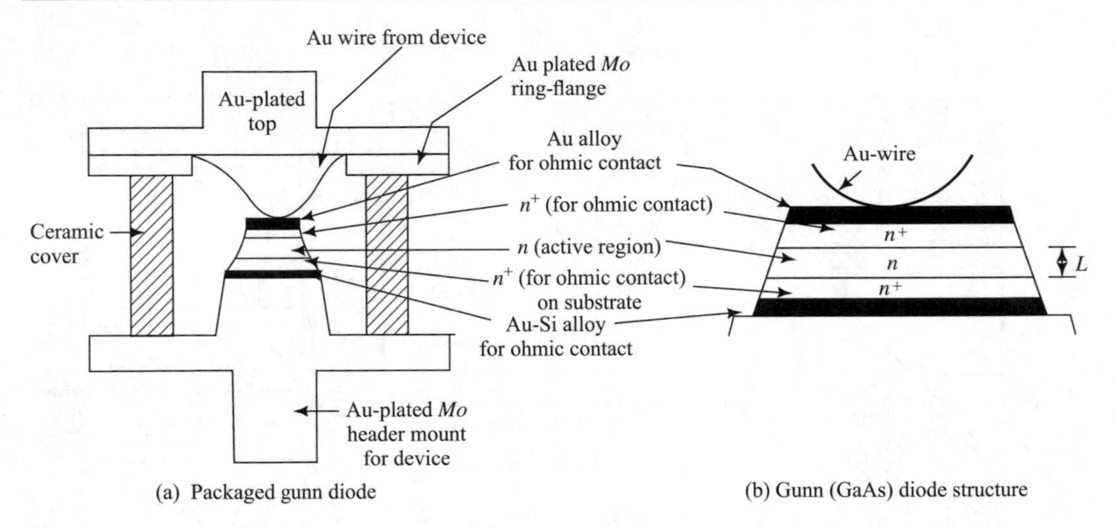

Fig. 6.23 **a** A typical packaged Gunn diode and **b** the GaAs device structure. Width of active region L depends on mode and frequency of operation. L = 50–100 μm

oscillation (Fig. 6.22). The frequency used is near transit time frequency (f_{tt}) and does not depend on the circuit, cavity, etc.

6.9.5 Diode Structure and Packaged Diode

Figure 6.23 gives the device structure as well as the packaged diode, and n^+ GaAs substrate, epitaxial layer of n-active layer is grown and then n^+ layer diffused on it for ohmic contact. The active n-layer is of around 150 × 150 μm cross-sectional area with width depending upon the desired mode and frequency of operation and is L = 5–100 μm. The doping density is of $N_d = 10^{14}$–10^{17}/cc. The gold alloy at the top and bottom is for (a) good contacts and (b) good heat transfer. This GaAs device is mounted on the header first then encapsulated with gold wire contact and top metal cover sealed with ceramic enclosure.

Note: Gunn diode is the only diode which has both (a) −ve dc resistance in I–V characteristic due to two valleys and (b) ac differential negative resistance due to dipole domain growth and moment.

6.9.6 The Gunn Diode −Ve Resistance Oscillator and Amplifier Circuits

The Gunn diode oscillator normally has a resonant cavity where the diode is mounted inside, along with (a) frequency tuner, (b) *r.f.* output coupling line, (c) *r.f.* choke with bias, (d) frequency tuner, (e) diode holder and heat sink.

The commonly used cavity in coaxial line and waveguide is shown in Fig. 6.24a–d, along with their equivalent circuit. The cavity lengths l_{ce} and l_w decide the resonant frequency f_0 and this can be changed a bit by the frequency tuner, normally a dielectric rod of sapphire. The tuning short plunger can give wider frequency tuning. The dc supply voltage, the cavity frequency $f_{0,}$ and the diode parameters (L and N_d) will decide the mode of the oscillation (Fig. 6.22). (e.g. transit tune, delayed mode, quenched mode, or LSA mode).

The Gunn diode amplifier circuit is similar to that of tunnel diode amplifier (Fig. 6.16), where a circulator is used. It is less popular as the gain of Gunn diode amplifier is very less (10 dB or so), while in tunnel diode amplifier gain is 30 dB or so, near X-band frequency.

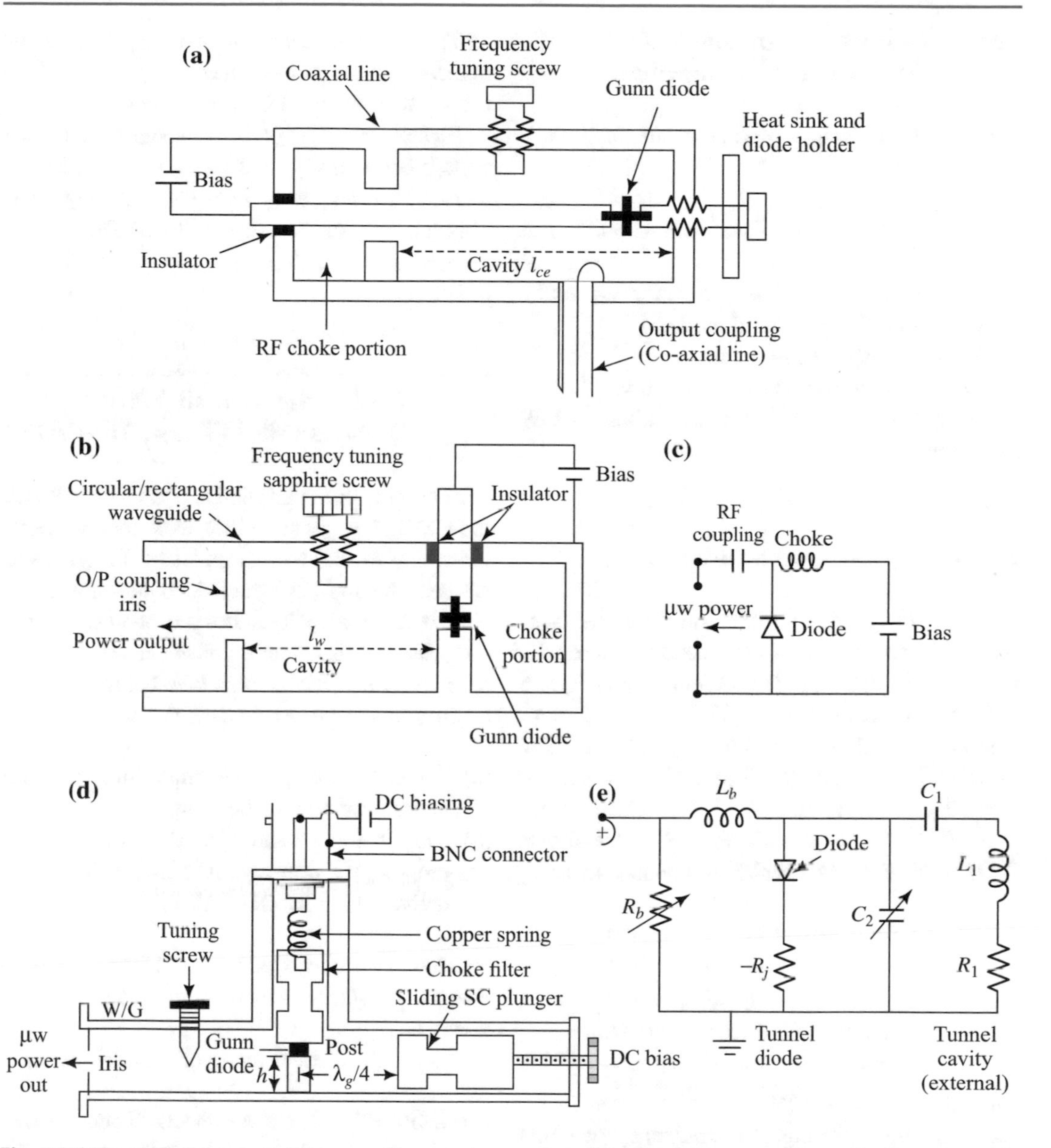

Fig. 6.24 Gunn diode −ve resistance oscillator units with small frequency tuning by tuning screw, **a** coaxial line, **b** rectangular or circular waveguide mount, **c** equivalent circuit, **d** rectangular or circular waveguide mount with tuning screw and sliding short also for wider frequency tuning, **e** equivalent circuit

6.9.7 Application of Gunn Diode Oscillators and Amplifiers

Gunn diode is mainly used as oscillator in various applications:

1. Medium power oscillator (1.0–100 GHz) in microwave receivers.
2. As a source in transponder for air traffic control (ATC) in pulsed mode.
3. Telemetry system in industry.
4. Pump source in parametric amplifier.
5. As a source in RADAR transmitter in CW mode.

6.9.8 Typical Characteristics

Voltage of operation depends on the active layer width. As the device has to operate between E_{th} (3.3 kV/cm) and E_{min} (11 kV/cm) say at $E_0 = 6$ kV/cm. Current density at 6 kV/cm is 400 A/cm^2. Therefore requirement of bias depends on active layer width, e.g. for $L = 5\ \mu\text{m}$, $V = 3$ V and for $L = 50\ \mu\text{m}$, $V = 30$ V. Therefore for $3 \times 10^{-8}\ \text{m}^2$ area, $I = 120$ mA. With the above, the characteristic of a typical diode may be as:

$$\text{(a) Power}\begin{cases} \text{CW} = 1\text{W}(1\text{GHz}) - 0.1\text{W}(2.5\text{GHz}); \\ \qquad 100\text{mW}(18-26\text{GHz}); \\ \qquad 40\text{mW}(26-40\text{GHz}); \\ \text{Pulsed} = 200\text{W}(1\text{GHz}) - 2\text{W}(10\text{GHz}) \\ \text{Pulsed(LSA)} = 500\text{W}(1\text{GHz}) - 20\text{W}(20\text{GHz}) \end{cases}$$

(b) Efficiency = 2–12%

(c) Frequency Tuning: The frequency of Gunn oscillator can be changed by mechanical tuning and electronic tuning.

Mechanical Tuning: As seen that the main control is by the size of resonant cavity, which can be varied by the short plunger as well as by the tuning screw (see Fig. 6.24a–d). The screw introduces susceptance in the cavity, thereby changing the frequency a little. The short plunger changes the size of the cavity, therefore wider change is possible.

The range of mechanical tuning can be up to ±5% of the central frequency, e.g. 1 GHz band tuning for a 10 GHz oscillator.

Electronics tuning: By changing the bias voltage between V_{max} and V_{min} (i.e. −ve resistance region), the frequency changes, but very little. This tuning is of the order of 3 MHz/V.

Normally mechanical tuning is always preferred.

6.10 Avalanche Transit Time Devices-IMPATT and TRAPATT

We saw that in tunnel diodes and Gunn devices, their DC. *I–V* characteristic exhibits −ve resistance due to different mechanism. In avalanche devices, the dc *I.V.* characteristic does not have a −ve resistance region. But at microwave frequencies, the current and voltage can become out of phase when the current lags behind voltage, leading to ac −ve resistances, due to:

(a) Avalanche delay due to finite time taken for charge and current build-up time.
(b) Transit time delay for the charge pulse crossing the drift region in IMPATT and the active region in TRAPATT.

These two delays add up to 180°, leading to −ve ac resistance, i.e. making it an ac signal generator.

These two distinct mode of oscillations, i.e.

(a) IMPATT (IMPact Avalanche Transit Time) has dc to RF conversion efficiency of 5–10%, can be made with drift region length suitable to operate up to 100 GHz, delivering up to 2 W (CW).
(b) TRAPATT (TRApped Plasma Avalanche Triggered Transit) has dc to RF conversion efficiency of 20–60%, can be made with active region length suitable to operate up to X-band frequency delivering 1–2 W (CW), while 1 kW pulsed at 1 GHz and 50 W pulsed at 10 GHz or so.

Avalanche multiplication is well known to be taking place when the electric field is above 400 kV/cm and electrons with their saturated velocity of 10^7 cm/s impacts and knocks off the outer most electrons, leading to a chains process of multiplication of electrons. In silicon, the drift velocity saturates to become 10^7 cm/s at 5 kV/cm, while avalanche multiplication starts above 400 kV/cm electric field.

6.10.1 IMPATT Diode, Read Diode Oscillator and Amplifier

The IMPATT diode consists of avalanche carrier generating p^+n thin junction followed by a long- and very-low-level-doped n^- region. The IMPATT diode structure, its doping profile, and field distribution when reversed biased just at breakdown are given in Fig. 6.25. This structure was proposed by Read in 1959 and is used even today.

At the end of the drift region, it has n^+ layer so that it makes ohmic contact with metal layer and not Schottky contact.

When a reverse voltage (corresponding to avalanche breakdown field of 400kV/EM) is applied, it results into avalanche multiplication of minority carriers around the p^+n junction. Generation of carriers starts at $t = t_1$. When field >400 kV/cm due to ac signal and starts reducing after t_2, when $E < 400$ kV/cm (Fig. 6.25b). This charge bunch then starts moving in the drift space and results into a current pulse at time t_3 in the n^+ region as output. Therefore time t_1 to t_3 is $T/2$. Thus we see that the phase difference of

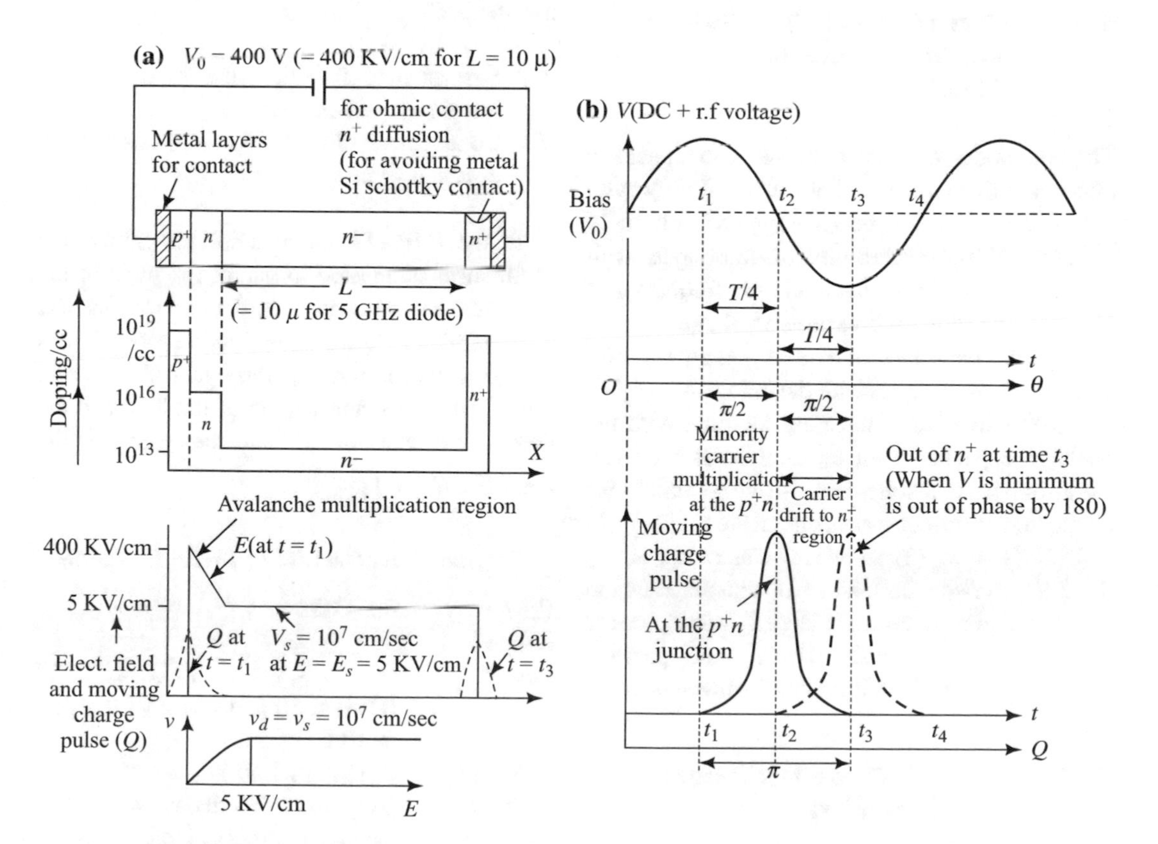

Fig. 6.25 a IMPATT diode structure, impurity doping, elect. field profile, and moving charge in the diode. In the drift region electron velocity $v_d = v_s$ (saturated velocity) = 10^7 cm/s at $E > 5$ kV/cc m. b Voltage and moving charge pulse, changing with time 't'. At the output end (n^+) the current pulse of charge reaches when voltage is minimum (i.e. π phase shift)

180° is there, between the *r.f.* voltage maximum and *r.f.* pulse current maxima at the output terminal. Therefore it has −ve dynamic resistance, i.e. becomes a source of microwave of frequency given by:

$$f = \frac{v_s}{2L} \quad (6.43)$$

where

L length of the drift space region

v_s carrier velocity in the drift region, which is saturated velocity, as the field here is >5 kV/cm

Because of the −ve dynamic resistance, the IMPATT diode is used both as an oscillator and as an amplifier.

6.10.2 Packaged IMPATT Diode and its Equivalent Circuit

The silicon device is encapsulated into a package after mounting it on a gold-plated metal alloy base. A typical package (Type S_4) is shown in Fig. 6.26 along with its equivalent circuit components, with values for a diode having operating frequency of 10 GHz, with bias at breakdown voltage.

The −ve resistance of diode R_D and its capacitance (C_D) are in series with the inductance (L_p) of the gold wire. These three are in shunt with the package capacitance of top and bottom covers. Three important features and also the conditions for oscillation of the equivalent circuit are (1) $|R_D| \ll |X_D|$, (2) $X_D \approx X_p$, (3) net −ve impedance $Z_{dp} \ll Z_0$, the characteristic impedance of the transmission line (waveguide or coaxial). Here X_D and X_P are the reactances of the diode ($1/\omega C_D$) and package ($\omega L_P - 1/\omega C_p$) which becomes the tuned circuit.

6.10.3 IMPATT Diode Oscillators and Amplifiers

When properly embedded in microwave cavity and circuit, IMPATT diode can generate microwave power. This circuit of Figs. 6.27a and 6.26b is similar to that used for Gunn diode oscillator (Fig. 6.24) and tunnel diode amplifier (Fig. 6.17).

In Fig. 6.27a, the diode is mounted centrally on the bottom broad wall of the rectangular waveguide, with a disc cap (placed in the H-plane) on the top of the packaged diode, for pressing it for electrical contacts. The waveguide is terminated by a slide short for mechanical frequency tuning.

Maximum power transfer conditions: Low pass filter (LPF) shorts the microwave signal for preventing them to reach the bias circuit. The disc cap acts as a quarter-wave impedance transformer between low −ve impedance of diode and high impedance of the waveguide, for maximum transfer of power to circuit.

Additional conditions for maximum power transfer are:

(i) Susceptance of diode = that of circuit (i.e. imaginary parts are equal).
(ii) Conductance of diode = that of circuit with opposite sign.

In the IMPATT diode amplifier circuit, the oscillator is connected to one of the ports of the circulator for using its −ve dynamic resistance. The amplified signal which comes out of it is passed to the output and through isolator as in Fig. 6.27b. If R_L and R_D are of the load and the diode −ve resistance, then the gain of the amplifier $G = \left(\frac{R_D - R_L}{R_D + R_L}\right)$.

Typical characteristic of IMPATT diode:

Frequency Range	0.5–GHz.
Power	0.5 W (CW) at 30 GHz, 10 W (CW) at 10 GHz, 100 W (pulsed) at 10 GHz.
Efficiency	5–15% (Si), 25% (GaAs).
Noise level	Very high (30 dB or so) as the avalanche process is noisy.
Tuning	As the frequency is decided by (v_S/2L), very less tuning is possible.

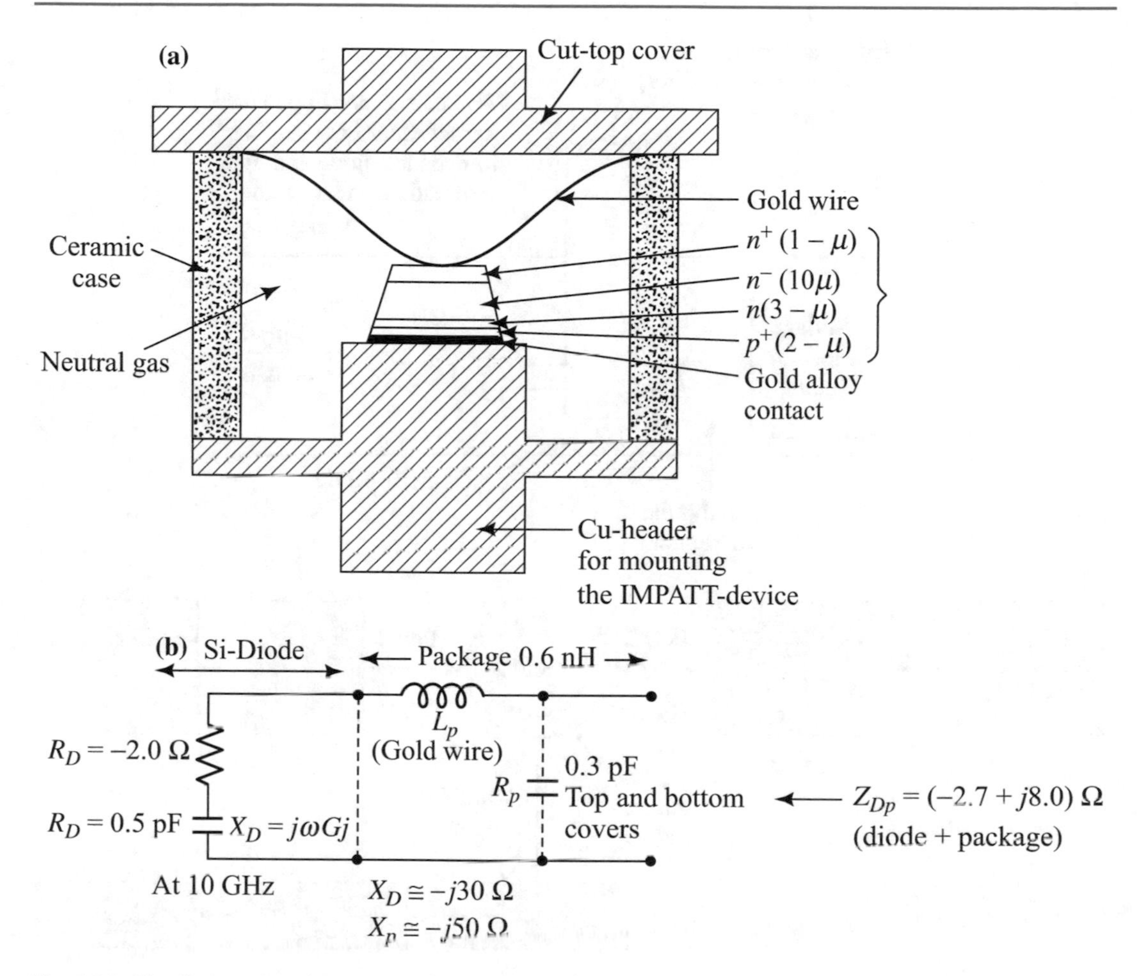

Fig. 6.26 The diode package (S_4-type) and their equivalent circuit, with typical values for 10 GHz diode

The advantage of Gunn diode/klystron tube is their broad frequency range, while here different IMPATT diodes will have to be used for different frequencies, which becomes its disadvantage.

6.10.4 Applications

(i) Microwave generator/local oscillator.
(ii) Pump in parametric amplifier.
(iii) CW Doppler radar transmitter.
(iv) FM telecommunication network.
(v) Transmission part of TV system.

6.10.5 TRAPATT Diode Oscillators

As we know that the abbreviation TRAPATT stands for TRApped Plasma Avalanche Triggered Transit mode of operation. It is high-efficiency microwave generator for 0.5–20 GHz range. The basic structure is p^+nn^+ (or n^+pp^+) with n (or p) active region varying from 2 to 15 μm. For delivering larger power in CW/pulsed mode, diameter is kept as large as 50–750 μm.

Mr. Liu and Risku of RCA Laboratory USA (Reference 8), in Oct 1969, noted that the IMPATT diode which with W_n = 5 μm, generates

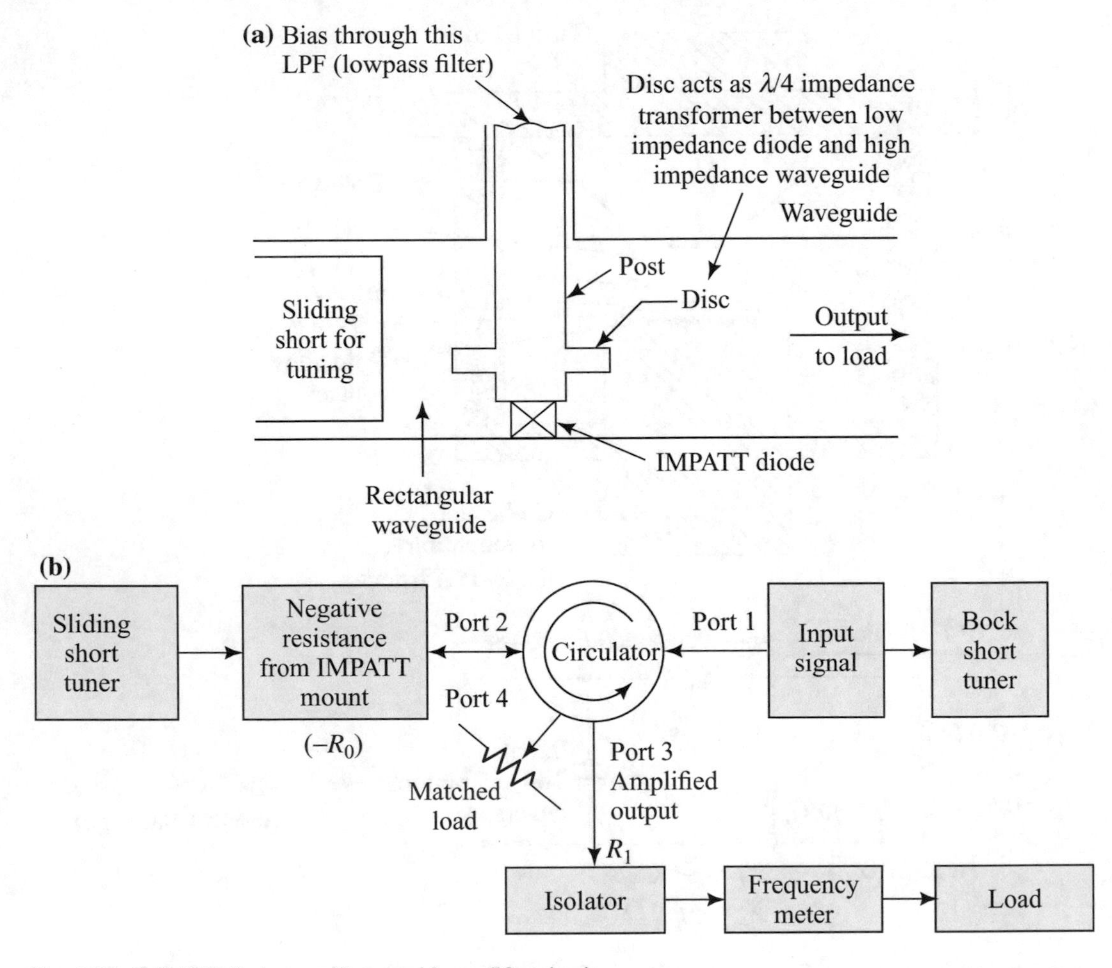

Fig. 6.27 IMPATT diode **a** oscillator and **b** amplifier circuit

1.5 or 2.0 GHz, can be made to oscillate at lower frequency, e.g. 1 GHz or so, as sub-harmonic mode of IMPATT by (i) placing it in a cavity resonator of 1 GHz and (ii) supplying dc voltage slightly above V_B, the breakdown voltage of that diode. They called this mode as anomalous mode, which was later given the name TRAPATT by DeLoach of Bell Telephone Laboratories (BEL), USA (Reference 9), who also gave the physics of its working.

This mode was found to give much higher power, at higher conversion efficiency.

The working of TRAPATT oscillators can well be understood from Fig. 6.28, which gives the structure, the transient moving electric field (E), current density (J), and power density ($p = J \cdot E$), in the active region of the diode. Figures 6.28 and 6.29 give the avalanche shock front (ASF) sweeping across the active region. As a result of the above, the current and voltage of the diode behave as given in Fig. 6.30.

To understand the process in detail, let us start with when the diode is reversed biased just at avalanche breakdown voltage V_A (Points A of Fig. 6.30 at $t = t_A$). The carrier density and field profile in the active region for different moments t_A to t_G are given in Fig. 6.31. At point A, voltage rises further (charging) up to point $B(t = t_B)$, where the avalanche multiplication of minority carriers takes place across the p^+n junction, where the electric field is very high (7×10^5 V/cm or so), i.e. higher than the normal avalanche breakdown field of 4×10^5 V/cm.

But thereafter E gradually reduces to $E_A = 4$ 10^5 V/cm where the multiplication is much less

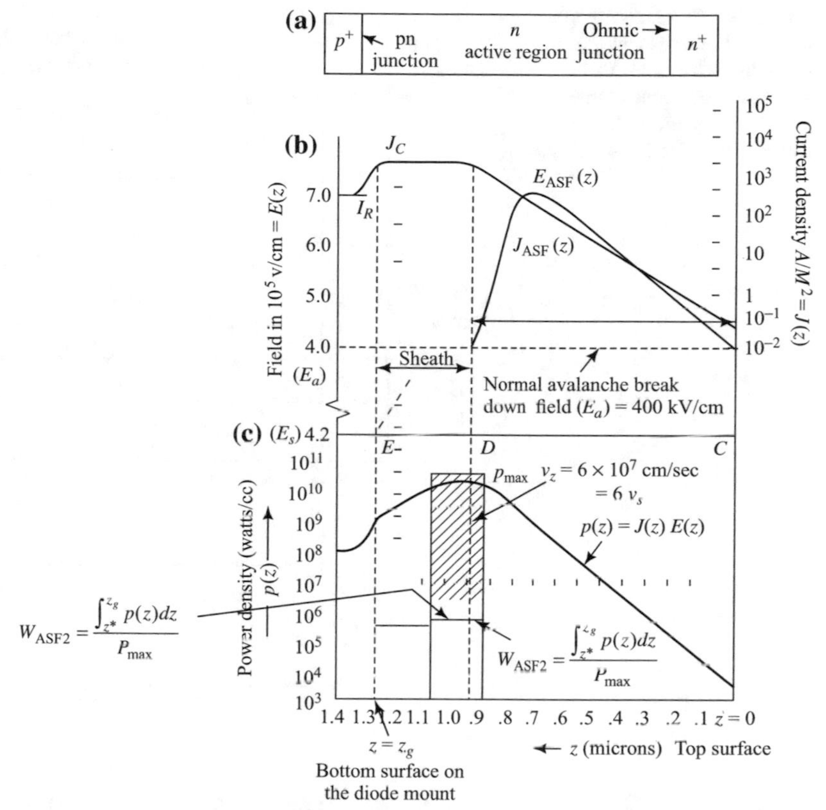

Fig. 6.28 Typical structure, current density, field, and power density profile during an avalanche zone (ASF) transit time in the depletion layer. *ASF* moves with velocity $v_z \approx 6 \times v_s$ (From Reference 10, Chaturvedi et al.)

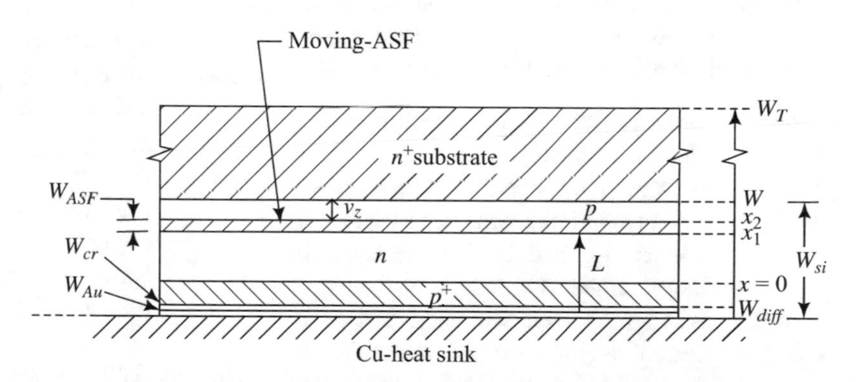

Fig. 6.29 Moving ASF in depletion region (From Reference 10) with velocity $v_z \gg v_s$

(Fig. 6.29b). This leads to a large current density region (J_C) near the junction having high electric field also. Thus this region of very high power density ($p = J.E. = 10^{11}$ W/cc) called avalanche shock front (ASF) starts moving away from the junction, leaving behind a high carrier region (sheath) as shown in Fig 6.28c. Once the ASF has sweeped fully to the right (Reference 10), the device is filled with large amount of carriers called plasma (see Fig. 6.30-BC region). This high-field (EM > Ea) region is called avalanche shock front, and it moves with a velocity $v_z > v_S$.

The generation and movement of carriers along with changing electric field inside the active region is explained in Fig. 6.31 with reference to Fig. 6.30.

The external voltage and current variation can be explained with reference to the above-explained variation of electric field and carriers in the active region (Fig. 6.31) during

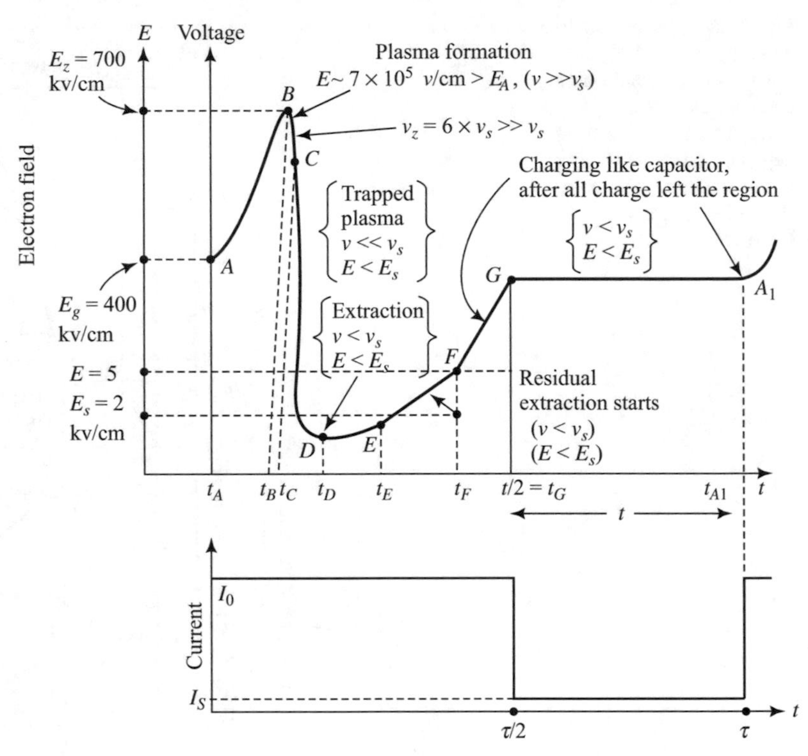

Fig. 6.30 External voltage, field, and current wave form changing with time. Field values for different moment $t = t_A$, t_B, t_D and t_E (From Reference 9, De Loach et al.)

$t = 0$ to $t = T/2$ of the complete cycle. This external voltage and current behaviour are given in Fig. 6.30 from = 0 to T, as explained below from the time points A to A_1:

Time A Only thermal carriers are present as electric field $< E_A$.

Time AB Charging, i.e. rise of voltage above avalanche breakdown (from V_A to V_B) voltage V_A, and carriers generate in large number (Fig. 6.31).

Time BC A very high electric field and high current region—the avalanche shock front (ASF) propagates through the diode at a speed v_z much larger than v_s, leaving behind whole of the n-active region with dense electron hole plasma (Fig. 6.31).

Time CD As the diode is filled with very high carrier density, the electric field falls much below E_s, causing their velocity to be $\ll v_s$. Because of its very slow movement, plasma is said to be trapped.

Time DE Some plasma get extracted during CD and extraction continues very slowly as $v < v_s$ here also.

Time EF: Now $v = v_s$ here as a result the residual extraction is fast, leading to rise in voltage (here $E = E_s$).

Time FG After full extracting, field and voltage rise very fast like a charging of capacitor and after full charging, the current leakage current I_s drops for ($v = v_s$ and $E > E_s$)

In Fig. 6.30, we see that from A to G, the current remains very high as a step current because of lot of carrier charges coming out of the terminal as shown in Fig. 6.31, but from G the voltages reach V_A (just the break down voltage) and current falls to just the leakage current I_s. By this time t_G, half of the time period of the wave time has lapsed and the voltage and current stay there for another half time period, before repeating the cycle at $t = t_{A1} = T = 1/f$ (f = frequency of TRAPATT), which has to

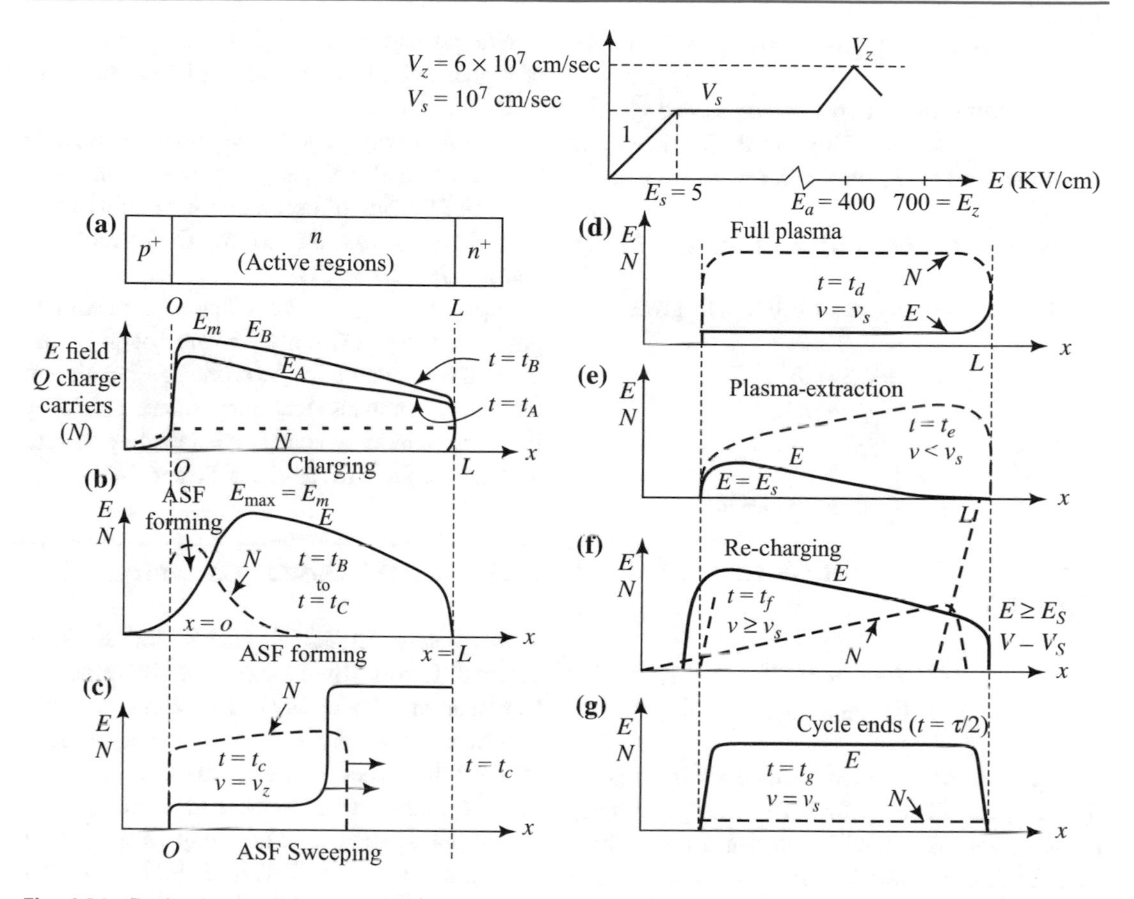

Fig. 6.31 Carrier density and field profile in the active region for different seven moments of time $t = t_A$ to t_g of $T/2$ period (Fig. 6.30). Velocity of electron-vs-field strength also given at the top right

match the resonant frequency of the external circuit.

6.10.6 Calculating v_Z the Velocity of ASF Region

As indicated in the beginning that this mode operates with voltage slightly higher than V_A (the avalanche breakdown voltage), where current also reaches very high due to very fast rise of electric field near the junction. The Maxwell's displacement currents for this region *AB* of Fig. 6.30 give:

$$J_0 = \varepsilon_s \frac{\partial E}{\partial t} \tag{6.44a}$$

(ε_S = permittivity of the semiconductor). Here the dramatic thing happens that the velocity of carrier of ASF becomes more than v_S and we calculate this velocity v_Z as follows:

For $E < E_A$ the Poisson equation for the active region is:

$$\frac{dE}{dx} = -eN_d/\varepsilon_s \tag{6.44b}$$

(N_d = doner density in n-region).

Integration (6.44a) and (6.44b) with the condition that at $x = 0$ (the junction point at $t = 0$) $E = E_a$ (the avalanche breakdown field):

$$E(x) = E_a - (eN_d/\varepsilon_s) \cdot x \tag{6.45}$$

$$E(x,t) = E(x,0) + J_0 t/\varepsilon_s \quad (6.46)$$

Combining these two Eqs. (6.45) and (6.46), we obtain the field to the right of the p^+n junction, when the field is E_A (at $x = 0$, $t = 0$), we get:

$$E(x,t) = E_a - (eN_a/\varepsilon_s)\cdot x + (J_0 t)/\varepsilon_s \quad (6.47)$$

Therefore $E = E_a$, for $x = 0$, $t = 0$, gives:

$$eN_d x = J_0 t$$

$$\therefore \quad \frac{dx}{dt} = v_z = J_0/eN_d$$

For $N_d = 2 \times 10^{15}$/cc, $J_0 = 2 \times 10^4$ A/cm^2 (Fig. 6.18).

$$v_z = 2 \times 10^4/(1.6 \times 10^{-19} \times 2 \times 10^{15}) \approx 6.25 \times 10^7 \text{ cm/s} \gg v_S$$

After this ASF sweeps across the active region 'n'; it gets filled with high density plasma, reducing the field $E < E_S$ and hence $v < v_S$ getting the plasma virtually trapped.

6.10.7 Power Output, Efficiency, and Frequency Limits

Once (a) the external circuit matches the dynamic −ve resistance of the diode with the load at the oscillating frequency of TRAPATT mode, (b) the cavity frequency matches with the TRAPATT mode, high power and efficiency is achieved. Power, frequency obtained in the labs, so far are given Table 6.5.

We get high power and efficiency at the cost of higher noise level in TRAPATT as compared to IMPATT diodes.

An effort was made to compute theoretically the power and efficiency. It was found that TRAPATT mode of oscillation is possible up to 20 GHz only (As proved by Chaturvedi and Khokle, Reference 11).

Figure 6.32 gives the efficiency, maximum power density (W/cm^3), power output, and internal temperature in CW mode as a function of frequency, showing that power and efficiency falls very rapidly beyond 2 GHz tending to zero near 25–30 GHz (References 11 and 14).

6.11 BARITT Diodes Oscillator

The acronym BARITT stands for BARrier Injected Transit Time. Diode was first made by Coleman and Sze in 1971. The structure is just two back-to-back diode p^+np^+, with very wide n-region sandwiched between two heavily doped regions (unlike transistors) and is used as a two-terminal device. The long drift region 5–10 μm is similar to IMPATT diode, but there is no avalanche mechanism but majority carriers (holes) are injected from the forward-biased p^+n junction thin barrier, which move into the drift region with saturated velocity ($v_s = 10^7$ cm/s in Si). These carriers get collected at the np^+ junction which is reversed biased (Fig. 6.33a). In this process of long transit time in the 'n' region, the output charge-pulse current is delayed from the applied voltage. This transit time is between (3/4 T to T/2) with the optimum.

Transit time being $\tau = W/v_s = 0.8T$ (where T is the time period of the microwave generated

Table 6.5 Typical peak pulsed power, average power and efficiency of TRAPATT diodes as a function of frequency

Frequency (GHz)	Peak pulsed power (W)	Average power (W)	Efficiency (%)
0.5	600	3	40
1.0	200	1	30
4.0	100	1	20
8.0	50	1	10

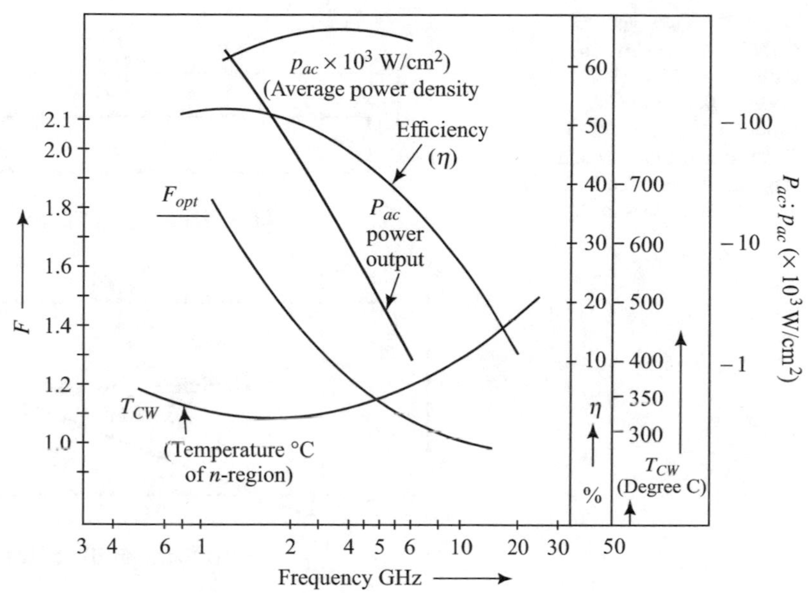

Fig. 6.32 Optimised values of efficiency (η) (output power density (p), output power (P_{ac} watts), active n-region, temperature (T_{CW}) with the corresponding values of F for 10-Ω negative resistance as a function of frequency (from Ref. 11, Chaturvedi et al.)

with $T = 1.25\tau$). Therefore the optimum frequency of oscillation is

$$f = 1/T = 0.8/\tau = 0.8\, \upsilon_s/L \qquad (6.48)$$

i.e. this optimal frequency is below the transit time frequency.

Because of this delay of charge-pulse current at output vis-a-vis the ac voltage, the diode exhibits dynamic −ve resistance and hence becomes capable of generating microwave power with time period around 1.25 of the transit time. Thus the microwave oscillations depend on two factors:

(a) The rapid increase of carriers injected caused by decreasing potential barrier of the forward-biased junction p^+n.
(b) The long transit time of the carrier to traverse the depletion region caused by the reverse-biased junction np^+.

A typical structure of BARITT diode could be (a) p^+np^+ (Fig. 6.33a, b) with n-region of 10 μm and 4 × 10^{14}/cc doping level (11 Ω cm resistivity) and (b) the two p^+ regions of 1 μm thickness and 10^{17}/cc doping level (0.2 Ω cm resistivity). (c) Instead of p^+ contacts, Schottky contacts of PtSi (thickness 0.1 μm) also can be used (Fig. 6.33a–c).

When a forward voltage is applied as in Fig. 6.33b, the voltage gets distributed into two portions:

(a) To overcome the p^+n potential barrier to turn on the forward bias by 0.7 V.
(b) To create depletion layer in the 'n' region due to reverse-biased np^+ junction (see Fig. 6.33b).

The full punch through voltage (for whole 'n' region) to get depleted can be obtained by double differential of the Poisson equation $dE/dx = qN_d/\varepsilon_s$ and will be the critical voltage:

$$V_C = \frac{qN_d L^2}{2\varepsilon_{si}} \qquad (6.49)$$

The average breakdown voltage is normally twice of V_c, and the corresponding breakdown fields are as:

$$V_{BD} = \frac{qN_d L^2}{\varepsilon_{si}} \qquad (6.50)$$

$$E_{BD} = \frac{qN_d L}{\varepsilon_{si}} \qquad (6.51)$$

For $N_d = 10^{15}$/cc, $L = 10$ μm, $V_c \approx 152$ V, $V_{BD} = 304$ V.

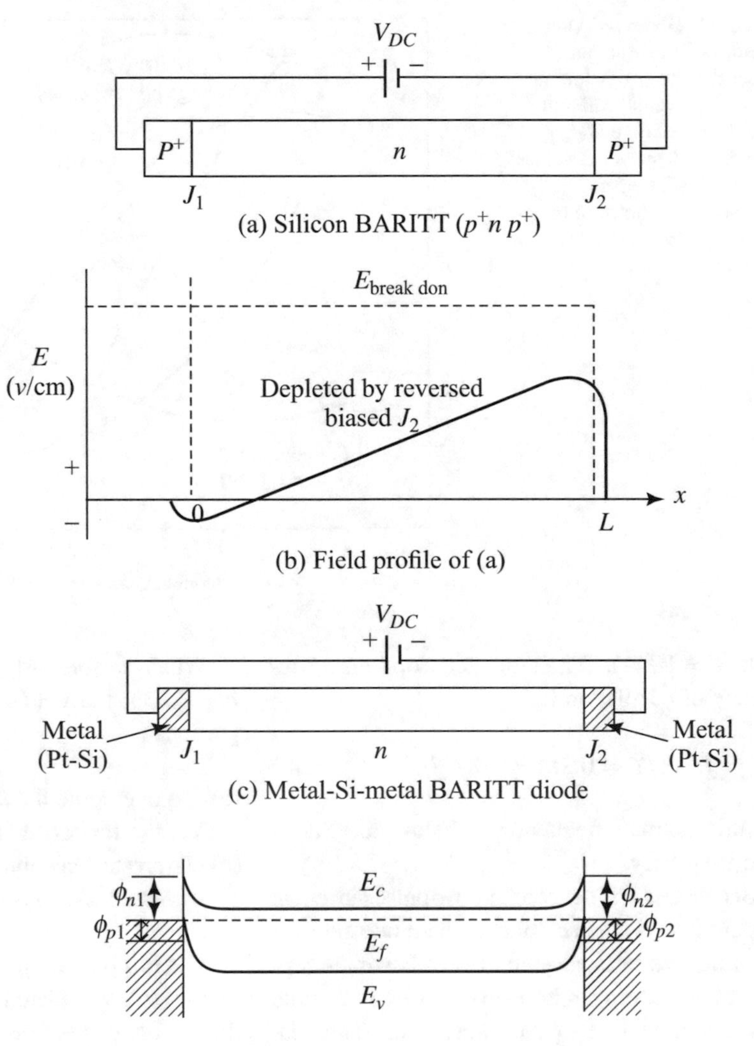

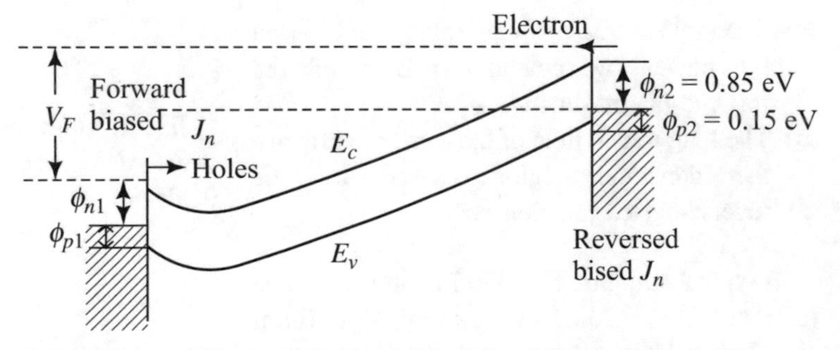

Fig. 6.33 Two types of BARITT diode structure **a** silicon p^+np^+ BARITT diode, **b** field profile of **a**, **c** metal-si-metal BARITT diode, **d** and **f** energy and diagram of **c** with no bias and under forward bias at J_1

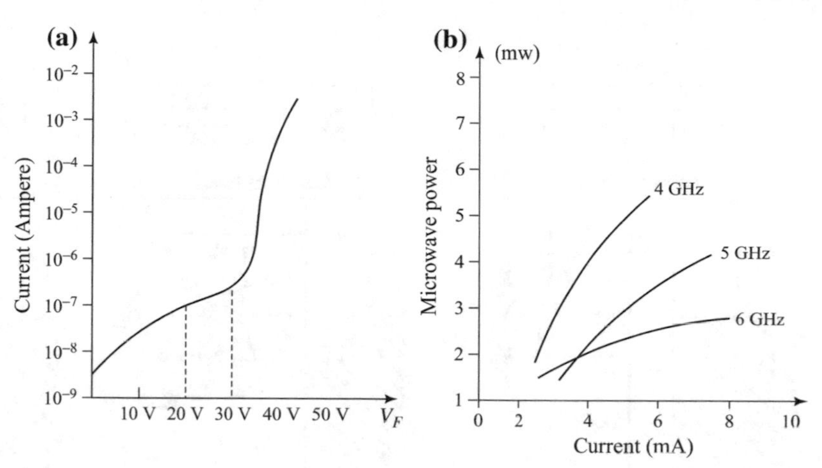

Fig. 6.34 **a** Current–voltage characteristics of a typical BARITT diode at room temperature with J_1 forward biased and **b** power output versus current of a BARITT diode

BARITT diodes are less noisy (<15 dB) than IMPATT diodes (and comparable Gunn-oscillators), as the holes injected from the forward-biased junction are responsible for increase of current and not avalanche multiplication, as obvious from the voltage shown in Fig. 6.34.

The disadvantages in its performance are its relativity narrowband width (0.2 GHz), low-power output (few milli watts to 50 mW CW at 4.9 GHz), small frequency range 4–8 GHz, and low conversion efficiency ($\approx$1.8%) (Fig. 6.34b).

Major advantages are of low cost, low power supply, low noise close to Gunn diodes, and highly reliable. The major applications of BARITT oscillator are as low-power local oscillators in RADAR, communication, etc.

6.12 Schottky Barrier Diodes (SBD) —As Detector and Mixer

We know that for metal contacts with semiconductor to be ohmic, the semiconductor has to be doped close to degenerate level (i.e. n^{++}, p^{++} making its conductivity close to that of metal by doping density $\geq 10^{19}$/cc). If the semiconductor doping is just below degenerate level than normally with metal, it makes a Schottky barrier rectifying contact diode. The ON-OFF or OFF-ON switching time is very small (in pico seconds range), being a majority carrier device as in the 'n'-type devices electrons enter the metal quickly. Therefore it is used as μW detector diode (see Sect. 4.12.2).

Let ϕ_m, ϕ_s be the work function energy (in joules or eV) of metal and semiconductor. Work function is the difference between fermi level and its energy levels of electron in free state in vacuum (i.e. zero level) in the metal or semiconductor (Fig. 6.35). Then the difference of work functions between metal and semiconductor contact will be:

$$W = (\phi_m - \phi_s) \qquad (6.52)$$

Following three conditions are there for having a Schottky barrier rectifying diode property.

(a) Doping density is low ($<10^{17}$–10^{18}/cc).
(b) Metal of type $\phi_m > \phi_s$ (i.e. W = +ve) for n-type semiconductor (Fig 6.35).
(c) Metal of type $\phi_m < \phi_s$ (i.e. W = −ve) for p-type semiconductor (Fig 6.36).

Reverse of any of the above conditions makes it ohmic contact, e.g. $n > 10^{18}$/cc or $\phi_m < \phi_s$ (in n-type) or $\phi_m < \phi_s$ (in p-type).

Work functions of platinum, palladium, nickel, gold are ≥ 5.1 eV and therefore are more suitable for n-type Schottky, while molybdenum, silver, aluminium, titanium, and tungsten have $\phi_m \approx 4.2$–4.6 can be used for p-type Schottky diode (Tables 6.6 and 6.7).

(i) ***n*-type semiconductor in contact with metal of type $\phi_m > \phi_s$, ($E_{fm} < E_{fs}$) making Schottky contact:**

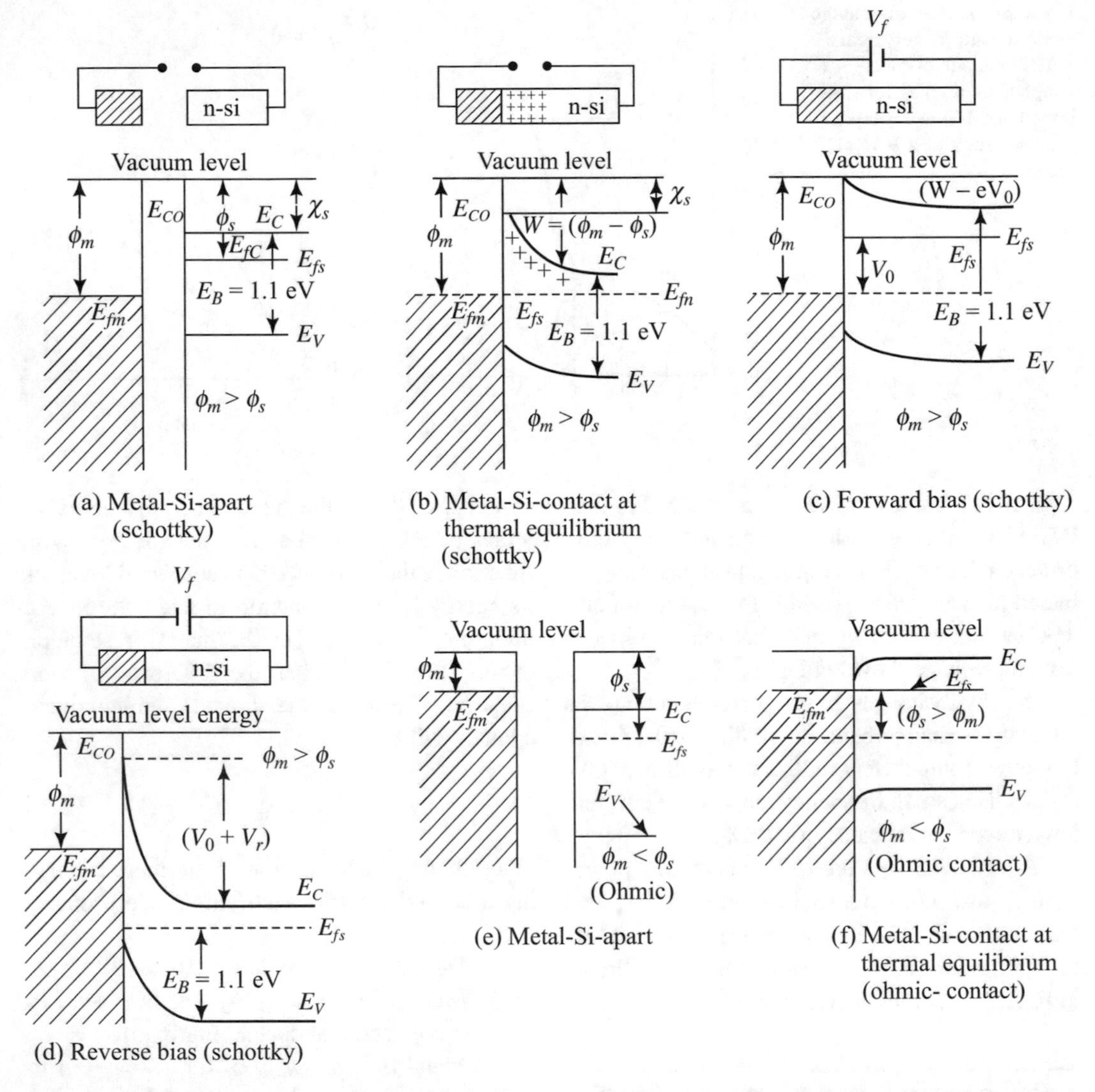

Fig. 6.35 Energy bond diagram for *n*-type semiconductor in contact with metals of two type (i) $\phi_m > \phi_s$ for Schottky barrier diode rectifier **a** for apart, **b** in contact, **c** forward bias (V_f), and **d** reverse (V_f) bias cases (ii) $\phi_m < \phi_s$ ohmic **e** apart **f** in contact. Here E_c. E_v, E_{fs} are conduction band level, valence band level, and fermi level of the *n*-semiconductor

When the metal and *n*-semiconductor are in contact (Fig. 6.35) such that work function ϕ_m of metal (energy needed to remove an electron from solid out to a point in the vacuum) is higher than ϕ_s in semiconductor, i.e. fermi energy level of electrons in semiconductor is higher than in metal, as a result electrons will flow from *n*-semiconductor to the metal, till the fermi-level equalisation takes place. This process makes the semiconductor +vely charged as it gets depleted of electrons, generating a potential barrier of $eV_0 = (\phi_m - \chi_s)$ Joules or V_0 electron volts, with a field which opposes further flow of electrons. Here χ_s is the affinity of electron in semiconductor.

Under forward bias (V_f), the potential barrier height reduces to $eV_0 = (\phi_m - \chi_s - eV_f)$ Joules (Fig. 6.35c) as a result more electrons gets injected from semiconductor into the metal. In reverse bias case, this barrier height increases (Fig. 6.35d) and electron flow almost stops. Further increase of reverse bias leads to breakdown.

Table 6.6 Electron affinity (χ) and work function of some semiconductors

Semiconductor	Electron affinity (χ_s)	[a]Work function of the semiconductors (ϕ_s) $\phi_s = [\chi_s + (E_c - E_f)]$	
		For n-type (ϕ_{sn})	For p-type ϕ_{sp}
Ge (eV)	4.13	4.23	4.66
Si (eV)	4.01	4.11	5.01
GaAs (eV)	4.07	4.17	5.27

[a]Here it has been assumed that $(E_e - E_f) - (E_e - 0.1)$ eV for p-type close to ≈ 0.1 eV for n-type close to degenerate

Table 6.7 Metals suitable for Schottky diodes

Suitable for n-type semiconductor metals with ($\phi_m > \phi_s$)		Suitable for p-type semiconductor metals with ($\phi_m < \phi_s$)	
Metals	ϕ_m (eV)	Metals	ϕ_m (eV)
Au	5.10	Ag	4.26
Ni	5.15	Al	4.28
Pd	5.12	Ti	4.33
Pt	5.65	Cr	4.50
		Mo	4.60
		W	4.55

The I–V characteristic of the diode action is shown in Fig. 6.37, and it acts as Schottky barrier rectifying diode, with turn-on voltage of 0.3 V or so, i.e. just half of normal diode as barrier potential is in semiconductor only.

(ii) ***n*-type semiconductor in contact with metal of type $\phi_m < \phi_s$, ($E_{fm} > E_{fs}$) making ohmic contact:**

With such metal, the band diagram without and with contact with semiconductor will be as given in Fig. 6.35e, f, respectively. The free electron negative charge in metal being at higher level ($E_{fm} > E_{fs}$) will flow to the semiconductor after contacting, increasing the −ve charge over there (see Fig. 6.35f), but no depletion region or barrier potential is formed. Therefore full current flows in both the directions (whether forward or reversed biased) and hence behaves as an ohmic contact.

(iii) ***p*-type semiconductor in contact with metal of type $\phi_m < \phi_s$, ($E_{fm} > E_{fs}$) making Schottky diode:**

Such a contact can be well explained by diagram Fig. 6.36a, b. The electron has higher energy in metal, therefore flows from metal to semiconductor, until fermi level is same thought. The excess −ve charge in semiconductor makes an electric field across the junction and hence a potential barrier, therefore it acts like a diode with rectifying behaviour.

(iv) ***p*-type semiconductor in contact with metal of type $\phi_m > \phi_s$, ($E_{fm} < E$) making ohmic contact:**

Such a contact is explained in Fig. 6.36c, d for metal and semiconductor separate and then in contact, respectively. Here the fermi level (E_{fs}) in semiconductor has higher energy level than of metal and therefore electrons flow from semiconductor to metal, making the semiconductor as positively charged, but no depletion or potential barrier. Therefore it behaves as ohmic contact, in forward as well as in reverse bias.

The Schottky barrier diode is sometimes called the hot electron diode because of electrons in semiconductor having higher energy level than in metal. In the forward current, it is all majority

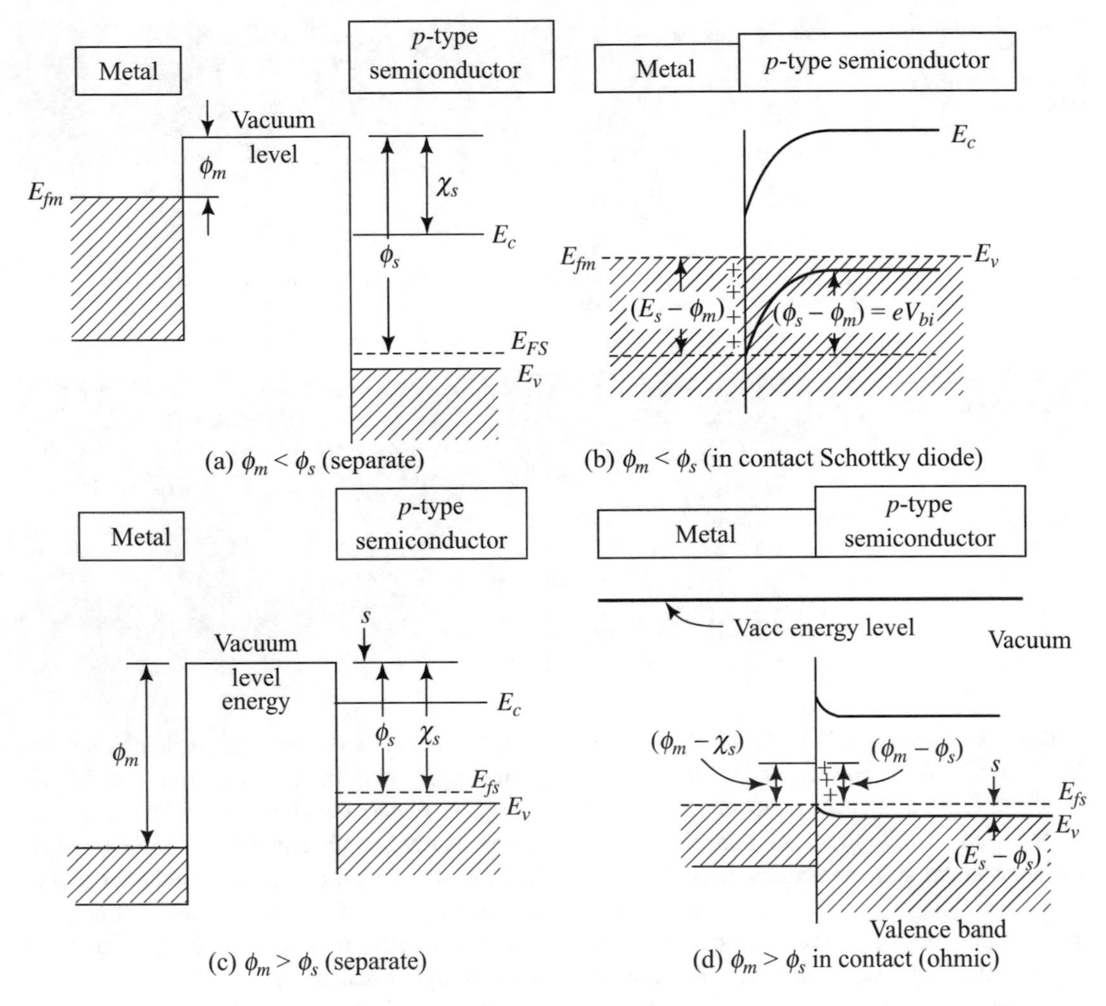

Fig. 6.36 Energy-level diagrams *p*-type semiconductors in contact with metal: **a** and **b** are for $\phi_m < \phi_s$. Contact (**b**) acts as a Schottky barrier diode rectifier. In **c** and **d** the contact is ohmic

carriers with minority initially absent. Therefore it has very short reverse recovery time as the storage capacitance is almost nil. Therefore Schottky diode switches from non-conducting stable to conductivity state very fast (in less than 100 p/s), whereas ordinary pn junction switching time is large (around 100 n/s).

Figure 6.38 gives the construction of Schottky diodes (a) the point contact type and (b) planar technology type. The latter has a n^+–Si-substrate, upon which a thin epitaxial *n*-layer of 2–3 μ thickness is grown. Schottky contact is from n-surface to Si-ohmic contact from n^+ is taken through a window opened and gold evaporated for Au–Al contact.

The second one can be manufactured in large scale easily, but can be used up to 100 GHz only due to larger metal contact area capacitance of 0.3–0.5 pF. Two more advantages of this type are:

(a) Lower forward resistance (<0.5 Ω)
(b) Lower noise generation (<4 dB)

The point contact type can be used for frequencies up to 1000 GHz due to very low shunt capacitance of the contact of 0.01 pF, but has high series resistance of 2–5 Ω.

The limitations of SBD are (a) low reverse breakdown voltage (<100 V) and (b) high

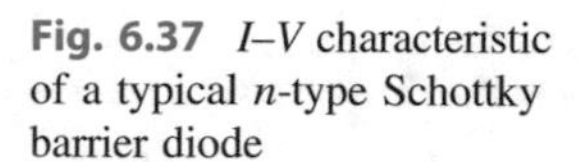

Fig. 6.37 I–V characteristic of a typical n-type Schottky barrier diode

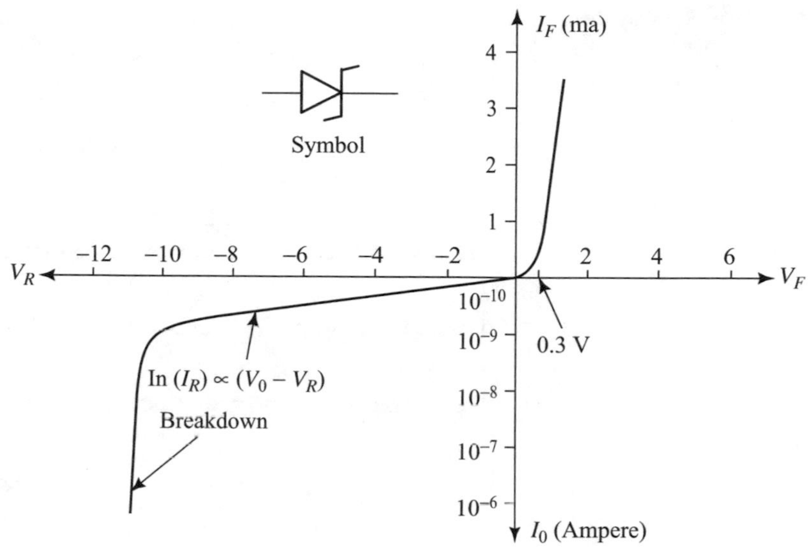

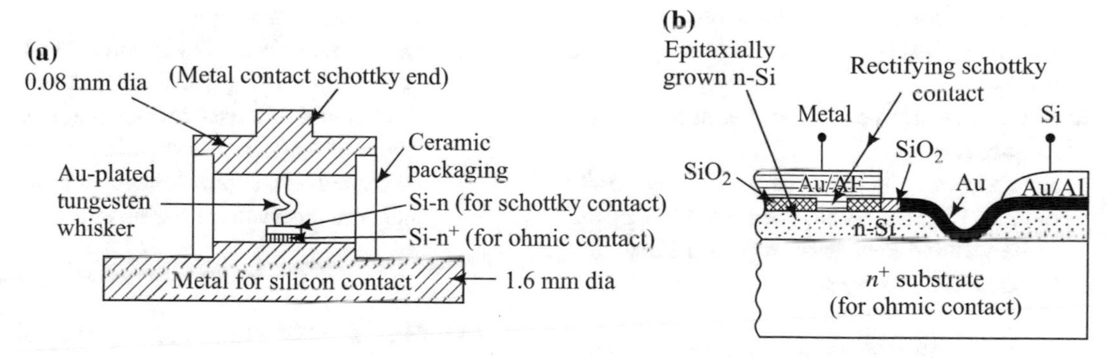

Fig. 6.38 Construction of Schottky barrier diode (SBD) **a** point contact Schottky **b** planar technology Schottky

reverse leakage current which increases with temperature causing thermal instability. The reverse bias leakage current is of the order of 10^{-6} A/cm^2 when compared to 10^{-11} cm^2 in conventional pn junctions.

For more properties in a circuit as a detector, see Sect. 4.12.2.

6.12.1 Application

The SBDs are used in the following circuits as microwave devices:

(i) Mixer in CW-RADAR

(ii) Microwave power detection as in Fig. 6.39 due to very small switching time $\approx 10^{-9}$ s. The rectified (detected) output is dc, and it goes to VSWR meter, etc.

6.13 PIN Diode for Switching/Controlling Microwave Power, Phase Shifting, Modulating etc.

PIN diode is very useful control element used at microwave frequencies as a (a) switch, (b) attenuator, (c) phase shifter, (d) power limiters,

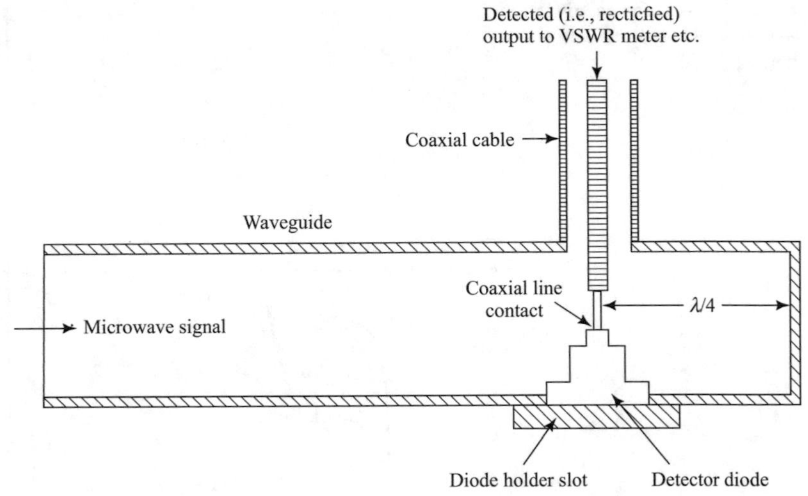

Fig. 6.39 Schottky diode detector mount assembly in a waveguide

(e) amplitude modulating element. This is because of its four important properties:

1. It can control very large microwave power of kW range, just by changing its bias by a small voltage.
2. It does not behave as a rectifier at microwave frequencies.
3. Breakdown voltage is very large generally over 500 V, as a result, even large μW power in its positive cycle cannot forward bias it.
4. Capacitance is very small.

Now we will discuss its structure, characteristics, and working:

(i) **Structure**: PIN Diode consists of heavily doped *p*- and *n*-regions separated by a high-resistivity *i*-region ($\approx$1000 Ω cm) (Fig. 6.40), nearly intrinsic. In fact, this so-called i-region is high-resistivity *p*-layer (called π-type) or high-resistivity *n*-type (called ν-type). The reverse bias resistance being very high, most of the reverse bias voltage is across it, fully depleting the region. Therefore the reverse breakdown voltage will be very high over 1000 V or so and the capacitance very small (0.2 pF or so). Therefore:

(i) Length L of the *i*-region is kept large (100 μm or so)

(ii) Doping level is kept very low (10^{12}–10^{13}), i.e. π or ν-type

(iii) Device capacitance is kept quite small (<1 pF by keeping small area). Moreover it remains constant with voltage, as whole of *i*-region gets depleted with a very small reverse bias voltage itself and higher reverse voltage has no effect.

$$C_s = (\varepsilon_s A/L) \tag{6.53}$$

(iv) For $L = 200$ μm, transit time of charge across *i*-region is approximately:

$$\begin{aligned}\tau &= L/\upsilon_s \\ &= (200 \times 10^{-4}/1.3 \times 10^{-2}) = 1.4 \times 10^{-9}\,\text{s}\end{aligned} \tag{6.54}$$

Figure 6.40 gives the (a) systematic diagram, (b) impurity concentration, (c) space charge, and (d) electric field distribution in fully depleted PIN diode. Figure 6.41 gives the equivalent circuit and RV and IV characteristics of the three biasing regions (before A, AB and after B). Figure 6.42 gives the actual PIN device and diode structures. It also gives the V_B *vs* doping level, forward bias resistance *vs* current and transit time *vs* length of the I region of the PIN diode.

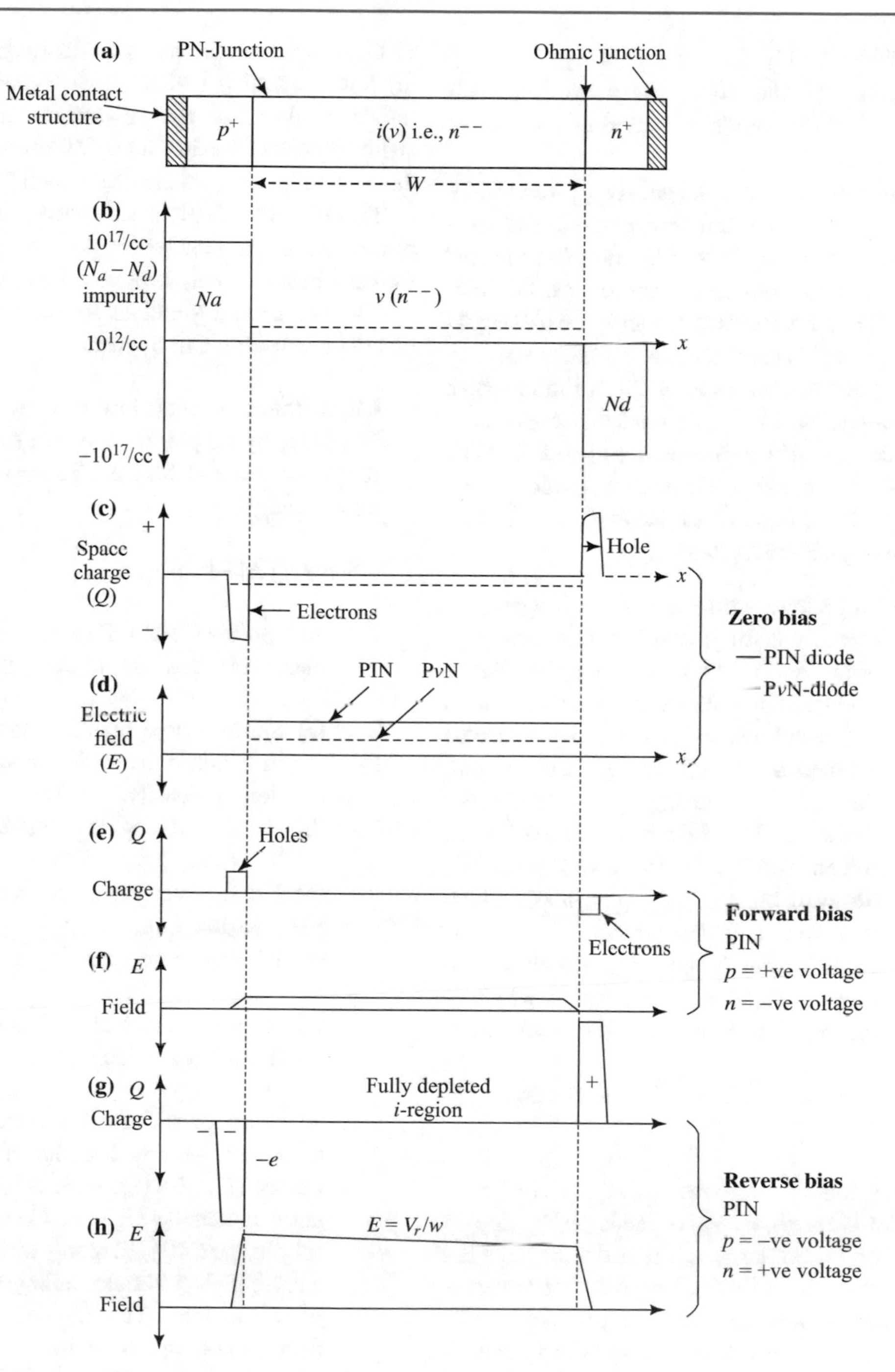

Fig. 6.40 PIN diodes **a** schematic diagram, **b** impurity distribution, **c** and **d** space charge and electric field for zero bias case, **e** and **f** space charge and electric field for forward bias case, **g** and **h** space charge and electric field for reverse bias case

(i) Characteristic

Because of the following three important properties of PIN diode, it is used as microwave switch;

1. **Can control/switch large μW power** of kW level or so, by a small change of bias from forward to reverse. This is by allowing the μW signal to flow through a very low resistance of 0.1 Ω or so in forward bias and stop μW signal in reverse bias resistance of 10 kΩ or so.
2. **Does not behave as a rectifier at microwave frequencies**, like other diodes like Schottky diodes, etc. At low frequencies up to 100 MHz, PIN diode behaves like other pn-diodes as rectifiers, but at higher frequencies the rectification property decreases due to

   (a) Large transit time across the *i*-region.
   (b) Large switching/recombination time (τ_{sw}) from ON to OFF, causing the carrier storage taking place in the *i*-region, which as a result acts as a variable resistance as a function of both voltage and current. This switching time/carrier re-combination time is $\tau_{sw} \approx 10^{-4}$ s for silicon and 10^{-9} s for GaAs from ON (forward bias) to OFF (rev. bias), which is ≫ than μW time period.
   (c) Large carrier lifetime τ_{sw}, which in turn leads to large diffusion length which is the length after moving, it recombines

$$L_d = \sqrt[2]{(D_{Si} \cdot \tau_{sw})} \approx 400\text{–}600\,\mu\text{m}$$

3. **Very high breakdown voltage V_{bd} of over 1000 V or so**, as V_{bd} depends on the doping and on *W*. For large V_{bd}, *W* had to be large but less than the diffusion length L_d for recombination to take place within *i*-layer thickness *W*, otherwise forward bias voltage drop across *i*-region as well as forward bias resistance will become high. For $W = 100$ μm, typical values of V_{bd} and L_d are as:

Semiconductor	Doping/cc	L_d (μm)	V_{bd} (V)
GaAs	10^{12}–10^{13}	10–20	150–250
Si	10^{12}–10^{13}	400–600	1000–6000

GaAs PIN diode has the disadvantages of (a) low V_{bd} and (b) three times higher thermal resistance, but has the advantages of smaller switching/recombination time < 10 ns, still Si-PIN diode is preferred specially due to high V_{bd}.

Therefore PIN diode is safe even at large μW power, as in its positive cycle also it cannot forward bias it, when kept at a reverse bias of 500 V. For handling still larger power, series–parallel connection can be made.

4. **Capacitance is very low due to** (a) large *W* and (b) by keeping/device area small. As a result the forward bias device impedance is also small.

(ii) Working Mechanism

1. At zero bias: The diffusion of holes and electrons across the junction causes;

   (a) Space charge region: The thickness of which is inversely proportional to doping density.
   (b) Fixed +ve space charge in the *n*-region.
   (c) Fixed −ve space charge in the *p*-region
   (d) No depletion region in ideal *i*-region

   ———

   (e) High resistance of PIN diode 10 kΩ or so Figs. 6.40c and 6.41.

2. At reverse bias: The space charge regions of *p* and *n*-layers become thicker and denser (Fig. 6.41b), with reverse resistance remaining high ≈10 kΩ and constant (Fig. 6.40g, h) along with uniform and high electric field, falling to zero in p^+, n^+ regions. This high resistance is equivalent to open circuit.
3. At forward bias: The hole and electron carriers get injected from both the junctions (n^+i and p^+i), respectively, into the *i*-region and this:

   (a) reduces the thickness as well as density of the space charge regions of *n* and *p* layers.

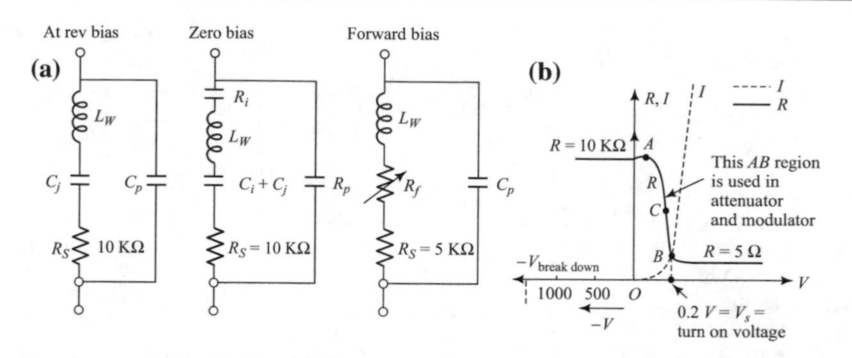

Fig. 6.41 PIN diodes **a** equivalent circuit of diode in package C_p = package cap; C_j = junction cap; C_{is}, R_i = Unswept intrinsic portion cap and res; R_s = series res; R_f = forward res; L_w = bonding Au-wire inductance and **b** *R–V* and *I–V* characteristic

(b) raises the carrier concentration in the *i*-layer above equilibrium.
(c) reduces the resistance of *i*-layers.
(d) causes fall of over all resistance of the diode, (Fig. 6.40c, f).
(e) causes virtual short circuit, as the switching time/carrier lifetime ($\approx 10^{-4}$–10^{-9} s) in this-*i*-region is ≫ period of microwave frequency. From high μW power to low power switching time is 40–1 ns.

6.13.1 PIN Diode Application in Circuits (as Switch, Attenuator, Phase Shifter, Limiter, and AM Unit)

PIN diodes are used in the circuit as (a) switch, (b) attenuator, (c) phase shifter, (d) power limiter, (e) amplitude modulating elements.

The property of short and open in forward and reverse bias is used as switch, and the switch property is used as phase shifter and limiter. The variation in resistance with bias voltage (Figs. 6.41b and 6.42) is used in attenuator and modulator.

(a) **PIN Diode as a Microwave Power Switch**: PIN diode is used in the series configuration or shunt configuration as in Fig. 6.43. The series configuration is more suitable for coaxial line while shunt one for waveguides. With the dc forward bias, the

  (i) series circuit transmission is ON.
  (ii) shunt circuit transmission is OFF.

In I-region impurity concentration has to be quite low, width $L > 20$ μm so that breakdown voltages remain large >500 V. This is because the microwave signal should not be able to forward bias it even in its +ve cycle, when it is reverse biased, i.e. OFF. The capacitances C_1, C_2 and RFC_1, RFC_2 have to be large to pass and stop the signal, respectively.

(b) **PIN Diode as Attenuator**: If we use *A*–*B* portion of the *I–V* characteristic in Fig. 6.41b, where the forward resistance decreases with bias along the points *A*, *C*, *B*, then PIN diode can function as attenuator using the same circuit of switch (Fig. 6.43). In this attenuator, the forward bias value will control the attenuation.

At point A-bias the series (shunt) configuration has max (nil) attenuation

At point B-bias the series (shunt) configuration has max (nil) attenuation

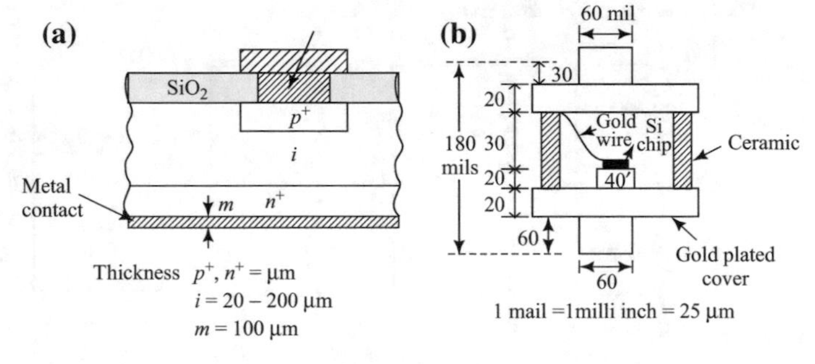

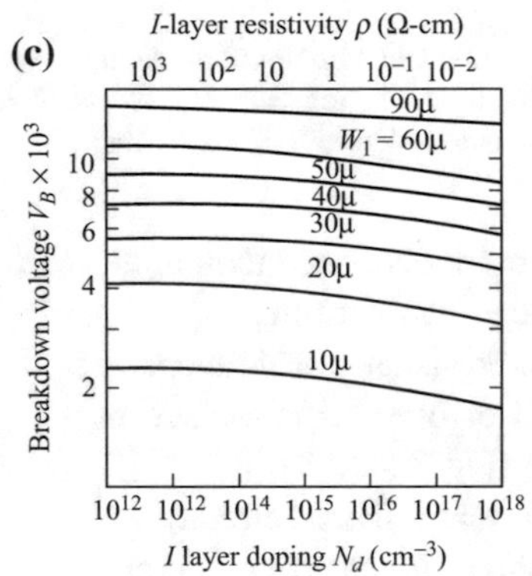

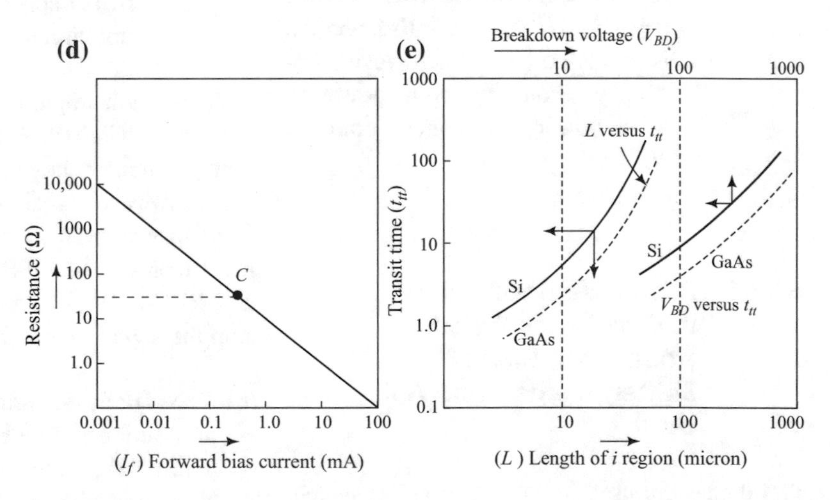

Fig. 6.42 PIN diode **a** construction of the planar diode device with measurements in mil and **b** a typical diode in package. **c** Breakdown voltage of PIN diode as a function of doping in I-layer for different width (W_I). Corresponding to the doping, the resistivity (ρ) in Si is also given at the top (from references 12 and 13, Chaturvedi et al.). **d** PIN diode as a variable resistance 0.1 Ω – 10 kΩ versus forward bias current 1 μA –100 mA, with centre point C at 50 Ω making Smith chart analyses. **e** Carrier transit time t_{tt} as a function of L and V_{BD}

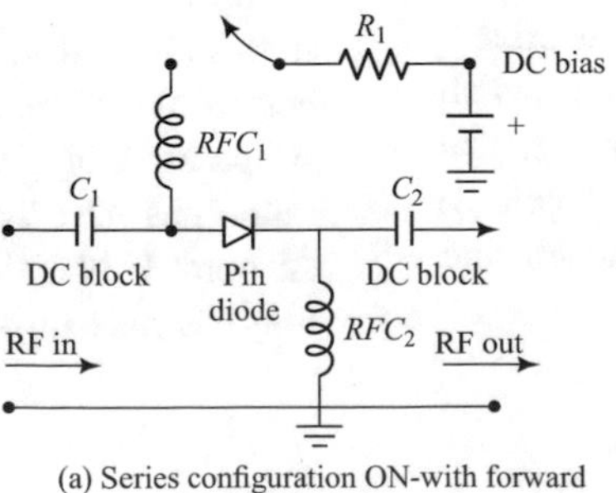

(a) Series configuration ON-with forward bias OFF-with reverse

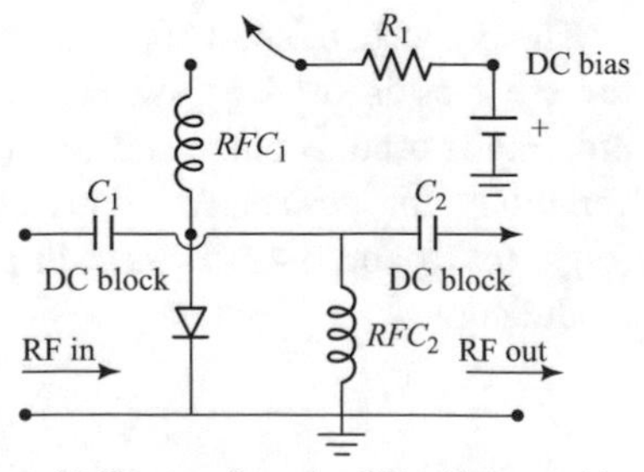

(b) Shunt configuration ON-with reverse bias OFF-with forward bias

Fig. 6.43 PIN diode as a switch

At point C-bias the series (shunt) configuration has medium (medium) attenuation

(c) **PIN Diode as Phase Shifter**: The switch property is indirectly used for introducing a phase shift in the signals (Fig. 6.44). The signal is made to travel a length of 2 (l_1 + l_2) at the port 2 when diode is OFF (to and fro path), while only length $2l_1$ when diode is ON i.e. short at diode. The phase shift $2l_2$ can be changed by moving the short plunger. The phase shift is

$$\phi = 2\pi \cdot l_2/\lambda \tag{6.55}$$

l_2 has to be fraction of λ. These phase shifters are used in phased array radars.

(d) **PIN Diode as Power Limiter**: PIN diodes are also used as microwave power limiters and act as short beyond a power limit (which is set p_{max} as per requirement). Here we make use of the diode property, i.e. beyond V_1 (the junction turn-on voltage) acts as short (Very low resistance) and microwave voltage above V_1 gets shorted (Fig. 6.45). Therefore external bias on diode is not required and microwave signal voltage acts as bias.

For increasing the power handling capacity, p_{max}, shorting current can be increased by having more number of diodes in shunt and can handle up to 100 kW.

Thus shorting the input power beyond p_{max} will not allow power to reach the output end beyond p_{max}, for protection of microwave system.

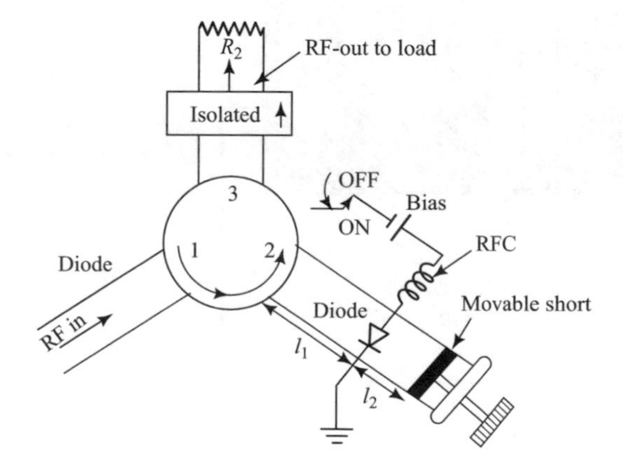

Fig. 6.44 Phase shifting using PIN diode as switch

(e) **PIN Diode as Amplitude Modulator**: PIN diode can be employed as an amplitude modulator as in Fig. 6.46, where the amplitude modulating (AM) signal of low frequency (f_m) and amplitude lower than the microwave carrier signal (t) are mixed. The diode is kept at a very low reverse (or zero) bias and in series with the modulating signal (f_m).

The f_m signal in its +ve cycle (point B) will make the diode forward biased and hence of very low resistance across PQ (Fig. 6.45a, b), allowing very low power to reach the output end. During the −ve cycle of f_m, (points A_1, A_2), more power will reach the output and hence we get AM output. The modulation could be by a square wave also. In this reference, the Gunn diode power supply could be referred (see Fig. 8.11 also).

6.14 Varactor Diode as a Variable Capacitor

The term varactor (or varicap, as it is so called) was coined from its property of variable reactance (or capacitor) of a pn junction with reverse bias, due to variation of its depletion layer width (acting as a dielectric) (Fig. 6.47).

It is a pn junction having I–V characteristic just like other pn junction diodes, but because of special impurity doping (Figs. 6.48 and 6.49) profile (abrupt or hyperabrupt), its properties differ as:

(a) Its capacitance vary in a nonlinear manner $\left(C\alpha 1/V_R^n\right)$ (Fig. 6.50) with the reverse bias voltage.

(b) It is fast enough to follow microwave frequency.

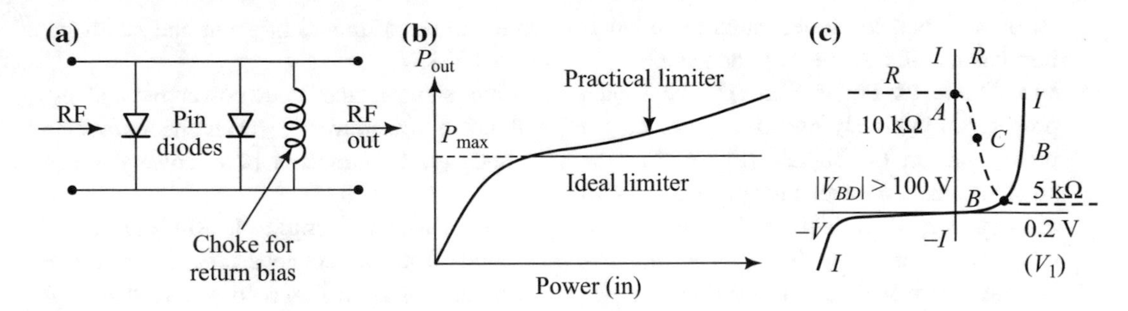

Fig. 6.45 **a** PIN diode limiter and **b** input–output characteristic of PIN-limiter **c** *I*–*V* and *R*–*V* characteristic and of PIN diode

Fig. 6.46 PIN diode as amplitude modulator **a** the circuit in principle, **b** mix of *dc* with f_m, f_c, **c** modulated outputs

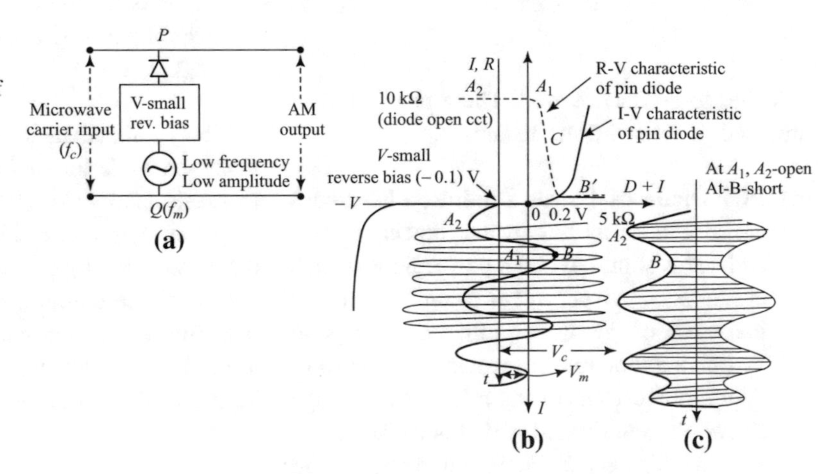

Fig. 6.47 Increase of reverse bias increases the depletion width and hence reduces capacitance ($C = \varepsilon A/d$)

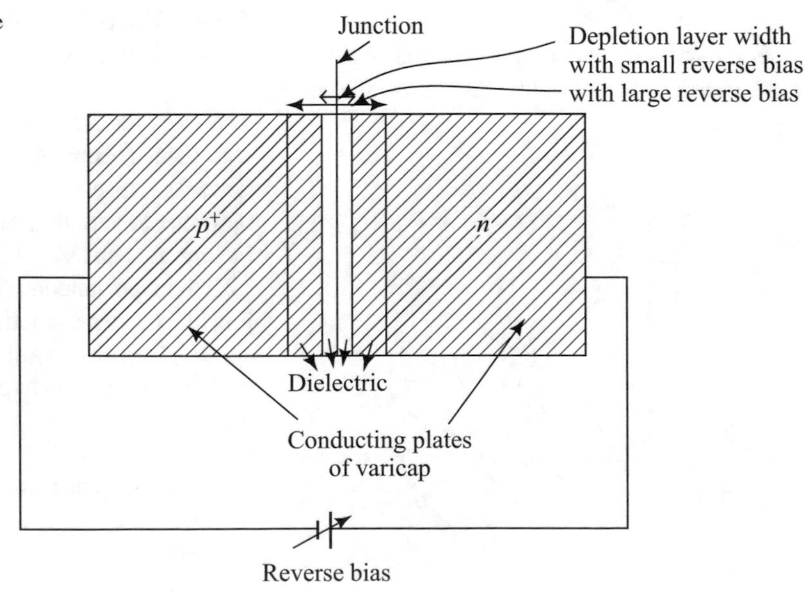

(c) The junction capacitance variation depends on the type of impurity profile linear or abrupt or hyperabrupt. The hyperabrupt gives best variation of C (Fig. 6.50). i.e. gradual change with voltage.

(d) It has negligible power loss, as the equivalent series resistance is very small (Fig. 6.51).

In an ordinary pn junction diode, the C–V curve does not show large variation but in varactor diode it does, as the depletion width vary as:

$$W \alpha (V_0 + V_s)^m \quad C \alpha \frac{1}{(V_0 + V_R)^m} \quad \begin{pmatrix} m = \frac{1}{3} \text{ for linearly graded Junction} \\ m = \frac{1}{2} \text{ for abrupt Junction} \\ m = 2 \text{ for hyper abrutpt Junction} \end{pmatrix} \tag{6.56}$$

6.14.1 The Device Structure

The three types of impurity profiles (linear, abrupt, and hyperabrupt) are given in Figs. 6.48 and 6.49 with their C–V characteristic in

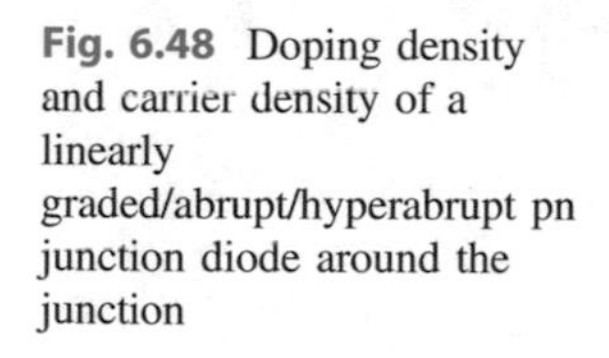

Fig. 6.48 Doping density and carrier density of a linearly graded/abrupt/hyperabrupt pn junction diode around the junction

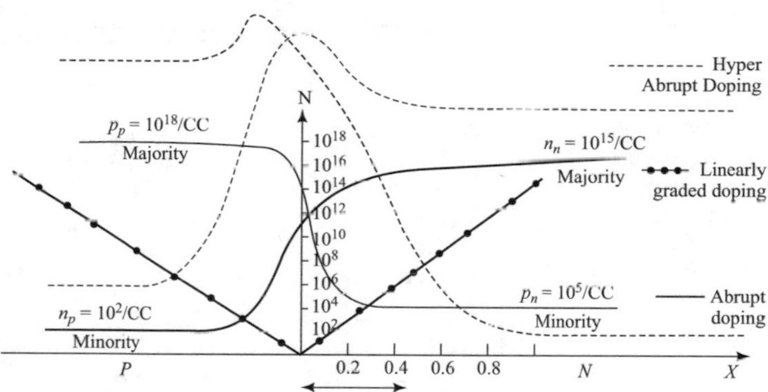

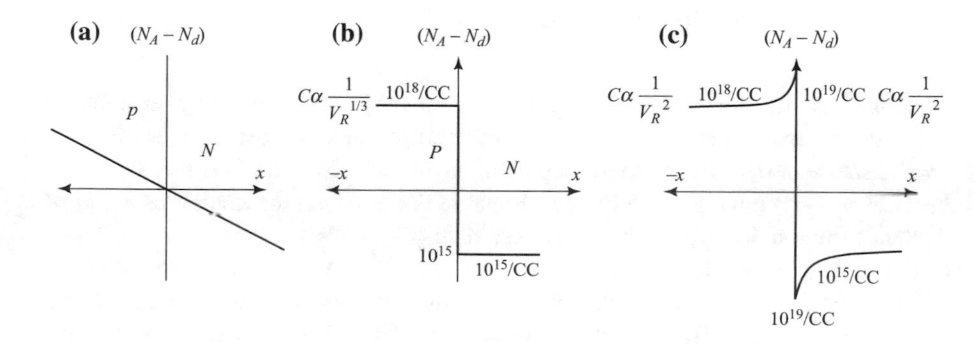

Fig. 6.49 Ideal doping profile of pn junction electrons (Jn) show by $N_a - N_d$ around the junction, **a** both sided linearly graded Jn (ordinary diode), **b** both sided abrupt Jn (varactor diode), **c** both sided hyperabrupt Jn (best varactor diode) with gradual variation of C with V_o

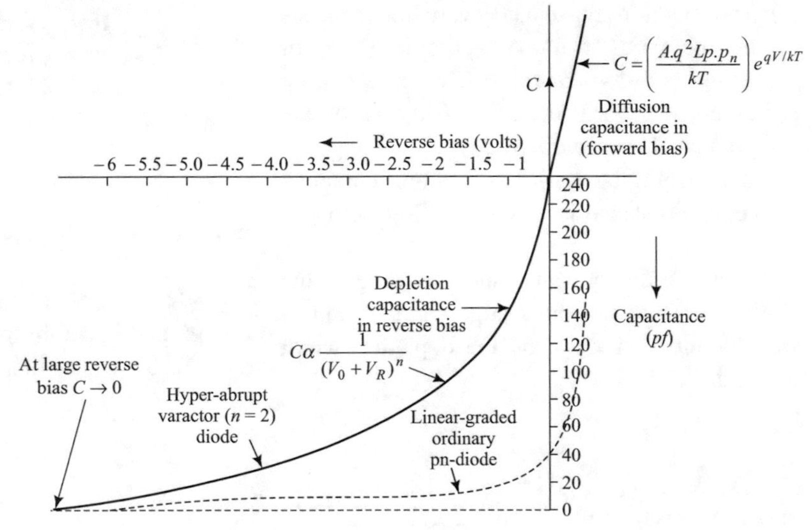

Fig. 6.50 C–V characteristic of a typical hyperabrupt varactor diode and ordinary diode

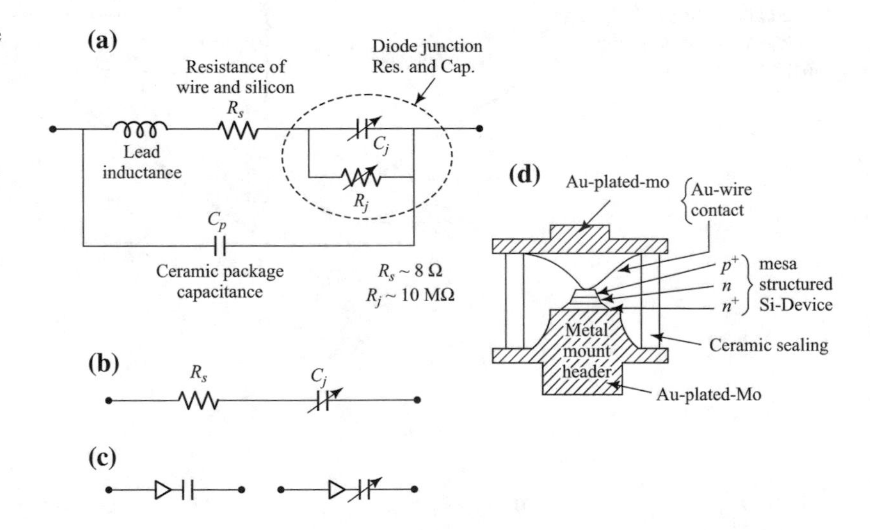

Fig. 6.51 Varactor diode **a** full equivalent circuit, **b** simplified equivalent circuit, **c** symbols used, **d** packaged device with mesa-structure (trapezium type)

Fig. 6.50. In the abrupt and hyperabrupt junction, the change of in purity from N_D to N_A takes place in a very small distance of 0.4 μm, while in linear graded it happens in 2–10 μm region. As a result of this, depletion region in hyperabrupt junction is very thin leading to higher capacitance value as well as gradual variation of C with reverse bias as shown in Fig. 6.50. The abrupt junction diode is fabricated (Fig. 6.51d) by starting with n^+ substrate over which epitaxial growth of n-layer is done and the np^+ is diffused. For hyperabrupt diode, on the n^+ substrate, two epitaxial layers of n-Si and p^+-Si are deposited, such that at the junction, the doping of n and p^+ is very high (Fig. 6.49c).

Individual Si devices are then given a mesa-structure (Trapezium Type) before dicing from the water. The device is then mounted on the header by T.C. bonding and then encapsulated with gold wire bonded on the top (p^+) (Fig. 6.51).

6.14.2 Characteristic

(a) **At zero bias** the pn junction has:
- Built-in voltage due to diffusion of charges $V_0 = \frac{kT}{e}\ln\left(\frac{N_A N_d}{n_i^2}\right)$
- Diffusion capacitance $C_d = \frac{A \cdot e^2 \cdot L_p \cdot p_n}{kT} \cdot e^{\frac{eV}{kT}}$

This C_d becomes very large, while V_0 depends on doping levels N_d, N_A on the two sides.

(b) **In the reverse bias** the variation of capacitance (C_j) is quite large in hyperabrupt junction diode Eq. (6.56) (Fig. 6.50).

$$C_j \alpha \frac{1}{(V_0 + V_R)^2} \quad (6.57)$$

(c) **For large reverse** bias $V_R \gg V_0$, therefore above equation reduces to an approximate equation as:

$$C_j \alpha \frac{1}{V_R^2} \quad (6.58)$$

6.14.3 Applications

Highly nonlinear $C_j - V_R$ characteristic of varactor diode permits it to be used as:

(a) Frequency converter/multiplier.
(b) Parametric amplification.
(c) Frequency tuner along with a fixed inductor.

6.14.4 Varactor as Harmonic Generator/Frequency Multiplier

As we know that $C = Q_0/V^n$, any variation in V for a fixed Q on it will lead to nonlinear variation of C. If an ac signal $V_p \sin(w_p t)$ is given, to a reverse-biased (V_R) varicap then its capacitance will be time varying as:

$$\begin{aligned} C_j(t) &= Q_0/\left(V_R + V_p \sin \omega_p t\right)^n \\ &= C_0/\left(1 + \frac{V_p}{V_R} \sin \omega_p t\right)^n \quad \text{where } C_0 = Q_0/V_R^n \end{aligned} \quad (6.59)$$

The expansion of $C_j(t)$ in harmonic series gives:

$$\begin{aligned} C_j(t) = C_0 + C_1 \sin \omega_p t + C_2 \sin(2\omega_p t) \\ + C_3 \sin(3\omega_p t) + \cdots \end{aligned} \quad (6.60)$$

This nonlinear behaviour of 'C_j' is shown in Fig. 6.52.

The diode ac current due to above varying, capacitance will be:

$$i(t) = I_1 \cos\left(\omega_p t\right) + I_2 \cos\left(2\omega_p t\right) + I_3 \cos\left(3\omega_p t\right) \quad (6.61)$$

Thus a single frequency ω_p gives rise to $2\omega_p$, $3\omega_p$, … etc., frequencies across the diode, i.e. acts as multiplier of frequency. The desired harmonic say $3\omega_p$ can be taken out by a putting a resonant circuit tuned to this harmonic ($3\omega_p$) as in Fig. 6.53 in these types of multiplier, the stable input frequency signal is fed from a crystal controlled signal of generator of frequency (f_b). There are three resonant circuits tuned to frequencies f_p = ($_p/2\pi$), $2f_p$, and $3f_p$. The input circuit 1 acts as isolater between source and the diode/output circuit. The tuned circuit 2 acts as heterodyne intermediate resonant circuit ($2f_p$) isolating input and output frequencies by bypassing it. The tuned circuit 3, tuned to $3f_p$, allows the third harmonic as output. Here it may be noted that as the third harmonic current is due to capacitance effect, very small power loss occurs, with very little noise. However the power of third harmonic ($3f_p$) is much smaller than main signal (f_p).

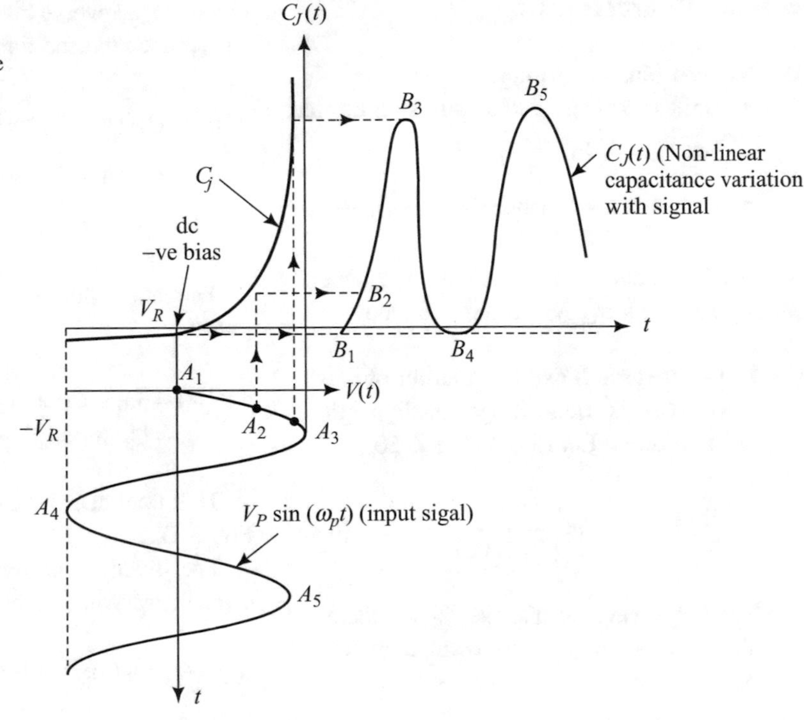

Fig. 6.52 Nonlinear variation of $C_1(t)$ with ac pump voltage correspondence of points A_1–A_5 of signal causes capacitances of value at B_1 … to B_5

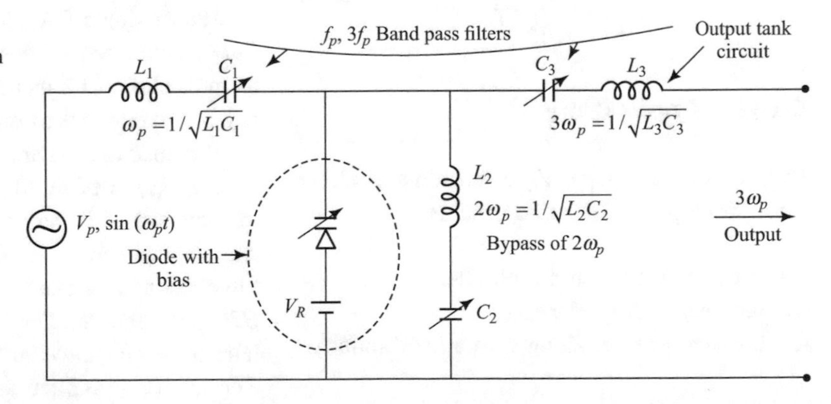

Fig. 6.53 Varactor diode multiplier tripler circuit, with three band pass filters

6.15 Parametric Amplifier: An Amplifier with Up/Down Conversion of Frequency

Parametric amplifiers use the variation of reactance for amplification purpose (rather than resistance as in normal amplifiers), therefore it is a very low-noise amplifier. Here input signal energy level of a wave (f_s) can be increased by another signal of different frequency (2, 3, … multiple higher), called pump (f_p). For understanding the mechanism, let us have a capacitor C (of varactor) charged to Q, so that the voltage across is

$$V = \frac{Q}{C}\left(\text{where } C = \frac{\varepsilon A}{\text{d}}\right). \tag{6.62}$$

We can increase the voltage by reducing C, by increasing the reverse bias in varactors, as Q is constant. The energy for increasing the reverse biasing is obtained from the pump signal f_p, which is normally some multiple of input signal frequency f_s). This way the energy $W = (1/2)\ C \cdot V^2$ in the capacitance increases. By partial differential of Eq. (6.62), we can prove that this increase in energy received from pump signal is

$$\Delta W = -\frac{1}{2} V \cdot (\Delta C)^2 \qquad (6.63)$$

Now let us study two types of paramps:

(a) Degenerate type for understanding the working of paramps ($f_p = 2f_s$).
(b) Non-degenerate type used in practice ($f_p \neq 2f_s$).

(a) Degenerate parametric amplifier
Let us have a pump signal with double the frequency of input signal ($f_p = 2f_s$) and in phase as in Fig. 6.54. As $V_p(t)$ and hence decreases from maxima to minima at t_1, t_3, t_5, t_7, therefore the energy of V_s increases during these moments of time.

This way energy is added to the signal (t_s) two times in one time period, which keeps on increasing its amplitude and hence, the amplification taken place. This amplification keeps on going, till energy added is equal to the energy dissipated in varactor, as it has equivalent resistance component also. A circuit for this amplifier is given at Fig. 6.55.

We see that in the above type of parametric amplification, the following two crucial conditions are there:

(i) **Frequency ratio**: $f_p = 2 \cdot f_s$
(ii) **Phase factor**: $(V_p)_{max}$ has to be exactly at the times when $V_s = (v_s)_{min}$ and $(V_s)_{max}$ occur.

This type of phase-sensitive amplifier with $f_p = 2f_s$ is called degenerate parametric amplifier. In practice the current relationship of phase and frequency between the pump and the input signal is difficult to have, owing to the lack of control over the input signal. Therefore this type is used just where higher frequency amplification is required, e.g. radars, space communication.

(b) Non-degenerate parametric amplifier
In practice the pump frequency (f_p) is normally other than twice of f_s and such a system is called non degenerate type and offers

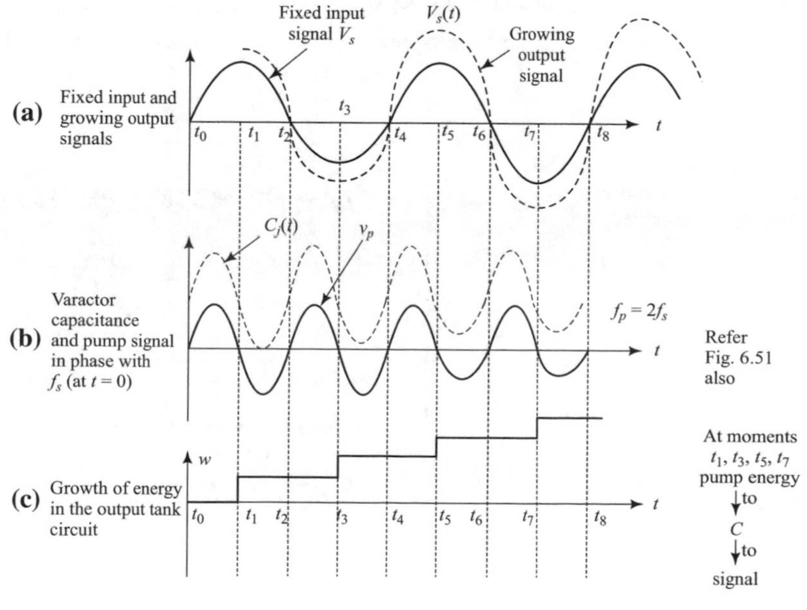

Fig. 6.54 Principle of parametric amplification (degenerate type): energy getting transferred from the pump (f_p) signal to input signal (f_S) at time moments t_1, t_3, t_5 (for $f_p = 2f_s$) when pump signal and hence C_j reduces from maximum to minima, as reduction in C only increases the energy in it [Eq. (6.63)]

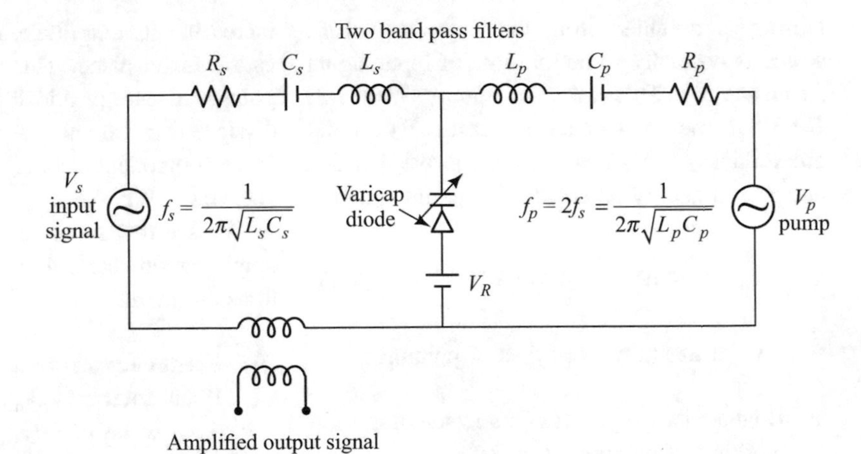

Fig. 6.55 Circuit of a degenerate parametric amplifier

greater flexibility in operation and power gain improvement. When we apply V_p and V_s signals to the varactor, we get amplified signal of sum and difference frequencies, besides other higher harmonics [i.e. $f_i = (mf_p \pm nf_s)$ where $n, m = 0, 1, 2, 3, \ldots, \infty$], at the output across R_i of idler circuit side (Fig. 6.56). Amplification mechanism is same as explained above in degenerate type and energy for amplification of V_s comes from the pump signal (V_P).

Thus out of these available harmonic frequencies, we can choose frequency $f_0 = f_i = \frac{1}{2\pi\sqrt{C_iL_i}}$, by choosing elements L_i, C_i the corresponding band pass resonant elements.

There are four cases of special interest in parametric amplifiers. These four cases are three of non-degenerate type (a, b, c) and one (d) of degenerate type as given below:

Non-degenerate type:

(a) If $f_0 > f_s$ the device is called parametric up converter.
(b) If $f_0 < f_s$ the device is called parametric down converter.
(c) If $f_0 = (f_p - f_s)$ the device is called −ve resistance parametric amplifier.

Degenerate type:

(d) If $f_0 = (f_p - f_s) = f_s$, the device is called −ve resistance parametric amplifier.

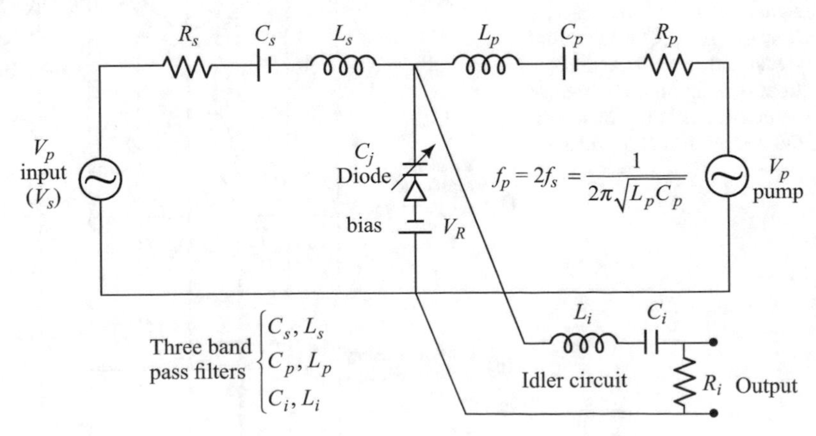

Fig. 6.56 Equivalent circuit of a non-degenerative parametric amplifier

All these special cases will be taken up again along with Manley–Rowe relation, which deals with harmonics frequencies generated when two signals are imposed on a nonlinear reactance, e.g. varactor.

6.15.1 Manley–Rowe Relation and Types of Paramps

A general set of relation between power flowing into and out of an ideal reactance (L or C), when two signals (f_s and f_p) are applied to it, was given by Manley and Rowe (M and R) in 1956. This relation is used for:

(a) Predicting whether power gain is possible in a para-amp.
(b) Predicting some characteristic of varactor circuit.

This Manley–Rowe (MR) relation is just power conservation equations when harmonics of f_s, f_p, their sum, and differences get generated across a varactor. These harmonics are separated by band pass filters and their power made to dissipate in separate resistive loads as in Fig. 6.57.

$$\sum_{n=0}^{\infty}\sum_{m=\infty}^{\infty}\frac{n\cdot P_{nm}}{(m\cdot f_p+n\cdot f_s)}=0 \quad (6.64)$$

$$\sum_{n=-\infty}^{\infty}\sum_{m=0}^{\infty}\frac{m.P_{nm}}{(m\cdot f_p+n\cdot f_s)}=0 \quad (6.65)$$

Here P_{nm} represents the average power at the output entering or coming out of nonlinear capacitor (varactor) at frequency $(m \cdot f_p + n \cdot f_s)$ Here f_p is pumping signal and f_s the input signal and n, m = integers (0 to ∞). Power flowing into the varactor is taken as +ve while that coming out of it and flowing into the load resistance are −ve.

To understand more, let us take some examples four of special cases:

(a) **$f_0 > f_s$ Case: Parametric Up Converter**: We can have a band pass filter of freq (f_p + f_s) at the output idler circuit, so that the above Eqs. (6.64) and (6.65) reduce to

$$\frac{P_{10}}{f_p}+\frac{P_{11}}{f_p+f_s}=0 \quad (6.66)$$

$$\frac{P_{01}}{f_s}+\frac{P_{11}}{f_p+f_s}=0 \quad (6.67)$$

$$f_0=f_p+f_s \quad (6.67a)$$

where P_{10} and P_{01} are powers of pump (f_p) signal and input signal (f_s) and they are taken as +ve (i.e. entering the varactor). The powers are the power flowing out of the varactor into the resistance load at frequency (f_p + f_s) and are therefore −ve.

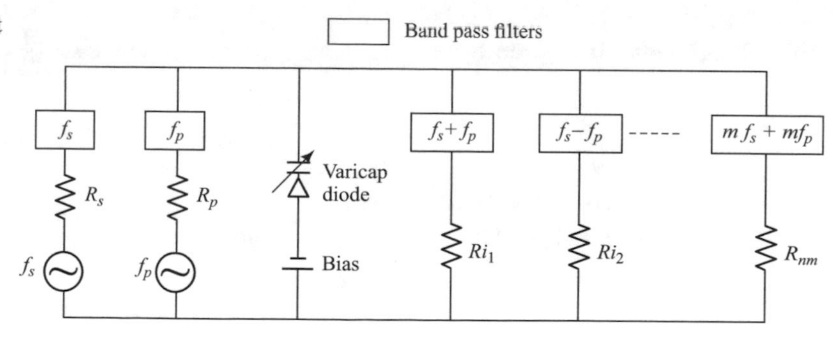

Fig. 6.57 Equivalent circuit for Manley–Rowe relation

Therefore the power gain (P_{11}/P_{01}) from MR relation is:

$$G_{MR} = \frac{P_{11}}{P_{01}} = \frac{f_p + f_s}{f_s} = \frac{f_0}{f_s} \tag{6.68}$$

Here (a) the output frequency $f_0 > f_s$ (b) the gain is >1 therefore it is called Parametric up converters (PUC) amplifier and has the gain, noise figure, and band width as per given below:

(i) *Maximum Power Gain*

$$G_{max} = \frac{f_0}{f_s} \cdot \frac{x}{\left(1+\sqrt{1+x}\right)^2}$$
$$7.4\,\text{dB for } f_0/f_s = 20 \tag{6.69}$$

Here

$$x = \frac{f_0}{f_s}(\gamma Q)^2$$

γQ Figure of merit of varicap.
Q Quality factor = $1/(2\pi f_s \cdot C_j \cdot R_j)$.
γ Merit factor.

Power gain given by MR relation [Eq. (6.68)] is lower than G_{max} [Eq. (6.69)].

(ii) *Noise Figure*: Optimum noise figure is expressed with T_d, T_0 as diode temperature and ambient temperature:

$$F = 1 + \frac{2T_d}{T_0}\left[1/\gamma Q + 1/(\gamma Q)^2\right]$$
$$= 0.9\,\text{dB for typical diode with } f_0/f_s = 10 \tag{6.70}$$

(iii) *Band width*: It is given by:

$$\text{BW} = 2\gamma\sqrt{\frac{f_0}{f_s}}$$
$$= 1.26,\ (\text{Typical diode with } f_0/f_s = 10 \text{ and } \gamma = 0.2) \tag{6.71}$$

(b) **$f_0 < f_s$ Case: Parametric Down Converter (PDC)**: If we choose the resonant circuit components of idler such that $f_0 < f_s$ then the corresponding harmonic only will resonate and the gain given by (which being <1 is actually loss):

$$G_{am} = \frac{f_s}{f_p + f_s}\ (\text{by MR-relation}) \tag{6.72}$$

$$= \frac{f_s}{f_0} \cdot \frac{x}{\left(1+\sqrt{1+x}\right)^2}\ (\text{By other methods}) \tag{6.73}$$

(c) **$f_0 = (f_p - f_s)$ Case: Negative Resistance Parametric Amplifier**: If the idler circuit frequency is $f_0 = (f_p - f_s)$ in Fig. 6.64, then from the Manley '*N*' Rowe relation Eqs. (6.66) and (6.67), power P_{11} is +ve, while P_{10} and P_{01} are −ve. In other words, capacitor delivers power to the input source (f_s) instead of absorbing. This can be proved analytically also that V_d (ac diode voltage) will be −ve.

Therefore the varactor diode becomes an oscillator for f_s and f_0 both and therefore the name of −ve resistance parametric amplifier. The gain, noise figure, and band width will be:

$$\text{Gain} = \frac{4f_0}{f_s} \cdot \frac{R_s \cdot R_i}{R_{Ts} \cdot T_{Ti}} \cdot \frac{a}{(1-a)^2}$$

Noise figure (same as up converter)

$$= 1 + 2\frac{T_d}{T_0}\left[1/(\gamma Q) + 1/(\gamma Q)^2\right]$$

$$\text{Bandwidth} = \frac{\gamma}{2}\sqrt{\frac{f_i}{f_s}}$$

where R_i = output idler resistance and R_{Ti}, R_{Ts} = are total resistance at f_i and f_s, respectively, and $a = \gamma^2/\left(\omega_s \omega_i c_j^2 R_{Ti}.R_{Ts}\right) = R/R_{TS}$ R = equivalent-veresistanceofvaractor $= \gamma^2/\left(\omega_s \omega_i c_j^2 R_{Ti}\right)$

$T_0 = 300°$K ambient temperature

$T_d =$ diode temperature in Kelvin.

(d) **$f_0 = (f_p - f_s) = f_s$ Case: Negative Parametric Amplifier-Degenerate**: Specific case of negative parametric amplifier is the degenerate paramps where $f_0 = (f_p - f_s) = f_s$, i.e. $f_p = 2f_s$, i.e. the output frequency is same as signal frequency. Here it becomes an oscillator with circuit given in Fig. 6.55.

6.15.2 Advantages, Limitations, and Application of Paramps in General

Advantages: Out of the above four types of paramps, the up converter paramps are the most useful due to the following advantages:

1. Stable amplification.
2. Power gain independent of changes in its source impedance.
3. The input impedance is +ve.
4. No circulator is required in the circuit.
5. Reasonable band width, 5% or so due to tuned circuit it can be increased by stagger tuning.
6. It has very low noise of 1–2 dB, as thermal noise is absent.

Limitations

1. *Frequency Limits*: As the frequency pump has to be much larger than that of the input signal, the upper frequency limit is due to difficulty in getting higher frequency pump source. If the available pump is of 200 GHz, then signal amplifier can be for highest frequency of 100 GHz or to for $f_p \approx 2f_s$. Therefore with −ve resistance paramps for higher frequencies, the up converter is non-practical. The lower frequency limit is set by the microwave components used due to their cut-off frequencies.
2. *Gain*: Gain is limited (up to 20–80 cB) due to (a) pump source and (b) quality of varactor.

Applications

1. In ground radar, radio telescopes, space communication applications as the requirement of frequency is above *x*-band, and noise to be very less. Therefore −ve resistance paramps become the choice.
2. For lower frequency application (below *x*-band), degenerate amplifier is used as it is simple to use, with low pump frequency.
3. In airborne radar systems, parametric amplifiers were used, but are getting replaced by GaAs-MESFET amplifiers due to its simplicity, fabrication, and production.

6.16 New Devices in Microwaves GaAs and GaN-HEMT, GaN-HEMT, and FINFET

The technology push for higher power at higher frequencies of operation with higher speed has brought new devices, like GaAs-HEMT, GaN-HEMT, and FINFET. HEMT stands for high electron mobility transistor, while FIN is prefixed to FET in FINFET as the gate protudes, i.e. comes out of the surface vertically upwards like a fin.

GaAs-HEMT had already reached maturity. Now GaN-HEMT, which gives highest power density and highest efficiency of all the semiconductor devices so far, has also reached maturity, with commercial devices available now. FINFET is yet to reach maturity.

6.16.1 GaAs-HEMT

There has been a continuous effort to find a microwave device which can give higher power at still higher frequencies than the existing devices. When MESFET came with frequency of operation up to 50 GHz, it surpassed MOSFET which was up to 30 GHz only. Then came GaAs-HEMT, which is a modified form of GaAs-MESFET, which can go even beyond 100 GHz due to its high mobility. The modification is that GaAs-HEMT is made of heterostructure of GaAlAs and GaAs materials. It has all the properties of GaAs-MESFET, therefore we can call it also as GaAs-HEM-MESFET.

In GaAs-MESFET the charge transport, i.e. carrier movement, takes place in highly doped GaAs, as a result lot of energy is lost due to coulombian scattering by the doped ions, leading to low mobility and hence low frequency. In GaAs-HEMT the conduction channel of electron carrier is not threedimensional broad channel region, but just two-dimensional plane-path along the thin interface of the two material, i.e. *n*-GaAlAs and GaAs (Fig. 6.58). This two-dimensional electron gas flow is called 2DEG (2D electron gas), which is in fact in the zero-doped GaAs region, having less coulombian scattering of electron by impurity ions (being absent). This leads to high mobility (up to 9000 cm^2/Vs as compared to 4000 cm^2/Vs in GaAs-MESFET) and short transit time along the channel, leading to higher frequency of operation. Normally 2DEG has a surface electron density of 10^{12}–10^{13}/cc.

Thus this high-frequency behaviour is due to shifting of electrons, i.e. polarisation from donor sites of *n*-GaAlAs to undoped GaAs due to difference of band gaps (E_g), equalisation of the fermi level takes place, with formation of a very small potential barrier across the interface of 0.2–0.4 eV (see Fig. 6.59b). Thus the 2DEG carrier gets confined to a very narrow layer (about 10 nm) (Table 6.8).

Thus the difference in the band gaps of GaAlAs and GaAs is exploited to:

(a) Reach frequencies beyond 100 GHz.
(b) Maintain low noise figure of that of GaAs-MESFET.
(c) Get high power rating specially due to high E_g and high thermal conductivity.
(d) Maintain no mechanical stress between the heterostructure (GaAlAs and GaAs) as both have nearly same lattice constant of 5.65 Å (Fig. 6.58).

For increasing the current and power as well, we can think of having higher doping density in *n*-GaAlAs for transfer of more number of electrons to 2DEG. But then the scattering of carriers by the impurity ion also increases, reducing the mobility. Therefore for reducing this coulombian scattering by the impurity ions near 2DEG, a spacer layer of 20–100 nm of undoped GaAlAs is grown between *n*-GaApAs and GaAs (see Fig. 10.11b). With this spacer layer, higher

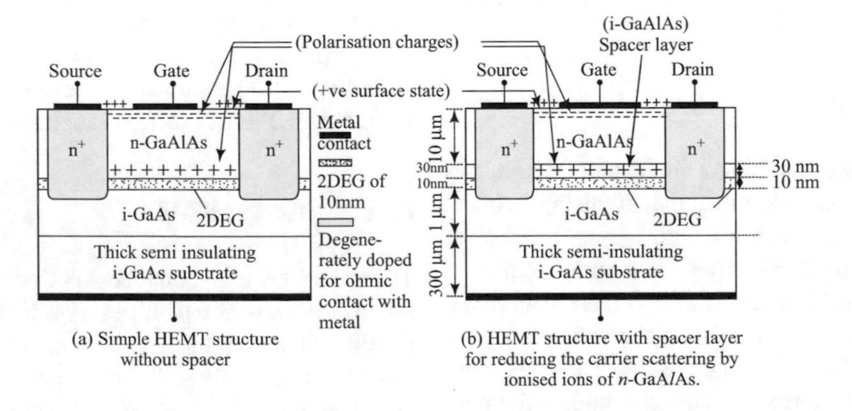

Fig. 6.58 Generic heterostructure of a depletion-mode HEMT, with and without spacer layer of intrinsic GaAlAs. Only gate has Schottky metallic contact while the rest (e.g. source/drain) have ohmic metal contact. The thick substrate (GaAs) has very high dielectric constant of 13 and high thermal conductivity of 6.54 W/cm K. 2DEG is formed by polarisation of carriers from *n*-GaAlAs to GaAs due to difference of fermi level, natural equalisation of electrons take place, leading to barrier potential of 0.2–0.4 eV across interface

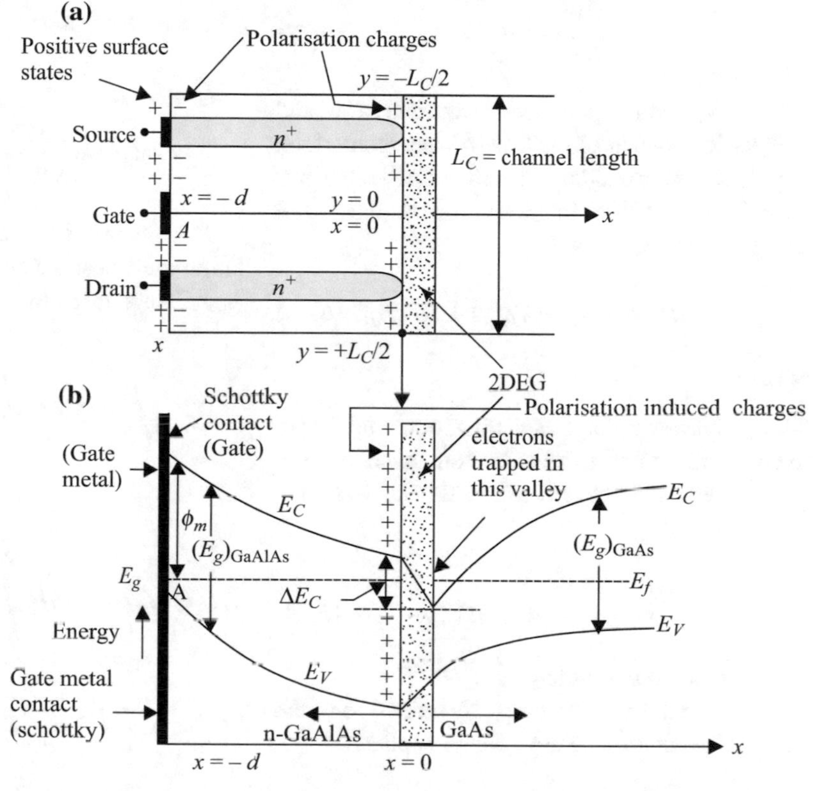

Fig. 6.59 **a** Simple Ga. As HEMT structure (shown without substrate) and formation of 2DEG and **b** energy band diagram with $x = -d$ at the gate and $x = 0$ at the interface of n-GaAlAs, GaAs. Electron carriers shift from doped region to undoped GaAs, i.e. polarisation for equalising the fermi levels of both sides, thereby developing very low potential barrier of $\Delta E_c = 0.2$–0.4 eV across the interface

Table 6.8 Band Gaps (E_g)

Si (eV)	GaAs (eV)	$Ga_x\,Al_{1-x}\,As$	GaN (eV)
1.1	1.43	1.43–3.4 eV (for $x = 0$–1)	3.4

doping in n-GaAlAs is possible for larger current and power. This spacer layer is grown through molecular beam epitaxy and is kept very thin (20–100 nm), so as to control the gate voltage (V_{gs}) for controlling the 2DEG through electrostatic mechanism.

As can be expected that the manufacture of HEMT is expensive compared to GaAs-MESFET, because of three reasons:

(a) Uses difficult to fabricate semiconductor material GaAlAs.
(b) Steep doping gradient requirement in n-GaAlAs.
(c) Precise control on all the thin structure required including the thin spacer GaAlAs.

The thick substrate of 300 μm or so of pure GaAs is an ideal material due to the following:

(a) Has very high dielectric constant of 13.
(b) Free intrinsic carrier density is very small, e.g. 1.8×10^6/cc as compared to Si of 1.5×10^{10}/cc. Thus GaAs substrate acts as a good insulator.
(c) Has crystal compatibility with GaAlAs as both have same lattice constant of 5.65 ÅU.

6.16.2 Drain Current Equation of GaAs-HEMT

The key point that decides the drain current in HEMT is the 2DEG narrow interface between GaAlAs and GaAs layers. For a simple calculation, we neglect the spacer layer of GaAlAs and study the energy band diagram (Fig. 6.59a) carefully. A mathematical model can be developed using the Poisson equation of the form

$$\frac{d^2V}{dx^2} = \frac{-qN_D}{\varepsilon} \quad (6.77)$$

where N_D and ε are donor concentration and dielectric constant of the GaAlAs heterostructure. Applying the boundary conditions of voltages $V(x = 0) = 0$ at the interface and at the metal surface $x = -d$, we get:

$$V(x = -d) = [-V_o = V_G(y) + \Delta E_c/q] \quad (6.78)$$

where

V_o barrier voltage (see Figs. 6.58 and 6.59)
ΔE_C difference of conduction band energy levels between n-doped GaAlAs and GaAs

$$V_G(y) = -V_{GS} + V(y) \quad (6.79)$$

V_{GS} gate-source voltage
$V(y)$ channel voltage drop at y point due to the V_{DS} source–Drain voltage applied

For finding the potential from Eq. (6.77), we integrate twice with the above boundary condition. We get at the metal–semiconductor interface (for $x = -d$), i.e. at the gate

$$V(-d) = \frac{qN_D}{2\varepsilon} \cdot d^2 - E_y(0) \cdot d. \quad (6.80)$$

where $E_y(0)$ = electric field at $y = 0$, $x = -d$. [At gate point A in Figs. 6.59a, b]

$$\therefore \quad V(-d) = V_P - E_y(0) \cdot d \quad (6.81)$$

where V_P pinch-off voltage $= \frac{qN_D}{2\varepsilon} \cdot d^2$

Now let us define HEMT threshold voltage as

$$V_{To} = \left(V_0 - \Delta E_c/q - V_p\right) \quad (6.82)$$

then above Eq. (6.80) can be written as:

$$E(o) = \frac{1}{d} \cdot \left[V_{gs} - V(y) - V_{To}\right] \quad (6.83)$$

Now the drain current using basic equation as:

$$\begin{aligned} I_d &= \sigma \cdot E_y \cdot A = (-q\mu_n N_D) \cdot E_y \cdot Wd \\ &= q\mu_n \cdot N_D \left(\frac{dV}{dy}\right) \cdot Wd \end{aligned} \quad (6.84)$$

For getting the surface charge density, we know the Gauss's law $Q_S = \varepsilon\, E_y(0)$.

∴ Integrating along the channel length L_C

$$\therefore \quad \int_0^{L_C} I_d \cdot dy = \mu_n \cdot W \cdot \int_0^{V_{DS}} Q_S dV \quad (6.85)$$

$$\begin{aligned} \therefore \quad I_d \cdot L &= \mu_n \cdot W\varepsilon \int_0^{V_{DS}} E_y(0) dS \\ &= \mu_n \cdot W \cdot \frac{\varepsilon}{d} \int_0^{V_{DS}} [V_{GS} - V(y) - V_{To}] dV \end{aligned}$$

$$\therefore \quad I_d = \frac{\mu_n \cdot W \cdot \varepsilon}{L_C \cdot d}\left[V_{DS}(V_{GS} - V_{To}) - \frac{V_{DS}^2}{2}\right] \quad (6.86)$$

The pinch off occurs when $V_{DS} \geq \left(V_{gs} - V_{To}\right)$ ∴ with this inequality I_d of Eq. (6.86) becomes

$$I_d = \frac{\mu W \varepsilon}{2L_C \cdot d} \cdot \left(V_{gs} - V_{To}\right)^2. \quad (6.87)$$

For depletion type of HEMT $V_{To} < 0$ or $\left(V_o - \Delta E_c/q - V_p\right) < 0$ ∴ for $V_p = qN_d \cdot d^2/(2\varepsilon)$, this inequality gives:

$$d > \left[\frac{2\varepsilon}{qN_d} \cdot (V_o - \Delta E_c/q)\right]^{1/2} \quad (6.88)$$

For d less than above, the HEMT operates in enhancement mode.

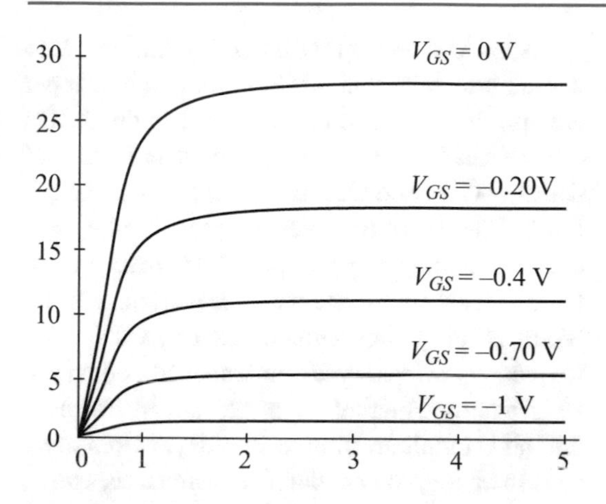

Fig. 6.60 Drain current in a typical GaAs-HEMT

Thus Eq. (6.87) shows that the I_d versus V_{ds} curve of HEMT is similar to that of MESFET (Fig. 6.60) and can be represented by the same electric circuit model. The HF performance is also similar to MESFET, with the transit time given by

$$T = \frac{L}{\mu_n E_y} = \frac{L^2}{\mu_n V_{ds}} \tag{6.89}$$

6.16.3 GaN-HEMT

After 1990s GaAs-MESFET got surpassed by GaAs-HEMT specially due to its high frequency of operation, with noise and power levels remaining the same. Then came GaN-HEMT in the year 2000, which is just a modified form of MESFET only, having high effective electron mobility of 9000 cm^2/VS in 2DEG as compared to that in GaAs-MESFET of 5000 cm^2/Vs and in Si-MESFET of 1900 cm^2/Vs. Here it may be noted as shown in Table 6.9 that in the bulk material, the electron mobilities are much less.

Now for still higher power at these high frequencies of 100 GHz or more, now GaN-HEMT has become the most promising device due to the following nine plus points:

(a) GaN has high breakdown voltage of 5000 kV/cm due to high band gap of 3.4 eV, therefore high power up to 100 W per device is possible (see Table 6.9). Because power density in GaN is 5–12 W/mm^2 as compare to 1.5 W/mm^2 in GaAs.
(b) GaN has much higher thermal conductivity and hence can dissipate much higher power.
(c) Higher mobility of electron is there in 2DEG and therefore has still higher frequency of operation is possible than GaAs-HEMT.
(d) The power efficiency of the GaN-HEMT is as high as 94%, i.e. very low energy consumption and hence low heating also.
(e) Intrinsic GaN has high electron density of 10^{13}/cc, vis-a-vis of Si 1.5×10^{10}/cc and GaAs 1.8×10^6/cc only.

Table 6.9 Properties of material used in HEMT

S. no.	Material	Coefficiency of expansion 10^6/K	Melting point (° C)	Band gap (E_g) (eV)	Thermal conductivity (or) W/cm K	Electron mobility (μ_e) cm^2/Vs	Breakdown field (E_C) (10^6 V/cm)	Saturated velocity of electron $v_s \times 10^7$ cm/s	Dielectric constt. (ε_r)
1	GaN	5.59	2500	3.43	1.5	1000	5	2.7	9.5
2	AlN	4.2	2275	6.2	3.2	1100	6.15	1.8	
3	InN	3.83	1925	0.7	0.8	2700	2.0	4.2	
4	SiC (4H)	3.08	2830	3.3	5	900	3.5	2.7	10
5	SiC (6-H)	4.2	2830	3.0	5	370	2.4	2.0	
6	Diamond	1.5	5000	5.43	1.5	1900	5.6	2.7	
7	Silicon	2.56	1415	1.12	1.57	1350	0.3	1.0	11.9
8	GaAs	6.8	1238	1.43	0.54	8500	0.4	2.0	12.6

(f) Good linearity in the gain of amplifier over a wide frequency range.
(g) Very low on-state resistance (R_{ON}) due of 2DEG.
(h) Is a very low-noise device.
(i) Very low parasitic capacitor, therefore useful is switch mode power supply.

Thus GaN-HEMT indicates a superb performance as a high-power device at high frequencies. To achieve power of 100 W, parallel configuration of multiple GaAs-MESFET is needed, while the same 100 W power can be achieved using a single GaN-HEMT device and that too with very low power consumption and heating as the efficiency is as high as 94%.

The transition frequencies (f_T) near THz, i.e. 1000 GHz, have been achieved in the laboratories with commercial device in the range of f_T = 200–400 GHz.

The shortcomings of GaN devices presently (2017) are:

(i) It has extremely low hole mobility of 40 cm^2/Vs at $N_d = 10^{16}$/cc at 300 °K circuit. This fact limits the designers to use *n*-channel GaN-HEMT only.
(ii) GaN material production cost is very high.
(iii) GaN devices are fabricated on different substrate, e.g. Si/SiC, increasing the cost further. This is special due to high material cost of GaN/higher thermally conductivity of Si/SiC, for enabling still higher power dissipation.
(iv) Because of the above suppliers of GaN devices are very few.
(v) It has a small market demand of only 0.85 billion dollars (2015) vis-a-vis of GaAs, which is around 130 billion dollars (2015). However gallium nitride high electron (GaN-HEMTs) has been available as commercial off-the-shelf devices since 2005.

As GaAs device growth rate is of around 25% per annum while that of GaN is around 50% per annum, the production cost as well as the device cost of GaN devices will come close to that of GaAs devices and then it is bound to overtake it fully. Based on the present growth rate, this overtaking may happen in 12–15 years or so. Global GaN semiconductor device market by products are power semiconductors, GaN radio frequency devices, opto-semiconductors devices. User agencies/market segments are in information and communication technology, automotive, consumer electronics, defence, aerospace, etc.

The GaN-HEMT technology as power amplifier can be used for mobile base station,

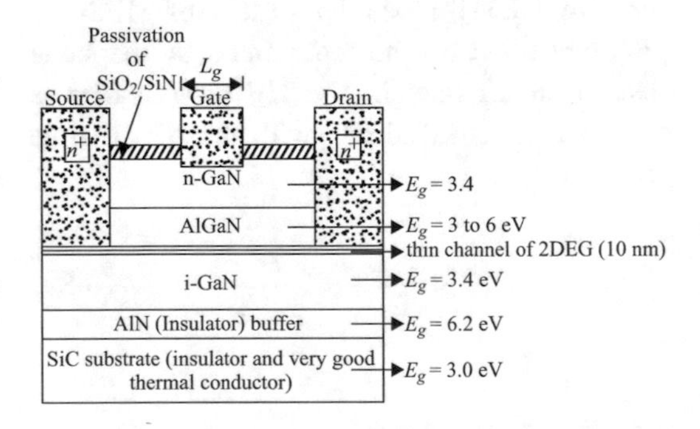

Fig. 6.61 A basic structure of GaN-HEMT

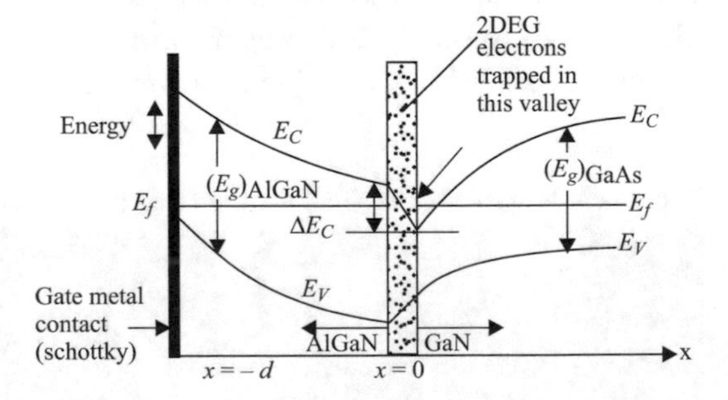

Fig. 6.62 Band diagram of simple GaN-HEMT (shown without substrate) and formation of 2DEG with barrier potential of ΔE_C = 0.2–0.4 eV across the interface due to shifting of electrons across the interface, i.e. polarisation for equalising the fermi level of both sides

RADAR sensor application equipments, power conversion, etc., where high microwave power is a desirable factor.

This technology will grow much faster due to the factors (a) to (i) listed above.

The structure of the device is given in Fig. 6.61, which is similar to GaAs-HEMT, including its working (Fig. 6.62).

The formation of 2DEG can be explained by the band structure given in Fig. 6.59 for GaAs-HEMT. However in fact, the piezoelectric effect and natural polarisation between the boundaries of GaN and n-ApGaN cause the accumulation of 2DEG (Fig. 6.11c). The I_d, V_{DS} characteristics are given in Fig. 6.63 for different V_{gs}. Which is similar to GaAs-HEMT except that current, voltages are high I_d increases with broader gate length L_g, as well as with charge density (N_S) in 2DEG which is in the range of 2–20 × 10^{12}/cc. This $N_s \propto V_{gs}$; $N_s \propto d_{AlGaN}$; N_S increases with Al mole concentration in $Al_x\, Ga_{1-x}As$. This makes it possible to have a very low ON-state resistance (R_{ON}). This low R_{ON} combined with high breakdown voltage (due to high E_g) and very high efficiency makes it possible to have a superb performance as a CW power source of very small size. The weight and size reduction is up to 10 times. M/S Fujitsu has shown commercially transmitter receiver amplifier on a single chip of 1.8 mm × 2.4 mm, with transmission loss of 1.1 dB in whole of the range 0–12 GHz, which they have claimed to be world's highest performance. They have also made a single chip with the functions of switch, transmitter amplifier, and a receiver, with an output of 6.3 W at 10 GHz. This chip size was 3.6 mm × 3.3 mm, which is 1/10th of conventional devices using multiple chips.

In addition to high power and high frequency, the amplifier system can be switched from band (4–8 GHz) to X-band (8–12 GHz) to Ku-band (12–18 GHz). M/S Fujitsu has shown a single GaN-HEMT amplifier having gain of 16 dB, 10 W power in the frequency range of 3–20 GHz, with a size of 2.7 mm × 1.2 mm on MMIC.

For comparing the advantages of materials of GaN-HEMT (which works on the lines of GaAs-HEMT) (Fig. 6.13), properties of a number of materials used are given in Table 6.9.

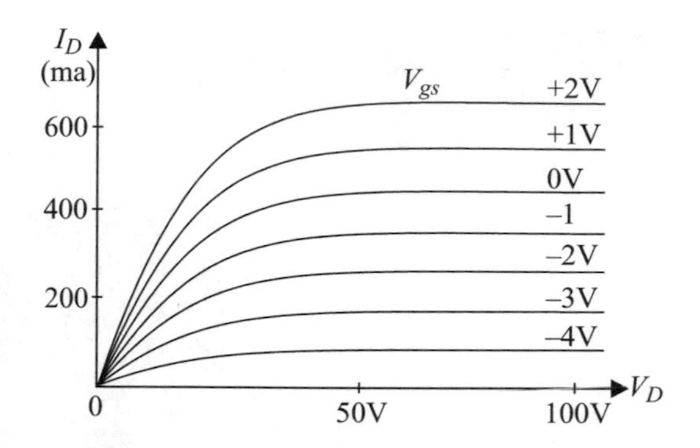

Fig. 6.63 A typical I_d versus V_D diagram of a GaN-HEMT

6.16.4 FINFET

Since the MOSFET was discovered, continuous zest for reducing the channel length has been the effort. So much so that it has nearly reached the scalable limit of 20 nm, as below this dimension of source and drain encroaches into the channel, leading to the following problems:

1. Increase of gate–drain junction tunnelling leakage current, making it very difficult to turn the transistor OFF completely. Therefore OFF state is not perfectly OFF. This problem in MOSFET becomes much more severe as due to scaling needs gate dielectric also has to be made thinner, thereby increasing further the gate–drain tunnelling leakage current.
2. Above degrades the circuit performance, power and noise margin.
3. Short channel effects comes into play.

For overcoming the above problems of scaling limits, wrap-around of the conducting channel by gate called FIN (as it protudes vertically above the substrate in 3D as a fin) was found to be most successful in going below this limit of 20 nm feature size. This type of gate (Fig. 6.64) has better electrical control (due to area

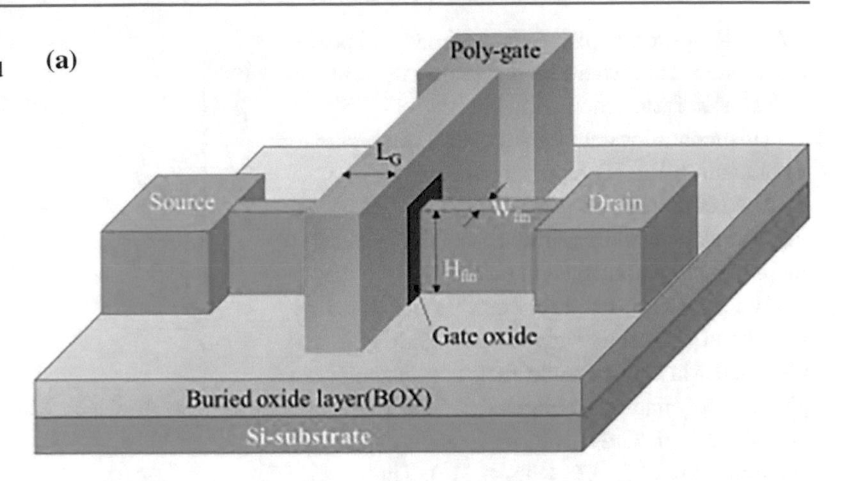

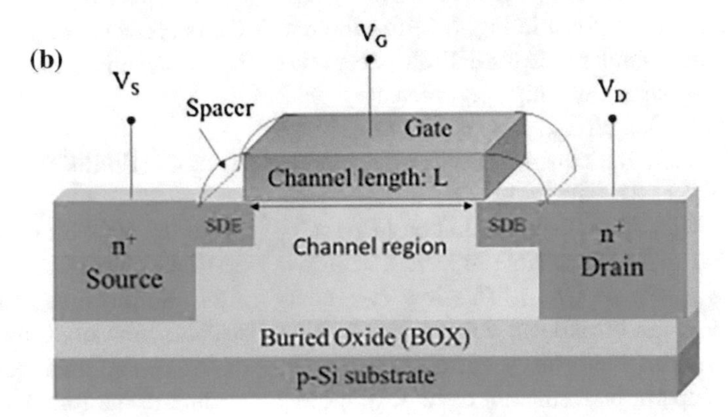

Fig. 6.64 Structure of FINFET channel surrounded by oxide and than gate

coverage) over the channel. This FET was there after called FINFET, and this technique was found to:

1. Have excellent control by the gate which is wrapped around the channel.
2. Have very low leakage current through the body when the device is in OFF state. In fact, it reduces by 90% or so.
3. Allows the use of lower threshold voltage, leading to higher switching speed by 30% or so.
4. Power consumption reduces by 50% or so.
5. Overcomes the short channel effect.
6. Have lower operating voltage say 0.9 V or so, requiring low power supply.
7. Allow feature size smaller than 20 nm which was thought as end point for planar FET's.
8. Be able to fulfil the Moore's law of device count/area doubling every 2 year and thereby can reach 5 nm feature size.
9. Have effective channel length larger than geometrical channel length. This is because thickness of the FIN contributes in determining the effective channel length as the FIN protudes above the surface as a fin.
10. Have little additional cost of 2–5% of fabrication of this 3D FINFET, as compared to the corresponding MOSFET.
11. Have now application in microwaves also specially due to its higher speed if not power with much cheaper technology than GaN-HEMT.

Thus we see that FINET with its 3D structure promises to rejuvenate the chip industry by rescuing it from the short channel effect that had put a threat of limit to the scalability faced by the planar transistor structure. Therefore we see that, out of the three parameters, e.g. (a) cost, (b) devices density/Area on a die, i.e. chip, (c) Power density of frequency whichever becomes the most important criteria the device technology, the devices that becomes important are respectively (a) MOSFET (b) FINFET (c) GaN-HEMT. This trend is likely to continue for the next decade or so, thereafter some new technology to go below 5 nm, will be required or else dead end will reach.

6.17 Solved Problems

Problem 1 For the Si-Jn-FET with the following specification, calculate pinch-off voltage and current (v_p and I_p), built-in voltage, drain current (I_d), and cut-off frequency. Parameters given are:

$N_d = 2 \times 10^{17}$/cc, (doping of channel), dielectric constant of $S_i = \varepsilon_r = 11.8$, channel height (a) =0.2 μm, channel length $L = 8$ μm, channel width $Z = 50$ μ, doping for junction diffusion $N_a = 10^{19}$/cc, mobility $\mu_n = 800$ cm²/Vs, drain voltage $V_d = 10$ V, gate voltage $V_g = -1.5$ V

Solution

$$V_p = \frac{eN_d \cdot a^2}{2\varepsilon_0 \cdot \varepsilon_r}$$
$$= \frac{1.6 \times 10^{-10} \times 2 \times 10^{17} \times (0.2 \times 10^{-4})^2}{2 \times 8.854 \times 10^{-14} \times 11.8}$$
$$= 6.1\,\text{V}$$

$$I_p = \frac{m_n e^2 \cdot N_a^2 \cdot z \cdot a^3}{L\varepsilon_0 \cdot \varepsilon_r}$$
$$= \frac{800 \times (1.6 \times 10^{-19})^2 \times (2 \times 10^{17})^2 \times (50 \times 10^{-4})(0.2 \times 10^{-4})^3}{8 \times 10^{-4} \times 8.854 \times 10^{-14} \times 11.8}$$
$$= 39.2\,\text{mA}$$

$$\text{Built in voltage} = V = \frac{\text{kT}}{\text{e}} \ln\left(N_d \cdot N_a / n_i^2\right)$$
$$= 26 \times 10^{-3} \cdot \ln \frac{(2 \times 10^{17} \times 10^{19})}{(1.5 \times 10^{10})^2} = 0.939\,\text{V}$$

Drain current I_d

$$= I_p \left[\frac{V_d}{V_p} - \frac{2}{3} \left(\frac{V_d + |V_g + V_0|}{V_p} \right) \right]^{3/2} + \frac{2}{3} \left(\frac{|V_g| + V_0}{V_p} \right)^{3/2}$$

$\therefore$

$$I_d = 39.2 \times 10^{-3} \left[\frac{10}{3.06} - \frac{2}{3} \left(\frac{10 + 1.5 + 0.94}{3.06} \right)^{3/2} + \frac{2}{3} \left(\frac{1.5 + 0.94}{3.06} \right)^{3/2} \right]$$
$$= 27.2\text{mA}$$

Cutoff frequency f_c

$$= \frac{2\mu_n \cdot \text{e} \cdot N_d a^2}{\pi \cdot \varepsilon_s L^2}$$
$$= \frac{2 \times 800 \times (1.6 \times 10^{-19}) \cdot (2 \times 10^{17})(0.2 \times 10^{-6})^2}{3.14 \times (8.854 \times 10^{-14} \times 11.8) \cdot (8 \times 10^{-4})^2}$$
$$= 9.6\text{GHz}$$

Example 2 In a bipolar Jn transistor, the t_c = charging time of depletion cap. of collector = 5×10^{-12} s, t_{tb} = transit time in base region of collector = 10^{-12} s, t_{tc} = transit time in collector depletion region = 5×10^{-11} s. Find the cut-off frequency.

Solution Cut-off frequency = $1/(2\pi\tau)$

$$\tau = t_c + t_{tb} + t_{tc}$$
$$= (0.5 + 0.1 + 0.5) \times 10^{-11} = 1.1 \times 10^{-11}$$

$$\therefore\ f_T = \frac{1}{2 \times 3.14 \times 1.1 \times 10^{-11}}$$
$$= \frac{100 \times 10^9}{2.2 \times 3.14} = \frac{100 \times 10^9}{6.908}$$
$$= 14.5\,\text{GHz}$$

Problem 3 For the following data of Si-junction-FET find V_{gs}, V_p.

$$\text{Given is } I_d = 5\,\text{mA}, I_{ds} = 10\,\text{mA}, V_{gsoff} = -6\,\text{V}$$

Solution

$$I_d = I_{ds}(1 - V_{gs}/V_p)^2 \text{ Here } V_p = |V_{gsoff}| = 6$$

$$\text{As } V_{gsoff} \text{ is same as } V_p; 5 = 10[1 - V_{gs}/6]^2$$
$$\therefore V_{gs} \cong 1.8\,\text{V} = V_p$$

Problem 4 In a Si-JFET and in GaAs-MESFET, find the pinch-off voltage if a = Channel height = 0.1 μm, $N_d = 10 \times 10^{17}$/cc = $10 \times 10^{23}/\text{m}^3$ Solution: For MESFET and JFET the pinch-off voltage is

$$V_p = eN_a a^2/2\,\varepsilon; \quad \varepsilon = \varepsilon_0 \varepsilon_r.$$

∴ For Si,

$$\varepsilon_r = 11.8$$

and for GaAs

$$\varepsilon_r = 12.1$$

In Si-JFET:

$$V_p = \left[\frac{1.6 \times 10^{-19} \times (10 \times 10^{23}) \cdot (0.1 \times 10^{-6})^2}{2 \times (8.854 \times 10^{-12}) \times 11.8}\right]$$
$$= 8.33\,\text{V}$$

In GaAs-MESFET

$$V_p = \left[\frac{1.6 \times 10^{-19} \times (10 \times 10^{23}) \cdot (0.1 \times 10^{-6})^2}{2 \times (8.854 \times 10^{-12}) \times 13.1}\right]$$
$$= 7.5\,\text{V}$$

Problem 5 In a Si-MESFET, $N_d = 8 \times 10^{17}$/cc, $\varepsilon_r = 13.1$, channel height (a) = 0. 1 μm, channel length (L) = 14 μm, channel width (Z) = 36 μm, $\mu = 0.08\ \text{m}^2/\text{Vs}$, $a = 800\ \text{cm}^2/\text{Vs}$, $v_s = 2 \times 10^7$ cm/s. Find V_p, I_p, for $V_g = 0$.

Solution

$$V_p = \frac{eN_d \cdot a^2}{2\varepsilon}$$

$$V_p = \frac{1.6 \times 10^{-19} \times 8 \times 10^{23} \times 10^{-14}}{2 \times 8.854 \times 10^{-12} \times 13.1} = 5.52\,\text{V}$$

For $V_g = 0$,

$$I_p = \frac{eN_d \mu a Z V_p}{2L}$$
$$= \frac{1.6 \times 10^{-19} \times 8 \times 10^{23} \times 0.08 \times 10^{-7} \times 36 \times 10^{-6} \times 5.52}{2 \times 14 \times 10^{-6}}$$
$$= 4.85\,\text{mA}$$

Problem 6 In a p-MOSFET; $N_a = 3 \times 10^{17}$/cc $(\varepsilon_r)_{Si} = 11.8$; $(\varepsilon_r)_{sio_2} = 4.0$; SiO_2 layer = 0.01/μm.

Calculate surface potential for strong inversion and oxide capacitance.

Solution

$$\text{Surface potential} = \frac{2\,kT}{e} \cdot \ln\left(\frac{N_a}{n_i}\right)\left(\text{Here } \frac{kT}{e} = 26\,\text{mV}\right)$$
$$= 2 \times (26 \times 10^{-3}) \ln\left(\frac{3 \times 10^{17}}{1.5 \times 10^{10}}\right)$$
$$= 0.87\,\text{V}$$

$$SiO_2 \text{ insulator capacitance} = \frac{\varepsilon_r}{d} = \frac{4 \times (8.54 \times 10^{-12})}{0.01 \times 10^{-6}}$$
$$= 3.54\,\mu\,\text{Fd/m}^2$$

Problem 7 A tunnel (diode) (or IMPATT diode or BARITT diode) amplifier is using circulator, find the gain if $R_j = -40\ \Omega$ and $Z_0 = 50\ \Omega$.

Solution Any amplifier using any device with −ve resistance as above will have the same result.

$$\text{Reflection coefficient}\ \Gamma = \frac{R_j - Z_0}{R_j + Z_0}$$

$$\therefore\quad \text{Power gain} = \Gamma^2 = \left|\frac{R_j - Z_0}{R_j + Z_0}\right|^2$$
$$= \left(\frac{-40-50}{-40+50}\right)^2 = \left(\frac{-90}{10}\right)^2 = 81$$

$$\therefore\quad \text{Power Gain(dB)} = 10\ \log(81)\text{dB} = 19.1\ \text{dB}$$

Problem 8 A tunnel diode has $R_j = -26\ \Omega$, $C_j = 5$ nF, $R_S = 1\ \Omega$ calculate:

(a) Resistive cut-off frequency.
(b) Gain if the diode is used as an amplifier with a load of 24 Ω in parallel to it.

Solution

$$f_{rc} = \frac{1}{2\pi R_j C_s}\sqrt{\frac{R_j}{R_s} - 1}$$
$$= \frac{1}{2 \times 3.14 \times 26 \times 5 \times 10^{-9}}\sqrt{\frac{26}{1} - 1} = 6.12\,\text{GHz}$$

$$\text{Gain with parallel load} = \frac{R_j}{R_j - R_L} = \frac{26}{26-24}$$
$$= \text{i.e. voltage gain} = 13$$

$$\therefore \text{Gain(dB)} = 20\ \log(13) = 22\,\text{dB}$$

Problem 9 A GaAs-Gunn device operates in transit time mode at 20 GHz. If $v_s = 10^7$ cm/s, find the length of the active region of the device.

Solution

$$f = \frac{v_S}{L} \therefore\quad L = \frac{v_S}{f} = \frac{10^5}{20 \times 10^9} = 5\ \mu\text{m}$$

Problem 10 In a n-type GaAs-Gunn diode at a certain bias, the electron density in the upper valley is $n_u = 10^8$/cc = 10^{14}/m^3 and in the lower valley is $n_l = 10^{10}$/cc = 10^{16}/m^3. If the doping density is 10^{18}/cc, calculate the conductivity in this −ve resistance region where $\mu_l = 8000$; $\mu_u = 180$ cm^2/Vs.

Solution

$$\text{Conductivity}\ \sigma = (n_l\mu_l + n_u\mu_u)$$
$$= 1.6 \times 10^{-19}$$
$$(10^{16} \times 8000 \times 10^{-4} + 10^{14} \times 180 \times 10^{-4})$$
$$= 1.28\ \text{milli mhos}$$

Problem 11 A GaAs diode has the following specifications:

n-type doping = 4 × 10^{14}/cc
Applied electric field = 3200 V/cm
Threshold electric field = 2800 V/cm
Active area length = 10 μm
Frequency of operation = 10 GHz

Calculate drift velocity, current density, and −ve mobility.

Solution

$$v_d = f \cdot L = 10 \times 10^9 \times 10 \times 10^{-6} = 10^5\ \text{cm/s}$$

$$J = ne\,v_d = 4.0 \times 10^{14} \times \left(1.6 \times 10^{-19}\right) \times 10^5 = 6.4\,\text{A/cm}^2$$

$$\mu = -\frac{v_d}{E} = \frac{-10^5}{3200} = -3.1\,\text{cm}^2/\text{Vs}$$

Problem 12 A GaAs-Gunn diode has drift length of 12 μm, calculate (a) natural frequency and (b) minimum voltage to start/initiate Gunn mode of oscillation, if the drift velocity = 2 × 10^7 cm/s.

Solution

(a) Natural frequency means transit time frequency = $v_d/L = \frac{2\times10^7}{12\times10^{-4}} = 16.66\,\text{GHz}$.
(b) As the field, i.e. voltage gradient (slope) has to be 3.3 kV/cm (the threshold field), the voltage for 12 μ, for Gunn oscillation will be:

$$= 3.3 \times 10^3 \times 12 \times 10^{-4} = 3.96\,\text{V}$$

Problem 13 Four Gunn diodes have the following characteristic.

Gunn Diode	A	B	C	D
Doping	5 × 10^{16}/cc	5 × 10^{16}/cc	2.5 × 10^{16}/cc	10^{15}/cc
L (μM)	15	12	10	50
f (GHz)	50	40	20	2.0

Determine the modes they are operating with external tuned circuit as per above frequencies.

Solution Calculate *f.L.*, $n_d \cdot L$ and N_d/f. for the four diodes:

Gunn Diode	A	B	C	D
fL	7.5 × 10^7	4.8 × 10^7	2 × 10^7	10^8
$N_d \cdot L$	7.5 × 10^{13}	6.0 × 10^{13}	2.5 × 10^{13}	5 × 10^{12}
N_d/f	10^6	0.8 × 10^6	1.25 × 10^6	5 × 10^4
Mode	Quenched	Gunn	Transit time	LSA

Comparing the data with Fig. 6.22, we infer that the modes for the four diodes are as given above.

Problem 14 IMPATT diode has a drift length of 2 μm. Saturated drift velocity = 2 × 10^7 cm/s, maximum operating voltage and current are 100 V and 200 mA, with the breakdown voltage = 90 V. If the efficiency is 15%, then calculate (a) operating frequency and (b) maximum CW output.

Solution

(a) $f = \frac{v_d}{2L} = \frac{2\times10^5}{2\times2\times10^{-6}} = 50\,\text{GHz}$
(b) $P_{CW} = \eta \times P_{dc} = 0.15 \times 100 \times 0.2 = 3\,\text{W}$

Problem 15 An IMPATT diode has the specification as C_D = 0.5 pF, L_p = 0.5 nH, and C_p is negligible. If the breakdown voltage is 110 V, bias current = 100 mA peak RF current = 0.9 A; load = 2 Ω, L = 4.8 μm, v_d = 10^7 cm/s. Find the resonant frequency of packaged diode, operating frequency, and efficiency.

Solution

(a)

$$f = \text{Resonant frequency of diode} = \frac{1}{2\pi\sqrt{L_p C_D}} = \frac{1}{2 \times 3.14\sqrt{0.5 \times 10^{-9} \times 0.5 \times 10^{-12}}} \approx 10\,\text{GHz}$$

(b)

$$\eta = \frac{P_{ac}}{P_{dc}} \times 100 = \left(\frac{I^2 R_L/2}{V_{BD} \cdot I_{dc}}\right) \times 100 = \left(\frac{(0.9)^2\times2/2}{110 \times 0.1}\right) \times 100 = 7.3\%$$

(c) f_0 = operating frequency $= f = \frac{v_d}{2L} = \frac{10^7}{2 \times 4.8 \times 10^{-4}} =$ 10.5 GHz

As $f_0 \approx f_r$ the design is a good design.

Problem 16 A TRAPATT diode has *n*-active region of doping = 5 × 10^{15}/cc. If the current density of a certain location in the ASF is 20 kA/cm², where the electric field is 7 × 10^5

V/cm, then find the velocity of the avalanche shock front (avalanche zone) v_z and the power density inside it. Why $v_z > v_{sat}$, explain.

Solution

(a) ASF velocity $v_z = J/(eN_d) = \frac{20\times10^3}{1.6\times10^{-19}\times2\times10^{15}} = 6.25\times10^7$ cm/s

(b) Power density inside ASF = $J.E.$ = $20\times10^3\times7\times10^5 = 1.4\times10^{10}$ W/CC

V_z is $>V_{sat}$ because the field $E_{ASF} > E_{avalanche}$. That is, electric field inside ASF-zone is > avalanche breakdown field.

Problem 17 A Si-BARITT diode has the specifications as $N_d = 2.5\times10^{15}$/cc; $L = 5$ μm. As the silicon has dielectric constant of 11.8. Find the breakdown voltage and corresponding electric field.

Solution

$$V_{bd} = eN_d.L^2/\varepsilon = \frac{1.6\times10^{-19}\times2.5\times10^{21}\times(5\times10^{-6})^2}{(8.854\times10^{-14})\cdot(11.8)} = 95.7\,\text{V}$$

$$E_{bd} = 95.7\text{V}/5\times10^{-6}\,\text{m} = 1.9\times10^7\,\text{V/m} = 1.9\times10^5\,\text{V/cm}$$

Problem 18 In an up converter parametric amplifier, the input signal (f_s), pump (f_p), and output frequencies (f_0) are f_s = 2 GHz, f_p = 16 GHz, f_0 = 18 GHz, figure of merit is = (γQ) = 10, merit factor γ = 0.4. Find the power gain and band width.

Solution By Manley–Rowe relation power gain

$$(G_{MR}) = \frac{f_p + f_s}{f_s} = \frac{f_0}{f_s} = \frac{18}{2} = 9.0$$

Gain in dB = 10 log (G_{MR}) = 9.54 dB

$$G_{max} = \text{maximum power gain} = \frac{f_0}{f_s}\cdot\frac{x}{\left(1+\sqrt{1+x}\right)^2}$$

Here

$$x = \frac{f_0}{f_s}(\gamma Q)^2 = \frac{18}{2}\times(10)^2 = 900$$

$$\therefore\; G_{max} = \frac{18}{2}\cdot\frac{900}{\left(1+\sqrt{901}\right)^2} = 8.42$$

∴ Maximum gain in dB = 10 log (G_{max}) = 9.2 dB

$$\text{Bandwidth} = 2\gamma\sqrt{f_0/f_s} = 2\times0.4\sqrt{18/2} = 2.4$$

Review Questions

1. Give comparison of microwave diode sources with their frequency range, CW power range, and its speciality.
2. If we want a simple oscillator to give 10 mW, with wide frequency turning of at least ±10%, which solid-state device will be chosen and why?
3. Why MESFET amplifier is replacing the very age-old proven parametric amplifier in aeroplane, etc?
4. How does a parametric amplifier differ with other diode or transistor amplifier?
5. Is BARITT an avalanche device? If not, explain the reasons.
6. In an IMPATT diode, the power supply is square pulsed at 10% duty cycle. The μW power gives 100 V peak voltage. If the dc current is 900 mA and the diode has an efficiency of 10%, then find the output power pulsed and average output power.

7. How can PIN diode be used as a microwave switch? Describe a single PIN switch in shunt and series configuration (UPTU 2003, 2004).
8. How is plasma trapped in TRAPATT diode? Why its operating frequency is lower than IMPATT give major merits and de-merits (UPTU 2003, 2004)?
9. What is the maximum frequency limitation in TRAPATT and why?
10. The electric field in the moving avalanche zone in TRAPATT is 7×10^5 V/cm while avalanche breakdown field is 4×10^5 V/cm. Explain how does plasma gets trapped due to this.
11. Explain the –ve ac resistance of Gunn diode mechanism by moving dipole domain formed by high mobility electron low mobility electron on the two sides.
12. Explain −ve dc resistance in Gunn diode by two-valley theory.
13. In *n*-Schottky diode we say that the two conditions make such *n*-diodes are $\phi_M >$ $_s$, and doping = 10^{18} to 10^{19}/cc. Explain.
14. Explain the difference between tunnel diode and ordinary PN junction diode (UPTU 2003–2004).
15. It is said that μω gets generated in IMPATT due to −ve ac differential resistance whole in tunnel diode or Gunn diode due to −ve dc differential resistance. Explain.
16. Explain the terms in Gunn diode

(i) Gunn effect	(ii) High-field domain theory	
(iii) Two-valley theory		(MDU 2006)

17. Write short notes on

(i) IMPATT	(ii) TRAPATT	
(iii) PIN diode	(iv) Parametric amplifiers	(MDU 2004)

18. Explain the structure of packaged μω-diode with reference to the IMPATT device.
19. In a μω point contact detector reverse saturation current is 1 μA. If the μω signal amplitude is 1V, then find the detector current.
20. A Gunn diode operates at 10 GHz in transit time mode. If $v_s = 10^7$ cm/s, then compute the device thickness.
21. In a varactor, 3 GHz is applied and it acts as tripler. Give the circuit and determine idler and output frequencies.
22. Explain the types of −ve resistances and operating frequencies (v_S/L or v_S/2L) in Gunn diode, tunnel diode, and IMPATT diode.
23. Explain the working of OFF-MOSFET and ON-MOSFET. Can we use OFF-MOSFET in depletion mode and why?
24. In which of the microwave solid-state (semiconductor) devices, only '2-terminal, 3-terminal, or 4-terminals are used'. List them out.
25. Which of the solid-state μW oscillators (a) do not have −ve resistance in dc *I–V* curve and (b) which have?
26. Which is μω semiconductor device which has −ve resistance in both the *I–V* characteristic of ac and dc?

(Ans: Only tunnel diode and Gunn diode have dc −ve resistance also along with ac)

Microwave Measurement: Instruments and Techniques

7

Contents

© Springer Nature Singapore Pte Ltd. 2018
P. K. Chaturvedi, *Microwave, Radar & RF Engineering*,
https://doi.org/10.1007/978-981-10-7965-8_7

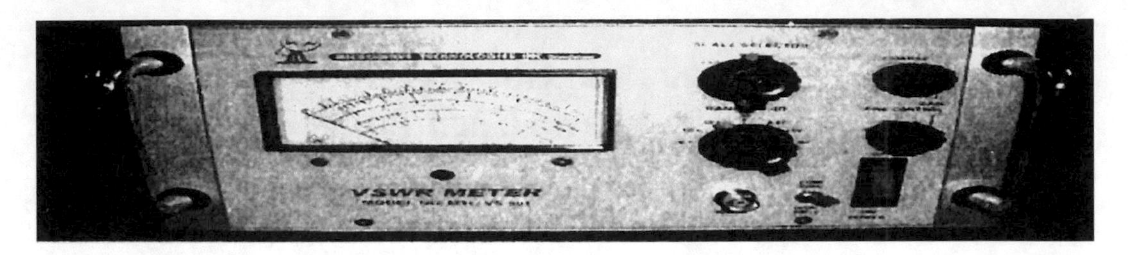

7.1 Introduction

As we have seen that measurement of voltage, current and resistance at microwave frequencies is not possible, as they vary with position and time along the transmission line. Therefore, the power being independent of location along a lossless line is measured. Many times, it is sufficient to know the ratio of two powers in dB, rather than their actual values. Similarly, for circuit elements (*R*, *L*, *C*) being distributed over the line and not lumped, measurement of line impedance suffices.

Besides power and impedance, the other properties/parameters of the circuit and that of the devices are also normally required to be known. These parameters which we normally measure are:

1.	Frequency	2.	Wavelength
3.	Power	4.	Voltage standing wave ratio (VSWR)
5.	Impedance	6.	Insertion loss
7.	Attenuation	8.	Return loss
9.	Q factor of a cavity	10.	Phase shift
11.	Dielectric constant	12.	Noise

For measurement of these parameters, some of the basic set-up and instruments required will be discussed.

7.2 Basic Microwave Bench

Normally, every teaching laboratory has a microwave bench set-up, the block diagram of which is given in Fig. 7.1. It consists of the microwave source, isolator, frequency meter, variable attenuator, slotted line section with detector and VSWR indicating meter with matched terminations. Normally in teaching laboratories, X-band (8.0–12.4 GHz) frequency devices and circuits are used, for which the waveguide has the inner dimension as $a \times b = 0.9'' \times 0.4''$.

1. The microwave source could be reflex Klystron (Chap. 5) or Gunn diode oscillator, (Chap. 6) with output of the order of few milliwatts. The output can be CW or pulsed power. The pulsed power is generally of 1 KHz frequency modulation, with duty cycle of 1% (refer Chap. 1 on CW/pulsed power).
2. The second component is an isolator (Chap. 4), isolating source from rest of the bench/equipment, so that signal reflected back from the circuit does not enter the source or else its power and frequency will get altered.
3. The third is a frequency meter in the form of a cylindrical cavity (absorption type) (Chap. 3), which can be adjusted to resonate. This resonance is observed at the VSWR meter to give reduced (i.e. a dip in the) output.

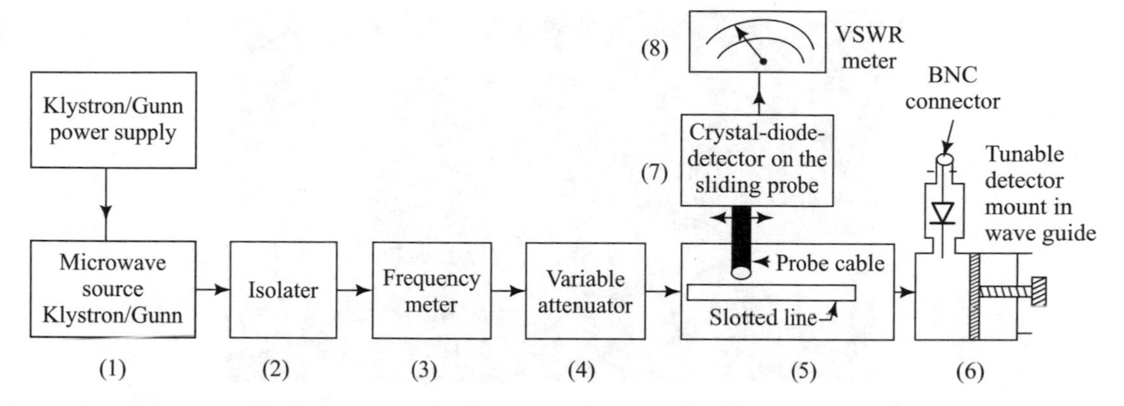

Fig. 7.1 Block diagram of basic microwave bench

4. The next is a variable attenuator (0–50 dB) for controlling the power to the slotted line and finally to the VSWR meter.
5. Power then enters the slotted line waveguide, which has a slot cut along the centre of the broad wall of the waveguide, (where the electric field is maximum), so that the moving E-probe of the detector can pick up maximum voltage.
6. At the end of the slotted line is the sliding short (Chap. 4) for this end to have a matched load for minimum reflection.
7. The movable E-probe has the square law crystal diode detector (Chap. 4), which acts as microwave signal rectifier, giving dc voltage to the VSWR meter.
8. VSWR meter amplifies the detector output and has a sensitive calibrated voltmeter for direct reading of VSWR. It has an amplifier gain control, for weak signal to be displayed on its corresponding scale.

Now, we will study various types of measuring devices used for measuring (a) frequency, (b) power, (c) VSWR, (d) impedance, (e) dielectric constant, (f) noise factor, (g) Q of a cavity, (h) insertion and attenuation loss.

7.3 Measurement Devices and Instruments

Some of the devices and instruments commonly used are Klystron tube on the mount/Gunn diode in the waveguide, power supplies of them, isolater, frequency meter (wave meter), power meter, spectrum analyser, etc.

7.3.1 Microwave Sources and Their Power Supplies

Normally in the laboratories, the Reflex Klystron and Gunn diode oscillators are used as sources. For waveguides, Klystron and its mount and Gunn diode mounts are as given in Fig. 7.2. Working of these devices is given in Chaps. 5 and 6, respectively.

Power supplies of these devices are different, and typical specifications are as given below,

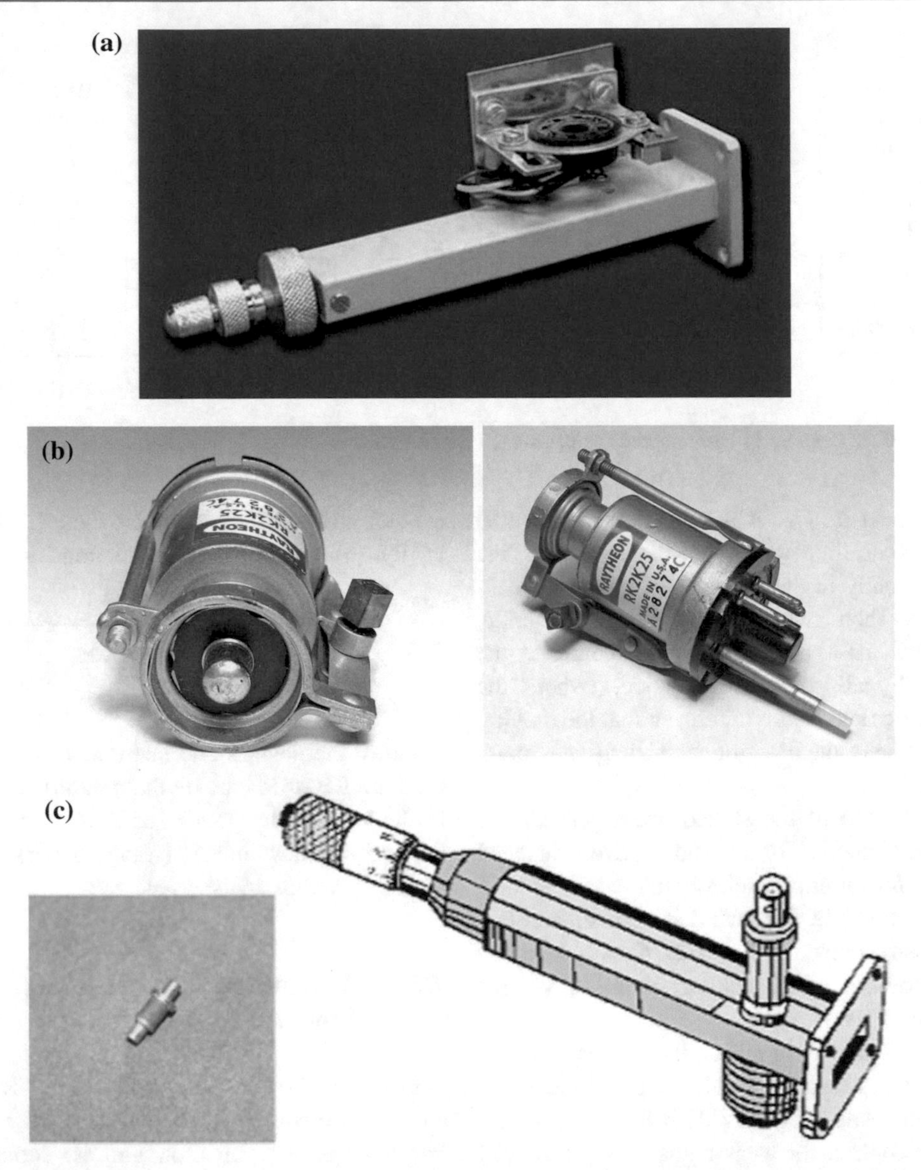

Fig. 7.2 **a** Klystron mount and tuner short slug. **b** Reflex klystron 2K25 (see Fig. 5.12 for more details). **c** Gunn diode mount and tuner short slug

while the CW and pulsed power supplies have been explained in Chap. 1.

(a) **Typical Klystron Power Supply**

1. Mains: 230 VAC + 10% (50 Hz).
2. Beam voltage (V_B): 240–420 V.dc variable knob.
3. Beam current (I_B) 50 mA (max).
4. Maximum, i.e. trip off beam current = 65 mA.
5. Repeller voltage (V_R) = −10 V to −270 V.dc (variable knob).
6. Filament supply = 6–3 V.dc.

7.

Modulation:	AM (square)	FM (sawtooth)
Frequency:	500–2500 Hz	150–300 Hz
Amplitude:	0–110 V	0–65 V

8. Single LCD display of V_B, V_R, I_B
9. Selector switch for V_B, V_R, I_B.
10. Connectors = BNC for external modulation source-8-pin octal socket.
11. Dimension of a typical instrument (cm) = (width × depth × height) = 32 × 36 × 29.

(b) **Typical Gunn Power Supply**

1. Mains: 230 V ± 10% (50 Hz).
2. Voltage supply output: 0 to 10 V.dc (variable knob).
3. Current 750 mA (max).
4. Modulation frequency: 800–1200 Hz.
5. Modulation modes are with square wave, audio frequency or P.C. data modulation.
6. Output connectors: BNC for Gunn bias.
7. Display: Two-line LCD display for voltage and current.

7.3.2 Isolator

The device has been explained in Chap. 4 (Sect. 4.14.1) along with its Fig. 4.29.

7.3.3 Frequency Meter or Wave Meter

This form an important device of the microwave bench. The shorting plunger is used to change the cavity length and hence its resonant frequency. This has already been discussed at length in Chap. 4 (Sect. 4.13), with Fig. 4.26 giving the various types of frequency meters.

7.3.4 Variable Attenuator

These devices have also been explained at length in Chap. 4 (Sect. 4.10) along with Fig. 4.20.

7.3.5 Slotted Line

In the waveguide, standing wave is normally formed; therefore, a probe moving (Fig. 7.3) along the longitudinal slot on the broad side of the waveguide where the electric field is maximum, will pick up voltage, in its coaxial line. Here we may note that such a slot does not radiate power outside in the dominant mode, as the electric field is perpendicular to the slot. A small portion of the probe is inside the waveguide and is on a carriage which can be moved on the top surface of the waveguide by rotating a knob (Fig. 7.3). The position of the probe can be read on a scale of the carriage. The probe is the extended central cable of a coaxial cable and is connected to crystal diode detector, and then to VSWR meter. The outputs of the detector (being a square law device), (Sect. 6.12), are proportional to the square of the input voltage at the position of the probe. As we move the probe, the output changes as per the standing wave pattern inside, which can be noted on the VSWR meter.

The position of the maxima and minima of VSWR can be read on the centimetre scale having vernier callipers also with a least count of 0.1 mm. The distance between two minima = ($\lambda_g/2$).

The slotted line is normally used for measuring impedance, reflection coefficient, SWR and frequency of the signal.

7.3.6 Tunable Detector and Probe System

Various types of tunable detectors (Fig. 4.25) have already been discussed in Chap. 4. The crystal diode inside is normally a Schottky barrier diode, which rectifies the signal as discussed in Sect. 4.12.2 in Chap. 4. Detectors are of the following types:

(a) Tunable coaxial detector (Fig. 4.25a).
(b) Fixed frequency-tuned coaxial detector (Fig. 4.25b).

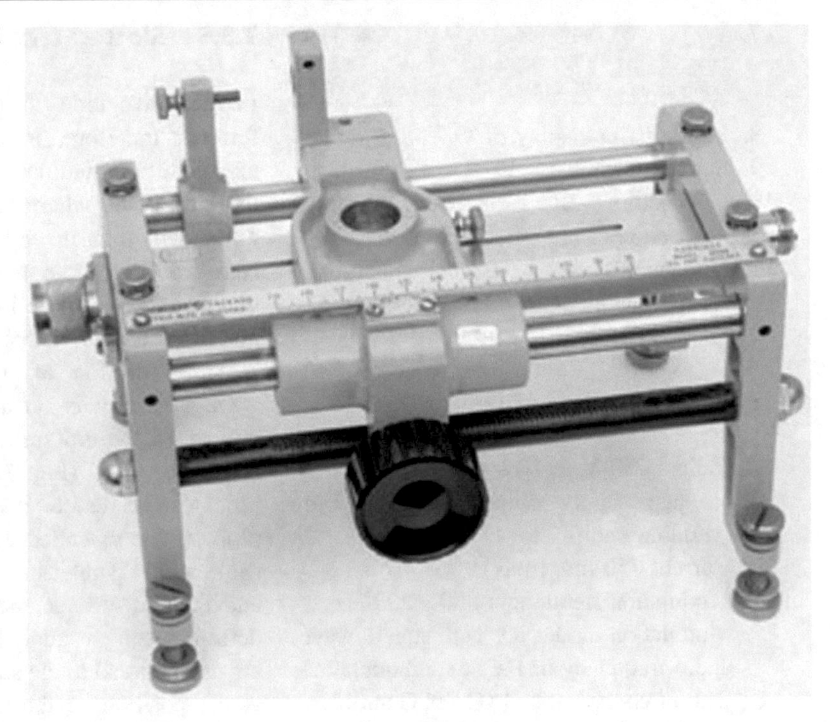

Fig. 7.3 Waveguide slotted line carriage

(c) Tunable waveguide detector (Fig. 4.25c)
(d) Fixed frequency-tuned matched detector (Fig. 4.25d).

The types (a) and (b) are mounted (by screwing in) on the moving carriage of the slotted line, while (c) and (d) are fixed on the waveguide of the slotted line (see Fig. 7.3).

7.3.7 VSWR Meter

A VSWR meter is just calibrated dc voltmeter detecting the rectified DC voltage from the crystal diode detector. It can measure VSWR as well voltage, as can be seen on its scale (Figs. 7.4 and 7.5). VSWR and the reflection coefficients (Γ) are very important parameters for

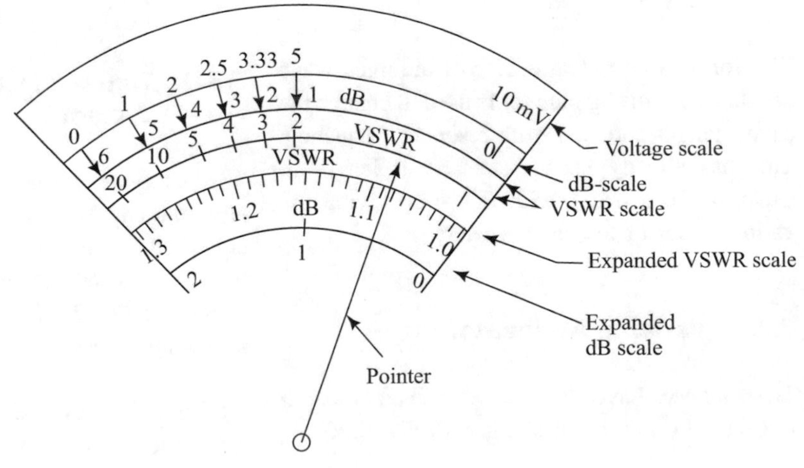

Fig. 7.4 VSWR meter scales

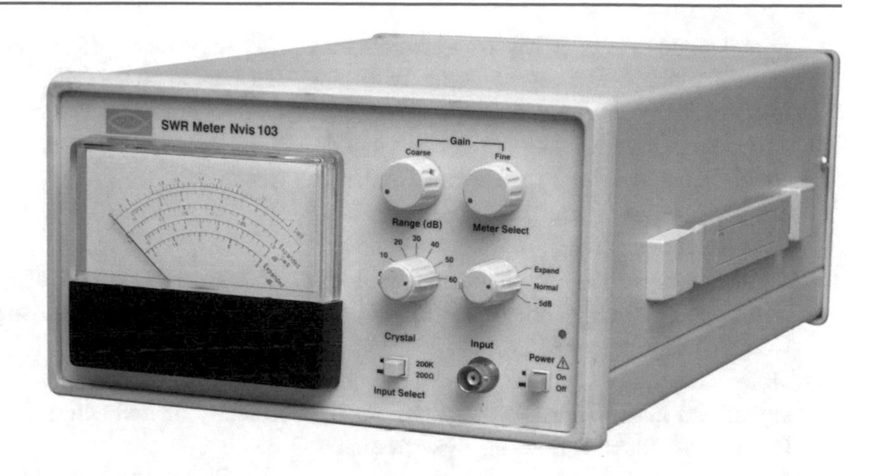

Fig. 7.5 VSWR meter

measuring load impedance and the degree of matching with the line. By moving the crystal diode detector carriage of the slotted waveguide, VSWR (S) can be measured by noting V_{max} and V_{min} on the VSWR meter, giving $S = V_{max}/V_{min}$. We can measure VSWR directly on the scale of the meter by calibrating it by adjusting the gain knob to show maximum scale deflection of unity VSWR at the carriage detector position of V_{max}, (see Fig. 7.4). Now we move the detector carriage to the minimum V_{min} reading on the meter, and here the reading on the VSWR scale of the meter gives directly the VSWR of the line in dB.

If we read the voltages at these two locations, the ratio of the V_{max}/V_{min} will be same as VSWR read on the VSWR meter. This can be noted just by seeing the two scales (voltage and VSWR) of the meter; e.g. if at the maximum scale deflection (VSWR = 1) the voltage is 10 mV then at $V = 5$ mV, on the scale voltage will correspond to $V_{max}/V_{min} = 10/5 = 2 =$ VSWR. For measuring very high VSWR, the techniques used are given in this Chapter itself in Sect. 7.4.

Actual VSWR meter is shown in Fig. 7.5. The overall gain is normally 125 dB, adjustable by a coarse and a fine knob. There are three scales on the VSWR meter:

(a) Normal VSWR for $S = 1$–10.
(b) Expanded VSWR for $S = 1$–1.3.
(c) dB scale for 0–2 dB, for measuring VSWR directly in dB.

An input selector switch is also there for different inputs for crystal diode, and this is for low current (4.5 mA) and high current (8.75 mA).

7.3.8 Power Meter

A microwave power meter equipment is quite expensive and is not available in normal teaching laboratories. In fact, it consists of power sensor (thermistor, thermocouple, etc.) which converts microwave power into heat indicator. Therefore, for measuring power in the laboratory, different heat sensing devices are used directly and are given in the measurement Sect. 7.4.

7.3.9 Spectrum Analyser

It is a frequency domain instrument which displays frequency spectrum of the signal, with frequency (*X*-scale)-versus-signal amplitude (*Y*-scale). The CRO part of it gives the signal amplitude (*Y*-Scale) as a function of time actual time variation. The frequencies contained in this time domain of CRO can be obtained by its Fourier transform.

Spectrum analysers are useful at RF and microwave frequencies for analysing the spectrum of:

(a) Signal source, e.g. a Klystron, which may have a fundamental signal along with side frequencies of low amplitudes.
(b) Any transmission line (e.g. waveguide, coaxial line antenna) to see amplitudes of different mode frequencies present, along with dominant mode.
(c) Signal analyser as a diagnostic tool for checking the compliance of EMI/EMC (see Chap. 1) requirements.
(d) Noise frequencies present along with main signal and hence signal to noise ratio.
(e) Output of filter for seeing its frequency response.
(f) Wave modulated output for determining its type, e.g. FM, AM.

7.3.10 Network Analyser

As we know that the slotted line can do measurements at a single frequency. For broadband testing, experiment at every frequency has to be carried out separately, which is time-consuming. A network analyser measures both, amplitude and phase of a signal over a wide frequency range, with reference to an accurate reference signal.

It is used for measurement of:

(a) Transmission characteristic and hence gain of a device.
(b) Reflection characteristic and hence impedance of the line.
(c) S-parameters of active and passive devices.
(d) Phase magnitude display.

7.4 Measurement Techniques in Microwaves

After having covered various measuring instruments, we now study the measurement techniques for measuring (1) frequency, (2) power, (3) SWR, (4) impedance, (5) insertion/attenuation, (6) Q factor, (7) phase shift, (8) dielectric constant and (9) noise factor. The first three can be measured directly, while the rest are measured by indirect methods.

7.4.1 Measurement of Frequency and Wavelength

There are three methods of measuring frequency directly or indirectly:

(a) **By Frequency Meter Directly**: As discussed in Chap. 4 (Fig. 4.41), we can measure frequency directly. **Normally, these frequency metres are cavity which takes power from the waveguide line through an iris (slot), and at resonance it absorbs maximum power**. Therefore, we change the length (i.e. cavity size) of the cavity by rotating the knob, and hence resonant frequency. Then, when the line signal frequency and cavity resonant frequency are equal, we get a dip in VSWR meter, showing maximum absorption of energy by the cavity. Here the calibrated drum or micrometer reading gives the frequency of the signal, within an accuracy of $\pm 1\%$.
(b) **By Slotted Line by Measuring λ_g (Maxima-Minima Method) and Waveguide Size**: The slotted line (Fig. 7.3) is connected with the circuit as in Fig. 7.1. The location of maximum voltage and minimum voltage due to standing wave can be noted. The difference between two consecutive maxima = $\lambda_g/2$, i.e. half of the guide wavelength. If the wider dimension of the waveguide is a, then the free space wavelength λ_0 is given by:

$$\frac{1}{\lambda_0^2} = \frac{1}{\lambda_g^2} + \frac{1}{(2a)^2} \quad (7.1)$$

(For dominant mode $TE_{10}, \lambda_c = 2a$)

$$\therefore 1/\lambda_0 = 1\sqrt{\left(\frac{1}{\lambda_g^2} + \frac{1}{4a^2}\right)}$$

$$\text{As } f_0(\text{GHz}) = 30/\lambda_0 \text{ (in cm)}$$

Therefore

$$f_0(\text{GHz}) = 30 \cdot \sqrt{\frac{1}{\lambda_g^2} + \frac{1}{4a^2}} \text{ (with } \lambda_g, a \text{ in cm)} \quad (7.2)$$

The accuracy here also is ±1%.

(c) **By Electronics Technique—Beats Method**: Electronics techniques are generally more accurate than the mechanical techniques (discussed above) but are expensive. These techniques are based on the mixing of the unknown frequency with harmonics of a known standard frequency. If one of the harmonic is close to the unknown frequency, beats will be formed. Such an instrument is known as transfer oscillator, which gives the frequency in the direct display.

The functional block diagram of the instrument is given in Fig. 7.6. Here the signal from a standard, stable variable frequency source is fed to a harmonic generator, which generates frequencies (which can be varied) in the desired microwave range. This standard signal frequency (f_1) has to be much multiple lower than unknown frequency signal (f_0). This harmonic output is mixed (using *E* or *H*-plane Tees) with the unknown signal and then passed through a detector (which allows only low beat frequency) to the null beat indicator (NBI).

The NBI could be a CRO also, where the beat-frequency signal can be observed. The frequency of the standard frequency (f_1) is measured from the null beat condition—as here the unknown frequency is an integral multiple of f_1, i.e. $f_0 = nf_1$.

Now the standard source frequency is increased very slowly till at f_2 the next null beat is noticed, and here $f_0 = (n - 1) f_2$ as f_0 is constant.

$$\therefore f_0 = nf_1 \quad \text{and} \quad f_0 = (n-1)f_2$$

eliminating *n*, we get

$$f_0 = \left(\frac{f_1 \cdot f_2}{f_2 - f_1}\right) \quad (7.3)$$

Since f_1 and f_2 can be measured very accurately, the accuracy of f_0 measured is within $<0.01\%$. Therefore, frequency meter scales are normally calibrated by this method.

7.4.2 Measurement of Power

Frequency as well as power can be measured directly. Most of the power measurement techniques use suitable calibrated power detectors, which convert the RF power into heat. Then by measuring the temperature rise by a temperature sensitive element, power is read directly or indirectly.

Power requirements of radars are quite high (in kW). While that of microwave solid state devices, it is quite low (few mW). Therefore, power measurement technique is different as per the power level and divided into the following four categories:

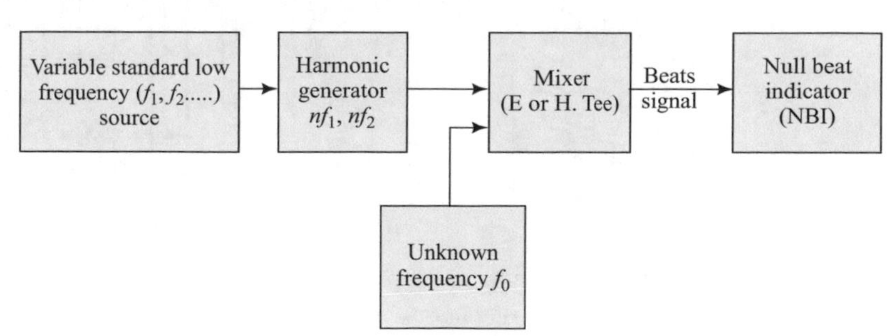

Fig. 7.6 Set-up for frequency measurement by null-beat electronic method

(a) Very low power (0.01–1 mW): Thermocouple technique.
(b) Low power (0.1–10 mW): Bolometer technique.
(c) Medium power (10 mW–10 W): Bolometer with directional coupler.
(d) High-power (10 W–1 kW) calorimetric watt meter.

(a) **Very low-power measurement (0.01–1 mW) Thermocouple method**: For this level of power, Sb-Bi thermocouples are used, where one of the junctions is connected to a conducting plate put inside the transmission line where the microwave is flowing. This plate gets heated, and e.m.f. gets generated between Sb and Bi, which is measured by suitably calibrated voltmeter. This way the voltmeter can directly read microwave power.

(b) **Low-power measurement (0.1–10 mW) Bolometer method**: For this, we use bolometers which are microwave power detectors, in which internal resistance changes (increase or decrease) with rise in temperature caused by microwave power dissipating inside them. They are of two types (Fig. 7.7a, b).

(i) **Barretters**: It has a resistance with positive temperature coefficient consisting of a fine platinum wire mounted in a cartridge like a fuse used in instruments (see Fig. 7.8a).
(ii) **Thermistors**: It has resistance with negative temperature coefficient, consisting of semiconducting material (see Fig. 7.8b). Both these bolometers are square law device, (like diode detector). Here the power (P) increases the

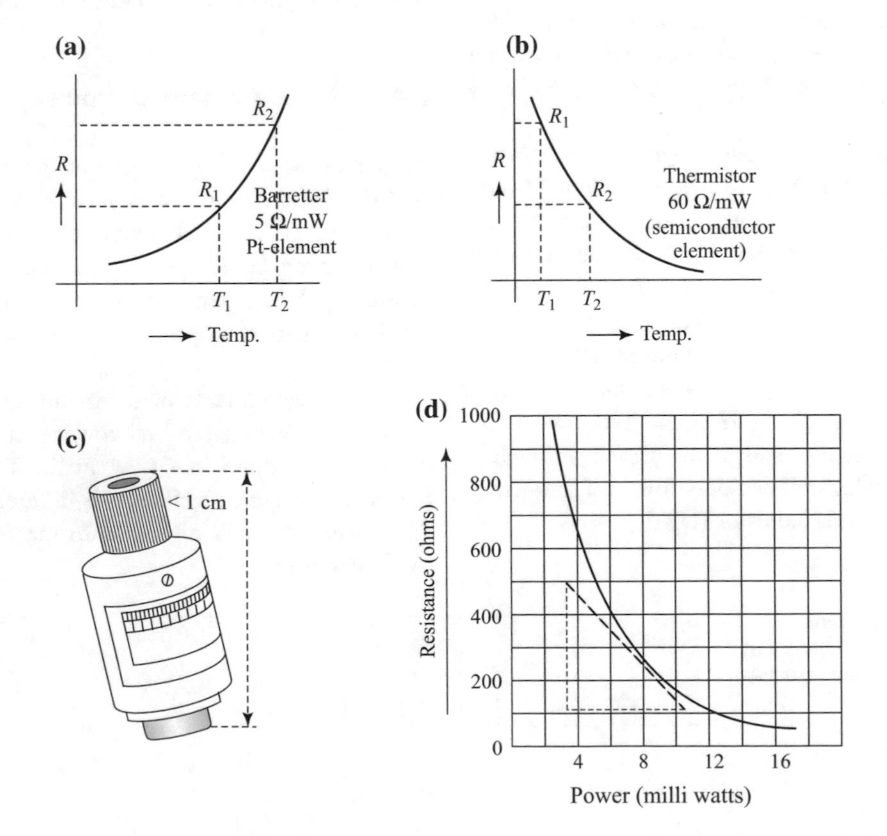

Fig. 7.7 Resistance-versus-temperature of the two bolometers. **a** Barretter. **b** Thermistor. **c** Look of these devices. **d** Curve of power-versus-resistance of a thermistor

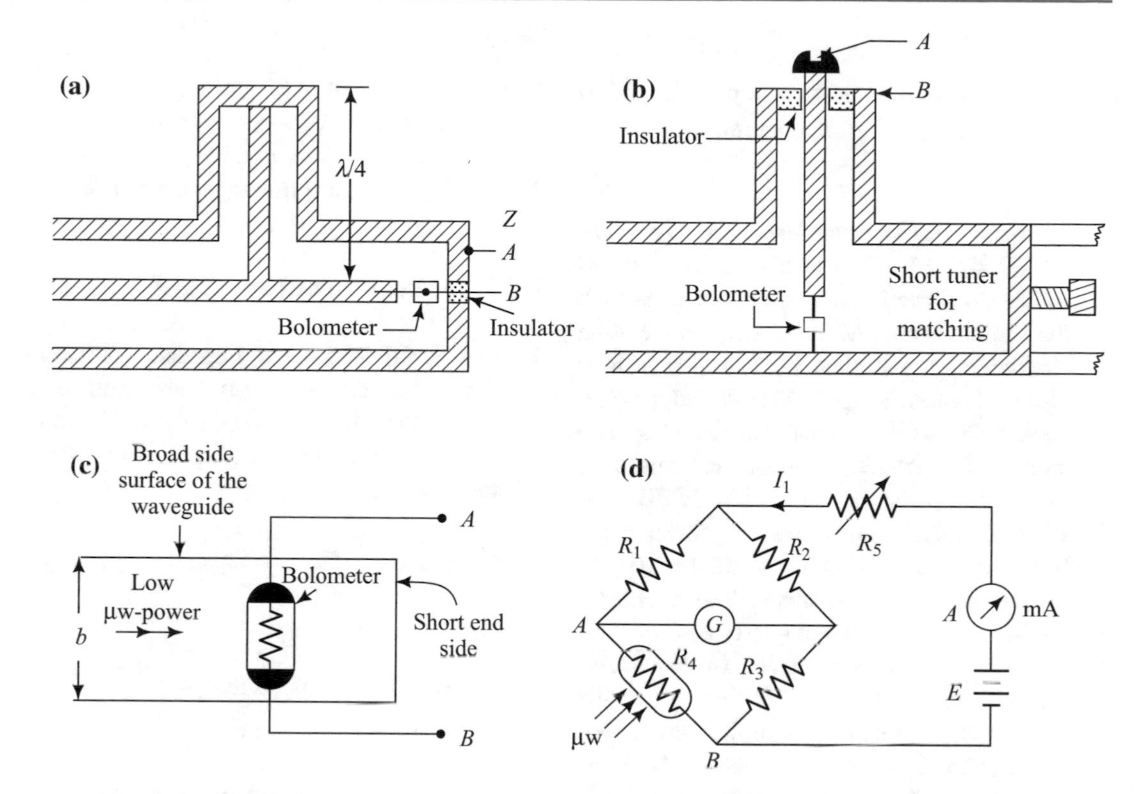

Fig. 7.8 **a** Bolometer mounted inside the coaxial line and **b** inside a waveguide near the shorted end. **c** Indicative figure. **d** Very low-power measurement by bridge: bolometer resistance as one arm (R_4) of the balancing bridge

temperature, which in turn increases/decreases the resistance by square law (Sect. 4.12.2) therefore the microwave powers $P \propto (R_1 - R_2)$

The bolometers are mounted inside the waveguide as in Fig. 7.8a and is used as a load to the microwave power for fully dissipating into heat. Because of this rise in temperature, its resistance changes from initial value of R_4 to R_4' with the difference $(R_4 - R_4')$ being −ve or +ve, its modulus is related to power:

$$\text{Power} \propto \left|R_4 - R_4'\right|^n (\text{with n-as a material index}) \tag{7.4}$$

Following steps may be followed:

1. This change in resistance is measured using a bridge technique. The bolometer becomes one of the arms of the bridge circuit as in Fig. 7.8d.
2. Now we balance the bridge without microwave powers where resistance of bolometer may be R_4 and current on ammeter as I_1, with the galvanometer reading null.
3. Microwave power is now applied to dissipate in the bolometer, and the rise in temperature changes the resistance, disturbing the balance.
4. Now for measuring microwave power, we can follow one of the following two methods:

   (i) The galvanometer G can be calibrated for directly displaying the microwave power on its panel.
   (ii) Vary R_5 (reduce or increase) so that the dc current I_1 to the bridge changes such that heating (I^2R) changes the value of R_5 and the bridge is balanced again showing null reading in G. In this case, the change in the ammeter current ($I_2 - I_1$) can also display microwave power if the ammeter is calibrated for it as:

$$\text{Power} \propto |I_2 - I_1|^m$$
$$\text{(where} = m \text{ index value of material used in bolometer} \quad (7.5)$$

(c) **Medium power measurement (10 mW–10 W) By extension of Bolometer method by 20 dB directional coupler (DC)**: As both the bolometers (–ve and +ve temperature coefficient) are limited in their power handling capabilities up to 10 mW only; therefore, a 20 dB directional coupler (Fig. 7.9) along with a 10 dB attenuator can be used, so that the power is down by 30 dB (i.e. 1/1000). Thus, it increases the power capability up to 10 W, as 10 W/1000 = 10 mW but at the cost of accuracy. The attenuator could be a precision variable type also.

(d) **High-Power Measurement (10 W–50 kW) Calorimeter Method**: This method is by calorimeter method, where the microwave power is dissipated in a perfectly absorbing load and then sensing the power level from the rise in temperature. It uses a circulating water being a bipolar absorber (also see Chap. 1 on μW heating) of microwave energy as given in Fig. 7.10. Here water is constantly flowing through a glass tube passing through a short-ended waveguide. Water will absorb microwave power and get heated. The water exit temperature will be higher than that of input water by ΔT. Then, the average power is given by:

$$P = 4.187 \cdot \mathrm{C} \cdot Vd\,\Delta T \text{ W}$$

where

$C =$ specific heat of water in cal/g.
$V =$ rate of flow of water in cc/s.
$d =$ density of water in g/cc.
$\Delta T =$ $(T_2 - T_1)$ = rise in temperature in °C.

7.4.3 Measurement of VSWR

A mismatched load (load impedance different than line impedance of 50 Ω) leads to reflected waves, resulting into standing waves. The ratio of maxima to minimum voltage gives the VSWR as shown in Fig. 7.11.

$$\text{VSWR} = S = \frac{V_{\max}}{V_{\min}} = \frac{1+\Gamma}{1-\Gamma} = 1 \text{ to } \infty \quad (7.6)$$

where

$$\Gamma = \text{reflection coefficient} = \sqrt{\frac{P_{\text{reflected}}}{P_{\text{incident}}}} = 0 \text{ to } 1$$

For measurement of VSWR, the microwave bench (Fig. 7.1) is required. As the crystal detector carriage moves, we can note the position and value of $V_{\max}$ and $V_{\min}$. Some special care has to be taken for measuring very low, medium and very high VSWR.

(a) **Very Low VSWR (5 = 1 to 2)—Averaging method**

When VSWR is very low, the minima and the maxima will be very broad and will be difficult to locate its position accurately. Therefore, we

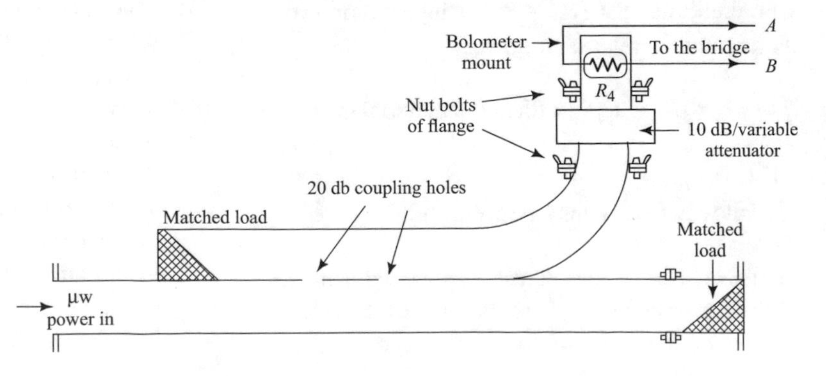

Fig. 7.9 Medium power measurement increasing the range of measurement of bolometer by directional Coupler + Attenuator

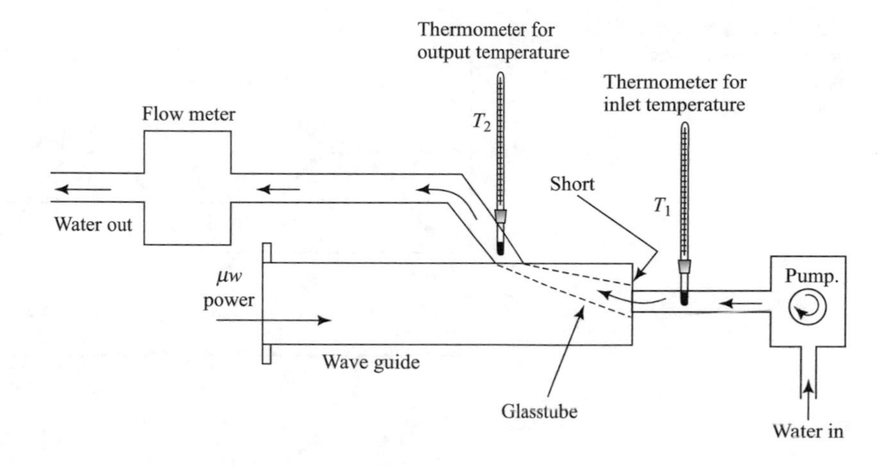

Fig. 7.10 High-power (>10 W) measurement by calorimetric method

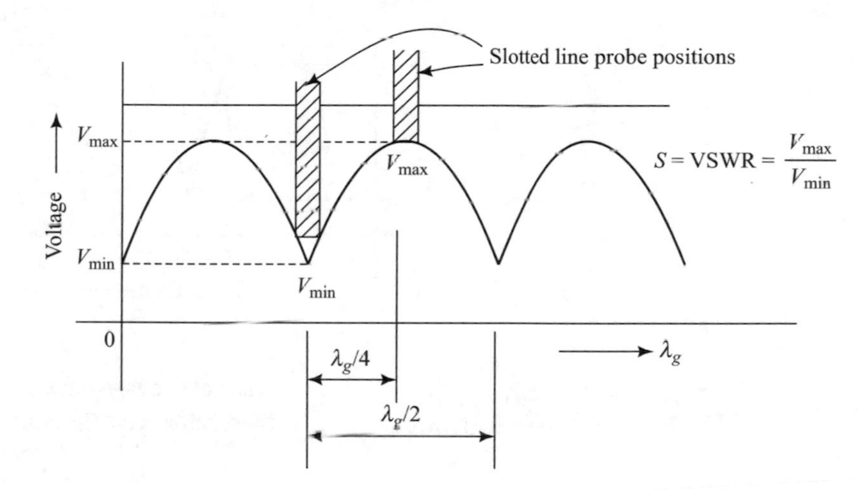

Fig. 7.11 Standing wave formation and VSWR

measure V_{max} and V_{min} at least at three locations each on the slotted line and average them to give (Fig. 7.12a):

$$S = \left(\frac{V_{max1} + V_{max2} + V_{max3}}{V_{min1} + V_{min2} + V_{min3}} \right)$$

(b) **Medium VSWR (S = 2 to 10)—Slotted Line Method**

As discussed in Sect. 7.3.8, the value of V_{max} and V_{min} can be easily measured to give accurate value of S, by slotted line method $S = V_{max}/V_{min}$.

(c) **Very High VSWR (S > 10)—Double Minima Method**

For VSWR > 10, we use the method known as double minima method (Fig. 7.12b). Here we use the slotted line (Fig. 7.1) to locate the minima point (d_0) and read the minimum voltage value (V_{min}). Then, we locate the position of the two points (d_1 and d_2) on the left and right of this V_{min}, which has $\sqrt{2} \cdot V_{min} = 1.414\ V_{min} = V_x$, value, i.e. double the power at point d_0, as:

$$P_{min} \propto V_{min}^2$$

$$\therefore 2P_{min} \propto V_x^2$$

$$\therefore V_x^2 = 2V_{min}^2$$

Then the empirical relation gives VSWR as:

Fig. 7.12 **a** V_{max} and V_{min} for very low VSWR. **b** Double minima method for very high VSWR

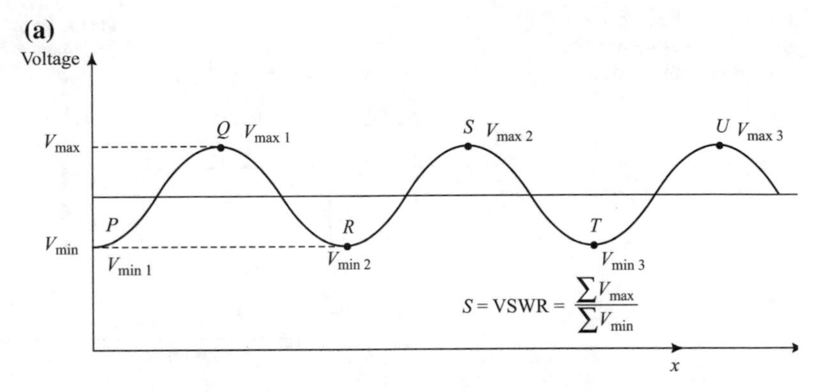

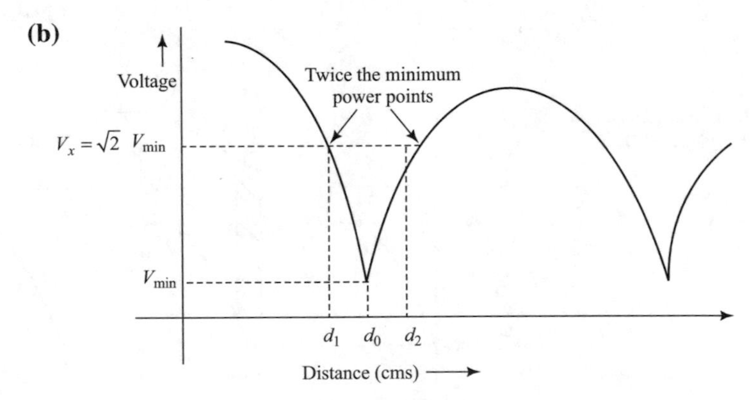

$$\text{VSWR} = \lambda_g/[\pi(d_2 - d_1)]$$

7.4.4 Measurement of Impedance

Impedance in a waveguide line can be due to some discontinuity, e.g. (posts, iris a device inside). Therefore, it could be capacitance or inductance or resistive or their combination (Sect. 4.11 and its figures). There are a number of methods for measuring it, and we will discuss three for resistive loads and two for complex loads:

(a) **Two Directional Coupler/Reflectometer Method (for Pure Resistive Load)**

The circuit set-up is given in Fig. 7.13, where two directional couplers are used to sample the incident power P_i and reflected power P_r from the load. Both the directional couplers are identical (20 dB) except their direction. The reflectometer directly reads the voltage reflection coefficient, from which reasonable accurate value of Z_L is computed:

$$\Gamma = \frac{V_r}{V_i} = \sqrt{\frac{P_r}{P_i}} = \frac{Z_L - Z_0}{Z_L + Z_0} \cdots \quad (7.7)$$

$$\therefore Z_L = Z_0 \cdot \left(\frac{1+\Gamma}{1-\Gamma}\right) = \frac{Z_0}{S} \quad (7.8)$$

This also gives an alternate method for measuring S (VSWR):

$$S = \frac{1+\Gamma}{1-\Gamma}$$

If a laboratory is not having reflectometer, we measure the VSWR and then the equation $Z_L = Z_0/S$ gives approximate value of load impedance.

(b) **Directly by VSWR meter (for pure resistive load)**

(c) **Magic tee and Null detector method (for Pure Resistive Load)**

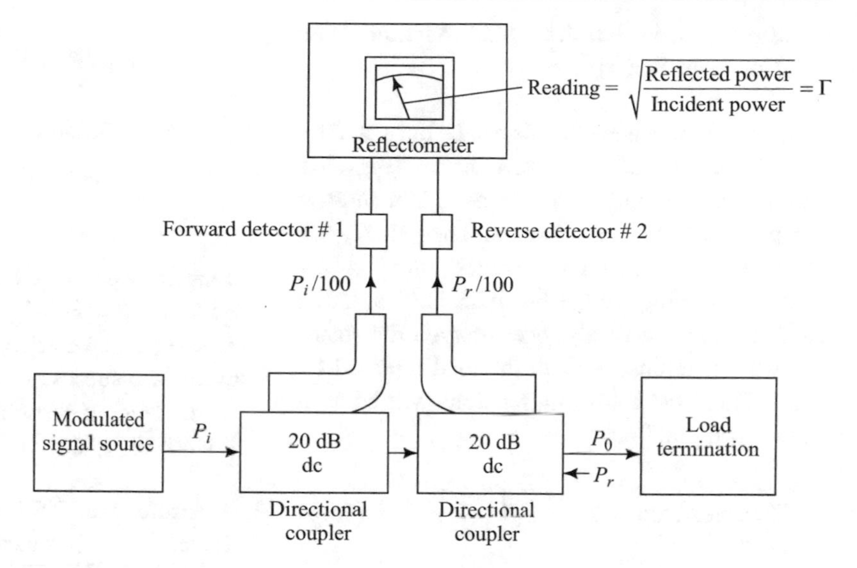

Fig. 7.13 Reflector set-up for measurement of impedance

A magic tee can be used to measured pure resistive load as per the circuit given in Fig. 7.14. The unknown load Z_2 and variable known load Z_1 are on the linear ports 1 and 2. Microwave source is at port 3 of H-Plane, while the null detector with a variable attenuator at port 4 (E-plane port). Null detector is nothing but a crystal detector followed by a galvanometer as shown in Fig. 7.14b.

The power source is at port 3 giving voltage level of a_3, which gets divided into two linear ports 1 and 2 of equal power with voltages as $a_3/\sqrt{2}$ and $a_3/\sqrt{2}$, respectively. The loads at these two ports 1 and 2 may not be equal to the characteristic impedance of Z_0 and therefore have reflection coefficients as Γ_1 and Γ_2, respectively. Therefore, the reflected microwave voltages will be $\Gamma_1 \cdot a_3/\sqrt{2}$ and $\Gamma_2 \cdot a_3/\sqrt{2}$, respectively, from ports 1 and 2, which will reach port 4 as

$$\frac{1}{\sqrt{2}}\left(\Gamma_1 a_3/\sqrt{2} - \Gamma_2 a_3/\sqrt{2}\right) = (a_3/2)(\Gamma_1 - \Gamma_2)$$

We now vary Z_1 till the null detector shows zero, and here $Z_2 = Z_1$; therefore, we get very accurate value of Z_2 directly.

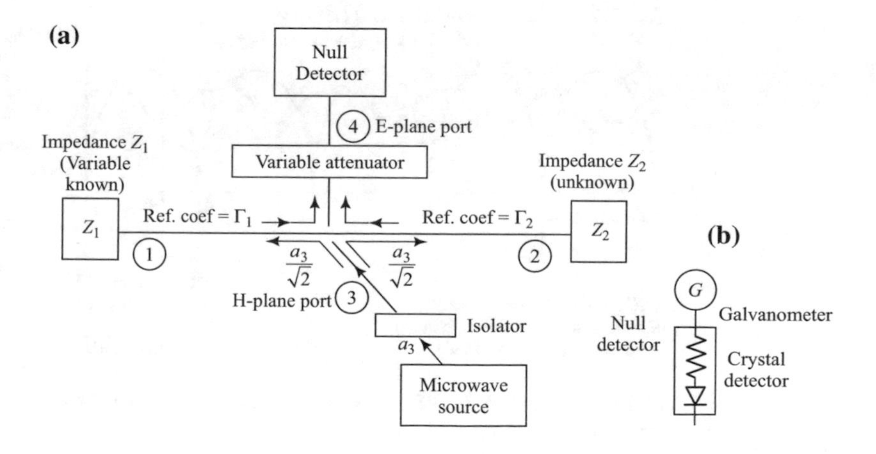

Fig. 7.14 **a** Impedance measurement by magic tee and **b** the null detector method

(d) **Slotted Line Minima Shift Method (For Complex Load)**

Complex impedance of loads can be measured by the slotted line of microwave bench (Figs. 7.1 and 7.15). The only thing we need is a short at the plane of the load location. Then **firstly** we observed the standing waves minima x_1 (Fig. 7.16) using slotted line with the load at its end, and then **secondly**, we observed the standing waves minima x_2 with the load replaced by short. Then, the following relations will be used in this computation:

$$\text{Complex load} = Z_L = Z_0\left(\frac{1+\Gamma_c}{1-\Gamma_c}\right) \quad (7.10)$$

$$\text{Complex reflection coefficient } \Gamma_c = \Gamma_0 \cdot e^{i\phi} \quad (7.11)$$

$$\text{VSWR} = S = \left(\frac{1+\Gamma_0}{1-\Gamma_0}\right)$$

Phase angle of load noted from shift of minima

$$\phi = [2\beta(x_1 - x_2) - \pi] \quad (7.12)$$

Imaginary part of the wave propagation factor $\alpha + i\beta$, is = $\beta = 2\pi\beta/\lambda_g$

Guide wavelength = λ_g = 2 × distance between two successive minima with load.

Thus for measuring/calculating Z_L, the following steps may be followed:

1. Measure the VSWR (S) = V_{max}/V_{min} and hence reflection coefficient $\boxed{\Gamma_0 = (S-1)(S+1)}$ with load Z_L.
2. Find the distance (d) between two successive minima with load (Z_L) to find $\lambda_g = 2d$ and $\beta = 2\pi/\lambda_g$.

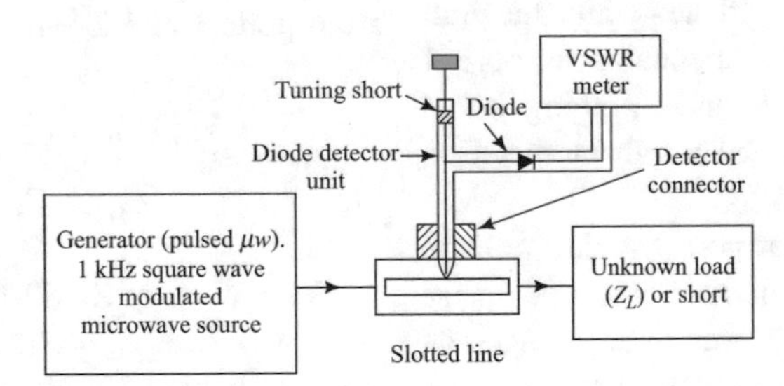

Fig. 7.15 Slotted line method for complex load

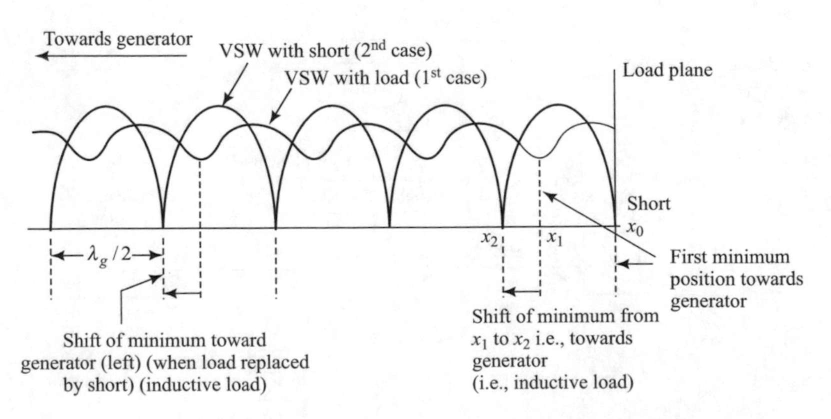

Fig. 7.16 Shifting of minima (x_1) of the VSWR pattern when the load to be measured is replaced by short on slotted line end (Fig. 7.1). The x_1 shifts to **a** left for inductive load **b** to the right for capacitative load and **c** for resistive, shift is exactly by $\lambda/4$ or no shift (i.e. zero shift)

3. Measure the position of first minima starting from load side towards generator, firstly with the load (x_1) then when it is replaced by short (x_2).
4. Calculate load phase angle using the shift of minima $\phi = [2\beta(x_1 - x_2) - \pi]$, and hence, $\Gamma_c = \Gamma_0 \cdot e^{i\phi} = \Gamma_0\,(\cos\phi + i\sin\phi)$.
5. Hence, we calculate:

$$Z_L = Z_0\left(\frac{1+\Gamma_c}{1-\Gamma_c}\right) \tag{7.13}$$

6. If the shift is:
   - to the left, then the load is inductive.
   - to the right, then the load is capacitative.
   - No shift or exactly $\lambda_g/4$ shift, then the load is resistive.

(e) **By Smith Chart (For Complex Load)**

We can use Smith chart also for calculating Z_L after getting VSWR and shift of minima.

7.4.5 Insertion Loss, Attenuation Loss and Return Loss

When an input power (P_i) is flowing in a lossless transmission line up to a matched load (non-reflection from it), and a device is inserted in between, then due to it (Fig. 7.17a):

(a) A portion of power P_r will get reflected back.
(b) Power P_{attn} will get attenuated/absorbed in the device.
(c) Remaining power (P_0) will go to the load.

Now by definition, the insertion loss in dB:

$$I = 10\log(P_0/P_i) \tag{7.14}$$

$$\text{Here}\quad P_0 = (P_i - P_r - P_{attn})$$
$$\therefore P_{attn} = (P_i - P_r - P_0)$$

$$\text{As}\quad \frac{P_0}{P_i} = \frac{P_i - P_r}{P_i} \times \frac{P_0}{P_i - P_r}$$

$$\therefore 10\log\left(\frac{P_0}{P_i}\right) = 10\log\left(1 - \frac{P_r}{P_i}\right) + 10\log\left(\frac{P_0}{P_i - P_r}\right) \tag{7.15}$$

Therefore, by definition of the losses:

$$\therefore \underset{(I)}{\text{Insertion loss}} = \underset{(R)}{\text{reflection loss}} + \underset{(A)}{(\text{attenuation loss inside the device})}$$

Because of insertion of the device in the line, the power lost is called insertion loss.

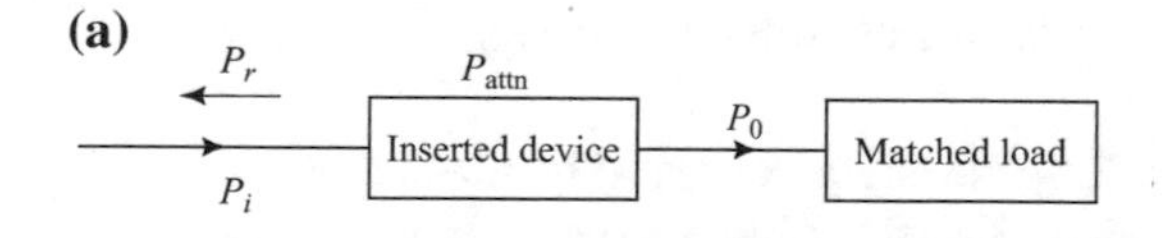

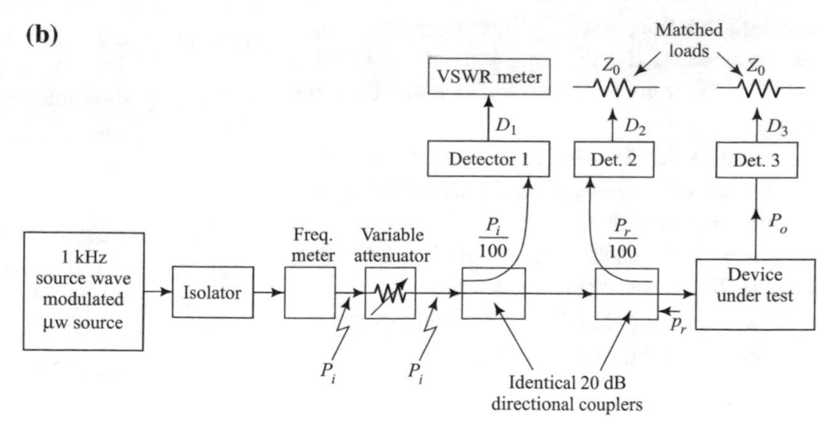

Fig. 7.17 **a** Line diagram of power flow with device in the line. **b** Set-up for measuring insertion loss and attenuation loss

(a) **The reflection loss (dB)**:

$$R = 10\log\left(\frac{1-\Gamma_r}{\Gamma_i}\right) = 10\log\left(1-\Gamma^2\right) \quad (7.16)$$

$$\therefore R = 10\log\left[\frac{4S}{(1+S)^2}\right] \quad (7.17)$$

$$\left[\left(\text{As } S = \frac{1+\Gamma}{1-\Gamma}\right), \Gamma = \left(\frac{S-1}{S+1}\right)\right]$$

(b) **The attenuation loss (dB)**:

$$A = 10\log\left(\frac{P}{P_i - P_r}\right) = 10\log\left(e^{2\alpha l}\right) = 8.686\,\alpha l \quad (7.18)$$

(c) **Return/reflection loss** is defined as the power lost due to reflected portion alone and

$$\therefore R = \text{Ret(dB)} = 10\log\left(\frac{P_r}{P_i}\right) = 20\log(\Gamma) \quad (7.19)$$

(d) **Insertion loss** is a measure of the energy loss through a transmission line as compared to direct transmission of energy without the transmission line, given by

$$\text{Insertion loss (dB)} = 10\log_{10}\frac{E_1}{E_2} \quad (7.20)$$

where E_1 is the energy received by the load when connected directly to the source without the transmission line, and E_2 is the energy received by the load when the transmission line is inserted between the source and the load, keeping the input energy constant.

The insertion loss is due to mismatch losses at the input and output plus the attenuation loss in the transmission line.

For perfect matching of the device in the line, $P_r = 0$, $R = 0$ and then $A = I = 0$.

For measuring A and I, the following steps can be followed, using the set-up as per Fig. 7.17b:

1. Keeping the variable attenuator to minimum, the VSWR at detector D_1 should read maximum. For this, either the modulation level of power source can be adjusted or the gain of the VSWR meter. At this, the maximum reading of VSWR should be = 1 (i.e. zero dB).
2. Read the frequency using the frequency meter when dip is observed in VSWR meter.
3. Now interchange the VSWR meter at D_1 with the matched loads at D_2. Here the VSWR meter will read (P_r/P_i) the return lost in dB. From this, reflection loss = 10log ($1 - P_r/P_i$) is calculated.
4. Now the variable attenuator is adjusted to give an attenuation of 20 dB, (i.e. equals that of directional coupler). Then, interchange the VSWR meter at D_2, with matched load at D_3, so that matched loads are at D_1 and D_2. The VSWR will give the reading of insertion loss (P_0/P_i) directly on its meter. The attenuation of the device under test can be obtained by subtracting reflection loss from insertion loss.
5. For measuring only attenuation A due to device, we can just measure power in the line with and without the device. Then for a matched line, in both these cases $P_r = 0$.

$$\therefore A_{\text{Matched line}} = 10\log\left(\frac{P_0}{P_i - P_r}\right) = 10\log(P_0/P_i) \quad (7.21)$$

7.4.6 Q-of a Cavity: Reflection and Transmission Types

As discussed in Sect. 3.5 (Chap. 3), the quality factor is a measure of frequency selectivity of a resonant or antiresonant circuit and is defined as:

$$Q = 2\pi\left(\frac{\text{Maximum in energy stored}}{\text{Energy dissipated per cycle}}\right) = \frac{\omega W}{P} \quad (7.22)$$

Thus, the electromagnetic energy has to be high, while energy dissipated through heat has to be as small as possible. Therefore for high Q, the circuit/cavity has to be made highly reactive (L or C) and minimum resistive (R).

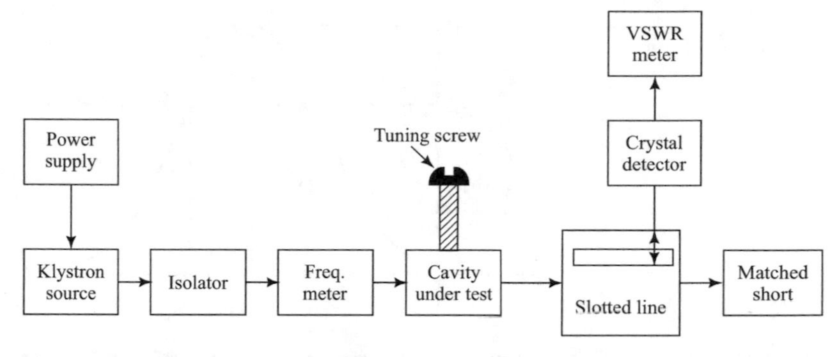

Fig. 7.18 Measurement of Q factor of a cavity

At resonant frequency, the electric and magnetic energies are equal in time quadrature. Also, when E-energy is maximum, H-energy is minimum and vice versa. There are two types of resonant cavities:

(a) Transmission type, e.g. in frequency meter, two-cavity klystron, reflex klystron.
(b) Reflection type, e.g. magnetron type, waveguide cavity.

In (a) type, the μW signal passes through the cavity, while in (b) type, it passes tangentially and a small port of power enters the cavity. Also, cavities put at the terminal end of the line are also of this type. (For details, chapter on cavity could be referred.)

(a) **Q-Measurement of Transmission Type Cavity by Slotted Line Method**

The set-up used is in Fig. 7.18 where the cavity under test is placed just before slotted line. Here the output signal is measured as a function of frequency resulting into a resonance curve (Fig. 7.19). Klystron frequency can be varied by the turning its cavity screw slowly and measuring every time the frequency and the power output on VSWR meter and plotting (Fig. 7.19).

Measure the half power band width frequency Δf and calculate Q by:

$$Q = \frac{f_0}{2 \cdot \Delta f} \tag{7.23}$$

Instead of changing the frequency of the klystron/source, the cavity also can be tuned a bit by a tuning screw on either side of resonant frequency, and power output noted in VSWR meter as per Figs. 7.18 and 7.19.

(b) **For Reflection-Type Cavity**

For cavity, like short plunger cavity, it can be placed at the end of slotted line and same experiment repeated.

7.4.7 Measurement of Phase Shift by Comparison with Precision Shifter

The phase shift introduced by a microwave network can be measured by using the set-up shown in Fig. 7.20. As it is not possible to distinguish between one-quarter wavelength and say, seven quarter wavelengths, we must have an approximate idea of the network's electrical length. Also, we know that each wavelength (λ_g), corresponds to a phase shift of 2 radians, knowing the approximate electrical lengths of the network, phase shift can be determined to a fairly accurate value as follows.

(i) The source with a 1 kHz sine wave amplitude modulation is split up into two equal parts using the H-plane Tee junction, one going to the unknown network whose accurate phase shift is to be measured and the other to the comparison adjustable precision phase shifter. (ii) Standard precision phase shifter is now adjusted until the two demodulated 1 kHz sine wave on the CRO are in phase, as shown by Fig. 7.21, and the relative phase shift of the two networks are now equal. (iii) Dial reading on the

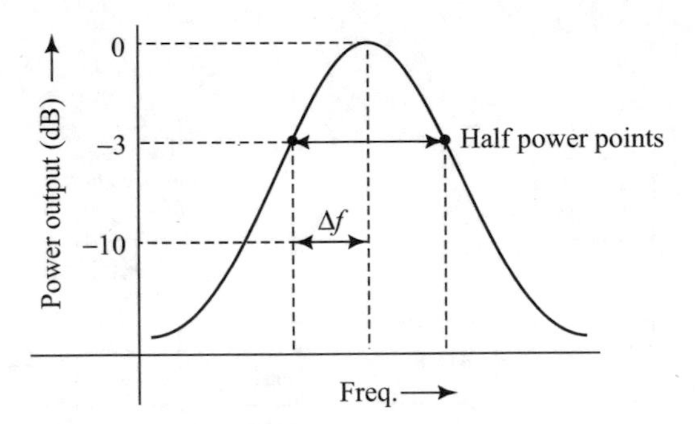

Fig. 7.19 Power versus frequency resonance curve and half power points (see Fig. 4.22 for band stop/pass cavities)

Fig. 7.20 Measurement of phase shift

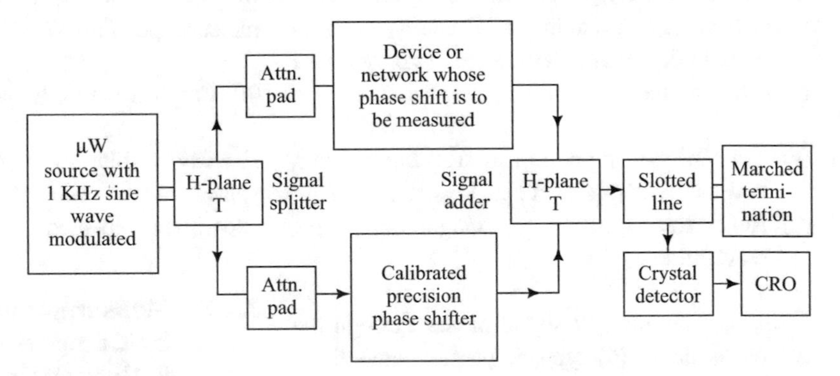

precision phase shifter now gives the phase shift offered by the device as shown in Fig. 7.22, to which the multiplies of 360° is to be added.

If from the preliminary measurement we get the phase shift of the unknown network to be in the vicinity of $4\lambda_g$ − 1440° and the reading on the calibrated precision phase shifter is 15°, then the total phase shift must be 1440° + 15°= 1455. If the reading is 310°, i.e. more than 180°, then the total phase shift must be [1080° (360 − 310)] = [1080° − 50°] = 1030°.

7.4.8 Measurement of Dielectric Constant (ε_r)-Minima Shift Due to Dielectric

(a) **The Dielectric Constants**

Dielectric constants (ε_r) is the relative permittivity ($\varepsilon_r = \varepsilon * / \varepsilon_0$) and is the measure of efficiency of transfer of electric lines of force just like the permeability (μ) has the ability of magnetic field to permeate and the conductivity (ρ) for conduction of charge. These three parameters (ε, μ and ρ) of a material appear in maxwell equations and play very important role in the analysis.

Dielectric material causes power loss/heating also. The permittivity is a complex quantity (ε^*) with imaginary component representing losses.

$$\varepsilon^* = \left(\varepsilon_1 - j\frac{\sigma}{\omega}\right) = \varepsilon_0(\varepsilon' - j\varepsilon'') = \varepsilon_0 \cdot \varepsilon_r \quad (7.24)$$

where

$\varepsilon'' = \frac{\sigma}{\omega\varepsilon_0}$ = measure of dissipation of energy
$\varepsilon' = \varepsilon_1^*\varepsilon_0$ = measure of ability to store energy
$\varepsilon_r = \frac{\varepsilon^*}{\varepsilon_0} = (\varepsilon' - j\varepsilon'')$ = relative permittivity = (complex dielectric constant)

$$\therefore \varepsilon_r = \varepsilon'(1 - j\tan\delta) \quad (7.25)$$

$\tan d = \varepsilon''/\varepsilon'$ = loss tangent
= ratio of power dissipated to the power stored per cycle when EM − wave propagates

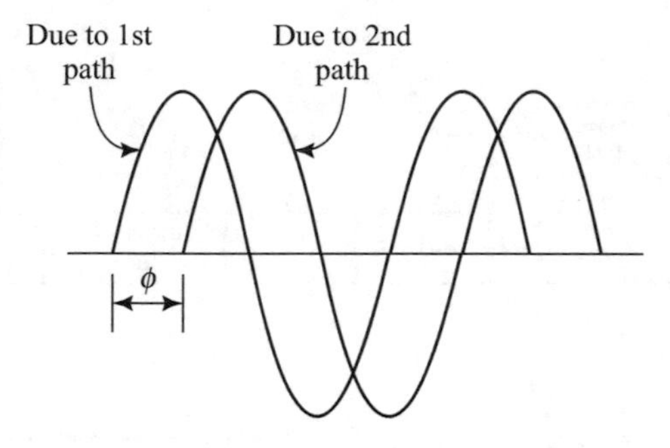

Fig. 7.21 Output in CRO of the two sine waves from two paths

Fig. 7.22 Dial of the standard precision variable phase shifter calibrated for the frequency being used

In general in measurements, the real part of ε_r suffices.

(b) **Measurements**: Here we measure the shift: (Δs) in the minima of standing wave, when the dielectric sample (in a waveguide) is not there and when it is placed between slotted line and the matched short as in Fig. 7.23. Approximate value of dielectric constant should be known before band. Following steps may be followed.

1. The frequency meter gives the frequency, and from this, λ_0 and λ_g can be computed. Alternatively, λ_g could be measured by slotted line.
2. The thickness (t) of the dielectric could be measured by a micrometer accurately.
3. Then, we calculate the value of a parameter (P):

$$P = \frac{\lambda_g}{t} \cdot \tan\left[\frac{2\pi(\Delta s + t)}{\lambda_g}\right] = \frac{\lambda_g}{t} \tan\theta \quad (7.27)$$

This P could be −ve or +ve, but we will use the modulus of P.

4. For another relation between P parameter N the number of wavelengths inside the dielectric over its thickness t is given by a transcendental equation in N. (Here it may be noted that λ_d, the guide wave length in the dielectric filled waveguide is greater than λ_g, the guide wave length with air filled). This transcendental equation in N= $\frac{t}{\lambda_d}$is given by:

$$|P| = \frac{\tan(2\pi N)}{N} \quad (7.28)$$

A plot of $|P|$ versus N is in Fig. 7.24, which we use for reading the first value of N (if we get two), from the value of P computed earlier by (Eq. 7.27).

5. Using this N, the final value of ε_r is computed from the following formula:

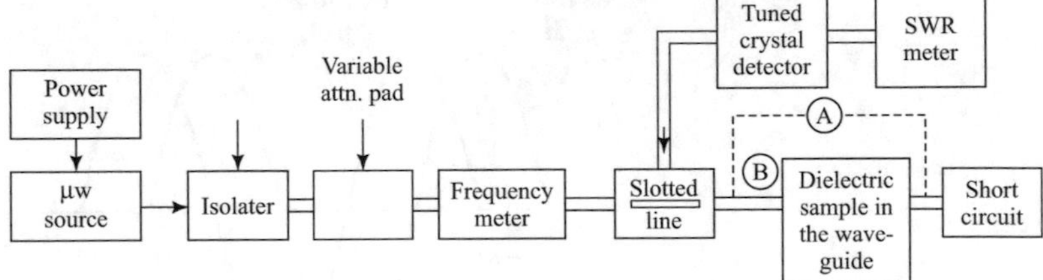

Fig. 7.23 Set-up for measuring dielectric constants

$$\varepsilon_r = 1 - \left(\frac{\lambda_0}{\lambda_g}\right)^2 + \left(\frac{\lambda_0 N}{t}\right) \tag{7.29}$$

6. The above procedure can be repeated at a slightly different frequency and ε_r computed again. The average of these two values will give more accurate value of ε_r.

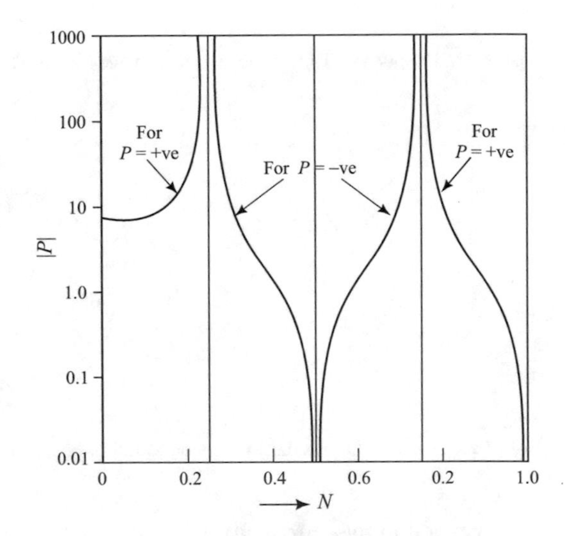

Fig. 7.24 N-versus-$|P|$ plot of transcendental equation $P = \frac{\tan(2\pi N)}{N}$. Once $|P|$ is known by Eq. (7.27), read N from above graph and put in Eq. (7.29) to get ε_r

7.4.9 Measurement of Noise Figure and Noise Factor by Standard Noise Source and Noise Meter

(a) What is Noise

Noise is an unwanted signal with frequencies from $t = 0$ to ∞ and is generated in the source, the transmission line and in the devices/components in between. The causes of noise include:

1. **Shot Noise**: It is due to random fluctuations of charge carriers in microwave tubes or solid state devices.
2. **Thermal Noise**: It is due to thermal agitation of bound charges.
3. **Flicker Noise**: It is due to random variation in the activity of electrons getting emitted from cathode in microwave vacuum tubes or solid state devices.
4. **Plasma Noise**: Due to random motion of electrons in an ionised gas, e.g. ionosphere or electric sparking contacts.
5. **Other Source**: Other causes are varying resistances, leaky insulation, stray field, vibration of tube and any circuit elements.

Noise finds its use in all types of oscillators, low/high frequency or microwave tubes/solid state device, without noise oscillator cannot generate signal. As oscillator is nothing but a +ve feedback amplifier, which amplifies only that noise frequency which is suitable to its tuned circuit.

(b) Measurement

Noise factor of a device in the lines is = Input signal to noise ratio/output signal to noise ratio. The noise figure (F) is in dB and is obtained by taking log of noise factor and can be defined as:

Noise figure of a device:

$$F = 10\log_{10}\left(\frac{\text{Input signal to noise ratio of the device}}{\text{Output signal to noise ratio}}\right) \tag{7.29}$$

A set-up for measurement is in Fig. 7.25, with the steps for measurement as:

1. The device under test is switched on with input side having matched termination, and output to the noise meter is noted (N_1).
2. The input of the device is now connected through a precision variable attenuator (with dB reading) to a standard noise source (normally argon discharge tube).
3. The attenuation is adjusted from zero dB (where noise of standard source is fully blocked onwards, till output noise (N_2) through the device to the power meter shows double the earlier reading (N_1) (i.e. $N_2 = 2N_1$). Then, the noise figure of the device is equal to that of the attenuator; therefore, the reading of the attenuator gives the noise figure in dB.

Alternatively we can get still move accurate value by avoiding the non-linearity of the detector, where noise source is also not required. For this we can put 3-dB attenuator (i.e. reduce noise signal to noise meter by half) before the meter, and then the precision attenuator adjusted to give the same reading as N_1. Then also the attenuator gives the noise figure directly.

Solved Problems

Problem 1 Two-directional couplers (DC) of 20 dB and 10 dB are used in a waveguide for sampling incident and reflected power in a line and outputs from the couplers at arm 3 is 25 mW and 50 mW, respectively. Find the VSWR and reflection coefficient of the line.

Solution The reflection coefficient = $\Gamma = \sqrt{\frac{P_r}{P_i}}$

The input-end coupler power $P_i/100 = 25$

The reverse coupler power $P_r/10 = 50$

$$\therefore P_i = 2500, P_r = 500$$

$$\therefore \Gamma = \sqrt{500/2500} = 1\Big/\sqrt{5} = 0.447$$

$$\therefore S = \frac{1+\Gamma}{1-\Gamma} = \frac{1.447}{0.553} = 2.79$$

Problem 2 In a double minima experiment for measurement of VSWR, find VSWR if the other details for TE_{10} mode are: frequency = 10 GHz, waveguide dimensions is 4 cm × 2.5 cm, $d_2 - d_1 = 1$ mm.

Solution For the TE_{10} dominant mode.

$$\lambda_c = 2a = 2 \times 4 = 8 \text{ cm}$$

$$\therefore \lambda_0 = c/f = 3 \times 10^8/10 \times 10^8 = 3 \text{ cm}$$

$$\therefore \lambda_g = \lambda_0 \Big/ \sqrt{1-(\lambda_0/\lambda_c)^2} = \frac{3}{\sqrt{1-(3/8)^2}}$$

$$= 3.236 \text{ cm}$$

For double minima method, the empirical formula gives:

$$\text{VSWR} = \frac{\lambda_g}{\pi(d_2 - d_1)} = \frac{3.236}{\pi(1 \times 10^{-1})} = 10.3$$

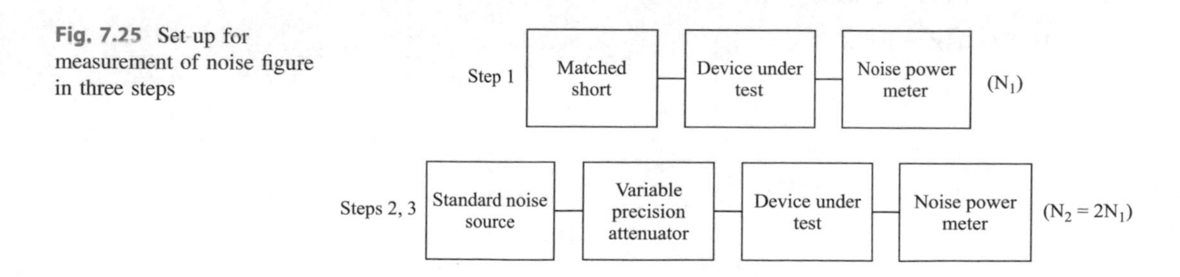

Fig. 7.25 Set up for measurement of noise figure in three steps

Problem 3 In a double minima experiment, distance between two maxima is 3.5 cm and distance between two double minima is 2.5 mm, find VSWR.

Solution

$$\lambda_g = 2 \times 3.5 = 7.0 \text{ cm}$$

$$(d_2 - d_1) = 2.5 \text{ mm} = 0.25 \text{ cm}$$

$$\therefore \text{VSWR} = \frac{7.0}{3.14(0.25)} = \frac{7.0}{0.785} = 8.9$$

Problem 4 In a waveguide of 2 cm × 1 cm inner cross section, distance between two successive minima is 1 cm in the standard wave pattern. Find the frequency of the signal in dominant mode.

Solution In dominant mode TE_{10}

$$\lambda_c = 2a = 2 \times 2 \text{ cm} = 4 \text{ cm}$$

$$\lambda_g = 2 \times 1 \text{ cm} = 2 \text{ cm}$$

$$\therefore \frac{1}{\lambda_0^2} = \frac{1}{\lambda_g^2} + \frac{1}{\lambda_c^2} = \frac{1}{(4)^2} + \frac{1}{(2)^2} = \frac{5}{16}$$

$$\therefore \frac{1}{\lambda_0} = \frac{\sqrt{5}}{4} = 0.559/\text{cm}$$

$$f_0 = \frac{c}{\lambda_0} = 3 \times 10^{10} \times 0.559 = 1.677 \times 10^{10} = 16.77 \text{ GHz}$$

Problem 5 In a dielectric measurement experiment, the dielectric thickness along the line is 5 cm and the shift of the first maxima of VSWR pattern shifts by 0.3 cm. The distance between two maxima = 4 cm. If the frequency of the wave is 10 GHz, find the dielectric constant using the plot of P-versus-N (Fig. 7.24).

Solution Given:

$$t = 5 \text{ cm}$$

$$\lambda_g = 4 \times 2 \text{ cm} = 8 \text{ cm}$$

$$\Delta s = 0.3 \text{ cm} \therefore \Delta s + t = 5.3 \text{ cm}$$

$$\lambda_o = c/f = 30 \times 10^9 / 10 \times 10^9 = 3 \text{ cm}$$

$$\therefore P; = \frac{\lambda_g}{t} \cdot \tan\left[\frac{2\pi(\Delta s + t)}{\lambda_g}\right]$$

$$; = \frac{8}{0.3} \cdot \tan\left(\frac{2 \times 3.14 \times 5.3}{8}\right)$$

$$; = \frac{8}{0.3} \cdot \tan(2.08)(\text{Here } 2.08 \text{ rad} = 119.24^\circ)$$

$$; = 26.67 \times \tan(119^\circ) = -1.8$$

$$\therefore \frac{\tan(2\pi N)}{N} = P = -1.8$$

$$\therefore |P| = 1.8$$

Reading from P-versus-N graph, we get first value of $N = 0.4$, (for −ve values of P).

$$\therefore \varepsilon_r = 1 - \left(\frac{\lambda_0}{\lambda_g}\right)^2 + \left(\frac{\lambda_0 N}{d}\right)$$

$$= 1 - \left(\frac{3}{8}\right)^2 + \frac{3 \times 0.4}{0.3}$$

$$= 1 - 0.14 + 4$$

$$= 4.86$$

Problem 6 For attenuation measurement of an attenuator in a line where the power flow = 26 mW, it is inserted and the line is matched. Then the power output from attenuator is 10 mW. Find the attenuation of the attenuator.

Solution As the load is matched

$$P_r = 0$$

$$\begin{aligned}\text{Attenuation} = A &= 10\log_{10}\left(\frac{P_0}{P_i}\right)\\ &= 10\log\left(\frac{10}{26}\right)\\ &= -10\log_{10}(2.6)\\ &= -10 \times 0.415\\ &= -4.15\ \text{dB}\end{aligned}$$

Review Questions

1. Explain the basic microwave bench and what are the measurement possible with it.
2. A waveguide is having signal of $\lambda_g = 5$ cm. Now it is filled with a powder of dielectric constant $\varepsilon_r = 4.0$. Find the new λ_g.
3. A waveguide of 1 cm × 1 cm, successive minima of standing wave is 1 cm, calculate the frequency.
4. Explain the methods of measuring power.
5. Why VSWR measurement is one of the most important types of measurement. List out the measurements of various parameters where it is used.
6. Explain the types of bolometers and how it is used for power measurement. How do we extend the range of its measurements.
7. What does complex impedance means? Explain a method to measure real impedance with no reactive component.
8. Name the devices which detect power, explain their properties, and its mounting in a microwave line.
9. A line has VSWR = 0.75 getting power from a source of 234 mW, through an isolator of 25 dB. Find the reflected signal power received by the source.
10. A CW-signal is modulated by a square pulse signal of 1500 Hz and pulse width 0.56 s. If the peak power is 1300 W, find the average power, i.e. CW power of the signal source and duty cycle.
11. If a signal is of frequency 2.4 GHz, how long a slot is required to measure the wavelength by a slotted line method using L-band waveguide of 10.9 cm × 5.5 cm.
12. Two identical directional couplers samples input and reflected power. The power level of P_r was 10 dB down than P_i find VSWR.

8 Microwave Propagation in Space and Microwave Antennas

Contents

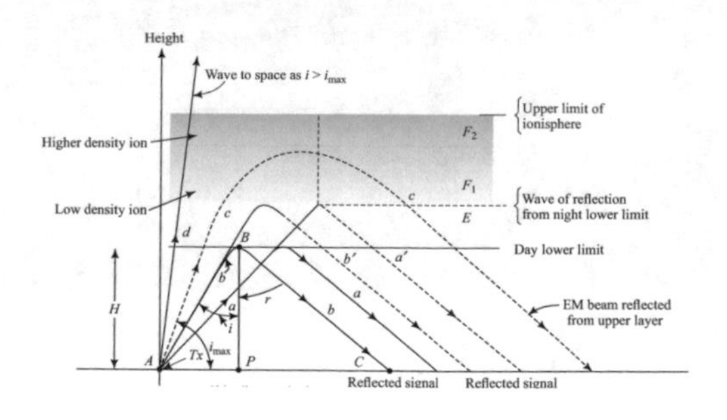

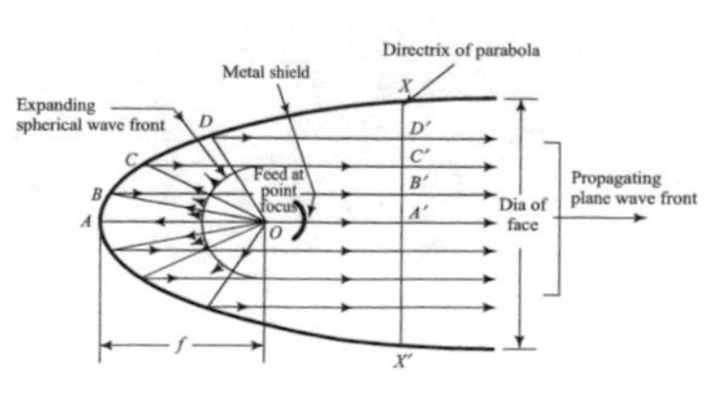

© Springer Nature Singapore Pte Ltd. 2018
P. K. Chaturvedi, *Microwave, Radar & RF Engineering*,
https://doi.org/10.1007/978-981-10-7965-8_8

8.1 Introduction

One of the most important applications of microwave is in communication, followed by RADAR. In RADAR, the $\mu\omega$ signal used has frequency range 0.5–300 GHz. The devices and circuit needed for generation of microwave become a part of the system along with antenna for transmission and receiving of signal. There are three parameters which decide the method of communication to be adopted, e.g.

(i) Distance of communication receiver.
(ii) Frequency in hand, i.e. the source and circuit.
(iii) Application.

Table 8.1 Nominal frequency bands, their names, and their applications

S. No.	Frequency range (name)	Application
1.	30–300 Hz (ELF)	Penetration into earth submarine (sonar)
2.	10–30 kHz (VLF)	Radio navigation, sonar aeronautical communication, point-to-point long-distance communication
3.	30 kHz–300 MHz (LF: 30–300 kHz; 300 kHz–3 MHz) (HF: 3–30 MHz)	Sound broadcasting, amplitude modulation (AM), ultrasonics, ship communication
4.	30 kHz–100 MHz	Telephony using single side band (SSB), police, aviation
5.	30–300 MHz (VHF)	Sound broad casting frequency modulation (FM), mobile communication, police, aviation, TV, navigation
6.	300 MHz–3 GHz (UHF)	TV, mobile communication, satellite communication, and space communication using RADARS
7.	3–30 GHz (SHF)	Satellite communication, exploratory space communication using RADAR, airborne RADAR
8.	30–300 GHz (EHF)	Same as SHF

S. No.	Distance on earth	Max. height of wave	Method of propagation	Freq. used
(i)	Up to 50–100 km	Ht of antenna	Line-of-sight (LOS) propagation	30–300 MHz
(ii)	Up to 600 km	10–16 km	**Troposcatter**: Tropospheric scattering, refraction, and reflection	30–300 MHz
(iii)	Up to 1000 km	<1 km	**Ground waves**: Duct propagation in troposphere	30 kHz–31 MHz
(iv)	Up to 1000 km	50–600 km	**Sky wave**: Ionospheric reflection and refraction	3–30 MHz
(v)	Above 1000 km	>1000 km	**Space wave**: For satellite communication—This is by refraction/reflection from ionosphere which stretches from 50-60 km above earth	1–300 GHz

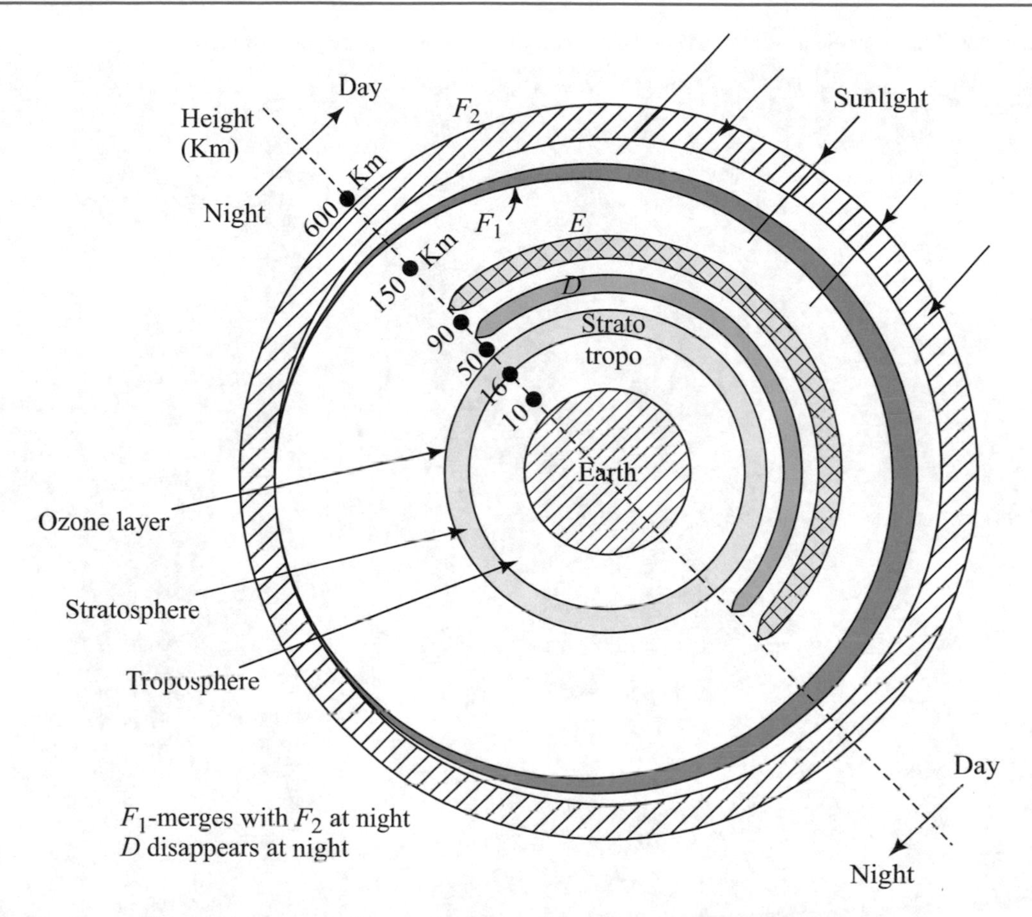

Fig. 8.1 Layers of troposphere, stratosphere, and ionosphere enclosing earth. The ozone layer in the stratosphere absorbs the ultraviolet rays and hence protects the earth from it

Nominal frequency bands (HF, VHF, UHF, L, S, etc.) are given in Table 8.1 along with the RADAR applications.

For long- and medium-range communication, method of communication becomes quite important because:

(a) Different frequency signals travel in different ways.
(b) For different distance transmission, method of propagation is different as:

For all the above types of communication, we need to study the medium of propagation, e.g. troposphere, stratosphere, ionosphere, and exosphere. Figure 8.1 gives these layers during day and night conditions along with their height from earth surface. The characteristics of these layers are summarised in Table 8.2.

8.2 Various Layers Enclosing Earth Acting as Medium for EM Wave Propagation

There are discrete two set of layers which enclose the earth:

(a) Troposphere (0–16 km) and stratosphere 16–50 km, with ozone layer at the uppermost portion of stratosphere.
(b) Ionosphere (50–600 km) having layers D, E, F_1, and F_2.

Table 8.2 Characteristic of various layers above earth: EM wave propagation point of view

Name of layer (thickness in km)	Height above earth km	Molecular density (per cc)	Ionic density (per m^3) (N)	Critical frequency $f_c = 9\sqrt{N}$	Content/activity in the layer
1. Troposphere (16 km thick)	0–16 km height (maximum 20 km at equator and 7 km near poles)	Normal air density falls with height 10^{23}–10^{20} per cc	Zero	>300 MHz in the duct propagation	• Has 75% of mass of air • Has 76% N_2; 21% O_2; 3% other gases • Has clouds of three types (i) 0–2 km height (ii) 2–4 km height (iii) 3–8 km height All the above change with season. The planes fly at different heights: • Air bus at 8 km or so • Small planes at 5 km
2. Stratosphere (35 km thick) Ozone layer (1 km thick)	16–50 km height	10^{20}–10^{16}	Zero	–	• Has ozone layer between 48–49 km height • Ozone stops ultraviolet rays of sun • Very dry, very little vapour • Polar stratosphere cloud (PSC) at 15–25 km is exception • PSC-form the infamous PSC-ozone hole at $T < -78$ °C
3. Ionosphere (4 layers) (a) D-layer (10 km thick)	50–600 km 50–90 km height	10^{16}–10^{14}	10^8–10^{10} (day–night)	100 kHz	• Shooting stars burn here • Disappears at sunset due to recombination of ions • It reflects VLF/LF but absorbs MF and HF • Temperature falls with height
(b) E-layer (25 km thick)	90–150 km height	10^{13}–10^{10} (typical value 6×10^{10})	10^{10}–10^{11} (day–night)	2.0 MHz or so	• Recombination is very fast after sunset and disappears • Temperature increases with height
(c) F_1-Layer (20 km thick)	150–250 km height	10^{12} or so	10^{11}–10^{12} (day–night)	1–3 MHz	• Maximum ionisation at noon • Reflects HF wave partly • F_1 forms with sunlight and merges with F_2 at night
(d) F_2-layer (300 km thick)	250–600 km height	10^{10} or so	10^{11}–10^{12} (day–night)	5–12 MHz	• Layer present whole of the 24 hrs • Ionisation atom density changes a lot from day to night
4. Exosphere (Infinite thickness)	>600 km height	Nearly zero	Nearly zero	All frequencies allowed to pass	• Temperature increases with height • Does not interfere or change the path of EM wave • Satellites are placed here

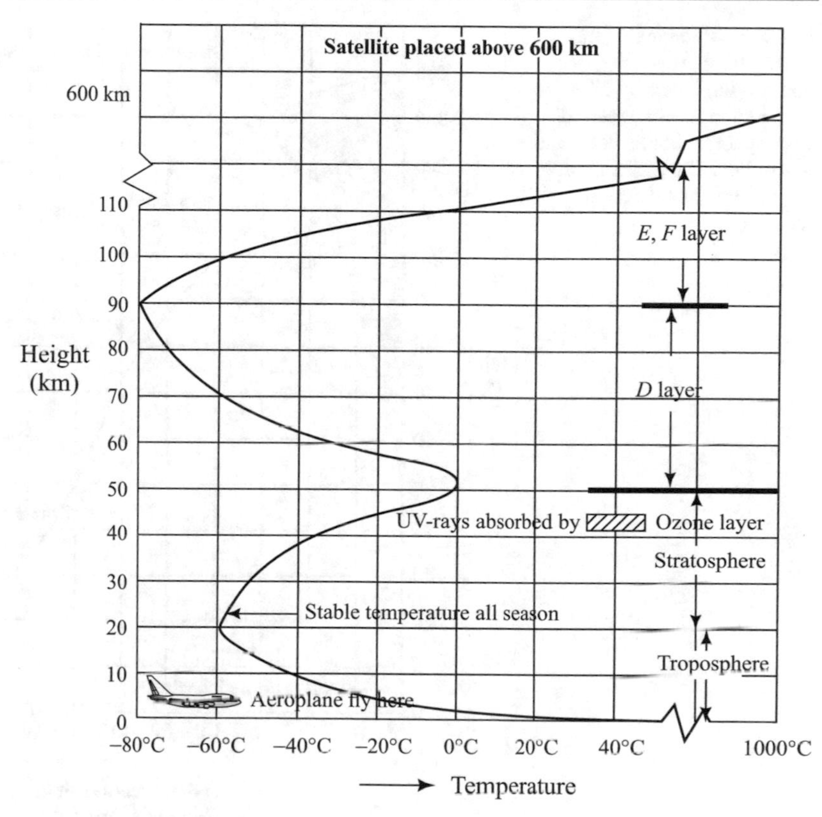

Fig. 8.2 Variation of temperature inside the layers of troposphere, stratosphere, and ionosphere. It is lowest (−80 °C) between D- and F-layers and very high (1000 °C) at 600 km height

All these layers are depicted in Fig. 8.1, and their properties/particulars are given in Table 8.1. Ionisation in ionosphere is greater in winter and falls in summer.

The temperature (as in Fig. 8.2):

- of troposphere falls with height
- of stratosphere increases with height
- of D-layer falls with height
- of E- and F-layers increases with height.

The low temperature of −60 °C is at the junction point of troposphere and stratosphere, while the lowest of −80 °C (like Antarctica) is in between D and E. In F, the temperature goes as high as 1000 °C (Fig. 8.2).

The gas density and atmospheric pressure keep a falling with height, while ionic density increases slowly from 10^8 to 10^{12} at 300 km height of ionosphere and falls thereafter (Fig. 8.3) beyond 300 km. In this figure, units of density may be noted.

8.2.1 Ionospheric Disturbances

The solar radiation (which causes the ionisation) can change its strength, and it will lead to change in the ionisation and hence change of:

(a) Critical angle of reflection due to change in refraction.
(b) The height of the point from where total reflection will take place.

Some of the changes in ionisation can be predicted, while some may be sporadic due to occurrence of sunspots. The causes of these changes are:

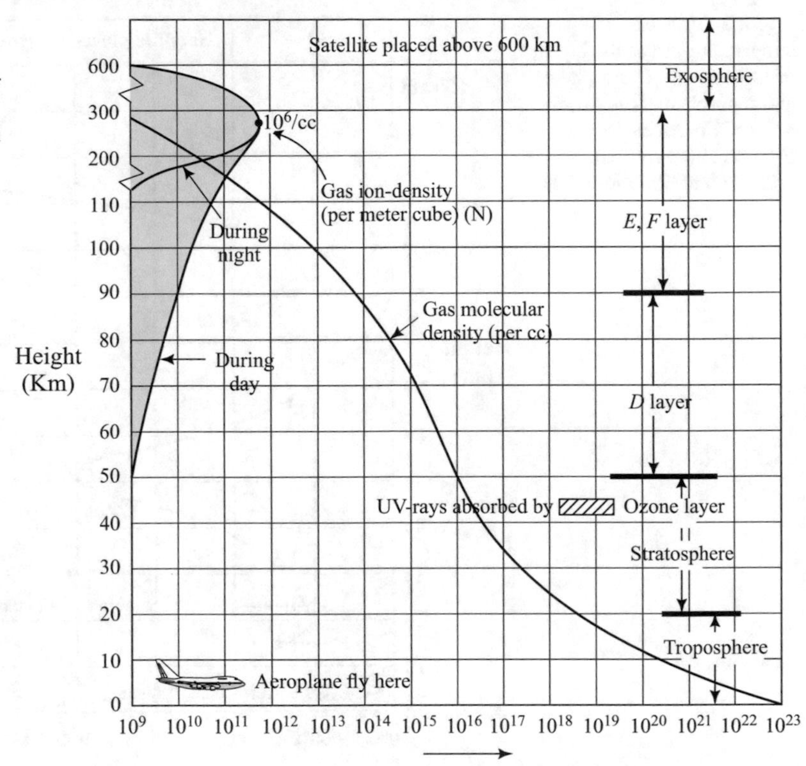

Fig. 8.3 Variation of gas molecular density in per cc and gas ionic density N in per metre cube, in the layers of troposphere, stratosphere, and ionosphere (D, E, F). During night, ionosphere region reduces

(i) *Diurnal variation*: This is due to daily rotation of earth around its axis.
(ii) *Seasonal variation*: This is due to annual rotation of earth around of the sun.
(iii) Eleven-*year sunspot cycle*: Sunspots are maximum every 11th year and then the radiation is also high, up to 2 times than normal. This leads to higher ionisation.
(iv) *Twenty-seven-day sunspot cycle*: Every 27th day, the fluctuation in F2-layer is maximum.
(v) *Different latitudes of earth*: As the radiation will be oblique at higher latitudes, ionisation is less. Therefore at the equator, ionisation is maximum.
(vi) *Ionospheric storm*: This is due to:
   (a) Disturbance in magnetic field of earth.
   (b) Disturbance from beyond solar system.

When the ionisation is less, then this thinly ionised layer refracts sky wave over a wider area. Therefore, the sky wave returns by total reflection from further away point at night as compared to day. Moreover, different frequency bands get affected differently.

The VLF (10–30 kHz) and LF (30–300 kHz) band do not get affected much in their day/night propagation, as they travel as ground wave. The MF (300 kHz–3 MHz) band waves travel as sky wave and therefore travel longer distance at night (Fig. 8.4).

Now we will discuss the different methods of wave propagation, as indicated in the introduc-

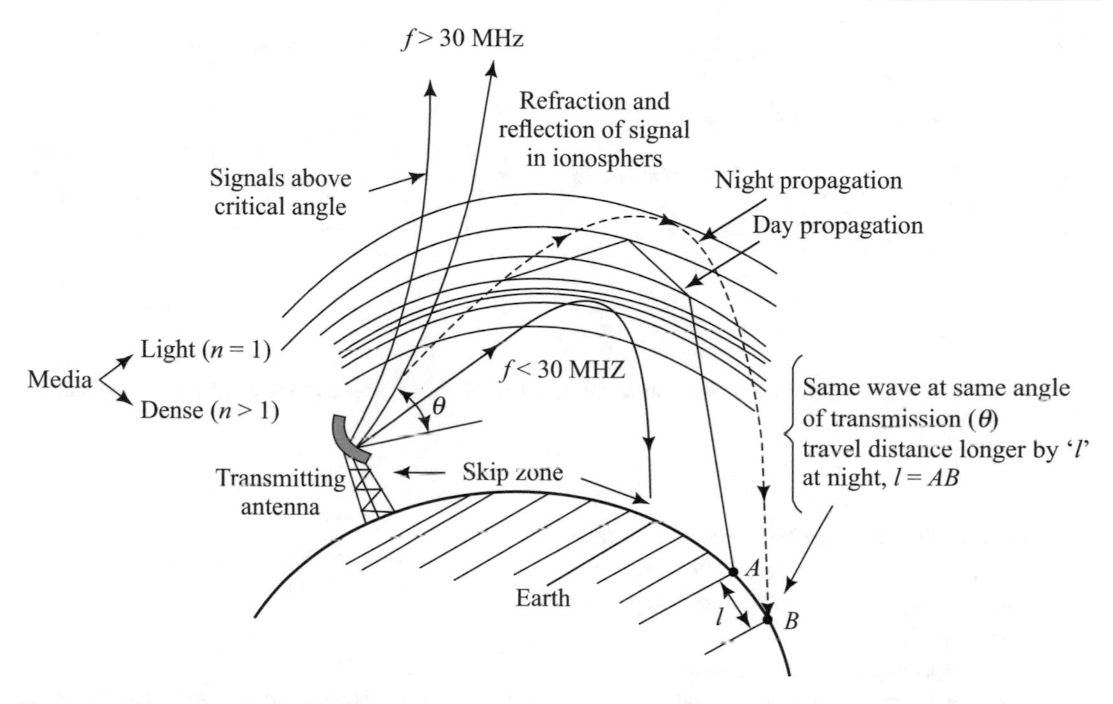

Fig. 8.4 Refraction through ionosphere and finally total reflection of wave below critical angle (θ). Above θ degree, the wave goes to space and gets lost hence loosing its energy. At night because of less ionisation, it travels longer by a distance 'l'

tion, e.g. line of sight, ground wave and ducted ground wave, space wave, sky wave, and finally satellite communication.

8.3 Different Methods of Wave Propagation

As per our need of distance and frequency being used, waves propagate by the following three distinct methods:

1. **Ground waves**: Propagation through line of sight or by getting reflected from earth, diffraction/reflection or ducted.
2. **Space wave**: Propagation by troposcatter from troposphere.
3. **Sky wave**: Propagation through refraction and reflection from ionosphere. These can be studied in detail now.

8.3.1 Ground Wave: LOS, Diffraction/Refraction, and Duct Propagation

There are three types of ground wave propagations:

(a) **Line of sight (LOS) (up to 50–60 km distance)**: In LOS propagation, the wave remains close to the ground (0–10 km height) and can travel up to 50–60 km distance only and is restricted by the curvature of earth (see Fig. 8.5a) and the height of antenna.

(b) **By diffraction/refraction (up to 4000 km distance)**: In case of frequencies 10 kHz–300 kHz, due to diffraction, the wave follows the curvature of earth and thereby can travel/propagate up to 4000 km, but at least 1000 km normally. Due to its diffraction

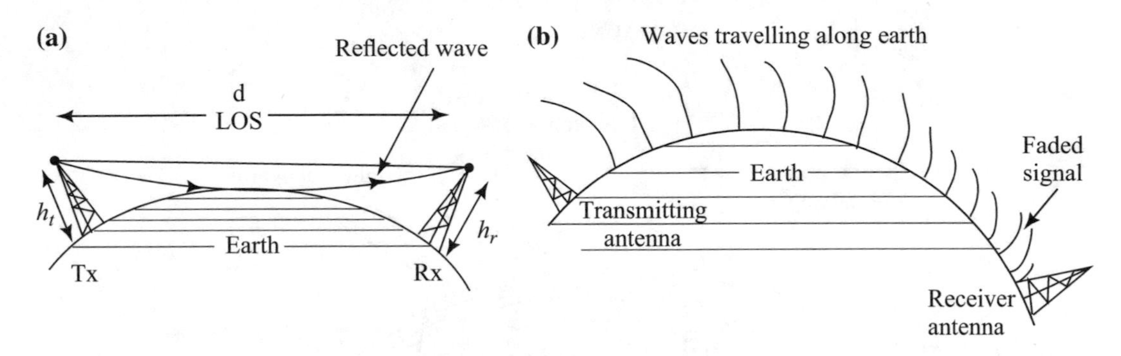

Fig. 8.5 Different methods of ground wave propagation: **a** line of sight (LOS) and wave reflected from earth and **b** diffraction and refractions

property, the wave can turn around obstacles like building, hills. A vertically polarised wave (see Fig. 2.5a) travels longer as ground wave (Fig. 8.5b).

(c) **Duct propagation**: Wave-propagating ducts are formed between the earth and height of 10–16 km, where dry and warm air is there.

Now, we will discuss each of these three ground wave propagations at length.

(a) **Line-of-sight (LOS) propagation**: This could be up to 100 km or so and depends on the:

(i) height of the two antennas and (ii) curvature of earth. If h_t and h_r are the heights of transmitter and receiver antenna, then the total distance of the tops of the two antennas will be (as proved below):

$$d = (d_1 + d_2) = \left(3.57\sqrt{h_t} + 3.57\sqrt{h_r}\right) = 3.57\left(\sqrt{h_t} + \sqrt{h_r}\right)\ \text{km} \tag{8.1}$$

where d, d_1, d_2 are in km and h_t, h_r in metres.

The strength of electric field reaching the receiver antenna will be given by:

$$E_R = \frac{88\sqrt{P}\cdot h_t \cdot h_r}{(\lambda d^2)}\ \text{V/m} \tag{8.2}$$

where

$P =$ power transmitted at wavelength λ in metres

$d =$ distance between transmitting/receiving antenna top points In addition to line-of-sight (LOS) wave, the wave which gets directly reflected from the ground also reach the receiving antenna (Fig. 8.5a).

For computing the LOS distance d, we use Fig. 8.6, where h_t, h_r as antenna height of transmitter and receiver, r the radius of earth, and d_1 and d_2 the two distances from horizon to antenna tops. This distance d_1 of transmitter is called radio horizon.

By ΔOPQ and ΔOQR

$$d_1 = \sqrt{(h_t + r)^2 - r^2} = \sqrt{h_t^2 + r^2 + 2h_t r - r^2} = \sqrt{2r \cdot h_t}\ \text{m} \quad (\text{as} h_t \ll \text{r})$$

Similarly, $d_2 = \sqrt{2r \cdot h_r}$

Therefore, the total distance d as r = 6370 km (radius of earth)

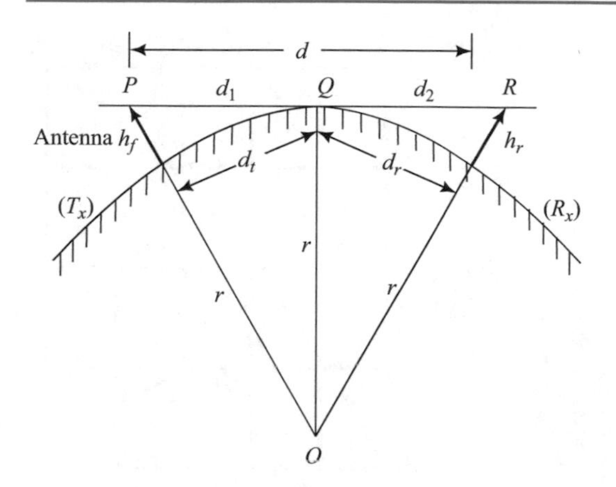

Fig. 8.6 Calculating LOS distance d

$$d = d_1 + d_2 = \sqrt{2r} \cdot \left(\sqrt{h_t} + \sqrt{h_r}\right)$$
$$d = \sqrt{2 \times 6370 \times 10^3}\left(\sqrt{h_t} + \sqrt{h_r}\right)$$
$$= 3570\left(\sqrt{h_t} + \sqrt{h_r}\right)$$

∴ Distance between the tops of antenna

$$d = 3.57\left(\sqrt{h_t} + \sqrt{h_r}\right) \text{ km} \tag{8.3}$$

Here, h_t, h_r are in metres.

In fact, that wave does not travel in a straight line PQR, but along a curved path along the surface of earth. This is because refractive index of atmosphere, which is greater than unity, is near the surface of earth and keeps on reducing towards unity where air density approaches to zero. Thus, radio wave bends towards earth (Figs. 8.4 and 8.6). As a result, the actual distance travelled by the wave along the earth is and is $\sqrt{4/3}$ times the d. Therefore, effective LOS distance also called radio horizon or distance horizon is:

$$d_{\text{eff}} = \sqrt{4/3} \cdot d = 4.12\left(\sqrt{h_t} + \sqrt{h_r}\right) \text{ km}$$
$$= (d_t + d_r) \tag{8.4}$$

(b) **By refraction and diffraction**: Ground wave propagation in troposphere beyond the LOS takes place due to continued refraction (Fig. 8.7a) along the tropospheric layer. This depends on the variation of refractive index with height (dN/dH), in the tropospheric layer at heights 0–16 km. Refractive index and dielectric constant of air are greater than units near the earth surface and becomes 1 at greater height, where air is very thin. In addition to refraction, the diffraction property of bending around obstacles like hills, high-rise buildings also helps in this propagation. This curve path of propagation has radius of curvature of R = 25,000 km, while the radius of curvature of earth is only 6370 km. Various forms of refraction are given in Table 8.3.

Average refractive index of air is taken as 1.000315. As its change with heights is very small, therefore for computing the radius of curvature (R) of the wave, we define a term refractivity (N) as:

$$N = (n - 1) \times 10^6$$

where

n = refractive index of air

e.g. for n = 1.000315; $N \approx 315$

n in terms of air pressure (P), temperature (T), and vapour pressure of water vapour (p_w) can be proved to be:

$$N = (n - 1) \times 10^6 = \frac{77.6}{T}(P + 4810 p_w/T) \tag{8.5}$$

Here, T is in K, while P, p in m bar (1 mm Hg = 1.3332 m bar). As P, p, T fall with height in troposphere, therefore it can be proved that N varies with height as:

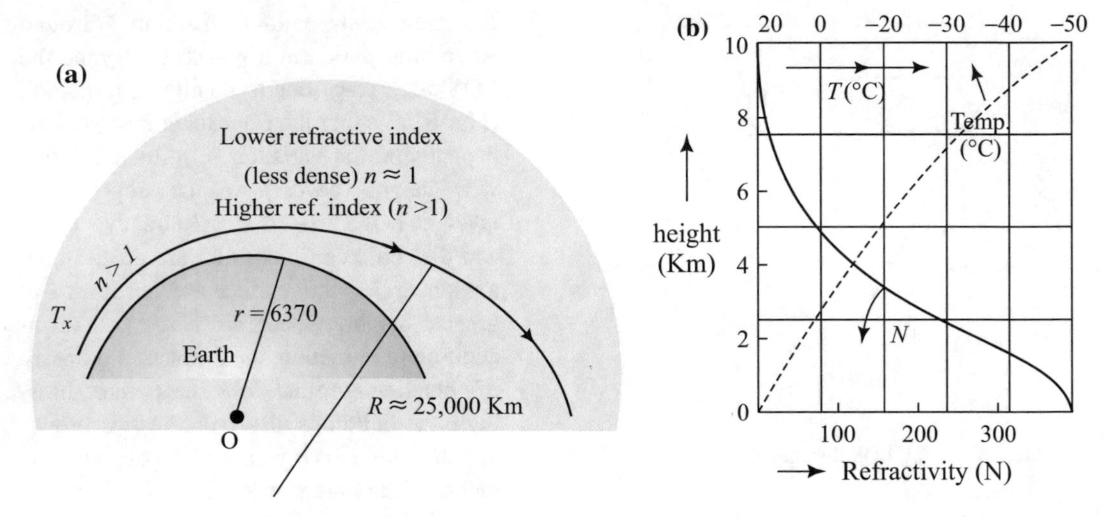

Fig. 8.7 **a** Wave propagation path bending due to reducing refractive index of air with height. **b** Refractively as well as temperature falls with height in troposphere

$$\therefore \frac{dN}{dH} = -4.3 \times 10^{-2} \quad \text{Units of refractivity per meter}$$

$$\therefore \frac{dN}{dH} = -43 \quad \text{N units/km} \tag{8.6}$$

From Fig. 8.8, $R \approx \frac{AB}{d\psi}$
From ΔABC, $AB = \frac{dH}{\cos(\psi + d\psi)} \approx \frac{dH}{\cos\psi}$

$$\therefore R = \frac{dH}{d\psi} \cdot \frac{1}{\cos\psi} \tag{8.7}$$

Apply law of refractive index as:

$$n \cdot \sin\psi = (n + dn)\sin(\psi + d\psi)$$

Expanding RHS and neglecting small terms, we get:

$$n \cdot \sin(\psi) = n \cdot \sin(\psi) + n \cdot \cos(\psi) d\psi + \sin(\psi) \cdot dn$$

$$\therefore \cos(\psi) \cdot d\psi = -\sin\psi \cdot \frac{dn}{n}$$

Putting this in Eq. (8.7), we get by $\sin(\psi) \approx 1$ and $n \approx 1$

$$R = \frac{n}{\sin \cdot \psi\left(-\frac{dn}{dH}\right)} \approx \frac{-dH}{dn} \times 10^6 \text{ m} \tag{8.8}$$

For standard atmosphere $\frac{dn}{dH} \approx -4 \times 10^{-2}$ N-units/metre (average in Fig. 8.6b)

$$R = +25{,}000 \text{ km} \quad \therefore |R| = 25{,}000 \text{ km} \tag{8.9}$$

The −ve sign is because n falls with height, giving propagation path as convex along the surface of earth.

(c) **Duct propagation**: As shown in Fig. 8.6, the value of refractivity N falls with height. Also, the temperature falls very fast in this 0–10 km height of troposphere. This temperature versus N relation can be defined by the empirical relation:

$$N = \frac{77.6}{T}(P + 4810\, p/T) \tag{8.10}$$

Here, T is in K, while P, p_w the air pressure and vapour pressure in m bar.

(1 mm Hg = 1.332 m bar)

Table 8.3 Wave propagation under various forms of refraction depending on d*N*/d*H*

Refraction	dN/dH	Actual path of wave	Types of wave propagation	Environmental conditions
1. Sub-refraction	0–40	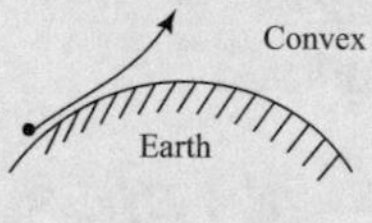 Convex Earth	Length of propagation is very small; by the time, it can be received by any antenna	In moist air over cool surface (e.g. sea) d*T*/d*H* = −ve (i.e. temperature falls with height)
2. No refraction (uniform atmosphere)	0	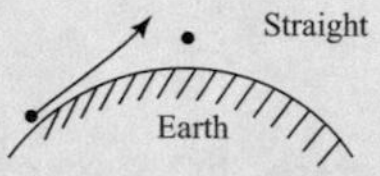 Straight Earth	—do—	Surface and air temperature are same and uniform
3. Standard refraction	−40	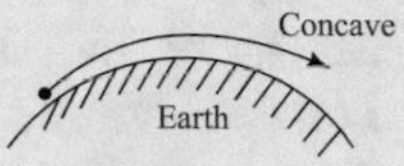 Concave Earth	d*N*/d*H* = −ve causes waves to curve down, which increases the range	–
4. Normal/standard refraction	0–79	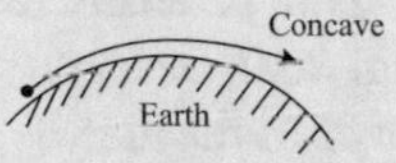 Concave Earth	—do—	–
5. Critical refraction (no ducting)	−79	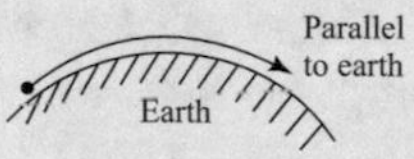 Parallel to earth Earth	• No ducting • Here the wave propagation is concave and it follows the earth's; curvature, resulting into long propagation up to 1000 km	This condition is there at night for some time only
6. Super-refraction (ducting)	−79 to −157	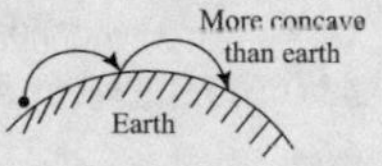 More concave than earth Earth	• Wave bends more • Bending is due to total internal reflection in the troposphere itself • Wave skips large distance due to hops of 50 m height by refraction from earth	• Temperature rises with height at the rate of 6.5 °C/km and finally reaches 50 °C at the top of the duct of 50 m height (where it is dry air). This hot dry air traps the cooler layer (e.g. air above sea gets trapped)
7. Trapping by refraction (ducting)	Above −157	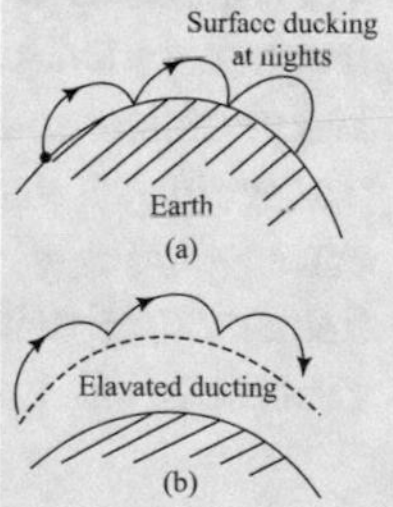 Surface ducking at nights Earth (a) Elavated ducting (b)	• Reach/propagate to large distance up to 1000 km (e.g. Singapore TV signals is noticed in eastern India during monsoon time) • Antenna height has to be less than duct height which is 50 m or so • Signal received by antenna is larger if placed inside the duct • $\lambda_{cutoff} = 0.084\ H^{3/2}$	• Temperature gradient inversion region, i.e. d*T*/d*H* = −ve instead of d*T*/d*H* = +ve

The refractive index *n* of air falls, and hence, N also falls with height as air pressure also falls. Therefore, (d*N*/d*H*) becomes an important parameter to decide the wave propagation.

Table 8.3 summarises the seven forms of refraction, based on the value of d*N*/ d*H*, and the corresponding types of wave propagation. Out of these seven types of conditions, the condition |d*N*/d*H*| > 79 leads to ducted propagation.

Ducted propagation means wave gets super-refracted (Fig. 8.9), i.e. bends more than curvature of earth due to refraction, and finally,

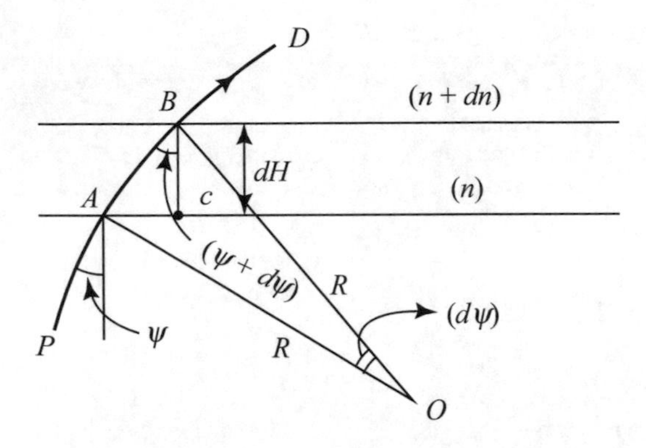

Fig. 8.8 Bending of wave due to change of refractive index n to $(n + dn)$ causes effective radius of curvature of wave propagation = R

total reflection turns it down to the earth, from where it gets reflected back to air. Thus, it hops and travels with hop heights of around h = 50–100 m or so.

Thus, the ducting of wave needs the following three conditions for it to happen:

(i) *Temperature gradient inversion* (dT/dH = +ve): Normally at heights, temperature falls, but for ducting, temperature should increase with height, making upper layer hot and dry.

(ii) $dN/dH = -79$ to -157 (i.e. refractive index reduces very fast with height).

(iii) *Frequency of signal has to be in VHF* range and above.

Modified refractivity (M): For describing the atmospheric property better, the height factor is included in it to represent M as another parameter:

$$M = (N + H/r) \times 10^6 = (n - 1 + H/r) \times 10^6 \tag{8.11}$$

Here, the unit of height H and earth radius r has to be the same. Ducting occurs when $dM/dH = -$ve and $dT/dH = -$ve.

This parameter M versus height is plotted in Fig. 8.10, for explaining the surface ducting and elevated ducting.

Some special points to remember on ducted propagation are:

1. **Surface ducting**: This type of propagation happens mostly at night with modified refractivity (M) having inversion gradient, i.e. $(dM/dH) = -$ve. This $-$ve gradient of M happens over sea at night and disappears in day.
2. **Elevated ducting**: It mostly happens near seacoast, at an elevated layer, where $|dT/dH)| > 79$ and also $dM/dH = -$ve; $dT/dH = -$ve.
3. **Waveguide-type propagation with a cut-off frequency**: Duct propagation is like waves in leaky dielectric-filled waveguide. Just like waveguide, it has a maximum wavelength, correspondingly the lowest allowed critical frequency, in which the duct type of propagation can take place. These two parameters

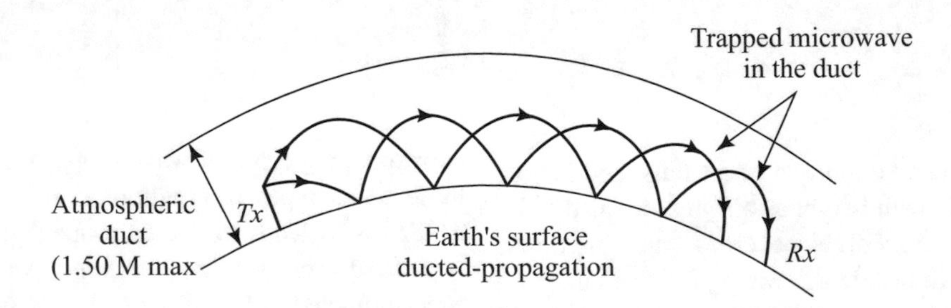

Fig. 8.9 Duct propagation through super-refraction in the atmosphere duct, where $dM/dH = -$ve between the transmitter and receiver

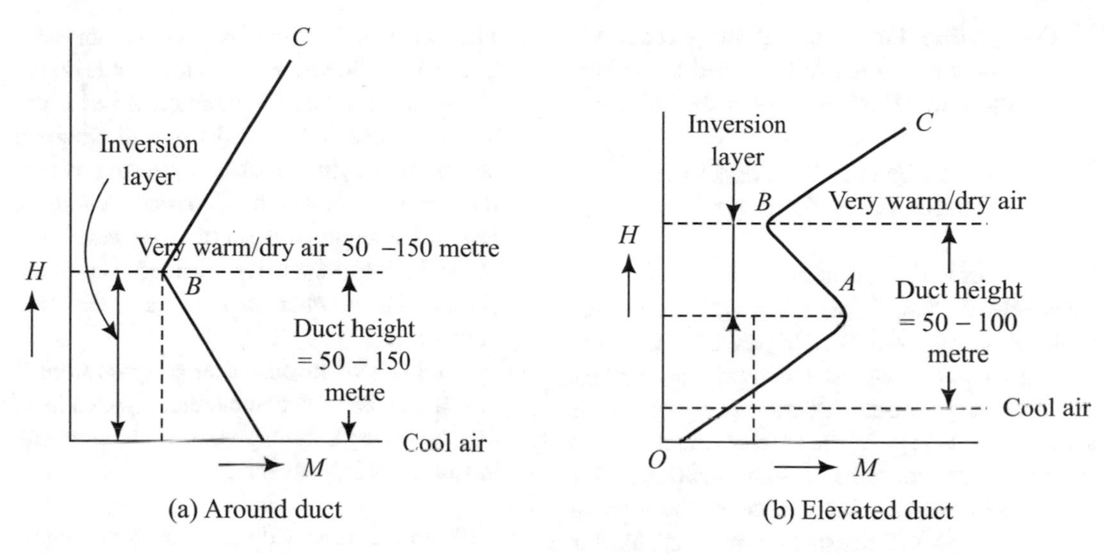

Fig. 8.10 **a** Ground duct d*M*/d*H* = −ve at *AB* region and **b** elevated duct d*M*/d*H* = –ve at *AB* region

(λ_c, f_c) are given by the following with h the duct height in metre:

$$\boxed{\lambda_c = 0.08\ h^{3/2}\ \text{cm}\ , f_c = 357/h^{3/2}\ \text{GHz}} \tag{8.12}$$

For h = 100 m, f_c = 0.357 GHz. Therefore, ducting is possible at UHF and microwave frequencies where $f > f_c$.

4. **Angle of transmission**: For better ducted propagation, the transmitter should send signal which is parallel to the duct within ±0.5° angle.
5. **Temperature inversion**: For ducted propagation, the temperature inversion, i.e. $(dT/dH) = +\text{ve}$, has to take place, while normally, $(dT/dH) = -\text{ve}$ is there in troposphere, see Fig. 8.2.

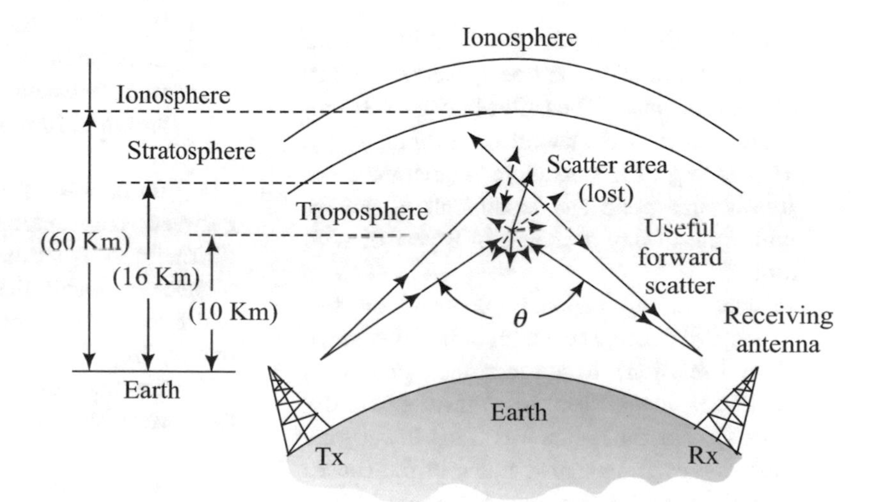

Fig. 8.11 Tropospheric scatter propagation

6. **For getting larger signal amplitude**: For this, the antenna has to be placed at the base of hop, rather than in between the hops.

8.3.2 Space Wave: Tropospheric Scatter Wave Propagation

By tropospheric scattering of UHF (30–300 MHz) band signal, by the gas molecules/irregularities of troposphere, a part of the signal goes towards the receiving antenna, while the major portion of wave energy is lost in space (Fig. 8.11), in undesired directions. In spite of major portion lost, some scattered wave can reach the receiving antenna as far as 1000 km from the transmitter. It is reliable but not cost-effective because of fading of signal.

Some of the features of troposcatter from troposphere (16–60 km height—Fig. 8.11) which need attention are:

1. **Field strength (*E*) falls drastically with distance (*d*)** as per the following relation:

$$E = k/d^{7-8} \tag{8.13}$$

Here,

$k =$ proportionally constant

Roughly, the field strength falls by around 0.3 dB/km or so in the free space value up to 200 km distance. Thereafter beyond 200 km, it falls faster at the rate of 0.8 dB/km.

This fading of troposcattered signal does not follow any trend and is difficult to predict with time of day or month of a year or with attitude.

In fact, fading definitely depends on frequency, distance, and length of hop (i.e. angle of transmission). In summer, the signal level is 10 dB lower than in winter. Similarly, midday (hot time) signal is 5 dB lower than morning–evening signal. Some of the causes of fading are absorption in medium, skips distance, polarisation, fading of down coming signal, etc.

2. **Phenomenon of troposcatter** is not very clear so far. However, it is felt that it is due to (a) inhomogeneous discontinuities and irregular refractive index or (b) due to stratified layers of varying thickness causing partial reflection. The amount of loss/attenuation is less, if the angle between the two waves, i.e. transmitted and the wave received, is as large (Fig. 8.11), i.e. when angle of transmission is very small.
3. **Advantages of troposcatter propagation**: It has a number of commercial applications, which outweigh the high cost and losses due to the following advantages:

   (i) For the same distance of communication, it requires lesser number (one-third to one-tenth) of repeaters stations, and hence, less staff required than with radio link LOS system.
   (ii) It can cross the territories of another political administration/country without the possibility of the signal being caught in between by spies. Therefore, it is very much useful for (a) multichannel tactical military field environment even for linking small distances of 50–350 km and (b) small-channel (16 Kbps) military links up to 1500 km.
   (iii) Provides reliable multichannel communication across large stretch of water/between islands/inaccessible terrain/hills/mountains, etc.

4. **The dominant mode of propagation between troposcatter and ground wave diffraction/refraction in propagation**: If the distance is small than we have to decide which to choose, based on the losses in two modes, troposcatter mode or ground wave diffraction/refraction mode. Properties of these waves are as:

   (i) The distance (*d*) at which troposcatter and diffraction/refraction losses are approximately equal depends mainly on frequency of operation as:

$$d_0 = 65\,(100/f)^{1/3}\ \text{km} \qquad (8.14)$$

where f = frequency of signal in MHz.
For $d < d_{0,}$ diffraction mode is dominant.
For $d > d_{0,}$ troposcatter mode is dominant.

(ii) For path having angular distance of 20 m radius or more, the path may be operating in troposcatter mode and diffraction mode may be neglected.

5. **Best frequency bands for propagation in troposcatter**/diffraction mode have been found with least loss/absorption as given below:
350–450 MHz
755–985 MHz
1700–2400 MHz
4400–5000 MHz.
6. **Disadvantages of troposcatter/diffraction**:

(i) Troposcatter system generally uses transmitter of 1–10 kW power with parabolic antenna having very large diameters of 4.5 m or 9 m or 18 m, increasing the cost.
(ii) The receiver has to be broadband FM receiver with front-end noise figure of 1–4 dB.
(iii) Troposcatter system needs much bigger investment than LOS microwave system.
(iv) Troposcatter path has much larger loss than radio link path.

8.3.3 Sky Wave Propagation in Ionosphere—Critical Frequency, Skip Distance, Etc.

As discussed in Sect. 8.2, above 50 km height is the ionosphere, which consists of ions, produced by ultraviolet rays, α-, β-, γ-ray as well as cosmic rays and meteors. The four sub-layers of ionospheres are D, E, F_1, and F_2 (Fig. 8.1) with different characteristics (Table 8.2). The ionisation temperature of these layers changes over the day, over the year, with sunspot cycle of 11 years, etc., as discussed in Sect. 8.2.

When an EM wave enters these layers, then it bends, due to gradual refraction, and then finally, total internal reflection takes place (Figs. 8.4 and 8.12), just like an optical ray. This is due to the fact that ion density of ionosphere keeps on increasing as we go deeper into space. The refractive index and the dielectric constant (ε_r) are function of ion density (N/cc). Besides these two parameters, there are some more parameters, which also depend on the value of ion density (N/cc), and these are:

- Minimum critical frequency (f_c), below which there is no reflection or refraction and the signal goes to space.
- Maximum usable frequency (f_{max}), lowest usable frequency, and optimum usable frequency.

We will discuss these parameters and derive expressions for them.

1. **Dielectric constant using currents (J_e, J_D) calculation**: Let the EM wave be represented by its electric fields:

$$E = E_m \sin(\omega t)\ \text{V/m} \qquad (8.15)$$

When this wave propagates through the ionosphere, then free electron (of N/m^3 density) experiences an electrostatic force (F-Newtons) due to this ac electric field, which will make it to move with an acceleration (dv/dt):

$$\therefore F = eE = m(dv/dt) \qquad (8.16)$$

where

e = electron charge (coulomb)
m = electron mass (kg)
v = velocity of electron (m/s)

Integrating Eq. (8.16) with respect to time, we get:

$$-e\int E\cdot \mathrm{d}t = \int m\cdot \frac{\mathrm{d}v}{\mathrm{d}t}\cdot \mathrm{d}t = m\int \mathrm{d}v$$

Putting Eq. (8.15) in above equation, we get:

$$-e\cdot E_m\int \sin(\omega t)\mathrm{d}t = m\int \mathrm{d}v$$

$$\therefore -eE_m\left(\frac{-\cos(\omega t)}{\omega}\right) = mv + K$$

If we set the constant of integration $k = 0$ then we get

$$mv = \frac{e\cdot E_m\cdot \cos(\omega t)}{\omega} \tag{8.17}$$

We can use this velocity expression for computing the electron current due to it as:

$$\begin{aligned} j_e &= -N\cdot e\cdot v = -N\cdot e\left(\frac{eE_m\cos(\omega t)}{m\omega}\right) \\ &= -\frac{N\cdot e^2}{m\omega}\cdot E_m\cdot \cos(\omega t) \end{aligned} \tag{8.18}$$

This shows that J_e the conduction current leads the electric field E by 90°.

Now because of varying electric field, the ionosphere will also have displacement currents (J_D) as:

$$J_D = \frac{\mathrm{d}D}{\mathrm{d}t} = \left(\frac{\mathrm{d}(\varepsilon_0\cdot E)}{\mathrm{d}t}\right) = \frac{\varepsilon_0\mathrm{d}(E_m\sin(\omega t))}{\mathrm{d}t}$$

$$\begin{aligned} \therefore J_D &= \varepsilon_0\cdot E_m\cdot \omega\cdot \cos(\omega t) \\ &= (E_m\cdot \omega)\varepsilon_0\cos(\omega t) \end{aligned} \tag{8.19}$$

This J_D leads electric field E by 90°. Total current in the ionised media is:

$$\begin{aligned} \therefore J_{\text{total}} &= (J_e + J_D) \\ &= E_m\cdot \omega\cdot \cos\omega t\left[\varepsilon_0 - Ne^2/m\omega^2\right] \end{aligned} \tag{8.20}$$

$$\begin{aligned} \therefore J_{\text{total}} = {} & E_m\cdot \omega \\ & \cdot \varepsilon_0\left[1 - Ne^2/\omega^2 m\varepsilon_0\right].\cos\omega t \end{aligned} \tag{8.21}$$

Comparing the RHS of (8.19) and (8.21), it can be said that the presence of displacement type of current changes the ε_0 to $\varepsilon_0[1 - Ne^2/\omega^2 m\varepsilon_0)$.

This is equivalent to saying that the new dielectric constant is:

$$\varepsilon = \varepsilon_0\left[1 - Ne^2/\left(\omega^2\cdot m\cdot \varepsilon_0\right)\right] \tag{8.22}$$

∴ Relative permittivity, i.e. dielectric constant, is less than one:

$$\varepsilon_\mathrm{r} = \frac{\varepsilon}{\varepsilon_0} = \left[1 - Ne^2/\left(\omega^2\cdot m\cdot \varepsilon_0\right)\right] \tag{8.23}$$

By putting $e = 1.6\times 10^{-19}$ c, $m = 9.1\ 10^{-31}$ kg, $\varepsilon_0 = 8.854\times 10^{-12}$ Fd/m we get

$$\boxed{\varepsilon_\mathrm{r} = \left(1 - 81\,N/f^2\right)} \tag{8.24}$$

where

$f =$ frequency of EM wave in c/s
$N =$ free electron density in number/m^3

Note: Here we have taken: (a) the number of free electron (N) = No. of ions (b) Ion in ionosphere being heavy, do not respond to the electric field [$E = E_m$. sin (ωt) do not contribute to the current J_D.

2. **Refractive index of medium**: It is well known that refractive index (n) changes with height (Fig. 8.12) and is related to dielectric constant as:

$$\frac{\sin(i)}{\sin(r)} = n = \sqrt{\varepsilon_r} \quad (8.25)$$

$$\therefore n = \left(1 - 81\ N/f^2\right)^{1/2} = \ <1 \quad (8.26)$$

3. **Critical frequency (f_c) (lower frequency limit)**: For reflection to occur, n has to be real, which is true after $f > f_c$ whereas f_c is given by $n = 0$ in Eq. (8.26) and angle $i = 0$ in Eq. (8.26)

$$\therefore f_{\min} = f_c = \sqrt{81N} = 9\sqrt{N} \quad (8.27)$$

Thus, f_c is the minimum frequency which can get reflected back by ionosphere with J_{Total} and angle i each = 0.

4. **Maximum usable frequency (upper frequency limit)**: By Eqs. (8.26) and (7.26):

$$n = \frac{\sin(i)}{\sin(r)} = \left(1 - 81N/f_{\max}^2\right) \quad (8.28)$$

For angle $r = 90°$ (maximum value), $f_{\max}$ computed from Eq. (8.28).

$$\therefore \frac{\sin(i)}{\sin(90)} = \left(1 - 81N/f_{\max}^2\right)^{1/2}$$

$$\therefore \frac{\sin(i)}{1} = \left(1 - 81N/f_{\max}^2\right)$$

$$\left(81N/f_{\max}^2\right) = 1 - \sin^2(i) = \cos^2(i)$$

∴ Using Eq. (8.27), i.e. $f_c = \sqrt{81N}$, above equation becomes:

$$\left(\frac{f_c^2}{f_{\max}^2}\right) = \cos^2(i) \quad (8.29)$$

$$\therefore \boxed{f_{\max} = f_c \cdot \sec(i) = 9\sqrt{N}\ \sec(i)}$$

This equation is called secant law, which relates the maximum usable frequency ($f_{\max}$) with the critical frequency f_c (i.e. lowest frequency which can get reflected from ionosphere). f_c is a function of N only, while $f_{\max}$ is a function of N and the incident angle i of the transmitted signal, therefore this angle 'i' can never reach 90°, even if the transmitter sends a signal grazing the earth, as clear from Fig. 8.12. From ΔOAB, where H = height of ionosphere, $i_{\max}$ is given by:

$$\sin\left(i_{\max}\right) = r/(r+H) \quad (8.30)$$

$$\therefore f_{\max} = 9\sqrt{N}\ \sec(i_{\max})$$
$$= 9\sqrt{N} \cdot (r+H)\Big/\sqrt{H^2 + 2rH} \quad (8.31)$$

As different, ionosphere layers have different ionic densities and height, $f_c = 9\sqrt{N}$, $i_{\max}$ as well as $f_{\max}$ will be different, e.g.

Thus, we can see from maximum (in day) and minimum (at night) ionic densities of these layers that it is not possible to reflect frequencies above 50 MHz and therefore cannot be used by ionosphere reflection propagation. Normally, E-layer reflection from its height of 100 km is used in communication.

5. **Virtual height of reflected wave**: As can be noted from Fig. 8.12, the height of reflection point of waves (after multiple refractions and finally total reflection) is different than virtual reflection point (after assuming linear propagation). Virtual height is measured by an instrument known as ionosonde. This instrument sends a 150 μ sec pulse-modulated radio wave, and the reflected back time (T_r) from ionosphere is noted. Then, the virtual height (with c = velocity of light = 3×10^8 m/s) is given by

$$H_v = \frac{cT_r}{2}\ \text{km}$$

Different frequencies will get reflected from different heights.

Layer	Height (km)	Ion density (N)	$f_c = 9\sqrt{N}$ (MHz)	$i_{\max}$	$f_{\max} = f_c\ \sec(I_{\max})$ (MHz)
E-layer	110	$6 \times 10^{10}/\text{m}^3$	2.2	79	11.5
F-layer	250	$10^{12}/\text{m}^3$	9	74	32.6

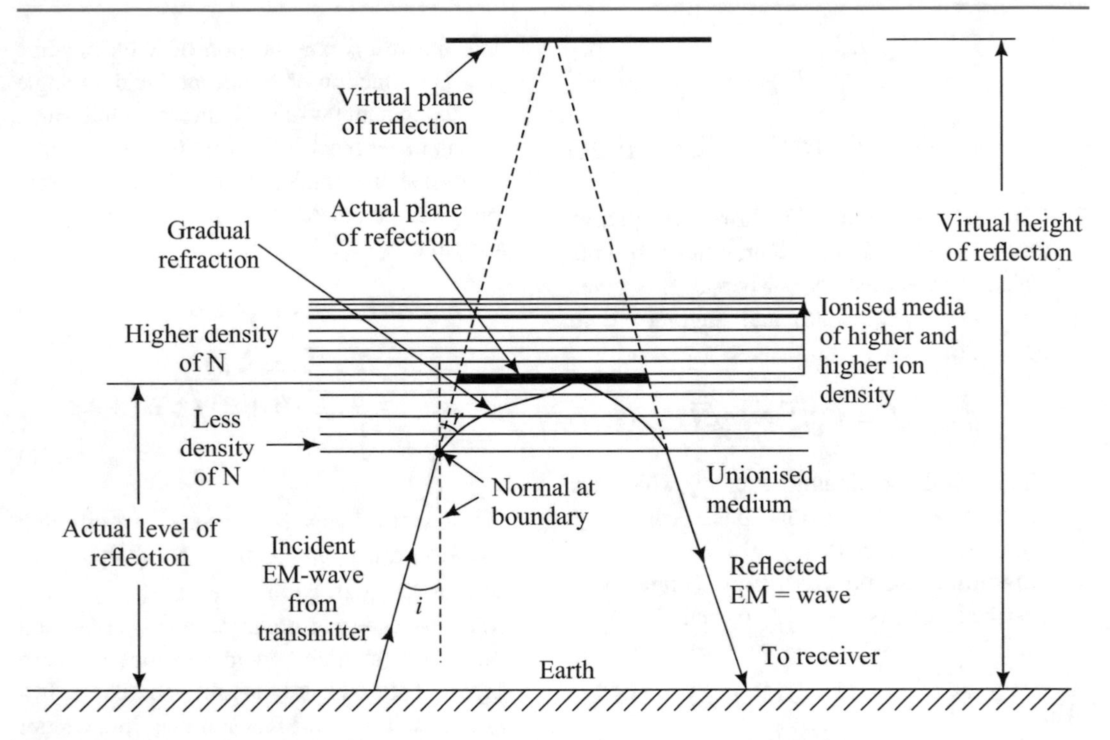

Fig. 8.12 Gradual refraction—bending and finally total reflection without much loss of energy

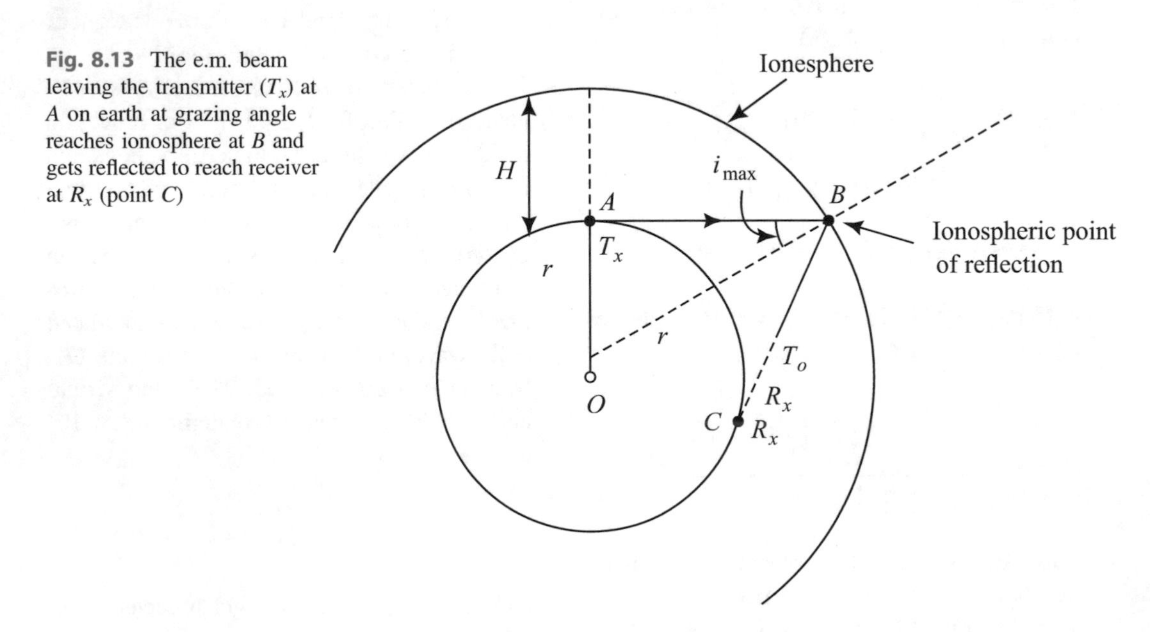

Fig. 8.13 The e.m. beam leaving the transmitter (T_x) at A on earth at grazing angle reaches ionosphere at B and gets reflected to reach receiver at R_x (point C)

6. **Skip distance**: It is taken as the shortest distance between transmitter position and the point where the signal is reflected back on earth as in Fig. 8.14. The region between is called skip zone, where signal is not there. As we reduce the angle (i), the reflected point is closer and closer, with smaller skip distance till 'b' beam. Beyond angle of beam 'b', it gets reflected from higher layer (beam 'c') and beyond that it disappears in space (beam 'd'). At the same time, we can note that at night, these ionised layers go to higher level

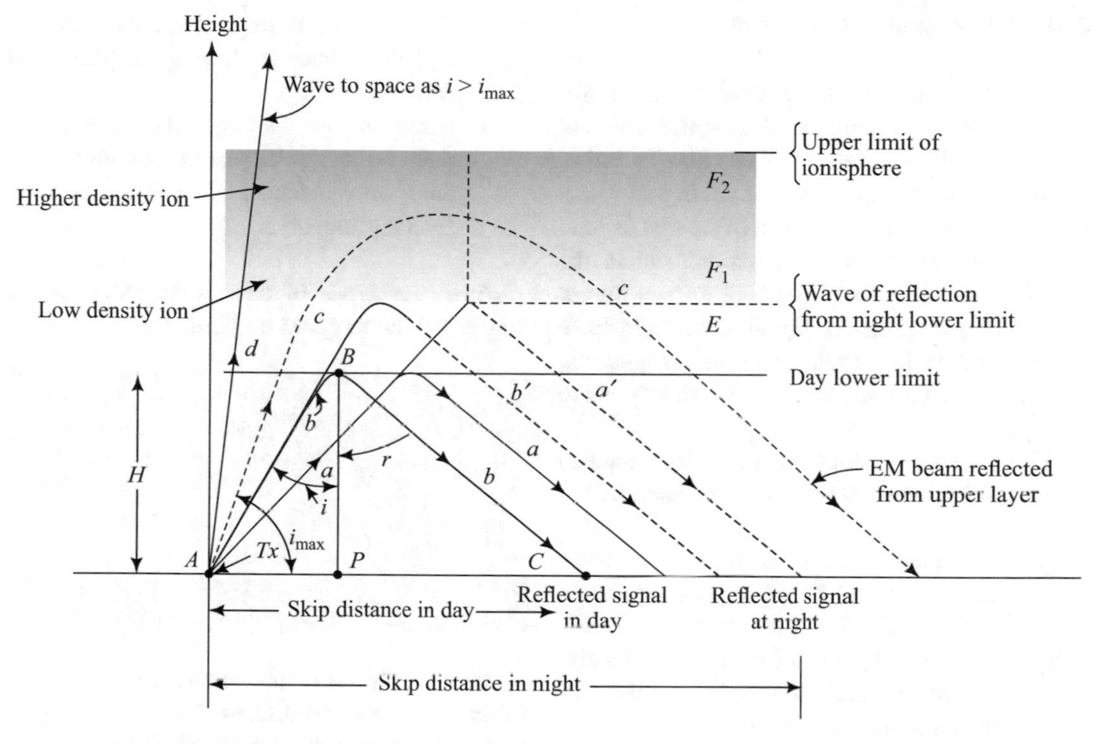

Fig. 8.14 Skip distance with **i** different angles of transmission and **ii** effect of day and night on propagation by reflection from E- and F-layers

and skip distance increases. Maximum skip distance is possible with the incident angle as to be $i = i_{max}$ (Fig. 8.14). This distance is around 4000 km, as compared to the circumference of earth $2\pi r$ = 40,000 km.

7. **Relation in f_{max} on skip distance (d_{skip}): From** Fig. 8.14

$$\cos(i) = \frac{BP}{AB} = \frac{H}{\sqrt{H^2 + (d_{skip}/2)^2}}$$

Also, $\cos(i) = \frac{f_c}{f_{max}}$ from Eq. (8.30).

∴ Equating the two RHS terms, we get:

$$\frac{f_c}{f_{max}} = \frac{H}{\sqrt{H^2 + (d_{skip}/2)^2}}$$

$$\therefore d_{skip} = 2H \cdot f_{max}^2/f_c^2 - 1 \quad (8.32)$$

This is with the condition of earth taken as flat. If earth curvature is also considered with radius = r, then above equation gets modified to a transcendental equation in d_{skip}:

$$d_{skip} = \left(2H + \frac{d_{skip}^2}{8r}\right) \cdot \sqrt{f_{max}^2/f_c^2 - 1} \quad (8.33)$$

d_{skip} has the maximum value of 4000 km with transmitter signal grazing with earth and getting reflected from highest layer (i.e. F_2) as in Figs. 8.13 and 8.14.

In all the above activities, we note that for transmitting and receiving microwave/VHF signals, direction of transmission is very important and therefore microwave antennas have to be highly directive. We will now study various types and properties of antenna.

8.4 Microwave Antennas

Microwave antenna in general is used for two-way point-to-point communication and not used for all directional broadcasts like TV/radio. Therefore, the antennas have to be highly directional, so that the angle of transmission is also controllable, as we saw the requirement in the previous articles. Also, it is well known that at higher frequencies, the signal generators have lower efficiencies (e.g. klystron, magnetron, IMPATTs); therefore, the gain of antenna has to be high.

Thus, the six reasons for microwave antenna to be highly directive can be summarised as:

(i) No broadcasting is done at microwave frequencies and therefore no need for an omnidirectional antenna.
(ii) To offset the effect of noise at the receiving end, the high power/high gain/ directional antenna is needed.
(iii) As the power and efficiency of microwave sources reduce with frequencies, high gain/directional antenna is required.
(iv) Microwave antenna is used to communicate with satellites or used in RADARs, which is very much directional.
(v) If the power transmitted is with a very small beam angle and highly directional, then very less power is lost, as with distance, wave front becomes bigger and bigger, thereby reducing the power density, i.e. power/m^2 (see Fig. 2.1a).
(vi) The microwave antenna has to act both as receiver and as transmitter, requiring high directivity.

Moreover at microwave frequencies, smaller antennas are adequate to have the desired (l/λ) ratio. Being small, it can be taken as a point source to emit spherical waves, which becomes plane wave front as it moves away. This plane wave front moves out in conical expanding shape, and it is preferred for the directional microwave communication (see Fig. 2.1a).

Generally, antenna could be categorised into the following three:

(i) Wire antenna having conduction current.
(ii) Aperture antenna having displacement current.
(iii) Array antenna having radiating elements with conduction/displacement current. The microwave antennas fall into the aperture antenna category.

As far as types of microwave antennas are concerned, these could be listed as:

1.	Dipole antenna	2.	Loop antenna
3.	Slot antenna	4.	Horn antenna
5.	Yagi array antenna	6.	Reflector antenna
7.	Lens antenna	8.	Microstripline antenna (patch or slot antenna)

Before we discuss the various types/structures of antennas, we will discuss some of the important parameters/properties of antenna.

8.4.1 Important Properties of Antenna

Following terms are used very often in the studies of antenna, which acts as receiver as well as transmitter:

(i) Radiation pattern: The power radiation pattern of a transmitting antenna with power from generation via waveguide is given and explained in Fig. 8.15.
The formation of lobes, i.e. wave from dipole antenna and its propagation, is shown in Fig. 8.16.
The vertical dipole gives vertically plane polarised wave, while horizontally placed dipole produces horizontally plane polarised wave (Fig. 2.1a) At the same time, the power density falls as we go away from axial (Fig. 8.17) wire. If the dipole size is increased from $l = \lambda/2$ to 2λ, then the radiation pattern and lobes change as in Fig. 8.18.

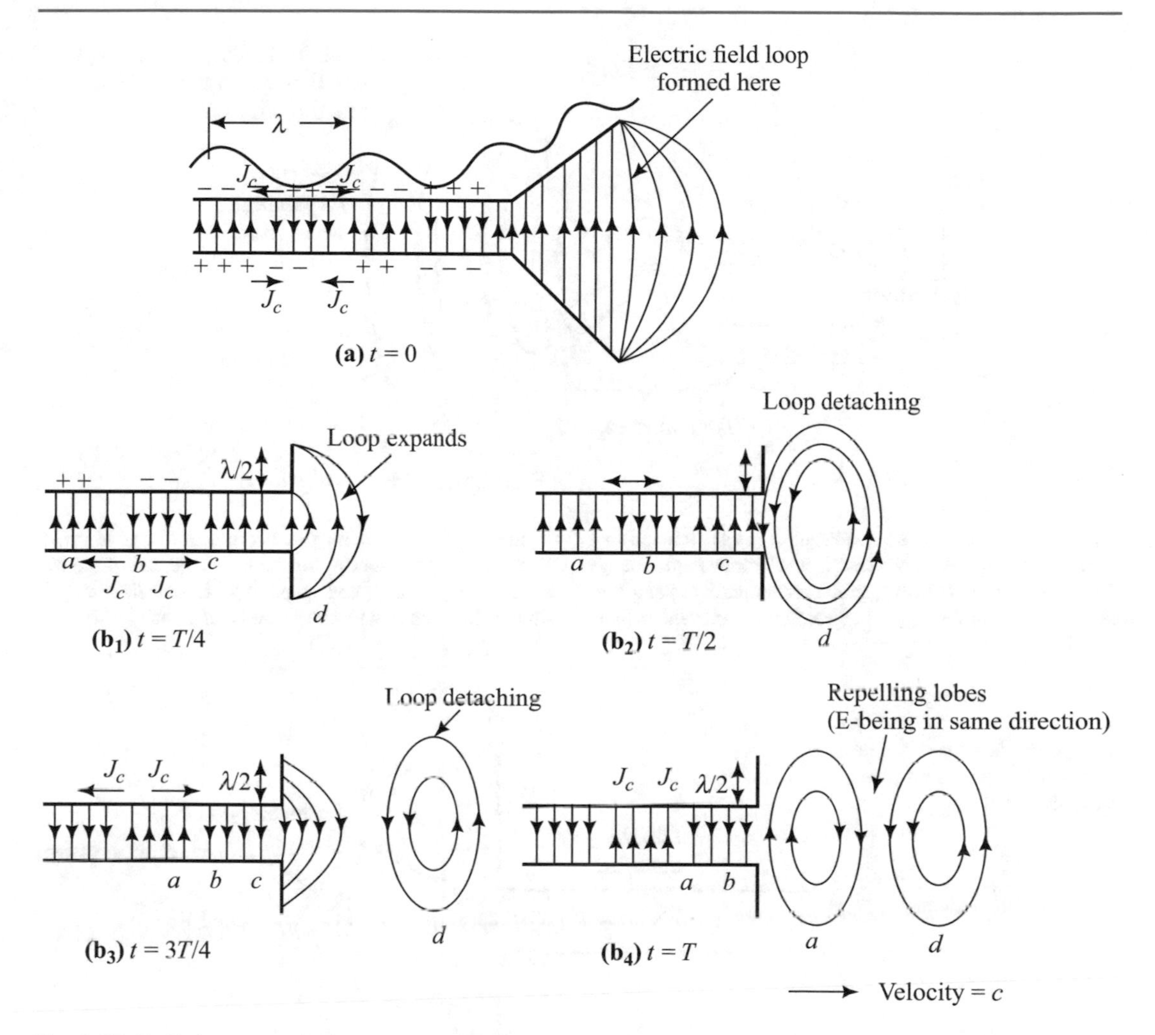

Fig. 8.15 Radiation pattern in free space wave by an antenna with directions of electric field, which will reverse after every half time period of signal. In this process, expanding and repelling E-loops get emitted into space with velocity c

(ii) *Power density at receiving antenna*: This is power per unit area received by the dish antenna, when it acts as receiver, and gets collected at the focal point (the feed point) and finally reaches the amplifier.

(iii) *Directivity of transmitting antenna*: It is defined as the ratio of radiation intensity of antenna in the direction of maximum radiation to the average radiation intensity of the antenna. If $U(\theta, \phi)$ is the radiation intensity in the direction (θ, ϕ), then the directivity D with the solid angle ($\mathrm{d}\Omega = \sin(\theta)\, \mathrm{d}\theta \cdot \mathrm{d}\phi$) will be:

$$D = \frac{4\pi U_{\max}(\theta, \phi)}{\int U(\theta, \phi)\mathrm{d}\Omega} \tag{8.34}$$

(iv) *Antenna gain*: It is defined as the ratio of maximum radiation intensity of an antenna to that from an omnidirectional (isotropic) antenna, with same power input. Thus, it is a measure of directive character of an antenna.

(v) *Impedance*: It is a complex quantity and is the ratio of voltage to the current across the antenna terminal, under no load condition.

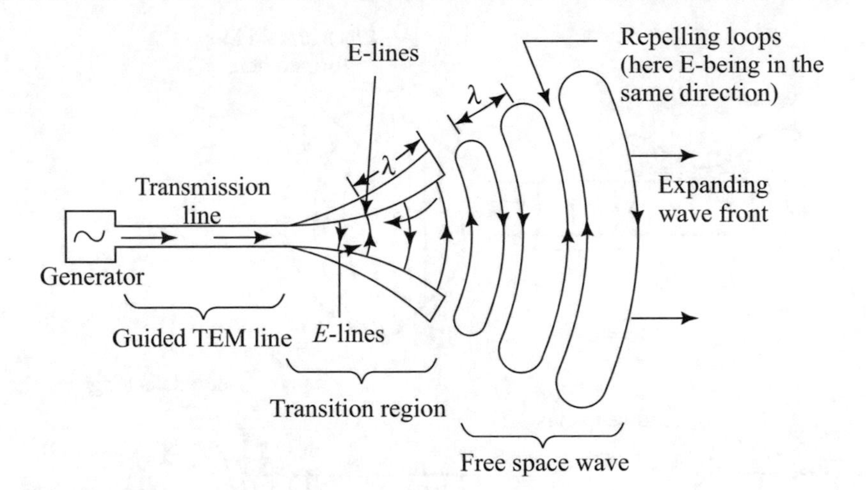

Fig. 8.16 A flared up waveguide (wg) transmitting waves into space. The transient electric field change direction energy T/2 time, with corresponding changes as its two ends on the wg. Thus conduction current is there on the wg, while switching of charges from +ve to -ve & vice versa. In the process, lobes form at the end of the wg, thereby expanding and repelling E-lobes/loops gets emitted into space with the velocity of light

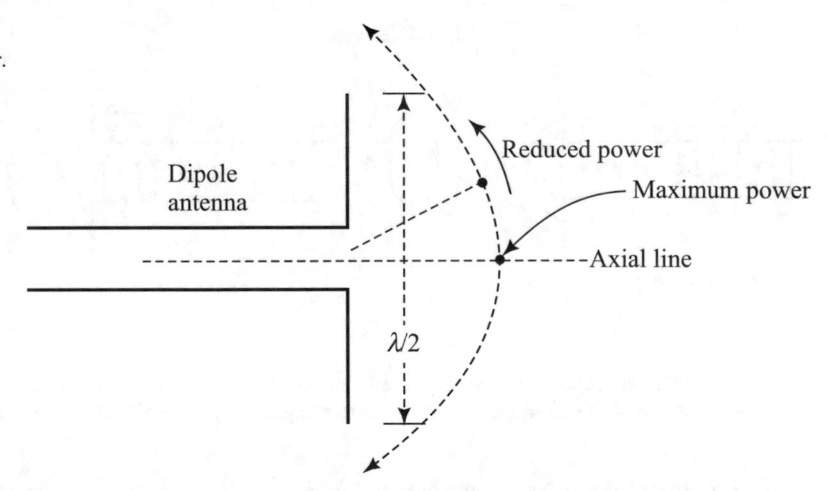

Fig. 8.17 A $\lambda/2$ dipole antenna acting as transmitter. The power density and field intensity fall as we go away from the axial line

(vi) *Effective aperture area* A_r *of receiving antenna:* This is for the receiving antenna and is the effective area A_r through which it can receive maximum of incident power which is given by:

$$A_r = \left(\lambda^2 g_r/4\pi\right)\ \text{m}^2 \qquad (8.35)$$

Here

$g_r =$ gain of the receiving antenna.

If the transmitting antenna is omnidirectional and transmits power P_t from generator, then power density P_d at a spherical distance R will be

$$P_d = P_t \cdot g_t/\left(4\pi R^2\right)\ \text{W/m}^2 \qquad (8.36)$$

Hence

$g_t =$ gain of transmitting antenna.

Then, the power received by the receiving antennas will be:

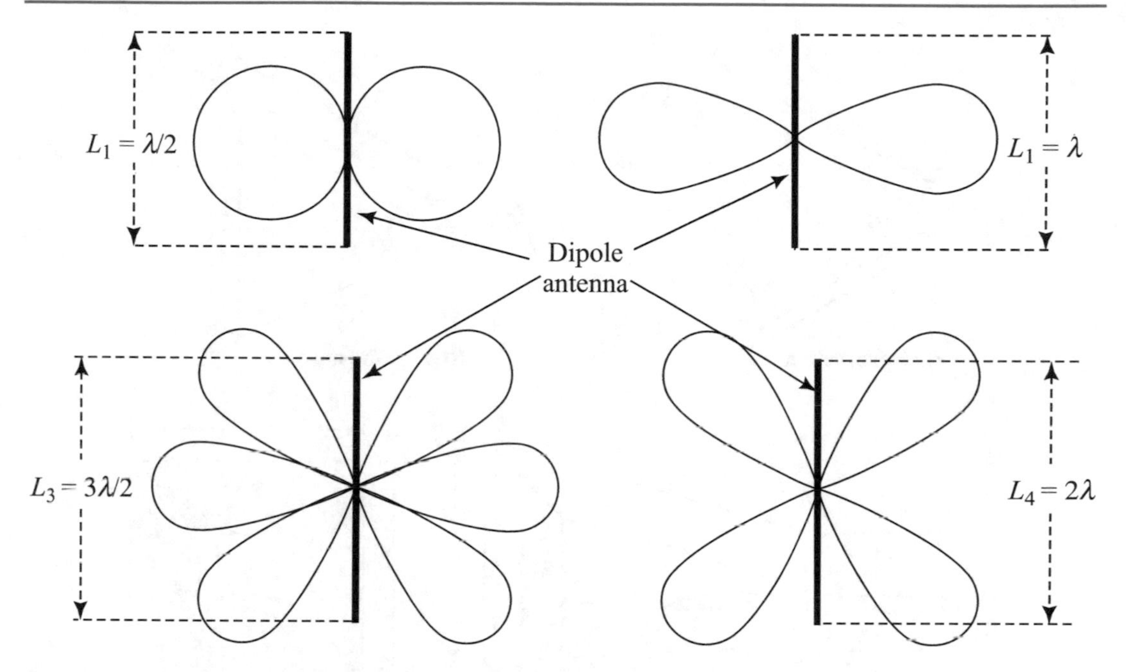

Fig. 8.18 Radiation pattern for different lengths of the dipole antenna

$$P_r = P_d \cdot A_r = \left(\frac{P_r \cdot g_t}{4\pi R^2}\right) \cdot \left(\frac{\lambda^2 g_r}{4\pi}\right) \text{ W}$$
$$P_r = P_t g_t \cdot g_r \left(\frac{\lambda}{4\pi R}\right)^2 \text{ W} \tag{8.37}$$

Now, we will discuss various types of antennas.

8.4.2 Horn Antenna

There is a variety of horn antennas, depending upon the direction of flare up, as well as its style, i.e. linear flared up or exponential flared up, as well as its style, e.g.

(a) H-plane flared up, H-plane horn (Fig. 8.19a).
(b) E-plane flared up, E-plane horn (Fig. 8.19b).
(c) E- and H-planes flared up pyramidal horn (Fig. 8.19c).
(d) Exponentially tapered pyramidal, E–H flared up (Fig. 8.19d).
(e) Exponentially tapered conical (front circular) (Fig. 8.19e).
(f) Linearly tapered elliptic front (Fig. 8.19f).

The flaring helps in:

(i) Matching of waveguide impedance with that of the free space impedance.
(ii) Providing greater directivity.
(iii) Smaller beam angle.
(iv) Smaller SWR, i.e. maximum energy radiation.

From Fig. 8.19g, $\cos\theta = \frac{L}{L+\delta}$; $\tan\theta = \frac{h/2}{L} = \frac{h}{2L}$

$$\therefore \theta = \tan^{-1}\left(\frac{h}{2L}\right) = \cos^{-1}\frac{L}{L+\delta} \tag{8.38}$$

Again from right-angled ΔOBC, we can write

$$(L+\delta)^2 = L^2 + (h/2)^2$$
$$\therefore L^2 + \delta^2 + 2L\delta = L^2 + h^2/4$$

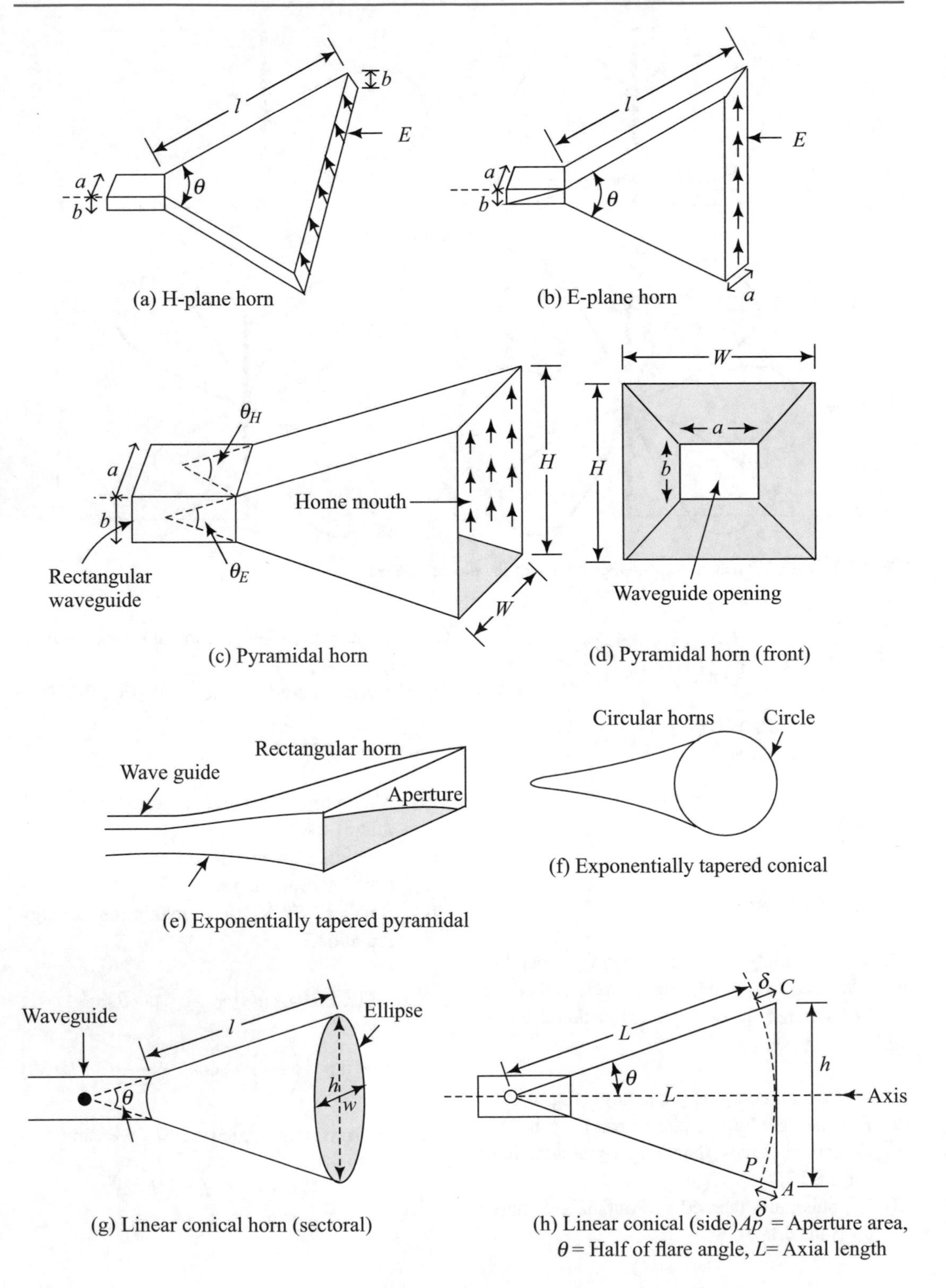

Fig. 8.19 Various types of horn antenna depending on the direction and style of flare of the front

Therefore, neglecting δ^2 being very small:

$$L = h^2/(8\delta) \quad (8.39)$$

These two Eqs. (8.38) and (8.39) are used as design equation of horn antenna. Following additional formulae are also used.

Beam width in degrees in E and H directions in pyramidal horn antenna (Fig. 8.19c)

$$\theta_E = 56\,\lambda/h \quad \text{and} \quad \theta_H = 67\,\lambda/W \quad (8.40)$$

Directivity

$$D = 7.5A/\lambda \quad (8.41)$$

Here

A = h.w. of mouth = Area
λ = wavelength of signal
h = height of mouth
here w = width of mouth

Power gain:

$$G_p = 4.5\,A/\lambda^2 \quad (8.42)$$

The directivity of parabolic antennas is better than horn antenna, as it:

(a) gives higher gain than 120 dB of horn antenna.
(b) size is smaller for the same gain requirement.

8.4.3 Paraboloidal Dish and Rectangular Aperture Antenna

One of the biggest advantage of a parabolic antenna is the fact that plane wave reaching it can converge to a point (focus) and also the reverse, i.e. a point source as transmitter at focus feed will make plane wave after reflection (see Fig. 8.20). We know from coordinate geometry that parabola is a curve formed by locus of a point, which moves so that sum of its distances from focus and a line (called directrix) is constant. The 3D version of this curve makes dish antenna as paraboloid.

$$\therefore \text{Path length } OBB' = OCC' = \text{etc. a constant} \quad (8.43)$$

(a) *Dish as transmitter*: In fact when it acts as transmitter, then the signal from focus spreads out as spherical wave front, which when reflected from paraboloid becomes plane wave, with good directivity and small beam angle. As a transmitter, signal from focus forms major lobe which goes to the dish antenna, but some minor lobes are also formed, which do not reach the dish antenna are lost. Therefore for minimising this loss, a metal shield is put (Figs. 8.20 and 8.21).
(b) *Dish as receiver*: Similarly when this antenna acts as receiver, the plane wave (i.e. parallel beam) converges to focus after

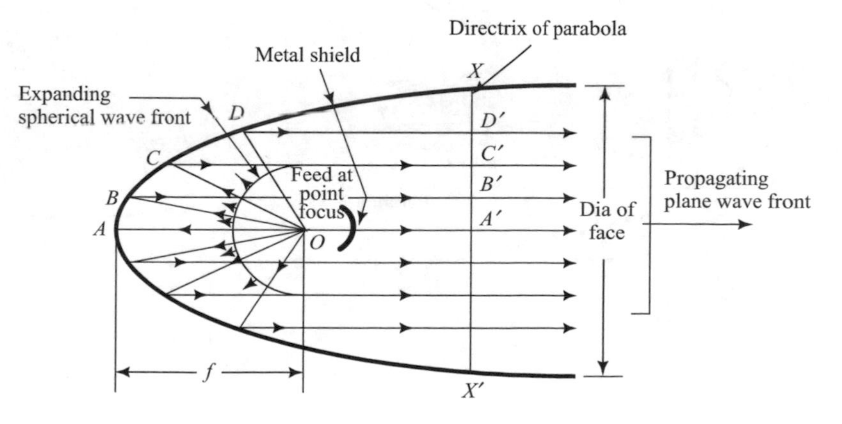

Fig. 8.20 Dish as transmitter. Due to parabolic reflector geometry, the spherical waves get reflected to become plane wave. The signal feed is at the focus of the parabola

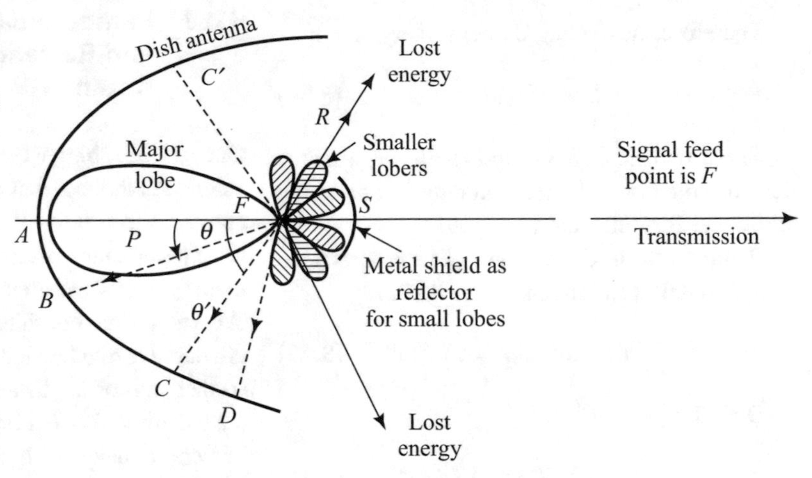

Fig. 8.21 As transmitter the reflections from lobes. From lobe 'R', the energy is lost, while from lobe 'S', the shield reflector takes care by reflecting back to dish for transmission along the main line

reflection as in Fig. 8.19. Some important parameters of the dish antenna, e.g. power gain, beam width, focal length diameter ratio, and field intensity around central axis, are as follows:

1. **Power gain**: Power gain of a parabolic dish antenna using Eq. (8.34) is given by:

$$G_{\mathrm{p}} = \frac{4\pi A_0}{\lambda^2} = \frac{4\pi \cdot KA}{\lambda^2} \tag{8.44}$$

where

A_0 = capture area, A = actual mouth area = $\pi D^{\,2}$

K = constant of efficiency depends on antenna feed type (for dipole feed $K = 0.65$)

D = diameter of the front of antenna

$$\therefore G_{\mathrm{p}} = \left(\frac{4\pi K}{\lambda^2}\right) \cdot \left(\frac{\pi D^2}{2}\right) \simeq 6(D/\lambda)^2 \tag{8.45}$$

This shows that power gain is a function of aperture ratio $(D/\lambda)^2$, and for $\lambda = 3$ cm (10 GHz) and $D = 1$ m, $G_{\mathrm{p}} \approx 6 \cdot \left(\frac{100}{3}\right)^2 \approx 6650$. The effective power transmitted by the antenna is the

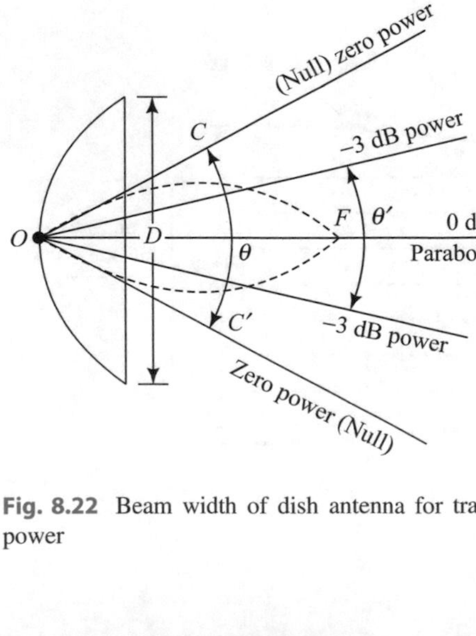

Fig. 8.22 Beam width of dish antenna for transmitting power

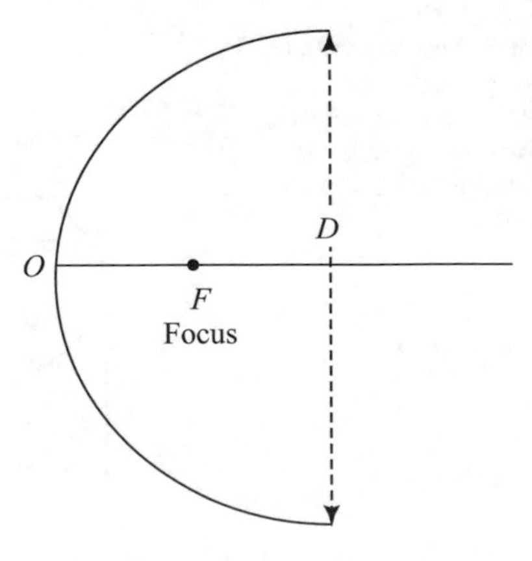

Fig. 8.23 Small FO/D ratio gives higher gain

product of power fed to the antenna and power gain, i.e. $P_{eff} = G_p\, P_{source}$. Power is maximum at the axes of the parabolic antenna and falls to zero at angle of $\pm\, \theta/2$ both sides around axe (Fig. 8.22).
If the $P_{source} \approx 1$ W, then P_{eff} = over 6 kW.

2. **Ratio of focal length to antenna aperture**: (i.e. FO/D in Fig. 8.23) It is an important characteristic, and its value varies between 0.25 and 0.5. This decides how much the focus is inside the dish antenna region. It has been found that $FO/D = 0.25$ gives maximum gain (Fig. 8.23).
3. **Field distribution around central axis**: As clear from Fig. 8.22, the dish antenna receives different intensities of field, as we move away from A to B to C to D. At C, it is nearly zero being no lobe point and again increases a bit at D. This fact has been given explicitly in Fig. 8.21 along with effect of obstacle (dipole/horn feed).
4. **Beam width** is an important parameter. Beam width between first nulls (BWFN) is the angle CC' as in Fig. 8.24 and angle $C'FC$ in Figs. 8.21 and 8.22. Value of BWFN in degree is given by:

$$\theta_r = \text{BWFN} = 115\,\lambda/D (\text{Rectangular antenna}) \tag{8.46}$$

$$\theta_c = \text{BWFN} = 140\,\lambda/D \text{ (Circular dish antenna)} \tag{8.47}$$

where

D = diameter of front aperture of antenna
L = length of the aperture of rectangular antenna.

Thus, circular dish antenna makes sharper beam than rectangular antenna.

5. **Band width**: Half power band width (HPBW) is the angle (degree) between the half power (3 dB) points on either side of axis. Also, it is clear from Fig. 8.24 and Table 8.4 that normally:

$$\text{HPBW} < \text{BWFN}/2$$

6. **Directivity = 4π Area/λ²** is given by:

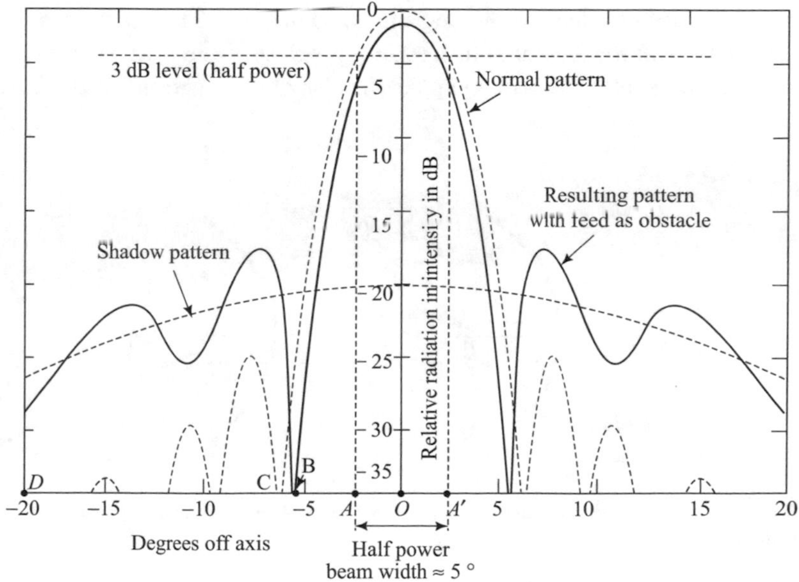

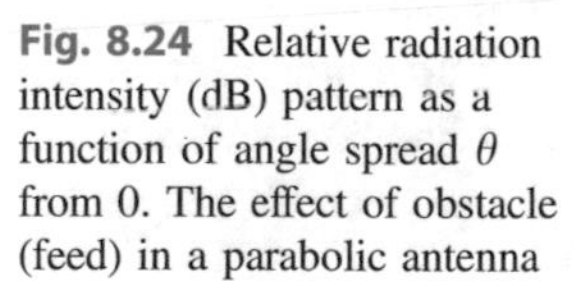
Fig. 8.24 Relative radiation intensity (dB) pattern as a function of angle spread θ from 0. The effect of obstacle (feed) in a parabolic antenna

Table 8.4 Different parameters for circular and rectangular apertures (antenna)

Parameter	Circular aperture	Square aperture of side L
Half power band width [HPBW (θ)]	58 (λ/D) degrees	51 (λ/L) degrees
Band width between first nulls: BWFN	140 (λ/D) degrees	115 (λ/L) degrees
Directivity	9.87 $(D/\lambda)^2$	12.6 $(L/\lambda)^2$
Gain	6 $(D/\lambda)^2$	7.7 $(L/\lambda)^2$ (square) or 4.5 A/λ^2 (rectangular)

$$D_r = 4\pi\ (l \times b/\lambda^2)(\text{Rectanglar aperture}) \quad (8.48)$$

$$D_c = 4\pi\ (\pi D^2/(4\lambda^2)) = 9.97\ (D/\lambda)^2(\text{Circular aperture}) \quad (8.49)$$

Thus, the circular dish antenna is preferred both from manufacturing and performance point of view.
Here, we note that at higher frequencies (e.g. at microwave frequencies), gain is higher and beam widths (θ_r, θ_c) are smaller, with higher directivity (D_r, D_c). Therefore, higher frequencies are preferred in RADAR antennas.

8.4.4 Feeds for Paraboloidal Dish Antenna

An ideal case of a feed (signal transmitter to the dish antenna at focal point of dish antenna) is the one, which radiates 100% of energy to the dish or receives 100% energy from dish, when acting as receiver.

There are different types of feeds, each of which has their own advantage:

1. **Waveguide horn feed**: Horn is the most common feed via waveguide (Figs. 8.25 and 8.26) as front feeding horn or rear side feeding horn.
2. **Dipole feed with reflector element**: A dipole is fed through coaxial line along with a reflector element for reflecting back the signal radiated by the dipole on the other side of dish antenna (Fig. 8.25a).
3. **Cassegrain feed antenna**: This antenna system is itself called cassegrain antenna (Fig. 8.26a, b, c), in which the feed is from the rear, into the vertex of main reflector, signal of which is reflected back to main reflector by a small hyperbolic sub-reflector. The two reflectors are placed such that their focal point is coplaced. The real focal point of the hyperbola is to be at the vertex of the main parabola.
The noise in a cassegrain antenna is much less, as there is no lossy line between receiver and the feed. There is aperture blockage due

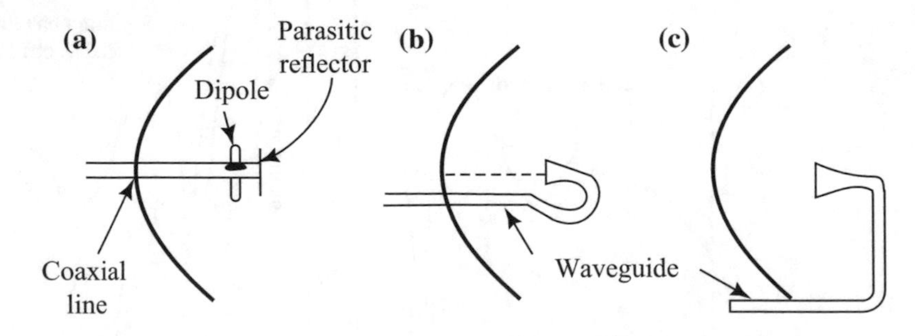

Fig. 8.25 Feeds of dish antenna: **a** rear feed via coaxial line using half-wave dipole and parasitic reflector, **b** rear feed using horn, and **c** front feed using horn

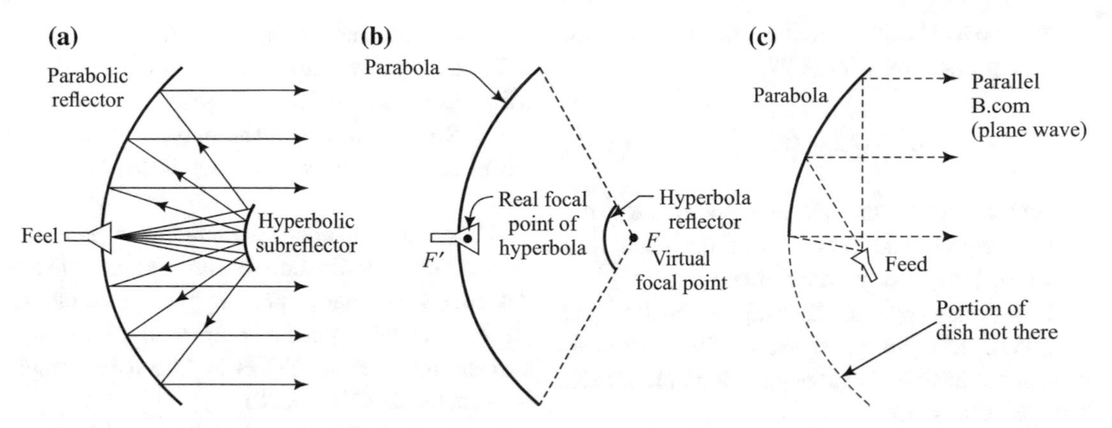

Fig. 8.26 **a** Cassegrain feed antenna showing the hyperbolic sub-reflector, with horn feed at the vertex (centre) of the main parabolic reflector antenna. **b** The geometrical position of horn and sub-reflector in cassegrain antenna. **c** Parabolic part (a section) reflector with offset feed

to sub-reflector, which can be reduced by having smaller sub-reflector, so that its blockage becomes equal to that of horn feed.

4. **Offset feed**: Normally, the signal is fed from the central axis of the parabolic dish; as a result, it obstructs the reflection of the main strong lobe of horn itself (see Fig. 8.26a, b). Therefore, this can be avoided/reduced by having an offset horn feed as in Fig. 8.26c, keeping the VSWR also low.

   Although the offset feed eliminates aperture blockage and mismatch of rear and front feeds, it introduces problem of its own. Its focal length divided by diameter (FID) ratio is greater than conventional parabola, thereby reducing the gain drastically.

$$\text{gain} \propto (F/D)^2 \tag{8.50}$$

8.4.5 Lens Antenna

For frequencies above 2 GHz, a convex lens like structure of dielectric is a preferred antenna. Here, a source at the focal point of the lens gives parallel beam out of it, just like optical rays. Here, the convex lens is made of polystyrene dielectric.

As the horn feed gives a powerful central lobe with small fringe lobes (which gets lost most of the time), there is no such loss here.

The refractive index 'n' as well as the dielectric constant of the lens is made to reduce

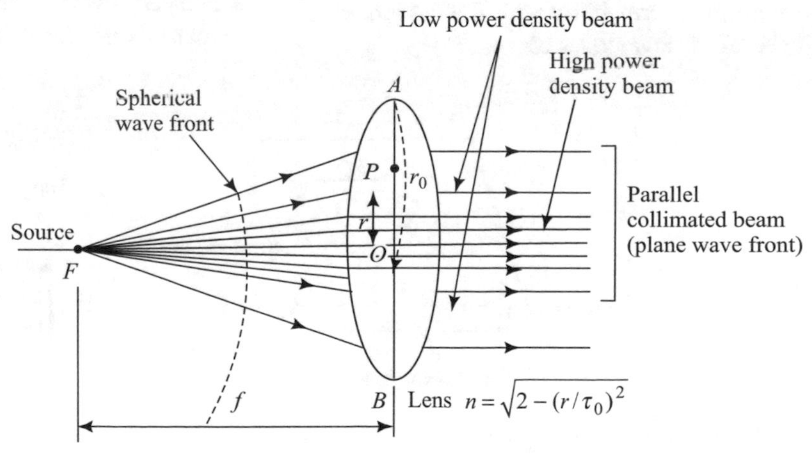

Fig. 8.27 Lens antenna with variable refractive index (n), which is maximum near 0 and minimum near A, B

as we move away from the centre as per the following equation (Fig. 8.27):

$$n = \sqrt{2 - (r/r_0)^2} \tag{8.51}$$

This increases the phase velocity near the central region, and hence from the edge ring region of lens, less power comes out.

This lens antenna is simple, small, light weighted, less lossy, specially for millimeter waves, but has not found application in RADAR. This is because of:

(a) Heat dissipation is a problem at high powers.
(b) Less efficient than reflector dish antennas due to unwanted reflections from its front and rear surface.

8.4.6 Microstrip Line Antenna—Patch Antenna

Microstrip line antennas have gained more importance after planar technology in semiconductor having gained speed, as it can be a part of microstrip line on which planar devices are mounted/soldered. With all the new technologies going for higher and higher frequencies, microstrip; antenna is becomming a promising candidate for applications in microwave and millimetre wave application. The additional reasons for its growing applications are simple design, low cost, ease of manufacture, low weight, conformability with the system, etc., in spite of disadvantages of:

(i) Narrow band width.
(ii) Low power capability.
(iii) Low gain ≈ 20 dB or so.
(iv) Radiates in half plane only.
(v) Loss of energy as surface wave, etc.

They are also called patch antennas, as it is a discontinuity/termination with some patch design, e.g. circular patch, ring patch, rectangular patch, triangular patch, at the end of the top surface line over the dielectric. The patch length is normally $\lambda/2$ (Fig. 8.28).

Because of electric field fringing out of the patch to the ground surface, which changes directions with μW frequency, radiation takes μ of the patch.

These antennas have very low efficiency and behave like cavity rather than a radiator. Wave gets reflected back with little power radiated.

Array of patch antenna is used for improving the directivity and power (see Fig. 8.29). A line of patches can be fed at the central point or at one of the end. Same is possible with a matrix of patches as in Fig. 8.29.

8.4.7 Waveguide Slot Antenna

When energy from a coaxial line is terminated across a slot of a conducting plane, then it becomes a radiating element and hence an antenna. The length of the slot has to be = $\lambda/2$ (see Fig. 8.30).

If one of the walls of the waveguide has a slot, it radiates power. There can be an array of slots to make it directional. These are normally used in

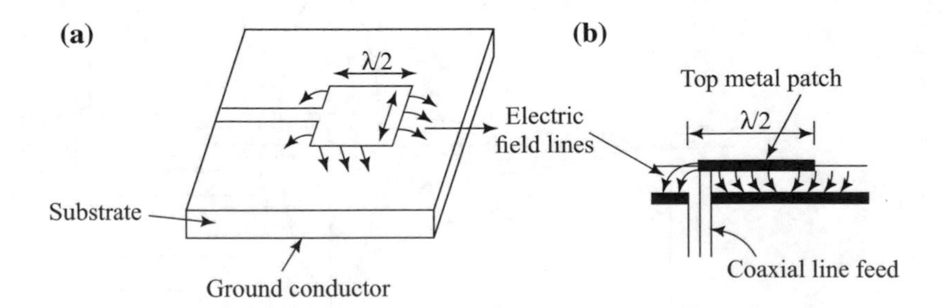

Fig. 8.28 Rectangular patch antennas: **a** signal fed from strip line itself and **b** signal fed from coaxial line

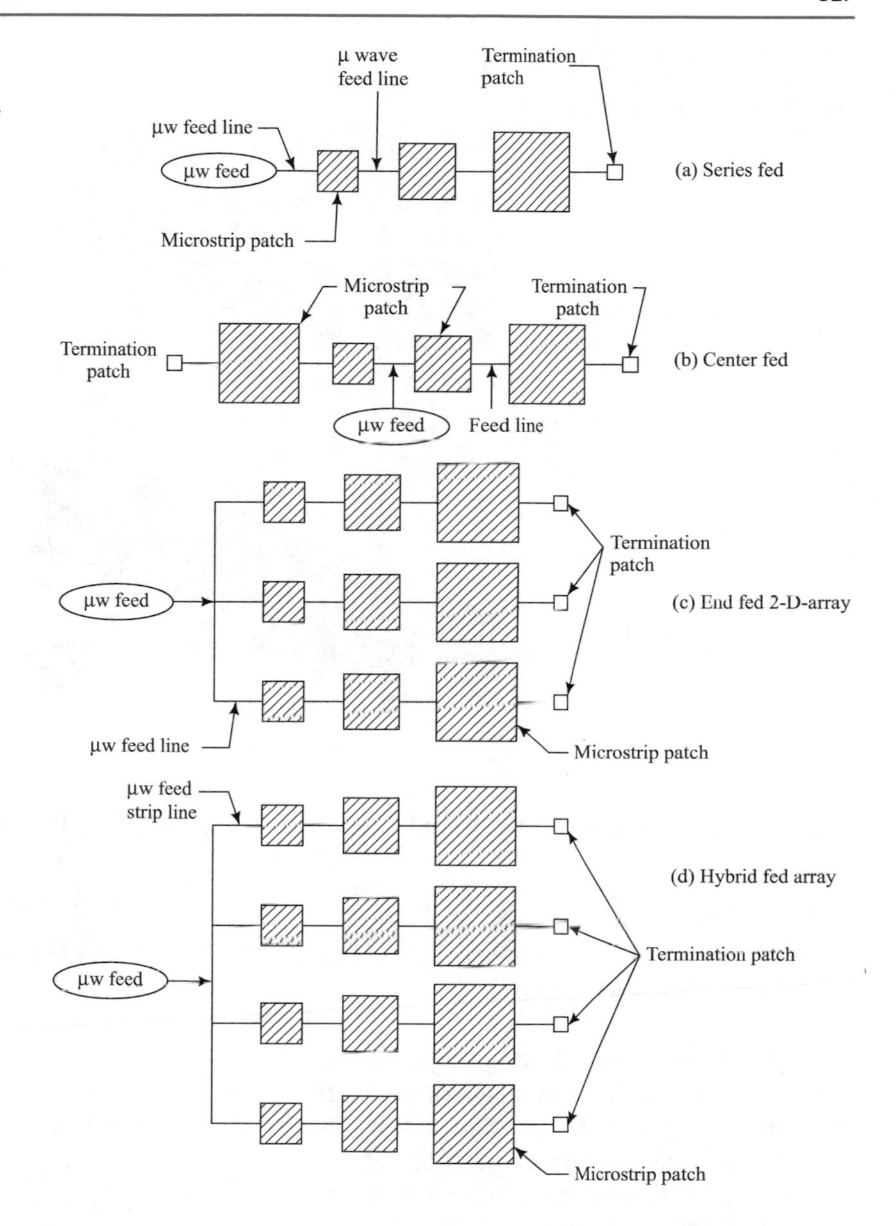

Fig. 8.29 **a** Series feed connection of patches. **b** Centre-fed series array. **c** End-fed shunt-type two-dimensional array. **d** Hybrid-fed array

high-speed aircraft in spite of having very low efficiency.

8.5 Other $\mu\omega$ Communication Systems—Satellite and Mobile

Satellite and mobile communication are also $\mu\omega$ communication systems, but their study is beyond the scope of this book, as each of them has grown very much and becomes a separate independent subjects.

8.6 Solved Problems

Problem 1 A transmitter antenna has 169 m height (being on a hill) and receiver antenna 16 m. What is the maximum distance the signal can be transmitted? What is the distance horizon, i.e. effective LOS of first antenna?

Solution The line-of-sight (LOS) effective distance:

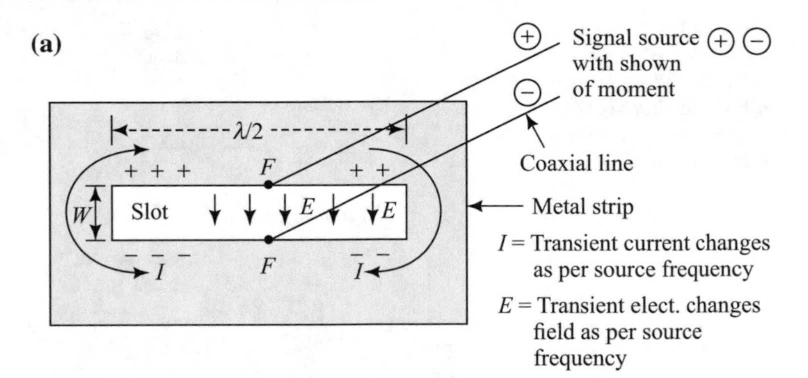

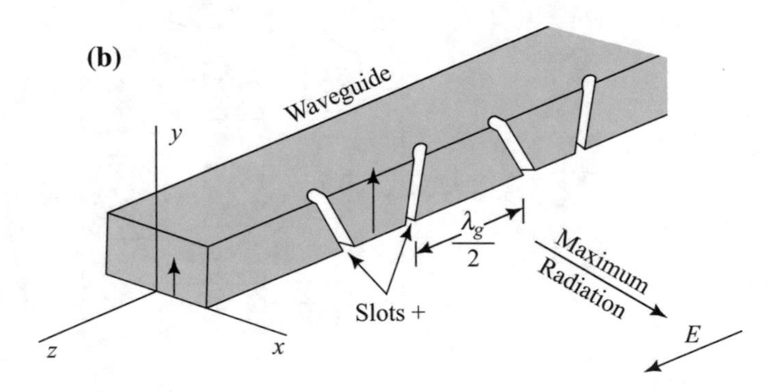

Fig. 8.30 Slot antenna: **a** A λ/2 in a large flat sheet. **b** Broad side array of slots in waveguides

$$d_{\text{eff}} = 4.12\left(\sqrt{h_\text{t}} + \sqrt{h_\text{r}}\right) = 4.12\left(\sqrt{169} + \sqrt{16}\right)$$
$$= 70.04 \text{ km}$$

Radio distance or horizon (d_t) = 4.12 $\sqrt{h_\text{t}} = 4.12 \times 13 = 53.56$ km.

Problem 2 An antenna of 50 m height radiates 10 kW power at 60 MHz, in all directions. (a) If the field strength of the receiver antenna is 1 mV/m, calculate its distance, and (b) find the field strength if the antenna is of 10 m height at 10 km distance.

Solution

(a) Field strength at D-distance if P is the power transmitted is:

$$E_\text{R} = \frac{88\sqrt{P} \cdot h_\text{t} \cdot h_\text{r}}{(\lambda D^2)} \text{ V/m}$$

$$10^{-3} = \frac{88\sqrt{10^4} \cdot 50 \times 10}{(5D^2)}$$
$$(\text{As } \lambda = 5\text{M}, \text{ for } f = 60 \text{ MHz})$$
$$\therefore D = 29.66 \text{ km}$$

(b) At D = 10 km; h_r = 10 m; h_t = 50 m

$$\therefore E_\text{R} = \frac{88\sqrt{10^4} \cdot 50 \times 10}{5 \times (10^4)^2} = 8.8 \times 10^{-3} \text{ V/m}$$

Problem 3 If radio horizon of an antenna is 41.2 km, what is its height?

Solution Radio horizon (d_t) = 4.12 × $\sqrt{h_\text{t}}$ = 41.2

$$\therefore \sqrt{h_t} = \frac{41.2}{4.12} = 10$$

$$\therefore h_t = 100 \text{ m}$$

Problem 4 Two aeroplanes at 4 and 6 km heights can send signal easily to each other up to what maximum distance?

Solution

$$d_{eff} = 4.12\left(\sqrt{h_t} + \sqrt{h_r}\right)$$

$$= 4.12\left(\sqrt{4000} + \sqrt{6000}\right) \text{ km}$$

$$= 4.12\,(63.24 + 77.45)$$

$$= 4.12 \times (140.7) = 579.66 \text{ km}$$

Problem 5 Two stations have LOS communication link with half-wave antenna at 100 km distance. If the transmitter transmits 5 kW power at 300 MHz, what is the maximum power received, if the maximum directive gain of each antenna is 1.64?

Solution

$$P_r = P_t \cdot g_t \cdot g_v \cdot (\lambda/4\pi R)^2$$

For 300 MHz frequency

$$\lambda = c/f = \left(3 \times 10^8 \text{ m/s}\right)/\left(300 \times 10^6\right) = 1 \text{ m}$$

$$P_t = 5 \times 10^3; R = \left(100 \times 10^3\right)$$

$$\therefore P_r = \left(5 \times 10^3\right)(1.64) \cdot \left[1/\left(4 \times 3.14 \times 100 \times 10^3\right)\right]^2 = 85.26 \times 10^{-6} \text{ W}$$

Problem 6 If for an EM wave, reflection takes place at 500 km height, what is the maximum ionic density of ionosphere if the refractive index is 0.8 at 10 MHz. Also find (a) the range for maximum usable frequency (MUF) of 10 MHz and (b) dielectric constant of the layer.

Solution

$$H = 500 \text{ km}, n = 0.8; f_{muf} = 10 \text{ MHz}$$

As $n = \sqrt{(1 - 81\,N/f^2)}$
For

$$f = 10 \text{ MHz} \quad 0.8 = \sqrt{1 - 81 \cdot N/(10 \times 10^6)^2}$$

$$\therefore \text{Ionic density} = N_{max} = 0.44 \times 10^{12}/\text{m}^3 = 44 \times 10^{10}/\text{m}^3$$

$$\therefore f_c = 9\sqrt{N_{max}} = 9 \times \sqrt{44 \times 10^{10}} = 5.96 \text{ MHz}$$

$$\therefore D_{skip} = 2h\sqrt{(f_{max}/f_c)^2 - 1}$$

$$= 2 \times 500\sqrt{(10/5.96)^2 - 1}$$

$$= 1349 \text{ km}$$

$$\text{Dielectric constant } (\varepsilon_r) = n^2 = \varepsilon_r = (0.8)^2 = 0.64$$

Problem 7 If the ionic densities of ionosphere of F_1-, F_2- and E-layers are 2.3×10^6, 3.6×10^6, and 1.7×10^5 electrons/cm^3, then find their cut-off frequencies.

Solution $f_c = 9\sqrt{N_{max}}$ c/s; (Here, N_{max} is in per m^3)

$\therefore$ For F_1-layer $f_c = 9 \times \sqrt{2.3 \times 10^6 \times 10^6} = 9 \times 1.5 \times 10^6$ c/s

$$= 13.7 \text{ MHz}$$

For F_2-layer $f_c = 9 \times \sqrt{3.6 \times 10^6 \times 10^6} = 9 \times 1.9 \times 10^6$ c/s

$$= 17 \text{ MHz}$$

For E-layer $f_c = 9 \times \sqrt{1.7 \times 10^5 \times 10^6} = 9\times\ 0.41 \times 10^6$ c/s

$$= 3.7\ \text{MHz}$$

Problem 8 If the D-layer has ionic density = 500/cc, and refractive index as 0.8, find the frequency of a wave which can propagate by reflection from D-layer. What is the dielectric constant of this D-layer?

Solution n = refractive index = $\sqrt{1 - 81\ N/f^2}$
and N = 500/cc = 500 × $10^6/\text{m}^3$

$$\therefore 0.8 = \sqrt{1 - 81 \times 500 \times 16/f^2}$$

$$\therefore f^2 = 112500$$

$$\therefore f = 335.4\ \text{kHz}$$

Problem 9 What is virtual height and actual height of reflection? If a pulse gets reflected back in 5 ms, find the virtual height.

Solution Virtual height = $C \cdot \Delta t/2$

$$= (3 \times 10^8\ \text{m/s})(5 \times 10^{-3})/2$$

$$= (15/2) \times 10^5\ \text{m} = 750\ \text{km}$$

Problem 10 If a layer has N = 3.24 × $10^4/\text{m}^3$ with refractive index (n) = 0.5, find the frequency of the EM wave that can propagate through this layer?

Solution

$$\therefore n = \sqrt{1 - 81\ \text{N}/f^2}$$

$$\therefore 0.5 = \sqrt{1 - 81 \times 3.24 \times 10^4/f^2}$$

$$\therefore f = 1.87\ \text{kHz}$$

Problem 11 If electron concentration at a height of 300 km is $10^{11}/\text{m}^3$, what will be the maximum angle of incident for 10 MHz frequency signal?

Solution As angle of reflection at total reflection is

$$90^\circ = r$$

$$\therefore \sin(r) = 1$$

$$\therefore n = \frac{\sin(i)}{\sin(r)} = \sqrt{1 - 81\ \text{N}/f^2}$$

$$\therefore \sin(i) = \sqrt{1 - 81 \times 10^{11}/(10 \times 10^6)^2}$$

$$= \sqrt{1 - 8.1 \times 100} = \sqrt{0.92} = 0.959$$

$$\therefore i = 66.9^\circ$$

Problem 12 If a transmitter transmits 100 kW power, find the field strength at a distance of 10 km.

Solution

$$E = \left(300\sqrt{P}/R\right)\ \text{mV/m}$$

where P is kW and R distance in km

$$E = 300 \times \sqrt{100/10} = 300\ \text{mV/m}$$

Problem 13 If front aperture (i.e. diameter) of a dish antenna = 5 m and frequency of signal = 600 MHz, then find the beam width between first nulls (BWFN) on the two sides of axis and half power band width (HPBW).

Solution For frequency of 600 MHz,

$$\lambda = c/f = 0.5\ \text{m} \quad \text{for} \quad f = 600\ \text{MHz}$$

$$\text{BWFN} = 140\lambda/D$$

$$= 140 \times 0.5/5$$

$$= 14^\circ$$

Similarly,
$\text{HDBF} = 58\lambda/D = 58 \times 0.5/5 = 5.8^\circ$

Problem 14 If a square horn antenna has a square aperture of 10 λ of a side, then calculate the power gain.

Solution In square horn antenna, the gain

$$G = 4.5A/\lambda^2 = 4.5 \times 10\lambda \times 10\lambda/\lambda^2$$
$$= 450$$

Problem 15 If a parabolic dish antenna with circular mouth has a gain of 1000 at λ = 10 cm; calculate its diameter and half power band width (HPBW).

Solution Gain of a dish antenna = $G = 6\left(\frac{D}{\lambda}\right)^2$

$$\therefore 1000 = 6 \cdot (D/\lambda)^2$$
$$\therefore D/\lambda = \sqrt{1000/6} = 12.9$$

$$\therefore D = 12.9 \times \lambda = 12.9 \times 0.1 = 1.29 \text{ m } (\lambda = 0.1 \text{ m})$$

$$\therefore \text{HPBW} = 50\lambda/D = 50/12.9 = 3.88^\circ$$

Problem 16 A parabolic dish antenna has $D = 20$ m, frequency = 6 GHz, illumination efficiency of 0.54, then calculate gain in dB.

Solution

$$\therefore \text{gain } (G) = 4\pi \cdot K \cdot A/\lambda^2$$

Hence K = efficiency; A = Area of face

$$\lambda = c/f = 3 \times 10^8/3 \times 10^9 = 0.1 \text{ m}$$

$$\therefore G = 4 \times 3.14 \times 0.54 \times 3.14(10)^2/(0.1)^2$$
$$= 8.52 \times 10^5$$

$$\therefore G(\text{dB}) = 10 \log(G) = 10 \log(8.5 \times 10^5)$$
$$= 10 \log(8.52) + 50 = 59.3$$

Problem 17 If BWFN of a dish antenna is 15° at 1.5 GHz, find the diameter of its face.

Solution

$$\text{At } f = 1.5 \text{ GHz } \lambda = c/f = 3 \times 10^8/1.5 \times 10^9$$
$$= 0.2 \text{ m}$$

$$\therefore \text{BWFN} = 15^\circ = 140\ \lambda/D$$

$$\therefore D = 140\ \lambda/15 = 140 \times 0.2/15 = 1.867 \text{ m}$$

Review Questions

1. Describe different layers above earth as media of propagation of EM waves.
2. How the ions get formed in the ionosphere and what is the highest density normally possible? Are the ionic density and electron density in these layers equal?
3. What is duct propagation and explain why it gets formed?
4. What are various types of feeds in dish antenna? Explain.
5. What are microstrip antennas and where do we use it?
6. Explain with figure the types of horn antennas?
7. What is the speciality of lens antenna?
8. Compare the types of propagation as ground wave, space wave, sky wave, showing the frequency range, ranges of propagation, and reliabilities of each.
9. Calculate the field strength of an EM wave at 10 km if the transmitter transmits 150 kW power [Hint: $E = 300\sqrt{P/d}$].
10. If the two half-wave antennas have directive gains of 1.64 each, the distance between them is 40 km and installed at a height of 20 m each, and the power of transmitter is 100 W at 150 MHz, then calculate the field strength of the field at the receiver antenna.

    Hint: $E_R = \left[88\sqrt{P} \cdot h_r \cdot h_t/(\lambda R)^2\right]$.
11. Calculate the maximum range in a single hop from E-layer on the earth; radius = 6370 km and E-layer height = 140 km.
    [Hint: $d = 2R$. $\theta = 2R \cdot \cos^{-1}$ [$R/(h_m + R)$] = 2645 km].
12. What is the function of ozone layer and where it is situated?

Radar

9

Contents

© Springer Nature Singapore Pte Ltd. 2018

P. K. Chaturvedi, *Microwave, Radar & RF Engineering*,
https://doi.org/10.1007/978-981-10-7965-8_9

9.1 Introduction

Radar, which is acronym of radio detection and ranging, is an electromagnetic system, which radiates energy in the space and detects the echo signal reflected back from an object or target. The frequency used is in VHF/UHF and microwave frequency range with different bands used for different type of radar application areas as given in Table 9.1, while the applications of radars in civilian and military are given in Table 9.2. The K-band and V-bands get absorbed by atmospheric water vapour, as resonance frequency of H_2O molecule lies in this band, and therefore normally not used for radar applications.

The high-power transmitter operating at GHz frequency range produce large reflection, enabling the radar system to detect the angle and distance of the target with good precision in every kind of weather condition, e.g. darkness, haze, fog, rain, snowfall, etc. but cannot recognise colours. Some radars are designed to detect the presence of all targets (with recognition of the type of object), while others (e.g. MIT) can only detect moving objects.

Nowadays, every general purpose radar, whether used for civilian or military application, has both:

(i) Moving target indicator (MTI) display.
(ii) Doppler shift indicator (DSI).

Besides above general purpose radar which is dealt in Sect. 9.7, tracking radar has also been discussed in Sect. 9.8.

9.2 Principle of Radar

Basic principle of radar is quite simple as explained in Fig. 9.1. The reflected back signal from the target reaches the display through the duplexer and receiver amplifier (Normally, a heterodyne receiver). Finally, this signal of receiver reaches display along with transmitting signal. Then comparison of these two signals gives the delay (t_d). As the velocity of EM wave = c, the range[1] or distance of the target at the moment of observation will be:

$$R = c \cdot t_d/2 \quad (9.1)$$

The radar system consists mainly of the following five units:

1. **Transmitter**: It consists of a microwave source (klystron, TWT, or magnetron) along with an amplifier and modulator. The modulated pulse of microwave train of duty cycle <1% is transmitted (for details of modulation pulse, duty cycle, etc., refer to Sect. 1.4). The average power maybe in hundreds of watts only, but the peak pulse power transmitted is normally 10 kW–10 MW.
2. **Duplexer**: This is a switch, which alternatively connects the antenna either with the transmitter or the receiver. Its purpose is also to protect the receiver from high μw power of transmitter. During the time of transmission

[1]**Note**: In radar studies, the range means the distance of the target from radar station, while echo means the signal received by the receiver after reflection from the target.

Table 9.1 Radar frequency bands and types of applications

Band names	Nominal frequency bands	IEEE-standard radar-frequency band	Application in radar
HF	3–30 MHz	–	Coastal radar systems, over the horizontal (OTH) radars
VHF	50–330 MHz	138–144 MHz 216–225 MHz	Very long range, ground penetrating
UHF	300–1000 MHz	420–450 MHz 820–942 MHz	Very long range (e.g. ballistic missiles early warning), ground penetrating foliage penetrating
L	1–2 GHz	1.215–1.4 GHz	Long range air traffic control, and surveillance; 'L' for 'long'
S	2–4 GHz	2.3–2.5 GHz 2.7–3.7 GHz	Terminal air traffic control, long-range weather, marine radar, 'S' for 'short'
C	4–8 GHz	5.25–5.925 GHz	Satellite transponders; a compromise (hence 'C') between X- and S-bands
X	8–12 GHz	8.5–10.78 GHz	Missile guidance, marine radar, weather, medium-resolution mapping and ground surveillance. Named 'X' for keeping secret in WW2
Ku	12–18 GHz	13.4–14.0 GHz 15.7–17.7 GHz	High-resolution mapping satellite altimetery; frequency just under K-band (hence 'u')
K	18–27 GHz	24.5–24.25 GHz	From German Kurz, meaning 'short'; limited use due to absorption by water vapour, so K_u and K_a were used instead for surveillance K-band is used for detecting clouds, by police for detecting speeding motorists. K-band radar guns operate at 24.150 + 0.100 GHz
K_a	27–40 GHz	33.4–36.0 GHz	Mapping, short range, airport surveillance: Photo radar, used to trigger cameras which take pictures of licence plates of cars running red lights, operates at 34.300 + 00.100 GHz. Frequency just above K-band
V	40–75 GHz	59–64 GHz	Very strongly absorbed by the atmosphere therefore used very less in RADAR
W	75–110 GHz	76–81 GHz	Used as a visual sensor for experimental autonomous vehicles, high-resolution meteorological observation, and imaging
mm	110–300 GHz	126–248 GHz	Radar and satellite communication

Table 9.2 Application area of radars

	Civilian application		Military application
1.	Navigational aid on ground, air or sea	1.	Detecting/ranging of enemy target
2.	For detecting height of plane above earth	2.	Aiming guns at aircraft/ships
3.	Instrument landing system (ILS) for landing aircraft under poor visibility, night or adverse weather conditions	3.	Bombing cities, ship, even in cloudy condition
4.	Satellite surveillance by airborne radar	4.	Early warning regarding ship/aircrafts
5.	MTI-for determining speed of moving target (ball speed etc. in sports), guided missile etc.	5.	Searching submarines
6.	Detecting speed of moving vehicles by police	6.	In guided missiles

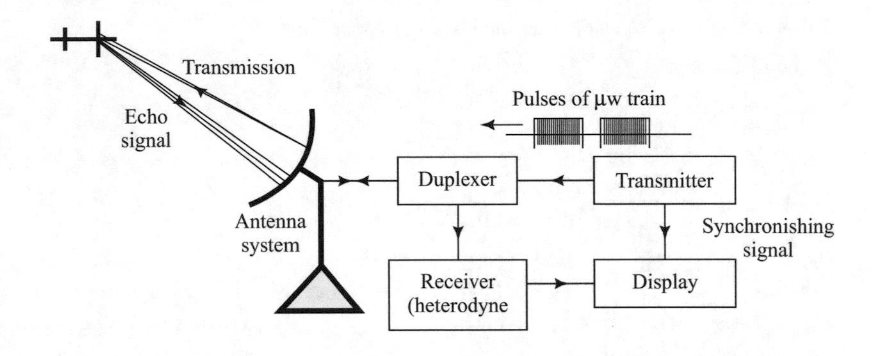

Fig. 9.1 A simple radar system

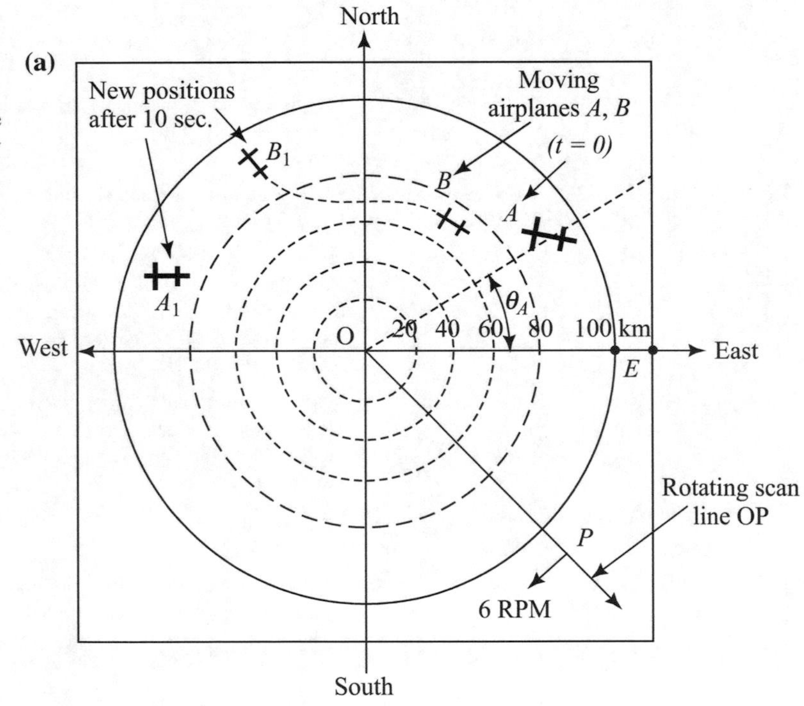

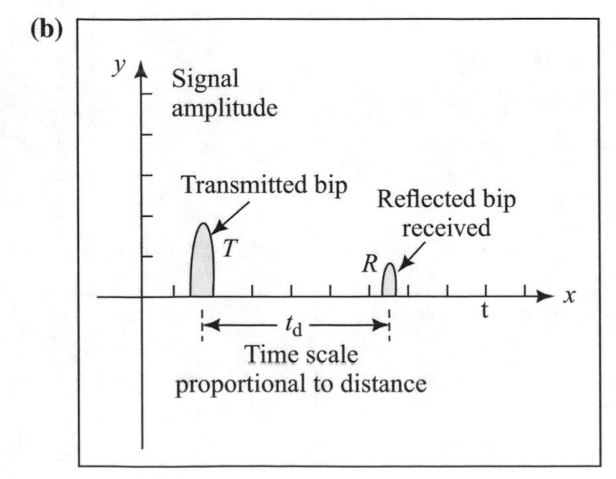

Fig. 9.2 **a** Plan position indicator (PPI) displays on CRO screen. The radar antenna and its scanning line OP rotate clockwise at the rate of 6 RPM. **b** A scope display of reflected pulse (R) along with transmitted pulse (T)

of the outgoing pulse μw train, the duplexer will connect the antenna to the transmitter. When the reflected (echoed) pulse is coming back, then it will connect antenna with the receiver.

3. **Antenna**: As discussed in the previous section, it transmits as well as receives echoed signal. The antenna must send a signal as a well focus small angle beam, so that the position of target is known accurately. Size of antenna has to be few wavelengths for sharper beam. The antenna can be mechanically steered to the desired azimuth and elevation angle.
4. **Receiver**: Normally, a superheterodyne receiver is used, which amplifies the echoed signal and cancels the noise, thereby giving high signal to noise ratio. This is because the carrier is converted into the intermediate frequency (30 or 60 MHz), and balanced mixer is used. Here a delay line is also required for cancelling clutter—the unwanted echoes.
5. **Displays**: Radar display is just like computer monitor screen. These are the following types where they display two of the parameters on the X- and Y-scales, out of four parameters like range, angle of elevation, azimuth angle, and height.

   (a) **Plan position indicator (PPI-Display)**: Here the screen presents the actual position of the target range around the radar station. The scanners cover 360° around the radar by revolving at the rate of 6-revolution per minute normally clockwise. This mechanical scanning of antenna is converted into electrical display by appropriately supplying the signal to the X and Y deflection of a CRT. This in turn gives an appearance of rotation axis at the rate of 6 RPM by a raster scan OP clockwise only (Fig. 9.2 a), that creates same visual effect. The objects A and B at distances of 90 and 70 km respectively get displayed on the screen, in the first scan at $t = 0$. Here the display system indicates both range and azimuth angle (θ_A) of the target (A) in polar coordinates. Azimuth angle is the angle measured in horizontal plane from a reference x-axis OE in Fig. 9.2a. This image of A and B persists on the screen for 15–20 s but gets faded. In the next scan after 10 s of OP, the targets may appear as the new locations A_1 and B_1 on the screen with faded images at A and B. The target illuminates on the phosphor coated screen by the echoed signal (i.e. reflected signal). Second scanning is done in 10 s by the rotating RADAR and shows the new position A_1, B_1 of the planes, while the first images A, B stay up to 15–20 s, but become dull.

   (b) **A-scope display**: It is deflection modulated rectangular display in which the vertical (V) deflection is proportional to the amplitude of receiver output signal and horizontal (X) coordinate is proportional to the time delay t_d or range (Fig. 9.2b) (which is proportional to the distance). The horizontal scale can be calibrated to give the distance between the forward signal bip and reflected signal bip. The term bip is used for the signal pulses on the screen i.e. transmitted & received.

   (c) **B-scope display**: Here, in the display system, the X-scale gives the azimuth (angle) while Y-scale gives the range (km) of the target. This is normally used in airborne military radar, where range and angle both are important.

   (d) **C-scope display**: Here both X- to Y-scales indicate angles. The X-scale indicates azimuth angle, while Y-scale the elevation angle. It is also used in airborne intercept radar and is similar to what a pilot might see when looking through windscreen.

   (e) **E-scope display**: Here X-scale gives the range, Y-scale the elevation angle.

   (f) **RHI-Display**: Here X-scales are for range and Y-scale for height attitude of the target from earth.

The above six types of radar display in a control room or aeroplane are summarised below:

S. no.	Display name	X-scale	Y-scale
1.	Plan position indicator (PPI)	Target distance along X-axis	Target distance along Y-axis
2.	A-scope	Actual range of target	Reflected signal amplitude
3.	B-scope	Azimuth angle (θ)	Actual range of target
4.	C-scope	Azimuth angle (θ)	Angle of elevation
5.	E-scope	Range of target	Angle of elevation
6.	RHI-display	Range of target	Height of target from earth

9.3 The Modulated Signal—Pulse Width Duty Cycle Etc.

The microwave signal used in radar transmitter is square pulse amplitude modulated, thereby it gives repetitive train of microwave in each pulse or we can say it gives busts of microwave power at intervals of the square pulse time period. It has been dealt in details in Sect. 1.4 of Chap. 1. Still with reference to Fig. 9.3, following are repeated.

- Duty cycle = PW/PRT and % duty cycle = (PW/PRT) × 100

$$\text{Pulse repetitive time (PRT)} = T_1 = t_{\text{on}} + t_{\text{off}} = (\text{PW} + \text{RT}) \tag{9.2}$$

$$\text{Pulse retititive frequency (PRF)} = 1/T_1 \tag{9.3}$$

$$\text{Average RF power} = \text{duty cycle} \times \text{peak pulse power.} \tag{9.4}$$

- Rest time RT = t_{off}

If microwave frequency is 1 GHz, modulating square wave frequency is 1 kc/s and if the pulse width is 1 μs, then:

PRF = 1 kc/s, PW = 1 μs; PRT = 1/1000 = 0.001 s = 1000 μs and t_{off} = 1000 − 1 = 999 μs;

$$\text{duty cycle} = \text{PW/PRT} = 10^{-6}/0.001 = 10^{-3}$$

$$\text{duty cycle } \% = 10^{-3} \times 100 = 0.1\%$$

If peak power = 500 kW, then

$$\text{Average power} = 500{,}000 \times 0.001 = 500\,\text{W}$$

In one pulse width, the number of waves of μw frequency will be PW/T = PW/(l/f) = $10^{-6} \times 10^{9}$ = 1000 waves and in one pulse repetitive time (PRT) the number of μw-power-waves will be PRT/T $\approx 10^{-3} \times 10^{9} \approx 10^{6}$ waves, i.e. out of 10^6 waves only 10^3 waves are on, at intervals of time (t_{on} + t_{off}) = T_1.

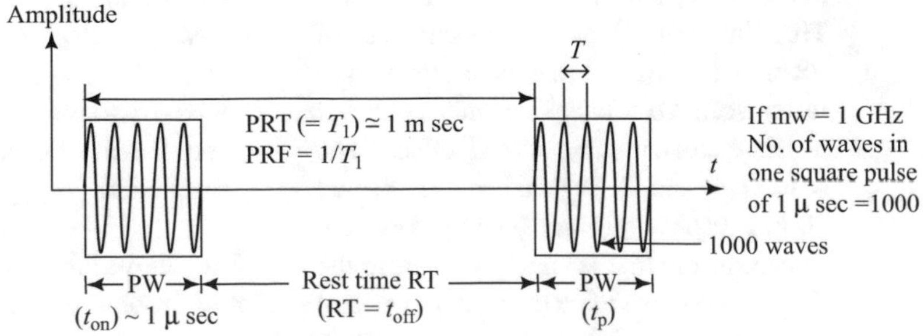

Fig. 9.3 Radar signal being transmitted

9.4 Range Accuracy (Resolution) and Range Ambiguity

The range of the target $R = C \cdot t_d/2$ needs to be accurate and non-ambiguous. Normally, accuracy depends on the width of the reflected pulse, while non-ambiguity on the pulse repetitive frequency (PRF) (Fig. 9.4).

(a) **Range Accuracy (Resolution)**

The pulse width transmitted and the pulse width of the reflected back echo may differ a bit due to the fact that, if the object is long, reflection may add from whole of its length. Moreover, the reflected pulse may not be perfectly rectangular because of the above and therefore its width PW_{ref} has to be taken from half amplitude points, as the reflection is not just from the central length of the target. Therefore taking at face value, the ability to accurately measure range is determined by the reflected pulse width. Thus, the uncertainty in the measure of range of the target is defined as range accuracy (i.e. resolution) as:

$$R_{ref} = \pm C \cdot PW_{ref}/2\,\text{m} \qquad (9.5)$$

Therefore for increasing the accuracy, we may reduce the pulse width of the transmitted signal such that $PW_{trans} \approx PW_{ref}$. This requires that PRF is to be increased for keeping both the average power and duty cycle same, which is the desired conditions in many cases. Increasing the frequency affects ambiguity of range, which we will study now.

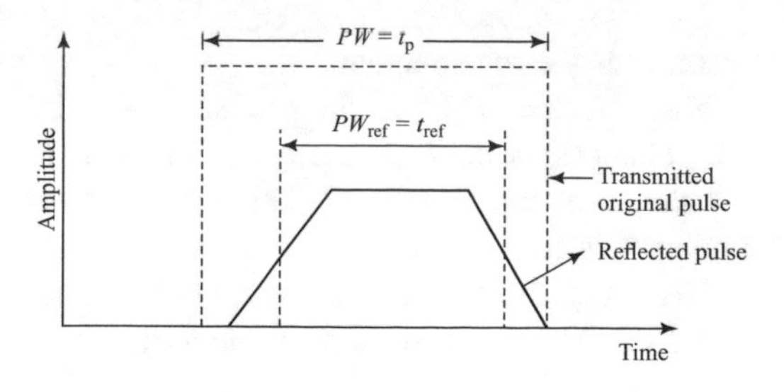

Fig. 9.4 Pulse width t_{ref} reflected back from target, along with original pulse overlapped, just for comparison

(b) **Range Ambiguity**

Figure 9.5a shows the snapshots of transmitted pulses as A_t, B_t, C_t, while their reflected pulses of reduced amplitudes received on the screen as A_R, B_R, while C_R comes still latter.

The reflected pulse should be received well before the next pulse is transmitted. The situation of Fig. 9.5a should not happen, because it will look as if the pulse A_R is the reflected wave of B_t transmitted and the reflected time might wrongly be taken as t_r' instead of t_r. Thus, the target range will look to be closer at $R' = Ct_r'/2$ instead of at $R = Ct_r/2$ (Fig. 9.5b).

Therefore, the pulse repetitive frequency has to be low, i.e. PRT to be high for range of a radar to be high. Thus, the maximum unambiguous range is given by seeing Figs. 9.3 and 9.5 as:

$$R_{max} = R_{un-amb} = C(t_{on} + t_{off})/2 = C/2(\text{PRF}) \qquad (9.6)$$

For example, If PRF = 1 Kc/s, PW = 1 μs; PRT = 0.001 s = 1000 μs

$$t_{on} = 1\,\mu s \quad \therefore \quad t_{off} = 1000 - 1 = 999\,\mu s;$$

$R = C/(2\text{PRF}) = 150$ km ∴ R could be much smaller ≅ 10 km

$$R_{un-amb} = 150\,\text{km}$$

If the two conditions of range accuracy and unambiguity are contradictory, then a balance is to be made, with giving more importance to ambiguity, as a result PRF is kept as low as possible.

9.5 Simple Radar Range Equation

(a) **Maximum range and effect of noise**: The radar range equation relates the range of a radar to the characteristics of transmitter, receiver, antenna, target, environment, etc. This equation gives maximum detectable distance of a target from a radar.

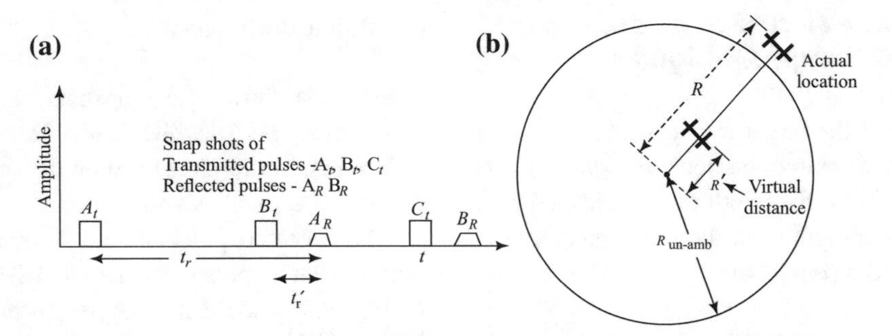

Fig. 9.5 **a** Pulses transmitted A_t, B_t, C_t, with pulse reflection received on the screen as A_R, B_R. **b** Ambiguous displayed range R' and actual range R. The maximum unambiguous distance of the given screen is shown in $R_{\text{un-amb}}$

Let power of an isotropic antenna (which radiates power uniformly in all directions) be P_t, then the power density at distance R (i.e. on the outgoing spherical wave of radius R) will be:

$$p_o = P_t/(4\pi R^2)\ \text{W/m}^2$$

Radars normally use directed power P_t in certain direction with a gain (say G), then the power density in that certain direction will be

$$p_{\text{trans}} = P_t \cdot G/(4\pi R^2)\ \text{W/m}^2 \tag{9.7}$$

If A_{tar} is the cross section of the target receiving the signal, then power received by the target is

$$p_{\text{tar}} = P_{\text{trans}} \times A_{\text{tar}} = P_t \cdot G \cdot A_{\text{tar}}/(4\pi R^2)\ \text{W} \tag{9.8}$$

The target reflects this power in all six directions as spherical wave therefore power density reaching the antenna will be:

$$\begin{aligned} P_i &= p_{\text{tar}}/(4\pi R^2) \\ &= P_t \cdot G \cdot A_{\text{tar}}/(4\pi R^2)^2\ \text{W/m}^2 \end{aligned} \tag{9.9}$$

If A_{rec} is the effective capture area of receiving antenna, then received echoed power is:

$$P_{\text{rec}} = P_i \cdot A_{\text{rec}} = P_t G \cdot A_{\text{rec}}/(4\pi R^2)^2 \text{W} \tag{9.10}$$

As per antenna theory, we know that for wavelength λ

$$\begin{aligned} \text{Antenna gain } &= G = \frac{4\pi A_{\text{rec}}}{\lambda^2} \text{ and} \\ A_{\text{rec}} &= G\lambda^2/(4\pi) \end{aligned} \tag{9.11}$$

Let

S_{mds} Minimum detectable signal of antenna and just the same has been received by the antenna then

S_{mds} P_{rec} in Eq. (9.10).

S_{mds} $P_t\left(\frac{4\pi A_{\text{rec}}}{\lambda^2}\right)(A_{\text{tar}}) \cdot A_{\text{rec}} \,/\, (4\pi R^2)^2 \ldots$

Solving for R which is $R = R_{\max}$ then:

$$\boxed{R_{\max} = \left(\frac{P_t \cdot A_{\text{rec}}^2 \cdot A_{\text{tar}}}{4\pi\lambda^2 \cdot S_{\text{mds}}}\right)^{1/4}} \tag{9.12}$$

Using Eq. (9.11) of A_{rec}, we get another expression for $R_{\max}$ as:

$$\boxed{R_{\max} = \left[\frac{P_t \cdot G^2 \cdot \lambda^2 \cdot A_{\text{tar}}}{(4\pi)^3 \cdot S_{\text{mds}}}\right]^{1/4}} \tag{9.13}$$

where G is the antenna gain.

Noise factor (F_n), i.e. in presence of noise $R_{\max}$ changes as the S_{mds} changes. For calculating this, we will use Eq. (9.12), which is not having G in it.

Here A_{tar}, the target front area is sometimes called RADAR cross section (Table 9.3).

(b) **Effect of noise on maximum range**: From the range Eqs. (9.12) and (9.13) of a radar, it

Table 9.3 Front cross section A_{tar} or RCS of some mobile objects as targets

Target	RADAR cross section[a] A_{tar} (m^2)
Birds	0.01
Missiles	0.5
A man	1.0
Small air force plane	2.0
Large air force plane	6.0
Cruiser boat	10.0
Large bomber	40.0
Large plane (737)	70.0
Car	70.0
Large plane (777)	100.0
Large truck	100.0

[a]Side cross section area maybe different, but for RADAR it is called radar cross section (RCS)

is clear that it can be increased by decreasing S_{mds}, which in turn depends on the sensitivity of the receiver and hence on its noise figure. As in Fig. 9.6, the input signal P_{si} and noise P_{ni} both get amplified along with noise generated at the receiver input side (P_{nri}). As a result, total noise at the output end becomes G_r (P_{si} + P_{nri}).

$$\text{Noise figure}(F_n) = \frac{\text{Input signal to noise power ratio}}{\text{Output signal to noise power ratio}}$$

$$= \frac{(P_{si}/P_{ni})}{(P_{so}/P_{no})} = \frac{P_{si}}{P_{ni}} \cdot \frac{P_{no}}{P_{so}}$$

$$\therefore \quad F_n = \frac{P_{si}}{P_{ni}} \cdot \frac{G_r(P_{ni} + P_{ri})}{G_r P_{si}}$$

$$= (1 + P_{nri}/P_{ni})$$

$$\therefore \quad P_{nri} = P_{ni}(F_n - 1)$$

$$= k \cdot T_o \cdot B \cdot G_r(F_n - 1) \tag{9.14}$$

Here P_{ni} depends on thermal noise $k \cdot T_O$, band width B and gain G_r of the receiver. Here k and T_o are:

k = boltzman constant

T_o = standard ambient temperature (°K)

The minimum detectable signal will be equal to the noise level of the receiver.

$$\therefore \quad S_{mds} = kT_oB \cdot G_r(F_n - 1) \tag{9.15}$$

Now putting Eq. (9.15) in Eq. (9.12) and not in (9.13), we get

$$R_{max} = \left[\frac{P_t \cdot A_{rec}^2 \cdot A_{tar}}{4\pi\lambda^2 \cdot kT_o \cdot B \cdot G_r(F_n - 1)}\right]^{1/4} \tag{9.16}$$

For parabolic antenna aperture–area is the receiver system area A_{rec} which uses around 65% of the face area (πR^2). Therefore, for D = distance of paraboloid face of dish antenna:

$$A_{rec} = 0.65(\pi D^2/4) \tag{9.17}$$

Putting all values of the constants k, π, T_o, and A_{rec} in Eq. 9.16, we get:

$$R_{max} = 48 \cdot \left[\frac{P_t \cdot D^4 \cdot A_{tar}}{B \cdot \lambda^2 \cdot (F_n - 1)}\right]^{1/4} \tag{9.18}$$

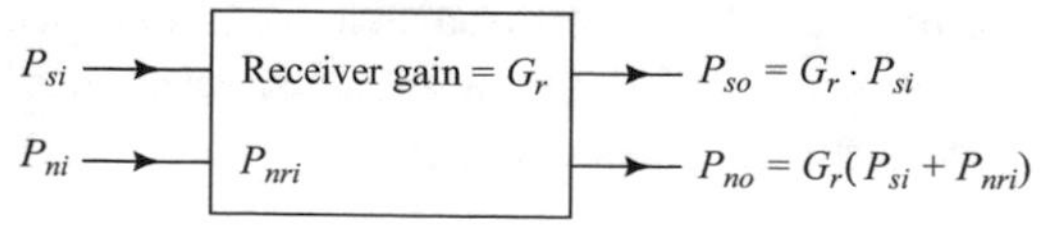

Fig. 9.6 Receiver signal and noise at input and output side

(c) **The threshold level of signal in receiver**: Normally, the input noise level P_{ni} of the receiver is equated to the minimum

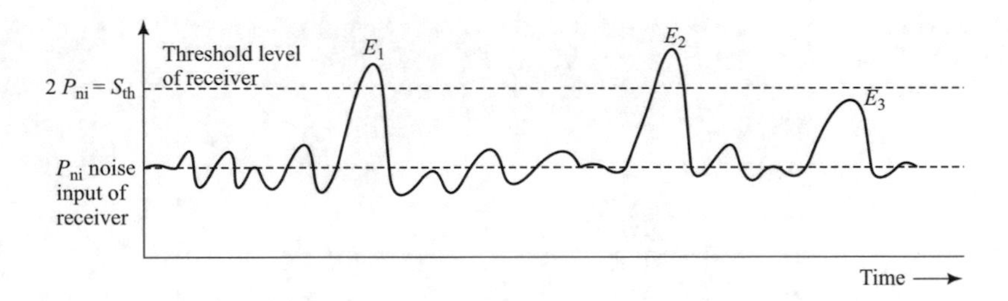

Fig. 9.7 Radar receiver input as a function of time. The signals reflected from targets, e.g. E_1, E_2 are detectable while E_3 is not

detectable signal (i.e. $P_{ni} = S_{mds}$); but it will be preferable to define a threshold level of S_{mds}, which should be at least twice of P_{ni} of Eq. (9.14), i.e. $S_{mds} = 2P_{ni} = S_{th}$

As in Fig. 9.7, the echo E_1 and E_2 are detectable while echo E_3 is not, if the threshold level is set at S_{th} which maybe 2 P_{ni}.

9.6 Some Special Parameters of Radar—Doppler Shift, Clutter, Jamming, Polarisation, Heterodyne Receiver, and Blind Speed

There are some five special parameters in the functioning of radars, which must be known, before we study various types of radars. These are Doppler shift, clutter, jamming, polarisation of signal, and blind speed. Besides these the receiver being of superheterodyne types, its knowledge is a must.

(1) **Doppler frequency shift**: Before the types of radar are taken up, we should study the concept of doppler frequency shift. When a signal of frequency f_t is transmitted towards a moving target, then the frequency of the reflected (echoed) signal is not f_t. It is lower if the target is receding and higher than f_t if the target as approaching. This change in frequency is called doppler shift and the difference frequency as doppler frequency. Let the target be at range R at the moment of study. The total number of wavelengths λ in the two-way path from radar to target and return will be $2R/\lambda$, which corresponds to a phase change of 2π for each of the wavelength. Therefore, the total phase change in the two-way propagation path is

$$\phi = 2\pi \times 2R/\lambda = 4\pi R/\lambda \tag{9.19}$$

As the target is moving, ϕ and R both are changing along with it. Differentiating the above ϕ w.r.t. time gives a relation between angular or doppler frequency $\omega_r = \mathrm{d}\phi/\mathrm{d}t$ and the radial velocity ($v_r = \mathrm{d}R/\mathrm{d}t$) as:

$$\omega_r = \omega_{dop} = \frac{\mathrm{d}\phi}{\mathrm{d}t} = \frac{4\pi}{\lambda}\cdot\frac{\mathrm{d}R}{\mathrm{d}t} = \frac{4\pi v_r}{\lambda} = 2\pi f_{dop} \tag{9.20}$$

$$\therefore \quad f_{dop} = (2v_r/\lambda) = (2\cdot f_t\cdot v_r/c) \tag{9.21}$$

where

f_t = frequency of signal transmitted
λ = wavelength of signal transmitted

If the target is moving with some velocity v_o at an angle θ with range R, then $v_r = v_o\cos(\theta)$ is the radial velocity in Eq. (9.21) above.

(2) **Clutter**: Clutter refers to the echo signals returning from targets which are not wanted by us, e.g. ground, sea, hills, tall buildings, birds, atmospheric turbulence, sea tides, sea ice, precipitation (metre or trails, snow, hail, ionospheric reflections, etc.

Some clutter maybe due to the long waveguide between radar trans-receiver and antenna.

Some sunbursts on the PPI radar screen are seen due to diffused transmitted pulses reflected by these waveguides before the signal leaves antenna.

In general, in application radars, these clutters are undesirable, but in weather meteorological radars, cloud, storm, hail maybe desirable clutter.

These are several methods of detecting and neutralising clutters. Most of these methods are based on the fact that clutter positions appear to be fixed between, first scan and subsequent scans and therefore can be removed. Sea clutter is removed by using horizontally polarised signal transmission, while effect of rain is reduced by circularly polarised signal transmission. The opposite is done in meteorological radars.

A method for removing clutter is discussed in MTI Radar Sect. 9.7.1 and Fig. 9.10.

(3) **Jamming**: Radar jamming refers to signals originating from sources outside radar, by transmitting signal to a radar at its own frequency for masking targets of interest by the enemy.

The jamming signal in such cases is directed to the radar and is much powerful. There are some electronic countermeasures also. Interference has now become a problem for 'C'-band 5.66 GHz meteorological radars due to 5.4 GHz band wi-fi equipments of Internet applications.

(4) **Polarisation**: Radars use horizontal, vertical, or circularly polarised signal to detect different types of material. Linearly, polarised echo indicates metal surface, random polarised indicate fractal surface, e.g. rocks, soils, etc. If one sense of circular polarised wave is sent to an aircraft, the echo has both left-hand and right-hand circular polarised wave (LHCPW and RHCPW).

Circularly polarised signal transmission is used to minimise the effect of rain, hail, etc., on echo signal.

(5) **Radar blind speed due to doppler effect**: The radial components of the velocity of an object (target) maybe towards the radar centre or away from the centre of radar, i.e. approaching or receding. In both the cases, doppler effect is observed where the pulse repetition frequency (PRF) reduces or increases and becomes different than the PRF of pulses transmitted. The difference of the two PRF transmitted and received back is called doppler PRF.

$$(\mathrm{PRF})_{\mathrm{rec}} - (\mathrm{PRF})_{\mathrm{trans}} = (\mathrm{PRF})_{\mathrm{doppler}}$$

When the speed of the target is quite large than a situation may reach where the doppler frequency becomes equals to transmitted frequency, i.e.

$$\begin{aligned}\left[(\mathrm{PRF})_{\mathrm{rec}} - (\mathrm{PRF})_{\mathrm{trans}}\right] &= (\mathrm{PRF})_{\mathrm{doppler}} \\ &= (\mathrm{PRF})_{\mathrm{trans}} \qquad (9.22)\end{aligned}$$

Under this situation, the radial signal component received from the target becomes zero and the situation is same as that of a stationary target, where the $(\mathrm{PRF})_{\mathrm{received}} = (\mathrm{PRF})_{\mathrm{transmitted}}$.

Therefore the object speed towards the radar as observed will be zero. This is called blind speed of the target. It may also happen when the speed of target is still high, doppler PRF maybe two times or higher than the transmitted PRF.

i.e.,

$$(\mathrm{PRF})_{\mathrm{doppler}} = n \cdot (\mathrm{PRF})_{\mathrm{transmitted}} \qquad (9.23)$$

where

$$n = 1,\ 2,\ 3,\ \ldots \text{ for blind speed}$$

This is also called doppler ambiguity.

Blind distances at blind speed can be known from the fact that if pulse width is t_{p}, then any signal returning within t_{p} time will not be indicated on radar, i.e. will not be seen. Therefore, blind spot distance from range equation gives:

$$R_{\text{blind}} = C \cdot t_{\text{p}}/2$$

e.g. if

$$t_{\text{p}} = 0.01 \text{ ms};$$

$$R_{\text{blind}} = (3 \times 10^8) \times (0.01 \times 10^{-3})/2 \cong 1.5 \text{ km}$$

9.6.1 Superheterodyne Receiver

This technique of radio receiver was devised in 1914 by an US army major Edwin Armstrong in France during the World War I for better detection of signals of enemy's ships of different frequencies. Latter in 1930, it became popular by getting it used in domestic radio receivers also, as here also large frequency range of radio stations was to be covered, e.g. medium wave (MW) (540–1640 Kc/s), short wave (SW), (3–18 Mc/s) and frequency modulated (FM) wave (88–108 Mc/s.)

As per the principle of superheterodyne receiver (Fig. 9.8), the signal of antenna (f_s) and that of local oscillator is mixed in the first detector to give its output as ($f_s \pm f_{LO}$), f_S. This goes to tuned intermediate frequency (IF) amplifier, which tunes and amplifies only ($f_S - f_{LO}$) = 455 kc (fixed) only. This is possible by a gang condenser having two sections each working for the *LC* tuners each of antenna signal and of oscillator (generator) signal. Therefore, radio station frequency tuning and oscillator tuning frequency happen together, such that the difference is always 455 kc/s. This IF signal can be easily be amplified with very high gain by a frequency tuned amplifier, which was not possible by any broadband amplifier for covering MW, SW, and FM. In fact, this has been the reason, because of which the technique of superheterodyne was discovered, where we need to amplify IF only using IF amplifier. The automatic gain and control (AGC) line in Fig. 9.8 is a −ve feedback line for stabilising audio output.

In radar, which operates in the microwave frequency band (0.3–300 GHz), superheterodyne receiver has IF = 30 or 60 MHz, with the audio amplifier and loudspeaker replaced by video amplifier and display.

9.7 The Three Types of MTI Doppler Radars

Almost all the radars of 1940–80 were of simple pulse system, which did not employ doppler effect. But today all military air defence radar and all civil air traffic control radar for detecting and tracking, depends on doppler frequency shift, for separating large clutter echoes from smaller echoes of moving targets. Many a times, the ratio of larger clutter echoes to the moving target echoes maybe as large as 70 dB (i.e. 10^7 times), but by using doppler effect technology, the moving target can be detected.

Thus nowadays every radar, whether used for civilian or military operation have:

1. Moving target indicator (MTI) display.
2. Doppler frequency shift indicator.

There are three types of radars in normal use, the difference among these three types are only in their different pulse repetitive frequency (PRF). These have been given the following three names:

(i) CW Doppler Radar: PRF = 0, i.e. no pulse, only CW
(ii) MTI Radar: PRF = very low, for avoiding the range ambiguities as $R = Ct_r/2$ (Sect. 9.4). Duty cycle is low $\approx 1.0\%$
(iii) Pulse Doppler Radar: $\therefore$ PRF = High, for avoiding the doppler ambiguities blind speed, but range ambiguities will be present. As PRF is high duty cycle is as high as 50%.

Now these three types of MTI Doppler Radar mentioned above will be discussed.

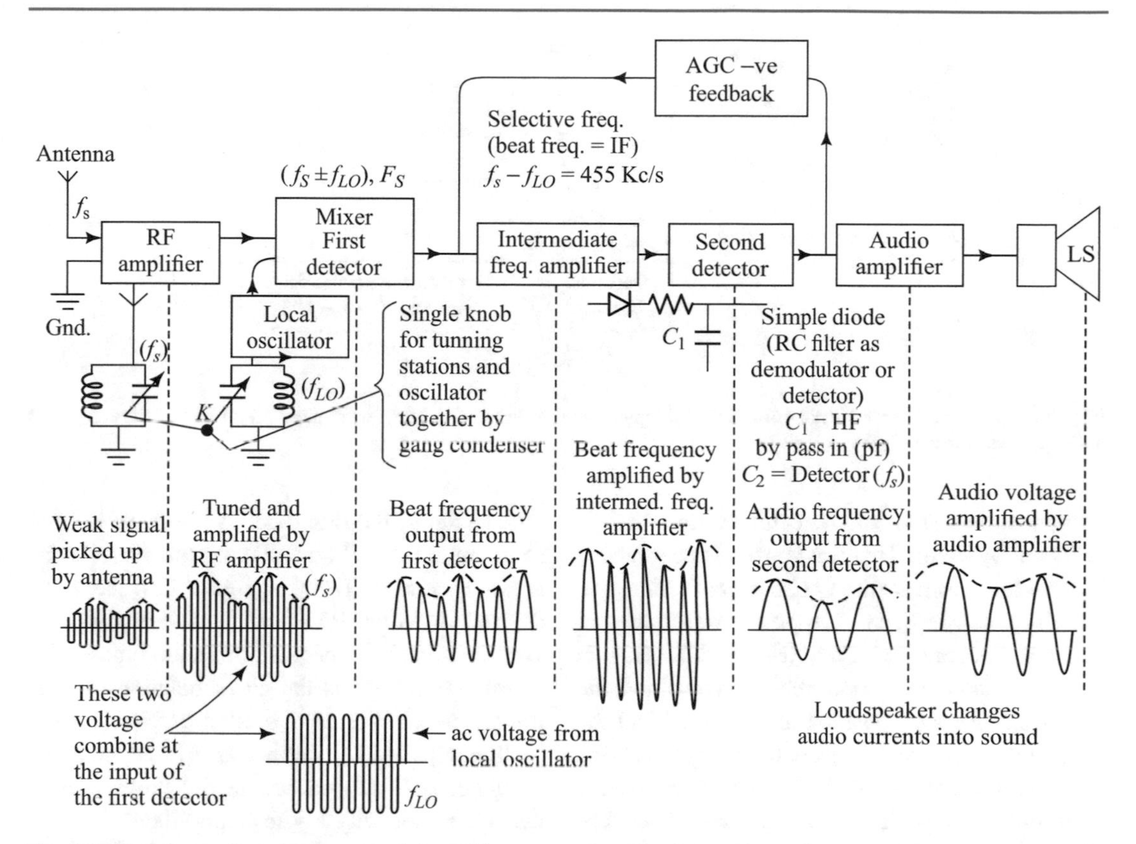

Fig. 9.8 Superheterodyne receiver used in radio sets: Difference of antenna signal frequency and frequency of oscillator has to remain constant = IF = 455 Kc/s and for this tuning circuit of both has to change the value of its C together in a gang condenser by a single tuning knob K. AGC is a −ve feedback for signal level stabilisation of audio signal output

9.7.1 CW Doppler Radar

Figure 9.9 gives the concept of CW doppler radar. The transmitter generates continuous (unmodulated) sinusoidal oscillators at frequency f_{trans} which is radiated by the antenna. On reflection (echoed) from the moving target, the transmitted signal is shifted by doppler effect by an amount $= \pm f_{dop}$ as per Eq. 9.21. The sign of f_{dop} decides that whether the target is approaching or receding the radar. So as to utilise the doppler shift, a radar must be able to recognise that the echo signal frequency is different (higher or lower) than the transmitted one. For this, a part of transmitter signal of frequency f_t is sent to the detector, along with the echoed signal of frequency $f_{dop} = (f_t - f_r)$ received by it. From the detector, the signal reaches doppler filter to give its output the signal of frequency f_{dop} after cutting off the clutter signals (Fig. 9.9).

CW radars are used where velocity information is of more interest, and actual range is not needed, e.g. (1) police radar for catching overspeeding vehicles, (2) for measuring water waves motion speed in sea, etc.

9.7.2 MTI Radar

The clutter has been a big problem for the civilian aircraft as well as for military plane, as our target may get obstructed by them or not visible at all due to clutter, many a times clutter maybe 100 times larger than the moving target. Most of the clutter signal is due to hills, large towers, etc., which are stationary, and there will not be any doppler shift

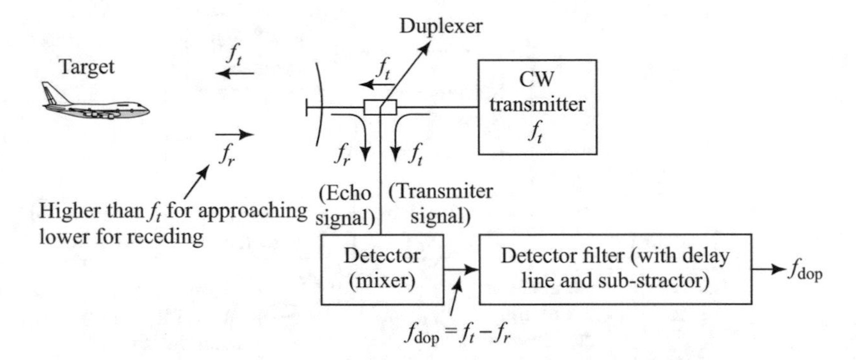

Fig. 9.9 A simple schematic diagram of CW-doppler radar for extracting doppler frequency shift from a moving target and reject stationary clutter echoes

or phase change in the transmitted signal when reflected by them. This fact is used for cancelling the clutters from desired echo signal using two successive sweeps, i.e. two pulses transmitted at intervals of one of PRT (Fig. 9.10). This is because, when the second pulse is transmitted, the approaching target position has changed and the amplitude of echo of target will be larger than the first sweep pulse. While for the stationary objects, amplitude and position will be same. Therefore after subtraction of these two successive echoes, the clutters get cancelled, while for the moving target echoes, the difference is not zero, but has some finite value (Fig. 9.10). As the second echo is late by one PRT, the first echo also has to be delayed by one PRT before subtracting, for keeping it in the same position in the timescale.

Figure 9.11 gives the block diagram of MTI Radar, which has two oscillator STable local oscillator (STALO) producing f_{lo} and coherent oscillator (COHO) producing f_{co}. Power amplifier is a multicavity klystron amplifier.

The IF stage is designed as a matched filter, as is usually the case in radar instead of an

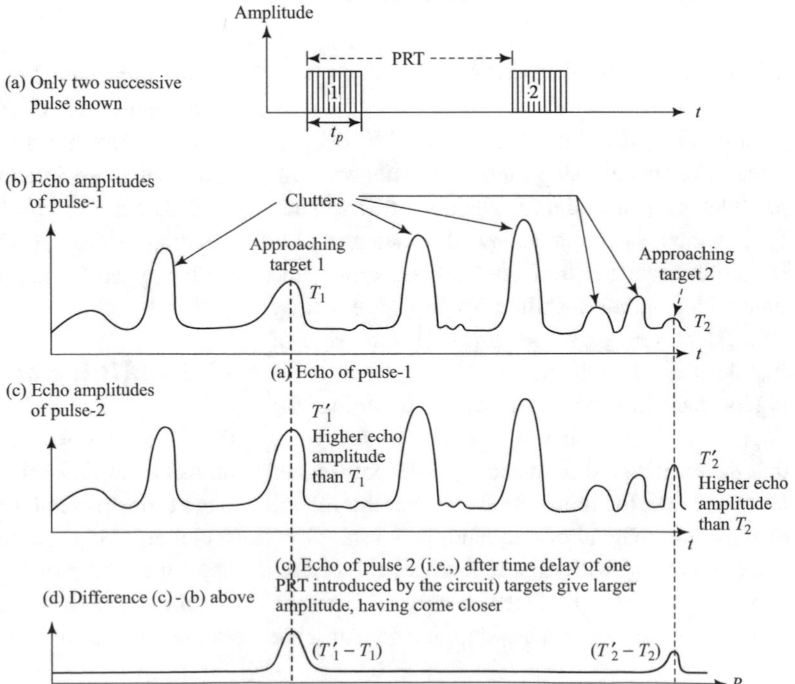

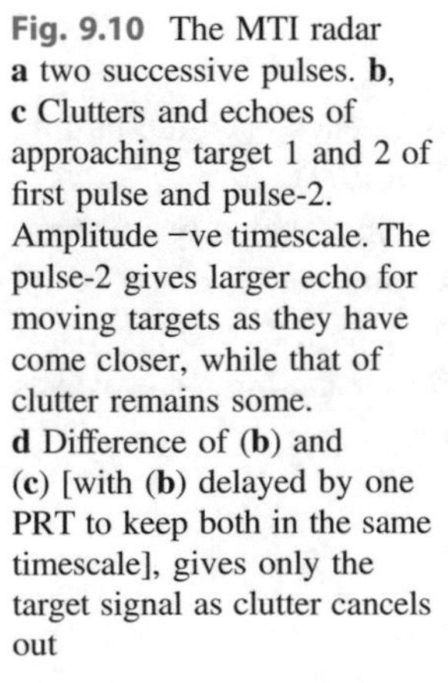
Fig. 9.10 The MTI radar **a** two successive pulses. **b**, **c** Clutters and echoes of approaching target 1 and 2 of first pulse and pulse-2. Amplitude −ve timescale. The pulse-2 gives larger echo for moving targets as they have come closer, while that of clutter remains some. **d** Difference of (**b**) and (**c**) [with (**b**) delayed by one PRT to keep both in the same timescale], gives only the target signal as clutter cancels out

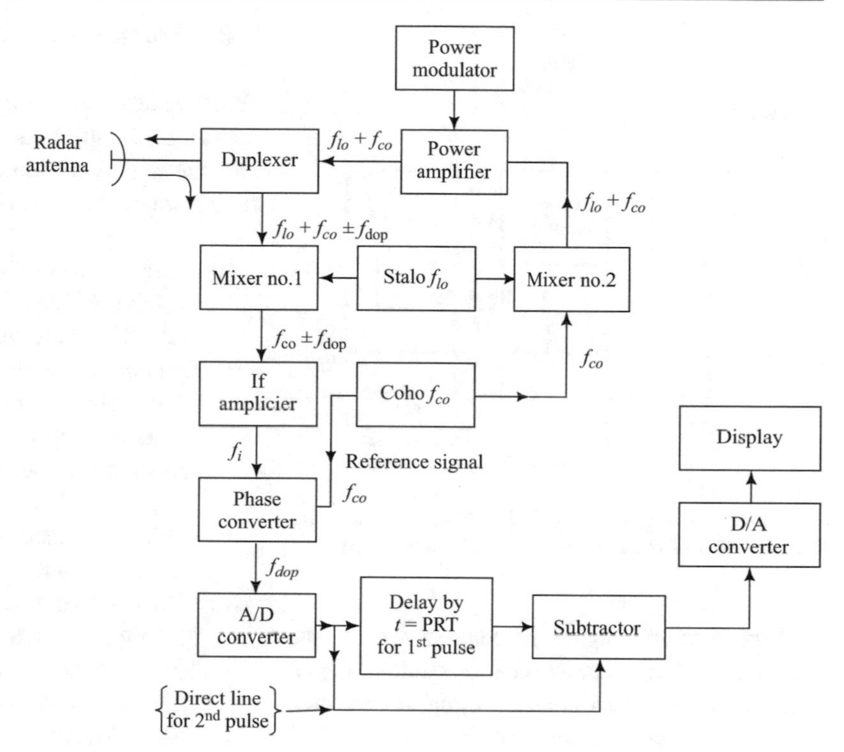

Fig. 9.11 MTI radar block diagram

amplitude detector there is a phase detector, followed by IF stage. This detector is a mixer type of device and combines IF signal (f_i) with reference signal (f_{co}) from COHO signal, producing a difference signal ($f_i - f_{co}$) = f_{dop}, which is nothing but the doppler frequency signal. The COHO oscillator is coherent and in phase with the transmitter signal (f_{tr}). Thus, the transmitter signal is the sum of STALO frequency f_{lo} and coherent frequency f_{co}, i.e. $f_{tr} = f_{lo} + f_{co}$. Output of the phase detector goes to the delay line and then to canceller or subtractor as in Fig. 9.11. Delay line acts as high pass filter to separate the doppler shifted echo signal of moving target from unwanted echoes of stationary clutters (see Fig. 9.10).

Practical MTI system can cancel the echoes of fixed targets having amplitude as high as 40 dB, i.e. 10^4 times or greater than the desired echoes of targets. Regarding differentiating the superimposed fixed and mobile targets, the mobile target could be as weak as 25 dB, i.e. $10^{2.5}$ times the fixed target and still it can be differentiated.

9.7.3 Pulse Doppler Radar

As discussed, this is called pulse doppler radar as (*i*) the pulse repetition frequency (PRF) is much larger than that it is in MTI doppler radar (*ii*) it avoids doppler ambiguities of Fig. 9.5, but increases the range ambiguities. Figure 9.12 gives the block diagram of a pulse doppler radar.

The output of the CW oscillator is amplified and turned ON and OFF (i.e. modulated) to generate high power pulse of high PRF.

Let the transmitted and received signal be $A_t \sin(w_t\ t)$ and $\sin[w_t\ (t - T_r)]$, with A_t and A_r as their amplitudes. If the object is moving towards the radar with radial velocity v_r, then the changing range and received signal at time 't' are:

$$R = (R_0 - v_r \cdot t) \tag{9.24}$$

$$V_{rec} = A_{rec} \cdot \sin[\omega_t \cdot (1 - 2v_r/c)t - 2\omega_t \cdot R_o/c] \tag{9.25}$$

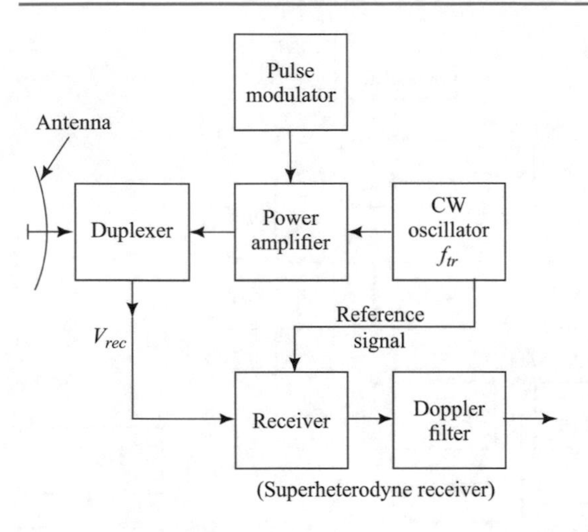

Fig. 9.12 A simple pulse radar which extracts the doppler shift of the echo signal from a moving target

The received signal is heterodyned in the receiver and referenced with transmitter signal (f_{tr}), and difference frequency signal is extracted and it will have amplitude as:

$$V_d = A_d \cdot \cos(\omega_d t - 4R_o/\lambda)$$

where

$$\omega_d = 2\pi f_d = 2\pi.(2v_r/\lambda)$$

As

$$f_{tr}.\lambda = c$$

Here more than one pulse is needed to recognise a change in echo frequency due to doppler effect. The high PRF reduces doppler ambiguities, but increases range ambiguities. Less or no doppler ambiguities means the number of blind speeds reduces to nearly zero in pulse doppler radar. The duty cycle is as large as 50%.

These radars are mostly airborne (i.e., fixed in the plane), e.g. airborne warning and control system (AWACS).

9.8 Tracking Radars

When a radar continuously observes a target over a time, such radars can predict its future course and are called tracking radars. These tracking radars are of the following four types:

(i) *Single Target Tracking (STT)*: It tracks continuously a single target at a fast data rate. These are very suitable for missiles.
(ii) *Automatic Detection and Tracking (ATD)*: This radar firstly detects and tracks its motion. Rate of tracking depends on the rotational velocity of antenna. It can track large number of targets at a time. This ADT facility is there in most of the radars stations.
(iii) *Phased Array Tracking Radar*: Here the rotation of beam OP (Fig. 9.2) is by using (the electronic method). Here a large number of targets can be detected at a time. Multiple targets are tracked on a time-sharing basis. The array of antennas can be electronically switched from one to another for covering the whole of 360°.
(iv) *Track While Scan (TWS)*: It tracks a small given angle only and can track more than one target on that angle. They are used in air defence system.

9.8.1 Methods of Scanning

Scanning is the method to keep the antenna moving in azimuth (horizontal circle) with or without elevation, for covering an area, which has the desired target. It sometime covers the complete hemisphere around the radar (Fig. 9.13).

Some of the typical scanning patterns are as follows:

(i) *Horizontal Scanning*: Pattern only for ship to ship type of scanning.

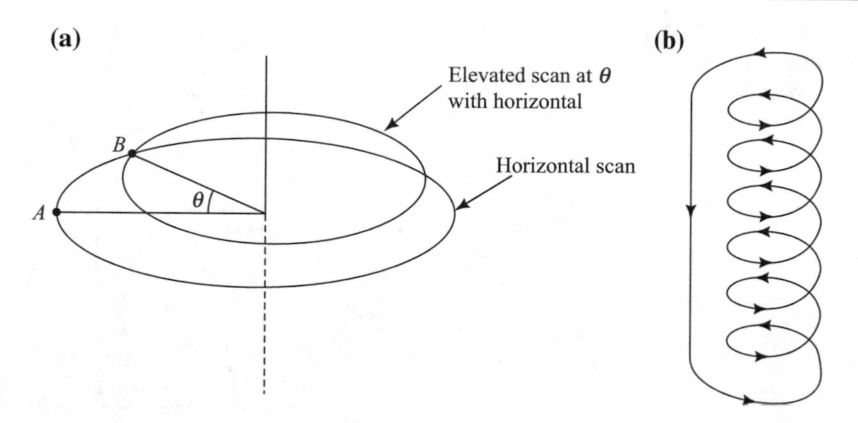

Fig. 9.13 **a** Horizontal (azimuthal) with Radar dish direction moving along the circle and elevated scan path of the point *B* along higher circle, i.e., rotating Radar dish antenna facing sky at an elevated angle θ. **b** Helical scan

(ii) *Elevated Scanning*: It covers horizontal as well as elevated areas.
(iii) *Helical Scanning*: As the antenna rotates, the angle of elevation also keeps on increasing. After completion of each scanning cycle, the antenna returns to the starting point (Fig. 9.13b).
(iv) *Spiral Scan*: If limited area of circular area is to be covered, spiral scan maybe used for covering horizontal as well as vertical plane.

9.8.2 Tracking

After scanning, when the target has been found, we have to track/find its path accurately, along with angle of elevation as well as range and this is called tracking. Automatic tracking is required the moving target which has to be aimed by gun and destroyed. A pencil beam antenna is not sufficient for accurate tracking with respect to (*a*) range, (*b*) velocity, (*c*) azimuth angle, (*d*) elevation. Therefore, additional three methods/techniques are used for tracking accurately:

(i) *Lobe Switching/Sequential Switching*: A narrow beam from the antenna switches very fast from position 1 to position 2, around the target which, may not be equidistant from the target (Fig. 9.14). The difference in the amplitude between the voltages obtained in the two switched positions is a measure of angular displacement (θ) of the target from the switching axis. When amplitudes of the received echoes are equal from both the positions, then only the target is on the switching axis. Figure 9.14a gives the two position/directions of the antenna, while Fig. 9.14b gives the amplitude of main lobe/side lobes for the two positions.
(ii) *Conical Scanning/Switching*: Also from Fig. 9.14a, we can see that the antenna can make a solid cone, if it rotates around the switching axis covering the two positions 1 and 2 also. While doing so, it is called conical scanning around the target. The target may not be on the axis. If the target is present within the solid angle θ, then the echo signal from the target will be frequency modulated at the frequency of rotation of the beam. The conical scan modulation is extracted from the echo signal and applied to the servo control system, which positions the antenna on the target both in azimuth (horizontal) and elevation planes. When this modulation becomes zero, target is on the axes of the beam and target location is known fully. The rotation speed has to be much lower than PRF.

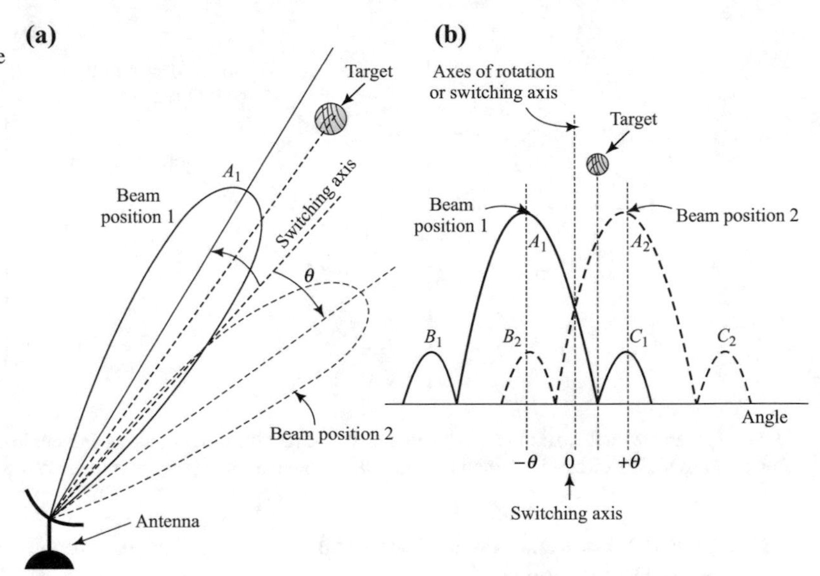

Fig. 9.14 Lobe switching between two positions **a** lobe position **b** signal amplitudes with angle

(iii) *Monopulse Tracking*: There are four disadvantages of the above two tracking systems:

(a) Additional servo mechanism is required as the motion of antenna is complex.
(b) Amplitude comparison is not accurate.
(c) Conical scanning is susceptible to enemy's electronic countermeasures (ECM), as the scan rate can be easily detectable by the enemy. Therefore, ECM can affect the working of the servo mechanism.
(d) For locating the target, a minimum of 4-pulses are required with conical scan one each for the four directions. Also if the target is changing its location, then its cross section (A_{tar}) will also change as our signal may cross the front or side in these locations, leading to error.

An ideal system will be which can give all information with one pulse alone (Fig. 9.15). Here in monopulse tracking four horn antenna feeds are used with one paraboloid reflector, which are placed around the axis of the parabola. The single transmitter feeds the horns simultaneously, so that the sum of four signals of same frequency and phase is transmitted. The echo signal is received by each horn, and the following three signals are generated by the receiver duplexer using hybrid ring (or rat race):

(*i*) A + B + C + D (Sum signal)
(*ii*) (A + B) − (C + D) (Horizontal difference signal)
(*iii*) (A + C) − (B + D) (Vertical difference signal)

If the target lies on the axis of the paraboloid antenna, then the difference signals =0. If the target is away from the axial line, then difference signals $\neq 0$ and the receiver has to process these three signals through three channels consisting of mixers, local oscillator, 3-IF-Amplifier, and 3-Detectors. The vertical and horizontal signals are used to drive a servo amplifier and a motor for positioning the antenna. Finally when the

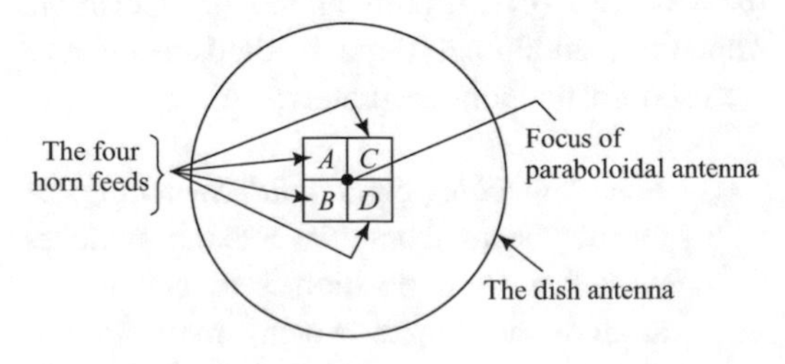

Fig. 9.15 Monopulse antenna: as seen from front: the four-horn antenna is placed around the focus of the paraboloid

differences are zero, the antenna is perfectly aligned with the target and keeps on tracking it.

Solved Problems

Problem 1

It a radar receives back the reflected signal from target after 200 μs find the distance of target.

Solution

$$\text{Range } (R) = c.\Delta t/2 = 3 \times 10^8 \times 200 \times 10^{-6}/2 \text{ m}$$
$$= 300 \times 10^2 \text{ m} = 30 \text{ km}.$$

Problem 2

Radar operates at 500 GHz, if its diameter = 2 m, calculate the beam width.

Solution

$\theta = 70\,\lambda/D.$ degrees.

$$\text{for 50 GHz; } \lambda = c/f = 3 \times 10^8/50 \times 10^9$$
$$= 6 \times 10^{-3}\text{m} = 0.6\,\text{cm}$$
$$\theta = 70 \times 6 \times 10^{-3}/2$$
$$\therefore \quad = 210 \times 10^{-3} = 0.21^\circ$$

which is very good because of very high frequency.

Problem 3

It peak power transmitted by a radar is 200 kW, with PRF = 1000 pulse/s; pulse width = 1 μs then find average power in dBW, maximum unambiguous range and minimum range.

Solution

$$P_t = 200 \text{ kW}; \text{PRF} = 1000;$$
$$t_p = 1 \times 10^{-6} \text{ s}$$
$$\therefore \quad \text{Duty cycle} = t_p \times \text{PRF}$$
$$= 1 \times 10^{-6} \times 1000 = 10^{-3}$$
$$= 0.1\%$$
$$\therefore \quad \text{Average power} = P_t \times \text{duty cycle}$$
$$= 200 \times 1000 \times 10^{-3}$$
$$= 200 \text{ W}$$
$$\text{dBW} = 10 \log\left(\frac{200}{1}\right)$$
$$= (10 \log 2 + 20)$$
$$= 20.3 \text{ dBW}$$

$$\text{Max. unambiguous range} = c/(2\,.\,\text{PRF})$$
$$= 3 \times 10^8/(2 \times 1000)$$
$$= 150 \text{ km}$$

$$\text{Also minimum. range} = c\,.\,t_p/2$$
$$= (3 \times 10^8 \times 10^{-6})/2$$
$$= 150 \text{ km}$$

Problem 4

If in a rectangular shaped radar, power transmitted = 200 W, frequency = 2.9 GHz, Front area = 9.0 m^2, Aperture efficiency = 0.6; minimum detectable signal (S_{mds}) = 1 pW; target cross section = 2 m^2; calculate the range.

Solution

$$\text{Range} = R_{max} = \left(\frac{P_t.A_{rec}^2.A_{tar}}{4\pi\lambda^2.S_{mds}}\right)^{1/4} \quad (9.12)$$

For frequency of 2.9 GHz; $\lambda = 3 \times 10^8/2.9 \times 10^9$

$$= 0.103\,\text{m}$$

$$A_{rec} = \text{effective area of radar}$$
$$= \text{efficiency} \times \text{front area}$$
$$= 0.6 \times 8 = 4.8\,\text{m}^2$$

$$R^4_{max} = \frac{200 \times (4.8)^2 \times 2}{4 \times 3.14 \times (0.103)^2 \times (1 \times 10^{-12})}$$
$$= \frac{9216}{0.133 \times 10^{-12}}$$
$$= 169137 \times 10^{12}\text{m}^4$$
$$R = 16.2 \times 10^3 = 16.2\text{ km}$$

Problem 5

What is the doppler shift when tracking a car moving away from two radars at speed of 100 km/h, if the two radars operate at 1 and 10 GHz.

Solution

(a)

$$1\,\text{GHz} \quad \therefore \quad \lambda = 0.33\,\text{m}; \; f_{\text{d}} = \left(\frac{2 \times v \cos\theta}{\lambda}\right)$$

$$v = 100\text{ km/h}$$
$$= 10^5\text{m}/(60 \times 60\,\text{s})$$
$$= 27.777\text{ m/s}$$
$$\theta = 0^\circ$$

$$\therefore \quad \text{Doppler shift}: \quad f_d = 2v\cos(0)/0.3$$
$$= 2 \times 27.77/0.33\text{ Hz}$$
$$= 168\,\text{Hz}$$

(b) At 10 GHz $\lambda = 0.03\,\text{m}$

$$f_{\text{d}} = 168 \times 10 = 1680\,\text{Hz}$$

Therefore, higher frequency is preferred as f_{d} increases with signal frequency.

Problem 6

A receiver at 1 GHz with 1 MW radar requires at least 0.001 W to detect a valid target properly; what is the radar-cross section (RCS) of the target, when the target is to be at 100 km range. Antenna gain is 40 dB.

Solution

R range has another expression [Eq. (9.13)]:

$$R = \left[\frac{P_{\text{t}}.G^2.\lambda^2.A_{\text{tar}}}{(4\pi)^3.s_{\text{mds}}}\right]^{1/4}$$

A_{tar} is also called radar cross section (RCS) of target.

As

$$G_{\text{dB}} = 40\,\text{dB} \quad \therefore \quad G = 10^4$$

As

$$f = 1\,\text{GHz} \quad \therefore \quad \lambda = 30\,\text{cm}$$

$$A_{\text{tar}} = \text{RCS (Assuming 100\% efficient antenna)}$$
$$S_{\text{mds}} = 0.001\,\text{W} = 10^{-3}\text{ W}$$
$$P_{\text{t}} = 10^6\text{ W}$$
$$R = 100{,}000 = 10^5\text{ m}$$
$$R = (10^5) = \left[\frac{10^6 \times (10^4)^2 \times (0.3)^2 \times (\text{RCS})}{(4 \times 3.14)^3 \times 10^{-3}}\right]^{1/4}$$

$$\therefore \quad 10^{20} = \frac{10^{14} \times 0.09 \times (\text{RCS})}{64 \times 30.96 \times 10^{-3}}$$
$$= 5.7 \times 10^{19} \times (\text{RCS})$$

$$\therefore \quad \text{Target area} = (\text{RCS}) = 10/5.7 = 1.75\text{ m}^2$$

Problem 7

A small jet plane can fool a radar system and made to think that a large plane is coming by adding radar reflector plates to it, for increasing the A_{tar} (i.e. RCS). If the small jet plane has RCS = 6 m^2 and target jet plane has RCS = 20 m^2, how much large flat plates are needed at 1 GHz for the small jet to look like large jet.

Solution

The plates required over the existing small plane will be:

$$20{-}6 = 14\ \text{m}^2$$
$$= 3.74\ \text{m} \times 3.74\ \text{m plate.}$$

Problem 8

If the doppler shift is f_d = 10 Kc/s when the target is approaching directly towards the radar. Find f_d when the same target turns by 45° in the direction and is at 45° to the line joining the radar.

Solution

$$f_d' = 2v\cos(\theta)$$

The approaching velocity (i.e. radial velocity) reduces by changing the direction.

$$\therefore \quad \text{new}\, f_d' = f_d\cos(45^\circ) = 10 \times 0.707$$
$$= 7.707\ \text{Kc/s}$$

Problem 9

If a pulse width of 0.01 s of a radar signal is transmitted, what are the blind speeds of target, at which the radar cannot easily see the target.

Solution

The blind speed of the target is given by

$$v_{\text{blind}} = nc \times t_p/2 = c \times t_p(n/2)$$

where t_p = pulse width; $n = 1, 2, 3 \ldots$

$$\therefore \quad \text{For } n = 1; \quad v_{\text{blind}} = 3 \times 10^8 \times 0.01 \times 10^{-3}/2$$
$$= 1.5\ \text{km/s}$$

$$\text{2nd blind speed} = 2 \times 1.5 = 3\ \text{km/s}$$
$$\text{3rd blind speed} = 3 \times 1.5 = 4.5\ \text{km/s}$$

Review Questions

1. If a radar has a band width of 3 MHz, calculate the highest resolution. (Ans. 50 m)
2. If a radar at 5 GHz has antenna diameter of 4 m, calculate peak power required to have a maximum range of 500 km with target area of 20 m square and minimum detectable signal of 10 W. (Ans. 8.94 MW)
3. A pulsed radar has duty cycle of 0.016 with resting time of 380 μs. What is the pulse width. (Ans. 6.18 μs)
4. Write three different types of expression of range equation.
5. What is clutter and how do we get rid of it?
6. What are the differences in the three basic types of radar, i.e. CW-doppler, MTI, and pulsed Doppler? List them out.
7. Define blind speed of target.
8. What are the factors that affect the range of a radar? [Hint: See all three expressions].
9. What is maximum unambiguous range?

RF Filter Design

10

Contents

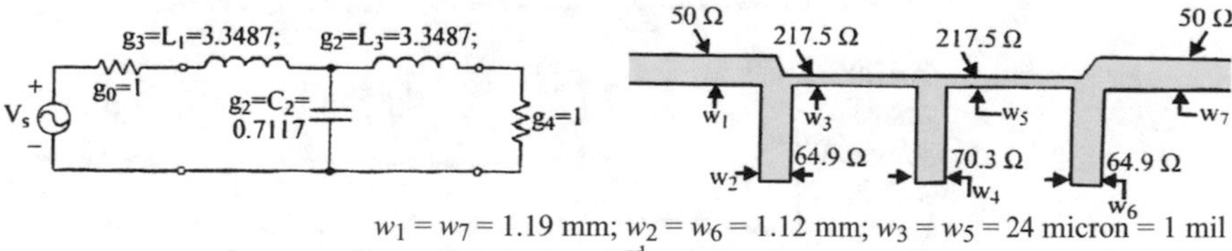

$w_1 = w_7 = 1.19$ mm; $w_2 = w_6 = 1.12$ mm; $w_3 = w_5 = 24$ micron = 1 mil
Low pass filter with 'g' values (3rd order) and corresponding μ-stripline design

© Springer Nature Singapore Pte Ltd. 2018

P. K. Chaturvedi, *Microwave, Radar & RF Engineering*,
https://doi.org/10.1007/978-981-10-7965-8_10

10.1 Introduction

The term radio frequency (RF) was used for the frequency range 300 kHz–30 MHz during 1960 and is now being used for covering also the microwave frequencies up to 300 GHz. Filters are required in most of the circuits in whole of the RF range. In fact, filter is a two-port, reciprocal, passive, and linear device, which attenuates heavily the unwanted signal frequencies, while allowing transmission of wanted signal frequencies.

Normally, filters are of four types, namely low pass (LS), high pass (HP), band pass (BP), and band stop (BS). Here the names itself indicate the frequency range which will be allowed to pass or stopped. The low pass filter allows low-frequency signals to be transmitted from its input and to the output with little attenuation. This attenuation (called insertion loss IL) increases significantly beyond cut-off frequency point. The opposite is true for high pass filter. The band pass filters allows the passing of signals to the output with low attenuation (i.e. low insertion loss, IL), between the given range of upper (f_u) and lower (f_L) cut-off frequencies. The reverse is true in the case of band stop filters.

All microwave filters are

1. Made by using reactive elements only, i.e. capacitance and inductance with sections of transmission lines having distributive impedance, as lumped elements cannot be used as microwave frequencies (Refer Chap. 1).
2. Assumed to the loss less with a very small resistance.
3. Designed to operate between generator impedance (Z_g) and load impedance (Z_L), with both resistive and normally equal to 50 Ω.

Figure 10.1 summarises the first-order basic four types of filters (LP, HP, BP, BS; their attenuation response and phase shift response, as a function of frequency. Table 10.1 and 10.2 gives their implementation as distributed components in waveguide, coaxial line andmicrostip line.

10.2 Basic Parameter of RF Filter

In filter, the cut-off frequencies are the primary parameters as shown in Fig. 10.1. The frequency versus IL plots normally are done with normalised frequency $f_n = (f/f_c)$. Here f_c is the

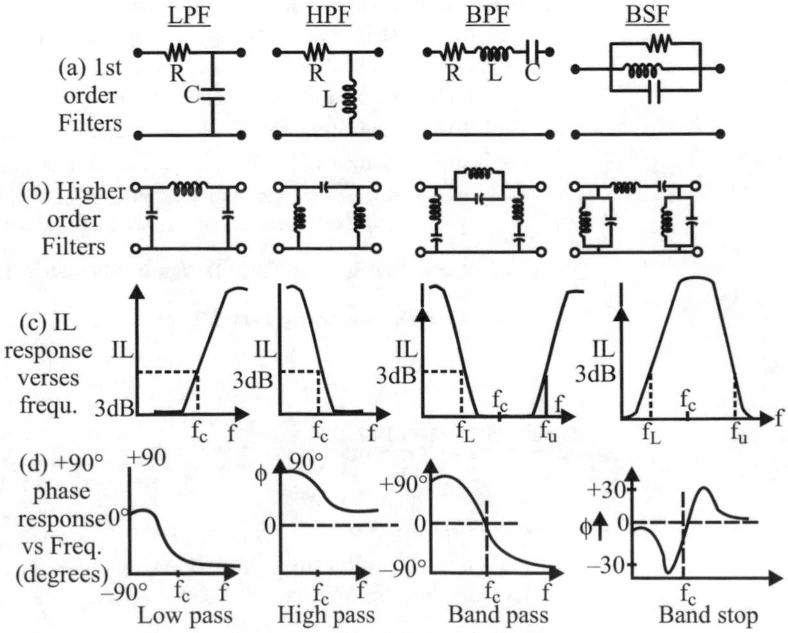

Fig. 10.1 Four basic filters LPF, HPF, BPF, and BSF; **a** first-order filters, **b** higher-order filters, **c** their typical attenuation (insertion loss IL) response in dB versus frequency. **d** Their typical phase shift response in degrees as a function of frequency

cut-off frequency in the case of low pass/high pass filters and the centre frequency of the band in the cases of band pass/stop filters (see Fig. 10.1). Therefore, this normalised frequency $f_n = 1$ is cut-off or centre frequency.

For analysing the performance/various trade-offs, the following eight parameters play key role.

1. **Insertion loss**: The basic parameter of filter design is insertion loss. The input power (P_i) given to the filter may not be equal to the output power (P_L) from the filter to the load. This difference is due to:

   (i) Power P_r reflected back due to mismatched load.
   (ii) Power P_h lost in resistive heating.

In normal cases where filter is made by using pure L and C, P_r is nearly zero. Therefore, IL in decibels may be defined as:

$$\begin{aligned}\text{IL(dB)} &= 10\log(P_i/P_L)\\ &= 10\log[P_i/(P_i - P_r)]\\ &= 10\log\left[1/(1-|\Gamma)^2\right]\end{aligned} \quad (10.1)$$

$$\text{Here}\ \Gamma = (P_r/P_i) = \text{reflection loss coefficient} \quad (10.2)$$

2. **Transmission phase change of signal (ϕ_t)**: The phase shift ϕ of signal when transmitted through a filter is different at various frequencies as depicted in Fig. 10.1 for the four filters. In low pass filters (LPF), ϕ decreases with frequency from 0° (lower frequency) to −90° after cut-off frequency, i.e. output lags the input (series component being inductance). In HPF, ϕ is 90° at lower frequency (i.e. leads) but tends to 0° beyond f_c. In BPF and BSF, The ϕ phase behaviour with frequency is given in Fig. 10.1.
3. **Group delay (T_{gd})**: It is the time it takes the information (or signal) to traverse the filter network. It is computed from the slope of the phase versus frequency curve by the equation

   $$T_{gd} = (1/2\pi)\cdot(d\phi_i/df) = (d\phi_i/d\omega)$$

   T_{gd} is in s, and it increases with the (*a*) slope of the skirt (*b*) higher order of the filter, i.e. number of L, C components. Therefore, T_{gd} is better (low) in Butterworth filter, then in the Chebyshev filter (Fig. 10.2).
   As it is desirable to have a filter with constant flat group delay, e.g. if $\phi = -A\omega$, then $(d\phi/d\omega) = A$ = constant. But $(d\phi/d\omega)$ is not constant for all frequencies; therefore, deviation from flat performance should be known.
4. **Band width (BW_{3dB})**: It is the difference between the upper cut-off frequency (f_u) and lower cut-off frequencies (f_L) in band pass and band stop filters. These two cut-off frequencies are those frequencies where the signal level has a difference of 3 dB with reference to that at centre band frequency (f_c) given in Fig. 10.3 (Chebyshev filter is discussed in Art 10.5).
5. **Transition range of skirt**: The IL response falls beyond the cut-off frequency between pass band and stops band region frequencies. This sloppy region is also termed as skirt.
6. **Ripple and ripple band amplitude**: In Chebyshev filter design, the output signal from the filter as a function of frequency has some ripple; however, shape factor is better than in other designs (see Fig. 10.2). Therefore, flatness of signal in the pass band filter can be quantified by specifying the ripple band amplitude in terms of dB, which is 3 dB given in Fig. 10.3.
7. **Shape factor (S_F)**: This describes the sharpness of the filter response and is the ratio of 60 dB band width to 3 dB band width (Fig. 10.3).

   $$S_F = \frac{BW_{60dB}}{BW_{3dB}} = \left(\frac{f_u^{60\,dB} - f_L^{60dB}}{f_u^{3\,dB} - f_L^{3\,dB}}\right) \quad (10.3)$$

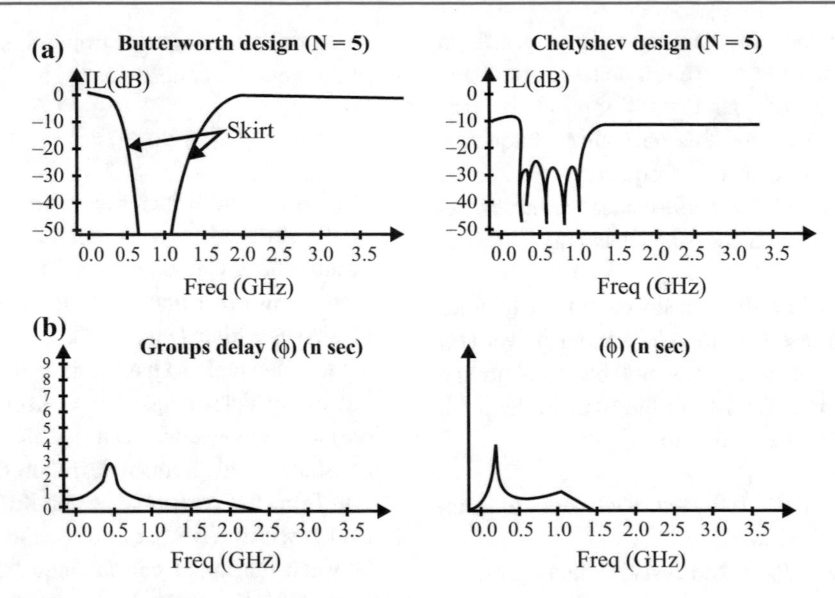

Fig. 10.2 **a** Insertion loss and **b** group delay in a typical band pass (0.5–1.5 GHz) filter of Butterworth and Chebyshev designs of order $N = 5$, discussed in detail in Art 10.5

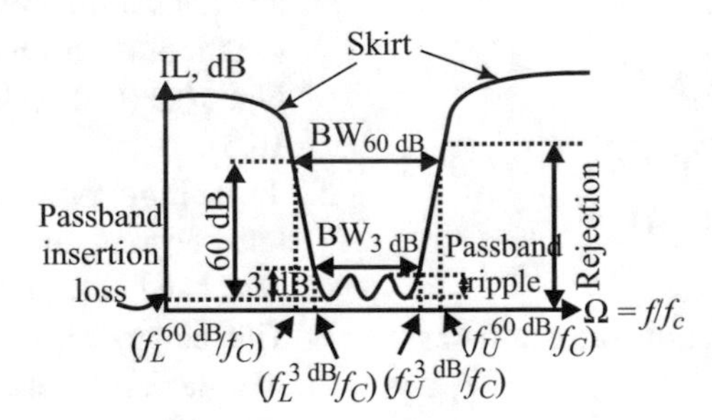

Fig. 10.3 Attenuation profile for a typical Chebyshev band pass filter (discussed in Art 10.5) as a function of normalised frequency

8. **Quality factor (Q)**: The frequency selectivity of a filter can be defined in terms of Q, the quality factor as the ratio of average energy stored per cycle to the energy lost per cycle at w_o.

$$Q = \frac{2\pi.\ \text{Average energy stored per cycle}}{\text{energy lost per cycle at } \omega_o}$$
$$Q = \omega \frac{\text{Average energy stored per cycle}}{\text{power loss}}$$
$$Q = \omega \cdot \left| \frac{W_{\text{stored}}}{P_{\text{loss}}} \right|_{\omega = \omega_c} \tag{10.4}$$

10.3 RF Filter Design Techniques

For getting the IL versus frequency profile of filters close to the Fig. 10.1, various design techniques are available. We will study only the three design techniques, e.g. the first order, and the two specialised designs of Butterworth and Chebyshev, which are with increasing level of performances in terms of better slope factor (i.e. sharpness of skirt). A comparison of the IL versus frequency profiles of these three techniques is given in Fig. 10.4, for comparison point of view.

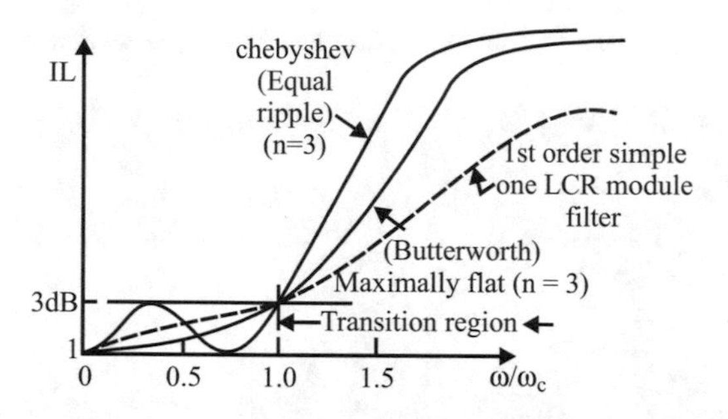

Fig. 10.4 Simple filter, Butterworth (Maximally flat) and equal-ripple low pass filter responses ($N = 3$) as a function of normalised frequency ω/ω_c $n = 3$ means three L, C modules

After having got the actual values of L and C from any of these three techniques, these components are physically realised by line sections of the waveguide or coaxial due or microstrips line, as given in Table 10.1.

In waveguides, low pass filter is not possible as waveguides itself are high pass filters.

For getting the actual filter, which has both L and C, these line sections of Figs. 10.5 and 10.6 have to be put in series. For better IL response, more number of filter sections (i.e. higher-order filter) with periodic structure of these filters are used. Possible L, C combination (but not used in practice) using the components of line section of Table 10.1 is given in Table 10.2. The practical simple structures of low pass and high pass are given in Figs. 10.5 and 10.6, respectively. **By putting low pass and high pass in series, we can get a single band stop filter.** Now we will discuss the design techniques.

(a) **First-order basic design techniques**: This is the lowest level of design of filter, consisting of only one module of L, C, R component values of which are computed by ABCD analyses.
(b) **Specialised filter design by insertion loss method**: This is done by using prototype low pass filter as the basic design. For better performance with sharper profile of IL versus frequency, we need to have more number of L, C components. For this, two design techniques are used:

   (i) Butterworth design for maximally flat or binomial.
   (ii) Chebyshev (or equal-ripple) design technique.

The first design technique of Butterworth gives a flat profile of IL versus frequency, while the Chebyshev design gives sharper IL versus frequency profile but with unavoidable ripples appearing in the pass frequency portion in all the filters (i.e. LPF, HPF, BPF, and BSF). See Figs. 10.2, 10.3 and 10.4. Detailed design will be discussed in Art 10.5

As a first order design we start with ABCD analysis of filters.

10.4 First-Order Filter Design by ABCD Analysis

10.4.1 ABCD Analysis

For analysis of a filter, its input and output parameters are represented in a matrix form called ABCD matrix as in Fig. 10.7. Here V_1, V_2 are input–output voltages and I_1, I_2 are currents entering from input or into the network from output side.

The input voltage (V_1) and current (I_1) are given by the following equation in terms of output voltage (V_2) and output currents (I_2), with A, B, C, D coefficients as:

$$V_1 = \mathrm{A}.V_2 - \mathrm{B}.I_2 \quad (10.5a)$$

$$I_1 = \mathrm{C}.V_2 - \mathrm{D}.I_2 \quad (10.5b)$$

$$\begin{bmatrix} \mathrm{A} & \mathrm{B} \\ \mathrm{C} & \mathrm{D} \end{bmatrix} \begin{bmatrix} V_2 \\ -I_2 \end{bmatrix} = \begin{bmatrix} V_1 \\ I_1 \end{bmatrix} \quad (10.6)$$

Here, these elements A, B, C, D of the matrix can be defined as follows:

Table 10.1 Basic L, C components realisation in waveguide, coaxial line, and microstrip line by using a section of these lines

Component	In waveguide	In coaxial line	In microstrip line
X_C Series C	b Z_1 Z_0 ω/g Narrow wall Sc (E-plane Tee) $\lambda_g/2 > l > \lambda/4$ $X_C = jZ_1.\tan(2\pi l/\lambda_g)$	$2b$ l Strip/coaxial line gap by a choke for a length $l < \lambda_g/4$ $X_C = jZ_1 \tan(2\pi l/\lambda_g)$ $= 1(\omega_c.c)_1$	Z_0 Z_0 Microstrip gap
X_L Series L	Z_1 Z_{in} Z_0 ω/g Narrow wall Sc stub (E-plane Tee) $l < \lambda_g/4$; $Z_{in} = jZ_1 \tan(2\pi l/\lambda_g) = X_L$	Z_0 l Z_0 Strip/coaxial line micro-strip line step down, for a length $l << \lambda_g/4$; $\omega_c L = X_L = Z_1 \tan 2\pi l/\lambda_g$	l Z_0 Z_0 $Z_1 >> Z_0$ Microstrip
C Shunt C	a b d w/g capacitive iris a b ω/g change in ht	C Z_0 Z_1 Z_0 l $Z_1 << Z_0$ Coaxial step up for a length $l << \lambda_g/4$; $1/\omega_c C = X_C$ $= Z_0 \cot 2\pi l/\lambda_g$; $Z_1 << Z_0$	Open Z_1 Z_0 Z_0 $Z_1 << Z_0$ Microstip
X_L Shunt L	x Z_1 $l < \lambda_g/4$ a Z_0 ω/g broad wall SC stub (H plane T) $l << \lambda_g/4$; $X_L = jZ_L \tan 2\pi l/\lambda_g$ a b d ω/g inductive iris	Z_0 l Z_0 Microstrip/coaxial–T, with SC termination with $l << \lambda_g/4$ to statisfy $\omega_c L = X_L = Z_0 \tan 2\pi l/\lambda_g$	Z_1 Z_0 Z_0 $Z_1 << Z_0$

Table 10.2 Possible shunt pairs of L, C components (generally not used in practice)

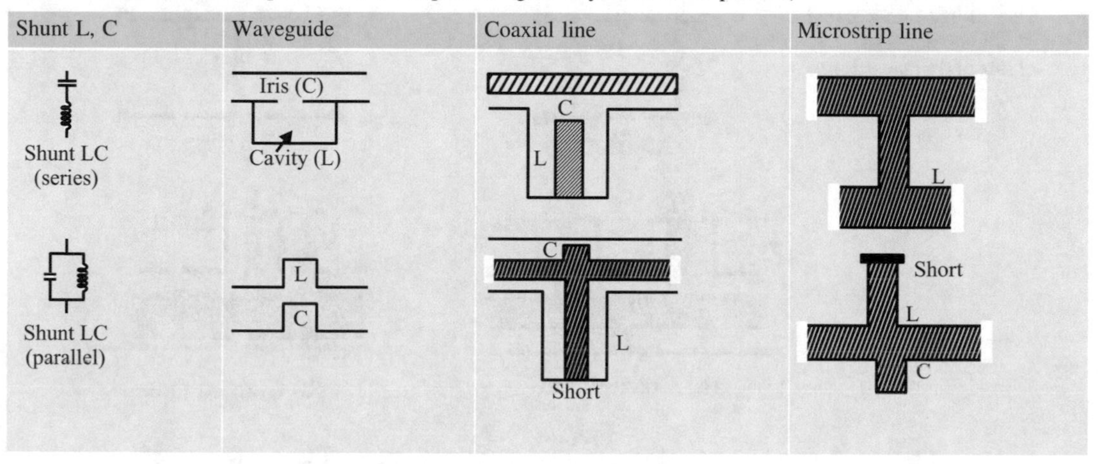

Shunt L, C	Waveguide	Coaxial line	Microstrip line
Shunt LC (series)	Iris (C) Cavity (L)	C L	L
Shunt LC (parallel)	L C	C L Short	Short L C

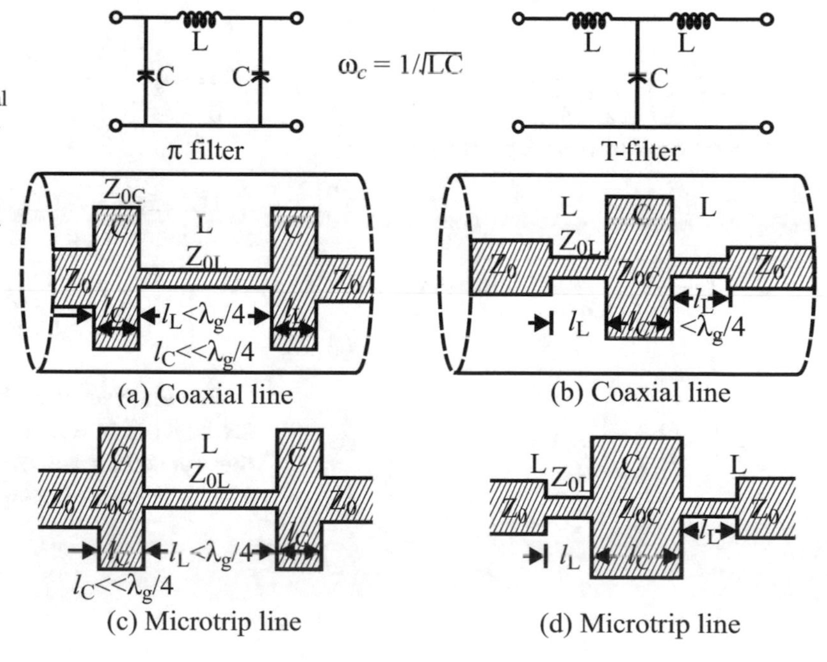

Fig. 10.5 Microwave low pass filters of first order: π section filter and T-section filters implemented in coaxial line and microstripline. In π-filters $\omega_c \cdot L = Z_{0L} \cdot \tan(\beta l_L)$ and in T-filters $1/(\omega_c \cdot C) = Z_{0C} \cdot \cot(\beta l_C)$, leading to required values of l_C, l_L

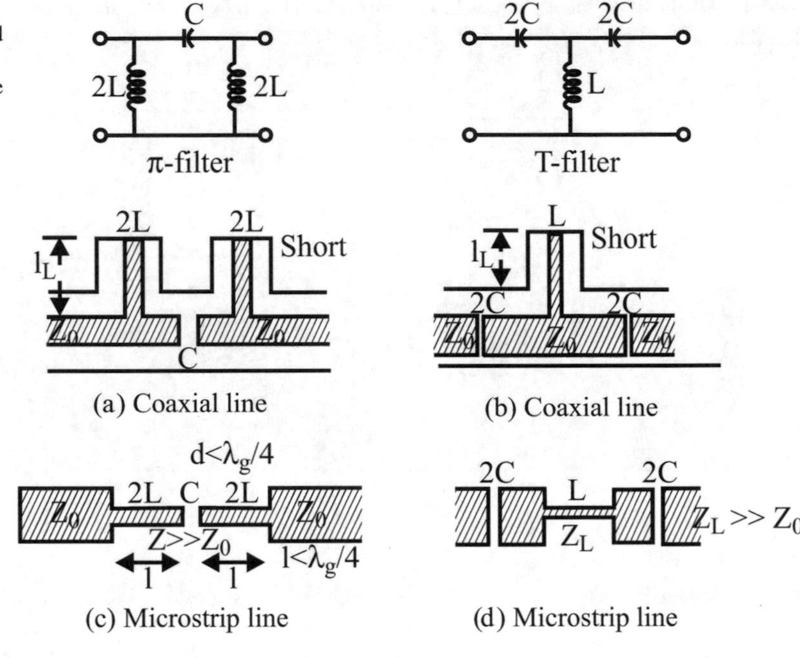

Fig. 10.6 Microwave high pass filter of first order: π and T types implemented in coaxial line and microstripline

$$\mathrm{A} = \left(\frac{V_1}{V_2}\right)_{I_2=0} = \frac{1}{\mathrm{A}_V}$$
$$= \text{reverse of voltage gain}(\mathrm{A}_V = 1/\mathrm{A}) \tag{10.7}$$

$$B = \left(-\frac{V_1}{I_2}\right)_{V_2=0} = Z_t = \text{transimpedance} \tag{10.8}$$

$$C = \left(-\frac{I_1}{V_2}\right)_{I_2=0} = Y_t = \text{transconductance} \tag{10.9}$$

$$D = -\frac{I_1}{I_2}\bigg|_{V_2=0} = -\frac{1}{A_I} = \text{reverse of current gain} \tag{10.10}$$

Therefore, ABCD matrix under the above condition will be:

$$\begin{bmatrix} \mathrm{A} & \mathrm{B} \\ \mathrm{C} & \mathrm{D} \end{bmatrix} = \begin{bmatrix} 1/A_V & Z \\ Y & -1/A_I \end{bmatrix} \tag{10.11}$$

For analysis of a filter, each of its elements are first put in terms of ABCD matrix; then all these matrices are multiplied to get the final ABCD matrix of complete filter. Therefore, let

Table 10.3 Some parameter of first-order band pass and band stop filters

	A band pass filter	A band stop filter
Parameter of the filter		
Impedance and admittance	$Z = R + j\omega L + \frac{1}{j\omega C_1}$	$Y = G + j\omega C_1 + \frac{1}{j\omega L}$
Resonant frequency (i.e. centre freq.)	$\omega_0 = \frac{1}{\sqrt{LC_1}}$	$\omega_0 = \frac{1}{\sqrt{LC_1}}$
Dissipation factor	$d = \frac{R}{\omega_0 L} = \frac{1}{R\omega_0 C_1}$	$d = \frac{R}{\omega_0 C_1} = G\omega_0 L$
Quality factor	$\frac{1}{d} = Q = \frac{\omega_0 L}{R} = \frac{1}{R\omega_0 C_1}$	$\frac{1}{d} = Q = \frac{\omega_0 C_1}{G} = \frac{1}{G\omega_0 L}$
3 dB band width	$BW = \frac{f_0}{Q} = \frac{1}{2\pi}\frac{R}{L}$	$BW = \frac{f_0}{Q} = \frac{1}{2\pi}\frac{G}{C_1}$
Voltage gain (V_2/V_1)	$\frac{Z_L}{Z_L + Z_g + R + j(\omega_L - 1/\omega C_1)}$	$\frac{Z_L}{Z_L + Z_g + 1/(G + j\omega C_1 - 1/\omega L)}$
Quality factor (internal)	$Q_{\text{int}} = \frac{\omega_0 L}{R}$	$Q_{\text{int}} = \frac{\omega_0 C_1}{G} = \frac{1}{G\omega_0 L}$
Quality factor (ext.)	$Q_{\text{ext.}} = \frac{\omega_0 L}{Z_L + Z_g}$	$Q = \frac{\omega L}{(Z_L + Z_g)}$
Quality factor loaded	$Q_{\text{loaded}} = \frac{\omega_0 L}{R + Z_L + Z_0}$	$Q = \frac{\omega_0 L}{R + Z_L + Z_g}$
Insertion loss of power (i.e. attenuation)	$2\ln(V_2/V_1)$ Np Or $-20\log(V_2/V_1)$ dB	$-2\ln(V_2/V_1)$ Np Or $-20\log(V_2/V_1)$ dB

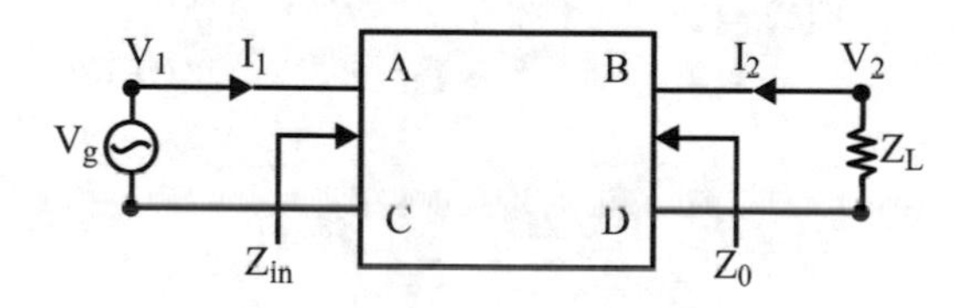

Fig. 10.7 Voltage, currents, and impedances at the input–output side for finding the ABCD matrix of a network

us see how a series element or shunt element is represented in ABCD matrix (Fig. 10.8).

(a) **ABCD matrix of a series load**: Here $I_1 = -I_2$, $V_1 \neq V_2$, $V_2 = V_1 - (I_1 - I_2)$z.

therefore

$$D = -\frac{I_1}{I_2}\bigg|_{V_2=0} = 1 \left(\begin{array}{l} \text{For O/P shorted } V_2 = 0 \\ \therefore \quad I_1 = I_2 \end{array} \right)$$

$$A = -\frac{V_1}{V_2}\bigg|_{I_2=0} = 1 \left(\begin{array}{l} \text{For open output } I_2 = 0 \\ \therefore \quad V_1 = V_2 \end{array} \right)$$

$$B = -\frac{V_1}{I_2}\bigg|_{V_2=0} = Z \left(\begin{array}{l} \text{For } V_2 = 0, \text{ i.e., O/P is shorted} \\ \therefore \quad I_2 = I_1 \end{array} \right)$$

and

$$C = -\frac{I_1}{V_2}\bigg|_{I_2=0}$$

$$= 0 \text{ (for } I_2 = 0, \text{ O/P open as } I_1 = I_2)$$

$$\begin{bmatrix} A & B \\ C & D \end{bmatrix} = \begin{bmatrix} 1 & Z \\ 0 & 1 \end{bmatrix} \tag{10.12}$$

(b) **ABCD of a shunt load**: From Fig. 10.8, we see that $V_1 = V_2$, therefore the ABCD matrix parameters will be:

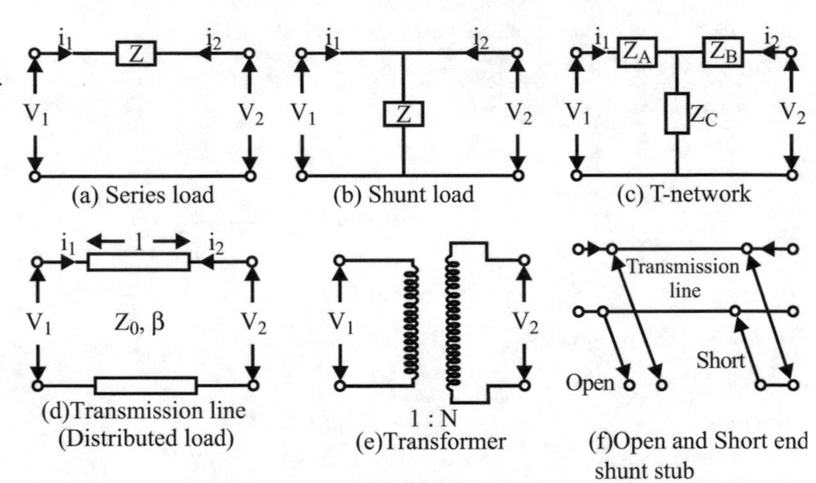

Fig. 10.8 Series, shunt, T network, transmission line, and the transformer loads for ABCD analysis

$$
\begin{aligned}
A &= \left.\frac{V_1}{V_2}\right|_{I_2=0} = 1 \quad (I_2 = 0 = \text{output open}) \\
B &= -\left.\frac{V_1}{I_2}\right|_{V_2=0} = 0 \quad (\text{As } V_2 = 0 = V_1, I_1 = I_2) \\
C &= \left.\frac{I_1}{V_2}\right|_{I_2=0} \cong \frac{I_1}{V_2} = Y \quad [I_2 = 0 \text{ means output open}] \\
D &= -\left.\frac{I_1}{I_2}\right|_{V_2=0} = 1 \quad [As\ V_2 = 0, \text{ means shorted output}, I_1 = I_2] \\
\therefore \begin{bmatrix} A & B \\ C & D \end{bmatrix} &= \begin{bmatrix} 1 & 0 \\ Y & 0 \end{bmatrix}
\end{aligned}
\tag{10.13a}
$$

(c) **ABCD of a T network**: For a T network, Fig. 10.8 can be taken as three networks as shown in Fig. 10.8a, b, c in series. Therefore, its ABCD matrix will be product of the three:

$$
\begin{aligned}
\begin{bmatrix} A & B \\ C & D \end{bmatrix} &= \begin{bmatrix} 1 & Z_A \\ 0 & 1 \end{bmatrix} \begin{bmatrix} 1 & 0 \\ 1/Z_C & 1 \end{bmatrix} \begin{bmatrix} 1 & Z_B \\ 0 & 1 \end{bmatrix} \\
&= \begin{bmatrix} 1 + \frac{Z_A}{Z_C} & \left(Z_A + Z_B + \frac{Z_A Z_B}{Z_C}\right) \\ \frac{1}{Z_C} & \left(1 + \frac{Z_B}{Z_C}\right) \end{bmatrix}
\end{aligned}
\tag{10.13b}
$$

(d) **ABCD of transmission line**: A transmission line (Fig. 10.8d) normally has distributed impedance (See art 2.2) at microwave frequency. For zero attenuation $\alpha = 0$ at the input of a short length 'l' of transmission line, the impedance with load Z_L is known from Eq. 2.1p to be:

$$Z_{in}(l) = Z_0 \left[\frac{Z_L + jZ_0 \tan(\beta l)}{Z_0 + jZ_L \tan(\beta l)}\right] \tag{10.13C}$$

The voltage and current can be written with the forward and reflected components as:

$$
\begin{aligned}
V(\ell) &= (V^+ \cdot e^{-j\beta l} + V^- \cdot e^{j\beta l}) \\
&= V^+ \cdot (e^{-j\beta l} + \Gamma_0 \cdot e^{j\beta l})
\end{aligned}
\tag{10.13c}
$$

$$
\begin{aligned}
I(\ell) &= \left(\frac{V^+}{Z_0} \cdot e^{-j\beta\ell} - \frac{V^-}{Z_0} \cdot e^{j\beta\ell}\right) \\
&= \frac{V^+}{Z_0} \cdot (e^{-j\beta\ell} - \Gamma_0 \cdot e^{j\beta\ell})
\end{aligned}
\tag{10.13d}
$$

where

$\Gamma_0 \;= V^-/V^+ =$ reflection coefficient.

For getting ABCD matrix of a transmission line, we will now study open-ended stub and short-ended stub.

Open-ended stub: It has $\Gamma_0 = +1$ as reflection coefficient (as $Z_L = \infty$, $I_2 = 0$)

$$
\begin{aligned}
\therefore\ & V(l) = 2V^+(\cos \beta l) \\
& I(l) = (2jV^+/Z_0) \sin \beta l \\
\therefore\ & Z_{in}(l) = -jZ_0 \cot(\beta l)
\end{aligned}
\tag{10.13e}
$$

Short-ended stub: It has $\Gamma_0 = -1$, as $Z_0 = 0$, $V_2 = 0$

$$V(\ell) = V^+(e^{j\beta l} - e^{-j\beta l}) = 2jV^+ \sin(\beta l) \quad (10.13f)$$

$$I(l) = (V^+/Z_0) \cdot (e^{j\beta l} + e^{-j\beta l}) = \frac{2V^+}{Z_0} \cos \beta l \quad (10.13g)$$

$$\therefore \quad Z_{insc} = jZ_0 \tan(\beta l) \quad (10.13h)$$

Now using the $V(l)$, $I(l)$ equations of open- and short-ended slubs, we can now compute the ABCD parameters of the transmission line (Fig. 10.8d). Here we may note that $i_2 = 0$ means open-ended stub and $v_2 = 0$ means short-ended stub where $i_2 = I_2(l)$ and $v_2 = V_1(l)$ are at the output end and $i_1 = I(l)$ and $v_1 = V(l)$ with $l = 0$ are at the input end.

Using open-ended stub equations ($i_2 = 0$)

$$A = (v_1/v_2)|_{I_2=0} = \frac{2V_0^+ \cos(\beta l)}{2V^+} = \cos(\beta l) \quad (10.13i)$$

$$= (i_1/v_2)|_{i_2=0} = C = (i_1/v_2)|_{i_2=0} \frac{(2jV^+/Z_0)\sin(\beta l)}{2V^+} = jY_0 \sin(\beta l) \quad (10.13j)$$

Using short-ended stub equations ($v_2 = 0$)

$$B = v_1/(-i_2)|_{v_2=0} = \frac{(2jV^+) \cdot \sin(\beta l)}{2V^+/Z_0} = jZ_0 \sin(\beta l) \quad (10.13k)$$

$$D = i_1/(-i_2)|_{v_2=0} = \frac{(2V^+/Z_0)\cos(\beta l)}{(2V^+/Z_0)} = \cos(\beta l) \quad (10.13l)$$

Therefore, for the transmission line of Fig. 10.8d:

$$\begin{bmatrix} A & B \\ C & D \end{bmatrix} = \begin{bmatrix} \cos(\beta l) & jZ_0 \sin(\beta l) \\ jY_0 \sin(\beta l) & \cos(\beta l) \end{bmatrix} \quad (10.13m)$$

Using all the ABCD parameters discussed above, we will now find their ABCD matrix of the following filters and then analyse their performance in terms of their insertion loss (IL) and transmission phase (ϕ) as a function of frequency:

- Low pass filter—RC type
- High pass filter—RL type
- Band pass filter—RLC in series
- Band stop filter—parallel RLC combination in series.

The relation between ABCD parameters of I–V relation and scattering matrix parameter S_{21} (received and reflected back power ratio) is related as:

$S_{21} = 2/A$, which will be used frequently.

(e) **ABCD of a transformer**: Without proof, we write the ABCD matrix of a transformer of winding ratio 1: N_1 (Fig. 10.8e) as:

$$\begin{bmatrix} A & B \\ C & D \end{bmatrix} = \begin{bmatrix} 1/N_1 & 0 \\ 0 & N_1 \end{bmatrix} \quad (10.13n)$$

and of a transformer of winding ratio N_2:1 is

$$\begin{bmatrix} A & B \\ C & D \end{bmatrix} = \begin{bmatrix} N_2 & 0 \\ 0 & 1/N_2 \end{bmatrix}$$

10.4.2 Low Pass Filter (LPF) of First Order

The first-order filter having generator impedance Z_g and load impedance Z_L is taken equal to the characteristics of line impedance Z_0 ($Z_g = Z_L = Z_0$), for simplifying the analysis (Fig. 10.9). This circuit can be best evaluated by cascading four ABCD networks with two series elements (Z_g, R) and two shunt elements (C, Z_L). ABCD matrices of these four elements can be written based on our previous article and multiplied to get the ABCD parameter of the whole filter circuit as:

$$\begin{bmatrix} A & B \\ C & D \end{bmatrix} = \begin{bmatrix} 1 & Z_g \\ 0 & 1 \end{bmatrix} \begin{bmatrix} 1 & R \\ 0 & 1 \end{bmatrix} \begin{bmatrix} 1 & 0 \\ j\omega C_1 & 1 \end{bmatrix} \begin{bmatrix} 1 & 0 \\ 1/Z_L & 1 \end{bmatrix} = \begin{bmatrix} 1+(R+R_g)(j\omega C_1+1/Z_L) & (R+Z_0) \\ (j\omega C_1+1/Z_L) & 1 \end{bmatrix} \quad (10.14)$$

Note: As C has been used as A, B, C, D parameters, C_1 is used for capacitance here.

Comparing Eqs. (10.11) and (10.14), we get voltage gain (A_V).

$$1/A = A_V = \text{voltage gain} = \left(\frac{V_2}{V_1}\right) = \frac{1}{1+(R+Z_g)(j\omega C_1+1/Z_L)} \quad (10.15)$$

Now we compute A_V for high/low frequency, attenuation, and phase shift for LPF.

(a) **At high frequencies** ($\omega \to \infty$):

$$A_V = 0 \quad (10.16)$$

(b) **At low frequencies** ($\omega \to 0$):

$$A_V = \frac{1}{1+(R+Z_g)/Z_L} = \frac{Z_L}{Z_L+R+R_g} \quad (10.17)$$

(c) **Power attenuation** (i.e. insertion loss):

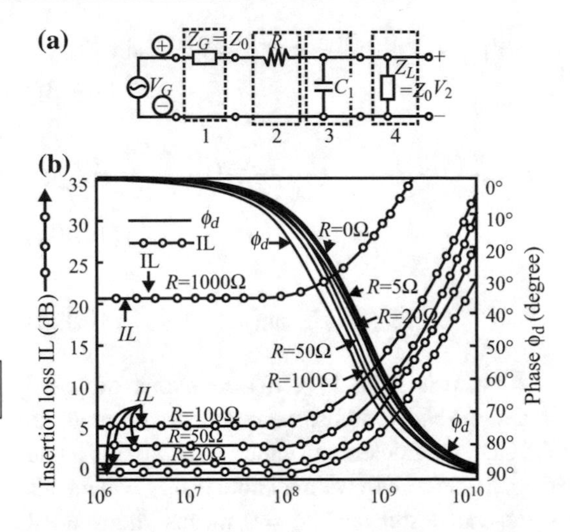

Fig. 10.9 **a** First-order RC low pass filter as cascaded four ABCD networks. **b** Its response: insertion loss IL (dB) and phase shift (in degrees) as a function of frequency for C = 12 pf and for different parasitic resistance R

$$\alpha = -\ln(V_1/V_2)^2 = -2\ln(V_1/V_2)(\text{In nepers}) \quad (10.18)$$

or

$$\alpha = \log(1/A_V) = -20\log(V_1/V_2)\ (\text{in dB}) \quad (10.19)$$

(d) **Phase shift** between input and output will be:

$$\phi_r(\omega) = \tan^{-1}\left[\frac{I_m(V_2/V_1)}{\text{Re}\,(V_2/V_1)}\right] \text{rad} \quad (10.20a)$$

i.e.

$$\phi_d(\omega) = (180/\pi)\phi_r\,^\circ. \quad (10.20b)$$

(e) **Group delay**

$$t_d = -\frac{d\phi}{d\omega} s \quad (10.21)$$

Normally, a filter is designed with linear phase delay, i.e. $\phi = -A_1\omega$, where A is some constant factor, then group delay $t_d = A$.

The parasitic resistance leads to unwanted loss and therefore ideal filter has to be with $R = O$, but R_L and R_g will always be there.

10.4.3 High Pass Filter (HPF) of First Order

Replacing capacitor with inductor, we get the first-order HPF (Fig. 10.10). Therefore, again getting the ABCD parameter of each of the component in cascade and then by multiplying them, we get the ABCD parameter of voltage–current relation as:

$$\begin{bmatrix} A & B \\ C & D \end{bmatrix} = \begin{bmatrix} 1 & R_g \\ 0 & 1 \end{bmatrix} \begin{bmatrix} 1 & R \\ 0 & 1 \end{bmatrix} \begin{bmatrix} 1 & 1 \\ 1/j\omega L & 1 \end{bmatrix} \begin{bmatrix} 1 & 0 \\ 1/R_L & 1 \end{bmatrix} = \begin{bmatrix} 1 + (R+R_g)\left(\frac{1}{j\omega L} + \frac{1}{R_L}\right) & (R_g + R_L) \\ \left(\frac{1}{j\omega L} + \frac{1}{R_L}\right) & 1 \end{bmatrix}$$

$$\text{And therefore}: A_V = \frac{1}{A} = \frac{V_2}{V_1} = \frac{1}{1 + (R+R_g)\left(\frac{1}{j\omega L} + \frac{1}{R_L}\right)} \quad (10.22)$$

The special cases/parameters for HPF will be:

(a) At high frequency ($\omega \to \infty$)

$$A_V = \frac{1}{1 + (R+R_g)/R_L} \quad (10.23)$$

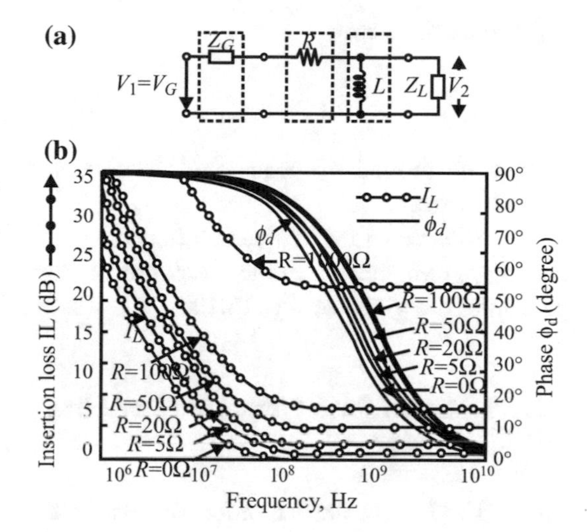

Fig. 10.10 **a** First-order RL high pass filter as cascaded four ABCD networks. **b** Its response: insertion loss IL (dB) and phase shift of signal (ϕ_d) in degrees as a function of frequency for L = 110 nH and different values of parasitic resistance R

(b) At low frequency ($\omega \to 0$); $A_V \to 0$ (the property of HPF).

(c) Insertion loss or power attenuation (in Np)

$$= -\ln\left(\frac{V_2}{V_1}\right)^2 = -2\ln\left(\frac{V_2}{V_1}\right) \quad (10.24)$$

(d) Insertion loss or power attenuation (in dB)

$$= -20\log\left(\frac{V_2}{V_1}\right) \quad (10.25)$$

(e) Transmission phase shift (in radians)

$$\phi_r = \tan^{-1}\left[I_m\left(\frac{V_2}{V_1}\right)/\text{Re}\left(\frac{V_2}{V_1}\right)\right] \quad (10.26)$$

(f) Transmission phase shift (in degrees)

$$\phi_d = (180/\pi) \cdot \phi_r{}^\circ \quad (10.27)$$

(g) Scattering matrix

$$S_{21} = 2/A \tag{10.28}$$

(can be proved by comparing S-matrix and ABCD matrix)

The insertion loss (IL) versus frequency and phase (ϕ_d) version frequency response of high pass filter are given in Fig. 10.10.

10.4.4 Band Pass Filter (BPF) of First Order

Figure 10.11 gives the first-order band pass filter, which can be analysed by taking the three LCR series components as one component $z\left(= R + j\omega L + \frac{1}{j\omega C_1}\right)$ for making the case simple (Fig. 10.11).

For complete ABCD matrix of the band pass filter (BPF) given in Fig. 10.11a, we get ABCD matrices of each of the partition and multiply them, e.g.

$$\begin{bmatrix} A & B \\ C & D \end{bmatrix} = \begin{bmatrix} 1 & Z_g \\ 0 & 1 \end{bmatrix} \begin{bmatrix} 1 & Z_1 \\ 0 & 1 \end{bmatrix} \begin{bmatrix} 1 & 0 \\ 1/Z_L & 1 \end{bmatrix} = \begin{bmatrix} 1 + (Z_1 + R_g)/R_L & (R_g + Z_1) \\ 1/R_L & 1 \end{bmatrix} \tag{10.29}$$

where

$$Z_1 = R + j\left(\omega L - \frac{1}{\omega C_1}\right) \tag{10.30}$$

$$\text{gain} = \frac{1}{A} = \frac{V_2}{V_1} = \frac{Z_L}{Z_L + Z_g + R + j(\omega L - 1/\omega C_1)} \tag{10.31}$$

Attenuation (or insertion loss), propagation phase shift, etc., due to filter will be:

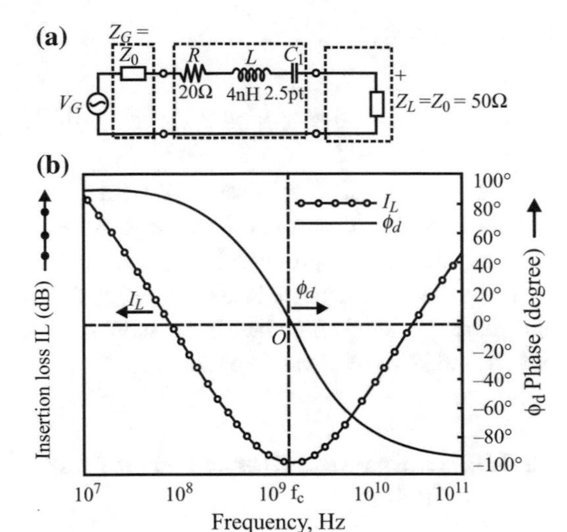

Fig. 10.11 **a** First-order band pass filter: series RLC components taken as one element Z = R + j (ωL − 1/ω C_1) for ABCD analysis. Thus, we have three cascaded ABCD networks. **b** Its response: insertion loss IL (dB) and phase shift of signal (ϕ_d) in degrees for $R = 20\ \Omega$, $L = 4$ nH and $C_1 = 2.5$ pF, $f_C = 1/(2\pi\sqrt{LC_1}) = 1.59$ GHz

(a) **IL**:

$$\text{IL(neper)} = \ln\left(\frac{V_1}{V_2}\right)^2 = -2\ln\left(\frac{V_2}{V_1}\right) \tag{10.32}$$

$$\text{IL(dB)} = 20\ \log\left(\frac{V_1}{V_2}\right) = -20\ \log\left(\frac{V_2}{V_1}\right) \tag{10.33}$$

(b) **Phase shift**:

$$\phi_r\,(\text{rad}) = \tan^{-1}\left[\frac{\text{Im}\,(V_2/V_1)}{\text{Re}\,(V_2/V_1)}\right] \tag{10.34}$$

$$\phi_d = \left(\frac{180}{\pi}\right) \cdot \phi_r\,^\circ \tag{10.35}$$

(c) **Cut-off frequency**:

$$f_c = \frac{1}{2\pi\sqrt{LC_1}} \quad \therefore \omega_c L = \frac{1}{\omega_c C_1} \tag{10.36}$$

(d) **Quality factor**:

As Impedance/Resistance External to the filter is $Z_E = (Z_g + Z_L)$

$$\therefore Q_{ext} = \frac{\omega_c L}{Z_E} = \frac{1}{\omega_c \cdot C_1 \cdot Z_E}; \; Q_{int} = \frac{\omega_c L}{R} = \frac{1}{\omega_c C_1 R} \tag{10.37}$$

$$\therefore \quad Q_{loaded} = \left(\frac{\omega_c \cdot L}{R + Z_E}\right) = \frac{1}{\omega_c \cdot C_1 \cdot (R + Z_E)} \tag{10.38}$$

$$\frac{1}{Q_{loaded}} = \frac{1}{Q_{ext}} + \frac{1}{Q_{int}} \tag{10.39}$$

(e) **Band width**:

$$\Delta f_{bw} = f_0 / Q_{loaded} \tag{10.40}$$

(f) **Power**: Taking $Z_L = Z_g = Z_0$

$$\text{Source power } P_{inmax} = V_g^2/(8Z_0) \tag{10.41}$$

$$\text{Power to load } P_L = V_g^2.Z_0/(Z_0 + 2R)^2 \tag{10.42}$$

10.4.5 Band Stop Filter (BSF) of First Order

In BSF, the series LCR of BPF is replaced by its parallel combination (Fig. 10.12), which has conductance $Y = G + j(j\omega - 1/\omega L)$; therefore, the three ABCD matrices of Z_g, Y, and Z_L will give total ABCD matrix as:

$$\begin{bmatrix} A & B \\ C & D \end{bmatrix} = \begin{bmatrix} 1 & Z_g \\ 0 & 1 \end{bmatrix} \begin{bmatrix} 1 & 1/Y \\ 0 & 1 \end{bmatrix} \begin{bmatrix} 1 & 0 \\ 1/Z_L & 1 \end{bmatrix}$$

$$\therefore \quad \begin{bmatrix} A & B \\ C & D \end{bmatrix} = \begin{bmatrix} \left(1 + \frac{Z_g + 1/Y}{Z_L}\right) & (Z_g + 1/Y) \\ 1/Z_L & 1 \end{bmatrix} \tag{10.43}$$

where $Y = G + j(\omega C_1 - 1/\omega L)$.

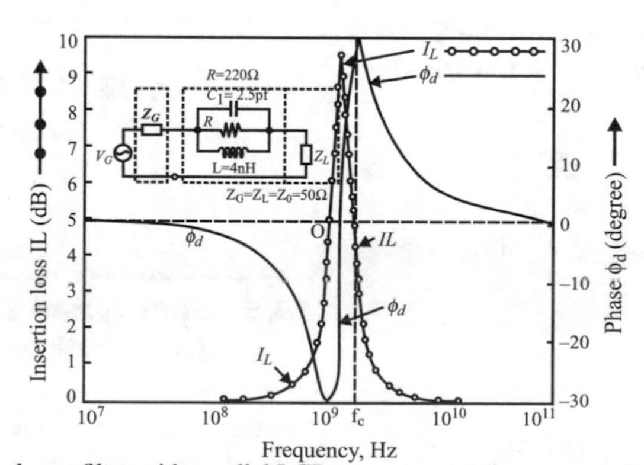

Fig. 10.12 **a** First-order band stop filter with parallel LCR components taken as one element $Y = G + j(\omega C_1 - 1/\omega L)$ for ABCD analysis. Thus, there are three cascaded ABCD networks. **b** Its response: insertion loss IL(dB) and phase shift of signal ϕ_d in degrees for R = 220 Ω, L = 4 nH and C = 2.5 pF; thus $f_c = 1/\left(2\pi\sqrt{LC_1}\right) = 1.59$ GHz.

The parameters of BSF are:

(a) **Resonance gain**:

$$A_V = \frac{V_2}{V_1} = \frac{1}{A} = \frac{Z_L}{\left[Z_L + Z_g + \frac{1}{G + j(\omega C_{-1}/\omega L)}\right]}$$
$$= \frac{Z_L}{Z_L + Z_g + 1/Y} = \frac{Z_L \cdot Y}{(Z_L + Z_g) \cdot Y + 1} \quad (10.44)$$

(b) **Resonant frequency**: $f_C = 1/\left(2\pi\sqrt{LC_1}\right)$

(c) I_L: **Power attenuation due to filter** (i.e. insertion loss) will be:

$$\text{IL (neper)} = -\ln\left(\frac{V_2}{V_1}\right)^2 = -2\ln(V_2/V_1) \quad (10.45a)$$

$$\text{IL(dB)} = -20\log(V_2/V_1) \quad (10.45b)$$

(d) ϕ: **Signal transmission phase shift** will be

$$\phi_r(\text{rad}) = \tan^{-1}\left[\frac{-(V_2/V_1)}{\text{Real}(V_2/V_1)}\right] \quad (10.46a)$$

$$\phi_d = (180/\pi) \cdot \phi_r \quad (10.46b)$$

Some of the parameters of first order Band Pass and Band stop filters are summarised in Table 10.3.

Problem In a BPF $Z_0 = 50\ \Omega$, $Z_g = Z_L = Z_0$, $R = 15\ \Omega$, L = 40 nH; $C_1 = 0.47$ pf; $V_g = 10$ V; Find Q_{loaded}, Q_{ext}, Q_{internal} (i.e. filter alone), P_{source}, P_{load}, f_u, f_L, IL.

Solution

$$f_c = \frac{1}{2\pi\sqrt{LC_1}} = 1.16\,\text{GHz}$$

$$\therefore \quad \omega_c = 2\pi f_c = 6.28 \times 1.16$$
$$= 7.29 \times 10^9 \text{Radian/s}$$

As $R_{\text{ext}} = R_g + R_L = Z_0 + Z_0 = 2Z_0$

$$Q_{\text{ext}} = \frac{\omega_0 L}{R_{\text{ext}}} = \frac{\omega_0 L}{2Z_0}$$
$$= \frac{7.29 \times 10^9 \times 40 \times 10^{-9}}{2 \times 50} = 2.9$$

$$\therefore \quad Q_{\text{Loaded}} = \frac{\omega_0 L}{R + 2Z_0}$$
$$= \frac{7.29 \times 40}{(15 + 100)} = 2.54$$

$$P_{\text{in}} = \frac{Vg^2}{8Z_0} = \frac{100}{8 \times 50}$$
$$= 0.250 = 250\text{mW}$$

$$P_L = \frac{Vg^2 \cdot Z_0}{(Z_0 + 2R)^2} = \frac{100 \times 50}{(50 + 30)^2}$$
$$= 0.781 = 781\text{mW}$$

$\Delta f_{3\text{dB}} = 3$ dB band width $= \frac{f_c}{Q_{\text{loaded}}} = \frac{1.16 \times 10^9}{2.54} =$ 456 MHz

$\therefore \quad f_u = (f_c + \Delta f_{3\text{dB}}/2) = 1.16 + 0.228 = 1.388\,\text{GHz}$
and $\quad f_L = f_C - \Delta f_{3\text{dB}}/2 = 1.16 - 0.228 = 0.932\,\text{GHz}$
$\therefore \quad \text{IL(dB)} = 10\log\left[\left(1 + \Delta^2 \cdot Q^2_{\text{Loaded}}\right)\left(Q^2_{\text{Loaded}}/Q^2_{\text{ext}}\right)\right]$

where Δ is the normalised frequency deviation $\Delta = (\omega/\omega_0 - \omega_0/\omega)$. At resonance $\Delta = 0$.

$$\therefore \quad \text{IL(dB)} = 20\log(Q_{\text{ext}}/Q_{\text{Loaded}})$$
$$= 20 \times 0.05767$$
$$= 1.15\,\text{dB}$$

10.5 Specialised Filter Design by Insertion Loss Method

Besides first-order filter design, the specialised designs, i.e. Butterworth and Chebyshev designs, are actually used in practice. These designs have sharper roll-off IL versus frequency profile at cut-off. Before analysing these designs, let us study the power loss ratio (P_{LR}) for matched source and load:

$$P_{LR} = \frac{\text{Power available from source}}{\text{Power delivered to the load}} = \frac{P_{in}}{P_L} = \frac{1}{1 - |\Gamma(\omega)|^2} = \frac{1}{|S_{12}|} \quad (10.47)$$

The insertion loss (IL) in dB is:

$$\text{IL} = 10\log(P_{LR}) = -10\log\left[1 - |\Gamma(\omega)|^2\right] \quad (10.48a)$$

Using the complex terms in $[\text{IL} = [Z_L(\omega) - Z_0]/[Z_L(\omega) + Z_0]$, it can be proved that $|\Gamma(\omega)|^2$ is an even function of ω [i.e. $\Gamma(-\omega) - \Gamma(\omega)$], therefore it can be expressed as a polynomial in ω^2 as:

$$|\Gamma(\omega)|^2 = \frac{M(\omega^2)}{M(\omega^2) + N(\omega^2)} \quad (10.48b)$$

where M and N are real polynomials in ω^2. Therefore, the Eqs. 10.47 and 10.48 become:

$$P_{LR} - 1 + \frac{M(\omega^2)}{N(\omega^2)} \quad (10.49a)$$

$$\text{IL} = 10\ \log\left[1 + \frac{M(\omega^2)}{N(\omega^2)}\right] \quad (10.49b)$$

Choice of this polynomial ratio M/N decides the type of response of IL versus frequency profile. The practical and useful responses are:

(i) Butterworth (called maximally flat)
(ii) Chebyshev (called equal-ripple)
(iii) Ecliptic function
(iv) Linear phase.

We will discuss the first two only, which have a number of common features of design. To start with, the final polynomial of P_{LR} for these two response is as follows with IL = 10 log (P_{LR})

$$P_{LR} = 1 + a^2\left(\frac{\omega}{\omega_c}\right)^{2N} \quad (10.50)$$

(Butterworth maximally flat design)

and $$P_{LR} = 1 + a^2 T_n^2\left(\frac{\omega}{\omega_c}\right) \quad (10.51)$$

(Chebyshev equal-ripple design).

In the process of designing any of the four filters, the starting point is the designing a prototype LPF (Table 10.4). The reactant of this LPF are in terms of a parameter called g_n values as given in the first column of Table 10.4. Here 'n' is the order (Number of components of that LPF) choosen by us.

As discussed in the previous article that for better performance of filter profile, i.e. sharper IL versus frequency, the number of L, C modules has to be larger. For this, the number of components is defined by the order 'n' of that filter. Also it has been observed that for the same number of components, the Chebyshev filter gives fast roll-off of IL after cut-off frequency, but at the cost of ripples in the IL versus frequency profile (Fig. 10.4).

For computing the normalised component values, called 'g' values of L and C components which are required in the distributed form of a transmission line of our choice (waveguide, coaxial or microstrip line), standard tables are available. (Table 10.5 for Butterworth and Table 10.6 for Chebyshev techniques) The order of 'g' values of the L, C components (i.e. 'n') is the number of L and C components chosen for the design. Higher is the order, the sharper is the IL profile.

In both, the design techniques (Butterworth and Chebyshev) following three steps are followed, for finally getting the de normalised values of L, C components from 'g' values and finally realising them on the transmission line sections.

(a) **Choose the filter design order 'n'**: This choice is done by the sharpness of the IL versus frequency profile expected, then get the 'g' values of 'L' and 'C' of the filter from Table 10.5 for Butterworth or Table 10.6 for Chebyshev. Next two steps are for scaling,

Table 10.4 Computing actual values of elements L_n, C_n from the prototype LPF elements L_{pn}, C_{pn} by impedance transformation and frequency transformation

Getting prototypic LPF elements L_{pn}, C_{pn} (by impedance transformation of g_n values from Tables 10.5 and 10.6)	Getting actual values of L_n (Henry) and C_n (Farad) of the four filters. (By frequency transformation using chosen f_c and BW) (All C_n in Farad and L_n in Henry)			
	LPF	HPF	BPF	BSF
1. Series arm $L_{pn} = g_n.Z_g$	L_n $L_n=\frac{L_{pn}}{\omega_c}$	C_n $C_n=\frac{1}{\omega_c.L_{pn}}$	$L_n C_n$ $L_n=\frac{L_{pn}}{(BW)}$ $C_n=\frac{(BW)}{\omega_c.L_{pn}}$	L_n C_n $L_n=\frac{(BW).L_{pn}}{\omega_c^2}$ $C_n=\frac{1}{(BW)L_{pn}}$
2. Shunt arm C_{pn} $C_{pn}=\frac{g_n}{Z_g}$	C_n $C_n=\frac{C_{pn}}{\omega_c}$	L_n $L_n=\frac{1}{\omega_c.L_{pn}}$	L_n C_n $L_n=\frac{(BW)}{\omega_c^2 C_{pn}}$ $C_n=\frac{C_{pn}}{(BW)}$	L_n C_n $L_n=\frac{1}{(BW)C_{pn}}$ $C_n=\frac{(BW).C_{pn}}{\omega_c^2}$

Band width (BW) = ($f_u - f_L$); $Z_L = Z_g = Z_0 = 50\ \Phi$

NB The above design rules are true for both Butterworth and Chebyshev designs with values g_n of Tables 10.5 and 10.6, respectively

Table 10.5 Butterworth filter: g_n values for maximally flat low pass filter x = 1–6

n	g_1	g_2	g_3	g_4	g_5	g_6	g_7
1	2.0000	1.0000					
2	1.4142	1.4142	1.0000				
3	1.0000	2.0000	1.0000	1.0000			
4	0.7654	1.8478	1.8478	0.7654	1.0000		
5	0.6180	1.6180	2.0000	0.6180	0.6180	1.0000	
6	0.5176	1.4142	1.9318	1.9318	1.4142	0.5176	1.0000

i.e. impedance or frequency transformation to the desired Z_0 and f_c values of our choice.

(b) **Impedance transformation for getting prototype of LPF components**: From the 'g_n' values of the prototype LPF, of order 'n', the normalised values of 'L' and 'C' (L_{pn}, C_{pn}) are obtained using formulas of Table 10.4 for the normalised frequency $f_n = f/f_c$ as per given below:

$$L_{p1} = g_1.Z_0; \quad C_{p2} = g_2/Z_0 \text{ etc.} \qquad (10.52)$$

Therefore, these prototype values L_n, C_n are valid for all frequencies.

(c) **Frequency transformation for getting actual components of the four filter (LPF, HPF, BPS, BSF)**: Here we choose the frequency and band width of our use, and then by using the prototype values of L_{pn}, C_{pn} of

Table 10.6 Chebyshev filter design: g_n values of L, C components of LPF for n = 1 to 6 for allowed ripples of **a** 3 dB **b** 0.5 dB. At the right are their loss (IL) versus frequency performances for n = 4

(a) g_n for 3 dB ripple LPF

n	g_1	g_2	g_3	g_4	g_5	g_6	g_7
1	1.9953	1.0000					
2	3.1013	0.5339	5.8095				
3	3.3487	0.7117	3.3487	1.0000			
4	3.4389	0.7483	4.3471	0.5920	5.3095		
5	3.4817	0.7648	4.5381	0.7618	3.4817	1.0000	
6	3.5045	0.7685	4.6061	0.7929	4.4641	0.6033	5.3095

(b) g_n for 0.5 dB ripple LPF

n	g_1	g_2	g_3	g_4	g_5	g_6	g_7
1	0.6986	1.0000					
2	1.4029	0.7071	1.9841				
3	1.5963	1.0967	1.5963	1.0000			
4	1.6703	1.1926	2.3661	0.8419	1.9841		
5	1.7058	1.2296	2.5408	1.2296	1.7058	1.0000	
6	1.7254	1.2479	2.6064	1.3137	2.4758	0.8696	1.9841

order n of LPF, we calculate the actual values of inductance (Henry) and capacitance (Farad) by Table 10.4 for the required filter (may be LPF or HPF or BPF or BSF). This is for order n, centre frequency (ω_c), and band width (BW) chosen by us.

(d) **Filter implementation**: We have got the lumped values of 'L_n' and 'C_n' of filter, which has to be realised in the transmission line by line segment of different structures (as shown in Figs. 10.5 and 10.6) as distributed circuit elements. This topic will be taken up in Art No. 10.6.

Here now we will discuss these two types of the filters of design techniques, i.e. Butterworth (maximally flat) and Chebyshev (equal-ripple) through the above-indicated steps.

10.5.1 Butterworth Filter Design (Maximally Flat or Binomial)

This type of filter is known as maximally flat filter since no ripple is permitted in its attenuation profile. For such low pass filters, the insertion loss (in absence of any internal resistance loss) is given by Eq. 10.51 and therefore by the following polynomial:

$$\text{IL(dB)} = 10\log\left[\frac{1}{1-|\Gamma(\omega)|^2}\right] = 10\log\left(\frac{P_{\text{in}}}{P_{\text{L}}}\right)$$
$$= 10\ \log(P_{\text{LR}})$$

We will get

$$\text{IL(dB)} = 10\ \log\left[1 + a^2(\omega/\omega_c)^{2n}\right] \quad (10.53)$$

where 'a' is a constant and we may select it to be a = 1, so that at (ω/ω_c) = 1, the value of IL = 10 log (2) = 3 dB at cut-off frequency $\omega = \omega_c$ at f_n = 1. Figure 10.4 gives the IL versus normalised frequency and the cut-off point for three types of filters of n = 3 for each. (i.e. for first order, Butterworth and Chebyshev filters) for showing the comparison.

The rate of increase of IL with 'ω' in equation No. 10.53 depends on the index $2n$, where n is the number of components.

Two possible realisations of π type and T type of generic normalised low pass filter are shown in Fig. 10.13.

The element values of L, C components in Fig. 10.13 are numbered as g_0 (for generate) to g_{n+1} (load). The elements in the circuit alternate between series inductance and shunt capacitance. These elements are defined as:

g_0 = Generator resistance in π-type filters (Fig. 10.13a) or generator conductance T–type filters (Fig. 10.13b)
g_{n+1} = Load resistance for π-type filter or load conductance for T-type filter
g_n = Inductance for series inductor of the filter or capacitance for shunt capacitor of the filter (n = 1, 2, 3, … N)

For example, from Table 10.5 for $n = 3$, $g_1 = 1.0$, $g_2 = 2.0$, $g_3 = 1.0$ and

g_{3+1} = normalised load resistance or conductance = 1.

Insertion loss in LPF of Butterworth design for different frequencies is given in Fig. 10.14

(a) $\omega < \omega_c$: It has a flat behaviour i.e. v.v. low loss here.
(b) For $\omega > \omega_c$, insertion loss increase is very steep with frequency.
(c) **For** $\omega \gg \omega_c$: (IL) $\cong 10\log (\omega/\omega_c)^{2n} \cong 20\, n\log (\omega/\omega_c)$

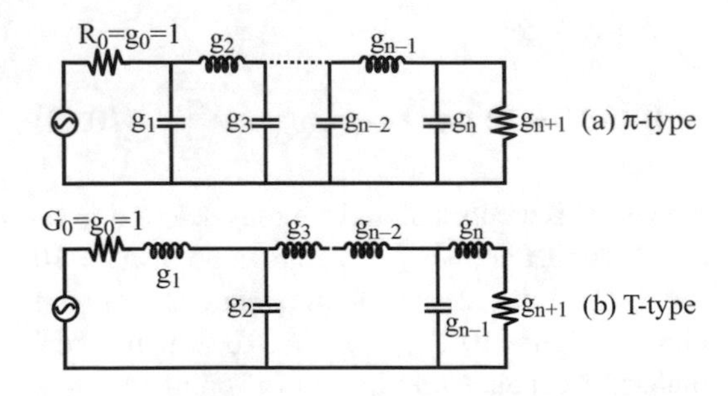

Fig. 10.13 Two types: **a** π type and **b** T type of realisation of low pass filter (LPF), with normalised elements values g_0, g_1, g_2, … g_n with g_0 as generator resistance or conductance and g_{n+1} the load impedance or conductance

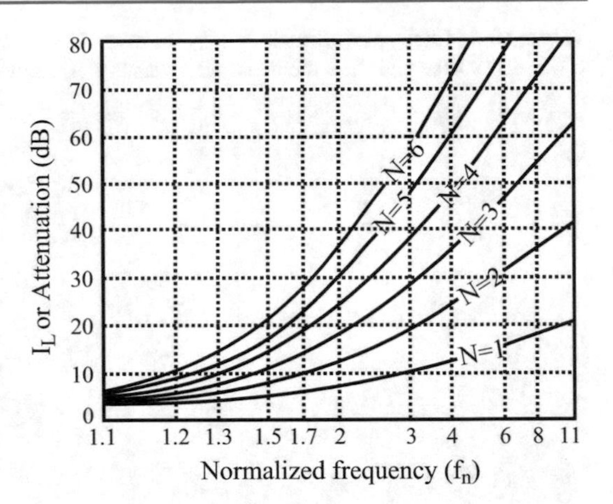

Fig. 10.14 Butterworth filter (maximally flat) filter. Insertion loss in a low pass filter versus normalised frequency (f_n) (for n = 1–6)

If $\omega = 10\,\omega_c$ (IL) = 20 $n$log 10 = 20 n dB i.e. IL increases at the rate of 20 n dB per decade of increase of frequency. For getting the final filter components by this technique, we follow the four designing steps (a) to (d) discuss in Article 10.5.

10.5.1.1 Demystifying the Origin of Design Values G_n

For this, let us study a filter for $n = 2$ case, i.e. only two L and C components for the case of Butterworth filter as in Fig. 10.15, with L, C, R as normalised values.

Here Z_{in} will be:

$$Z_{in} = (j\omega L) + \left(\frac{1}{j\omega c}\right) || R_L = \left(j\omega L + \frac{R_L}{1 + j\omega R_L C}\right) \quad (10.54)$$

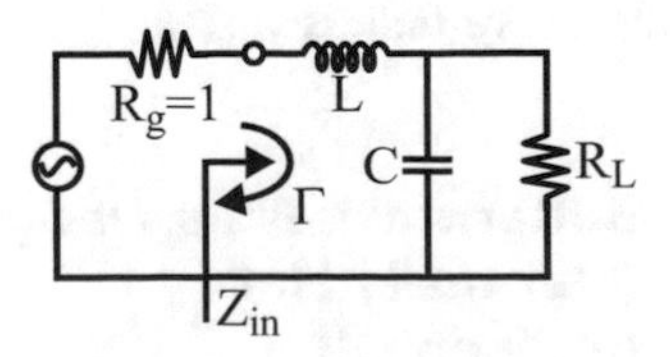

Fig. 10.15 Butterworth filter of second order (n = 2)

and reflection coefficient will be:

$$\Gamma = \left(\frac{Z_{in} - 1}{Z_{in} + 1}\right). \quad (10.55)$$

(where Z_{in} = normalised input impedance).

∴ Putting this in the equation of power loss (P_{LR}), we get:

$$P_{LR} = \left[1 - |\Gamma|^2\right]^{-1} = \left[1 - \left|\frac{Z_{in} - 1}{Z_{in} + 1}\right|^2\right]^{-1}$$
$$= \{1 - [(Z_{in} - 1)/(Z_{in} + 1)][(Z^*_{in} - 1)/(Z^*_{in} + 1)]\}^{-1}$$
$$\therefore \quad P_{LR} = \frac{|Z_{in} + 1|^2}{2(Z_{in} + Z^*_{in})} \cdot \left[\text{where } (Z_{in} + Z^*_{in} = \frac{2R_L}{1 + \omega^2 R_L^2 C^2}\right]$$

By putting the value of Z_{in} from Eq. 10.54 in above, we get a polynomial expression in ω^2.

$$P_{LR} - 1 + (1/4R_L)[(1 - R_L)^2 + (R_L^2 C^2 + L^2 - 2LCR_L^2)\omega^2 + L^2 C^2 R_L^2 \omega^4] \quad (10.56a)$$

Also take the Eq. (10.50), by $a = 1$ and at cut-off frequency, i.e. $\omega_c = \omega$, it becomes

$$P_{LR} = 1 + a^2 \left(\frac{\omega}{\omega_c}\right)^{2N}$$

Considering the simple case of two-element LPF Butterworth prototype as in Fig. 10.15, L, C, R_L are normalised then

$$P_{LR} = 1 + (\omega)^4 \quad (10.56b)$$

Comparing 10.56a and 10.56b of P_{LR}, we infer that coefficient of $\omega^4 = 1$ and of $\omega^2 = 0$, therefore for $R = 1$ we get:

$$C^2 + L^2 - 2LC = 0 \quad \text{i.e. } (C - L)^2 = 0 \quad \therefore \quad L = C$$

and

$$\frac{1}{4}L^2C^2 = \frac{1}{4}L^4 = 1 \quad \therefore \quad L = \sqrt{2}$$
$$\therefore \quad L = C = \sqrt{2} = 1.414$$

Therefore, in normalised scale it means

$$g_1 = g_2 = 1.414 \quad (10.56c)$$

This is the same value as in Table 10.5. Similarly for higher values of n, Table 10.5 can be obtained.

10.5.2 Chebyshev Equal-Ripple Filter

This type of filter is based in the insertion loss behaviour given by the polynomial in Eq. 10.51, i.e.

$$\text{IL(dB)} = 10\log(P_{LR}) = 10\log[1 + a^2 \cdot T_n(\omega/\omega_c)] \quad (10.57)$$

where the polynomial $T_n(\omega/\omega_c)$ has equal-ripple of amplitude 'a^2' due to its sinusoidal function for $\omega < \omega_c$ as following.

$$a^2 T_n(\omega/\omega_c) = a^2 \cos\left[n \cdot \cos^{-1}(\omega/\omega_c)\right] \text{ for } |\omega/\omega_c| < 1 \quad (10.58)$$

$$a^2 T_n(\omega/\omega_c) = a^2 \cos h\left[n \cdot \cosh^{-1}(\omega/\omega_c)\right] \text{ for } |\omega/\omega_c| > 1 \quad (10.59)$$

$$\cong (1/2)[2\omega/\omega_c]^N [\text{for } a = 1 \text{ and } (\omega/\omega_c) \gg 1] \quad (10.60)$$

This equal-ripple amplitude for $a = 1$ and for different frequency (for $\omega < \omega_c$) as well as for

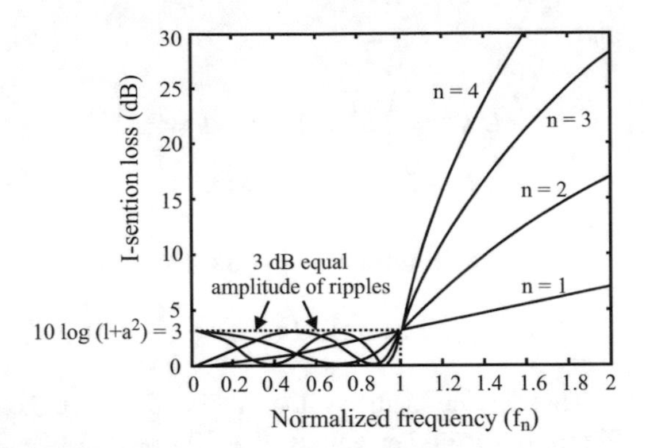

Fig. 10.16 Chebyshev equal-ripple filter frequency dependence of insertion loss of the low pass filter (for n = 1–4 and a = 1)

different 'n' can be seen in band pass region of the IL versus normalised frequency response curve in Fig. 10.16.

At $(\omega/\omega_c) = 1$ in Eq. 10.57, we note that IL = 10 log (1 + a^2)

$$\begin{aligned} &= 10\log 2\ [\text{for}\ a = 1] \\ &= 10 \times 0.3 = 3\,\text{dB} \end{aligned}$$

Also for

$$\begin{aligned} a &= 0.5 \\ IL &= 10\log 1.25 \\ &\approx 0.97 \approx 1.0\,\text{dB} \end{aligned}$$

Thus, we see that the magnitude of the ripple can be controlled by suitably choosing the value of 'a' of Eqs. (10.57)–(10.60), through the following equation:

$$a = \sqrt{10^{(\text{ripple in dB}/10)} - 1} \tag{10.61}$$

For example, for getting a good IL versus frequency profile with a small ripple of 0.5 dB, we have to select $a = \sqrt{10^{0.5/10} - 1} = 0.349$, using this value of a, the associated profile for the n = 1–6 is shown in Fig. 10.16. Comparing Figs. 10.17 and 10.18, we see that higher is the amplitude of the ripple, the sharper is the IL profile for the same n = 1 with these ripple of 3 dB and 0.5 dB; the corresponding values of g_n coefficients of L, C components are given in Table 10.6a, b.

Unlike butterworth filter, the Chebyshev filter provides steeper pass band/stop band transition. For $\omega \gg \omega_c$, from Eq. 10.60 we infer that this filter has improvement in attenuation of roughly $2^{2n}/4$ over Butterworth.

Problem 1

(a) A third-order Chebyshev low pass filter is to be designed with maximum allowed ripple of 3 dB with cut-off frequency of 2.1 GHz on 50 Ω line.

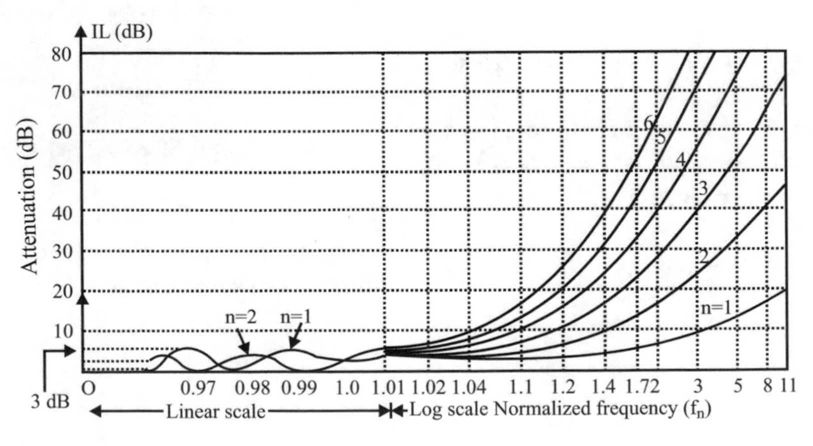

Fig. 10.17 Attenuation response for 3 dB Chebyshev design of LPF. It has larger ripple of 3 dB in $\omega < \omega_c$, but sharper profile for $\omega > \omega_c$

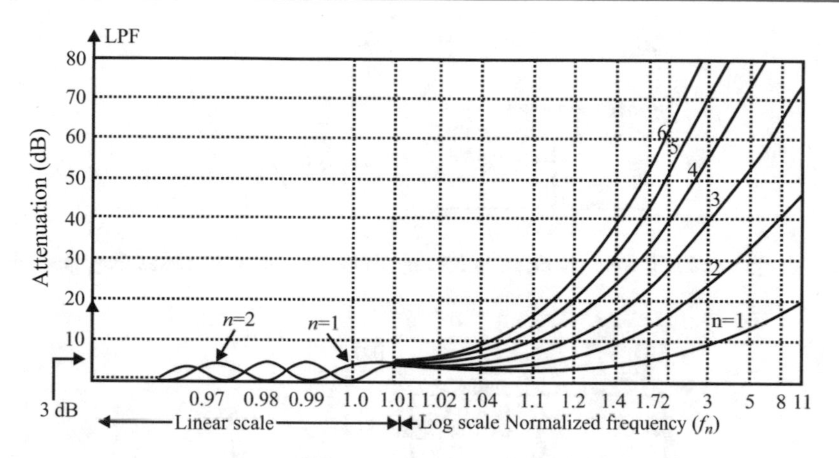

Fig. 10.18 Attenuation response for 0.5 dB Chebyshev design LPF. It has smaller ripple (0.5 d) in $\omega < \omega_c$ but less sharper profile for $\omega < \omega_c$. For higher n attenuation ripple amplitude variations at lower frequency at $<\omega_c$ are more

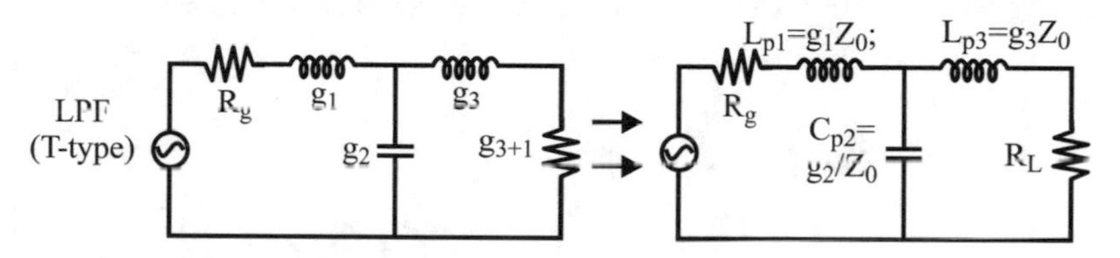

Fig. 10.19 Prototype LPF from g_n values

(b) Extend this design for band pass filter with centre frequency of 2.5 GHz and ±10% band width (i.e. 20%).
(c) Draw the IL performance of each.

Solution First we find the g values of L_{pn} and C_{pn} for LPF prototype and then actual filter by impedance and frequency transformation (Fig. 10.19).

Step 1: Third-order filter g values: we use, $n = 3$, Table 10.5.

$$(g_1 = 3.3487 = g_3; g_2 = 0.7117; g_{3+1} = 1.)$$

We choose

$$R_g = R_L = 50\ \Omega \quad \text{(For maximum power transfer)}$$

$$\therefore \quad g_0 = g_{n+1} = 1.0$$

Step 2: Prototype LPF by impedance transformation.

$$L_{p1} = L_{p3} = g_1 \times R_g = 3.348 \times 50 = 167.435\ \text{H}$$

$$C_{p2} = \frac{g_2}{R_g} = \frac{0.7117}{50} = 14.234\ \text{F}$$

Step 3: Frequency transformations:

(3a) getting actual values of L and C for LPF from proto LPF values for f_C = 2.1 GHz.

For f_c = 2.1 GHz; $\omega_c = 2\pi f_c = 13.188 \times 10^9$ rad/s.

By Table 10.4 (Fig. 10.20)

$$L_1 = \frac{L_{p1}}{\omega} = \left(\frac{167.435}{13.118 \times 10^9}\right) = 12.76\,\text{nH}$$

$$C_2 = \frac{C_{p2}}{\omega_c} = \frac{14.234}{13.118 \times 10^9} = 1091.9\,\text{pf}$$

(3b) **BPF**: Getting actual values of L and C of BPF from proto LPF values for f_0 = 2.5 GHz

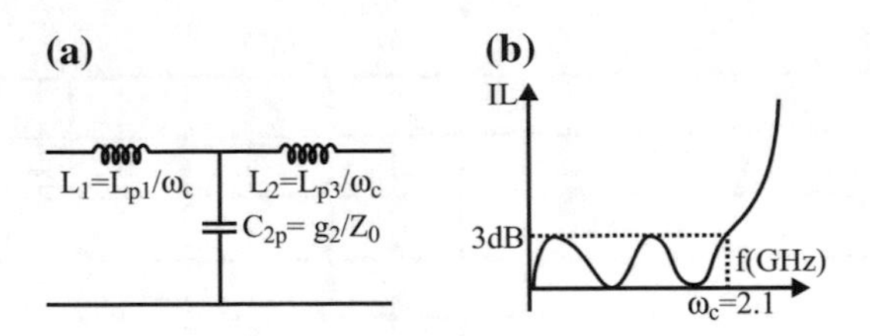

Fig. 10.20 **a** Getting actual LPF from prototype LPF using Table 10.4 and **b** IL versus frequency performance

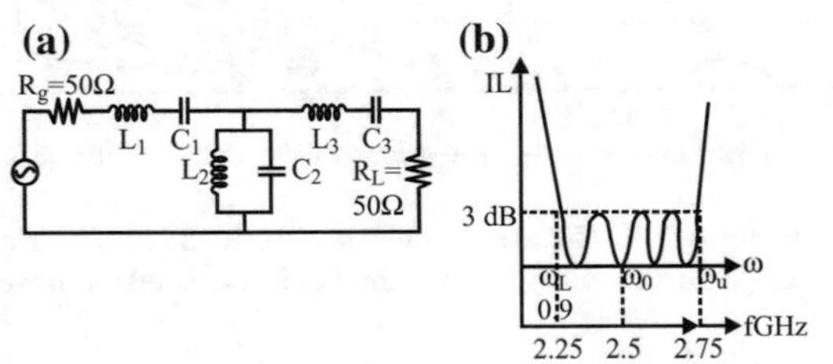

Fig. 10.21 **a** Getting actual HPF from proto LPF and **b** IL versus frequency performance

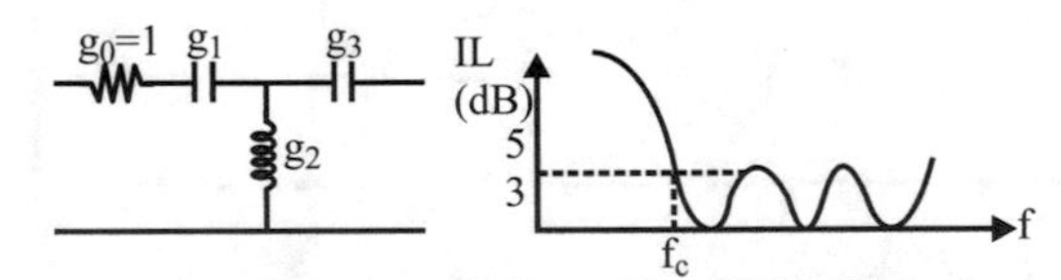

Fig. 10.22 High pass filter of third order (n = 3) with three components of T-type circuit

$\therefore \quad \omega_0 = 2.5 \times 2\pi = 15.7 \times 10^9$

For 10% up : $\omega_u = \omega_{upper} = 1.1 \times 2\pi \times 2.5 \times 10^9 = 17.27 \times 10^9$ rad/s

For 10% down : $\omega_L = \omega_{lower} = 0.9 \times 2\pi \times 2.5 \times 10^9 = 14.13 \times 10^9$ rad/s

$\therefore \quad (\omega_u - \omega_L) = \Delta\omega = \Delta W = 3.14 \times 10^9$ rad/s

Note: If ω_L, ω_u are given and not ω_0, then $\omega_0 = \sqrt{\omega_L.\omega_u}$.

∴ By Table 10.4

$$L_1 = L_3 = \left(\frac{L_{p1}}{\Delta\omega}\right) = \frac{167.4}{3.14 \times 10^9} = 53.3\,\text{nH}$$

$$C_1 = C_3 = \frac{\Delta\omega}{\omega_0^2 L_{p1}} = \frac{3.14 \times 10^9}{(15.7^2 \times 10^{18} \times 167.4)} = 0.0771\ \text{pf}$$

$$L_2 = \frac{\Delta\omega}{\omega_0^2 C_{p2}} = \frac{3.14 \times 10^9}{(15.7 \times 10^9)^2(14.234)} = 0.855\,\text{pH}$$

$$C_2 = \frac{C_{p2}}{\Delta\omega} = \frac{14.234}{3.14 \times 10^9} = 4.53\,\text{nF}$$

(3c) **HPF**: Actual values of L and C for HPF from prototype LPF values by Table 10.4.

The HPF circuit of order 3 (n = 3) from Table 10.6 is given in Figs. 10.21 and 10.22 along with its IL versus frequency profile. By Table 10.4 we get for f_c = 2.1 GHz

$$\text{Series capacitance}(C_1 = C_3) = \frac{1}{\omega_c.L_{p_1}} = \frac{1}{15.7 \times 10^9 \times 167.435}\ \text{F} = 0.380\,\text{pF}$$

$$\text{Shunt inductance } L_2 = \frac{1}{\omega_c.C_{p2}} = \frac{1}{15.7 \times 10^9 \times 14.234} = 4.47\,\text{pH}$$

3(d) **BSF**: Actual values of L and C from proto LPF from Table 10.4: Circuit is given as per Fig. 10.23 along with its IL profile.

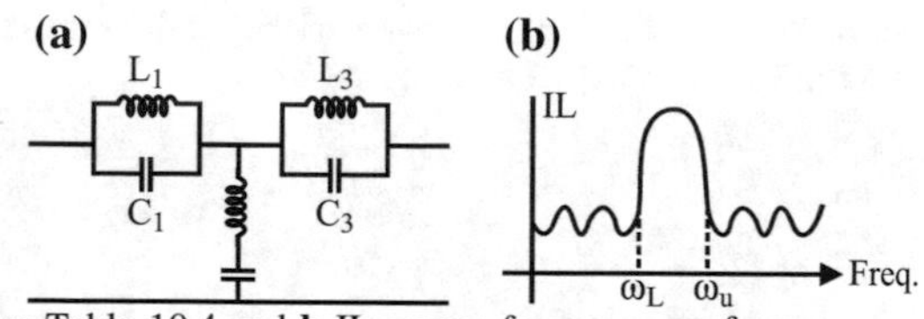

Fig. 10.23 **a** BSF circuit from Table 10.4 and **b** IL versus frequency performance

From BPF data above, BW = $\Delta\omega$ = 3.14 10^2, $\omega_C = 15.7 \times 10^9$

$$L_1 = \frac{(\text{BW}) \cdot L_{p1}}{\omega_c^2} = L_3 = \frac{(3.14 \times 10^9) \cdot 50}{(15.7 \times 10^9)^2} = 645\,\text{pH}$$

$$C_1 = \frac{1}{(\text{BW}) \cdot L_{P_1}} = C_3 = \frac{1}{3.14 \times 10^9 \times 50} = 6.37\,\text{pF}$$

$$L_2 = \frac{1}{(\text{BW}) \cdot C_{p2}} = \frac{1}{3.14 \times 10^9 \times 14.234} = 22.37\,\text{pH}$$

$$C_2 = \frac{(\text{BW})C_{p2}}{\omega_c^2} = \frac{(3.14 \times 10^9) \times 14.234}{(15.7 \times 10^9)^2} = 196\,\text{pF}$$

Problem 2 Solve the above problem for Butterworth design.

Solution Here all the calculations are same except g_n values for $n = 3$. Butterworth filter designs are from Table 10.5:

$$g_1 = 1.000 = g_3; g_{3+1} = 1 = R_\text{L}$$
$$g_2 = 2.000$$

Now we use Table 10.4 for the following:

(a) Prototype LPF:

$$L_{p1} = g_1 * R_g = 50$$
$$C_{p2} = g_2/R_g = 2/50 = 0.04$$

(b) Actual LPF:

$$L_1 = L_3 = \frac{L_{p1}}{\omega_c} = \frac{50}{15.7 \times 10^9} = 3.14\,\text{nF}$$

$$C_2 = \frac{C_{p2}}{\omega_c} = \frac{0.04}{15.7 \times 10^9} = 2.5\,\text{pF}$$

(iii) Actual BPF:

$$L_3 = L_1 = \frac{l_{p_1}}{\Delta\omega} = \frac{50}{3.14 \times 10^9} = 3.14 \times 10^{-9} = 318.5\,\text{pH}$$

$$C_1 = C_3 = \frac{\Delta\omega}{\omega_0^2 L_{p_1}} = \frac{3.14 \times 10^9}{(15.7 \times 10^9)^2 50} = 0.25\,\text{pF}$$

$$L_2 = \frac{\Delta\omega}{\omega_0^2 C_{p_2}} = \frac{3.14 \times 10^9}{(15.7 \times 10^9)^2 0.04} = 318.5\,\text{pH}$$

$$C_2 = \frac{C_{p_2}}{\Delta\omega} = \frac{0.04}{3.14 \times 10^9} = 12.7\,\text{pF}$$

Filter type	LPF (f_c = 2.1 GHz)		BPF (f_0 = 2.5 GHz; Δf = 0.25 GHz)			
Component	$L_1 = L_3$	C_2	$L_1 = L_3$	$C_1 = C_3$	L_2	C_2
Butter worth	12.76	1091.9	53.3	0.077	0.85	4500
Chebyshev	3.14	2.5	318.5	0.25	318.5	12.7

Note '*L*' in nH; *C* in pF

Table 10.7 Lumped equivalence of λ/8, λ/4, λ/2 lines is open-ended or short-ended stubs put as series or shunt stub. This table is just an extension of Fig. 2.24, in Chap. 2 and Fig. 10.24

	Transmission (open-/short-ended series shunt stubs)	Lumped equivalence in the transmission line sections of		
		λ/8 line section	λ/4 line section	λ/2 line section
(a) short-ended series stub	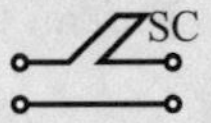 SC	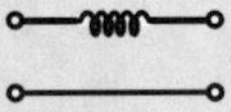	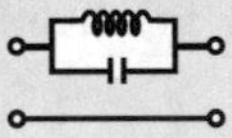	
(b) open-ended series stub	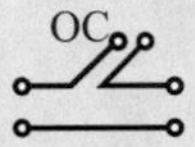 OC	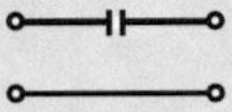	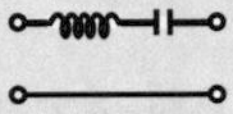	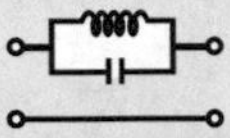
(c) open-ended shunt stub	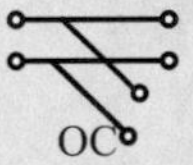 OC	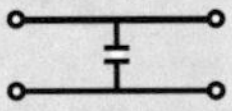		
(d) short-ended shunt stub	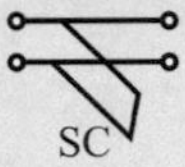 SC		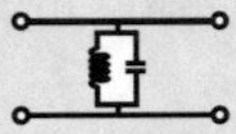	

Comparison of L, C component values for the two designs (i.e. Butterworth and Chebyshev) for same cut-off for LPF ($f_c = 2.1$) and 2.5 GHz as centre frequency for BPF, and same band width $\Delta f = (\pm 10\%\text{ of } f_0) = 0.25$ GHz, for $n = 3$, is given below:

10.6 Filter Implementation on Microstrip Line

The lumped components arrived at the previous articles cannot be used as discrete components beyond 1 GHz, as the size of the filter dimensions becomes comparable to the wave length. Therefore, for getting a practical filter, the lumped components of art 10.5 need to be converted into distributed elements of transmission line, which could be a section of waveguide, coaxial line, or microstrip line. The filter design could normally be chosen out of Butterworth or Chebyshev designs.

Here in the book, we will discuss only the microstrip line as our transmission line. In order to get the values of lumped series or shunt stub elements (C and L) in the distributed form in transmission line sections, we use

(a) **Richard's transformation**: This is by using a short- or open-ended transmission line segments, which may be put in series or in shunt.

(b) **Kuroda's identities**: The Kuroda identities are used for converting 'difficult to implement design to simpler implementable design'. For example, a series inductance (See Table 10.7) which is made by a short-ended stub line segment (difficult to fabricate) is replaced by an open-ended shunt line segment (simple to fabricate).

Using these two approaches [as (*a*) and (*b*) above], only low pass (LPF) and band stop filters (BSF) are designed, **as the Kuroda identities are not useful for designing**

high pass filter (HPF) and band pass filters (BPF). Therefore, for HPF and BPF, coupled line filters are used (Art 10.7). Before we discuss the Richard's transformation and Kuroda's identities, let us study how a short-ended/open-ended transmission line segment stub of $\lambda/2$ or $\lambda/4$ or $\lambda/8$ length behaves as L, C components (see Fig. 2.24). but differently. We will try to prove in the next section the equivalence given in Table 10.7 which is an extension of Fig. 2. 24 of Chap. 2.

10.6.1 Half-Wave and Quarter-Wave Section Lines as LCR Resonators-Proof

Consider a lossy transmission line, for which it has been proved (Eq. 2.1p) that:

$$Z_{in} = Z_0 \left[\frac{Z_L + Z_0 \tanh(\gamma l)}{Z_0 + Z_L \tanh(\gamma l)} \right] \quad (10.62)$$

where

$$\gamma = \text{Propogation constant} \\ = \sqrt{(R + j\omega L)(G + j\omega C)} = \alpha + j\beta$$

With attenuation constant $\alpha = (1/2)\left(R\sqrt{C/L} + G\sqrt{L/C}\right)$ and phase constant $\beta = \omega\sqrt{LC}$.

The reflection coefficient Γ is a complex quantity.

$$\Gamma = (Z_L - Z_0)/(Z_L + Z_0) = (\Gamma + j\Gamma i)$$

Now we prove the following of Fig. 10.24:

(a) **Short-ended $\lambda/2$ line as series LCR resonator**: The input impedance for this line ($Z_L = 0$) will be

$$Z_{in} = Z_0 \cdot \tanh[(\alpha + j\beta) \cdot l] \quad (10.63)$$

Using exponential expression of tanh we can prove that:

$$Z_{in} = \left[\frac{\tanh(\alpha l) + j\tan(\beta l)}{1 + j \cdot \tanh(\alpha l) \cdot \tan\beta l} \right] \quad (10.64)$$

For loss line $\alpha = 0$, $\tanh(\alpha l) = 0$

$$Z_{in} = jZ_0 \tan(\beta l)$$

But in practice most of the line has some small finite loss; therefore, we can take $\alpha l \ll 1$, so that $\tanh(\alpha l) \approx \alpha l$. Let $\Delta\omega = (\omega - \omega_0)$ an upper deviation from resonant frequency ω_0 and that the wave is TEM in the line where $\beta = \omega/v_p = 2\pi/\lambda$ and v_p the phase velocity, then for $l = \lambda/2 = \pi\, v_p/\omega_0$.

$$\beta l = \frac{\omega l}{v_p} = \left[\frac{\omega_0 l + \Delta\omega l}{v_p} \right] \\ = \frac{\omega_0 \lambda}{2v_p} + \frac{\Delta\omega.\lambda}{2v_p} = \left[\pi + \frac{\Delta\omega.\pi}{\omega_0} \right]$$

Then

$$\tan(\beta l) = \tan\left(\pi + \frac{\Delta\omega \cdot \pi}{\omega_0}\right) = \tan\left(\frac{\Delta\omega\pi}{\omega_0}\right) \approx \frac{\Delta\omega \cdot \pi}{\omega_0}$$

Using these results in Eq. 10.64, we get

$\therefore$

$$Z_{in} = Z_0 \left[\frac{\alpha l + j\Delta\omega \cdot \pi/\omega_0}{1 + j(\Delta\omega \cdot \pi/\omega_0)\alpha l} \right] \\ \approx Z_0(\alpha l + j\Delta\omega \cdot \pi/\omega_0) \quad (10.65)$$

as $\Delta\omega \cdot \alpha l/\omega_0 \ll 1$

This equation is of the form

$$Z_{in} = R + 2jL \cdot \Delta\omega \quad (10.66)$$

In a simple series LCR resonator

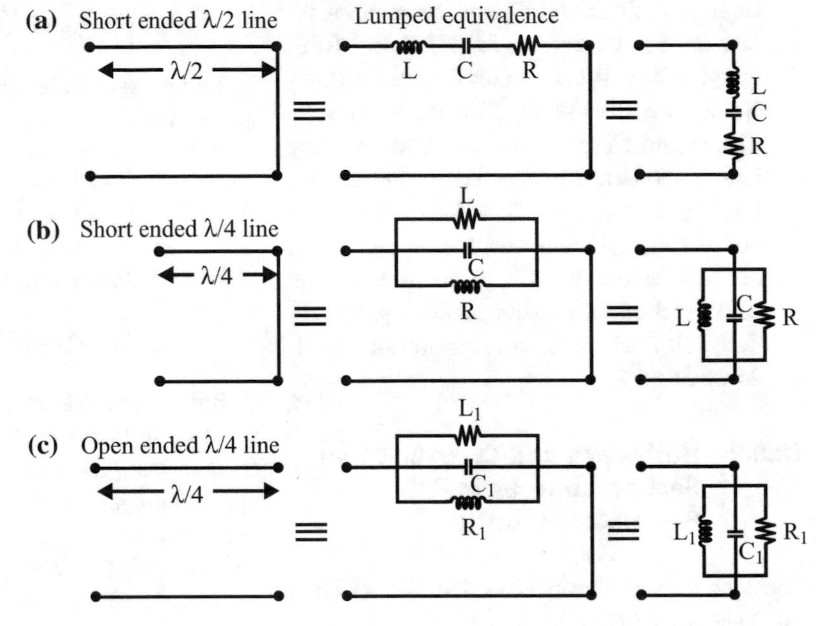

Fig. 10.24 Stubs of transmission line sections in microwave frequencies and their lumped equivalence as proved in Sect. 10.6.1 (see Table 10.7)

$$
\begin{aligned}
Z_{\text{in}} &= \left(R + j\omega L + \frac{1}{j\omega C}\right) \\
&= R + j\omega L\left(1 - \frac{1}{\omega^2 LC}\right) \\
&= R + j\omega L\left(1 - \frac{\omega_0^2}{\omega^2}\right)\left(\text{as } \omega_0^2 = 1/LC\right) \\
&= R + j\omega L\frac{(\omega - \omega_0)(\omega + \omega_0)}{\omega^2} \\
&= R + j\omega L \cdot \Delta\omega \times \frac{2\omega}{\omega^2}(\text{with } \omega + \omega_0 \approx 2\omega) \\
\therefore \quad Z_{\text{in}} &= R + j2L\Delta\omega
\end{aligned}
\tag{10.67}
$$

Therefore, we see that Z_{in} of Eqs. 10.66 and 10.67 are of the same form. Proving that shortened $\lambda/2$ line behaves as series L, C circuit with equivalent values of R, L, C components (Fig. 10.24) as:

$$
\begin{aligned}
&R = Z_0\alpha l; \\
&L = Z_0\pi/2\omega_0; C = 1/\left(\omega_0^2 L\right) = 2/(\pi Z_0\omega_0)
\end{aligned}
\tag{10.68}
$$

The line is resonant at all multiple of $l = \lambda/2$ also, i.e. $l = n\ \lambda/2$, $n = 1, 2, 3, \ldots$

Also at resonance $\Delta\omega = 0 \quad \therefore (Z_{\text{in}})_{\omega=\omega_0} = R = Z_0\,\alpha \cdot l$

(b) **Short-ended $\lambda/4$ line as parallel LCR resonator**: From Eq. 10.64 which is for short-ended line ($Z_L = 0$), we get by writing $j \tan \beta l = -1/j \cot \beta l$

$$
\begin{aligned}
Z_{\text{in}} &= Z_0 \tanh(a + j\beta)l \\
&= Z_0\left[\frac{\tanh(\alpha l) + j\tan(\beta l)}{1 + j\tan(\beta l)\cdot\tanh(\alpha l)}\right] \\
&= Z_0\left[\frac{1 - j\tanh(\alpha l)\cdot\cot(\beta l)}{\tanh(\alpha l) - j\cot(\beta l)}\right]
\end{aligned}
\tag{10.69}
$$

Now for $l = \lambda/4$ is exactly at $\omega = \omega_0$, while for normal condition let $\omega = (\omega_0 + \Delta\omega)$, then for TEM wave $\beta = \omega/v_p$.

$$
\begin{aligned}
\therefore \quad \beta l &= (\omega_0 l/v_p + \Delta\omega_e l/v_p) \\
&= (\omega_0 \cdot \lambda/4v_p) + \Delta\omega \cdot \lambda/4v_p \\
&= \pi/2 + \pi\Delta\omega/(2\omega_0) \quad \text{As}\left(\frac{\lambda}{v_p} = \frac{1}{f} = \frac{2\pi}{\omega_0}\right) \\
\therefore \quad \cos\beta l &== \cos(\pi/2 + \pi\Delta\omega/2\omega_0) \\
&= -\tan(\pi\Delta\omega/2\omega_0) \cong -\pi\Delta\omega/2\omega_0
\end{aligned}
$$

As angle of tan is very small, $\tanh(\alpha l) \approx \alpha l$ for low attenuation.

∴ Eq. 10.69 becomes:

$$Z_{\text{in}} = Z_0 \times \frac{1 - j\alpha \cdot l \cdot \pi \cdot \Delta\omega/2\omega_0}{\alpha \cdot l + j\pi \cdot \Delta\omega/2\omega_0} \approx \frac{Z_0}{\alpha l + j\pi \cdot \Delta\omega/2\omega_0}$$
$$(\text{As}\, \alpha l\pi \cdot \Delta\omega/2\omega_0 \ll 1) \tag{10.70}$$

Also impedance of a parallel RLC circuit is:

$$Z_{\text{in}} = \left(\frac{1}{R} + \frac{1}{j\omega L} + j\omega C\right)^{-1} = \left[\frac{1}{R} + j\omega C\left(1 - \omega_0^2/\omega^2\right)\right]^{-1}$$
$$\left(\frac{1}{R} + j\omega L(\omega - \omega_0)(\omega + \omega_0)/\omega^2\right)^{-1} [\text{As}\, \omega + \omega_0 \approx 2\omega_0]$$
$$\therefore \quad Z_{\text{in}} = (1/R + 2j \cdot \Delta\omega \cdot C)^{-1} = \frac{1}{1/R + 2j\Delta\omega C} \tag{10.71}$$

This is similar to the Eq. 10.70; therefore, a short-ended $\lambda/4$ line is equivalent to parallel LCR circuit put in shunt (Fig. 10.24), with the values of RCL as:

$$R = Z_0/\alpha l; \; C = \pi/(4\omega_0 Z_0); \; L = 1/\left(\omega_0^2 C\right)$$
$$= 4Z_0(\pi\omega_0)$$

At resonance $\Delta\omega = 0$.

$$\therefore \quad (Z_{\text{in}})_{\omega=\omega_0} = \text{R} = \text{z}_0/\alpha\text{l}$$

(c) **Open-ended λ/2 line as parallel LCR resonator**: Here $Z_L = \infty$, therefore Eq. 10.62 gives:

$$Z_{\text{in}} = Z_0 \operatorname{Coth}(\alpha + j\beta)l$$
$$= Z_0 \left[\frac{1 + j\tan(\beta l)\tanh(\alpha l)}{\tanh(\alpha l) + j\tan(\beta l)}\right] \tag{10.72}$$

Again for $l = \lambda/2$ exactly at $\omega = \omega_0$, while in practice $(\omega = \omega_0 + \Delta\omega)$ then for TEM wave as proved earlier $\beta = \omega/v_p; \lambda/v_0 = 1/f = 2\pi/\omega_0$. Therefore, we get

$$\beta l = \pi + \pi\Delta\omega/\omega_0$$
$$\therefore \quad \tan(\beta l) = \tan(\pi\Delta\omega/\omega_0) \cong \pi \cdot \Delta\omega/\omega_0$$
$$\text{Similarly} \quad \tan h(\alpha l) \approx \alpha l$$

∴ Eq. 10.71 becomes:

$$\therefore \quad Z_{\text{in}} = \frac{Z_0}{\alpha l + j(\Delta\omega \cdot \pi/\omega_0)} \tag{10.73}$$

This is similar to Eq. 10.70. Therefore, open-ended $\lambda/2$ lines also behave like a parallel LCR resonant circuit with equivalent but different values of components LCR (Fig. 10.24) as:

$$R = Z_0/\alpha l; \; C = \pi/(2\omega_0 Z_0); L = 1/\left(\omega_0^2 C\right)$$
$$= 2Z_0/\pi\omega_0$$

Here it may be pointed out that in microstrip line circuits, the practical resonator is an open-ended $\lambda/2$ lines or its multiples as short-ended lines are difficult to fabricate.

Now we study the Richard's transformation of lumped component values of L and C into its equivalent short-ended or open-ended shunt transmission line segments as discussed above. All the equivalent circuits for stub of $\lambda/2$, $\lambda/4$, $\lambda/8$ length in series or shunt, which may be open or short ended, have already been summarised in Fig. 10.24. **Here this λ is the cut-off wave length $\lambda_c = c/f_c$. In case of microstrip line having dielectric in between, $\lambda_g = \lambda_c/\sqrt{\varepsilon_{\text{eff}}}$ has to be used, and accordingly $\lambda_g/2$ or $\lambda_g/4$ or $\lambda_g/8$ lines, as the wavelength reduces to λ_g therein.**

10.6.2 Richard's Transformation for Low Pass and Band Stop Filters

Here we assume that the transmission lines of very small segment, i.e. stubs, are lossless ($R = 0$) and the lines have purely reactive reactances with distributed L and C. The short- and open-ended lines as seen are purely reactive as:

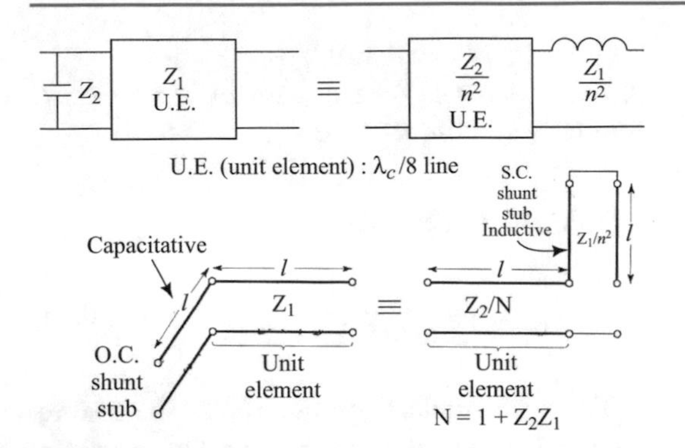

Fig. 10.25 Example of first identity of Kuroda's identities (Table 10.8), for stub conversion: initial circuit (left) with its equivalent (right) for S = tan (π/4· f/f$_c$) = 1 for f = f$_c$

$$Z_{insc} = jZ_0 \tan(\beta l) = j\omega L \text{ (i.e. inductive)}$$

$$Z_{inoc} = -j\left(\frac{1}{Z_0}\right) \cot \beta l = \frac{1}{j\omega C} \text{ (i.e. capacitive)}$$

Here arbitrarily we will use $l = \lambda/8$ for low pass filters and $l = \lambda/4$ for band stop filters, as these choices will meet their attenuation profiles as is clear from their equivalent circuits (Fig. 2.24). Moreover the above Z_{insc}, Z_{inoc} have tangent function and therefore have periodic behaviour after every $\beta l = (\beta l + n\pi)$, where $n = 1, 2, 3, \ldots$. Therefore, such filters cannot be regarded as broadband.

Now taking $l = \lambda_0/8$ we get

$$(\beta l) = \beta \cdot \lambda/8 = \frac{2\pi}{\lambda} \cdot \frac{\lambda_0}{8} = \frac{\pi}{4} \cdot \frac{\lambda_0}{\lambda} = \frac{\pi}{4} \cdot \left(\frac{f}{f_0}\right) = \frac{\pi}{4}\Omega \tag{10.74}$$

where $\Omega = (f/f_0)$ = normalised frequency.

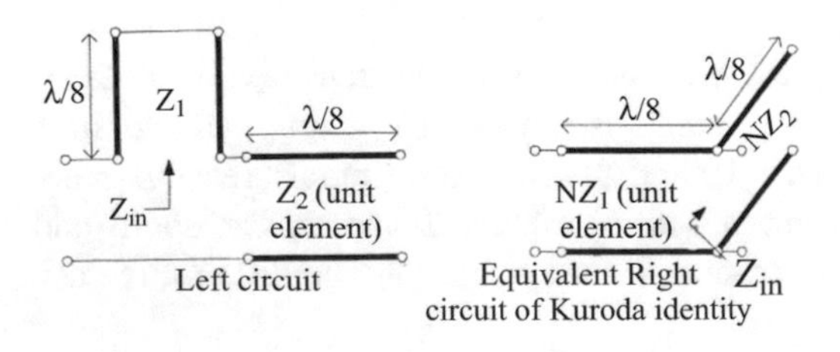

Fig. 10.26 Example of Kuroda's second identity (Table 10.8) as transmission lines equivalents in performance with $N = \left(1 + \frac{Z_2}{Z_1}\right)$.

(i) **For short-ended λ/8 segment:**

$$\begin{aligned} \therefore\quad Z_{insc} &= j\omega L = jZ_0 \cdot \tan\left[\left(\tfrac{\pi}{4}\right)\left(\tfrac{f}{f_0}\right)\right] \\ &= jZ_0 \tan\left[\left(\tfrac{\pi}{2}\right) \cdot \Omega\right] \\ \therefore\quad Z_{insc} &= S \cdot Z_0 \end{aligned} \tag{10.75}$$

Here $S = j\tan\left[\left(\frac{\pi}{4}\right) \cdot \Omega\right]$

is termed as Richard transform (10.76)

(ii) **For open-ended λ/8 line segment (i.e. stub).**

$$\begin{aligned} \therefore\quad Z_{inoc} &= \tfrac{1}{j\omega C} = j\left(\tfrac{-1}{Z_0}\right) \cdot \cot\left(\tfrac{\pi}{4}\Omega\right) \\ \text{i.e.}\quad Y_{inoc} &= jY_0 \tan\left(\tfrac{\pi}{4}\Omega\right) = SY_0 \end{aligned} \tag{10.77}$$

where

$Y_0 = 1/Z_0$ and $S = 1$ for $\Omega = f/f_c = 1$ i.e. $f = f_c$

$$\therefore \text{ with S} = 1 \text{ the } Z_{in} \text{ values becomes}$$
$$Z_{inoc} = Z_0 \text{ and } Y_{inoc} = Y_0$$

i.e. normal line impedance and this justifies the choice of taking $l = \lambda/8$ line length. Then at $\omega_0 = 2\,\omega_C$ line length will be $l = \lambda/4$ and another attenuation pole occurs.

10.6.3 Unit Elements

When the lumped elements get converted into transmission line section, we need to keep them separate specially for (*a*) achieving practically realisable configuration and (*b*) avoiding EM coupling. Therefore, a transmission line segment of length $l = \lambda/8$ is inserted in between them which is called unit elements (UE). This $\lambda/8$ is arbitrary but convenient in measurement etc. A open ended $\lambda/8$ line has capacitative impedance. These UEs have unit electrical length of $\beta l = \frac{\pi}{4}.(f/f_0)$ with characteristics impedance Z_{UE}. Therefore, ABCD parameter of this unit element will be from Eq. (10.13n) (Fig. 10.8d).

$$\begin{bmatrix} A & B \\ C & D \end{bmatrix}_{UE} = \begin{bmatrix} \cos(\beta l) & jZ_{UE}\sin(\beta l) \\ \frac{j\sin(\beta l)}{Z_{UE}} & \cos(\beta l) \end{bmatrix}$$

$$= \cos\beta l \begin{bmatrix} 1 & jZ_{UE}\tan(\beta l) \\ \frac{j\tan(\beta l)}{Z_{UE}} & 1 \end{bmatrix}$$

$$= \frac{1}{\sqrt{1-S^2}} \begin{bmatrix} 1 & Z_{UE}.S \\ \frac{S}{Z_{UE}} & 1 \end{bmatrix} \tag{10.78}$$

Where the Richards Transform is : $S = i\tan(\pi/4)\,\Omega$, where $\Omega = f/f_c$ (10.79)

All these lines are normally $\lambda/8$ length transmission lines and therefore called commensurate lines.

10.6.4 Kuroda's Identities

As indicated earlier in Art 10.6 that series inductance implemented by **short-ended line segment is difficult to fabricate specially because the short means a conductor through the dielectric to the base of a microstrip line**. Therefore by Kuroda's identity, it is replaced by open-ended shunt line segment giving the same IL versus frequency performance (Table 10.8). Similarly, other three transformations of the identities are used.

Table 10.8 Kuroda's identities

Initial circuit of λ/8 or λ/4 lines with characteristic impedance given there	Equivalent Kuroda's identity with characteristic impedance given there
(a) 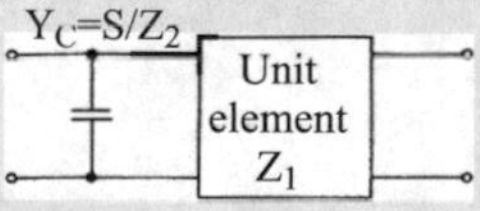 $Y_C = S/Z_2$; Unit element Z_1	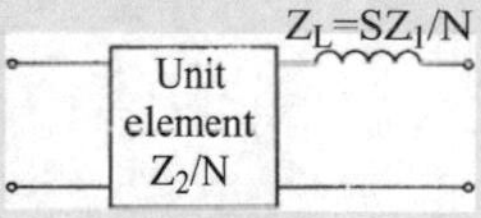 $Z_L = SZ_1/N$; Unit element Z_2/N
(b) $Z_L = Z_1 S$; Unit element Z_2 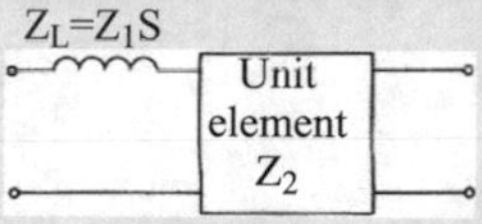	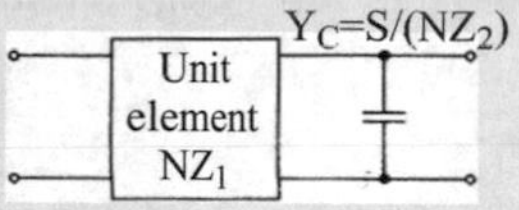 $Y_C = S/(NZ_2)$; Unit element NZ_1
(c) $Y_C = S/Z_2$; Unit element Z_1 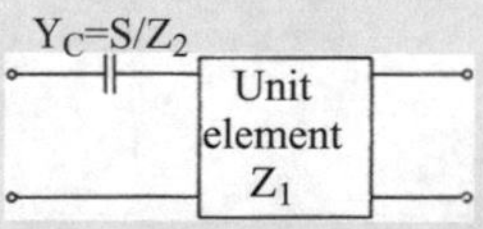	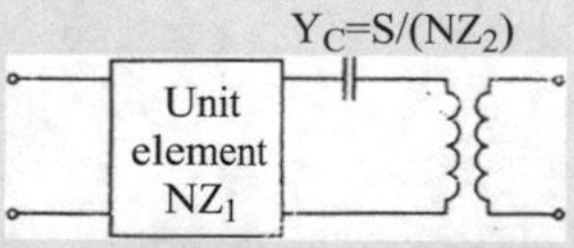 $Y_C = S/(NZ_2)$; Unit element NZ_1
(d) 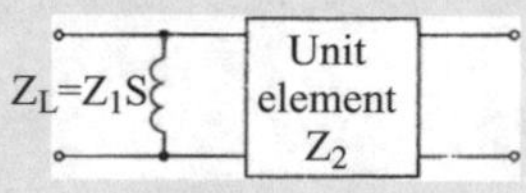 $Z_L = Z_1 S$; Unit element Z_2	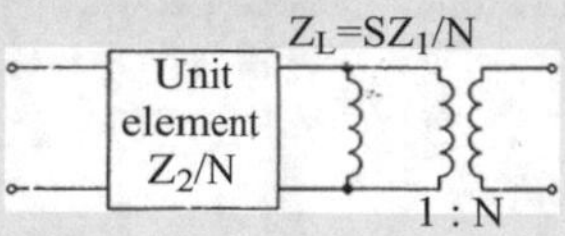 $Z_L = SZ_1/N$; Unit element Z_2/N; 1 : N

$N = 1 + Z_2/Z_1$; $S = \tan(\pi/4 \cdot f/f_c) = 1$ (for $f = f_c$)

Kuroda's identities can do the following operations

- Physically separates transmission line stubs by unit elements of different Z_0's.
- Transforms series stubs into shunt stubs, or vice versa.
- Change impractical characteristics impedances into more realisable ones. Only useful for LPF and BSF for f $\leq$ 10 GHz and not for HPF and BPF at all, as design size for f > 10GHz becomes very small and not practicable.

Note

(i) In implementation of Kuroda's identities, all the line elements are to be commensurate, i.e. equal in length, which is kept $\lambda/8$ normally with $\lambda = c/f_c$ and f_c = cut-off in LPF, HPF and centre frequency in BSF, BPF.

(ii) The IL versus frequency performance remains same with Kuroda's transformation.

(iii) The filter line segments have bilateral performance; therefore, putting unit element in the left or right of the open-ended shunt stub does not change the performance. Therefore, the unit elements are shifted for using them as buffer between two open-ended shunt stubs (i.e. capacitors).

(iv) Many a times the unit element which is placed on the right side of the element in the initial circuit goes to the left side of the new element in the Kuroda's changed circuit. However in the changed circuit, it can be placed on the right also as we will see in the examples, Kuroda's identities being bilateral in properties (see Fig. 10.25).

For proving these identities, let us first prove the ABCD matrix of the fourth identity to be equal for initial circuit and for Kuroda's circuit (Table 10.8).

Here we will use the ABCD matrix of series load, shunt load, transformer and of the unit element given by Eqs. 10.12, 10.13a, 10.13n, and 10.78, which are reproduced below for convenience.

$$\begin{bmatrix} A & B \\ C & D \end{bmatrix}_{\text{shunt load z}} = \begin{bmatrix} 1 & 0 \\ 1/Z & 1 \end{bmatrix} \quad (10.13a)$$

$$\begin{bmatrix} A & B \\ C & D \end{bmatrix}_{\text{Series load z}} = \begin{bmatrix} 1 & Z \\ 0 & 1 \end{bmatrix} \quad (10.12)$$

$$\begin{bmatrix} A & B \\ C & D \end{bmatrix}_{\text{1:N-Trans}} = \begin{bmatrix} N & 0 \\ 0 & 1/N \end{bmatrix} \quad (10.13n)$$

$$\begin{bmatrix} A & B \\ C & D \end{bmatrix}_{\text{Unit E}} = \frac{1}{\sqrt{1-S^2}} \begin{bmatrix} 1 & Z_{\text{uE}} \cdot S \\ S/Z_{\text{uE}} & 1 \end{bmatrix} \quad (10.78)$$

(a) **Proving the fourth identity**:

$$\text{Left side ABCD} = \begin{bmatrix} A & B \\ C & D \end{bmatrix}_{\text{shunt Load } SZ_1} \times \begin{bmatrix} A & B \\ C & D \end{bmatrix}_{\text{UE}}$$

$$= \begin{bmatrix} 1 & 0 \\ \frac{1}{SZ_1} & 1 \end{bmatrix} \cdot \frac{1}{\sqrt{1-S^2}} \begin{bmatrix} 1 & Z_2 S \\ \frac{S}{Z_2} & 1 \end{bmatrix}$$

$$= \frac{1}{\sqrt{1-S^2}} \begin{bmatrix} 1 & Z_2 S \\ \left(\frac{1}{SZ_1} + \frac{S}{Z_2}\right) & \left(1 + \frac{Z_2}{Z_1}\right) \end{bmatrix} \quad (10.79)$$

$$\text{Right Side ABCD} = \begin{bmatrix} AB \\ CD \end{bmatrix}_{UE} = \begin{bmatrix} AB \\ CB \end{bmatrix}_{\text{Shunt-load}(SZ_2/N)} \cdot \begin{bmatrix} AB \\ CD \end{bmatrix}_{\text{Transformer}}$$

$$= \frac{1}{\sqrt{1-S^2}} \begin{bmatrix} 1 & \frac{Z_2 S}{N} \\ \frac{SN}{Z_2} & 1 \end{bmatrix} \begin{bmatrix} 1 & 0 \\ \frac{1}{(SZ_2/N)} & 1 \end{bmatrix} \begin{bmatrix} 1/N & 0 \\ 0 & N \end{bmatrix}$$

$$= \frac{1}{\sqrt{1-S^2}} \begin{bmatrix} \frac{1}{N}\left(1 + \frac{Z_2}{Z_1}\right) & Z_2 S \\ \left(\frac{S}{Z_2} + \frac{1}{SZ_1}\right) & N \end{bmatrix}$$

$$\left(\text{By } N = 1 + \frac{Z_2}{Z_1}\right) \text{ this} = \frac{1}{\sqrt{1-S^2}} \begin{bmatrix} 1 & Z_2 S \\ \left(\frac{1}{SZ_1} + \frac{S}{Z_2}\right) & \left(1 + \frac{Z_2}{Z_1}\right) \end{bmatrix} \quad (10.80)$$

These Eqs. 10.79 and 10.80 being same the fourth identity gets proved. Similarly other identities can be proved.

(b) **Proving the second Kuroda's identity**: The left-hand circuit and RHS circuits of Table 10.8 are shown as in Fig. 10.26.

From this circuit

$$Z_{in} = jZ_1 \tan \beta l = jZ_1 S$$

where $S = \tan \beta l$; Z_{in} is an unnormalised value because of Z_1.

Cascading ABCD matrices for this LHS circuit we get:

$$\begin{bmatrix} A & B \\ C & D \end{bmatrix}_{LHS} = \begin{bmatrix} 1 & SZ_1 \\ 0 & 1 \end{bmatrix} \cdot \begin{bmatrix} 1 & jSZ_2 \\ \frac{jS}{Z_2} & 1 \end{bmatrix} \frac{1}{\sqrt{1+S^2}}$$
$$= \frac{1}{\sqrt{1+S^2}} \begin{bmatrix} 1 - S^2 \frac{Z_1}{Z_2} & jSZ_2 + j\Omega Z_1 \\ \frac{jS}{Z_2} & 1 \end{bmatrix} \quad (10.81)$$

For the right circuit in row 2 of Table 10.8, i.e. second Kuroda's identity, we have

$$Z_{in} = -jZ_0 \cot \beta l = \frac{Z_0}{jS}$$

or

$$Z_{in} = \frac{NZ_2}{jS} \quad \therefore Y_{in} = \frac{jS}{Z_2}$$

∴ Cascading ABCD matrices for this RHS circuit:

$$\begin{bmatrix} A & B \\ C & D \end{bmatrix}_{RHS} = \frac{1}{\sqrt{1+S^2}} \begin{bmatrix} 1 & SNZ_1 \\ \frac{jS}{NZ_1} & 1 \end{bmatrix} \cdot \begin{bmatrix} 1 & 0 \\ \frac{jS}{NZ_2} & 1 \end{bmatrix}$$
$$= \frac{1}{\sqrt{1+S^2}} \begin{bmatrix} 1 - S^2 \frac{Z_1}{Z_2} & SN \cdot Z_1 \\ \frac{jS}{NZ_1} + \frac{jS}{NZ_2} & 1 \end{bmatrix} \quad (10.82)$$

If we assume $N_2 = (Z_1 + Z_2)/Z_1$, then RHS of Eqs. 10.81 and 10.82 is same. This proves the second identity.

10.6.5 Microstrip Line Implementation of Low Pass and Band Stop Filters and Examples

We will discuss only the design of low pass filter (LPF) and band stop filter (BSF) implementation by Richardson's transformation followed by Kuroda's identities. This is because by this shunt and series shunt method of achieving a filter response, it is very difficult to design high pass and band stop filters. Therefore for this, coupled filters are more useful, which will be discussed later in Art. 10.7. ***Also for frequencies >10 GHz, i.e. λ/8 < 7.5 mm stub type of LPF or HPF is not preferred due to very small size required, therefore coupled line filters are used here also.***

Now for practical realisation of LPF and BSF using stub lines, we need to follow the following steps (as per Art 10.5):

(a) **Choose the order of the filter**: This is by the need of the sharpness (dB per GHz fall) of IL versus frequency response curve required.

(b) **Replace the lumped elements**: Replace inductance and capacitance by equivalent series or shunt $\lambda/8$ lines for LPF and $\lambda/4$ lines for BSF. The reason for this is clear from Table 10.7.

(c) **Insert unit element (UE) before applying Kuroda's identities**:
- Insert UEs (matched to the load and source) at the start and end of the total filter. Being matched line, UEs do not affect the performance of the filter, but are used for isolating two devices for avoiding EM coupling. This is clear if we see Fig. 10.46, the final filter.
- Insert UEs between series stub (L) and shunt stub (C) for isolation in case they become adjacent.

(d) **Apply Kuroda's identities**: Convert series stub line into shunt stub line using Kuroda's identities.

(e) **Denormalisation of Z_0, C, L**: In all these calculations, we use normalised values of Z_0, L, and C, till this stage. Here we de-normalise to get characteristic impedances of all the commensurate line segments lengths $\lambda_c/8$, where λ_c corresponds of cut-off frequency.

(f) **Compute Width of Lines**: Using the characteristic impedances of all commensurate lines, compute their width, length, and phase velocity using given formulas:

(f1) The line width to the dielectric thickness ratio (w/h) depends on Z_0, ε_r through a transcendental equation as given below:

For $w/h \leq 2$

$$\frac{w}{h} = \left(\frac{8e^A}{e^{2A} - 2}\right) \tag{10.83}$$

where

$$A = \left[2\pi \cdot \left(\frac{Z_0}{Z_f}\right) \cdot \sqrt{\frac{\varepsilon_r + 1}{2}}\right] + \left[\frac{\varepsilon_r - 1}{\varepsilon_r + 1} \cdot \left(0.23 + \frac{0.11}{\varepsilon_r}\right)\right]$$

$$Z_f = \text{characterstic impedance of free space} = 120\pi \approx 377\Omega$$

For $w/h \geq 2$

$$\frac{w}{h} = \frac{2}{\pi}\left[B - 1 - \ln(2B - 1) + \frac{\varepsilon_r - 1}{2\varepsilon_r} \cdot \left(\ln(B - 1) + 0.39 - \frac{0.61}{\varepsilon_r}\right)\right] \tag{10.84}$$

where $B = \frac{Z_f \pi}{2Z_0\sqrt{\varepsilon_r}}$ and $Z_f = 120\pi = 377\ \Omega$

(f_2) **All the line lengths are** $\lambda_g/8$, (i.e. commensurate lines)

where

$$\lambda_g = \frac{\lambda_0}{\sqrt{\varepsilon_{eff}}} \tag{10.85}$$

$$\begin{aligned}\varepsilon_{eff} &= \frac{\varepsilon_r + 1}{2} + \frac{\varepsilon_r - 1}{2} \cdot \left[\left(1 + \frac{12h}{w}\right)^{-1/2} + 0.04\left(1 - \frac{w}{h}\right)^2\right] \text{for} \frac{w}{h} \leq 1 \\ &= \frac{\varepsilon_r + 1}{2} + \left(\frac{\varepsilon_r - 1}{2}\right) \cdot \left[\left(1 + \frac{12h}{w}\right)^{-\frac{1}{2}}\right] \text{for} \frac{w}{h} \geq 1\end{aligned} \tag{10.86}$$

(f_3) **Phase velocity**: We know that

$$\frac{v_0}{\lambda_0} = \text{freq} = \frac{v_p}{\lambda_g} \quad \therefore \quad \frac{v_p}{v_0} = \frac{\lambda_g}{\lambda_0}$$
$$\therefore \quad v_p = \frac{v_0}{\sqrt{\varepsilon_{eff}}} = \frac{c}{\sqrt{\varepsilon_{eff}}} \tag{10.87}$$

A mix of first and second identities is given in Fig. 10.27, which will be used frequently, where characteristic impedance changes as follows:

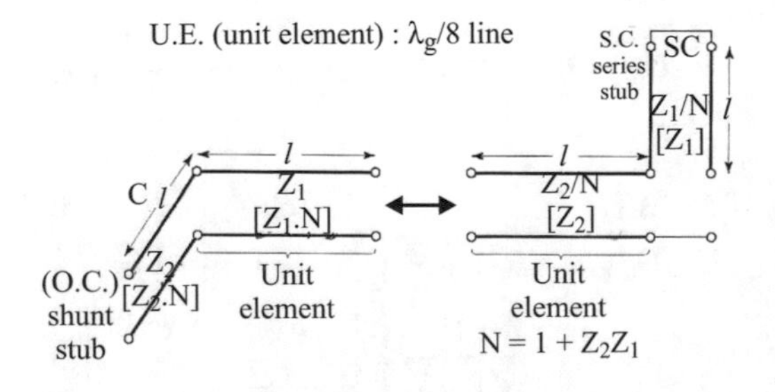

Fig. 10.27 Key identity for converting shunt stub to series stub left to right, i.e. first identity and vice versa (in bracket), e.g. shunt to series for second identity right to left

In shunt to series: Shunt $Z_2 \rightarrow (\text{UE})Z_2/N$ and $Z_1(\text{UE}) \rightarrow$ series Z_1/N;

In series to shunt: Shunt $Z_1 \rightarrow (\text{UE})Z_1 \cdot N$ and $(\text{UE})Z_2 \rightarrow$ series $Z_2 \cdot N$.

10.7 Some Examples of Filter Design in Microstrip Line

Example 1 Design a third-order stub low pass Chebyshev filter on microstrip for f_c = 4 GHz, a 50-Ω system impedance, and 3 dB equal-ripple in the pass band region. The strip line dielectric laminate is 32 mil thick with $\varepsilon_r = 3.55$.

Solution The dielectric thickness is $h = 32$ mil = 32 × 25 = 800 μ. Also from Table 10.6 we get $g_1 = g_3 = 3.3487$, $g_2 = 0.7117$ and $g_4 = 1$. The low pass prototype is then as per Fig. 10.28.

Next we synthesise the stub elements to provide the equivalent reactances equal to the lumped elements at the centre frequency. Using Fig. 10.24, the circuit of Fig. 10.28 becomes circuit of Fig. 10.29 by applying $(Z_0)_L = L_2$ and $(Z_0)_C = 1/C_2$.

We cannot actually get this circuit operated correctly in the laboratory because there is no physical separation between the stubs. If we were to build it with a 'small' separation between them, there would be extensive coupling between the stubs.

Therefore, we use Kuroda's identities to transform this impractical circuit into an equivalent, and more practical, one. For applying Kuroda's identity on this circuit, we can add 'unit elements' to either end of the circuit without affecting the power loss factor P_{LR}, provided that their normalised characteristic impedances are

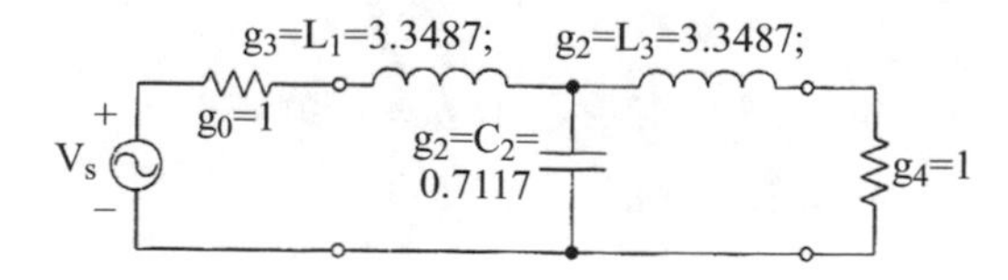

Fig. 10.28 Simple LPF and the 'g' values of its L and C

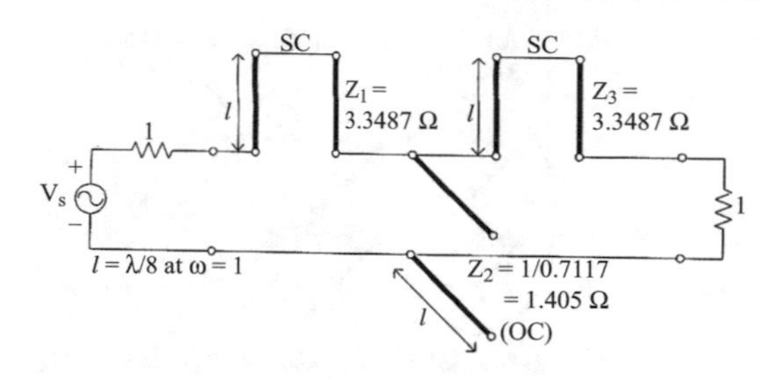

Fig. 10.29 Richard's transformation: replacing inductor and capacitor by series and shunt stubs with normalised Z_0's as $(Z_0)_L = 3.3487$, $(Z_0)_C = 1/0.7117 = 1.405$

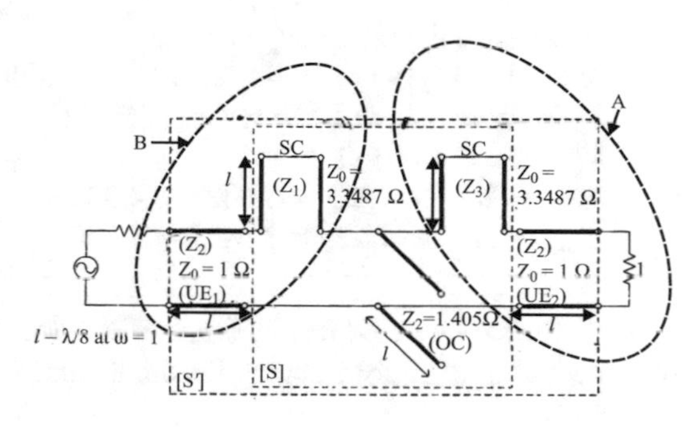

Fig. 10.30 UE added to the two ends of the filter of unit Z_0 each

$Z_0 = 1$. Adding unit elements (UEs) on each end gives Fig. 10.17b becomes (Fig. 10.30).

These UEs do not affect the filter performance because their characteristic impedances are matched to the system impedances at the source and load ends. This is an accurate statement as the two filter of Fig. 10.29 and 10.30 can be described by the same *S*-parameters ($S = S'$). As the phase planes are moved along the UEs towards the source and load, only the phases of the *S*-parameters will change, not the magnitudes, which are mainly of interest in many filters, including low pass, high pass, band pass, and band stop filters. Therefore, the two UEs that is added on each end of this low pass filter prototype will not alter the overall filter performance as measured by the power loss ratio P_{LR} or *S*-parameters.

Now we apply Kuroda's second identity on the right block 'A' and left block 'B' of Fig. 10.30, and we get Fig. 10.31, with new blocks *A'* and *B'*:

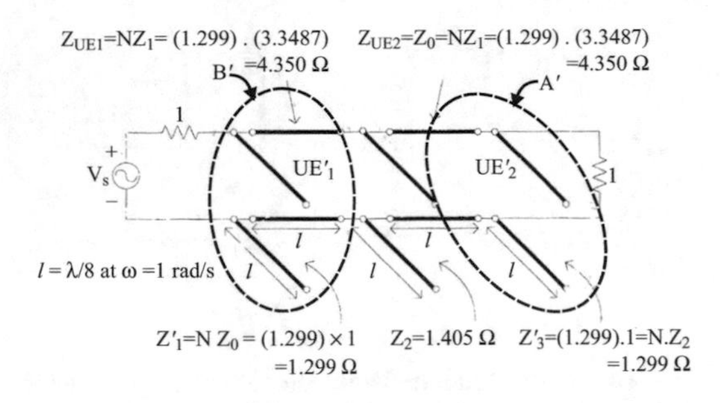

Fig. 10.31 Applying second Kuroda's identity on Fig. 10.30 gives this Figure

$$
\begin{aligned}
\text{As}\quad & Z_1 = Z_3 = 3.3487 \text{ and } Z_2 = 1.405 \\
\therefore\quad & N = (1 + Z_2/Z_1) = (1 + 1/3.3487) = 1.299 \\
\therefore\quad & Z_1' = Z_3' = NZ_2 = 1.299 \times 1 = 1.299\ \Omega \\
& Z_2' = Z_2 = 1.405\ \Omega \\
& Z_{\mathrm{UE}_1} = NZ_1 = 1.229 \times 3.3487 = 4.350 \\
& Z_{\mathrm{UE}_2} = NZ_1 = 1.229 \times 3.3487 = 4.350
\end{aligned}
$$

Here it may be noted that UE_1' and UE_2' have been kept around Z_2 for isolating it from the new Z_1' and Z_3'.

The final step is the impedance scaling and frequency scaling of the circuit. To do this, we multiply all impedances by 50 and scale the TLs to $\lambda/8$ at 4 GHz using ε_r of the microstrip line (Fig. 10.32).

Using the above Z_0's, we find the line width and length of each of the line segments (Fig. 10.33). Using the formula of w/h, λ_g, and $\varepsilon_{\mathrm{eff}}$ given in Eqs. 10.85 and 10.86, depending upon $w/h \le 2$ or $w/h \ge 2$. This can be checked from curves of Figs. 2.20 and 2.21.

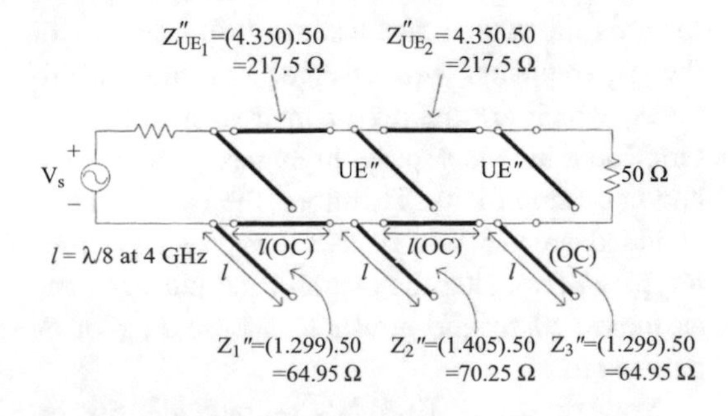

Fig. 10.32 Impedance and frequency scaled new values of character impedances of unit elements (UE) and stubs of Fig. 10.31

for $\varepsilon_r = 3.55$; $Z_f = 377\ \Omega$; $h = 800\ \mu$

We now compute line widths:

(a) For $Z_0 = 217.5\ \Omega$: We will see that $w/h \le 2$ for $\varepsilon_r \approx 3.5$

∴ Use 'A' formula of Eq. 10.83

$$
\begin{aligned}
\frac{w}{h} &= \left(\frac{8e^A}{e^{2A}-2}\right) \\
A &= 2 \times 3.14 \times \frac{217.5}{377} \times \sqrt{\frac{3.55+1}{2}} + \frac{2.55}{4.55}\left(0.23 + \frac{0.11}{3.55}\right) \\
&= 5.435 + 0.134 = 5.569. \\
e^A &= 262.17;\ e^{2A} = 68733 \\
\therefore\quad w &= \frac{8 \times 262.17}{68733} \times 800\ \mu \\
&= 24\ \mu
\end{aligned}
$$

(b) **For $Z_0 = 64.95\ \Omega$ and 50 Ω**: we will see that $w/h \ge 2$ for $\varepsilon_r \approx 3.5$

∴ Use 'B' formula of Eq. 10.84

$$
\begin{aligned}
\therefore\quad B &= Z_f\pi/\left(2.Z_0 \cdot \sqrt{\varepsilon_r}\right) \\
&= 377 \times 3.14/\left(2 \times 64.95 \times \sqrt{3.55}\right) \\
&= 4.837 \\
\therefore\quad \frac{w}{h} &= \frac{2}{3.14}\{4.837 - 1 - \ln(2 \times 4.837)\} \\
&\quad + \left(\frac{3.55-1}{2 \times 3.55}\right)[\ln(3.837) + 0.39 + 0.61/3.55] \\
&= 0.637\{3.837 - 2.269 + 0.359[1.345 + 0.39 + 0.17]\} \\
&= 1.43896 \\
\therefore\quad w &= 1.43896 \times 800\ \mu = 1151.2\ \mu = 1.15\ \text{mm}
\end{aligned}
$$

(c) **Also for $Z_0 = 50\ \Omega$:**

$$
\begin{aligned}
B &= Z_f\pi/\left(2.Z_0\sqrt{\varepsilon_r}\right) = 6.283 \\
\therefore\ \frac{w}{h} &= 0.637\{5.283 - \ln(11.466) \\
&\quad + \frac{2.55}{2 \times 3.55} \cdot \left[\ln(5.283) + 0.39 - \frac{0.61}{3.55}\right]\} \\
&= 0.637\{5.283 - 2.438 + 0.359 \\
&\quad [1.665 + 0.39 - 0.172]\} \\
&= 2.243 \\
w &= (2.243 \times 800)\ \mu = 1794.4 \\
\mu &= 1.19\ \text{mm}
\end{aligned}
$$

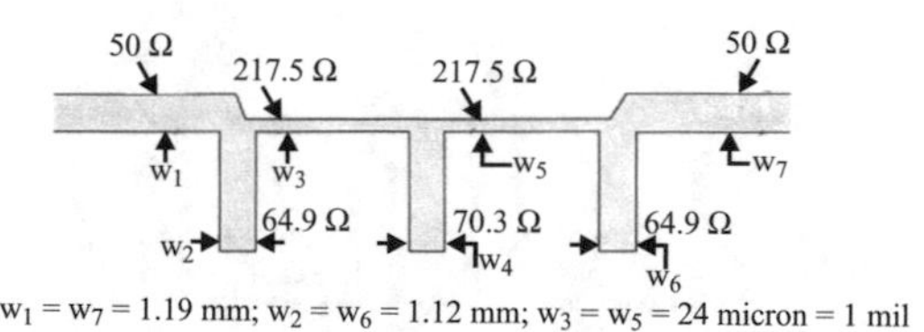

Fig. 10.33 Microstrip fabrication of final low pass filter. The 'w' calculated from Eqs. 10.83 and 10.84

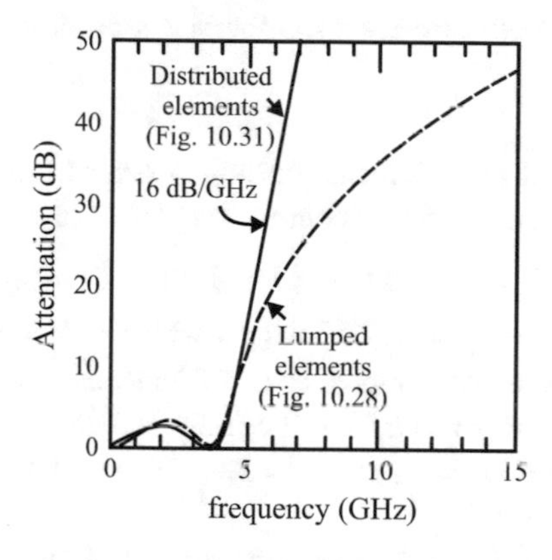

Fig. 10.34 Attenuation performance of third-order and final microstrip line low pass filter (LPF) with cut-off frequency of 4 GHz as compared to LPF with lumped elements. The attenuation profile of the former is 16 dB/GHz

Alternatively *w*/*h* can be read from Figs. 2.20 and 2.21 directly also, for knowing the approximate value of *w*/*h*, to see whether $w/h \geq 2$ or ≤ 2. This final strip line filter is shown in Fig. 10.33, with attenuation performance in Fig. 10.34.

This example illustrates one of the chief disadvantages of the stub filters: they often require physically unrealistic strip widths, e.g. W3, W5 in Fig. 10.33. Even for a large 50% band width, the band pass filter in this example would be extremely difficult to manufacture. Therefore, coupled microstrip line filters are preferred.

Example 2 Design a microstrip line LPF of order 5 with $f_c = 3.0$ GHz equiripple of 0.5 dB. If $\varepsilon_r = 3.5$ and $h = 800\ \mu$. Find the phase velocity also.

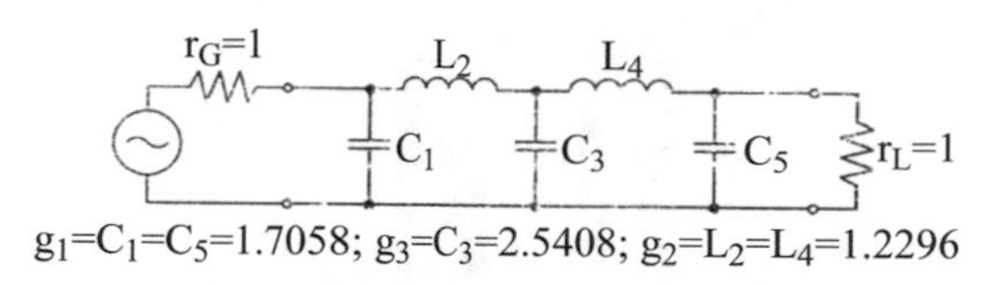

Fig. 10.35 Normalised low pass filter of order $N = 5$

Solution

(a) **Commensurate length segment $\lambda_g/8$ and the phase velocity (v_p).**

For $\varepsilon_r = 3.5$ and $Z_0 = 50\ \Omega$

The *w*/*h* is approximately = 2.6 from Figs. 2.20 and 2.21, i.e. $w/h \gg 2$, therefore we use B-formula of Eq. 10.84:

$$\therefore \quad \varepsilon_{\text{eff}} = \frac{\varepsilon_r + 1}{2} + \frac{\varepsilon_r - 1}{2}\left(1 + 12.\frac{h}{W}\right)^{-1/2}$$
$$= 2.25 + 1.25(1 + 4.5)^{-1/2} = 2.78$$

$$\text{Guide wave length} = \lambda_g = \lambda_0/(\sqrt{\varepsilon_{\text{eff}}}) = \frac{c}{(f\sqrt{\varepsilon_{\text{eff}}})}$$
$$= 3 \times 10^{10}/(3 \times 10^9 \times 1.667)\ \text{cm}$$
$$= 6.0\,\text{cm}$$

$$\text{Commensurate length} = \lambda_g/8 = 6/8 = 0.75\,\text{cm}$$
$$= l_{cl}$$
$$\text{Phase vel.} = \frac{c}{\sqrt{\varepsilon_{\text{eff}}}} = 3 \times 10^{10}/1.667$$
$$\approx 1.8 \times 10^{10}\,\text{cm/s}$$

(b) **Getting the '*g*' values**: The '*g*' values of the components of Fig. 10.35 from Table 10.5b are:

$$g_0 = g_6 = 1.0;\ g_1 = g_5 = 1.7058;\ g_2 = g_4 = 1.2296;\ g_3 = 2.5408.$$

(c) **Replacing *L*, *C* by series, shunt line segments**: The inductance and capacitance of Fig. 10.35 are replaced by short-ended series

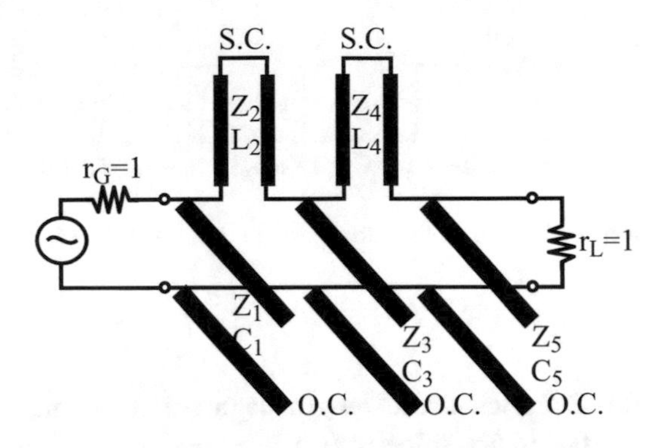

Fig. 10.36 Richard's transformation: Replacing inductors and capacitors by series and shunt stubs (oc = open--circuited line, sc = short-circuited line) with characteristic impedances $Z_1 = Z_5 = 1/C_1 = 1/1.7058 = 0.5862$; $Z_3 = 1/C_3 = 1/2.5408 = 0.3926$, $Z_2 = Z_4 = L_2 = L_4 = 1.2296$

stub and open-ended shunt stub, respectively, with normalised characteristic impedances computed by $Z_\mathrm{L} = L$ and $Z_\mathrm{C} = 1/C$ in Fig. 10.36.

(d) **Inserting unit elements and applying Kuroda's identities**: For making realisable filter of fifth order, we need to apply the first and second Kuroda's identities, for converting all the short-ended series stub into open-ended shunt stubs. This requires the following steps.

(d_1) Adding unit elements on both side of the filter.

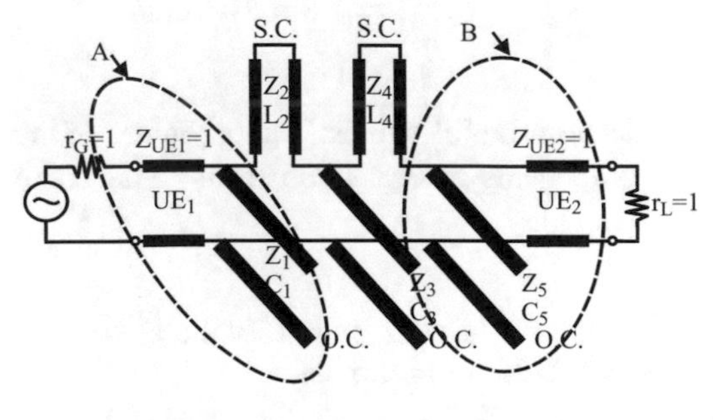

Fig. 10.37 For making practical filters add first set of unit elements (UE = unit element). For both the pairs, i.e. UE_1 & Z_1 and UE_2 & Z_5

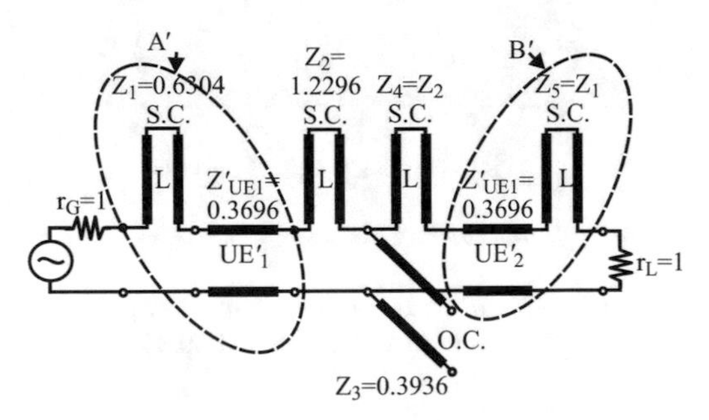

Fig. 10.38 Converting shunt stubs to series stubs

Add unit elements on both sides of the filter then apply first identity on '*A*' and '*B*' of Fig. 10.37, using $N = \left(1 + \frac{Z_{C1}}{Z_{UE1}}\right) = \left(1 + \frac{Z_{C5}}{Z_{UE2}}\right) = \left(1 + \frac{0.5862}{1}\right) = 1.5862$ we get with $S = 1$, the new characteristic impedances of unit elements as $Z'_\mathrm{UE} = Z_\mathrm{c}/N = 0.5862/1.5862 = 0.3696$ and of the series shunt as $Z'_\mathrm{L} = Z_{\mathrm{UE}_1}/N = 1/1.5862 = 0.6304$.

This leads to new structure of *A* and *B* of Fig. 10.37 as *A*′ and *B*′ of Fig. 10.38. Note that UE'_1 and UE'_2 have been placed after Z'_1 and before Z'_5 for isolating the two stubs. This now requires two more new unit elements UE_3&UE_4 for isolating source and load from the filter of Fig. 10.38.

(d_2) Add two more UEs: For isolating load and source from filter in Fig. 10.38 to get Fig. 10.39 with two more UEs. Introduction of UEs does not affect the performance as they are matched to source and load.

(d_3) Apply Kuroda's second identity to each pairs of A'', B'', C, and D of Fig. 10.39.

$$\begin{aligned}
\text{For } A'', B'' \text{ pairs}: N &= \left(1 + \frac{Z_{\mathrm{UE}_3}}{Z_\mathrm{L}}\right) = \left(1 + \frac{1}{0.6304}\right) \\
&= 2.5862 \\
Z'_{\mathrm{UE}_4} &= N \cdot Z_\mathrm{L} = (2.5862)0.6304 \\
&= 1.6304 = Z_{\mathrm{EU}_3} \\
Z''_1 &= N \cdot Z_{\mathrm{UE}_3} = (2.5862) \cdot (1) \\
&= 2.5862 = Z''_5
\end{aligned}$$

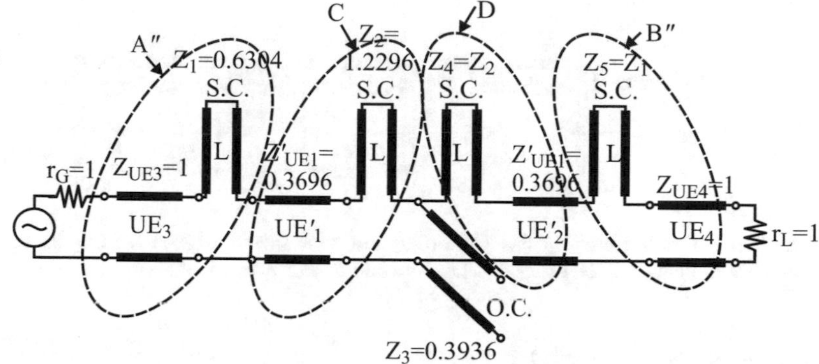

Fig. 10.39 Inserting the second set of unit elements UE_3, UE_4 before and after the filter of Fig. 10.38

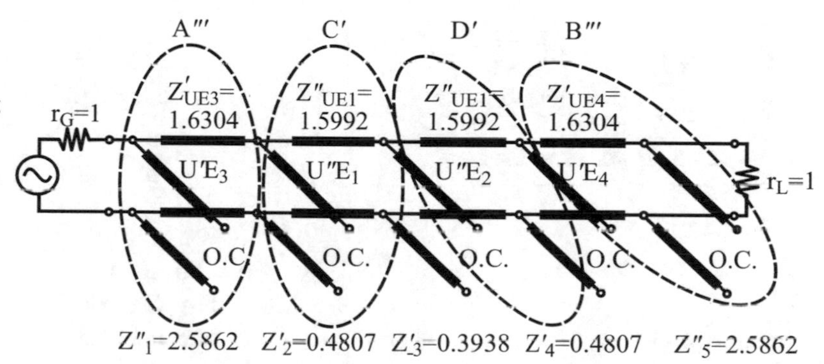

Fig. 10.40 Final realisable filter circuit obtained by converting series stubs into open-ended shunt stubs using Kuroda's identities in Fig. 10.39

For C, D pairs using second identities:

$$N = \left(1 + \frac{Z'_{\mathrm{UE}_1}}{Z_2}\right) = \left(1 + \frac{0.3696}{1.2996}\right) = 1.3006$$

$$Z''_{\mathrm{UE}_1} = Z_2 \cdot N = 1.2996 \times 1.3006 = 1.5992 = Z''_{\mathrm{UE}_2}$$

$$Z'_2 = Z'_{\mathrm{UE}_1} \cdot N = 0.3696 \times 1.3006 = 0.4807 = Z'_4$$

The new unit elements are to be kept to the inner side for isolating the other stubs. Thus, the blocks A'', C, D, B'' of Fig. 10.39 become new blocks A''', C', D', B''' in Fig. 10.40.

(c) **De-normalising the impedance**: We now de-normalise multiplying the normalised impedance of Fig. 10.40 by 50 Ω. Then we compute the width of the microstrip line segments by using Z_0 of these lines and formula of Eqs. 10.83 and 10.84 as given in Fig. 10.41.

$$(Z''_1)_0 = 129.3\ \Omega;$$
$$(Z'_{\mathrm{UE3}})_0 = 81.5\ \Omega;$$
$$(Z'_2)_0 = 24.0\ \Omega;$$
$$(Z''_{\mathrm{UE1}})_0 = 80.0\ \Omega;$$
$$(Z'_3)_0 = 19.7\ \Omega;$$
$$(Z''_{\mathrm{UE1}})_0 = 80\ \Omega;$$
$$(Z'_4)_0 = 24\ \Omega;$$
$$(Z'_{\mathrm{UE4}})_0 = 81.5\ \Omega;$$
$$(Z''_5)_0 = 129.3\ \Omega;$$

For Z_0 = 80, 81.5: Being larger than 50 Ω, then normally $w/h < 2$, therefore for $\varepsilon_r = 3.5$ we use 'A' formula of Eq. 10.83 ∴ with $h = 800\ \mu$ we get:

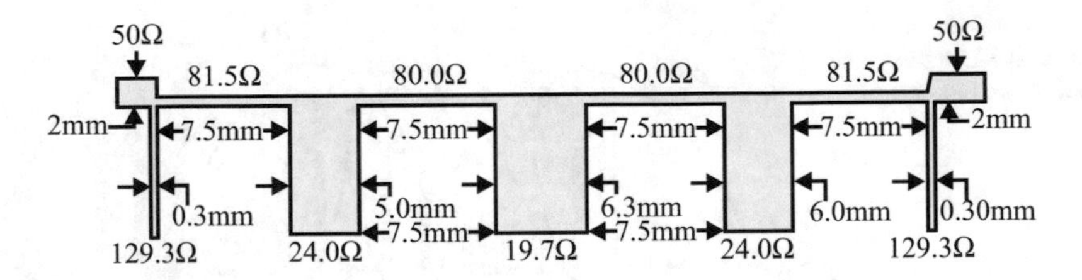

Fig. 10.41 Final microstrip line (5° order) low pass filter designed for 0.5 dB ripple Chebyshev method of $\lambda_g/8 = 7.5$ mm elements. The substrate has $h = 800\ \mu$ with $\varepsilon_r = 3.5$. Approx size is 5 cm × 1 cm

For 80 Ω: $\frac{w}{h} = \left(\frac{8e^A}{e^{2A}-2}\right)$

where

$$A = 2\pi\left(\frac{Z_0}{Z_f}\right)\cdot\sqrt{\frac{\varepsilon_r+1}{2}} + \left[\frac{\varepsilon_r+1}{\varepsilon_r-1}\cdot\left(0.23 + \frac{0.11}{\varepsilon_r}\right)\right]$$

$$\therefore\ A = 2\times 3.14\times\frac{80}{377}\cdot\sqrt{\frac{3.14+1}{2}} + \left[\frac{2.5}{4.5}\cdot 0.23 + \frac{0.11}{3.5}\right]\ (1)$$

$$= 80\times 0.025 + 0.1452 = 2.144$$

$$\therefore\ e^A = 8.5335$$

$$\therefore\ \frac{w}{h} = 0.4687$$

$$\therefore\ w = 800\times 0.4687 = 375$$

For 81.5 Ω

$$A = (81.5\times 0.25 + 0.1452) = 2.173$$

$$\therefore\ e^A = 8.8463 \because e^{2A} = 78.257$$

$$\therefore\ \frac{w}{h} = 0.452 \because w = 800\times 0.452 = 361.6\,\mu m$$

For $Z_0 = 24$ **Ω**, 19.7 **Ω**: Being less than 50Ω, normally $w/h > 2$; therefore, we use 'B' formula (Eq. 10.84).

$$\therefore\ \text{Here } B = Z_f\cdot\pi/(2Z_0\sqrt{\varepsilon_r}) = 13.1824\ \mu\Omega$$

$$= 16.06 \text{ for } 19.7\ \Omega$$

$$\therefore\ \frac{w}{h} = \frac{2}{\pi}\left\{(B-1) - \ln(2B-1) + \frac{\varepsilon_r - 1}{2\varepsilon_r}\left[\ln(B-1) + 0.39 - \frac{0.61}{\varepsilon_r}\right]\right\}$$

$$= 6.188 \text{ for } 24\ \Omega; = 7.9 \text{ for } 19.7\Omega$$

$$\therefore\ w = 4.951,\ \mu = 0.495 \text{ mm (for24 } \Omega)$$

$$= 6.350,\ \mu = 0.632 \text{ mm (for } 19.7\ \Omega)$$

Alternatively w/h can be obtained from Figs. 2.20 and 2.21 for a given Z_0 and ε_r

For Z_0= **129 Ω** & $\varepsilon_r = 3.5$ As $\frac{w}{h} < 2$, We apply 'A' formula

$$A = (129\times 0.025 + 0.1452) = 3.37$$

$$\therefore\ e^A = 25.15; e^{2A} = 632.7$$

$$\therefore\ \frac{w}{h} = \left(\frac{8e^A}{e^{2A}-2}\right) = \frac{201.2}{630.7} = 0.319$$

For $h = 800\,\mu \therefore w = 255\ \mu = 0.255\text{ mm} \approx 0.3\text{ mm}$

For Z_0 **= 50 Ω**: w = 2 mm by 'A' formula.

The final microstrip line LPF of Fig. 10.41 has a very sharp attenuation (33 dB/GHz) profile (see Fig. 10.42) beyond 3 GHz with 0.5 dB ripple. The sharpness of the skirt of the profile can be compared with that of LPF of third order of 5 dB ripple in Fig. 10.34. Figure 10.42 has an attenuation sharpness of 33 dB/GHz while

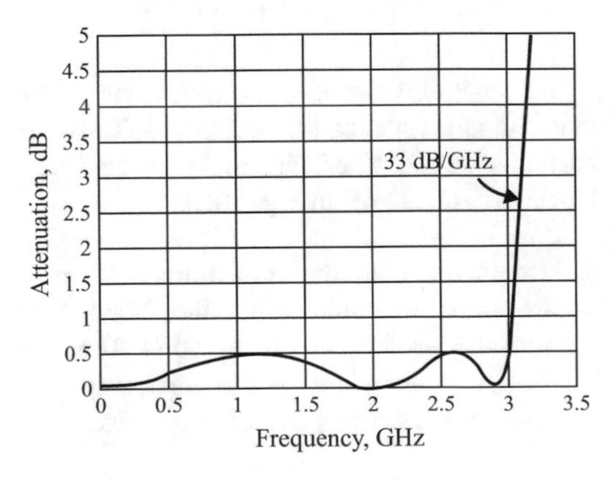

Fig. 10.42 Attenuation versus frequency response of final fifth-order microstrip line 3 GHz Chebyshev low pass filter for 0.5 dB ripple

Fig. 10.34 had only 16 dB/GHz. Also because $\lambda_g = \lambda_0/\sqrt{\varepsilon_{\text{eff}}}$, therefore by choosing a dielectric substrate of double ε_r, the size of the filter gets reduced to nearly half the size and *w*/*h* also reduces, showing the importance of dielectric constant of substrate. Figures 2.20 and 2.21 shows the relation of $Z_0 - \text{versus} - w/h$ and $\varepsilon_{\text{eff}} - \text{versus} - w/h$ for different values of ε_r.

Example 3 Design a third-order band stop filter with centre frequency of 5 GHz, with band width of 50%. The maximally flat (Butterworth) design could be used with the dielectric constant of the substrate as 3.5.

Solution For band stop filter design, we have to remember that the series short-ended element stub (inductance) should give very large of, i.e. infinite impedance, while the shunt open-ended stub (capacitor) should give zero/minimum impedance at resonant centre frequency f_0. This is possible only if we use λ/4 line as is clear from the properties of λ/4 lines (Fig. 10.24 and Table 10.7). Also we know that Richardson's transform gives $S = \tan\ (\beta l) = \tan\ (2\pi/\lambda)\ l = \tan\left(\frac{2\pi}{\lambda}\cdot\frac{\lambda}{8}\right) = \tan\left(\frac{\pi}{4}\right) = 1$ and not maximum value for $l = \lambda/8$, while for $l = \lambda/4$ line, $S = \infty$. In addition to this requirement, we require that the cut-off frequency $\Omega = 1$ of the low pass prototype filter should be transformed into this band stop filter as upper and low cut-off frequency. This can be a simple technique by introducing a so-called band width factor ($\Delta\omega_f$) by which the Richard's transforms need to be multiplied and hence also the *L* and *C* normalised values of the filter of Table 10.5.

For proving the above, let us assume the '$\Delta\omega_f$' as:–

$$\Delta w_f = \cot\left(\frac{\pi}{4}\cdot\frac{\omega_L}{\omega_0}\right) = \cot\left[\frac{\pi}{4}\cdot\left(1 - \frac{\omega_U - \omega_L}{\omega_0}\right)\right]$$
$$= \cot\left[\frac{\pi}{4}(1 - \Delta\omega_n)\right]$$

where $\Delta\omega_n = (\omega_U - \omega_L)/\omega_0 =$ normalised band width. For $\lambda/4$ line, we can prove that for $\omega = \omega_L$ the product, $S\ \Delta\omega_f$ will be $= +\ 1$ and for $\omega = \omega_U$, $S\cdot\Delta\omega_f = -1$. This is corresponding to the normalised frequency $\Omega = 1$ and -1 of a band stop filter as

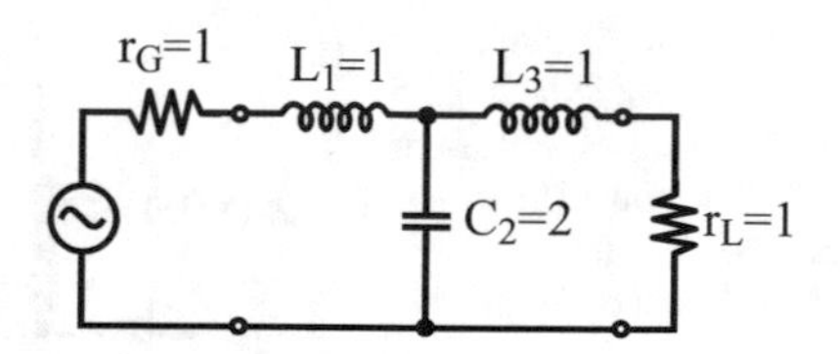

Fig. 10.43 Normalised third-order prototype low pass filter

$$S\cdot(\Delta\omega_f)|_{\omega=\omega_L} = \cot\left(\frac{\pi}{2}\cdot\frac{\omega_L}{\omega_0}\right)\cdot\tan\left(\frac{\pi}{2}\cdot\frac{\omega_L}{\omega_0}\right) = 1$$

$$S\cdot(\Delta\omega_f)|_{\omega=\omega_u} = \cot\left(\frac{\pi}{2}\cdot\frac{\omega_L}{\omega_0}\right)\cdot\tan\left(\frac{\pi}{2}\cdot\frac{\omega_u}{\omega_0}\right)$$
$$= \tan\left[\frac{\pi}{2}\cdot\left(\frac{2\omega_0 - \omega_L}{\omega_0}\right)\right] = -1$$

With the above, we follow the six steps, for using modified *g* values as $g' = g\cdot\Delta\omega_f$ for computing impedance values of the line segments and their dimensions.

(a) **The '*g*' values**: For normalised low pass prototype filter (Fig. 10.43) the '*g*' values of the low pass filter of third order to be chosen from Table 10.5 as g_0 (source) = 1; $g_2 = 2.0$; $g_3 = 1$; g_4(load) = 1] leads to LPF as in Fig. 10.44.

(b) **Replace *L*, *C* by stubs and their '*Z*' values**: The inductance and capacitance of Fig. 10.43 to be replaced by series short stub and shunt open ended stub respectively, with '*g*' values multiplied by "band width factor $\Delta\omega_f$", to get Fig. 10.44 with Z_1, Z_3, Y_1 as:

$$\Delta\omega_f = \cot\left(\frac{\pi}{4}\cdot\frac{\omega_L}{\omega_0}\right) = \cot\left[\frac{\pi}{4}\cdot\left(1 - \frac{\Delta\omega_n}{2}\right)\right]$$

For 50% band width $\Delta\omega$ at 5 GHz centre frequency ω_0:

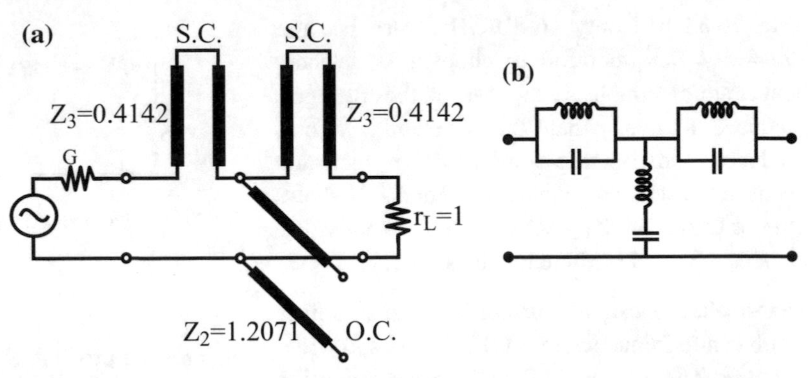

Fig. 10.44 a Replacing inductors and capacitors of Fig. 10.43 by series and shunt of in λ/4 line, the LPF becomes BSF. b Its equivalent lumped circuit

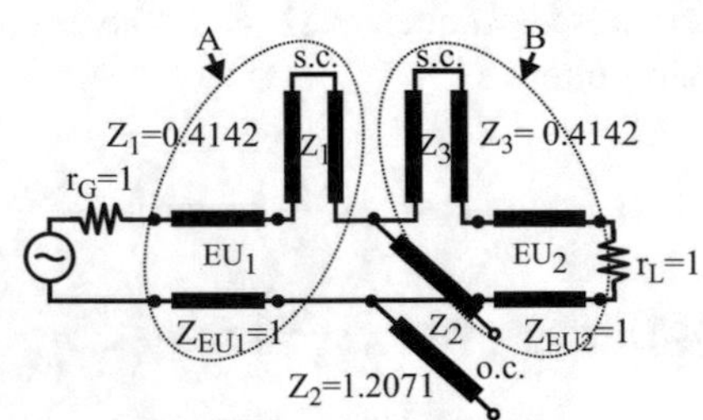

(a) Unit elements at source and load sides

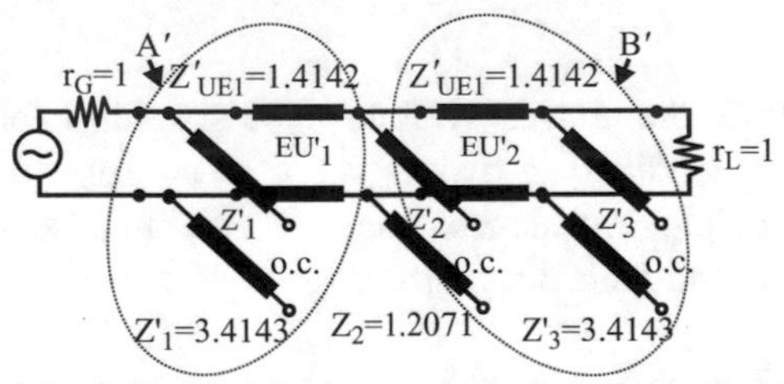

(b) Conversion from series to shunt stubs & isolating them

Fig. 10.45 Introducing unit elements and converting series stubs to shunt stubs

$$\Delta\omega_n = \frac{\omega_U - \omega_L}{\omega_0} = 0.5;$$

$$\therefore \quad \Delta\omega = (\omega_U - \omega_L) = 0.5 \times 5 = 2.5\text{ GHz}$$

$$\therefore \quad \Delta\omega/2 = 2.5/2 = 1.25\text{ GHz}$$

$$\therefore \quad \omega_L = 5 - 1.25 = 3.75\text{ GHz}$$

$$\omega_u = 5 + 1.25 = 6.25\text{ GHz}$$

$$\therefore \quad \Delta\omega_f = \cot\left[\frac{\pi}{4}\left(1 - \frac{0.5}{2}\right)\right] = \cot\left(\frac{\pi}{4}\cdot\frac{3}{4}\right) = 0.4142$$

$$\therefore \quad Z_1 = Z_3 = \Delta\omega_f . g_1 = 0.4142 \times 1 = 0.4142$$

$$Y_2 = \Delta\omega_f . g_2 = 0.4142 \times 2 = 0.8284$$

$$\therefore \quad Z_2 = \frac{1}{Y_2} = 1.2071$$

With these Z values, the prototype LPF of Fig. 10.42 with λ/4 sections become BSF in Fig. 10.44a. This is because by Fig. 10.24 and Table 10.7, a short-ended series stub is equivalent to a series of parallel combination of lumped L and C, while an open-ended shunt stub is equivalent to a shunted series of L and C as shown in Fig. 10.44b.

(c) **Inserting unit elements**: Unit elements of λ/4 lines are inserted for isolating the filter from source and load (Fig. 10.45a).

(d) **Apply Kuroda's identity**: Apply Kuroda's second identity to each of the two pairs Z_{UE_1}, Z_1 and Z_{UE_2}, Z_3 separately and use the new unit elements UE_1' and UE_2' to isolate the three open-ended shunt stubs Z_1', Z_2 and Z_3' (Fig. 10.45b). Thus, blocks A, B become A', B'.

Here

$$N = \left(1 + \frac{Z_{UE_1}}{Z_1}\right) = \left(1 + \frac{1}{0.4142}\right)$$

$$= 3.4143 = \left(1 + \frac{Z_{UE_2}}{Z_3}\right)$$

$$\therefore \quad Z_{UE_1}' = N \cdot Z_1 = 3.4143 \times 0.4142 = 1.4142 = Z_{UE_2}'$$

$$\text{and} \quad Z_1' = NZ_{UE_1} = 3.4143 \times 1 = 3.4143 = Z_3'$$

(e) **Denormalisation**: De-normalise all the impedances by multiplying by $Z_0 = 50\ \Omega$ to get

$$Z_1'' = Z_1' \times 50 = 170.7\ \Omega = Z_3'$$

$$Z_2' = Z_2 \times 50 = 60.4\ \Omega$$

$$Z_{UE_1}'' = Z_{UE_1}' \times 50 = 70.7\ \Omega = Z_{UE_2}''$$

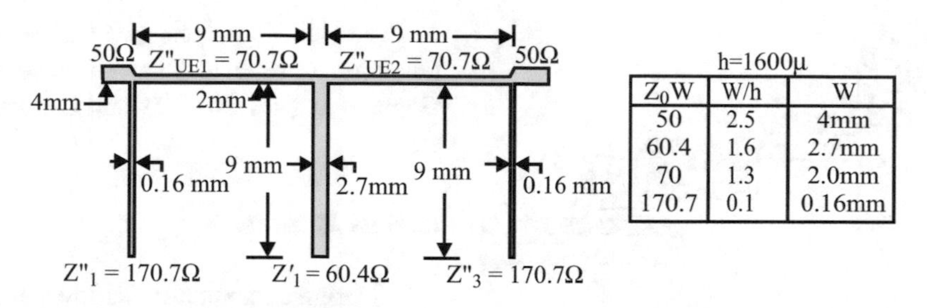

Z_0W	W/h	W
50	2.5	4mm
60.4	1.6	2.7mm
70	1.3	2.0mm
170.7	0.1	0.16mm

Fig. 10.46 Characteristic impedances of final microstrip line implementation of band stop filter design for substrate of $\varepsilon_r = 3.5$ and $h = 1.6$ mm. Approximate size of the filter area = 1.25 × 2.5 cm²

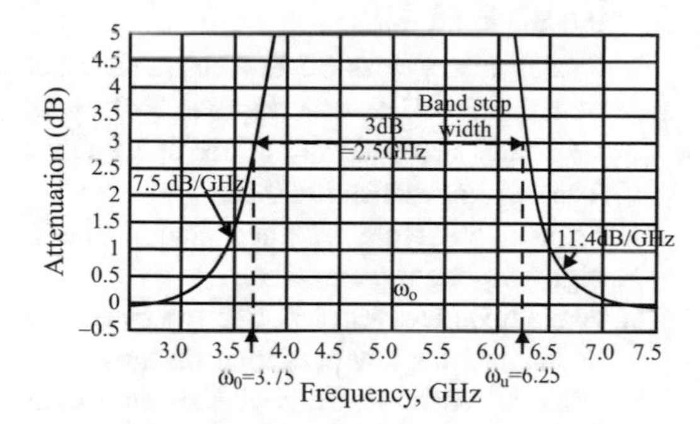

Fig. 10.47 Attenuation versus frequency response for microstrip line third-order band stop filter using maximally that (Butterworth) design

(f) **Line length the of microstrip line elements**. All the line lengths are $\lambda_g/4$, while $\lambda_g = \lambda_0/\sqrt{\varepsilon_{\text{eff}}}$; $f = 5$ GHz and $\varepsilon_r = 3.5$, for $Z_0 = 50\,\Omega$, $w/h \approx 2.6$ by Eq. 10.83 or by Fig. 2.20 directly.

$$\therefore\ \varepsilon_{\text{eff}} = \frac{\varepsilon_r + 1}{2} + \frac{\varepsilon_r - 1}{2}\left(1 + 12\frac{h}{w}\right)^{-\frac{1}{2}} \approx 2.78;$$

Therefore $\lambda_0 = 6\,\text{cm}\ \lambda_g = \lambda_0/1.5 = 3.6\,\text{cm}$ and $\lambda_g/4 = 0.9$ cm = 9 mm.

(g) Line width of microstrip line elements: As all the 'Z' are > 50 Ω therefore for $\varepsilon_r = 3.5$ and $w/h < 2$, we use the 'A' formula, i.e. Equation 10.83:

$$\frac{W}{h} = \frac{8e^A}{e^{2A} - 2};$$

where

$$A = 2\pi\left(\frac{Z}{Z_f}\right) \cdot \sqrt{\frac{\varepsilon_r + 1}{2}} + \left[\frac{\varepsilon_r + 1}{\varepsilon_r - 1} \cdot \left(0.23 + \frac{0.11}{\varepsilon_r}\right)\right]$$

Or also approximate values of line width w/h can be read from Fig. 2.20 of Z_0—versus w/h for different ε_r.

Attenuation versus frequency response of this filter is given at Fig. 10.47. Thus, the attenuation per GHz here being between 7 and 11 dB/GHz is much sharper than in Chebyshev design.

All these filters can also be designed by using commercial software packages which give also the attenuation response.

10.8 Coupled Microstrip Line Filters

As discussed earlier that for designing band pass and high pass filters, the stub approach design is not preferred and coupled line filters are preferred. In coupled line filters, the two lines are placed close to each other for allowing capacitative and inductive coupling. Here it may be noted that maximum coupling occurs over a $\lambda/4$ overlap region. Therefore, the band pass filter with top and bottom lines performance is obtained by having an coupling length (i.e. overlap length) of each segment $\lambda/4$ of, i.e. $\beta l = \pi/2$ (Fig. 10.48), as a result the total length

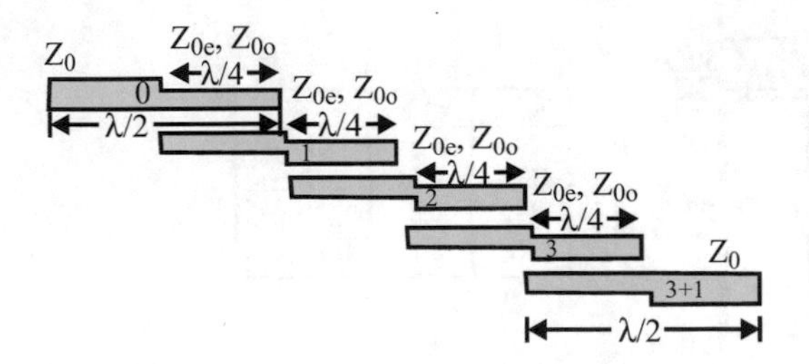

Fig. 10.48 A third-order coupled microstrip line band pass filter

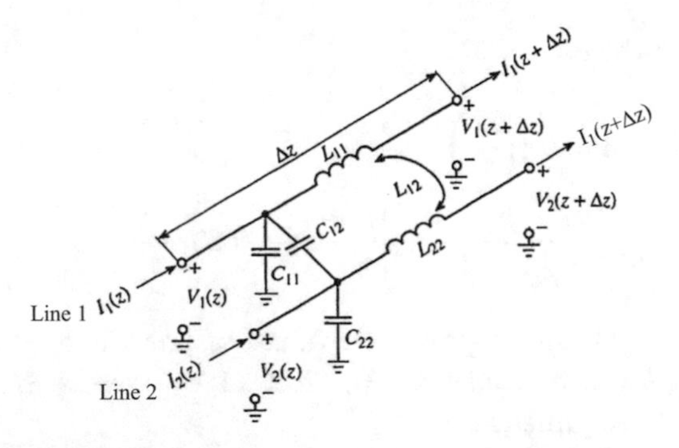

Fig. 10.49 Equivalent circuit diagram with appropriate voltage and current for a system of two lossless coupled microstrip lines

of each line segment = $\lambda/2$ at the centre frequency f_0.

The currents and voltages in the two adjacent coupled strip lines (I_1, I_2, V_1, V_2) can be in the same direction or in opposite directions (Fig. 10.49), therefore the concept of even and odd mode current has been introduced, which are

$$I_e = (I_1 + I_2)/2;\ V_e = (V_1 + V_1)/_2 \quad (10.88)$$

$$I_o = (I_1 - I_2)/2;\ V_o = (V_1 - V_1)/_2 \quad (10.89)$$

Using this concept, odd and even characteristic impedance Z_{oe}, Z_{oo} of a line has got evolved.

As usual we start with the 'g' coefficient of components of LPF of either Butterworth design or Chebyshev design. Then we choose ω_U, ω_L, centre frequency ω_0 and finally compute Z_0 of odd mode and even mode, i.e. Z_{oe}, Z_{oo} of a pair of line, e.g. $(Z_{oo})_{0,1}$, $(Z_{oe})_{01}$, $(Z_{oo})_{12}$, $(Z_{oe})_{12}$. Finally using these Z_o, the width of the strip lines and the spacing are computed.

For detailed analysis and design of these filters, we can refer to Das & Das (Ref. No. …), Gupta et al. (Ref. No. 17), and Ludwig (Ref. 18), etc.

Review Questions

1. Design a maximally flat low pass microstrip filter having cut-off frequency 2.5 GHz and insertion loss of 30 dB at 4 GHz. The dielectric constant of substrate is 9.0 and thickness 0.6 mm.
2. Design a low pass 50 Ω Chebyshev filter with f_c = 2.5 GHz with ripple of 3 dB, with stop band attenuation of 30 dB at 3.5 GHz. The I/O impedance is 50 Ω.
3. Explain the types of filters and its basic eight important parameters.
4. Explain Butterworth and Chebyshev design techniques for low pass filter design.
5. What is ABCD analysis of first-order filter design?
6. Write the ABCD matrix of a (*a*) low pass filter of first order. (*b*) Band stop filter.
7. Write the principle of filter design by insertion loss method.
8. Explain the principle of Chebyshev filter design of equal-ripple. How does reduction of ripple helps in filter performance.
9. Design a fourth-order Chebyshev filter having 3 dB ripple with cut-off frequency of 5 GHz on 50 Ω line.
10. For the above data, design a band pass filter with centre frequency = 6 GHz and Band width of ±10%.
11. Solve the above two questions for Butterworth design and give a comparison.
12. What is Richard's transformation and Kuroda's identities? Explain where do we use them.
13. How does a $\lambda/4$ line acts as single component *L* or *C* and under what condition, $\lambda/4$ line becomes series *L*, *C* resonant circuit and of what type? Explain the conditions.
14. Design a microstrip line LPF of order 4, with f_c = 5 GHz, equal-ripple allowed is 3 dB. If ε_r = 3.5 and height of the dielectric

of the microstrip line is 800 μm, compute the phase velocity also.

15. Design a low pass Butterworth filter that provides at least 25 dB attenuation at frequencies above $2f_o$.
16. Repeat the above for Chebyshev for 3 dB ripple factor allowed.
17. Prove the first three Kuroda's identities using ABCD matrix.
18. Design a microstrip line Chebyshev low pass filter with 3 GHz cut-off frequency, using FR4 substrate of $\varepsilon_r = 4.6$ and $h = 25$ mil. Get the length, width of each of the segment. Here attenuation of 25 dB should he there at 1.5 times the cut-off frequency.

RF Amplifiers, Oscillator, and Mixers 11

Contents

© Springer Nature Singapore Pte Ltd. 2018
P. K. Chaturvedi, *Microwave, Radar & RF Engineering*,
https://doi.org/10.1007/978-981-10-7965-8_11

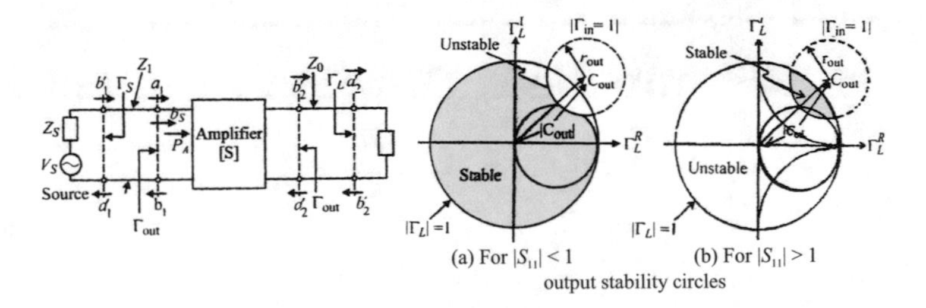

11.1 Introduction

In nearly each and every electronic circuit, an amplifier is one of the essential devices. As far as microwave circuits are concern, earlier microwave amplifiers and oscillators relied on tubes, such as klystron, TWT or on solid state amplifiers based on negative resistance characteristics of tunnel or varactor diodes, etc. Now due to dynamic improvement and innovations in solid state technology since 1970, most RF and microwave today use transistor devices, such as Si or Si–Ge BJTs, GaAs HBTs, GaAs or InP-FETs, or GaAs-HEMTs. The microwave amplifiers are rugged, low cost, small in size, less noisy, broad band width, reliable and can easily be integrated in both hybrid and monolithic IC. These can be used above 100 GHz as well.

Although microwave tubes are still required for very high power and/or very high-frequency applications, continuing improvement in the performance of microwave transistor is steadily reducing the dependence on microwave tubes.

Amplifier designs at RF differ very much from the conventional low-frequency circuit approach. This is because of the fact that voltage and current impinge upon the active device, e.g. transistor, which in turn requires matching with the transmission line, etc., for reducing the VSWR. In addition to this, amplifier sometime also tend to oscillate with certain source and load, requiring stability analysis in conjunction with gain and noise figures. This stability analysis is done with the help of Smith chart. Here in this book, we will deal with simple concepts only.

11.2 Amplifiers

Our discussion on the transistor amplifier design will rely on the terminal characteristic of transistor (e.g. S-parameters.) and will begin with some general definition of two-port power gain which is useful for amplifier design.

A simple single-stage amplifier embedded between input and output matching network is shown in Fig. 11.1. The key parameters in terms of performance specification are given below:

- Gain
- Operating frequency and band width
- Output power
- Power supply requirements (in V and A)
- Input and output reflection coefficients as in Fig. 11.1.
- Noise figure (dB).

The main amplifier has the S-parameters as defined by the voltage signals b_1, b_2, a_1, a_2 as:

$$\begin{bmatrix} b_1 \\ b_2 \end{bmatrix} = \begin{bmatrix} s_{11} & s_{12} \\ s_{21} & s_{22} \end{bmatrix} \begin{bmatrix} a_1 \\ a_2 \end{bmatrix} \quad (11.1)$$

The S-matrix has its usual properties (square matrix, unitary, symmetry, complex conjugate, lossy/lossless network, shifting of reference plane, etc.), with each element of 'S' matrix defined as:

(a) The two reflection coefficients:

$$S_{11} = \Gamma_{\text{in}} = \left.\frac{b_1}{a_1}\right|_{a_2=0}; \; S_{22} = \Gamma_{\text{out}} = \left.\frac{b_2}{a_2}\right|_{a_1=0}; \quad (11.2)$$

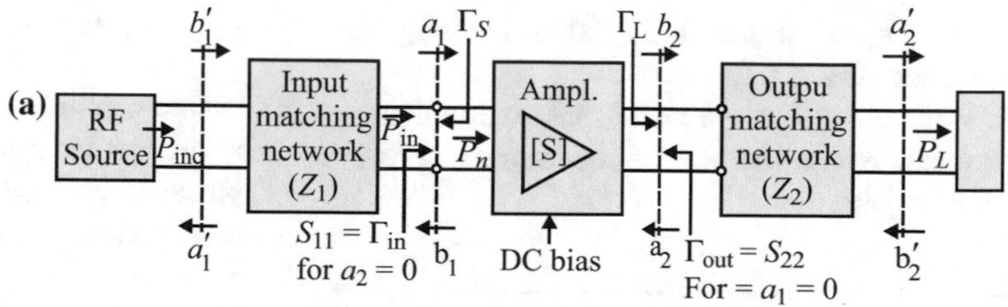

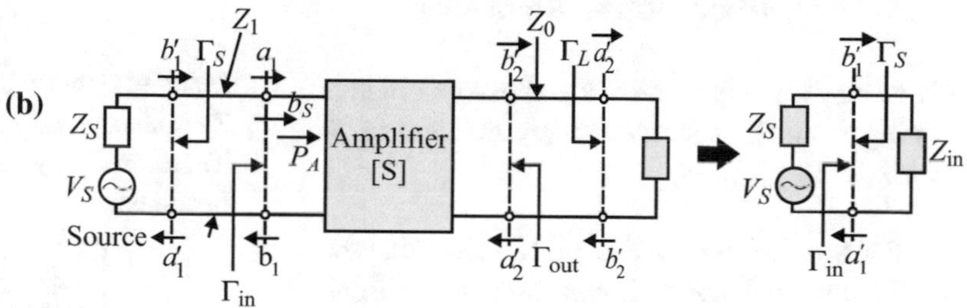

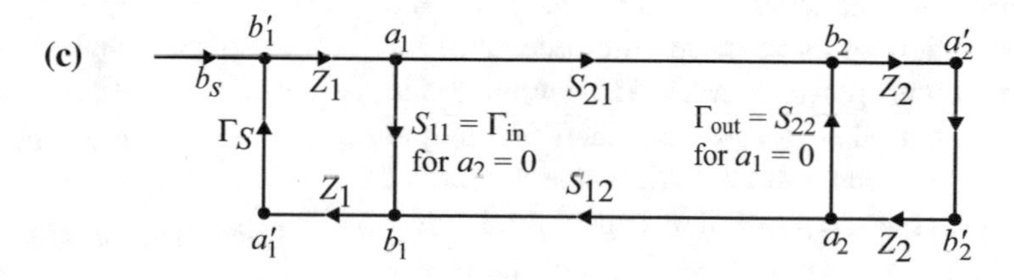

Fig. 11.1 **a** Simplified one-stage amplifier with input and output matching networks, input and output signals of each block along with reflections. **b** Further simplified one-stage amplifier. **c** Signal flow diagrams where Z_1, Z_2 as bilateral impedances of the two matching networks

(b) The two attenuation coefficients:

$$S_{12} = \left.\frac{b_1}{a_2}\right|_{a_1=0}; \; S_{21} = \left.\frac{b_2}{a_1}\right|_{a_2=0};$$

The reflection coefficients of signal reflected by source and load are:

$$\Gamma_S = \frac{Z_s - Z_0}{Z_s + Z_0}; \Gamma_L = \frac{Z_L - Z_0}{Z_L + Z_0} \tag{11.3}$$

Also, the VSWR at the input and output side of the amplifier will be

$$(\text{VSWR})_{in} = \frac{1 + |S_{11}|}{1 - |S_{11}|}; (\text{VSWR})_{out} = \frac{1 + |S_{22}|}{1 - |S_{22}|} \tag{11.4}$$

It can also be proved that Γ_{in}, Γ_{out}, Γ_L are related to S-parameters and power to the non-matched load as:

$$\Gamma_{in} = \left(S_{11} + \frac{S_{21}\,S_{12} \cdot \Gamma_L}{1 - S_{22} \cdot \Gamma_L}\right) = \left.\frac{b_1}{a_1}\right|_{a_2=0} \tag{11.5}$$

$$\Gamma_{out} = \left(S_{22} + \frac{S_{12}\,S_{21} \cdot \Gamma_r}{1 - S_{11} \cdot \Gamma_r}\right) = \left.\frac{b_2}{a_2}\right|_{a_1=0} \tag{11.6}$$

$$P_L = \frac{1}{2}|b_2|^2\left(1 - |\Gamma_L|^2\right) \tag{11.7}$$

$$b_2 = \frac{S_{21}\,a_1}{1 - S_{22}\Gamma_L} \tag{11.8}$$

$$b_s = \left[1 - \left(S_{11} + \frac{S_{21}S_{12}\Gamma_L}{1 - S_{22}\Gamma_L}\right)\right] \cdot a_1 \tag{11.9}$$

Here b_s is the net signal values entering the amplifier (Fig. 11.1b).

With all the above definitions, we will now establish some definitions of various power relations, etc.

11.3 Amplifier Power Relations

We will now study some of the above parameters starting from RF source, its representation as S-matrix, etc.

(a) **RF Source**: As per Fig. 11.1, we see that the RF source is normally connected to the input matching network and then only to the amplifier, so as to reduce the reflection, for maximum power transfer. The output of the amplifier also goes to the load via output matching network, for the same reason of improving the power flow capability. In spite of these matching networks, reflections still exist and these are depicted in Fig. 11.1 as Γ_s, Γ_{in}, Γ_L, and Γ_{out}.
From Fig. 11.1a, the signal voltage entering the amplifier can be written as $\Gamma_{in} = a_1'/b_1'$:

$$b_s = (b_1' - a_1'\Gamma_s) = b_1'(1 - \Gamma_{in} \cdot \Gamma_s) \tag{11.10}$$

Also

$$b_s = a_1 - \Gamma_{in}\, b_1 \tag{11.11}$$

(b) **Incident power (P_{inc}), power input (P_{in}) to amplifier**: In terms of power, we can say that power associated with b_1' (i.e. normalised power coming out of the source) will be:

$$\boxed{P_{inc} = \frac{|b_1'|^2}{2}} \tag{11.12}$$

(With maximum power transfer conditions of $Z_S = Z_0$.)
Therefore, by using Eq. (11.5) above becomes

$$\boxed{P_{inc} = \frac{1}{2}\frac{|b_s|^2}{|1-\Gamma_{in}\cdot\Gamma_s|^2}} \tag{11.13}$$

Also the actual power to the amplifier is

$$P_{in} = P_{inc}\left(1 - |\Gamma_{in}|^2\right)$$

Using this Eq. (11.7) becomes:

$$P_{in} = \frac{1}{2}\cdot\frac{|b_s^2|}{|1 - \Gamma_{in}\Gamma_s|^2}\cdot\left(1 - |\Gamma_{in}|^2\right) \tag{11.14}$$

(c) **Available Power (P_A)**: For maximum power transfer, the condition of $Z_{in} = Z_s^*$ has to be satisfied which is equivalent to $\Gamma_{in} = \Gamma_s^*$ in terms of reflection coefficient. Therefore, we define power available to the amplifier (see

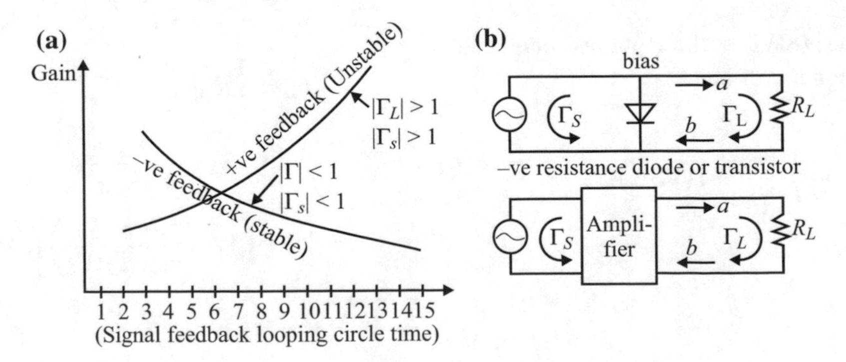

Fig. 11.2 Reflection coefficients Γ_s, Γ_L in an amplifier decide its stability as it leads to +ve or −ve feedback loop to the input of the amplifier. **a** Gain versus fb looping time, **b** negative resistance diode or transistor amplifier

Fig. 11.1a, b) with maximum power transfer condition as:

$$P_A = P_{in}|_{\Gamma_{in}=\Gamma_s^*}$$

$$P_A = \frac{1}{2} \cdot \frac{|b_s|^2}{|1-\Gamma_{in}\Gamma_s|^2}\Bigg|_{\Gamma_{in}=\Gamma_s^*} \cdot \left(1-|\Gamma_{in}|^2\right)$$

$$\boxed{P_A = \frac{1}{2} \cdot \frac{|b_s|^2}{1-|\Gamma_s|^2}} \quad (11.15)$$

(d) **Transducer Power Gain (G_T)**: The gain of an amplifier placed between source and load is called transducer power gain (G_T).

$$G_T = \frac{\text{Power delivered to load } (P_L)}{\text{Power available from source } (P_A)} \quad (11.16)$$

∴ Using Eqs. (11.7), (11.8), (11.9) we get P_L, and using P_A from Eq. (11.17) we get

$$\boxed{G_T = \frac{(1-|\Gamma_L|^2)\cdot|S_{21}|^2\cdot(1-|\Gamma_s|^2)}{|1-\Gamma_L\Gamma_{out}|^2\cdot|1-S_{11}\Gamma_s|^2}} \quad (11.17)$$

(e) **Unilateral Power Gain (G_{TU})**: The power gain, when feedback of amplifier is neglected (i.e. $S_{12} = 0$) then above equation becomes:

$$\boxed{G_T = \frac{(1-|\Gamma_L|^2)\cdot|S_{21}|^2\cdot(1-|\Gamma_s|^2)}{|1-\Gamma_L S_{22}|^2\cdot|1-S_{11}\Gamma_s|^2}} \quad (11.18)$$

(f) **Available Power Gain (G_A)**: The available power gain for load-side matching is given as:

$$G_A = \frac{\text{Power available from the amplifier}}{\text{Power available from the source } P_A} = G_T|_{\Gamma_L=\Gamma_{out}} \quad (11.19)$$

$$\boxed{G_A = \frac{|S_{21}|^2\cdot\left(1-|\Gamma_s|^2\right)}{\left(1-|\Gamma_{out}|^2\right)\cdot|1-S_{11}\cdot\Gamma_s|^2}} \quad (11.20)$$

(g) **Operating Power Gain (G)**: The operating power gain or simply power gain is defined as the ratio of power delivered to load, to the power supplied to the amplifier.

$$G = \frac{\text{Power delivered to load}}{\text{Power supplied to amplifier}} \quad (11.21)$$

$$G = \frac{P_L}{P_{in}} = \frac{P_L}{P_A} \cdot \frac{P_A}{P_{in}}$$

$= G_T \cdot \frac{P_A}{P_{in}}$ by Eq. (11.18)

By Eqs. (11.15), (11.17), (11.19)

$$\boxed{G = \frac{\left(1-|\Gamma_L|^2\right)\cdot|S_{21}|^2}{\left(1-|\Gamma_{in}|^2\right)\cdot|1-S_{22}\,\Gamma_L|^2}} \quad (11.22)$$

11.4 Stability Consideration of Amplifiers

Most important consideration of a quality of an amplifier is its stable performance over the entire frequency range. This is because in an RF circuit the amplifier has the tendency to oscillate depending upon the operating frequency and termination. If $|\Gamma| > 1$, then the return voltage in every feedback loop cycle increases in magnitude (positive feedback), causing instability. This feedback is always there indirectly due to reflection from load (self-feedback loop) (see Fig. 11.2). If $|\Gamma| < 1$, then the diminished return voltage in every feedback loop cycle keeps reducing the voltage and gain (negative feedback) (Fig. 11.2).

An amplifier can be taken as a two-port network characterised by S-parameters with external terminations (load and source) as Γ_L and Γ_s. Then for the stability, the reflection coefficients have to be less than unity. Using Eqs. (11.6) and (11.7) we rewrite as:

$$|\Gamma_{in}| = \left|\frac{S_{11} - \Gamma_L \cdot \Delta}{1 - S_{22}\,\Gamma_L}\right| \leq 1 \quad (11.23)$$

and

$$|\Gamma_{\text{out}}| = \left|\frac{S_{22} - \Gamma_S \cdot \Delta}{1 - S_{11}\,\Gamma_S}\right| \leq 1 \quad (11.24)$$

where

$$\Delta = (S_{11}S_{22} - S_{12}S_{21}) \quad (11.25)$$

As we know that S-parameters are fixed for a given frequency, therefore Γ_L and Γ_s decides the stability of the amplifier.

11.4.1 The Stability Circles

It we substitute the complex terms in above Eqs. (11.24) and (11.25) with real and imaginary parts as:

$$S_{11} = S_{11}^R + jS_{11}^I; S_{22} = S_{22}^R + jS_{22}^I; S_{21} = S_{21}^R + j \cdot S_{21}^I, S_{12} = S_{12}^R + jS_{12}^I$$

$$\Delta = \Delta^R + j\Delta^I; \Gamma_L = \Gamma_L^R + j\Gamma_L^I; \Gamma_S = \Gamma_S^R + j\Gamma_S^I$$

Then the above equations with $|\Gamma_{\text{out}}| = 1$ and $|\Gamma_{\text{in}}| = 1$ gives the circles of stability by a little algebraic operation as plots in Γ_L and Γ_S planes on Smith charts as:

(a) **Output Stability Circle**:

$$\left(\Gamma_L^R - C_{\text{out}}^R\right)^2 + \left(\Gamma_L^I - C_{\text{out}}^I\right)^2 = r_{\text{out}}^2 \quad (11.26a)$$

where radius

$$r_{\text{out}} = \frac{S_{12}S_{21}}{\left(|S_{22}|^2 - |\Delta|^2\right)} \quad (11.26b)$$

Centre of circle

$$C_{\text{out}} = C_{\text{out}}^R + jC_{\text{out}}^I = \frac{\left(S_{22} - S_{11}^*\Delta\right)^*}{\left(|S_{22}|^2 - |\Delta|^2\right)} \quad (11.26c)$$

(b) **Input Stability Circle**:

$$\left(\Gamma_S^R - C_{\text{in}}^R\right)^2 + \left(\Gamma_S^I - C_{\text{in}}^I\right)^2 = r_{\text{in}}^2 \quad (11.27a)$$

where radius

$$r_{\text{in}} \frac{S_{12}S_{21}}{\left(|S_{11}|^2 + |\Delta|^2\right)} \quad (11.27b)$$

Centre of circle

$$C_{\text{in}} = C_{\text{in}}^R + jC_{\text{in}}^I = \frac{\left(S_{11} - S_{22}^*\Delta\right)}{\left(|S_{11}|^2 - |\Delta|^2\right)} \quad (11.27c)$$

These two Eqs. (11.26a–11.26c) and (11.26a–11.27c) when plotted in the complex planes Γ_L and Γ_S, respectively, give output-port stability and input-port stability circles as in Figs. 11.3 and 11.4. The circle is for a given frequency, Z_L, Z_S and will change if these change.

(c) **Output Stability**: If $\Gamma_L = 0$ then by Eq. (11.24), $|\Gamma_{\text{in}}| = |S_{11}|$, then two cases arise depending upon whether $|S_{11}| < 1$ or $|S_{11}| > 1$. For $|\Gamma_{\text{in}}| = |S_{11}| < 1$, the origin (the point of $\Gamma_L = 0$, no reflection at load side) is a part of stable region (Fig. 11.3a). While $|S_{11}| > 1$ means $|T_{\text{in}}| = |S_{11}| < 1$ then, $\Gamma_L = 0$ results in $|\Gamma_m| = |S_{11}| > 1$, i.e. the origin is a part of unstable region. In this latter case, the stable region is the one which is common to $|\Gamma_{\text{in}}| = 1$ and $|\Gamma_L| = 1$ circles (Fig. 11.3b).

(d) **Input Side Stability**: Let us assume it $\Gamma_s = 0$ then by Eq. (11.25) $\Gamma_{\text{out}} = S_{22}$. Then here also two cases arise depending upon whether $|S_{22}| > 1$ or $|S_{22}| < 1$. For $|S_{22}| < 1$, the $\Gamma_s = 0$ is a centre which is stable. Therefore, the region where the centre lies for $\Gamma_s \neq 0$ will be stable region (shaded in Fig. 11.4a). In case $|S_{22}| > 1$, the region where the centre is there will be unstable region, while the stable is as shaded in (Fig. 11.4b), i.e. common region of $|\Gamma_s| = 1$ and $|\Gamma_{\text{out}}| = 1$ circles.

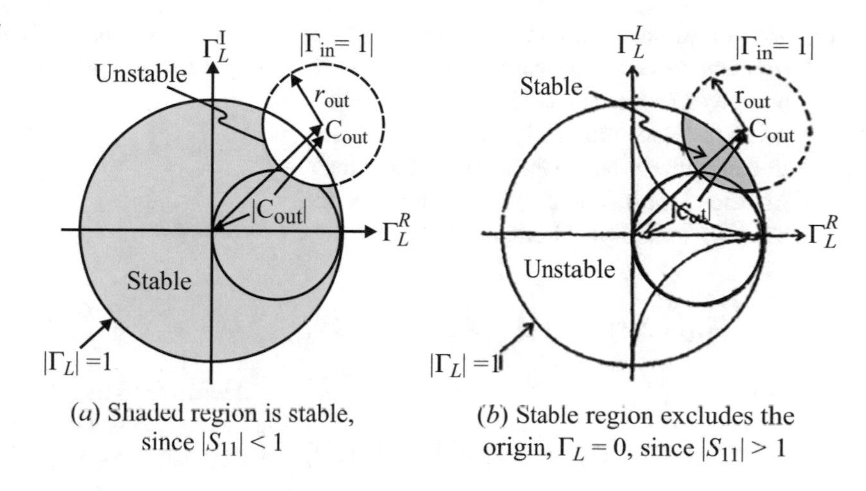

(*a*) Shaded region is stable, since $|S_{11}| < 1$

(*b*) Stable region excludes the origin, $\Gamma_L = 0$, since $|S_{11}| > 1$

Fig. 11.3 Output stability circles with stable and unstable regions

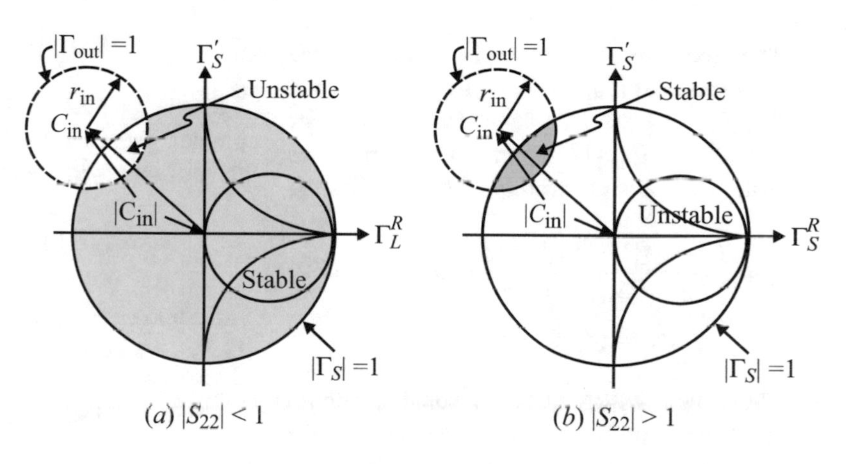

(*a*) $|S_{22}| < 1$

(*b*) $|S_{22}| > 1$

Fig. 11.4 Input stability circles with stable and unstable regions

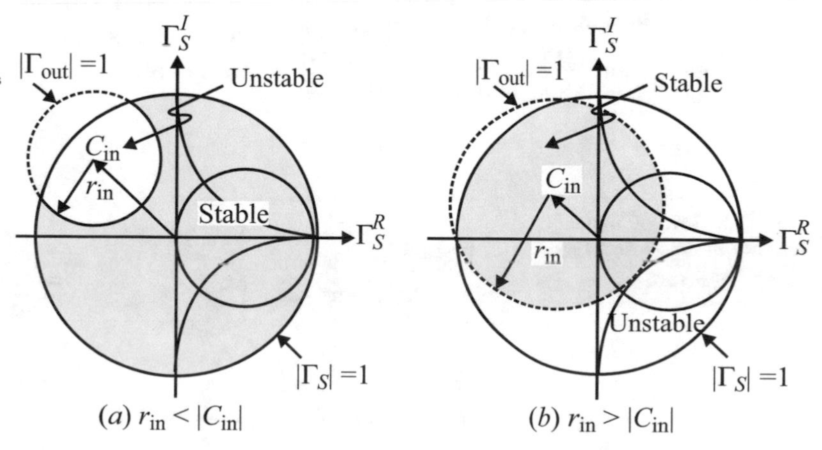

(*a*) $r_{in} < |C_{in}|$

(*b*) $r_{in} > |C_{in}|$

Fig. 11.5 Different input stability regions for $|S_{22}| < 1$ depending on ratio between r_s and $|C_{in}|$

Figures 11.3 and 11.4 are the cases when $|\Gamma_L| = 1$ or $|\Gamma_s| = 1$ circles have radius smaller than C_{in}, C_{out}. But it could be other way round also. A case of the input stability circles for $|S_{22}| < 1$ with two possibilities of $r_{in} < |C_{in}|$ or $r_{in} > |C_{in}|$ is given in Fig. 11.5. Similar cases can be there for the output and stability also.

(e) **Unconditional Stability**: If an amplifier remains stable for any passive source or load, at a given frequency and bias conditions, it is said to be unconditionally stable. In this case, the amplifier is stable throughout the domain of the Smith chart. This statement for $|S_{11}|<1$ and $|S_{22}|<1$ can be stated as:
For

$$|S_{11}|<1 : ||C_{in}| - r_{in}| > 1 \quad (11.28)$$

For

$$|S_{22}|<1 : ||C_{out}| - r_{out}| > 1 \quad (11.29)$$

Therefore, the stability circles have to reside fully outside the $|\Gamma_s| = 1$ and $|\Gamma_L| = 1$ circles.

For $|\Gamma_s| = 1$, circle is shown in Fig. 11.6. For this condition, Eq. (11.28) can be used to prove the following term of stability or Rollett factor k:

$$k = \frac{1 - |S_{11}|^2 - |S_{22}|^2 + |\Delta|^2}{2|S_{12}|\,|S_{21}|} > 1 \quad (11.30)$$

$$|\Delta| = (S_{11}S_{22} - S_{12}S_{21})$$

The above Rollett factor of stability applies for both input and output ports. Other than the above, the unconditional stability can also be in terms of the behaviour of Γ_s in the complex plane $\Gamma_{out} = \Gamma^R_{out} + j\Gamma^I_{out}$. Here, the $|\Gamma_s| \le 1$ domain has to be fully within the $|\Gamma_{out}| = 1$ circle Fig. 11.6b. Plotting $|\Gamma_s| = 0$ in plane Γ_{out}, we get a circle whose centre is located at

$$C_S = S_{22} + \frac{S_{12}S_{21}S^*_{11}}{1 - |S_{11}|^2} \quad (11.31)$$

with radius:

$$r_s = \frac{|S_{12}\,S_{21}|}{1 - |S_{11}|^2} \quad (11.32)$$

Here naturally, the condition $(|C_S| + r_s) < 1$ has to be true and if we put this in Eq. (11.31) we get by using $|S_{12}S_{21}| \le |S_{22} - S^*_{11}| + |S_{12} \cdot S_{21}|$:

$$|S_{12}S_{21}| < |1 - |S_{11}|^2| \quad (11.33)$$

Similar analysis can show that for Γ_L in the complex plane Γ_{in} for $|C_L| = 0$ and $r_s < 1$, from the corresponding equations of C_L and r_L:

$$|S_{12}S_{21}| < \left(1 - |S_{22}|^2\right) \quad (11.34)$$

Addition of the above Eqs. (11.33) and (11.34), we get

$$2|S_{12} \cdot S_{21}| < 2 - |S_{11}|^2 - |S_{22}|^2 \quad (11.35)$$

As we know that $|ab - cd| < (|ab| + |cd|)$ therefore from the definition of Δ (Eq. 11.25a), we can also write:

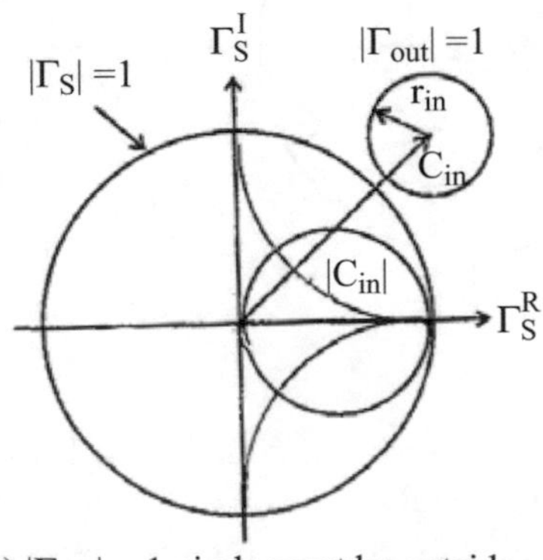

(a) $|\Gamma_{out}| = 1$ circle must be outside

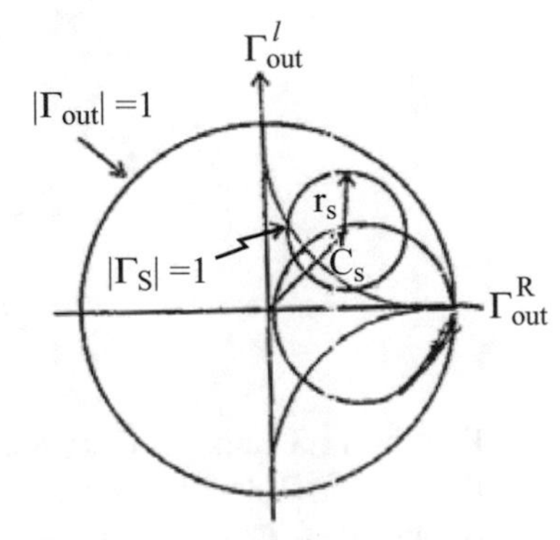

(b) $\Gamma_{out} = 1$ circle must be inside

Fig. 11.6 Unconditional stability in the T_S and Γ_{out} planes for $|S_{11}| < 1$

$$|\Delta| = S_{11} \cdot S_{22} - S_{12} \cdot S_{21}| \leq (|S_{11} \cdot S_{22}| + |S_{12} \cdot S_{21}|)$$

$\therefore$ Using a Eq. (11.35) for $|S_{12} \cdot S_{21}|$, we get

$$\begin{aligned} |\Delta| &\leq |S_{11} S_{22}| + 1 - \frac{1}{2}\left(|S_{11}|^2 + |S_{22}|^2\right) \\ &\leq 1 - \frac{1}{2}\left(|S_{11}|^2 + |S_{22}|^2 - 2|S_{11} S_{22}|\right) \\ &\leq 1 - \frac{1}{2}(|S_{11}| - |S_{22}|)^2 \end{aligned}$$

As for stability $|S_{11}|$ and $|S_{22}|$ both are ≤ 1.

$$\therefore \frac{1}{2}(|S_{11}| - |S_{22}|)^2 < 1$$

$$\therefore |\Delta| < 1 \qquad (11.36)$$

$\therefore$ It is always better to see that both $|\Delta| < 1$ and $k > 1$ conditions are satisfied to ensure unconditionally stable design at that frequency. At a frequency if it is unstable, then at some other frequency it may be stable.

11.4.2 Stabilisation Methods by Loading

Sometimes the operation of an amplifier (whether FET or BJJ, etc.) is not stable, as they have $|\Gamma_{in}| > 1$ and $|\Gamma_{out}| > 1$, then an attempt can be made to stabilise the transistor. As we know that this condition of unstability, i.e.

$$\begin{aligned} |\Gamma_{in}| &= \left|\frac{Z_{in} - Z_0}{Z_{in} + Z_0}\right| > 1; \text{and} \\ |\Gamma_{out}| &= \left|\frac{Z_{out} - Z_0}{Z_{out} + Z_0}\right| > 1; \end{aligned} \qquad (11.37)$$

leads to the conditions:

$$\text{Re}\,(\Gamma_{in}) < 0 \quad \text{and} \quad \text{Re}\,(\Gamma_{out}) < 0$$

For stabilising the input side or output side, we can add a series resistance R_{in}, R_{out}, or shunt conductances (G_{in}, G_{out}), so that the above inequality condition changes (see Figs. 11.7 and 11.8). This loading in conjunction with Re(Z_s) will compensate the −ve contribution of Re(Z_{in}), thus requiring:

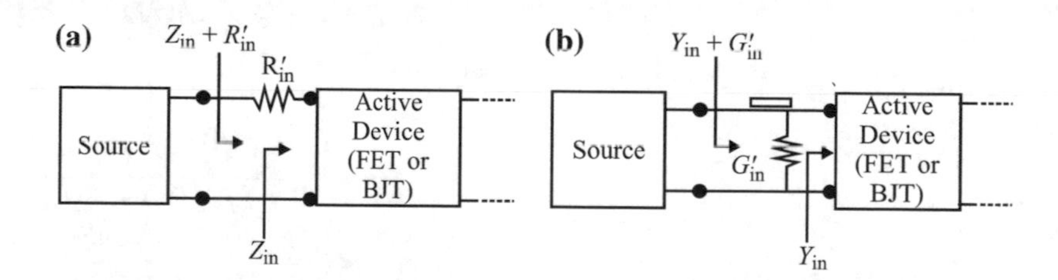

Fig. 11.7 Stabilisation of input port by loading **a** series resistance or by **b** shunt conductance

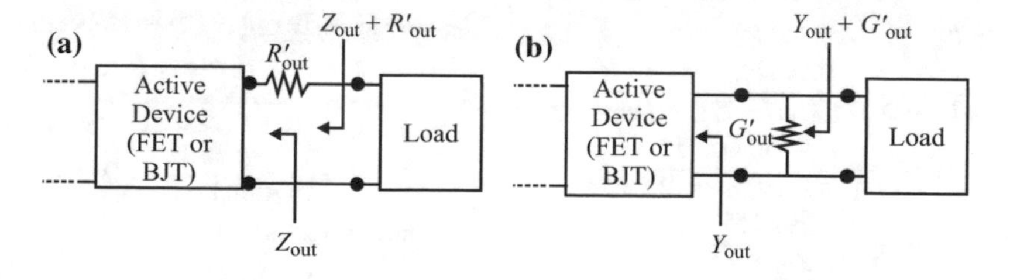

Fig. 11.8 Stabilisation of output port by loading **a** series resistance or by **b** shunt conductance

$$\mathrm{Re}(Z_{in} + R'_{in} + Z_s) > 0 \text{ or } \mathrm{Re}\,(Y_{in} + G'_{in} + Y_s) > 0 \quad (11.38)$$

With similar logic for the output port, we can prove that:

$$\mathrm{Re}(Z_{out} + R'_{out} + Z_L) > 0 \text{ or } \mathrm{Re}\,(Y_{out} + G'_{out} + Y_L) > 0 \quad (11.39)$$

where Z_s, Y_s are impedance and conductance of source while Z_L, Y_L that of load.

Example 1 An FET power amplifier has the following parameters.

$$S_{11} = 0.5\angle-70^\circ, S_{21} = 0.6\angle-10^\circ, S_{12} = 4.5\angle 85^\circ,$$
$$S_{22} = 0.4\angle-45^\circ$$
$$\Gamma_L = 0.187, \Gamma_i = 0.187, Z_L = 73\,\Omega, Z_S = 73\,\Omega, Z_0 = 50\,\Omega$$

Find transducer gain G_T, unilateral gain G_{TU}, available gain G_A, operating gain G, power to the load P_L, power available P_A, and incident power P_{in}.

Solution First, we find the source and load reflection coefficients, assuming $Z_0 = 50\ \Omega$.

$$\Gamma_S = \frac{Z_S - Z_0}{Z_S + Z_0} = \frac{40-50}{40+50} = \frac{-10}{90} = -0.1111$$

$$\Gamma_L = \frac{Z_L - Z_0}{Z_L + Z_0} = \frac{73-50}{73+50} = 0.187$$

$$\Gamma_{in} = S_{11} + \frac{S_{21}\,S_{12}\,\Gamma_L}{1 - S_{22}\,\Gamma_L}$$
$$= 0.5\angle-70^\circ + \frac{(0.6\angle-10^\circ)(4.5\angle 85^\circ)(0.187)}{1+0j-0.4\angle-45^\circ(0.187)}$$
$$= 0.5\angle-70^\circ + \frac{0.54049\angle 75^\circ}{1+0j-0.0748\angle-45^\circ}$$
$$= 0.5\angle-70^\circ + \frac{0.54049\angle 75^\circ}{1+0j-0.0748[\cos(-45^\circ)+j\sin(-45^\circ)]}$$
$$= 0.5\angle-70^\circ + \frac{0.5049\angle 75^\circ}{1+0j-0.0528+0.05289j}$$
$$= 0.5\angle-70^\circ + \frac{0.5049\angle 75^\circ}{0.947.\angle+3.19^\circ}$$
$$= 0.5\angle-70^\circ + 0.533\angle 71.91^\circ$$
$$= 0.5[\cos(-70^\circ)+j\sin(-70^\circ)] + 0.533[\cos(71.91^\circ)+j\sin(71.91^\circ)]$$
$$= 0.171 - 0.469j + 0.1669 + j0.5069$$
$$= 0.03379 + 0.0379j$$

Similarly,

$$\Gamma_{out} = S_{22} + \frac{S_{12}S_{21}\Gamma_S}{1 - S_{11}\Gamma_S}$$
$$= 0.4\underline{-45^\circ} + \frac{(0.6\angle-10^\circ)(4.5\angle 85^\circ)(-0.111)}{1-(0.5\angle-70^\circ)(-0.111)}$$
$$= 0.4\angle-45^\circ - \frac{-0.2997\angle 75^\circ}{1+0.0555\angle-70^\circ}$$
$$= 0.4\angle-45^\circ - \frac{0.2997\angle 75^\circ}{1+oj+0.0555[\cos(-70^\circ)+j\sin(-70^\circ)]}$$
$$= 0.4\angle-45^\circ - \frac{.29997\angle 75^\circ}{1+oj+0.0189-0.052j}$$
$$= 0.4\angle-45^\circ - \frac{0.29997\angle 75^\circ}{1.0189-0.052j}$$
$$= 0.4\angle-45^\circ - \frac{0.29997\angle 75^\circ}{1.0189.\angle-2.92^\circ}$$
$$= 0.4\angle-45^\circ - 0.2941\angle 77.92^\circ$$
$$= -0.4[\cos(-45^\circ)+j\sin(-45^\circ)] - 0.2941[\cos(77.92)+j\sin(77.92^\circ)]$$
$$= 0.2828 - 0.2828j - 0.06143 - 0.2876j$$
$$\Gamma_{out} = 0.2212 - 0.5704j$$

$$G_T = \frac{\left(1-|\Gamma_L|^2\right)|S_{21}|^2\left(1-|\Gamma_S|^2\right)}{|1-\Gamma_L\Gamma_{out}|^2|1-S_{11}\Gamma_S|^2}$$
$$= \frac{[1-(0.187)^2](4.5)^2[1-(0.111)^2]}{|1-(0.187)(0.221)^2||1-(0.5)(0.111)|^2}$$
$$= \frac{0.965\times 20.25\times 0.987}{(0.918)(1.052)} = \frac{19.28}{0.979}$$
$$G_T = 19.87$$
$$G_T(\text{In dBm}) = 10\log(19.87) = 12.98\,\text{dBm}$$

$$G_{TU} = \frac{(1-\Gamma_L)^2|S_{21}|^2\left(1-|\Gamma_S|^2\right)}{|1-\Gamma_L S_{22}|^2\,|1-S_{11}\Gamma_S|^2}$$
$$= \frac{19.28}{(0.855)(1.052)} = 21.43$$
$$G_{TU}\text{ in dBm} = 13.31\,\text{dBm}$$

$$G_A = \frac{|S_{21}|^2\left(1-|\Gamma_S|^2\right)}{\left(1-|\Gamma_{out}|^2\right)(1-S_{11}\Gamma_S)^2}$$

$$= \frac{20.25 \times 0.987}{(0.9510)(1.052)} = \frac{19.986}{1.00052}$$

$$G_A = 19.976$$

$$G_A(\text{in dBm}) = 10\log(19.976)$$

$$G_A = 13.005\,\text{dBm}$$

$$G = \frac{\left(1 - |\Gamma_L|^2\right)|S_{21}|^2}{\left(1 - |\Gamma_{in}|^2\right)(1 - S_{22}\Gamma_L)^2}$$

$$= \frac{0.965 \times 20.25}{0.885 \times 0.555} = \frac{19.541}{0.756} = 25.82$$

$$G(\text{in dBm}) = 14.120\,\text{dBm}$$

$$P_{inc} = \frac{|b_S|^2}{2|1 - \Gamma_{in} \cdot \Gamma_S|^2}$$

$$= \frac{|V_S|^2 \times Z_0}{2 \times (Z_0 + Z_S)^2(1 - \Gamma_{in} \cdot \Gamma_S)}$$

$$= \frac{25 \times 50}{2(8100)(1.037)}$$

$$= \frac{1250}{16799.4} = 0.074407\,\text{W}$$

$$P_{inc} = 74.407\,\text{mW or } 18.716\,\text{dBm}$$

$$P_A = \frac{|b_S|^2}{2\left(1 - |\Gamma_{in}|^2\right)}$$

$$= \frac{|V_S|^2 \times Z_0}{2(Z_0 + Z_S)^2\left(1 - |\Gamma_{in}|^2\right)}$$

$$P_A = \frac{25 \times 50}{2 \times 8100 \times .8858} = 0.871059\,\text{W}$$

$$P_A = 87.1059\,\text{mW} = 19.40\,\text{dBm}$$

$$P_L = G_T \cdot P_A$$

$$= 19.87 \times 87.05$$

$$= 1730.77\,\text{mW}$$

$$P_L(\text{in dBm}) = G_T\text{dBm} + P_A\text{dBm}$$

$$= 12.981 + 19.40$$

$$= 32.38\,\text{dBm}$$

Example 2 A MESFET operator at 5.0 GHz and has *S*-parameters as:

$$S_{11} = 0.5\angle -60^\circ, S_{12} = 0.02\angle 0^\circ, S_{21} = 6.5\angle 115^\circ \text{ and } S_{22} = 0.6\angle -35^\circ$$

Determine whether it is unconditionally stable.

Solution Stability test is $k > 1$ and $|\Delta| < 1$

$$\Delta = |S_{11}S_{22} - S_{12} \cdot S_{21}| = 0.42$$

$$\therefore$$

$$k = \frac{1 - |S_{11}|^2 - |S_{22}|^2 + |\Delta|^2}{2|S_{12}| \cdot |S_{21}|}$$

$$= \frac{1 - (0.5)^2 - (0.6)^2 + (0.42)^2}{2.(0.02)(0.05)} = 2.12$$

Therefore, the transistor is unconditionally stable and no need to draw stability circles.

Example 3 The *S*-parameters for a transistor are as given below. Check the stability and draw the input and output stability circles using Smith chart.

$$S_{11} = 0.385\angle -53^\circ, S_{12} = 0.045\angle 90^\circ, S_{21} = 2.7\angle 78^\circ \text{ and } S_{22} = 0.89\angle -26.5^\circ$$

Solution

$$\boxed{\Delta = S_{11}S_{22} - S_{21}S_{12}}$$

$$= (0.385\angle -53^\circ \times 0.89\angle -26.5^\circ) - (0.045\angle 90^\circ \times 2.7\angle 78^\circ)$$

$$= 0.343\angle -79.5^\circ - 0.122\angle 168^\circ$$

$$= 0.0625 - j0.337 - 0.1193 - j0.02536$$

$$\therefore |\Delta| = |-0.0568 - j0.3625| = 0.1389;\ |\Delta|^2 = 0.0193$$

$$\phi = \tan^{-1}\left(\frac{0.3625}{0.0568}\right) = \angle 81.01^\circ$$

$$\Delta = 0.3189\angle 81.01^\circ \text{ and } |\Delta| = 0.1389$$

$$k = \frac{1 + |\Delta|^2 - |S_{11}|^2 - |S_{22}|^2}{2|S_{12}S_{21}|}$$

$$= \frac{1 + 0.0193 - 0.1474 - 0.792}{0.243} = 0.3329$$

$\therefore |\Delta| < 1$ and $k > 1$; it is the case of a potentially unstable amplifier.

Input stability circle parameters:

$$
\begin{aligned}
C*_{\text{in}} &= S_{11} - \Delta^* S_{22} \\
&= 0.385\angle-53^\circ \\
&\quad - (0.1389\angle 81.01^\circ \times 0.89\angle-26.5^\circ) \\
&= 0.385\angle-53^\circ - 0.1236\angle 54.91^\circ \\
&= (0.2316 - j0.3073) \\
&\quad - (0.0372 - j0.1173) = 0.1944 - j0.4248
\end{aligned}
$$

$$|C*_{\text{in}}| = (0.0379 + 0.01846)^{1/2} = 0.4673$$

$$\phi = \tan^{-1}\left(\frac{-0.4248}{0.1944}\right) = \angle-65.4^\circ$$

$$
\begin{aligned}
C_{\text{in}} &= \frac{C^*_{\text{in}}}{|S_{11}|^2 - |\Delta|^2} = \frac{0.1577\angle 14.15^\circ}{0.1482 - 0.0193} \\
&= 3.98\angle-65.4^\circ
\end{aligned}
$$

$$r_{\text{in}} = \frac{|S_{12}S_{21}|}{|S_{11}|^2 - |\Delta|^2} = \frac{0.1215}{0.1289} = 0.9426$$

Parameters for output stability circle:

$$
\begin{aligned}
C^*_{\text{out}} &= S^*_{22} - \Delta^* S_{11} \\
&= 0.89\angle 26.5^\circ \\
&\quad - (0.1389\angle 81.01^\circ \times 0.385\angle-53^\circ) \\
&= 0.89\angle 26.5^\circ - 0.0536\angle-134.01^\circ \\
&= (0.7964 + j0.3970) \\
&\quad - (-0.0368 - j0.0812) = (0.8332 + j0.4351)
\end{aligned}
$$

or

$$|C^*_{\text{out}}| = \left[(0.8332)^2 + (0.4351)^2\right]^{1/2} = 0.9399$$

$$\text{and } \phi = \tan^{-1}\left(\frac{0.4351}{0.8332}\right) = \angle 27.5^\circ$$

$$C_{\text{out}} = \frac{C^*_{\text{out}}}{|S_{22}|^2 - |\Delta|^2} = \frac{0.9399\angle 27.5^\circ}{0.7921 - 0.0193} = 0.1216\angle 27.5^\circ$$

$$r_{\text{out}} = \frac{|S_{12}S_{21}|}{|S_{22}|^2 - |\Delta|^2} = \frac{0.1215}{0.7720} = 0.1572$$

C_{in} and C_{out} is given in the Smith Chart (Fig. 11.9).

11.5 Oscillators

A stable harmonic oscillator is a key requirement for a signal source in radio, mixer, trans-receiver, radar, etc. For low-frequency requirements, an active nonlinear device, e.g. diode or transistor along with required passive components will suffice. For its frequency stability, crystal resonator can be used. For microwave frequencies, diodes and transistors biased to a negative operating point can be used with cavity transmission line, etc., to produce ac signal up to 100 GHz as seen in Chap. 6. 'Frequency multipliers also can be used by nonlinear devices, like varactor diode, to produce power at millimeter waves frequency (refer Sects. 6.14 and 6.15)'. Rigorous analysis and design of oscillator circuit being a difficult task, it is usually done by advanced CAD tools available commercially (Fig. 11.9).

We will first discuss the low-frequency Hartley and crystal-controlled oscillators, for developing the fundamental concept of positive feedback oscillators which differ (a) due to different transistor characteristics and (a) the ability to make practical use of negative resistance devices.

Some of the important considerations of oscillators at RF and microwave frequency in a system include:

(a) Study of its negative resistance and positive feedback mechanism.
(b) Loading by the parasitic components.
(c) Loading by the subsequent circuit.
(d) Frequency stability of the device with temperature.
(e) Frequency turning range.
(f) Harmonics generation (lower or higher).
(g) Noise.

11.6 Basic Oscillator Models

11.6.1 Feedback Oscillator Model and Source of Signal

An oscillator converts DC power to ac power specially due to its positive feedback mechanism. Each and every device (passive or active) while passing dc current generate thermal noise voltage due to its resistance by random motion of electrons with collisions of atoms in its material. This voltage is nothing but flicker in the dc current, i.e. noise with infinite number of ac components

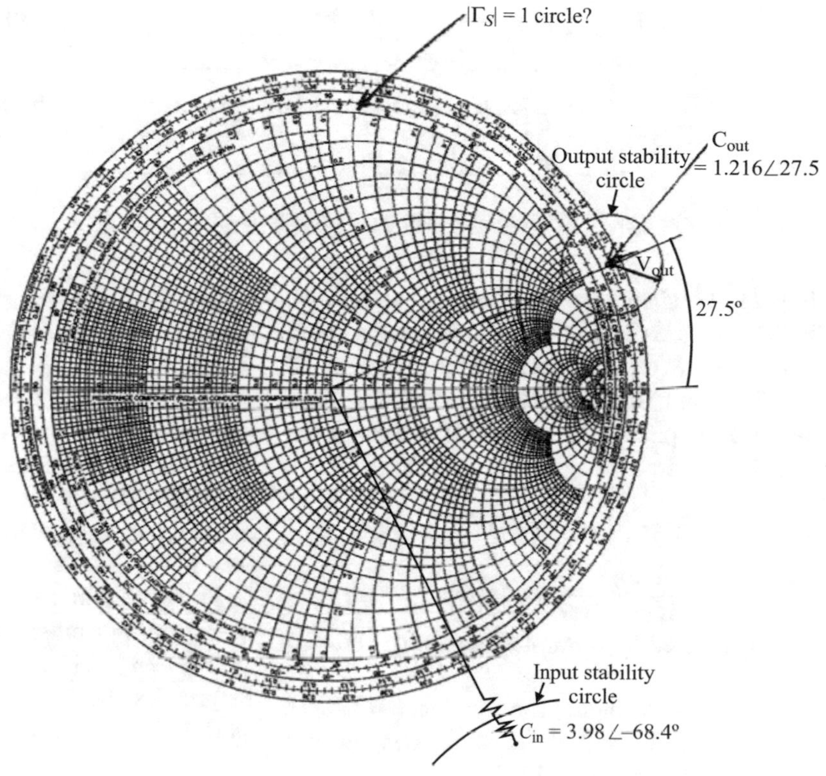

Fig. 11.9 Input and output stability circles

with frequencies ranging from 0 to ∞. If an amplifier of single stage, an input signal is given then it amplifies it and the output has a phase shift of 180°, then if this signal is feedback through a network which is able to phase shift by another 180° at this frequency only, then the amplifier becomes an oscillator due to this situation of regenerative amplification of a signal. This additional 180° phase shifting is possible by the feedback network which depends on its *L, C, R* values (i.e. its resonant frequency) and phase shift response. Now if no input signal is present, then the amplifier will take that noise signal frequency suitable to this feedback network which can cause 180° phase shift and regenerative amplification of this frequency only will start leading to signal output of this frequency.

Noise signals of all the other frequencies will remain untouched and will show off as noise in the signal generated by this oscillator. Thus, at the core of any oscillator circuit is a loop that causes positive feedback of a self-selected frequency, at which it gets overall 360° phase shift in the total circuit. If we change the *L, C* values, frequency of oscillation changes and we call it tuning.

Figure 11.10a shows the generic closed loop system of an amplifier with feedback while

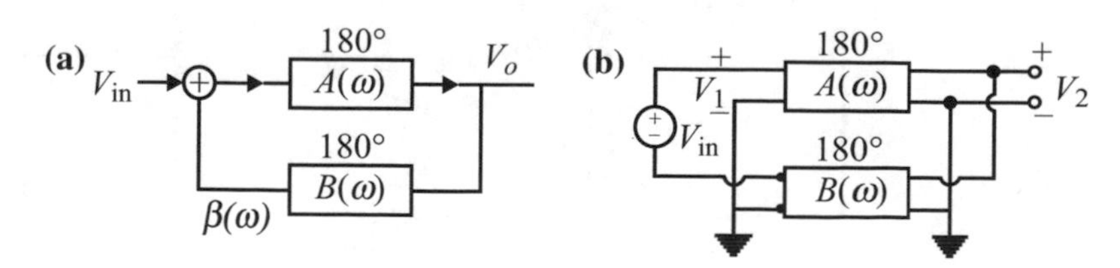

Fig. 11.10 Basic oscillator feedback system of an amplifier

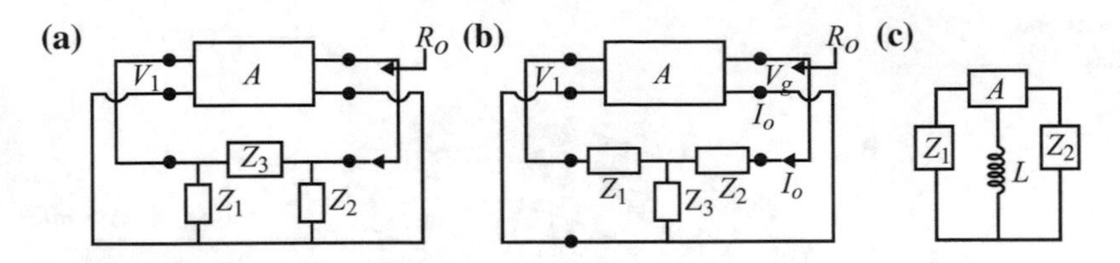

Fig. 11.11 Pye type, T-type, and induction types of feedback

Fig. 11.10b shows the two-port normal network in which V_{in} is internal feedback voltage. Here $A(\omega)$ and $B(\omega)$ are the transfer functions of amplifier and the feedback network, with each causing a phase shift of 180°. This leads Fig. 11.1a with:

$$V_0 = [V_{in} + V_0 \cdot \beta(\omega)]A(\omega)$$

$$\frac{V_0}{V_{in}} = \frac{A(\omega)}{1 - A(\omega) \cdot \beta(\omega)} = A_{cl}(\omega)$$
$$= \text{(closed loop transfer function)} \quad (11.40a)$$

As there is no input in an amplifier working as an oscillator, V_{in} = 0, therefore the denominator of Eq. (11.1) will be = 0.

$$A(\omega) \cdot \beta(\omega) = 1 \quad (11.40b)$$

This condition is called loop gain equation and also known as Nyquist or Barkhausen criteria for oscillation. Gain of the amplifier is a real quantity, i.e. $A = A_r$.

$\beta(\omega)$ generally being the transfer function for feedback network which has to cause a phase shift of another 180°, consists of inductance and capacitance and therefore is a complex quantity say $\beta = \beta_r + j\beta_x$.

This feedback loop could be a π type of network or a T type or inductance type as in Fig. 11.11.

As soon as we switch on an oscillator, due to regenerative feedback mechanism voltage V_0 keeps on increasing with time, i.e. loop gain function $A_r\beta_r > 1$ and soon it stabilises of saturation figure at V_Q with arrest of the fall of A_r as seen in Fig. 11.12a.

Figure 11.12b shows that noise signal frequency also stabilises at a certain resonant frequency ωQ, as for all other noise signal frequencies the 180° phase shift by the β-network is not there and $A\,\beta \neq 1$.

11.6.2 Negative Resistance or Conductance Oscillators Model

An oscillator, one can treat as a passive resonator ($z = R + jX$) coupled to an active device $Z_a = R_a + jX_a$ that creates negative resistance. This negative resistance exactly cancels the resonator's equivalent internal resistance, allowing oscillations to continue at the resonance frequency (Fig. 11.13).

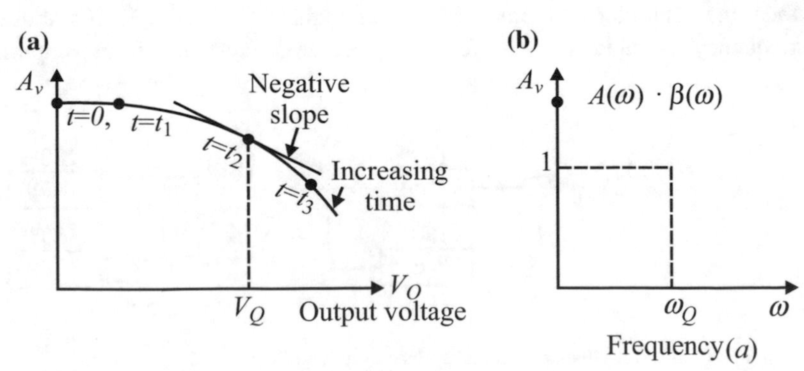

Fig. 11.12 **a** Gain versus output voltage changing with time. **b** Loop gain versus frequency

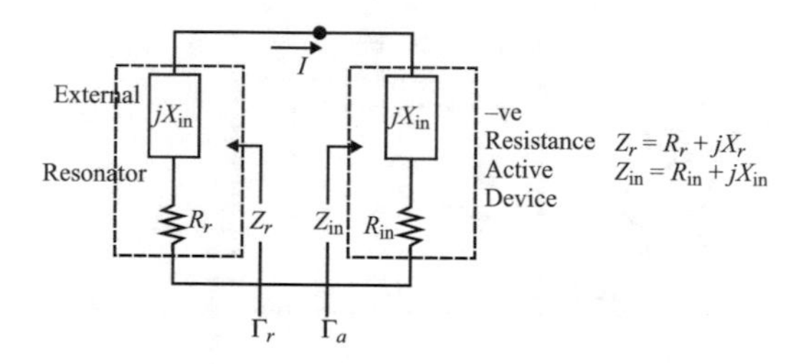

Fig. 11.13 Generic negative resistance oscillator

Applying Kirchhoff's voltage law in the loop for no ac supply but with a finite ac current I, we get:

$$(Z_r + Z_{in})I = 0 \tag{11.41}$$

This requires the real and imaginary parts of the above equation to be zero separately, i.e.

$$R_{in} + R_{out} = 0; (X_r + X_{in}) = 0$$

That is, the negative resistance of device and resonator is of equal and opposite sign. Same is true for the two reactances, for a sustained oscillation, $R_r = -R_{in}$; $X_r = -X_{in}$.

As soon as the oscillation sets in $(R_r + R_{in}) > 0$ then with the increasing voltage of oscillation R_{in} increases, and finally stabilises at $R_r + R_{in} = 0$, i.e. $|R_r| = |R_{in}|$. At the start of the oscillation $|R_{in}| > |R_r|/3$ (at least).

We also know that as the signal of the oscillator increases due to internal regeneration, the resistive part of the device (R_{in}) either increases or decreases with the oscillator voltage amplitude, and then accordingly we call oscillator as negative resistance oscillator or negative conductance oscillator. The oscillator's voltage amplitude level growth soon stabilises and leads to sustained oscillations. Thus, the practical rule which should be there are:

(a) Negative resistance of the active device has to be three times the external series resistance (Fig. 11.14a).
(b) Starting negative conductance of the active devices has to be three times the external shunt conductance (Fig. 11.14b).

Now we may note that Eq. (11.24) implies that $Z_r = -Z_{in}$ for a steady state oscillation: (Eq. 11.26)

From Fig. 11.13, we see that

$$\Gamma_r = \frac{Z_r - Z_0}{Z_r + Z_0} \text{ and } \Gamma_{in} = \frac{Z_{in} - Z_0}{Z_{in} + Z_0} \tag{11.42}$$

$\therefore$ by Eq. (11.41) we get $Z_r = -Z_{in}$

$$\Gamma_r = \frac{Z_r - Z_0}{Z_r + Z_0} = \left(\frac{-Z_{in} - Z_0}{-Z_{in} + Z_0}\right) = \frac{Z_{in} + Z_0}{Z_{in} - Z_0} = \frac{1}{\Gamma_{in}}$$

$$\Gamma_r \cdot \Gamma_{in} = 1 \tag{11.43}$$

These oscillators have been discussed in Chap. 6, where we have dealt in length the following.

- Tunnel diode −ve resistance oscillators.
- Gunn diode −ve resistance oscillators.
- IMPATT diode oscillators.
- TRAPATT diode oscillators.
- BARITT diode oscillators.

11.7 Oscillator Noise

As discussed in Sect. 11.6.1, all oscillators, etc., have noise signal associated with the output signal. There are specially two types of noise in the signal output (a) random variation of frequency (or phase) called jitter (b) random amplitude variation. In oscillators because of the signal amplitude feedback mechanism, the amplitude variation part of the noise gets attenuated, but not phase noise. Therefore, phase noise which dominates in oscillators is from thermal noise, shot, and $1/f$ noise.

In an ideal oscillator, we can assume that the parallel L.C. resonator is completely noise free, while its conductance generates white thermal noise, which is normally written as Norton equivalent current source as:

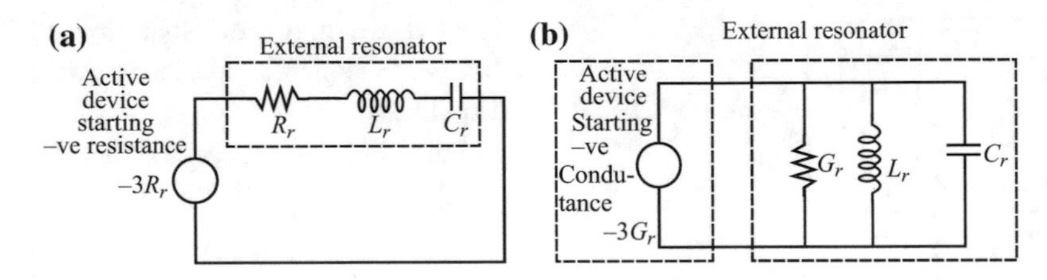

Fig. 11.14 **a** Negative resistance oscillator **b** Negative conductance oscillator

$$\frac{I_n^2}{\Delta f} = 4kTG \tag{11.44}$$

where k is Boltzmann's constant, T is the temperature, G is the resonator conductance, and Δf the noise band width.

While measuring noise by a spectrum analyser, it cannot differentiate between these two noises and the measured noise can be called as the phase noise.

The pure reactive impedance of the LC parallel resonator can be written as:

$$Z(\omega) = \frac{1}{j\omega c + 1/j\omega L} = j \cdot \frac{1}{G.Q\left(\frac{\omega_0}{\omega} - \frac{\omega}{\omega_0}\right)} \tag{11.45}$$

where Q is the quality factor of the original resonator with resistance also. If we consider noise close to the resonant frequency, then a small change $\Delta\omega \ll \omega_0$. in frequency will lead to new impedance at $(\omega_0 + \Delta\omega)$ as:

$$Z_\omega = j\frac{1}{GQ\left[\frac{\omega_0}{\omega_0 + \Delta\omega} - \frac{\omega_0 + \Delta\omega}{\omega_0}\right]} \approx -j\frac{\omega_0}{2GQ(\Delta\omega)} \tag{11.46}$$

$\therefore$ We can write noise voltage square as

$$\frac{V_n^2}{\Delta f} = \frac{I_n^2 \cdot Z^2(\omega + \Delta\omega^2)}{\Delta f} = 4kTG\left(\frac{\omega_0}{2GQ.\Delta\omega}\right)^2 = \frac{4kT}{G} \cdot \left(\frac{\omega_0}{2Q\Delta\omega}\right)^2 \tag{11.47}$$

Due to feedback mechanism, this mean square of noise is halved, i.e. noise power band width

$$\frac{P_n}{\Delta f} = \frac{V_n^2}{\Delta f}.G = 2kT\left(\frac{\omega_0}{2Q \cdot \Delta\omega}\right)^2 \tag{11.48}$$

Thus, the phase noise strongly depends on the frequency and inversely to the Q factor of resonator, and hence effective way to reduce is to increase the Q factor of the resonator.

It has been found that the short noise being dependent on collector current in a BJT, it varies considerably within one oscillation cycle also, being highest in its positive cycle.

11.8 Basic Feedback Low-Frequency Oscillators

We discussed in Sect. 11.8.1, various types of feedback mechanism (Fig. 11.11). In case of Fig. 11.2a, the feedback network is:

$$\beta(\omega) = \frac{V_1}{V_e} = \frac{Z_1}{Z_1 + Z_3} \quad (11.49)$$

In the device, amplifier voltage gain is μ_v with its output resistance as R_0 then loop equation will be:

$$\mu_v V_1 + I_0 R_0 + I_0 Z_c = 0 \quad (11.50)$$

where

$$Z_C = \frac{1}{1/Z_2 + 1/(Z_1 + Z_3)} \quad (11.51)$$

$$\therefore V_0 = I_0 Z_C$$

By applying the loop gain of this π type of feedback amplifier, it can be proved that all the three impedances Z_1, Z_2, Z_3 have to be reactive and also that their sum equal to zero:

$$X_1 + X_2 + X_3 = 0 \quad (11.52)$$

If we say $X_3 = -(X_1 + X_3)$ therefore there are two possibilities:

$$X_1 = \omega L_1; X_2 = \omega L_2; X_3 = 1/\omega C_3 \text{ (Hartley oscillator)} \quad (11.53)$$

$$X_3 = -(X_1 + X_2)\, X_1 = 1/\omega C_1; X_2 = 1/\omega C_{2;} X_3 = \omega L_3 \text{ (Colpitt oscillator)} \quad (11.54)$$

Besides Hartley and Colpitt oscillators, Fig. 11.15, other configurations are of standard common source, common gate, and common drain analogue to CE, CB, and CC in BJT (Fig. 11.16). The frequency of oscillation of these circuits are function of h-parameters of the transistor, besides other passive components.

11.9 High-Frequency Oscillators

As the frequency approaches GHz range, the wave nature of voltages and currents comes into play, requiring study of reflection coefficient, S-parameters, etc., for the study of circuit's functionality. This requires to re-exam in the previous Sects. 11.4.1 and 11.4.2 from transmission line point of view. The Barkhausen criteria need to be re-formulated in the above context.

Starting from the amplifier circuits of Fig. 11.10a, in absence of the signal and the two matching networks, with amplifier replaced by a negative resistance oscillator, we get a new circuit, as in Fig. 11.17, with source replaced by a resonator and load by a terminating network. Therefore, we replace Γ_L by Γ_T and Γ_S by Γ_r in Eqs. (11.5) and (11.6) to give:

$$\Gamma_{\text{in}} = S_{11} + \frac{S_{21} \cdot S_{12}\Gamma_L}{1 - S_{22}\Gamma_L} = \left(\frac{S_{11} - \Delta\Gamma_T}{1 - S_{22}\Gamma_T}\right) \quad (11.55)$$

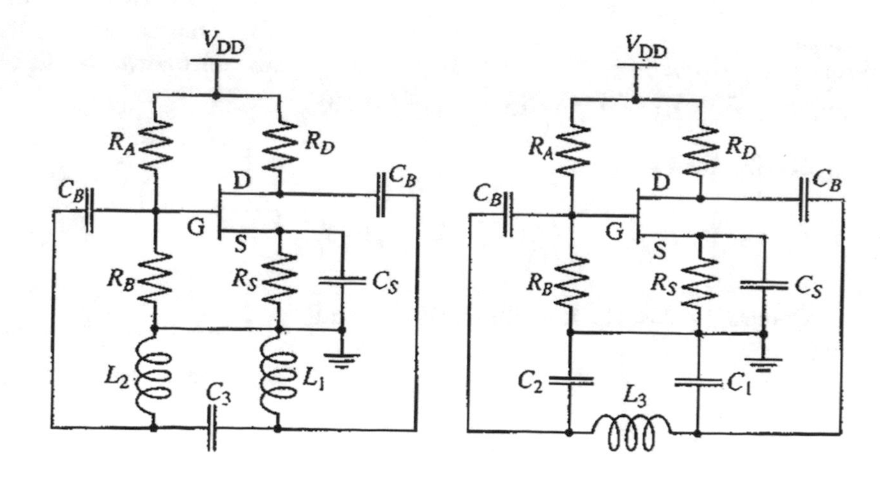

Fig. 11.15 Hartley and Colpitts oscillators, with feedback network

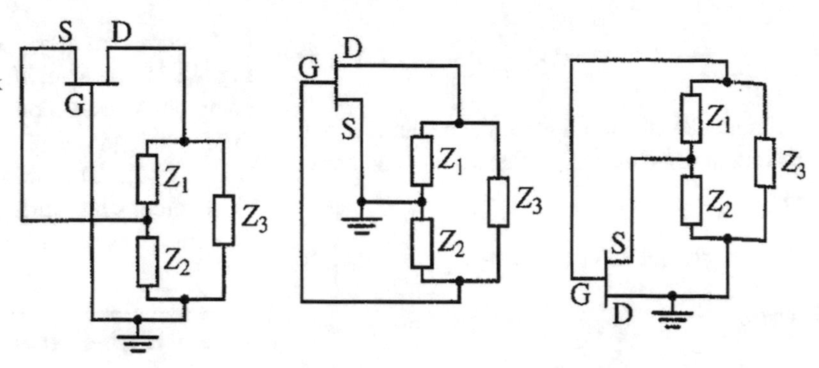

Fig. 11.16 Common gate, source, and drain configurations, with feedback network

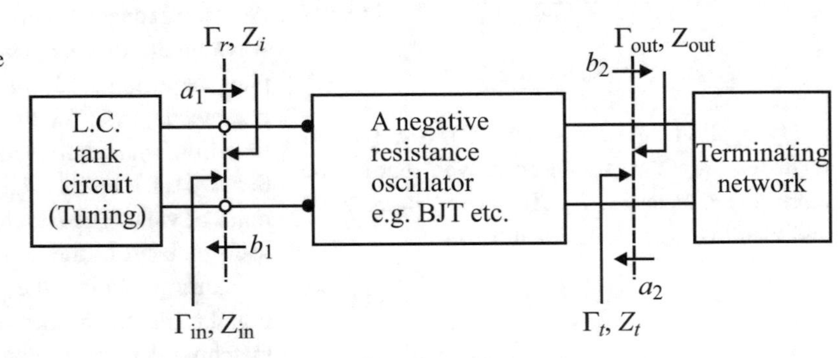

Fig. 11.17 Negative resistance oscillator with the resonator (tank circuit), as source

And

$$\Gamma_{out} = S_{22} + \frac{S_{12}S_{21} \cdot \Gamma_S}{1 - S_{11}\Gamma_S} = \frac{S_{22} - \Delta\Gamma_r}{1 - S_{11}\Gamma_r} \quad (11.56)$$

For steady state oscillations, we know from Eq. (11.43) that $\Gamma_{in} \cdot \Gamma_r = 1$, therefore Eq. (11.55) becomes

$$\frac{1}{\Gamma_r} = \frac{S_{11} - \Delta\Gamma_T}{1 - S_{22}\Gamma_T} \quad (11.57)$$

where

$$\Delta = S_{11} \cdot S_{22} - S_{12}S_{21} \quad (11.58)$$

Solving for Γ_T we get

$$\Gamma_T = \frac{1 - S_{11}\Gamma_r}{S_{22} - \Delta \cdot \Gamma_L} \quad (11.59)$$

Comparing Eqs. (11.59) with (11.50) we find that:

$$\Gamma_{out} \cdot \Gamma_T = 1 \quad \text{and} \quad \text{hence } Z_T = -Z_{out} \quad (11.60)$$

Thus, the condition for oscillation in the terminating network is also satisfied.

Various types of positive feedback mechanism were given in Fig. 11.11. The example of Fig. 11.11c is given below with a BJT as the active device, with base inductor giving positive feedback (Fig. 11.18), as it appears at the input as well as in the output. Once the S-parameters of this circuit are known, the stability analysis of this microwave oscillator can be done. For

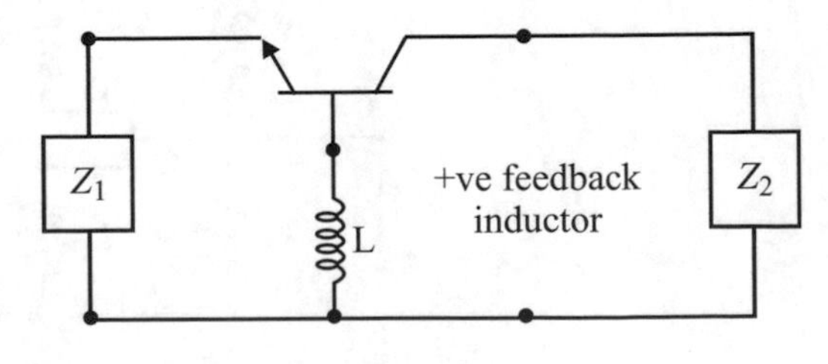

Fig. 11.18 A typical BJT oscillator

detailed analysis, we can see the references (17) and (18).

For oscillators with devices like tunnel diode, Gunn diode, IMPATT diode, TRAPATT diode, BARITT diode, we can refer to Chap. 6.

11.10 Mixers in Heterodyne Receiver

A mixer is a three-port device which with a nonlinear element uses two different frequency (f_1, f_2) inputs at the two parts and produces an output at the third port which consists of sum and difference of these two frequencies $(f_1 \pm f_2)$, besides f_1 to f_2.

11.10.1 Origin of Mixer and Heterodyne Receiver

The need started before 1930 when domestic radio receiver was supposed amplify signals of all the radio stations transmitting at different AM and FM frequencies from medium waves (MW = 540–1640 kc/s), short wave (SW = 3–18 mc/s) to frequency modulated wave (FM = 88–108 mc/s). There is no single amplifier possible, which can be of so much wide band, that it covers all the frequencies MW to FM of the radio station. But at the same time, a single oscillator is possible which can be tuned to generate signal over that wide frequency range as above. This fact of the oscillator was used to remove the above difficulty of amplifier by mixing the carrier signal (f_c) and local oscillator f_{eo} such a way that tuning of f_{eo} and f_c is done simultaneously by a single tuning control by a twin set of variable capacitors called gang condenser. This way the difference frequency after mixing remained constant during tuning also $= (f_c - f_{eo}) = f_{\mathrm{IF}} =$ intermediate frequency (Fig. 9.8). This frequency f_{IF} has been 455 kc/s, which is much lower that the main RF carrier, and thus the mixing has down converted the RF carrier frequencies to a single frequency. Amplification of this single frequency f_{IF}, which already has the signal (audio/data) in it is easily possible. This type of receiver (Fig. 9.8) is called superheterodyne receiver as given in Sec. 9.6.

At RF also the heterodyne receiver has similar concept. The RF carrier signal (f_c) received from the antenna is pre-amplified by LNA and send to a mixer along with local oscillator frequency (f_{eo}). The output of this mixer will be $(f_c \pm f_{lo})$. This is sent to a band filter to pass only $f_c - f_{lo} = f_{\mathrm{IF}}$, the intermediate frequency (Fig. 11.19). This IF is not 455 kc/s but in the MHz range. The mixer has two parts the combiner and detector.

Combiner is normally a 90° or 180° directional coupler while the detector is a single diode as it has the nonlinear behaviour/$I(r) = I_0$ ($e^{V/VT}$$-1$).

Other than the diode, BJT and MESFET mixer with low noise figure and high conversion gain can also be used. This is because the BJT and MESFET also have a nonlinear behaviour of:

$$I(V) = I_{DSS}(1 - V/V_{TO})^2 \tag{11.61}$$

Output from an ideal mixer is given by the product of the carrier RF signal and local oscillator signal:

If $v_c = v_1 \cos(\omega_c t)$

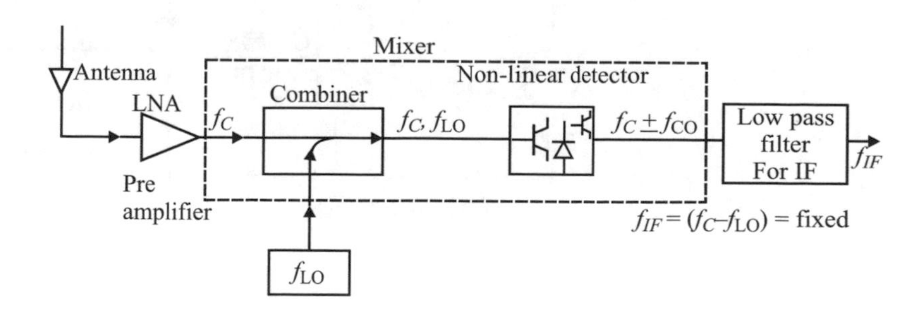

Fig. 11.19 Heterodyne receiver at RF

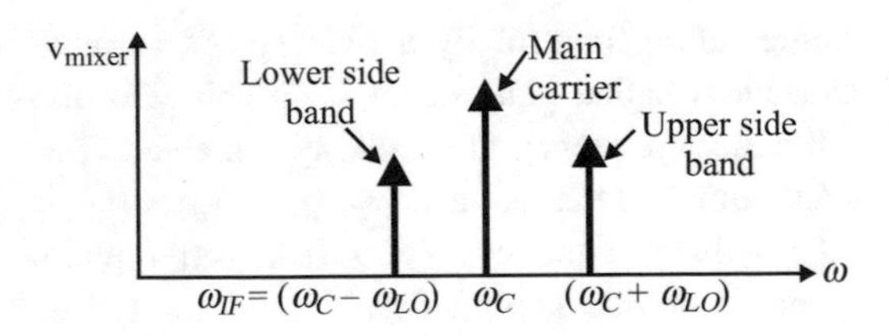

Fig. 11.20 Spectral representation of the output of mixer

$$v_{\mathrm{LO}} = v_2 \cos(\omega_{\mathrm{LO}} t)$$

Then $v_{\mathrm{mixer}} = v_1 v_2 \cos(\omega_C t) \cdot \cos(\omega_{\mathrm{LO}} t)$

$$v_{\mathrm{mixer}} = \frac{v_1 v_2}{2}[\cos(\omega_C - \omega_{\mathrm{LO}})t + \cos(\omega_C + \omega_{\mathrm{LO}})t] \tag{11.62}$$

The output of the mixer (Fig. 11.20) will have four frequencies: the original carrier signals ω_C and ω_{LO} plus two signals $(\omega_C \pm \omega_{\mathrm{LC}})$ as per Eq. (11.62). This is clear from the spectral representation in Fig. 11.20, e.g. if f_C = 2 GHz, then f_{LO} = 1.8 GHz, then $(f_C + f_{\mathrm{LO}})$ = 3.8 GHz, and $f_C - f_{\mathrm{LO}}$ = 0.2 GHz = 200 MHz = f_{IF}. This IF remains fixed by simultaneously tuning of the load oscillator and antenna input carrier by a gang condenser (refer Sect. 9.6 and Fig. 9.8).

Thus, we see that mixing has caused a down conversion of GHz (RF signal) to MHz (IF signal), handling of which is easier as the modulation information of RF carrier gets transferred to the IF with the same modulation, e.g. radio speech, songs. As normally, we will be using only $f_{\mathrm{IF}} = (f_C - f_{\mathrm{LO}})$, these circuits are called single-ended mixers.

11.10.2 Important Parameters of a RF Mixer

Some of the important parameters of RF mixer can be listed as:

(a) **Port Isolation (LO to RF, LO to IF, and RF to IF)**: All these isolations decide which type of mixer we need to use. The double-balanced mixer is the best in this regard. In fact, the LO and RF carrier input can pass to the detector, resulting into a DC and $2f_o$ component in the IF output. Also the higher harmonics of LO lowers the power output of IF and therefore need to be reduced.

(b) **Operating Frequency Range**: This is as per our need of RF carrier signal.

(c) **Conversion Gain**: The active device like transistor gives gain (output IF to input RF power Ratio), in the range of +10 dB, while passive devices (e.g. diodes) will give −ve gain.

(d) **Noise Figure**: Mixer also adds its own noise to the IF output. The presence of harmonics, RF, LO, side bands, etc., due to less isolation, also becomes the sources of added noise. Thus, better isolation reduces noise as well.

Thus, if we use a double-balanced mixer with four FET in a ring instead of four diodes (Fig. 11.24) it becomes the best mixer.

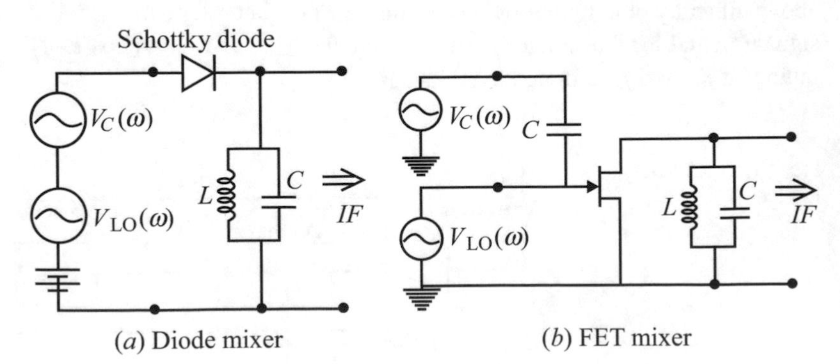

Fig. 11.21 Indicative single-ended mixers circuits

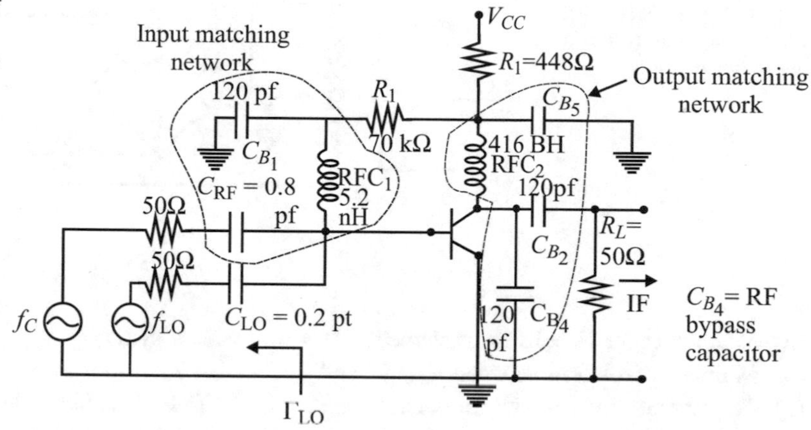

Fig. 11.22 A typical type of BJT mixer for f_{LO} = 1.7 GHz, f_C = 1.9 GHz we get f_{IF} = 200 MHz

11.10.3 Simple Mixer Circuits

Normally, all the mixers are single ended with I.F. = $(f_C - f_{LO})$, *i.e.* down conversion of frequency (from RF-GHz to IF-MHz) takes place, for ease of further operation. Indicative simplest circuit of diode and transistor mixed is given in Fig. 11.21.

In both the cases, the combined RF and LO signal is subjected to a nonlinear device (Schottky diode and FET) followed by a LC band pass filter for getting IF signal only. As here the RF carrier and LO signals are not electrically isolated, LO signal can interfere with RF reception from the antenna. Unlike diode mixer, the FET mixer provides gain in the IF signal also.

Figure 11.22 gives a typical circuit of a BJT single-ended mixer for f_C = 1.9 GHz f_{LO} = 1.7 GHz; f_{IF} = 0.2 GHz.

Here C_{LO} is sufficiently low so that RF carrier signal should not get coupled into the LO source.

C_{B1} and RFC$_1$ together make series resonator for IF, rest is clear from Fig. 11.22 itself.

This circuit has so many components which become difficult to implement and also has the difficulty of maintaining f_{LO}, f_C, and f_{IF} isolated. This problem is removed in balanced mixers.

11.10.4 Single-Balanced Mixer

A balanced mixer has a hybrid coupler to which the f_{RF} and f_{LO} is provided, which in conjunction with dual-diode or dual-transistor generates the required IF as in Fig. 11.23.

In the hybrid coupler, input at port *A* gives output at port *C* and *D* at 90° phase difference. The input at port *B* gives output at port *D* and *C* with 90° as phase difference.

Thus, the advantages of a balanced mixer are as follows:

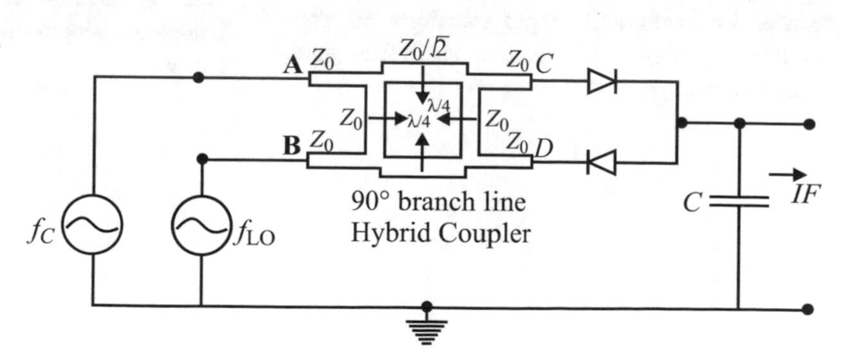

Fig. 11.23 Single-balanced mixer using hybrid coupler

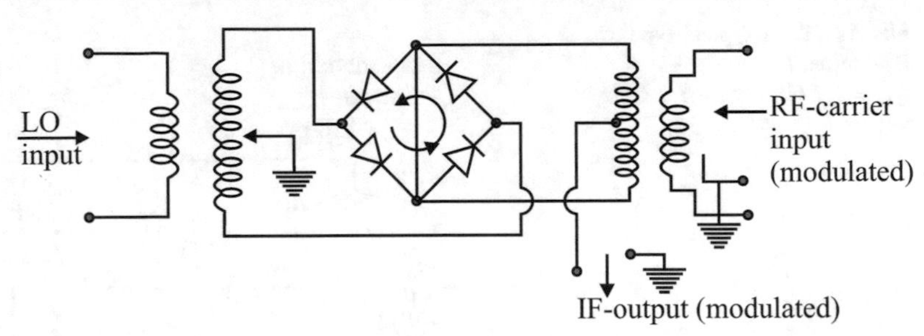

Fig. 11.24 Double-balanced mixer using four cyclic diodes and a transformer

(a) Much simpler circuits to implement.
(b) Maintains isolation between f_{LO}, f_C, and f_{IF}
(c) Can operate over a wide frequency band.
(d) It provides noise suppression because of opposite diode arrangement in conjunction with 90° phase shift.
(e) It also suppresses spurious product rejection. The spurious product arises out of mixing of higher harmonics of RF carrier and LO as well as within IF band.
(f) It has excellent VSWR.
(g) It has the best linearity.

11.10.5 Double-Balanced Mixer

For still better isolation of RF carrier, LO and IF signals with better suppression of spurious harmonics of LO and RF signals much more effectively, double-balanced mixer, are required. It is constructed by using four diodes in a cyclic configuration Fig. 11.24. The only disadvantages are the higher LO drive power requirement, higher noise figure, and higher conversion loss. Here all the three signal paths are de-coupled, with symmetric mixing of *RF* signals f_{LO} and f_{RF} carrier by input and output transformers. The modulatory signal information of f_c(GHz) gets transferred to f_{IF} (MHz).

Problems

1. Define incident power, power input, power available, and power to the load in block diagram. Also explain transducer gain, available gain, operating gain.
2. What is stability of an amplifier? What makes it unstable explain the mechanism of instability?
3. Draw input and output stability circles on smith chart.
4. *S*-parameters of a BJT for certain frequency and condition are as:
   $S_{11} = 0.60\angle 57°$; $S_{21} = 2.18\angle 61°$;
   $S_{12} = 0.09\angle 77°$; $S_{22} = 0.47\angle -29°$;
   Check the stability. How can we stabilise it and suggest the value of the resistances and check the stability again.
5. Explain the origin of oscillation in oscillators. From where does the ac signal comes when only dc is given?
6. What is heterodyne receiver explain. How information (modulatory signal) gets transferred from so many different carriers of GHz range to a fixed single intermediate frequency of MHz range, explain.
7. What is balanced mixer. Explain the important parameters of mixer. Which mixer is best and why?

Simple Laboratory Experiments and Laboratory Manual

12

Contents

© Springer Nature Singapore Pte Ltd. 2018

P. K. Chaturvedi, *Microwave, Radar & RF Engineering*,
https://doi.org/10.1007/978-981-10-7965-8_12

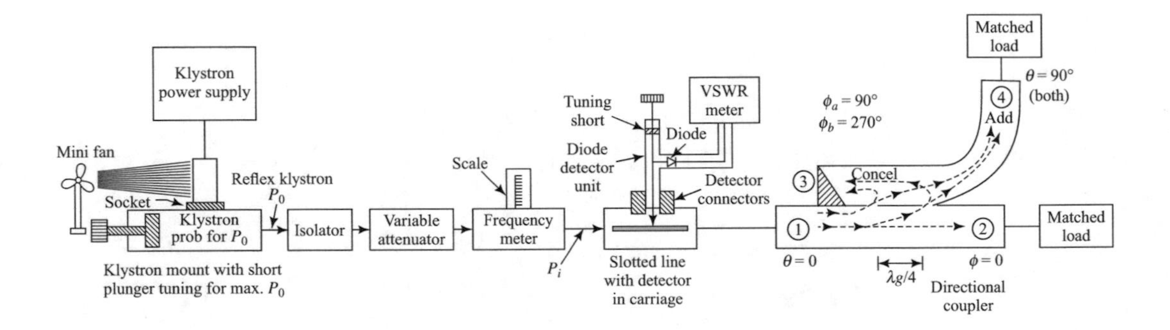

In this chapter, simple experiments on the study of microwave devices and measurement techniques, which can be performed in academic environment, have been given. **It gives the laboratory manual in the conventional form with required tables and some sample readings, besides viva and quiz questions after each experiment**. It will therefore help the students and teachers in bridging the gap between theory and practice, as well as make their foundation strong, for their career in academics, industry, and research.

12.1 Experiment No. 1: Reflex Klystron Characteristics—Modes for Power and Frequency with Repeller Voltage Using Electronics Tuning

Aim

(i) To study the power output and frequency characteristic with changing repeller voltage V_R (electronic tuning).
(ii) Find electronic tuning sensitivity (ETS).
(iii) To find the mode number and the transit time at a given repeller voltage.

Equipments Required

Reflex Klystron with mount and cooling fan, klystron power supply, variable attenuator, frequency metre, tuned detector with its slotted waveguide unit, waveguide stands, and VSWR metre/microammeter.

Theory

Reflex klystron is vacuum tube microwave oscillator, the operation of which depends on the principle of electron velocity modulation. Normally the reflex klystron model used in the laboratories is 2K25, as given in figure of Chaps. 7 and 5 (Figs. 7.2 and 5.12).

Reflex klystron has already been dealt in detail in Chap. 5. The important parameters are mode number, transit time, power frequency characteristic, frequency, repeller voltage characteristics, mechanical tuning, electronic tuning, power efficiency, etc.

The working of reflex klystron oscillator (where electron gets accelerated due to the firstly coming anode cavity positive voltage and then gets repelled beyond that due to –ve repeller voltage) it can be summarised in the following 6 steps. The klystron generates RF signal from the dc voltages given to it, by way of amplifying only that frequency signal from the white noise (frequency 0 to ∞), which is favourable to it, just like any oscillator.

The working of reflex klystron oscillator can be summarised in the following six points:

1. Electron ejection: The electron coming out of heated cathode gets accelerated towards the perforated buncher cavity grid as anode with +ve voltage. Due to momentum it crosses the buncher anode.
2. Electron bunching: Thereby its density gets modulated; i.e. bunching takes place after crossing the perforated anode cavity, due to its RF voltage accross its walls. This is becuase those electrons accelerate which face +ve RF cycle across the cavity while crossing, while other electrons retard which face -ve RF cycle while crossings.
3. Beyond bunching cavity: Due to momentum, the electron continues to move in the −ve dc repeller field between anode cavity and repeller. The repeller reduces its velocity and finally repels back to the cavity, and by this time the modulation (bunching) becomes still more sharper, i.e. denser.
4. Energy transfer by electron bunches to cavity after getting repelled back: If this concentrated bunch of electron in the return path re-enters the cavity at that moment when the perforated anode cavity's first wall has +ve RF voltage cycle, then it loses its energy by getting retarded, thereby transferring energy to the RF field, and amplifies it (+ve feedback process).
5. Signal Tapping: This amplified signal (microwave) in the cavity can be tapped out using tapping probe of coaxial line.
6. Frequency of oscillation: The frequency of oscillation is primarily determined by:

(a) The dimension of the cavity of klystron, which can be changed a little by rotating the knob/screw on the tube. This is called mechanical tuning, which varies the frequency up to 1 GHz in 2K25 tubes, i.e. ±5% around central frequency of oscillation.
(b) The repeller voltage, which changes the frequency within a given mode marginally, is called *electronic tuning*. It has a small range of 60 MHz tuning or so (A to B of Fig. 12.2) in $1\frac{3}{4}$ mode, i.e. ±3% (max) around the central frequency of oscillation.
(c) The distance of repeller from the cavity is fixed and it cannot be changed.

Note: Power metres are normally expensive and are not used in such laboratories of academic institution. In place of power measurement directly, we measure the detected voltage on VSWR metre or current (which are directly proportional to power) on a microammeter, with its other terminal grounded.

Procedure

1. Assemble the equipments as per Fig. 12.1.
2. Set the power supply to CW position.
3. Switch on the cooling fan and then klystron power supply. Wait for 3–4 min for stabilisation of the supply and the klystron.
4. Set the attenuator to a certain level say 3 dB initially, which should not be changed in whole of the experiment.
5. Keep the frequency metre to one end fully.
6. Set the beam voltage at 250 V with beam current <30 mA.
7. Increase the −ve voltage of the reflector (V_R) in small steps of 5 V, go up to −250 V, and see that some power output is shown in the VSWR micrometre which shows that klystron has started oscillating.
8. Adjust the short plunger tuners of klystron mount, of detector and of the detector mount, for getting maximum power output in VSWR metre.
9. If the VWSR metre goes out of scale, go to its higher scale (e.g. 0–10 dB to 10–30 dB scale). See that detectable output is there.
10. At each step, measure/note the:

    (a) Frequency by frequency metre corresponding to a dip in the output power.
    (b) Voltage–current reading in the VSWR metre/microammeter which is directly proportional to the μw power.

11. Plot the readings of power/current output and frequency with V_R, and note the different modes after calculating it (see Fig. 12.2 and calculations).

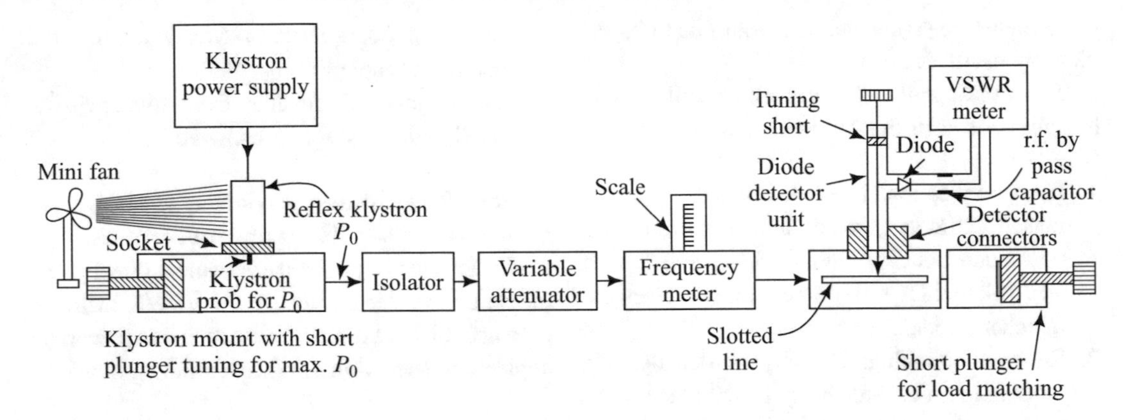

Fig. 12.1 Microwave bench set-up for reflex klystron characteristic

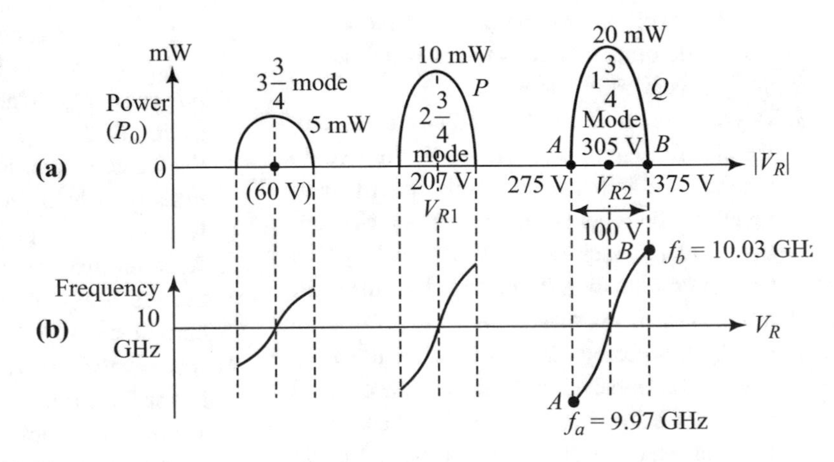

Fig. 12.2 **a** Power output and **b** frequency tuning with $|V_R|$ (modulus of V_R)

12. Calculate the mode number transit time and electronic tuning sensitivity (ETS), i.e. frequency tuning for per volts change in V_R.
13. Change the power supply setting from CW to modulated/pulsed mode of square wave of 1 Kc/s, and repeat all the above steps 5–12. As the power output is mainly controlled by V_R, the modulation is superimposed on it inside the power supply.

Results and Observation Calculations (a sample case):

1. **Mode number**: From the locations of the two consecutive maxima power in the V_R versus power plot, read the values of repeller voltages V_{R_1}, V_{R_2} Fig. 12.2. We now compute 'n' from the following equation, which gives the mode numbers N_1 and N_2 (Table 12.1).

$$\frac{V_{R_1}}{V_{R_2}} = \frac{N_2}{N_1} = \frac{n+1+3/4}{n+3/4}$$

(For N_1, N_2 two consecutive modes with n = an integer)

Example 1 If V_{R_1} and V_{R_2} are −305 and −207 V at 10 GHz.

Then

$$\frac{n+1.75}{n+0.75} = \frac{-305}{-207} = 1.57$$

Table 12.1 Reflex klystron characteristic

A. Fixed settings for an experiment are:

- Attenuation setting …… dB
- Beam voltage …… 300 V or so
- Beam current …… mA

B. CW power supply case:

S. no.	Repeller voltage (V_R)	Voltage in VSWR metre/microammeter reading (μA) $\propto P_o$	Frequency (GHz)
1 2 3 : 15			

C. Square wave modulated/pulsed power case:

S. no.	Repeller voltage (V_R)	Microammeter reading (μA) $\propto P_o$	Frequency (GHz)
1 2 : 15			

D. Power modes from the plot:

S. no.	Mode no.	Central freq. of	Maximum power in terms of mV in VSWR metre	Electronics tuning range		Electronics tuning sensitivity
				f_{min} (GHz)	f_{max} (GHz)	
1	$1\frac{3}{4}$	10 GHz	10	9.97	10.03	0.6
				($\lvert V_R \rvert$ = 275)	($\lvert V_R \rvert$ = 375)	MHz/V
2	$2\frac{3}{4}$					
3	$3\frac{3}{4}$					

$$\therefore \quad n = \frac{1.75 - 1.57 \times 0.75}{1.57 - 1} \cong 1$$

Therefore

V_{R_1} −305 V gives maximum power at mode number $1\frac{3}{4} = N_1$

V_{R_2} −207 V gives maximum power at mode number $2\frac{3}{4} = N_2$

2. **Transit time**: The transit time of electron to return back to the cavity resonator in $\left(n + \frac{3}{4}\right)$ mode is given by:

$$t = \frac{\left(31 + \frac{3}{4}\right)}{f_0}$$

Example 2 At 10 GHz for mode number $N_2 = (2 + 3/4)$, $N_1 = (1 + 3/4)$

$$\therefore \quad t_2 = \frac{2+\frac{3}{4}}{10}\,\text{ns} = 275\,\mu\text{s}$$

$$\text{and} \quad t_1 = \frac{1+\frac{3}{4}}{10}\,\text{ns} = 175\,\mu\text{s}$$

3. **Electronics tuning sensitivity (ETS)**: In a given mode $\left(n+\frac{3}{4}\right)$ when V_R is increased from mode end (zero power point) to another end (zero power point) (see Fig. 12.2b, A and B), then ETS in that mode is:

$$\text{ETS} = \frac{f_b - f_a}{|V_{R_b}| - |V_{R_a}|}$$

Example 3 It in $2\frac{3}{4}$ mode; then $V_{R_a} = (-275)\,\text{V}, V_{R_b} = -375\,\text{V}$ (see Fig. 12.2)

$$f_a = 9.97\,\text{GHz}, f_b = 10.03\,\text{GHz}$$

then for $1\frac{3}{4}$ mode

$$\text{ETS} = \frac{10.03 - 9.97}{|-375| - |-275|} = \frac{0.06}{100} = \text{GHz/V}$$
$$= 0.6\,\text{MHz/V}$$

Precautions

1. An isolator and attenuator avoids reverse reflected signal loading of klystron and therefore must be used in the circuit.
2. While measuring frequency, the metre frequency should start from one end and should be de-tuned each time.
3. Normally repeller voltage should be applied first fully before anode voltage. Therefore start with zero anode voltage of power supply.
4. For klystron, starting setting could be as:
   - Mini fan is ON.
   - Mode switch—CW/AM.
   - Reflection voltage full.
   - Beam voltage zero.
   - Amplitude of modulation full.
   - Frequency of modulation at mid-point.
5. In VSWR metre starting setting could be as:
   - VSWR indicating metre—mid-range.
   - Input switch—low impedance point.
   - dB range from 0 to 20.
   - Gain control centred—at maximum.
6. Cooling fan should be used to avoid overheating of klystron.
7. The mechanical screw of klystron should be rotated very carefully or else it may get spoiled.
8. Never allow your body parts to come in front of the end of waveguide, when the power is ON, as the microwave radiation is over there, as it is cancerous.

Quiz/Viva Questions

1. Why conventional tubes cannot be used at microwave frequencies.
2. Explain bunching in reflex klystron.
3. How does beam focusing takes place in klystron.
4. Mechanical tuning has a higher span range (e.g. maximum 5% of central frequency). While electronic tuning has lower span range (e.g. ±2% of central frequency), still electronic tuning is preferred for application why?
   (**Hint**: If mechanical tuning is done frequently, it will spoil the Klystron and it will shift the modes also.)
5. What are velocity and density modulation?
6. With higher V_R, the mode number reduces but power output increases and electronic tuning range decreases. Explain this phenomenon with reference to bunching and return of electrons back to the cavity.
7. What are CW and pulsed power? What is duty cycle?
8. What are electronic tuning sensitivity (ETS) and electronic tuning range (ETR)?
9. Why short-circuit plunger is provided with: (a) klystron mount, (b) detector probe point in waveguide, (c) detectors tuning.
10. What is the normal range of power and efficiency of a reflex klystron?

11. Output power of a reflex klystron depends on V_R or beam voltage or reflector distance?
12. The perforated cavity of reflex klystron is said to be re-entrant type, explain what is re-entering there also explain cavity and its mechanism of mechanical tuning.
13. If $Z_L = 200\,\Omega$, $Z_0 = 100\,\Omega$, find the values of $\Gamma = \ldots\ldots$, and $S = \ldots\ldots$
14. What is the purpose of isolator?
15. Write the range of reflection coefficient (Γ) and VSWR(s) in general.
16. What does free space velocity (v_0), group velocity (v_g), and phase velocity (v_p) means? What is the relation connecting them? $\left(\text{Ans.}: v_0^2 = v_p \cdot v_g\right)$
17. The detector is said to follow square law, explain.
    Hint: *I–V* relation of detector is as follows: $I_{det} = K \cdot V_{guide}^2$ and $\therefore \frac{V_{max}}{V_{min}} = \sqrt{\frac{I_{max}}{I_{min}}}$.
18. In a two-cavity klystron, how do we find the number of bunches travelling in the drift space?
19. In VSWR the detector current shown is related to the voltage in waveguide as $i_{det} = KV_{W\ guide}^2$ [square law detector]; therefore it represents power as $P \propto V_{guide}^2$; therefore, $P \propto i_{det}$. Explain $I \times V$ characteristic of a diode.
20. What is the purpose of waveguide flange?
21. Why the waveguides are made of metal?
22. What is the surface inside a waveguide? Why it is highly polished and Ag-plated (**hint**: for better reflection). What is meant by 50 micron polish?
23. Why waveguide is air filled only.
24. How many cavities a reflex klystron has?
25. In one line, state on which principle any klystron works.
    (Ans.: Velocity modulation of electron).
26. Explain how different modes of oscillation a reflex klystron can have with the same central frequency but different transit time.
27. Why dc to ac conversion efficiency of reflex klystron is only 20–30% and where does the remaining energy go?
28. Explain how in a reflex klystron single cavity acts as both buncher and catcher cavity.
29. What is the range of *X*-band and why it is used in academic laboratories?
30. Why in reflex klystron modulation is required on the repeller voltage, when the indicator is VSWR metre?
31. Why reflex klystron oscillates only in certain intervals of repeller voltage? (**Hint**: Here the phase of the returning electron bunch matches with that of cavity.)
32. Which reflex klystron is used in the laboratories, give its model number? (**Ans.**: 2K25)
33. When we repeat the same experiment again in microwave, same readings do not come easily, explain the reasons.
34. In electronic tuning, we get different modes with same central frequency (f_0), but in mechanical tuning we get different modes with no central frequency. Explain why.

12.2 Experiment No. 2: Calibration of Mechanical Tuning Screw of a Reflex Klystron

Aim

Calibration of mechanical tuning screw of a reflex klystron for different modes.

Equipments Required

As in Experiment No. 1

Theory

Same as Experiment No. 1.

Procedure

1. Assemble the equipments as per Fig. 12.1.
2. Set the power supply to CW position.

3. Switch on the cooling fan and then klystron power supply. Wait for 3–4 min for stabilisation of the supply and the klystron.
4. Set the attenuator to a certain level (say 3 dB) initially, which should not be changed in whole of the experiment.

14. Plot frequency versus angle of rotation of mechanical tuning screw.

Observation and Results

See Table 12.2.

Table 12.2 Observation and results of mechanical tuning of Reflex Klystron

S. no.	I-mode		II-mode		III-mode	
	Tuning screw angle	Frequency by freq. metre	Tuning screw angle	Frequency by freq. metre	Tuning screw angle	Frequency by freq. metre
1	0°	8.5 GHz	0°	10.5 GHz	0°	11.0 GHz
2	45°	:	:	:	:	:
3	90°	:	:	:	:	:
		:	:	:	:	:
		:	:	:		
6	450°					

V_{beam} = … V; $V_{Repeller}$ = … V

5. Keep the frequency metre to one end fully.
6. Set the beam voltage at 250 V with beam current <30 mA.
7. Increase the −ve voltage of the reflector (V_R) in small steps of 5 V, go up to −250 V, and see that some power output is shown in the VSWR micrometre which shows that klystron has started oscillating.
8. Adjust the short plunger tuners of klystron mount, of detector and of the detector mount, for getting maximum power output in VSWR metre.
9. If the VSWR metre goes out of scale, go to its higher scale (i.e. 0–10 dB to 10–30 dB scale). See that detectable output is there.
10. Start the mechanical tuning screw from maximum anticlockwise position.
11. Set the V_R at one of the modes where it gives maximum power, i.e. at the peak point (Fig. 12.2).
12. Vary the screw clockwise slowly, in steps of 45°, and note the frequency using frequency metre. If the frequency changes are large, then reduce the steps of angle from 45° to 30° (Fig. 12.3).
13. Repeat 11 and 12 above for the next mode.

Precautions

1. The repeller and the beam voltage should not be changed while mechanical tuning.
2. An isolator and attenuator avoid reverse reflected signal loading of klystron and therefore must be used.
3. While measuring frequency, the metre should start from one end and should be de-tuned each time.
4. Normally repeller voltage should be applied first fully before anode voltage. Therefore start with zero anode voltage of power supply.
5. Starting setting could be as follows:
   - Voltage and current indicating metre at OFF.
   - Mode switch—CW/AM.
   - Reflection voltage at full.
   - Beam voltage at zero.
   - Amplitude of modulation at full.
   - Frequency of modulation at mid-point.
6. In VSWR metre starting setting could be as:
   - Voltage, current and dB indicating metre at mid-range.
   - Input switch—low impedance point.
   - Gain control—at maximum.

7. Cooling fan should be used to avoid overheating of klystron.
8. The mechanical screw of klystron should be rotated very carefully or else it may get spoiled.
9. Never allow your body parts to come in front of end of the waveguide when the power is ON, as the μW radiation is over there.

Quiz/Viva Question
Same as Experiment No. 1.

12.3 Experiment No. 3: Study Mode Characteristics of Reflex Klystron on CRO

Aim

To study modes of operation of reflex klystron on CRO by varying reflector voltage.

Equipments

As per Experiment No. 1 (i.e. Fig. 12.1) except CRO in place of VSWR metre and 1 Kc/s. modulating sawtooth signal from klystron power supply to the x-sweep external input to the CRO.

Theory

As per Experiment No. 1.

Procedure

1. Assemble the setup as per Fig. 12.4:

   (i) CRO in place of VSWR metre.
   (ii) Mode selector of power supply at FM MOD.
   (iii) CRO set with external sweep.
   (iv) Modulating signal output of klystron power supply to X-sweep of CRO.

2. Note the output on CRO, and if the CRO signal is low, tune the short plungers of klystron mount, detector mount, and the detector unit.
3. To 10 same as Experiment No. 1.
4. Finally see the modes directly on the screen as in Fig. 12.2a. Please see Table 12.3 for results and observations.

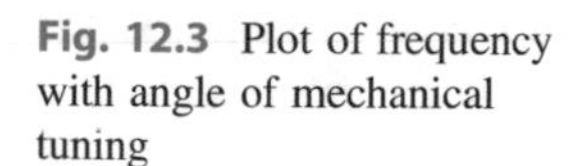
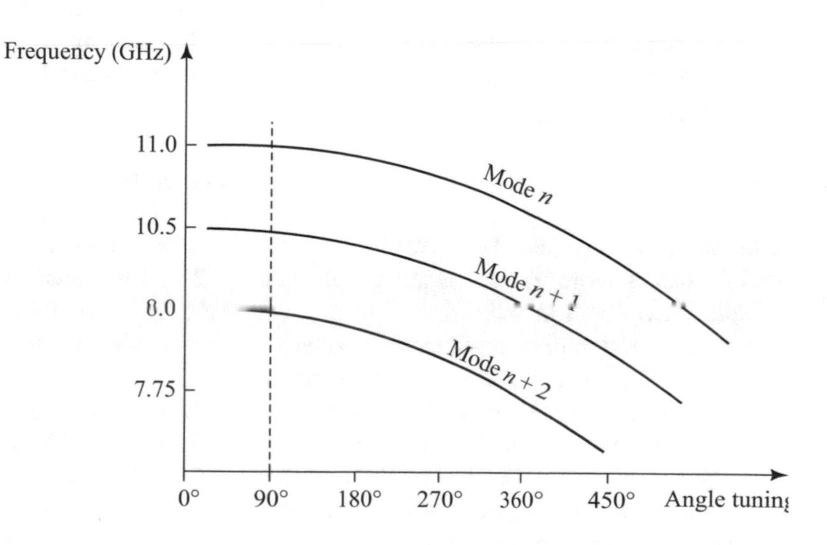

Fig. 12.3 Plot of frequency with angle of mechanical tuning

Table 12.3 Observations and results of mode characterisation by CRO

S. no.	Repeller voltage from klystron power supply	Voltage output as seen in CRO = power output	Frequency metre measurement reading
1			
2			
:			
15			

Precautions

As in Experiment No. 1.

Quiz/Viva Questions

1 to 32—same as Experiment No. 1

33—the sweep signal amplitude should be set to give reasonable readable power output—modes, explain.

12.4 Experiment No. 4: To Determine Frequency, Wavelength, and VSWR in a Rectangular Waveguide Using Slotted Line

Aim

To measure guide wavelength (λ_g), frequency (f_0), and VSWR using slotted line.

Equipments as per Fig. 12.5

Theory

Even when a waveguide has a small mismatch load (which is more or less always there), then standing waves are formed by the reflected wave. The maxima and minima positions at its voltages (V_{max} and V_{min}) can be noted by the carriage position on the slotted line. The distance in between may be measured as $(y_{max} - y_{min}) = \lambda_g/4$, giving the value of λ_g. The frequency (f_0) is measured by the frequency metre, and λ_0 can also be computed. We know that

$$f_0 = c/\lambda_0$$

where $c = 3 \times 10^{10}$ cm/s

$$\lambda_c = 2/\sqrt{\left(\frac{m}{a}\right)^2 + \left(\frac{n}{b}\right)^2} \quad \text{and} \quad \frac{1}{\lambda_0} = \sqrt{\frac{1}{\lambda_g^2} + \frac{1}{\lambda_c^2}}$$

Therefore for the dominant mode TE_{10}, $\lambda_c = 2a$.

$$\therefore \quad f_0 = c \cdot \sqrt{\frac{1}{\lambda_g^2} + \frac{1}{(2a)^2}}$$

We can then compare the computed and measured frequency.

In *X*-band waveguides $a = 2.286$ cm and $b = 1.016$ cm.

Also VSWR = V_{max}/V_{min}.

Detailed theory can be seen in Chap. 4, for waveguides and Chap. 6 for klystron.

Procedure

1. Set up the apparatus as per Fig. 12.5.
2. Set the attenuator at zero dB.
3. Switch on the power supply with modulator at sine wave or sawtooth wave, and the beam voltage at 250 V.
4. Tune the short of klystron for $(P_0)_{max}$.
5. Set the VSWR metre to 10–50 dB scale.
6. Adjust the repeller voltage V_R and modulation voltage V_m, and the VSWR metre shows some power output.

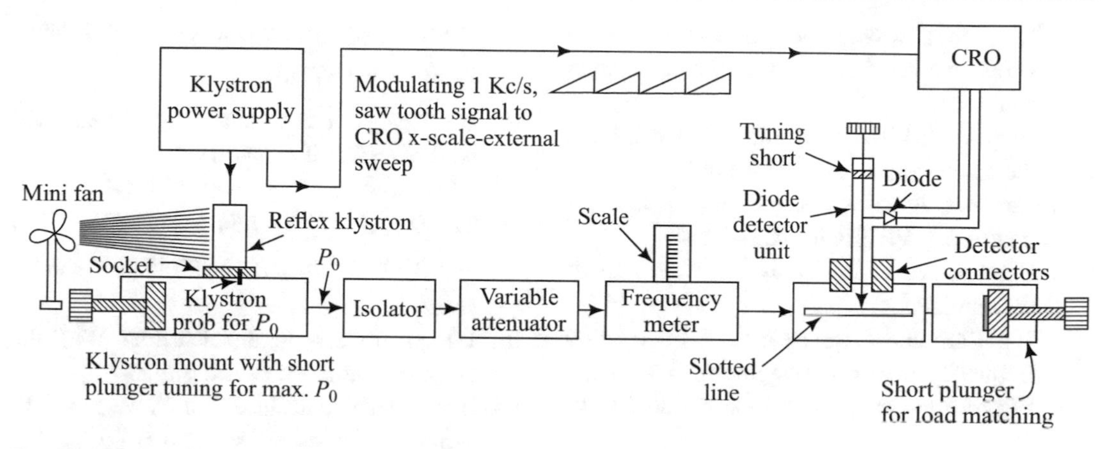

Fig. 12.4 Set-up for mode characteristics using CRO with external sweep

7. Move the detector carriage of the slotted line to read the positions of two consecutive minima d_1 and d_2 or two consecutive maximum output shown in VSWR metre, and note the positions as well as the voltages, i.e. V_{min}, V_{max} $\lambda_g = 2(d_2 - d_1)$ and $S = V_{max}/V_{min}$.

$$\text{TE}_{10}\,\text{mode} = \lambda_C = 2a \cong 2.3\,\text{cm} \times 2 = 4.6\,\text{cm}\,(\text{in}\,X\text{−band})$$
$$\lambda_0 = 3\,\text{cm}\,(\text{for}\,10\,\text{GHz})$$
$$\lambda_g = \lambda_0/\sqrt{1 - (\lambda_0/\lambda_C)^2} = 3.96\,\text{cm}$$

See Table 12.4.

Table 12.4 Observations and results of finding f, λ and VSWR

S. no.	Frequency setting and frequency measured (f_0)	d_1 first min.	d_2 second min.	$\lambda_g = 2.(d_2 - d_1)$ cm	$f_0' = c\sqrt{\frac{1}{\lambda_g^2} + \frac{1}{(2a)^2}}$ computed	Difference error ($f_0 - f_0'$)
1	$f_0 = 10.0\,\text{GHz}$, $\lambda_0 = 3\,\text{cm}$, $l_C = 2a = 4.6\,\text{cm}$, $\lambda_g = 3.96\,\text{cm}$	45	65	4.0	9.9 GHz	0.1 GHz
2	:	:	:	:	:	:
:	:	:	:	:	:	:

8. Measure the frequency by frequency metre.
9. Compute the frequency by the formula given above.
10. Compute VSWR = V_{max}/V_{min}.
11. Change the frequency by changing V_R (electronic tuning) or by klystron knob (mechanical tuning), and repeat for next frequency.

Observation and Results

Cutoff wavelength, etc., for dominant mode:

Precautions

Same as Experiment No. 1.

Quiz/Viva Questions

1. How does the slotted line probe pick up power and how the detection takes place by the detector. The detector is a square law detector, explain.
2. Why is the slot at the centre of the broad a side of the waveguide and not at the b side?

3. Why in a waveguide, the wavelength of wave changes from λ_0 to λ_g, which is larger and why?
4. Why is VSWR never infinity? (**Hint**: Even the space impedance $\neq \infty$ but 377.50.)
5. Can we have a coaxial slotted line for measuring VSWR in coaxial line.
6. Explain what is λ_g for a wave inside waveguide.
7. What are cutoff wavelength (λ_c) and cutoff frequency (f_c) in a waveguide?
8. Why a frequency metre is also called wave metre?
   (**Hint**: As in a cavity resonator, wavelength and frequency both are important, and a frequency metre is nothing but a cavity resonator.)
9. What does TE_{12} means, explain for a rectangular waveguide with sketch.
10. While measuring frequency in a wave metre, what does a dip mean in VSWR metre. What happens here?
    (**Hint**: At the resonance of frequency metre cavity, energy is absorbed by it and less energy reaches VSWR metre.)
11. What is dominant mode and what are degenerate modes?
12. If only the slotted line portion is filled with a dielectric with $\varepsilon_r = 5$, what will be the new value of λ_g and f_c higher or lower? Will the frequency change?
    [**Hint**: $\lambda_{c-new} = \sqrt{\varepsilon_r}\,\lambda_{c-old}$]
    As

$$\lambda_g = \left(\frac{1}{\lambda_0^2} - \frac{1}{\lambda_c^2}\right)^{-1/2}$$

$$\therefore \quad \lambda_{g\,new} = \left(\frac{1}{\lambda_0^2} - \frac{1}{\varepsilon_r \lambda_{c-old}^2}\right)^{-1/2}$$

$\therefore\ \lambda_{g\,new} < \lambda_{g\,old}$, i.e. lower wavelength, while frequency cannot change, because λ_c, f_c as well as group velocity also change.

13. What is the free space impedance for EM wave? Can VSWR be infinite in open waveguide circuit cases?
    (**Hint**: Open-space characteristic impedance $Z_0 = \sqrt{\frac{\mu_0}{\varepsilon_0}} = 377\ \Omega$)
14. If a waveguide is open at the end, what will be the VSWR if $Z_0 = 50\ \Omega$?
    (**Hint**: As Z_L = open-space impedance = $377\ \Omega \therefore S = \frac{377}{50} = 7.54$. As VSWR = $S = \frac{Z_L}{Z_0} \therefore$ VSWR can never be ∞ in open-circuit cases.
15. Which diode is in the detector? Why the detector unit has a shorting tuner?
16. If in a waveguide V_{max} = 5 V, V_{min} = 3 V. What will be the power ratio P_{max}/P_{min}?
17. If in a VSWR metre, current at voltage max and min are 20 and 13 μA, what will be the P_{max}/P_{min} and V_{max}/V_{min} inside the waveguide?
    (**Hint**: $P_{max}/P_{min} = \frac{20}{13} = \frac{V_{max}^2}{V_{min}^2}$ as the detector follows square law $P_0\ \alpha V_{guide}^2 \alpha i_{det}$)
18. The VSWR, Γ, Z_0, and Z_L are related as

$$\Gamma = \frac{S-1}{S+1}; S = \frac{1+\Gamma}{1-\Gamma}; \Gamma = \frac{Z_L - Z_0}{Z_L + Z_0}; S = \frac{Z_L}{Z_0}$$

    For perfect open circuit ($Z_L = \infty$), Γ = +l; for short circuit (Z_L = 0), Γ = −l.
19. Explain the meaning of Γ = −1 and Γ = +1.
    [**Hint**: Γ = 1 means reflection is in phase with the signal, and Γ = −1 means reflection is out of phase with the signal.]

12.5 Experiment No. 5: To Determine High Voltage Standing Wave Ratio (VSWR), Using Slotted Line Double Minima Method

Aim

To determine VSWR and reflection coefficient of a wave travelling in a waveguide. Using slotted line for (a) matched load, (b) shorted, and (c) open-circuit end.

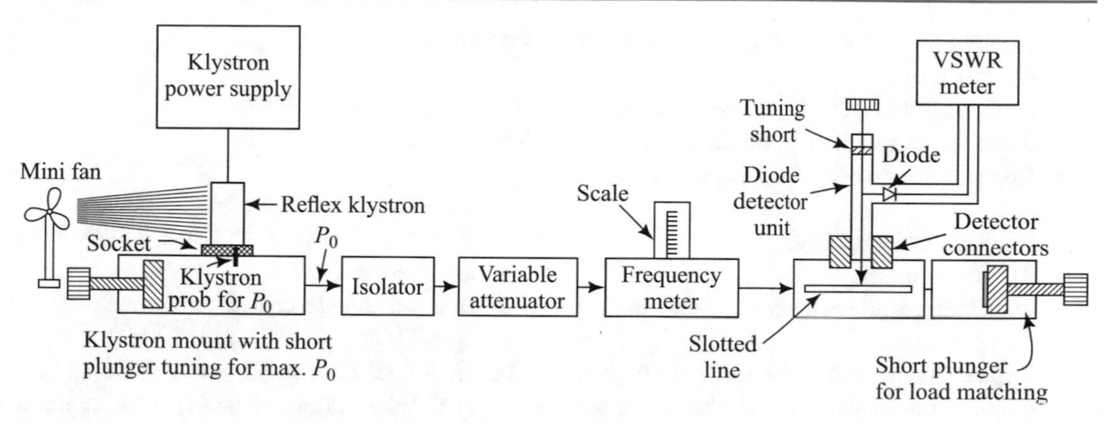

Fig. 12.5 Set-up for measuring frequency, λ_g, and VSWR using slotted line

Equipments

Same as per Fig. 12.5 of experiment No. 4.

Theory

For low VSWR: As discussed in Chap. 7, it is just by computing $S = (V_{max}/V_{min}) = \sqrt{I_{max}/I_{min}}$ on the slotted line, as $I \propto V^2$ square law detector. Here the voltage (V) is of wave in the waveguide, while current (I) is of VSWR metre.

For high VSWR: Here the V_{min} and V_{max} cannot be observed on the same scale of VSWR metre or CRO, being very much different in values and changing the scale will not give the correct value. Therefore the measurement is done using V_{minima} only. This is because of nonlinear relation (square law) being followed by the detector.

$$i_{det} = KV_{wg}^2$$

where

i_{det} is the detector current given on VSWR scale.

V_{wg} waveguide voltages.

Note the minima point d_0 on the slotted line scale, and read the minimum voltage V_{min} on VSWR metre (Fig. 12.6). Then locate the position of two points d_1 and d_2 on the left and right of V_{min}, which has $\sqrt{2} \cdot V_{min} = 1.414\, V_{min}$ value;

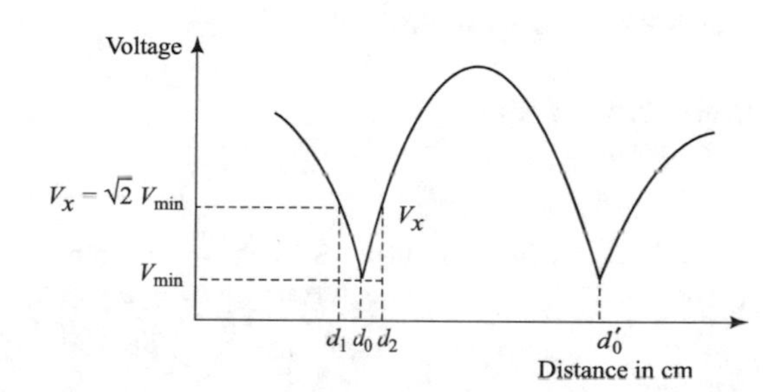

Fig. 12.6 Double minima method for measuring very high VSWR

i.e. double the power at d_0 (i.e. double the power at minima):

$$P_{min} \propto V_{min}^2$$
$$2P_{min} \propto V_x^2$$
$$\therefore \quad V_x^2 = 2V_{min}^2$$
$$\therefore \quad V_x = \sqrt{2}\, V_{min}$$

Then the empirical relation of VSWR gives:

$$\text{VSWR} = \lambda_g/[\pi(d_2 - d_1)]$$

Here λ_g is obtained from the position of two consecutive minima d_0 and d_0':

$$\lambda_g = 2(d_0' - d_0)$$

Procedure

1. Set up the microwave bench as in Fig. 12.4.
2. Same as 2–7 of Experiment No. 4.

8. Compute λ_g by $\lambda_g = 2\ (d_0 - d'_0)$, two minima positions.
9. Compute VSWR= $\lambda_g/[\pi \cdot (d_2 - d_1)]$, two double power minima positions (Table 12.5).
10. Repeat all the above for the three cases:

    (a) Matched load end.
    (b) Shorted end.
    (c) Open-circuit end after slotted line.

11. Repeat all the above for different frequencies by mechanical tuning the reflex klystron by rotating its tuning screw.

Observation and Result

High VSWR Cases

See Table 12.5.

Precautions

Same as Experiment No. 4.

Viva and Quiz Questions

1. Same as Experiments No. 1 and 4.
2. How many scales are there in VSWR metre and their ranges?
3. In a waveguide can '*S*' = ∞ with its end open? [Ans. No as Z_{opra} = 377 Ω]

3a. How does the tunable probe picks up voltage? Why the slotted line is at the centre of the width?

4. What is the size of the waveguide we use in the laboratories and its frequency band. (For *X*-band *a* = 0.9"; *b* = 0.4", size is convenient to use.)
5. Explain what are *X*-band, *K*-band, and *L*-band.

Table 12.5 Finding very high VSWR by double minima method

S. no.	Frequency (GHz)	d_0 (cm)	d'_0 (cm)	$\lambda_g = 2(d_0 - d'_0)$ cm	d_1 (cm)	d_2 (cm)	VSWR = $\lambda_g/[\pi\ (d_2 - d_1)]$
(a) Matched load end							
1	f_1						
2	f_2						
3	f_3						
(b) Short-circuit end							
1	f_1						
2	f_2						
3	f_3						
(c) Open-circuit end							
1	f_1						
2	f_2						
3	f_3						

6. What are the methods of impedance matching?
   (**Ans**: (i) Stub matching method.
   (ii) Half- and quarter-wavelength method.
   (iii) Screw tuner method.
   (iv) Iris method.
   (v) Reactance of load should be equal and opposite in sign to reactance of source.
7. Define guide wavelength.
8. What are the minimum and maximum values of VSWR and when it can be there?
9. Why for a slotted line open at the end will have a VSWR of 7.6 and not more, while a short can give VSWR much large also? What will be the reflection coefficient? [**Hint**: $S = Z_L/Z_0$ and $\Gamma = (Z_L - Z_0)/(Z_L + Z_0)$ as open are $Z_L = 377\,\Omega$ ∴ for $Z_0 = 50$, $S_{max} = 7.6$]
10. What is the VSWR of a perfectly matched line load?
11. Why VSWR is formed? Explain.
12. Can we have a coaxial waveguide slotted line?
13. How does the tunable probe picks up voltage? What does its turning short do?

12.6 Experiment No. 6: Computing Unknown Impedance Using Slotted Line by Measure of Minima Shift Method

Aim

To measure an unknown impedance using slotted line.

Equipments

As per Fig. 12.7.

Theory

Impedance of any terminal device of a transmission line, where the signal is to be send, should be known before hand; *e.g.* if a signal is fed to an antenna, its impedance should be known to us. Since impedance is a complex quantity, we need to know its modulus as well as phase angle, and we use the following relation for computation:

$$\text{Complex load} = Z_L = Z_0 \frac{1-\Gamma_c}{1-\Gamma_c}$$

$$\text{Complex reflection coefficient} = \Gamma_c = \Gamma_0 e^{j\phi}$$

$$\text{Phase angle of load} = \phi = [2\beta(x_1 - x_2) - \pi]$$

$$\beta = \text{Imaginary part of wave propagation constant} = 2\pi/\lambda_g \ (\text{where } \gamma = \alpha + j\beta)$$

$$\text{Waveguide length } \lambda_g = 2 \times \text{distance between two successive maxima on the slotted line}$$

Procedure

1. Connect the circuit as in Fig. 12.7 with load Z_L at the end of the slotted line.
2. For getting λ_g and Γ_0 the modulus of the reflection coefficient of load Z_L measures the location and values of V_{max} and (Fig. 12.8) V_{min} in the slotted line $\lambda_g = 4.\ (x_{min} - x_{max})$; $(S' = V_{max}/V_{min})$. Also note the position of one of the minimum (x_1) near the centre of slotted line, when the load is present.
3. Replace the load by short, and note the shift of minima to x_2 position.
4. Use the shift of minima from x_1 to x_2 to compute the load phase angle

$$\phi = [2\beta(x_2 - x_1) - p];\ \beta = 2\beta/\lambda_g$$

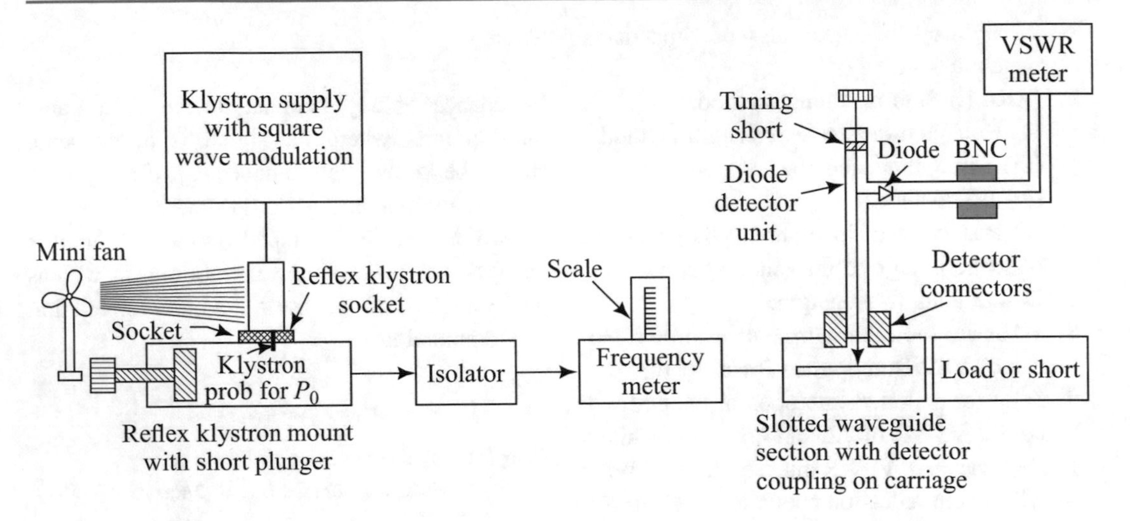

Fig. 12.7 Slotted line method for complex load

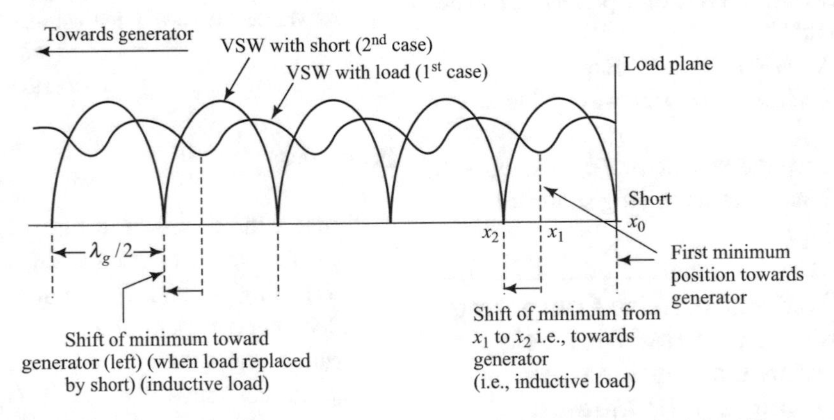

Fig. 12.8 Shifting of minima of VSWR pattern when short replaces the load. The minima with load are x_1 which shift left towards generator to x_2 by short. Left shift means inductive load and right shift means capacitative load. For resistance load, shift is exactly $\lambda_g/4$ or zero shift

and hence $\Gamma_c = \Gamma_0 e^{j\phi} = \Gamma_0 (\cos \phi + i \sin \phi)$

$$S = \frac{V_{\max}}{V_{\min}}; Z_L = Z_0\left(\frac{1-\Gamma_c}{1+\Gamma_c}\right); \Gamma_c = \frac{S-1}{S+1} = \frac{V_{\max} - V_{\min}}{V_{\max} + V_{\min}}$$

5. If the shift is:
   - to the left, then the load is inductive + resistive.
   - to the right, then the load is capacitative + resistive.
6. No shift or exactly $\lambda_g/4$ shift, then the load is purely resistive. Repeat the experiment by changing the frequency by mechanical tuning.

Observation and Results

See Table 12.6.

Table 12.6 Setting unknown impedance by minima shift method

S. no.	Freq. (GHz)	$\Gamma_0 = \frac{V_{max}-V_{min}}{V_{max}+V_{min}}$	First minima with load Z_L (x_1 cm)	Immediate next minima with load Z_L (x_1' cm)	$\lambda_g = 2(x_1' - x_1)$ (cm)	Position of first minima when Z_L replaced by short (x_2 cm)	ϕ in radians (and degrees)	Γ_c	$Z_L = \left(\frac{1-\Gamma_c}{1+\Gamma_c}\right)$ (Z_0 = 50)
1	8.8	0.33	5.23	11.20	5.97	4.87 (i.e. left shift)	0.757 (43.37°)	0.24 + 0.23i	43.5 + 22.5i inductive
2									
3									
4									

Precautions

1. Switch on the fan first before the klystron power supply.
2. Tune the short of the klystron mount and the tuner of the probe of the shorted line for maximum power indicated on VSWR metre.
3. Read the vernier calipers reading of the slotted line carefully, as precision of this will give correct value of Z_L.
4. All the precautions of Experiment No. 1.

Viva and Quiz Questions

1. List out the possible error in calculation of Z_L.
2. What does complex load means? What type of load will the screw tuners and iris give?
3. What is important while measuring Z_L?
4. Should we tune the reflex klystron by mechanical tuning or electronic tuning, for changing of frequency and why?
5. How do we measure the frequency by frequency metre? What happens inside the frequency metre and SWR metre at resonance?
6. What is wave modulation? What is CW wave or pulsed wave?
7. In measurements, the minima in the standing wave pattern are generally used and not maxima. Why? (**Hint**: Minima are sharp point and not maxima.)
8. Can impedance depend on frequency if yes or no, then why?
9. Why does the minima shift with load?

12.7 Experiment No. 7: To Study (a) Gunn Diode dc Characteristic (b) Gunn Diode Oscillator Power and Frequency Versus Its Bias (c) Modulation Depth of μw Signal by Using PIN-Diode Modulator

Aim

To study the following characteristics of Gunn diode:

(i) I–V characteristic.
(ii) Power output and bias voltage relation.
(iii) Variation of frequency of oscillation with bias voltage.
(iv) Square wave modulation using PIN diode.

Components

Gunn oscillator unit, Gunn power supply, PIN diode modulator unit, isolator, frequency metre, variable attenuator, detector mount, detector unit, and VSWR metre. Slotted line is not required (Fig. 12.9).

Theory

As discussed in detail in Chap. 7, the Gunn diode, oscillator is based on the following properties of GaAs:

1. GaAs has −ve differential conductivity as bulk material.

2. The reason is the two valleys/levels/states in the conduction band. The electrons in the upper valley (L_2) have less mobility (l/44)th than the electrons in the lower valley (L_1). [μ_2 = 180 cm^2/Vs; μ_1 = 8000 cm^2/Vs; μ_0 = 8500 cm^2/Vs]
   Here μ_1, μ_2, μ_0 are mobilities in lower level, upper level, and normal electron in GaAs. The upper valley has larger number of electrons for $E > E_{th}$ causing −ve conductivity region. Here μ_0 is electron mobility in +ve conductivity ohmic region ($E < E_{th}$).
3. The difference of energy levels between L_1 and L_2 is greater than that of thermal agitation energy kT = 26 mV at room temperature.
4. Therefore at room temperature, not enough electrons are at L_2 the upper level.
5. With voltage applied, more and more electrons go to the upper valley of lower mobility and they do not contribute much to the current ($J = \mu E$), leading to lowering of current (Fig. 12.10), causing −ve differential (dc) resistance.
6. *Domain theory*: A disturbance at the cathode end of the bulk GaAs gives rise to high-field dipole domain, which travels towards the anode, giving a voltage pulse at the output. Immediately after that another high-field domain is formed and it moves to anode to give another voltage pulse at the output. The time gap between these two pulses is of microwave frequency time period which is controlled by external circuit. This leads to −ve ac resistance and hence μW power output.
7. This microwave signal is allowed to resonate in a cavity, the size of which finally decides the frequency of oscillation.
8. This Gunn diode with DC bias supply with $E_{min} > E > E_{th}$ gives CW microwave power output.
9. Diode can be directly pulse modulated, but this will be accompanied itself by frequency modulation as the frequency changes with pulse voltage. Therefore **Gunn diodes are never modulated directly**. Therefore once the μW signal has got generated and goes to waveguide, it is modulated by PIN diode modulators circuit (Fig. 12.9).
10. The action of microwave signal modulation by PIN modulator is shown in Fig. 12.11. The *R–V* characteristics show very high resistance (10 kΩ or so) for $V < 0.2$ V and very low resistance (5 Ω) for $V > 0.2$ V. If the bias on PIN diode is set at 0.1 V and square wave superimposed, then for ON region, i.e. +ve square wave, the PIN diode

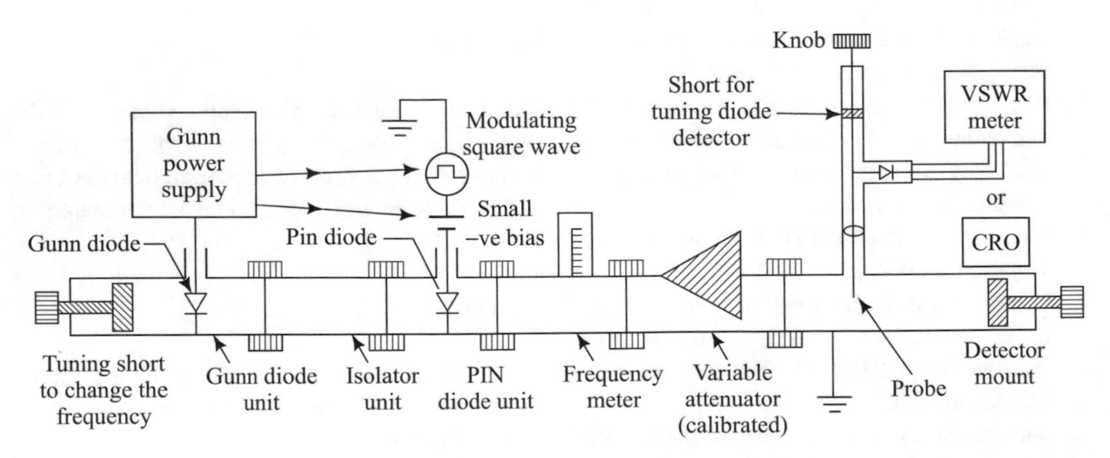

Fig. 12.9 Set-up for characteristic of Gunn diode. The PIN diode dc bias and its modulating square wave signal are coming from the Gunn power supply

is forward biased; therefore the μ*w* output is nearly short (i.e. very small resistance) with no μ*w* powers output (during the time BC or DE in Fig. 12.11).

11. This is modulated μ*w* signal, where the modulation depth is defined as V_{on}/V_{off}; see Fig. 12.11c. It is controlled by reverse bias of PIN diode.
12. Gunn diode tuning can be done by:

    (a) ***Mechanical tuning is*** by micrometre knob at the left end of the cavity which is just a short plunger in the Gunn diode mount. Normally it tunes up to 5% around the central frequency, i.e. up to 10% band width = 1 GHz for f_0 = 10 GHz.

    (b) ***Electronics tuning*** is by variation of the bias in the −ve resistance region, i.e. 6–10 V for the Gunn diode of $W = 20$ μm thickness. Normally this tuning region is very small 3 MHz/V, i.e. for 6–10 V region $\Delta f = 3 \times 4 =$ 12 MHz only, i.e. 0.01% only.

Procedure

1. ***I–V* characteristic**

    (i) Assemble the circuit as in Fig. 12.9.
    (ii) Keep the knob of voltage control of Gunn power supply to lowest position and switch it on. Keep the PIN −ve bias to −2 V so that it is open circuit for microwave signal to pass.
    (iii) Increase the Gunn bias voltage in steps of 0.2 V up to 10 V only, and note the current from the Gunn power supply unit.
    (iv) Plot *I–V* characteristic and measure the threshold voltage (V_{th}) for I_{max}.
    (v) Calculate the width of GaAs device by

$$W(cw) = \frac{V_{th}}{E_{th}} = \frac{V_{th}(\text{in kV})}{3.3\,\text{kV/cm}}$$

2. **Power P_0 versus voltage and f_0 versus voltage characteristics of Gunn diode**:

    (i) Keep the voltage control of Gunn power supply to minimum and switch it on.
    (ii) Keep the attenuator at minimum.
    (iii) Decrease PIN diode bias to −2.0 V, so that it acts as open circuit to transfer full power to the VSWR metre.
    (iv) Slowly increase the Gunn bias voltage (V) in 0.5 V step, and note power P_0 in VSWR metre and frequency f_0 in frequency metre, at each step.
    (v) Plot the P_0 versus *V* and f_0 versus *V*, and we note that both f_0 and P_0 increase with Gunn bias voltage.

3. **Modulated μW output**

    (i) Increase Gunn supply slowly up to 10 V.
    (ii) Connect CRO in place of VSWR metre.
    (iii) Tune the PIN diode dc bias voltage and its modulating square wave frequency for maximum voltage on the CRO.
    (iv) Coincide the bottom of the square wave in CRO to some arbitrary level to be taken as reference level, and note the calibrated variable attenuator reading.
    (v) Vary the variable attenuator, so as to coincide the top of the square wave and read the new reading of attenuator.
    (vi) Now replace CRO by VSWR metre back again, and for these two positions of the variable attenuator, read the two dB position in VSWR metre.
    (vii) The difference of these two readings of VSWR metre gives modulation depth of the PIN diode modulator for the given bias.
    (viii) Change the tuning micrometre knob of Gunn oscillator unit for changing the frequency, and repeat all above again.

Observations and Results

1. *I–V characteristics* (Table 12.7)
   Plot of the graph as per Fig. 12.10, and react the V_{th} = … volts
   ∴ Width of GaAs device = w (cm) = $\frac{V_{th}(\text{in volts})}{3300\,\text{V/cm}}$
2. *Power and frequency with Gunn bias* (Table 12.8)
3. *PIN diode modulated* μw *power* (Table 12.9)

Precautions

(i) Isolator should be put after the Gunn oscillator to save it from heavy μw power reflections from −ve biased PIN diode.
(ii) The Gunn diode bias voltage should not be increased beyond 10 V for standard Gunn diode.
(iii) Gunn diode should not be kept at max. voltage, i.e. critical voltage (V_{th}) for more that 1/2 min as it may burn.

Table 12.7 I.V. characteristics of Gunn diode

S. no.	Gunn bias volts (V)	Read current from power supply (I)
1	0.2	
2	0.5	
3	1.0	
12	5.0	

Table 12.8 Power and frequency of Gunn diode oscillator for different bias voltages

S. no.	Gunn bias volts	Detected current in VSWR metre representing power	Frequency measured
1	0.5		
2	1.0		
.	1.5		
.	2.0		
.	3.0		
.	4.0		
10	5.0		

Table 12.9 Measure of modulation depth in μw power from Gunn diode modulated by PIN diode modulator

S. no.	Tuning micrometre reading of Gunn oscillator	Frequency measured	Attenuator reading for lower upper level of square mod. output on CRO		VSWR dB reading for lower and upper level (dB) of μW detected		Modulation $(b - a)$ dB
			Lower	Upper	Lower (a)	Upper (b)	
1 2 3							

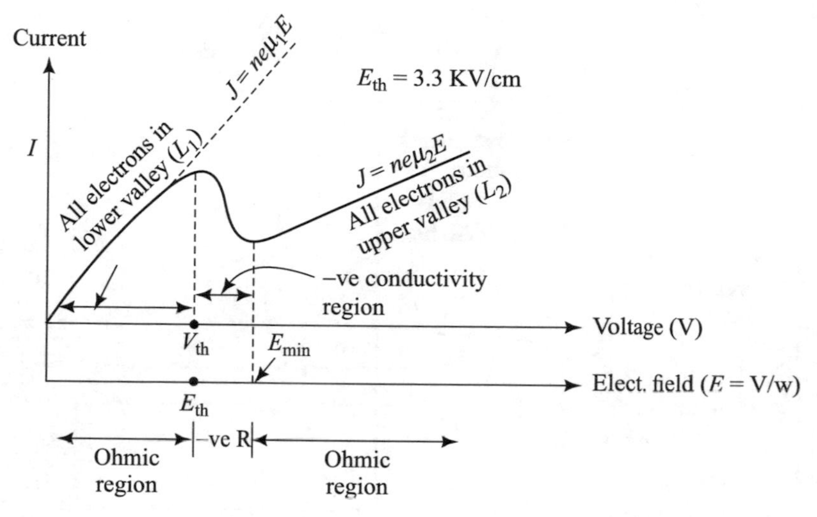

Fig. 12.10 *I–V* characteristics of Gunn diode

Quiz and Viva Questions

1. Why do we need PIN diode modulator instead of directly modulating the power supply of Gunn diode, for getting a modulated output. [Hint: Gunn diode voltage being very small needs to remain fixed, as the modulating signal will change this voltage and hence the frequency also.]
2. Why do we need modulation at all? Can't we get CW–μW power from Gunn diode without PIN diode modulator.
3. For changing frequency, mechanical tuning of Gunn diode is preferred over electronic tuning. Explain why and also the mechanism of tuning happening inside Gunn diode.
   [**Ans.**: Mechanical tuning is easy as it is by external cavity tuning.]
4. Write the values of the parameters of GaAs, i.e. n_i, E_g, density, threshold voltage—V_{th}, m_{lower}, m_{higher}, m_{lower}, m_{higher}.
5. What is Gunn effect and what is the material used.
6. Define −ve differential resistance and what happens in this region.
   [**Hint**: Differential resistance = $\frac{\partial V}{\partial I}$ = −ve; i.e. ∂V and ∂I are out of phase by 180°.]
7. Any special advantage of Gunn diode.
   [**Ans.**: Very less noisy, very wide tuning possible can be used up to 1000 GHz even]
8. Frequency in Gunn diode is mainly decided mainly by the cavity and not by diode size. Explain.
9. Any disadvantage of Gunn diode.
   [**Ans.**: Very much temperature dependent and delicate to bias voltage. It burns out with higher bias voltage.]
10. Describe various modes in which Gunn oscillator operates (e.g. LSA, etc.).
11. Which factors determine the frequency of oscillation of a Gunn diode?
12. What is −ve differential conductivity in Gunn diode? Give the reason for this.
13. Applications of Gunn diode are as an oscillator only.
14. Compare the mechanical tuning in reflex klystron and Gunn diodes. Explain why it is preferred in Gunn diode and not in reflex klystron.

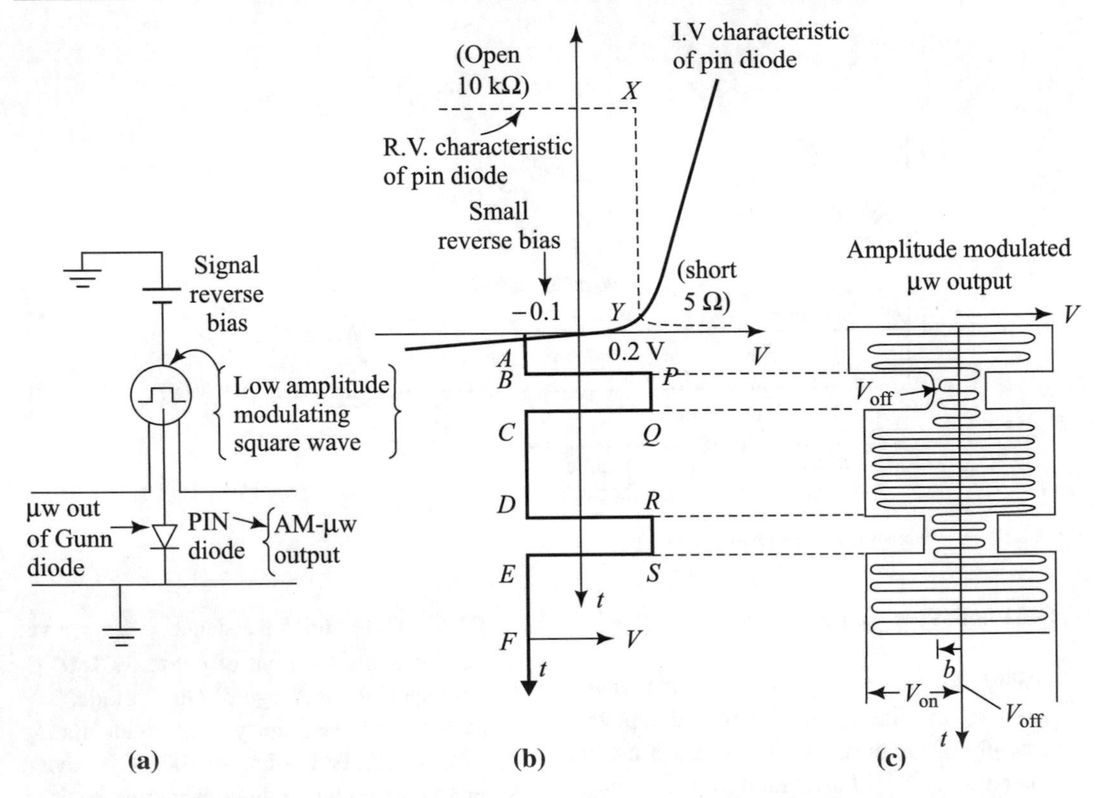

Fig. 12.11 **a** PIN diode modulator unit. **b** The PIN diode I–V and R–V characteristics plus the square wave modulating signal of small amplitude to cover the x–y region. During PQ, RS, the PIN diode is forward biased, hence μw signal gets shorted, i.e. very very small output signal is there. **c** Modulated μw power output

12.8 Experiment No. 8: Study of E- and H-Plane Tee Characteristics Isolation and Coupling Coefficients

Aim

To study E-plane and H-plane tee: (a) The characteristic plane of a tuning plunger, (b) coupling coefficients, and (c) isolation between arms.

Equipments

As per Fig. 12.12, the paths A B C D, and E may be connected or disconnected for getting different configurations of the circuit needed for getting different parameters, e.g. coupling, isolation.

Theory

1. Detailed theory has already been discussed in Chap. 3, still it is summarised in Table 12.10.
2. *The characteristic plane*: If power is fed to arm 1 (in E-tee or H-tee), shorting plunger is attached to arm 3 [connection E and not D] and power is detected out of arm 2. [connection B and not A] Here C is open circuited, and then the following is observed:
   - The portion of power coming out of arm 3 will get reflected back and get divided equally into arms 1 and 2 of the tee.

- Therefore power out of arm 2 will be sum of power from input arm 1 and half of reflected power from arm 3.
- As E line is connected, a position of short plunger of arm 3 will give maximum power at arm 2 [detected by the detector via path B], as the plunger position behaves as open circuit, reflecting back whole of the power in phase.
- Another position of short plunger P_2 of arm 3 will give minimum power at arm 2 as the plunger position will behave like a short, and the reflected power is in opposite phase.
- These two positions of this short plunger P_2 of arm 3 are called characteristic planes. At positions in between these two planes the load due to this plunger is complex (i.e. inductive or capacitative).

3. The properties of H- and E-plane tee are summarised as follows Table 12.10.
4. *Isolation and coupling coefficient*: Isolation or attenuation between two arms of a T-junction is the ratio of power supplied from a matched generator to one of the arms, to the power going to a matched detector through any other arm. The remaining third arm is terminated in a matched load (Fig. 12.13).

Let the circuit be set up as Fig. 12.13a, and then attenuation, i.e. isolation from port 1 to port 2, will be:

$$\text{Isolation } \alpha_{12} = 10 \log (P_1/P_2)$$

$$\text{coupling } C_{12} = 10^{-\alpha_{12}/20}$$

For α_{13}, C_{13}, the load and detector can be interchanged. See Fig. 12.13b.

Note: (1) The slotted line and A line are used only for finding VSWR and λ_g.

(2) The short plunger P_1 is used only for maximising the signal detected via line B, i.e. as matched load mount for detector.

Procedure

(i) Assemble the circuit as in Fig. 12.12 with B and C lines connected, with A D and E disconnected. Here the tee becomes out of circuit.

(ii) Switch on the power supply of generator to give maximum on VSWR metre by adjusting attenuation and tuner of signal source mount. Set the calibrated attenuator

Table 12.10 Properties of H- and E-plane Tee summary

H tee		E-tee	
Input at port	Output	Input at port	Output
a_3	$b_1 = a_3/\sqrt{2}$ $b_2 = a_3/\sqrt{2}$	a_3	$b_3 = 0$ $b_1 = \left(a_3/\sqrt{2}\right)$ $b_2 = \left(-a_3/\sqrt{2}\right)$
$a_1 = a_2 = a$	$b_2 = 0$ $b_2 = \frac{a_1}{\sqrt{2}} + \frac{a_2}{\sqrt{2}} = a\sqrt{2}$	$a_1 = a_2 = a$	$b_3 = \left(\frac{a_1}{\sqrt{2}} - \frac{a_2}{\sqrt{2}}\right) = 0$ (cancellation) $b_1 = a$ (reflection) $b_2 = a$ (reflection)
$a_1 = a$	$b_2 = -a/2$ $b_3 = -a/\sqrt{2}$ $b_1 = a/2$ (reflection back)	a_1	$b_1 = \frac{a_1}{2}$ (reflection) $b_2 = \frac{a_1}{2}$ $b_3 = -\frac{a_1}{\sqrt{2}}$

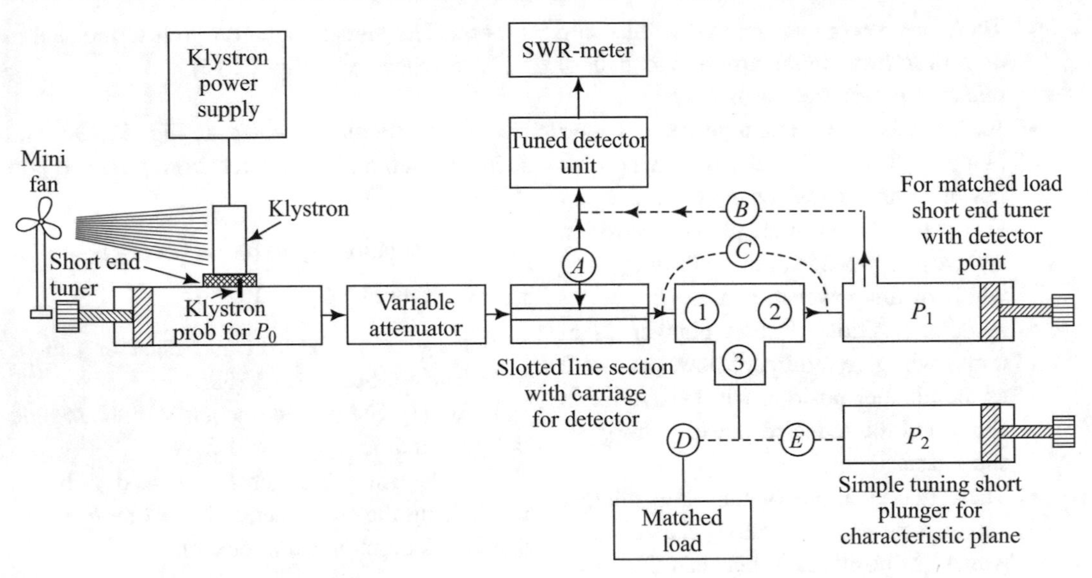

Fig. 12.12 Set-up for measuring parameters of tee

dB to around 30 dB. This is the power output from slotted line to port 1 of tee.

(iii) Now insert the tee as in figure such that the, *B*, *D* lines are connected and *A*, *C*, *E*, disconnected.

(iv) Decrease the attenuation, such that VSWR reading gives the same reading as before. This decrease in attenuation (in dB scale) is isolation α_{12} between ports 1 and 2, then compute C_{12} by the formula $C_{12} = 10^{-\alpha_{12}/20}$.

(v) Above (i) to (iv) can be repeated by interchanging ports 2 and 3 to get α_{13}, and compute $C_{13} = 10^{-\alpha_{13}/20}$.

Observations and Results

(a) *Characteristic plane at port 2* (Fig. 12.12) Here line *B* and *E* are connected and *A C D* are disconnected (Table 12.11).

(b) *Isolation and coupling set* (Table 12.12)

Precautions

(i) To start with, the VSWR metre has to be kept at normal scale and never at expanded scale.

(ii) Fan of klystron should be put on prior to the power of klystron.

(iii) Klystron reflector voltage is switched on prior to beam voltage.

Viva and Quiz Questions

(i) Where do we use E-tee and H-tee?

(ii) Draw the electric field diagram, inside the E-tee and H-tee, when a wave is travelling inside.

(iii) From the characteristic plane of a simple shorting slug, what information do get.

(iv) Why at the two points of characteristic planes the short plunger behaves as open or short circuit?
[**Hint**: The length of the cavity is $(2n + 1)\ \lambda/4$ or $(2n + 1)\ \lambda/2$]

(v) Show the wave form inside at these two situations. [**Hint**: See Fig. 2.2b.]

(vi) Why the current in VSWR metre directly represents power of the waveguide?
[**Hint**: $I_{\text{detector}} = KV^2_{\text{waveguide dc}}$ α power as the detector is a square law detector due to its nonlinear response.]

(vii) Write in a table of I/O of H-plane for the three ports.

Table 12.11 Characteristic plane locations in E- and H-plane tee

S. no.	Frequency	Detector of 'B' line		
		Maxima locations of short plunger with line *E*	Minima locations of short plunger with line *E*	Difference
1 2				

Table 12.12 Finding isolation (α_{12}) and Coupling (C_{12}) between ports of Tee junctions

Input at port (figure used)	Initial attenuator reading *x* dB output of slotted line end	Final attenuator reading *y* dB output from tee	Decrease in attn. $(x - y)$ dB	Isolation, i.e. attn. $\alpha = (x - y)$ dB	Coupling $C = 10^{-\alpha/20}$	VSWR by slotted line V_{max}/V_{min} by line *A*
Port 1 Figure 12.13a				$\alpha_{12} = 3$ dB (approx.) $\alpha_{13} = 3.5$ dB (approx.)	$C_{12} = 0.7$ $C_{13} = 0.6$	
Port 2 Figure 12.13b				$\alpha_{32} = \ldots$ $\alpha_{31} = \ldots$	$C_{32} - \ldots$ $C_{31} = \ldots$	

12.9 Experiment No. 9: Study of Magic Tee Characteristic—Isolation and Coupling Coefficients Between Various Pair of Ports

Aim

To study the isolation and coupling between its various ports.

Equipments

As per Fig. 12.14.

Theory

The magic Tee is a combination of H- and E-plane arms, while shunt arm 3 is H-arm and series arm (4) is E-arm. The properties have been discussed at length in Chap. 4, which is summarised as:

(a) **Power input–output**

1. Power fed into port 3 gets divided equally into ports 1 and 2 in phase, with no power out of port 4. Reverse of this is also true.
2. If power is fed into port 4, it gets divided equally into ports 1 and 2 out of phase, with no power out of port 3. Reverse of this is also true.
3. Thus if equal power in same phase is fed into at ports 1 and 2, then these powers get added at port 3 (H-arm), while it gets subtracted at port 4 E-(arm) to give zero output (Fig. 12.15). This is understood by the directions of transient oscillating field and charge.

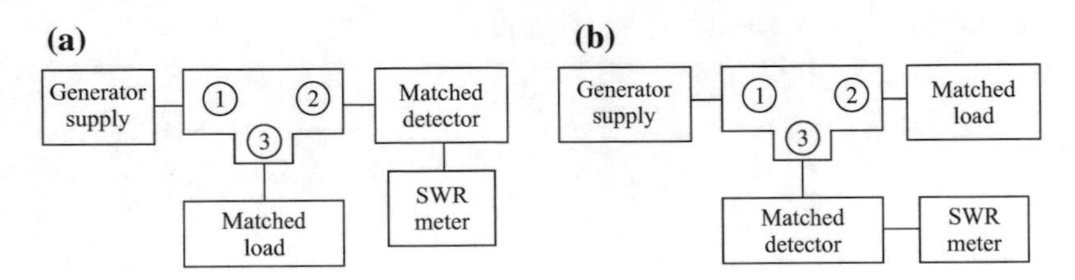

Fig. 12.13 Circuits for tee for measuring isolation and coupling **a** for α_{12}, c_{12}, and **b** for α_{13}, c_{13}

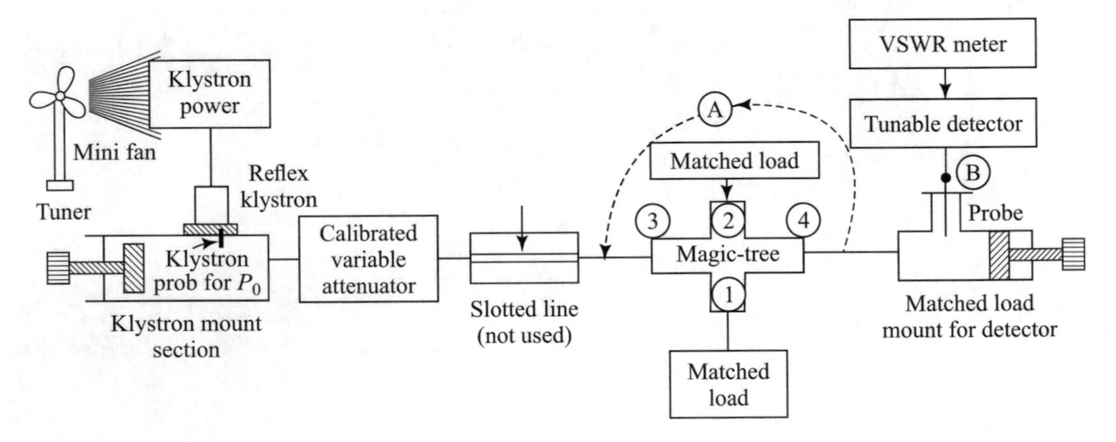

Fig. 12.14 Test bench for studying magic tee isolation and coupling coefficients

4. If power is fed into port 1 only, then no power comes out of port 2, but from ports 3 and 4 we get half power each.

(b) **Other properties**

5. If the H-arm and E-arm ports are matched, then the magic tee is matched, and at the same time these two ports are non-coupled ports. **Because of above two points 4 and 5, it is called magic tee**.
6. If three ports have matched loads, then the input at the fourth port will give the VSWR of that port. Thus each port has different values of VSWR.
7. Isolation or attenuation between H- and E-ports is the most important parameter. It is defined as power detected (P_3) at H-arm (port 3) with power input (P_4) at E-arm (port 4), with collinear port (1, 2) terminated with matched load.

$$\alpha_{34} = 10\ \log_{10}\left(\frac{P_4}{P_3}\right)\text{dB}$$

Similarly isolation between any other pair of ports can be computed.

8. *Coupling*: Coupling coefficient between port *i* and *j* is defined with all the remaining ports matched as:

$$c_{ij} = 10^{\alpha_{ij}/20}$$

where α_{ij} is isolation or attenuation between *i* and *j* ports.

$$\alpha_{ij} = 10\ \log\left(\frac{P_i}{Pj}\right)\text{dB}$$

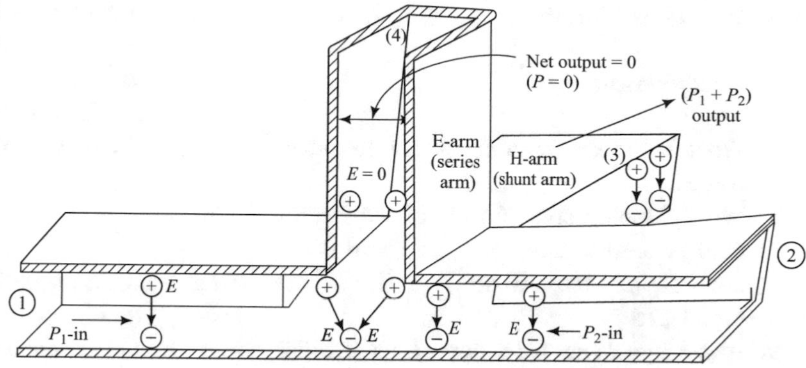

Fig. 12.15 Cutaway view of a magic tee, with equal and in phase inputs at ports (1) and (2). Transient electric fields and charge are also shown, which change with μW frequency

Procedure

(i) Assemble the circuit as in Fig. 12.14 with *A* path connected without the magic tee.

(ii) Switch on the fan and then power supply and see that maximum power is shown in VSWR metre, by tuning the short tuners of klystron, detector, and the matched mount of detector.

(iii) Adjust the attenuator at a middle point of its range say 30 dB, note this, as this is the power out of the slotted line.

(iv) Now insert the magic tee as in figure with *A* path disconnected, for observing the α_{14}, and C_{34}.

(v) Read the new VSWR reading.

(vi) Now decrease the attenuation, such that VSWR reading gives the same reading as before (say 30 dB). This decrease in reading is the attenuation, i.e. isolation between 3 and 4.

(vii) Compute α_{34} and C_{34}.

Observations and Results

See Table 12.13.

Table 12.13 Study isolation (α) and coupling in a magic tee, between various pairs of ports

S. no.	Orientation of magic tee		Variable attenuator reading		Isolations	Coupling
	Input	Output	Without magic tee (a) (dB)	With magic tee (b) (dB)	$\alpha =$ $(a-b)$ (dB)	$C = 10^{-\alpha/20}$
1	3	4	20	60	$I_{34} = 30$	$C_{31} = 0.647$
2	3	1	20	13.8	$I_{31} = 3.8$	$C_{32} = 0.661$
3	3	2	20	23.6	$I_{32} = 3.6$	$C_{41} =$
4	4	1			$I_{41} =$	$C_{42} =$
	4	2			$I_{42} =$	$C_{43} =$
	4	3			$I_{43} =$	$C_{12} =$
	1	2			$I_{12} =$	$C_{13} =$
	1	3			$I_{13} =$	$C_{14} =$
	1	4			$I_{14} =$	$C_{21} =$
	1	1			$I_{21} =$	$C_{23} =$
	2	3			$I_{23} =$	$C_{24} =$
	2	4			$I_{24} =$	

Viva and Quiz Questions

1–6 of Experiment No. 8.

7. What is the magic in magic tee, to get its name?
8. Why phase change of 180° of the electric field is observed in series tee (E-arm) and not shunt tee (H-arm)? (Hint: See Fig. 12.15.)
9. If two signals are fed at ports 1 and 2 which are equal in phase, what will be the output at E- and H-ports? (**Hint**: See Fig. 4.11.)
10. Why does power input to E- or H-arm gets divided equally into collinear arms?
11. How do the lines of force of electric field and magnetic field look like inside the magic tee, show along with directions?
12. How the powers in the magic tee are coupled, show with field diagram for various cases?
13. When the electric field changes the direction with μW frequency, what happens to the corresponding charges?

12.10 Experiment No. 10: To Study the Characteristic of Directional Coupler—Isolation and Coupling Coefficient

Aim

To study the (a) coupling factor (c); (b) insertion loss (I_n); (c) directivity (D) and; (d) isolation (I_s) in a directional coupler.

Equipments

As per Fig. 12.16.

Theory

1. It is a four-port device commonly used for coupling to output port 4 by a known fraction (20 dB or 30 dB) of the microwave input power of the port 1. This coupling many times is used for taking a sample for measuring power flow in the main line port 1–2.

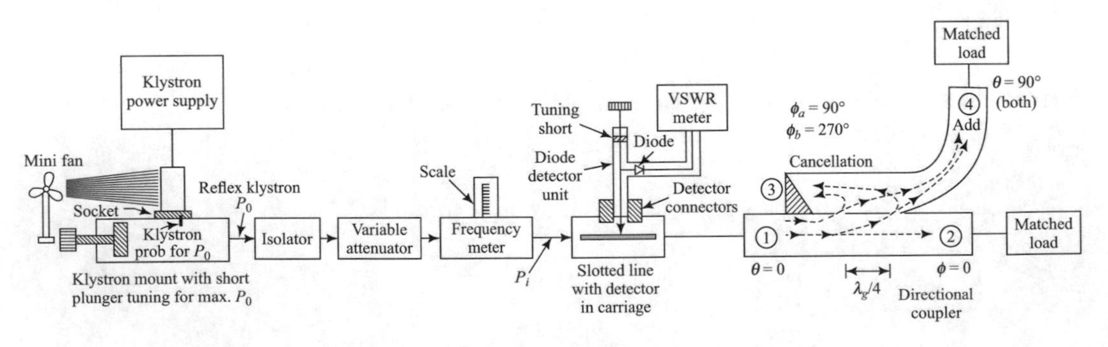

Fig. 12.16 Microwave bench for measure of VSWR of main line of directional coupler

2. The port 3 has perfect absorbent match termination for any power if at all coming to it. The ports 1 to 2 form the main line (Fig. 12.16).
3. As is clear from Fig. 12.16 that at port 3, the two signals reach from the two holes and they cancel due to being out of phase, while at port 4 two portions of signal get added, being in phase. However normally only 20 dB (1/100) or 30 dB (1/1000) power reaches port 4 and the remainder to port 2.
4. Frequency range of couplers are specified by the manufacturer, is that of the coupling from arms 1 to 4. The main arm (1–2) frequency response is much wider. If the specification says 2–4 GHz, then the main arm (1–2) can be operated at 1–5 GHz.
5. However we should also remember that coupler response is odd number periodic. For example, a $\lambda_g/4$ directional coupler will also have same response as a coupler for $(2n + 1)$ $\lambda_g/4$ wavelength. This is because, the separation of holes will give same phase shift so that signals cancel at port 3 and add at port 4 for $n = 1, 2, 3, \ldots$

For more details Chap. 4 can be referred, however the important parameters of directional coupler as given below.

Parameters of Directional Couplers

(a) *Coupling Factor (C)*

It is defined as the ratio of input (P_1) to output power (P_4) at port 4 (Fig. 12.17a).

$$C = 10\log\left(\frac{P_1}{P_4}\right)\text{dB} = 20\log\left(\frac{I_1}{I_4}\right)\text{dB}$$

Here, I_1, I_4 are currents detected by the VSWR metre proportional to powers (P_1, P_4) entering the detector, as it is a square law detector.

(b) *Insertion Loss (I_{ns})*

It is defined as the ratio of power input (P_1) to the power output (P_2) on the main lines (Fig. 12.17b):

$$I_{ns} = 10\log\left(\frac{P_1}{P_2}\right) = 20\log\left(\frac{I_1}{I_2}\right)$$

For above C and I_{ns} we should remember that $P_1 = P_2 + P_4$ (as P_3 is nearly zero).

(c) *Directivity (D)*

It is defined as the ratio of coupled power output (P_4) to the power reflected back at port 3 (P_3), both on the auxiliary line (Fig. 12.17c$_1$ and c$_2$);

$$D = 10\log\left(\frac{P_4}{P_3}\right) = 20\log\left(\frac{I_4}{I_3}\right)$$

It has to be as high as possible.

(d) *Isolation (I_{so})*

It is defined as the ratio of power lost in reflection at port (P_3) to the input power given (P_1). Here P_3 can only be computed indirectly by $P_3 = P_1 - (P_2 + P_4)$ as the port 3 is sealed:

$$I_{so} = 10log(P_3/P_1)$$

This has to be very high for good couplers.

All the above three parameters are frequency dependent as the coupling holes are separated by a distance $\lambda_g/4$, which is fixed for a coupler. For wider band couplers, more numbers of coupling holes with larger tolerances are made.

The normal values of these parameters are $C = 20$ dB, $D = 30$ dB, $I_{so} = 50$ dB, and $I_{ns} = 0.5$ dB.

Procedure

1. Assemble the equipment as per Fig. 12.16.

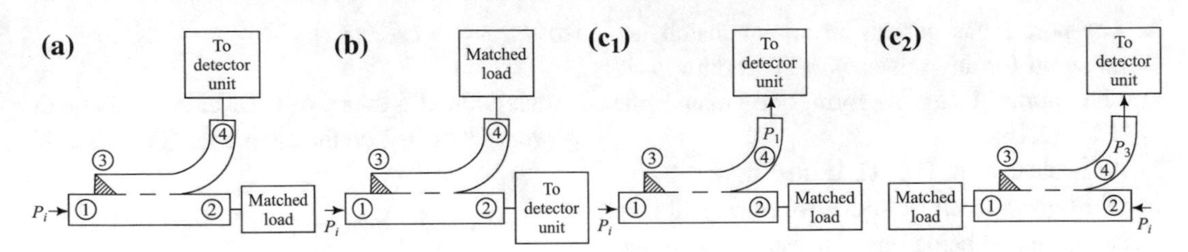

Fig. 12.17 **a** Set-up for measuring coupling factor. **b** Set-up for measuring insertion loss. **c_1, c_2** Set-up for measuring directivity

2. As per Experiments No. 4 and 5 find the VSWR of the line. This VSWR is that of the main line of directional coupler.
3. Change the frequency and again repeat as above.
4. Restructure the connection of directional couplers, matched load, and detector system as per Fig. 12.17a, b, c_1, c_2, and we note the detected output. Then compute C, I_{ns}, D, and I_{so} as given in theory.

Observations and Results (with a sample)

Square law detector current/voltage shown in VSWR metre is proportional to power; therefore the current or voltage is measured. Some of the samples are shown Table 12.14.

For isolation, normally P_3 cannot be measured as the end of port 3 is normally sealed by the manufacturers, with lossy and matched load, but P_4 is very small but never zero. Therefore we take $P_3 = P_1 - (P_2 + P_4)$.

Normally isolation (dB) = directivity (dB) + coupling (dB). As P_3 cannot be measured directly, but computed indirectly, to get I_{so}.

Precautions

Same as Experiment No. 1.

Viva and Quiz Questions

1. If coupling factor comes to be 20.5 dB, what does it mean?
2. What is the purpose/application of directional coupler in μ*w*?
3. If the main line is terminated at port 2 as short, how much power will appear in coupling auxiliary arm.
4. If VSWR comes out to be 1.2, what do we infer?
5. All the parameters of a coupler are frequency dependent, why?
[**Hint**: Mainly because of distance between holes.]
6. For making a wider band directional coupler, what techniques are used?
[**Hint**: Multiholes and larger holes.]
7. If specification of a coupler is C_{14} (coupling between port, 1, 4) = 20 dB ± 0.5 dB. What does it mean?

Table 12.14 In directional coupler getting coupling (C) coefficient, Insertion loss (I_{ns}), Directivity (D) and Isolation (I_{so})

S. no.	Frequency	P_1	P_2	P_3	P_4	S	C	I_{ns}	D	I_{so}
1	10 GHz	15 μA	14.8 μA	0.05 μA	0.15 μA	1.15	20 dB	0.86 dB	4.7 dB	24.8 dB
2										
3										

8. A directional coupler operates well at 2 GHz, will it operate on the same way at 6, 10, 14… GHz or not and why? Explain.
9. What is the material of a directional coupler?
10. Isolation (I_{so}) of a coupler is 40 dB, what does it mean?
11. For a simple directional coupler following is a normal specification:

$$\begin{aligned}\text{Frequency} &= 8.2-12.4\,\text{GHz}(\text{X-band})\\ \text{Coupling} &= 20\,\text{dB} \pm 0.8\,\text{dB}.\\ \text{Directivity} &= 35\,\text{dB}\\ \text{VSWR}(\text{Main line}) &= 1.1\\ \text{VSWR}(\text{Auxiliary line}) &= 1.2\\ \text{Insertional loss}\ (I_{\text{ns}}) &= 0.04\,\text{dB}\end{aligned}$$

Explain all the above parameters.
12. In practice it has been found that as we increase the coupling factor C, the insertion loss falls logarithmically, i.e. $\left(C\alpha\frac{1}{\log I_{ns}}\right)$

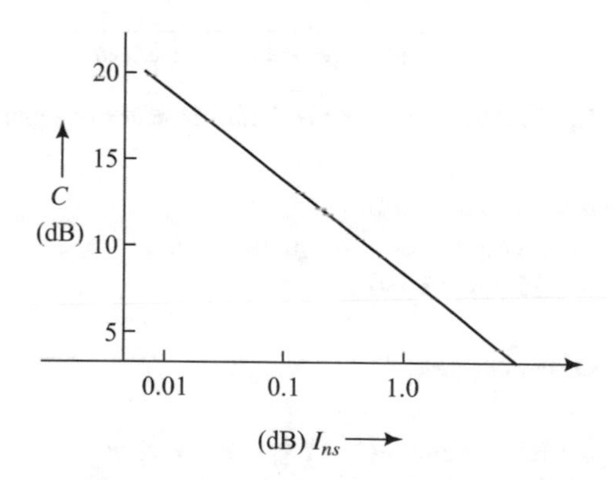

Explain the reason for the above.
13. What is the distance between two coupling holes of a directional coupler? [**Ans.**: Odd multiple of $\lambda_g/4$]
14. For more coupling to port 4, we have more than two holes. How does it help?
[**Ans.**: It gives (a) more coupling and (b) broad banding.]
15. When two signals reach from port 1 to port 3, how/why their phase shifts are 90° and 270%? Explain.

12.11 Experiment No. 11: Calibrating an Attenuator Using VSWR Metre

Aim

Calibrate (a) a fixed attenuator and (b) a variable attenuator.

Equipments

Full microwave bench with klystron power supply, klystron unit, isolator, frequency metre, slotted line, matched load, the given attenuators, and detector unit (Fig. 12.18).

Theory

1. An attenuator is a device which reduces power of a signal (from P_i to P_0) when passes through it (more details in Sect. 4.10), with attenuation in dB as:

$$\alpha = 10\ \log_{10}(P_i/P_0)$$

2. It could be fixed attenuator or continuously variable attenuator.
3. The attenuation will change with frequency; therefore an attenuator will be specified like 25 dB at 8–9 GHz with ±5% accuracy. Fixed types are normally of 3, 10, 20 dB, etc., but those too with given frequency range.
4. The fixed type uses tapered edge of resistive vane made of loss material.
5. The variable type has tapered resistive cards parallel to the electric field, depth of which decides the attenuation.
6. Attenuation is frequency sensitive. Some phase shift is also introduced by it. Therefore choose a material with maximum attenuation with lowest phase shift.
7. Attenuators are also called pad as it absorbs power and itself get heated.

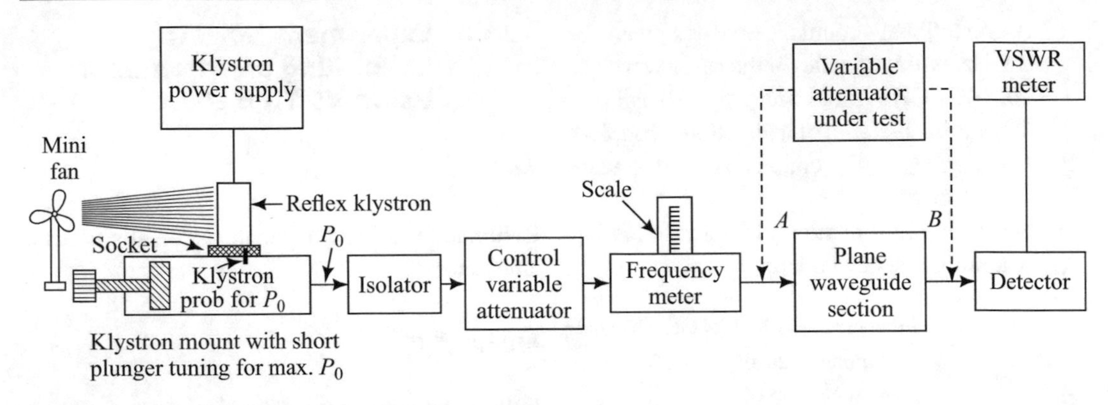

Fig. 12.18 Set up for calibrating an atternuator using VSWR meter

Procedure

1. Assemble equipment as per Fig. 12.18. Without the attenuator under test see that maximum power is detected in VSWR metre by tuning the short of klystron mount and also the detector tuner. Here the control variable attenuator in the line is just for controlling excess power.
2. Set 0 dB in VSWR metre by the variable attenuator by the main microwave bench.
3. Now insert the fixed or variable attenuation under test in place of waveguide section at AB.
4. Note the dB in VSWR metre; this is the attenuation of the attenuator.
5. For variable attenuator start from lowest micrometre reading of attenuator, read attenuation on VSWR metre, keep on changing the micrometre reading, and note dB on VSWR metre.
6. Plot a graph between micrometre reading and VSWR (dB) reading.

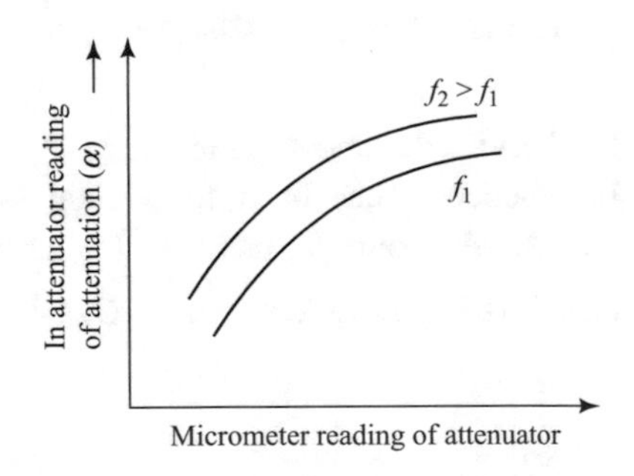

Fig. 12.19 Attenuator's performance calibrating graph

7. Change the frequency and repeat all (Figure 12.19).

Observations

Variable attenuator: Frequency = f_1 = ……
See Table 12.15.
Frequency = f_2 = ……
See Table 12.16.

Table 12.15 Calibration of attenuator using VSWR meter at frequency…… GHz

S. no.	Micrometre reading of attenuator	Attenuation on VSWR metre
1	–	–
2	–	–
3	–	–
10	–	–
	–	–

Table 12.16 Calibration of attenuator using VSWR meter at frequency…… GHz

S. no.	Micrometre reading of attenuator	Attenuation on VSWR metre
1	–	–
10	–	–
	–	–

Procedure

Same as Experiment No. 1.

Viva and Quiz Questions

1. In an attenuator, where do the microwave power go?
2. Why do the phase shift also takes place in attenuators, explain?
   [**Hint**: Medium is dielectric.]
3. Will the VSWR change with two positions of variable attenuators and why?
   [**Hint**: Due to it line impedance will change and hence reflections increase; therefore VSWR also increases.]
4. How do a ferrite attenuator differ from carbon pad? [Higher absorption]
5. Can we have a broadband attenuator, how?
6. When we repeat the experiment most of the time same reading is not there. List out the reasons for it.
7. Why attenuators are also called pad? (e.g. names like 3 dB pad are normally used for fixed attenuators)

12.12 Experiment No. 12: Measurement of Dielectric Constant and Phase Shift by Minima-Shift by Its Insertion

Aim

To find dielectric constant of dielectric sample placed inside a waveguide section and the phase shift in the wave caused by it.

Equipments

As per Fig. 12.20.

Theory

Dielectric constant (ε_r) is the relative permittivity [$\varepsilon_r = \varepsilon^*/\varepsilon_0$ $(\varepsilon' - \varepsilon''_0)$] and is the measure of efficiency of permitting electric lines of force in the material. Dielectric constant causes power loss/heating. In fact the permittivity (ε^*) is a complex quantity, so is the dielectric constant, ($\varepsilon_r = \varepsilon' - j\varepsilon''$). The real part ($\varepsilon'$) represents ability to store energy, while the imaginary part (ε'') is the measure of dissipation of energy.

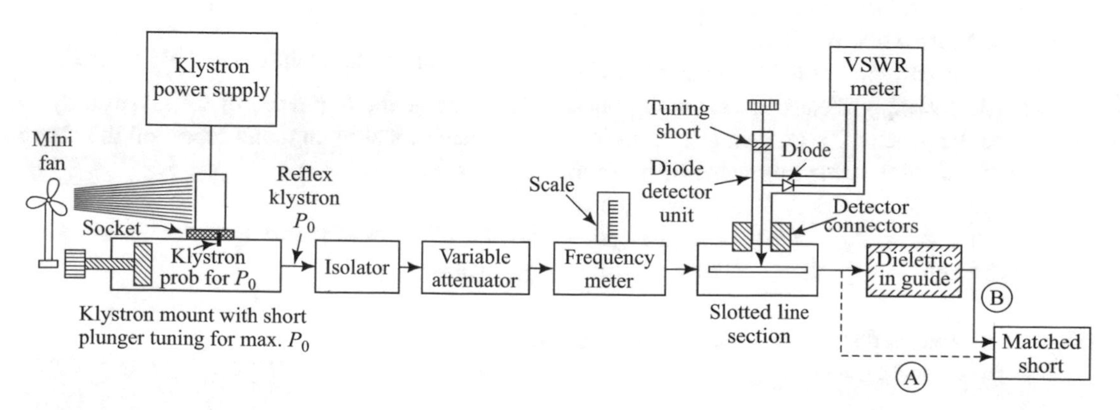

Fig. 12.20 Set-up for measuring dielectric constant

Normally measuring the real part only suffices, by computing from the shift of a reference minima when the dielectric section is not there (like A path in Fig. 12.20) and then when the dielectric section inserted (like B path in Fig. 12.20).

For detail Sect. 4.8 can be referred.

Procedure

1. Assemble the equipment as in Fig. 12.20 with link A, i.e. without dielectric waveguide section.
2. Switch on the fan, the microwave source, and then the power supply.
3. Maximise the power (a) by tuning the short of source and detector (b) with variable attenuator at minimum.
4. Measure the frequency by the metre and compute λ_0.
5. Measure λ_g by slotted line as double the distance between two minima, set the VSWR metre at one of the minima, and call it reference minima.
6. Note the thickness (t) of dielectric of the section by mechanical measurement.
7. Insert the dielectric section link B, and note the shift of the reference minima by Δs (refer Fig. 12.8).
8. Calculate the value of a parameter P as:

$$P = \frac{\lambda_g}{t} \cdot \tan\left[\frac{2\pi(\Delta s + t)}{\lambda_g}\right]$$

   This P could be −ve or +ve, but we will use the modulus of P as $|P|$.
9. Now we compute N the number of wavelength ($N = t/\lambda_d$) dielectric material which could be fraction also, using a transcendental equation using this value of $|P|$ from computed above (Here it may be pointed out that $\lambda_d < \lambda_g$, where λ_d is the guide wavelength in the dielectric filled waveguide instead of air). This transcendental equation is:

$$\frac{\tan(2\pi N)}{N} = |P|$$

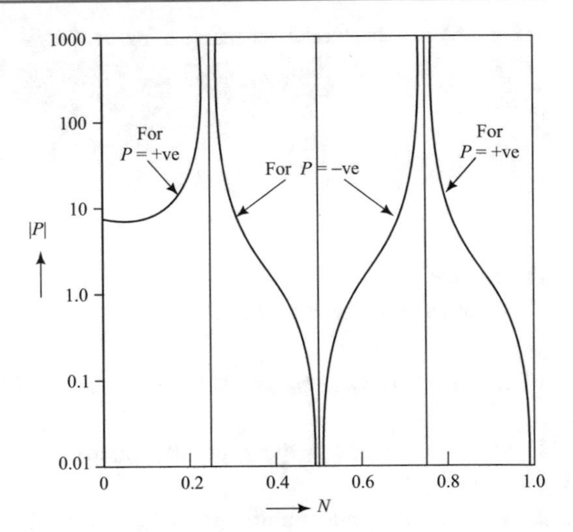

Fig. 12.21 N versus $|P|$ plot of transcendental equation $|P| = \frac{\tan(2\pi N)}{N}$. Once $|P|$ is known by equation of item 8 above, read N from above graph and put in the equation of item 10, as procedure to get ε_r

   For this we can use the readily available plot between N and $|P|$ of Fig. 12.21, and read N corresponding to the $|P|$, computed earlier.
10. With the N, λ_g, λ_0, t, we now compute ε_r as:

$$\varepsilon_r = 1 - \left(\frac{\lambda_0}{\lambda_g}\right)^2 + \left(\frac{\lambda_0 N}{t}\right)^2$$

11. Calculate phase shift $\phi = \left(\frac{\Delta s}{\lambda_g}\right) \times 360°$.
12. Change the frequency of the klystron by its mechanical tuning, and repeat all the above again.

Observations and Results

See Table 12.17.

Table 12.17 Finding dielectric constant by shifting minima by its insertion

$t = \ldots$ cm								
S. no.	Frequency	λ_0	λ_g	P	$\|P\|$	N	ε_r	f
1								
2								
3								

Precautions

1. When we have chosen one of the minima as reference minima with link *A*, none of the settings should be changed before inserting the dielectric section (link *B*).
2. For second set of reading with the frequency, all the settings of maximisation of power in VSWR have to be done again, with variable attenuator at minimum.
3. Follow all precautions of Experiment No. 1.
4. When we insert another device in the circuit (e.g. dielectric line), the nuts of the flanges of the waveguide should be tight, or else a small disturbance will change the setting, because even a small gap will be a discontinuity leading to reflections.

Viva and Quiz Questions

1. What is a dielectric?
2. What happens to the elect. field inside the dielectric?
3. A slotted line is filled with dielectric, what will happen to:

   (a) λ_g will it increase or fall. [**Ans**.: Fall]
   (b) Maxima and minima voltage [**Ans**.: Fall].
   (c) VSWR [**Ans**.: Increase].

4. If the screws of the flanges of waveguide are loose, what will happen?
   [**Ans**.: More reflections, high VSWR, and instability]
5. If two waveguides are filled with air and then with dielectric constants $\varepsilon_r = 5$, and $\varepsilon_r = 15$, which will have larger λ_g and why?
   [**Ans**.: Air filled]
6. If a two-hole 20 dB directional coupler is filled with $\varepsilon_r = 10$, which of its parameters will change, increase, or decrease and how much?
   [Ans.: I_{so}, dc will decrease, I_{ns} will increase by $\sqrt{10}$ times, λ_g reduces, and λ_c increases.]
7. Can a dielectric change the frequency of the wave, when it crosses it?
   [**Ans**.: Never]

12.13 Experiment No. 13: Study of the Ferrite Devices—Isolator and Circulator

Aim

To study VSWR, insertion loss (forward loss), and isolation in isolator and circulator.

Equipments Required

As per Fig. 12.22.

Theory

The isolator is a two-port device with small insertion loss of signal in the forward direction but very large attenuation of the signal in the reverse direction. Here the Faraday rotation of plane of polarisation of the microwave takes place, in the presence of magnetic field. Here the reflected wave from the circuit enters the isolator, becomes out of phase (180° rotation) with the main signal at the input end, and hence gets cancelled. For details, Chap. 4 on components could be referred.

The circulator is a multiport device, where the input power given to port 1 goes to port 2 only, input to port 2 goes to port 3 only, and so on. Normally a three-port circulator is used in the laboratories. This device also uses the property of Faraday rotation of the plane of polarisation of microwave. Here the input signal at port 1 reaches the port 2 from two paths in phase (add) but reaches port 3 from two paths out of phase (cancels). For detailed study, Chap. 4 on components may be referred.

Insertion loss (I_{ns}) or forward loss is the ratio of power by a source to the input of the device to the output power detected from the device at its output.

Isolation (I_{so}) is the ratio of power fed to a device under test, to the power detected in the arm not supposed to be coupled, e.g.

In circulator $I_{so} = 10 \log_{10} (P_i/P_{i+j})$ dB (where $i \neq 1$)
In isolator $I_{so} = 10 \log_{10} (P_{in}/P_{ref})$
Where P_{ref} = Power reflected back to the input

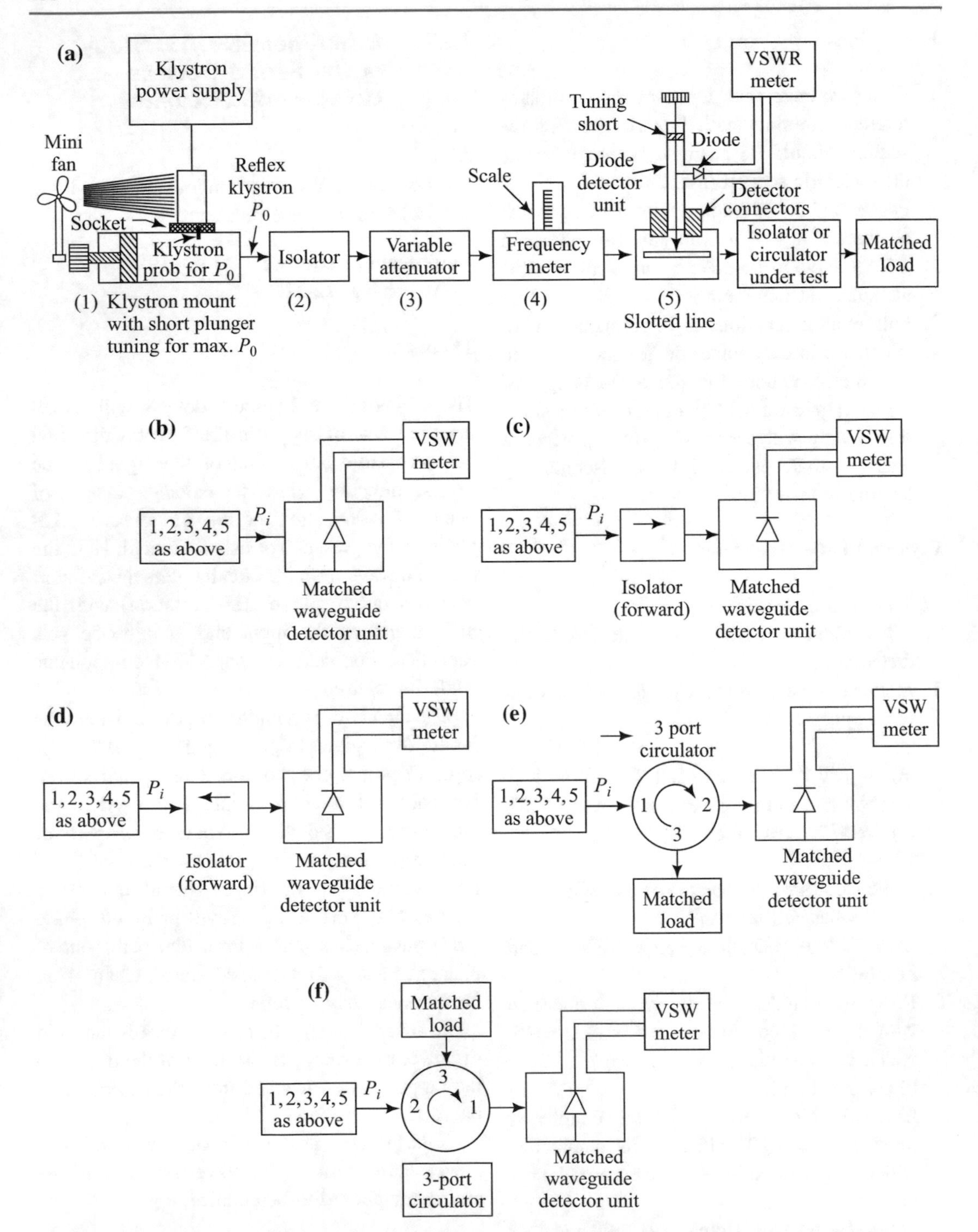

Fig. 12.22 Set-up for measurements in **a** input VSWR measurement, **b** power (P_i) for input to isolator/circulator, **c** insertion or forward loss in isolator, **d** isolation (I_{so}) in isolator, **e** insertion loss (I'_{ns}) in circulator, **f** isolation (I'_{so}) in circulator

Input VSWR: It is the ratio of maximum to minimum voltages detected at the input end of the device under test, by the slotted line.

As the waves get reflected by the device, it will form VSWR, even when the output end of the device is terminated with matched load.

Input Power (P_i): For all the tests, we have to keep the power from the slotted line (i.e. P_i) same (fixed).

Procedure

1. Set up the instruments as in Fig. 12.22a.
2. Switch on the cooling fan and then the power supply of the source.
3. The VSWR of isolator and circulator can be found by measuring the V_{max} and V_{min} on the VSWR metre or CRO connected through the tuned diode detector Fig. 12.22a.
4. Remove the probe mount from the slotted line, and put a matched detector waveguide section at the output side of slotted line. Connect the detector line to a VSWR metre (Fig. 12.22b).
5. Adjust the attenuator so that (a) reasonable power in the VSWR metre is shown. Record this input power P_i (in the scale of current in the VSWR meter as $V_{guide} \propto I_{det}$) and now attenuator is not to be changed for whole of the experiment.
6. For measuring insertion loss (I_{ns}) of isolator, insert it between output of the slotted line and the matched detector (Fig. 12.22c). Now read the P_{o1} on VSWR metre and calculate $I_{so} = 10 \log_{10} (P_i/P_{o1})$.
7. For measuring isolation loss (I_{so}) of isolator, reverse the isolator (Fig. 12.22d) and measure the output power (P_{O2}) in VSWR metre. In ideal condition this power should be zero, but some small amount may be there. Calculate

$$I_{so} = 10 \log_{10}(P_i/P_{o2}).$$

8. For measuring insertion loss of circulator (I'_{ns}) replace isolator with circulators (Fig. 12.22e), and measure the output power (P_{o3}). Calculate $I'_{ns} = 10 \log (P_i/P_{o3})$.
9. For measuring isolation loss (I'_{so}) of circulator, reverse its connection (Fig. 12.22f), i.e. port 2 with slotted line and port 1 with the matched load detector unit. Measure the power in VSWR metre as P_{o4}. Here in ideal condition, the power output should be zero, but in practice some power may be there. Calculate $I'_{so} = 10 \log_{10} (P_i/P_{o4})$.

All the above power can be measured in dB also directly by VSWR metre.

Observation and Results

See Table 12.18.

- **Precautions**

Follow all the precautions of Experiments No. 1 and number 4.

Quiz/Viva Questions

1. What is the basic property of ferrite when μW passes through it in the presence of magnetic field?
2. Why does relative phase change gets introduced by a ferrite in LCPW and RCPW? Does it depends on the direction of magnetic field?

Table 12.18 Finding Insertion loss (I_{ns}) and Isolation (I_{so}) in Isolator and Circulator

S. no.	Frequency GHz	VSWR		Input power Fig. 12.22b Pi	Isolator				Circulator			
		Isolator	Circulator		Forward output	Reverse output	I_{ns}	I_{so}	Forward output	Reverse output	I_{ns}	I_{so}
1	9.1	1.05	1.1	30	27	2.2	30 − 27 = 3	30 − 2.2 = 27.8	28	1.5	30 − 28 = 2	30 − 1.5 = 28.5
2												
3												

3. Explain Faraday rotation.
4. Ferrite devices are non-reciprocal, Explain.
5. List out the ferrite materials.
6. What are ideal values of insertion loss and isolation of isolator and circulator?
7. Can we convert a three-port circulator to an isolator by terminating one of its ports by matched load? (**Ans.**: Yes)
8. Does Faraday rotation angle increase with the thickness of ferrite?
   (**Ans.**: Yes by 100°/cm approximately at 10 GHz)

It does not depend on mag. field strength, after saturated field.

(As seen from input, direction of rotation remains same (e.g. clockwise) whether microwave moves forward or back, so far as magnetic field is in the same direction.)

12.14 Experiment No. 14: Measure of Q Factor of Resonant Cavities-Reflection Type and Transmission Types

Aim

Measure of the Q factor of resonant cavities—reflection type and transmission type.

Equipment

As per Figs. 12.23 and 12.24).

Theory

Power between waveguide and the cavity resonators is coupled either by a slot between or by a probe or loop. The cavities are mainly of two types (refer Chap. 3).

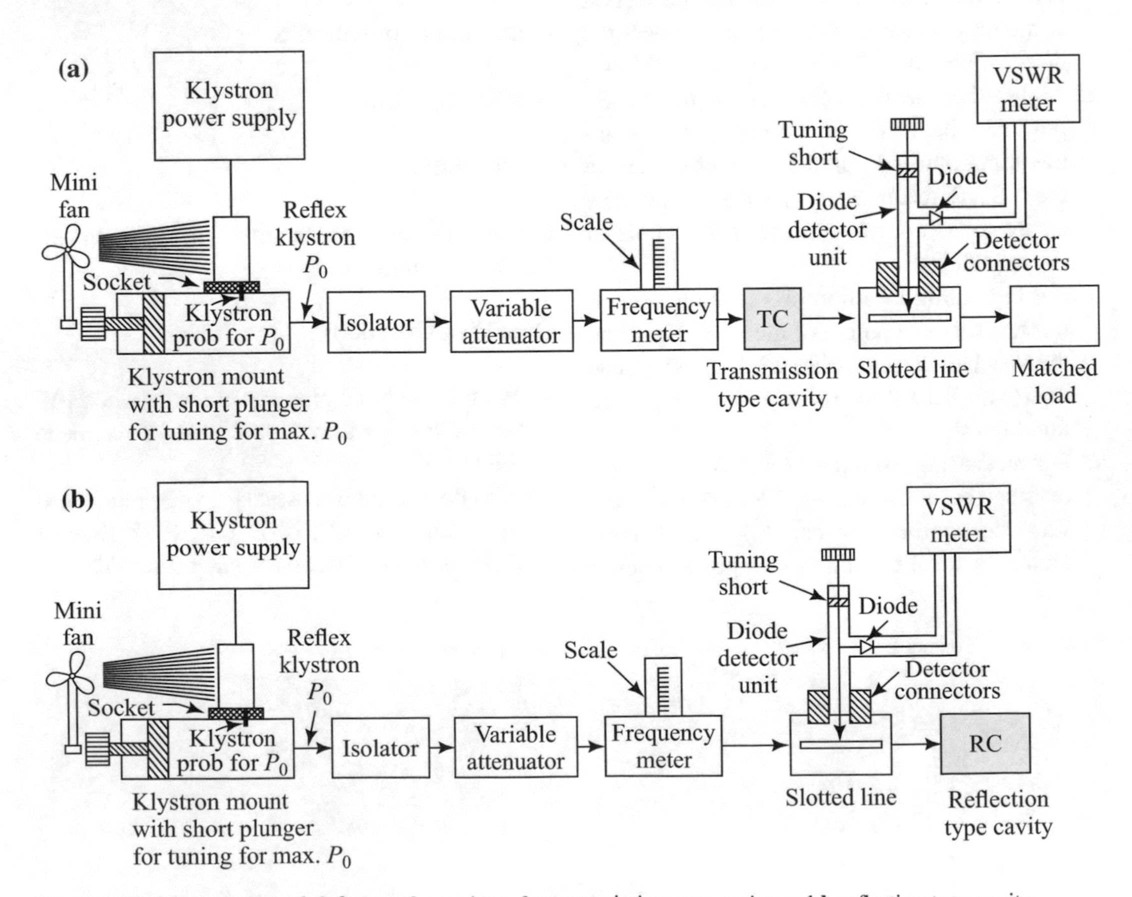

Fig. 12.23 Measurement of Q factor of a cavity **a** for transmission type cavity and **b** reflection type cavity

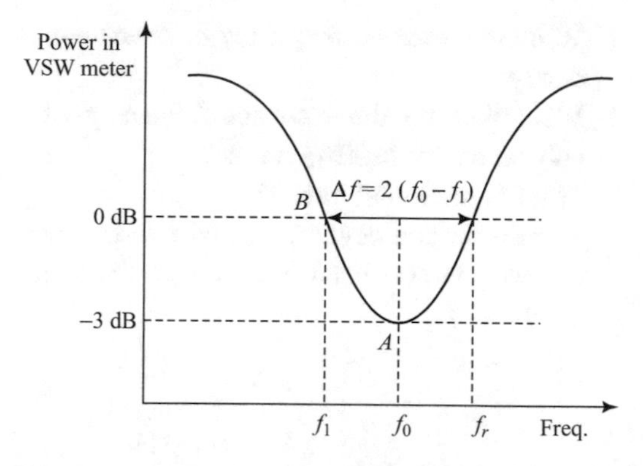

Fig. 12.24 Power detected in the VSWR metre. At the minima point A the VSWR reading has to be made 3 dB by attenuator and then frequency of klystron varied by changing V_0 for getting point B (i.e. 0 dB)

(i) *Transmission type* through which the main power flows. This cavity is connected with the main line by a slot on the main line which takes power from the main line at resonance, e.g. frequency metres. This being in series absorbs maximum power at resonance (Fig. 12.24).

(ii) *Reflection type* into which the μW signal goes and gets reflected back, e.g. terminal resonator. Reflection type gives maximum load at the terminal point at resonance.

The Q_0 factor of a cavity is the measure of a frequency selectivity defined as:

$$Q_0 = 2\pi\left(\frac{\text{maximum power } (W) \text{ stored during a cycle}}{\text{average power dissipated per cycle}}\right)$$
$$= \frac{\omega_r \cdot W}{P_0}$$

where

P_0 average power loss
$\omega = 2\ \pi/T$ the resonant angular frequency.

Above is called unloaded, while with the external circuit which adds to the loss by P_i, the loaded Q factor is defined as:

$$Q_L = \frac{\omega_r \cdot W}{P_0 + P_i}$$

To a very good approximation, the Q_L can be defined with the half power band width (Δf) around the resonant frequency as:

$$Q_L = (f_r/\Delta f)$$

Procedure

Set the equipment as in Fig. 12.23.

(a) **For Transmission Type Cavity**

1. Vary the frequency of klystron by varying V_R, then by short plunger tuning knob gets a point of maximum power, and this is resonance frequency.
2. Measure f_0 by frequency metre and detune the metre.
3. Adjust the attenuator such that VSWR metre reads 0 dB.
4. Without disturbing the attenuator, change the frequency, so that we get 0 dB in VSWR metre.
5. Measure this frequency f_1 by the frequency metre.
6. Compute $Q_0 = \frac{f_0}{2(f_0 - f_1)}$.

(b) **For Reflection Type Cavity**

1. Vary the frequency of the klystron by varying V_R, then by tuning the short plunger we get a point of maximum power, and this is resonant frequency (f_0).
2. Measure this by frequency metre and detune the frequency metre.
3. Adjust attenuator such that VSWR metre reads 0 dB (i.e. max power).
4. Without disturbing the attenuator now change the frequency, so that we get 3 dB, power in VSWR metre.
5. Measure this frequency as f_1.
6. Compute $Q_0 = \frac{f_0}{2(f_0 - f_1)}$.

NB: In case the cavity is tunable then after step 3 above, the cavity can be tuned to change the resonant frequency till we get the 1/2 power point *B* of Fig. 12.23. Here size and hence *Q* of cavity change a bit, but the result is within accuracy limits.

Results and Observation

See Table 12.19.

Table 12.19 Getting *Q*-factor of a resonant cavity

Transmission cavity			Reflection cavity		
f_0	f_1	$Q = f_0/[2(f_0 - f_1)]$	f_0	f_1	$Q = f_0/[2(f_0 - f_1)]$

Precautions

1. Most important precaution is that the resonant frequency of the cavity has to be within the range of frequency variation of the klystron.
2. Start with zero in the variable attenuator.
3. Remaining precautions as in Experiment No. 1.
4. After measurement of frequency the frequency metre should be de-tuned sufficiently to be much beyond f_1.

Quiz/Viva Questions

1. Define *Q* of a cavity, what does its value represent.
2. What type of cavities is in:

   (a) Frequency metre.
   (b) Diode detector tuner.
   (c) Tunable short and matched load.
   (d) Klystron.
   (e) Reflex klystron.
   (f) Magnetron.
3. Write the resonant frequency of a rectangular cavity.
4. What will be the resonant frequency of a cubical cavity in TE_{10}, mode?
5. What is re-entrant cavity?
6. In transmission cavity made by two iris, how can we have (a) band pass and (b) band stop cavities.

12.15 Experiment No. 15: Study of the Radiation Pattern and Gain of a Waveguide Horn Antenna

Aim

To plot the polar radiation pattern and compute gain of waveguide horn antenna.

Equipments

Microwave power source (klystron or Gunn diode oscillator), isolator, frequency metre, variable attenuator, two horn antennas, turntable, detector unit, detector mount, VSWR metre, and accessories, e.g. stand etc. (Fig. 12.25).

Theory

If any transmission line, e.g. waveguide carrying microwave power, is left open at the other end, then it will radiate power in all directions, which

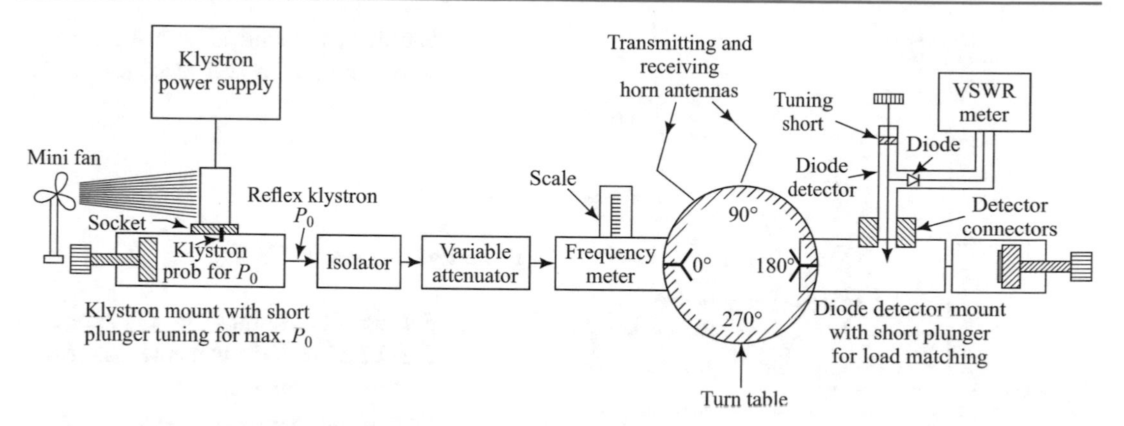

Fig. 12.25 Set-up for studying field radiation pattern of horn antenna and measurement of its gain

may not be uniform. It will get reflected also due to discontinuity/mismatch of impedance of waveguide (50 Ω) and free air (377 Ω). The match will improve with less reflection, and more power can be radiated if impedance changes gradually from 50 to 377 Ω using a horn antenna, which flared up waveguide in H-plane, i.e. a side) or E-plane (i.e. '*b*' side). (For details refer to Chap. 7).

Normally in the laboratories the H-plane horn antenna is used for studying the radiation pattern along *x*-plane using a turntable. Here the radiation is in *y*-plane also but comparatively less. The radiation pattern consists of several lobes of electric field, namely main front lobe, two side lobes, and one back lobe (Fig. 12.26).

The major power is concentrated in the main front lobe. Let us now study gain, beam width, and required distance between antennas.

1. ***Gain*: Gain as a word normally gives a feeling of amplification**, but gain of an antenna is defined as the power intensity at the maximum of the front lobe, as compared to the power intensity achieved from an imaginary omnidirectional antenna (radiating equally in all directions) with the same power fed to that antenna (Fig. 12.26).

As the gain of an antenna remains same, (a) whether it is acting as receiver or transmitter and (b) whether power level is high or low. Therefore we will use this property for finding the gain.

If the transmitting antenna of gain G_{tr} transmits power P_{tr} watts, then power intensity at distance will be $P_{tr} \cdot G_{tr}/(4\pi r^2)$ W/m^2. Also the capture area of the receiving antenna is G_{rec}/λ_0^2 4π met^2; therefore power (W) received by it will be

$$P_{rec} = \left[P_{tr}G_{tr}/(4\pi r^2)\right]\left[G_{tr}/\lambda_0^2/4\pi\right]$$

i.e.

$$P_{rec} = P_{tr}G_{tr}G_{rec}(\lambda_0/4\pi r)^2$$

As the two antennas are identical $G_{tr} = G_{rec} = G$ (say), then above equation gives gain as:

$$\boxed{G = (4\pi r/\lambda_0) \cdot \sqrt{P_{rec}/P_{tr}}}$$

(b) **Beam width**: Angle between the two points on the *x*-plane around the central axis of the main front lobe, where the power intensity is half (i.e. 3 dB down) the maximum power intensity on the axis.

The beam widths in degrees are found to be inversely proportional to the front aperture widths, i.e. a' and b' (in cm) as:

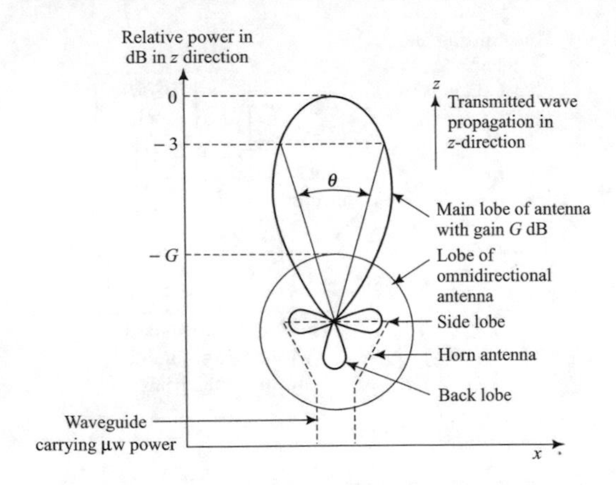

Fig. 12.26 Horn antenna (dotted line) and the radiation pattern of its electric field

$$\boxed{\theta_H = 80\lambda_0/a' \text{ and } \theta_E = 53\lambda_0/b' \text{ degrees}}$$

Here θ_{H} is beam width along H-plane (x-plane) and θ_{E} is beam width along E-plane (y-plane). If a' (flared up) = 10 cm, b' (non-flared) = 1.0 cm, λ = 3 cm (10 GHz), then θ_{H} = 24° and θ_{E} = 160°. Thus horn H-plane antenna is quite directional.

(c) **Distance between transmitting and receiving antenna**: The accuracy of the gain measurement depends upon the selection of proper distance (r) between the two antennas. When the aperture of the two antennas is different, then there will be phase lag between the electric fields at central part and peripheral of the aperture (face) of the antenna, causing an error in gain measurement. This error can be minimised by having the distance as given below:

(i) *Unequal size antenna*: If D is the aperture of larger antenna, and if the size of the two is different, then it has been found that error is minimised if the distance between is kept as:

$$r_{\min} > 2D^2/\lambda_0$$

(ii) *Identical antenna*: Further to avoid mutual reflection and interaction between two identical horn antennas the separation (r) for best result should be:

$$r_{\min} > 2a^2/\lambda_0$$

Procedure

(a) *Radiation Pattern Plotting*

1. Set up the equipments as shown in Fig. 12.25 with the two antennas face to face on the same axis.
2. Switch on the power supply of the klystron oscillator as well as its fan.
3. Set the beam voltage near 300 V. So that beam current is below 30 mA.
4. Change the modulation voltage and its frequency to get maximum deflection in VSWR metre with attenuation at minimum.
5. Set the tuners of klystron, detector mount, and detector for maximum power.
6. Check the frequency and detune the metre.
7. Set the VSWR metre to read maximum signal and if required change the scale.
8. Turn the table in steps of 10° left and right, and see the VSWR readings.
9. As the VSWR reading is directly proportional to the power (as the diode is a square law detector), these VSWR reading gives the plot of power (Fig. 12.26 and Table 12.20).
10. The half angle between the half power points on the two sides of the central line is the beam width.

(b) *Gain Measurement*

1. Keep both the antennas fully opposite to each other (i.e. 0° and 180°).
2. Keep the dB switch of the VSWR metre at 50 dB, with gain control knob at full.
3. Energise the klystron to give power so that VSWR metre gives full scale deflection by using variable attenuator. Let this reading be 6 dB on 20 dB full deflection scale.

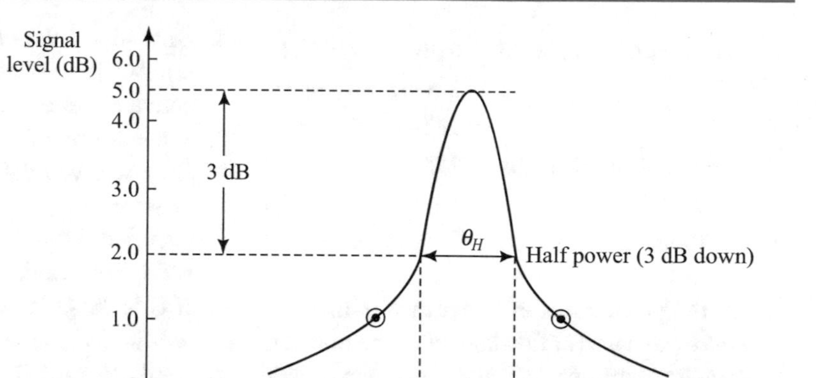

Fig. 12.27 Plot of signal level versus angle between the two antenna

4. Replace the horn antenna system by directly connecting the detector unit, and without changing the gain of VSWR metre, change the range.
5. Calculate the difference in power measured in dB at steps 3 and 4 above, and this will be the ratio P_t/P_r in dB.
6. Convert this dB into normal ratio to get the value of (P_t/P_r).
7. Compute gain $G = (4\pi r/\lambda_0)\ \sqrt{P_t/P_r}$, where r = distance between antenna λ_0 = wavelength of signal.
8. Convert G value to dB units by $G_{dB} = 10 \log_{10} (G)$ (Fig. 12.27).
9. *Example*: By step 5, $P_t/P_r = 50 - (20 + 6) = 24$ dB = 251 (normal ratio).

Let the distance $r = 140\,\text{cm}$

frequency $f = 10\,\text{GHz}$

$$\therefore \quad \lambda_0 = 3.0 \times 10^{10}/10 \times 10^9 = 3\,\text{cm}$$

$$\therefore$$

$$G = (4 \times 3.14 \times 140/3)\sqrt{251} = 9284 \equiv 39.7\,\text{dB}$$

Observation and Results

Reflex klystron	: Beam voltage	…… V
	: Reflector voltage	…… V
	: Beam current	…… mA

See Table 12.20.

Table 12.20 Observation of radiation pattern of a horn antenna

Left of axis		Right of axis	
Angle	Signal in VSWR metre (dB)	Angle	Signal in VSWR metre (dB)
180	3.9	180	3.9
175	3.6	185	3.6
170	3.2	190	3.2
165	–	195	–
160	–	200	–
155	–	205	–
145	–	210	–
140	–	215	–
135	–	220	–
130	–	225	–
125	–	230	–
120	–	235	–
		240	–

From the plot beam width = 3 dB down (left) to 3 dB down (right)
= 225 − 195 = 30°

From theoretical calculation = 24°

Precautions

1. All the precautions of Experiment No. 1.
2. Microwave power flowing out of the horn can damage retina of eye; therefore one should not see directly into the horn antenna. Avoid keeping the body parts also on the line of transmission as in general microwave is cancerous.
3. Around the experimental set-up, lossy/absorbing material should be placed and not reflecting type, i.e. not metal sheet, etc., or else radiation pattern will get disturbed due to reflected signal.

Quiz Questions and Viva

1. What does an antenna radiate?
2. How does the wave get emitted from a dipole? If they are loops, how do they get converted into wave front?
3. What is wave front and what it consists of?
4. How will you define an antenna based on its beam width?
5. Can we have vertically linearly polarised wave front from a dipole antenna and how?
6. If G is the gain of transmitting antenna, what will be its gain as a receiving antenna?
7. If 100 W microwave power is radiated by an omnidirectional antenna, what will be the power density at a point 1 km away?
   (**Ans.**: $100/4\pi r^2 = 7.9\ \mu W/m^2$)
8. If 100 W microwave power is emitted from a circular horn antenna at beam angles of 30° and if whole of the power goes to this conical beam, find the power density at 1 km distance.
   (**Hint**: $p = 100/\pi r^2\ W/m^2$ where distance $d = r/\sin 45°$ $p = 100/[\pi \sin^2 (15). (1000)^2] = 471.3\ \mu W/m^2$)

Printed in the United States
By Bookmasters